Probability and Stochastics Series

Advances in Queueing

Theory, Methods, and Open Problems

Edited by
Jewgeni H. Dshalalow
Department of Applied Mathematics
Florida Institute of Technology
Melbourne, Florida

CRC Press
Taylor & Francis Group
Boca Raton London New York

CRC Press is an imprint of the
Taylor & Francis Group, an **informa** business

CRC Press
Taylor & Francis Group
6000 Broken Sound Parkway NW, Suite 300
300 Boca Raton, FL 33487-2742

First issued in paperback 2019

ISBN 13: 978-0-367-44891-2 (pbk)
ISBN 13: 978-0-8493-8074-7 (hbk)

Visit the Taylor & Francis Web site at
http://www.taylorandfrancis.com

and the CRC Press Web site at
http://www.crcpress.com

Library of Congress Card Number 95-31086

Library of Congress Cataloging-in-Publication Data

Advances in queueing : theory, methods, and open problems / edited by
Jewgeni H. Dshalalow.
p. cm. -- (Probability and stochastics series)
Includes bibliographical references and index.
ISBN 0-8493-8074-X (alk. paper)
1. Queueing theory. I. Dshalalow, Jewgeni H. II. Series.
T57.9.A37 1995
519.8'2—dc20 95-31086
CIP

Probability and Stochastics Series

Edited by Richard Durrett and Mark Pinsky

Published Titles

Chaos Expansion, Multiple Wiener-Ito Integrals and Applications, Christian Houdre and Victor Perez-Abreu
Topics in Contemporary Probability, J. Laurie Snell
Linear Stochastic Control Systems, Guanrong Chen, Goong Chen, and Shih-Hsun Hsu

Forthcoming Titles

A First Course in Stochastic Calculus, Richard Durrett
Frontiers in Queueing Systems: Models and Applications in Science and Engineering, Jewgeni Dshalalow

Preface

Originally founded by a Danish scientist, A.K. Erlang (1878-1929), for the first telecommunication systems, the theory of queues evolved as an interdisciplinary science driven by a tremendous number of problems arising in virtually all areas of engineering, technology, operations research, management, and physics. By now, there are many thousands of research articles and about 500 text books and research monographs on queueing and related topics. *Queueing* has been adopted as a regular (undergraduate and graduate) course in many curricula of academic institutions, and it commands much attention in a large number of scientific conferences in operations research, stochastic processes, statistics, applied mathematics, and industrial engineering. This tremendous push for new results forced more and more academic journals to publish articles in queueing and even open new sections. In 1986, Baltzer Verlag, AG launched a new academic journal entitled *Queueing Systems* (edited by N.U. Prabhu), which is devoted entirely to queueing. Many other journals, such as *Stochastic Models* and the *Journal of Applied Probability*, publish articles on queueing extensively.

The flow of new theories and methodologies in queueing has become very hard to keep up with. Surveys on the hottest topics in queueing and related areas are scattered over a large variety of scientific magazines. A sort of manual would be desirable, indicating where to find the hottest topics and where to concentrate one's efforts should queueing become one's interest. An attempt to write such a comprehensive monograph or textbook would meet serious, if not insurmountable obstacles. Because of the above reasons, it is difficult for one or even a few coauthors to be proficient enough in so many areas simultaneously and to be able to complete such a book in a reasonable amount of time. A careful elaboration of major themes would take many years of work after which the results would then be outdated!

However, if a substantial number of top experts in their respective fields were to be assembled, the formidable task of writing an all-encompassing book in queueing could be accomplished. This idea is not entirely new. In 1983 and 1984, Akademie Verlag, in the former German Democratic Republic, published a handbook of queueing in two volumes, edited by Boris Gnedenko and Dieter König and written by a team of more than thirty renowned authors from what were formerly East Germany and the Soviet Union. The book was very informative and treated many basic and contemporary topics. However, it took a while to finish it, and it was published in German and never translated into English. Initially, Dieter König approached me and proposed to update and translate the book. The project did not work. Among other obstacles, over a period of years we could not reach a compromise with the former publisher about the copyright, and the translation of more than 1100 pages alone promised to become a very tedious task. Then, Prof. König proposed an alternative project to write an encyclopedia/monograph, which is more or less preserved in the current form. He initially agreed to co-edit, but later abandoned the project because of other obligations. However, he has been helping me with the choice of topics and various other tasks.

The diversity of topics in queueing suggested that it would be best to split them into a monograph in queueing methodology with a more theoretical focus, and a book of contemporary queueing models and more applied topics. Both monographs are totally independent and complement each other. Both are divided in chapters, each surveying a topic followed by a discussion of open problems and future research directions.

The current book consists of 19 chapters written by 25 renowned scientists in the area of probability and queueing. All of them contain the most significant recent methods in the theory of queues, which are introduced, surveyed, and demonstrated on a large number of examples of queueing models. Consequently, methods are primary goals, whereas queueing models are secondary in this book. [The reader wanting to become acquainted with models is referred to the second book, *Frontiers in Queueing: Models and Applications in Science and Engineering.*] The methods in this book are divided into stochastic methods, analytic methods, and generic methods such as approximations, estimates and bounds, and simulation.

The main goal of the book is to introduce the reader to the major up-to-date methodologies in queueing. The first chapter presents an overview of classical queueing methods from the birth of queues through the seventies. It also contains a comprehensive bibliography of books on queueing and telecommunications. Each of following 18 chapters surveys a method or methods applied to classes of queueing systems or processes related to queueing. The last section of each chapter contains a discussion of open problems and future research directions (which can be used for research ideas and even topics for doctoral theses), followed by an extensive bibliography of pertinent books and articles on the subject. The book can be used for independent studies, seminars, workshop lectures, or as a text for upper-level graduate courses in queueing, applied mathematics, operational sciences, industrial engineering, and operations research. Graduate level courses in probability and stochastic processes are prerequisites for most of topics. Some chapters may require, additionally, advanced level real and complex analysis courses.

I thank Professor Dieter König for his inspiration, Dr. Benjamin Melamed for his valuable suggestions on the selection of topics, and Professor Hideaki Takagi who kindly shared with me his huge bibliography of books on queueing. I also thank my colleague, Professor Jay Yellen, and my editorial assistants, Mr. Gary Russell and Dr. Lotfi Tadj (my former Ph.D. student) for their extensive help in editorial proofreading, suggestions, and critical remarks. Special thanks are due to Ms. Donn Miller-Harnish for typing this manuscript in a very short period of time and for always being available for all changes and corrections. Finally, I greatly thank all the contributors for their willingness to share their expertise in this volume.

– Jewgeni H. Dshalalow
Melbourne, Florida

Contributors

Søren Asmussen, Institute of Electronic Systems, The University of Aalborg Fr. Bajesvej 7, DK-9220 Aalborg, Denmark

Bartłomiej Błaszczyszyn, Mathematical Institute, Wrocław University, pl.Grundwaldzki 2/4, 50-384 Wrocław, Poland

Sid Browne, Graduate School of Business, Columbia University, 402 Uris Hall, New York, NY 10027, USA

Erhan Çinlar, Program in Statistics and Operations Research, Department of Civil Engineering and Operations Research, Princeton University, Princeton, NJ 08544, USA

Jos H.A. de Smit, Faculty of Applied Mathematics, University of Twente, P.O. Box 217, 7500 AE Enschede, The Netherlands

Jewgeni H. Dshalalow, Department of Applied Mathematics, Florida Institute of Technology, 150 W. University Blvd., Melbourne, FL 32901, USA

Alexander M. Dukhovny, Department of Mathematics, University of San Francisco, 2130 Fulton Street, San Francisco, CA 94117-1080, USA

Muhammad El-Taha, Department of Mathematics and Statistics, University of Southern Maine, 96 Falmouth Street, Portland, Maine 04103, USA

Vladimir V. Kalashnikov, Institute for System Analysis, Prospect 60 Let Oktyabrya, 9, 117312 Moscow, Russia

Charles Knessl, Department of Mathematics, Statistics, and Computer Science M/C 249, University of Illinois at Chicago, 851 South Morgan Street, Box 4348, Chicago, IL 60607-7045, USA

Igor N. Kovalenko, STORM, University of North London, Holloway Road, London N7 8 DB, England

Benjamin Melamed, Rutgers University, RUTCOR, P.O. Box 5062, New Brunswick, NJ 08903, USA

Isi Mitrani, Computing Laboratory, University of Newcastle-Upon-Tyne, Newcastle-Upon-Tyne, NE1 7RU, England

Marcel F. Neuts, Department of Systems and Industrial Engineering, University of Arizona, Tucson, Arizona 85721, USA

António Pacheco, Departamento de Matemática, Instituto Superior Téchnico, Av. Rovisco Pais, 1096 Lisboa Codex, Portugal

N.U. Prabhu, Operations Research and Industrial Engineering, Cornell University, 338 Upson Hall, Ithaca, NY 14853, USA

Tomasz Rolski, Mathematical Institute, Wrocław University, pl. Grundwaldzki 2/4, 50-384 Wrocław, Poland

Reuven Y. Rubinstein, Faculty of Industrial Engineering and Management Technion Institute of Technology, Haifa 32000, Israel

Volker Schmidt, Department of Stochastics, University of Ulm, Helmholzstrasse 18, D-89069 Ulm, Germany

Richard F. Serfozo, School of Industrial and Systems Engineering, Georgia Institute of Technology, Atlanta, GA 30332-0205, USA

Shaler Stidham, Department of Operations Research, University of North Carolina at Chapel Hill, 3180 Smith Building, Chapel Hill, NC 27599-3180, USA

Lajos Takács, Department of Mathematics and Statistics, Case Western Reserve University, Cleveland, OH 44106, USA

Charles Tier, Department of Mathematics, Statistics and Computer Science M/C 249, University of Illinois at Chicago, 851 South Morgan Street, Box 4348, Chicago, IL 60607-7045, USA

Ward Whitt, AT&T Bell Laboratories, (2C-178), 600 Mountain Avenue Murray Hill, NJ 07974, USA

David D. Yao, Department of Industrial Engineering and Operations Research Columbia University, New York, NY 10027-6699, USA

Contents

1 An Anthology of Classical Queueing Methods 1
Jewgeni H. Dshalalow

I **Stochastic Methods** 43

2 Queueing Methods in the Theory of Random Graphs 45
Lajos Takács

3 Stationary Distributions via First Passage Times 79
Søren Asmussen

4 An Introduction to Spatial Queues 103
Erhan Çinlar

5 Sample-Path Techniques in Queueing Theory 119
Shaler Stidham, Jr. and Muhammad El-Taha

6 Markov-Additive Processes of Arrivals 167
António Pacheco and N.U. Prabhu

7 The ASTA Property 195
Benjamin Melamed and David D. Yao

8 Campbell's Formula and Applications to Queueing 225
Volker Schmidt and Richard F. Serfozo

9 Excess Level Processes in Queueing 243
Jewgeni H. Dshalalow

II **Analytic Methods** 263

10 Matrix-Analytic Methods in the Theory of Queues 265
Marcel F. Neuts

11 Explicit Wiener-Hopf Factorization for the Analysis of Multi-Dimensional Queues 293
Jos H.A. de Smit

12 Applications of Singular Perturbation Methods in Queueing 311
Charles Knessl and Charles Tier

13 The Spectral Expansion Solution Method for Markov Processes on Lattice Strips 337
Isi Mitrani

14 Applications of Vector Riemann Boundary Value Problems to Analysis of Queueing Systems 353
Alexander Dukhovny

III Approximation, Estimates, and Simulation of Queues 377

15 Light-Traffic Approximation in Queues and Related Stochastic Models 379
Bartłomiej Błaszczyszyn, Tomasz Rolski, and Volker Schmidt

16 Quantitative Estimates in Queueing 407
Vladimir V. Kalashnikov

17 Steady State Rare Events Simulation in Queueing Models and Its Complexity Properties 429
Søren Asmussen and Reuven Y. Rubinstein

18 Piecewise-Linear Diffusion Processes 463
Sid Browne and Ward Whitt

19 Approximations of Queues via Small Parameter Method 481
Igor N. Kovalenko

Index 509

Chapter 1

An anthology of classical queueing methods

Jewgeni H. Dshalalow

CONTENTS

1.1	What queues are about	1
1.2	The anatomy of queueing systems	2
1.3	Historical notes	4
1.4	Chapters 2 through 19	15
	Acknowledgement	15
	Bibliography	15

This introductory chapter presents a brief historical overview of fundamental methods in queueing theory from its inception through the seventies; the rest of the book is devoted to the most significant methods for contemporary analyses of queues. In preparation for the remainder of the book, this chapter also includes an up-to-date bibliography of most books on queueing and related topics, and is organized in several sections. All references appearing in this chapter will be denoted by $[\mathbf{x},y]$, where $\mathbf{x}$ is the section number and y is the reference number in this section.

1.1 WHAT QUEUES ARE ABOUT

A queueing node consists of an input stream of single units (called *customers*) or batches thereof, a waiting room or buffer (finite or infinite), a servicing facility consisting of one or more *servers* (or *channels*), and a set of disciplines and policies specifying the rules for queueing and servicing customers. *Queueing networks* are formed by routing customers among multiple queueing nodes. If a queueing network consists of a single node it is referred to as a *queueing system.*

The word *queue* has a somewhat negative connotation stemming from its association with waiting, a familiar experience of everyday life, such as that encountered in banks, postal offices, airports, gas stations, automobile traffic, mail backorders, telephone traffic, dental and medical offices, hospitals, and organ transplants. In many of these real-world systems, it is desirable to reduce waiting by better scheduling, improved disciplines, and allocation of service intensities. In some queueing systems, ordinary waiting is replaced by blocking. For instance, in telecommunication systems, incoming calls (which we interpret as customers) arrive at a group of parallel servers (trunk lines), so that if all lines are busy, an arriving call is blocked and lost.

A complete and accurate description of a *queueing system* is elusive. Usually, a queueing system is characterized by an appropriately interpreted servicing process and an input stream of discrete-valued batches of customers. ("Continuous-valued

0-8493-8074-x/95/$0.00+$.50

batches" are associated with "fluid systems," such as dams or inventories.) However, even this notion has recently undergone a variety of modifications and extensions. (The concept of *routing* should be added to characterize queueing networks.) The following section will describe the structure of queueing systems.

1.2 THE ANATOMY OF QUEUEING SYSTEMS

Over the past eighty years or so, a large variety of real-world queueing systems have been successfully modeled and investigated by many hundreds of researchers. These activities gave rise to an enormous literature and the creation of a huge collection of analytical queueing models. A select set of methods used to analyze queueing systems and networks will be presented in this book. Queueing research consists of three phases: formulation of a model, identification of some process of interest, and devising a solution method for the requisite statistics. Consequently, queueing problems can be classified according to model, process, or method.

Model description. To describe a queueing model, one uses the following basic terminology:

1. Input (arrival) process
2. Queueing discipline
3. Buffer size (waiting room capacity)
4. Number of servers and their capacities
5. Service discipline and service process
6. Vacation discipline and vacation process
7. Network configuration (routing).

A succinct notation for the above terms was proposed by David Kendall [**V**,39] in 1953 and has been widely adopted in the contemporary literature on queueing. This notation has the form $A/B/m/C_w/C_s/Q/(q,N)$, where

A characterizes the the arrival process. Common arrival processes are:

a) Renewal process:

M - orderly Poisson process

D- deterministic

E_n-n-Erlang

H_n-hyperexponential of type n

PH-phase type

GI-general renewal process

b) SM-semi-Markov process

c) $BMAP$-batched Markov arrival process

A^X or alternatively BA stands for bulk arrivals

B characterizes the service process. Common service time distributions are:

M - exponential

D - constant

E_n - n-Erlang

H_n - hyperexponential of type n

PH - phase type

G - general

SM - semi-Markov

B^Y or BB stands for batch service

m denotes the capacity of the servicing facility, i.e. the number of servers (or channels) which is either 1, >1 and finite, or infinite (each of the three types requires a separate classification)

- C_w gives the capacity of the waiting room, according to which queueing models are classified for C_w being zero, any positive integer, or infinity
- C_s is the server capacity (≥ 1), which is the maximal number of customers in a servicing batch
- Q stands for a queueing discipline; common disciplines are *FIFO* (first in, first out), *LIFO* (last in, first out), *RSS* (random selection of service), and *PR* (priority)
- (q,N) denotes the busy period disciplines which specify the rules by which the system starts and ends a busy period. Under these rules, the server stops processing new customers when the queue length drops below level q (≥ 1), and it resumes service when the queue length reaches or exceeds level N (≥ 1). Busy periods alternate with periods during which the server suspends service. In the latter case, the server either idles, or takes "vacations," or takes vacations and idles. If the arrival stream is a bulk arrival process or if the server leaves on vacations, the queue length at the beginning of a new busy period need not be exactly N. The presence of q in the above notation is frequently associated with the term *q-quorum* or just *quorum*. The rule of entering a busy period with a specified value N is referred to as the *N-policy*.

Processes. The basic processes of most interest in the research of queueing systems are:

1. The queueing process $\{Q(t)\}$, where
   $Q(t)$ is the number of customers in the system (including those being in service) at time $t \geq 0$.
2. The actual waiting time process $\{W_n\}$, where
   W_n is the waiting time of the nth customer in the line (excluding his service time).
3. The Takács process (virtual waiting time process) $\{W(t)\}$, where
   $W(t)$ is the cumulative time needed to process all customers present in the system at time t.
4. The busy period process, which
   denotes the time during which the server continuously processes (primary) customers. Busy periods alternate with idle or vacation periods. In the former case, the server idles as long as the system has fewer than N (≥ 1) customers. In the latter case, the server *leaves* the system for *vacations*, which may be related to maintenance or service of secondary customers. A *vacation* or, more precisely, *vacation cycle* consists of one or more vacation periods (or segments). If, during a vacation period, the system hits (i.e. reaches or exceeds) level N, the server does not return from vacation prematurely. However, a vacation cycle ends once the period is completed, and the system starts a new busy period.
5. The output process, which
   refers to the flow of completely processed customers.

Each of these processes can be investigated in "transient" or "stationary" modes, and analyses conventionally target their one-dimensional distributions. A time-dependent distribution, is referred to as the *transient distribution*, as opposed to the *steady state* or *stationary distribution* in which case the underlying process is said to be in its (*statistical*) *equilibrium*. Although stochastic processes usually approach steady state quite rapidly, mathematical steady state holds only asymptotically (unless the process starts in steady state). Other special characteristics may be of interest in conjunction with the above processes or independently, such as various statistical data and optimization.

Methodology. A major portion of any research work is based on mathematical, probabilistic and statistical methods applied to processes of interest. Typically, some part of the analysis includes classical methods (differential, integral and integro-differential equations, Laplace and Fourier transforms, complex and functional analysis), boundary value problems (Wiener-Hopf factorization, Riemann-Hilbert boundary value problems, Banach algebras), numerical analysis, matrix calculus (standard methods, spectral analysis, matrix-geometric methods), stochastic analysis (fluctuation theory, embedded Markov chains, supplementary variables, Markov, semi-Markov and semi-regenerative processes, level-crossing analysis), approximation, statistical inference, estimates and bounds, or simulation.

1.3 HISTORICAL NOTES

The inception of **Queueing Theory** may be traced to a sequence of papers on telecommunications by the remarkable Danish engineer, statistician, and mathematician, Agner Krarup Erlang (1878-1929), during his appointment at the Copenhagen Telephone Company in the period 1909-1922. In 1909, Erlang was introduced to F. Johannsen, the managing director of that telephone company. Johannsen had published two papers entitled *Waiting times and number of calls* (on delay problems for incoming calls in manual telephone exchange switchboards) in 1907, and *Busy* (which analyzes the frequency of "busy" reports by subscribers of one or more lines) in 1908, as a prelude to queueing. These works employed some probability arguments and were essentially correct for practical use, though not mathematically rigorous.

As an employee of the company, in 1909 Erlang published the paper *The theory of probabilities and telephone conversations*, widely regarded as marking the birth of queueing theory. Particularly notable was Erlang's rationale for the input stream of telephone calls to obey the Poisson law. In 1917, Erlang published his most significant paper entitled *Solution of some problems in the theory of probabilities of significance in automatic telephone exchanges.* Here he assumed that the input is Poisson (based on his arguments in the 1909 paper) and that service times are exponential. Among other things, he studied the queueing processes in the $M/M/1/\infty$ system and loss model $M/M/\mathrm{m}/0$ in statistical equilibrium using "birth-and-death" representations. Erlang obtained the stationary distribution of the queue length and his famous loss formula, both of fundamental importance in the theory of telephone traffic.

Erlang's impressive contributions to teletraffic have stimulated and continue to stimulate an enormous volume of work in queueing. Besides Erlang, two other most notable pioneers and contributors to queueing are Félix Pollaczek in the thirties through sixties, and Lajos Takács (the most versatile and prolific contributor in the history of queueing) from the fifties to the present. The apparent significance of queueing was also appealing to Emil Borel, David Kendall, A.Y. Khintchine, and Andrey Kolmogorov in their historical works on queueing [**V**,8,38-41,44,

45]. Having been initiated by Erlang, the theory of teletraffic continues to dominate in telecommunications and has considerable overlap with queueing. Vazlav Beneš and Ryszard Syski are among the most notable queueing theorists who have also contributed to the theory of teletraffic. In the past twenty years, a large body of contributors from diverse areas of engineering and technology the world over have made queueing theory a towering and active discipline.

Markov processes. The origin of Markov processes derives from two independent sources. One source is the mathematical studies of Brownian motion initiated in 1900 by H. Poincare's student, L. Bachelier [**V**,3], and by A. Einstein in 1905. The other source is A.A. Markov himself who in 1907 introduced and studied Markov chains with finitely many states using probabilistic analysis. Besides Erlang's queueing models, birth-and-death processes were introduced and studied by G. Yule in 1924 as pure birth processes. This particular birth-death process is referred to as a *Yule process*. A general theory of Markov processes was developed by A. Kolmogorov in 1931 [**V**,44] and 1936 [**V**,45] with important follow-up work by J. Doeblin, J.L. Doob, W. Feller, and P. Lévy in the thirties through fifties.

Various types of Markov processes have been used in queueing. These include Markov chains with denumerable or continuum state space, Markov processes with continuum state space, and n-dimensional Markov processes.

A sequence $X = \{X_0, X_1, \ldots\}$ of random variables valued in a countable set I is said to form a *homogeneous Markov chain* if

$$\mathbf{P}\{X_{n+k} = j \mid X_0, \ldots, X_{n-1}, X_n = i\} = \mathbf{P}\{X_{n+k} = j \mid X_n = i\}$$

$$= \mathbf{P}\{X_k = j \mid X_0 = i\} = p_{ij}^k,\ i, j \in I,\ k = 0,1,, \ldots,\ n = 0,1, \ldots$$

The matrix $P^k = (p_{ij}^k; i, j \in I)$ is the *k-step transition probability matrix* (TPM) of X. For $k = 0$, $P^0 = \mathbf{I}$ is the identity matrix; for $k = 1$, $P^1 = P$ is called the *TPM*; and for $k \geq 1$, by the Chapman-Kolmogorov property, P^k is the kth power of P. A Markov chain is said to be *ergodic* if it is *irreducible*, *aperiodic* and *recurrent-positive*. For the relevant definitions see, for instance, Çinlar [**IV**,13]. The irreducibility and aperiodicity requirements are almost always met in queueing applications. The third condition can be established by various criteria (*cf.* Chung [**IV**,12]), such as *Foster's lemma* (a most prominent tool in queueing that frequently requires the use of Tauberian theorems) and its modifications which apply to special classes of Markov chains. (See for instance, Abolnikov and Dukhovny [**I**,1].) Given the intensity λ_i of the input process (which includes arrival rate and mean arriving batch size) and λ_s of service processes (which includes the number of servicing channels and processed batch sizes) in a particular queueing system, ergodicity is often equivalent to the condition $\lambda_i < \lambda_s$.

If an ergodicity condition for the Markov chain X is met, then the limit for $k \to \infty$ of each row vector $(p_{ij}^k; j \in I)$ of P^k exists (independent of i) and is called the *stationary distribution of* X; this limiting vector is denoted by $p = (p_j; j \in I)$. In this case, p equals the unique positive solution of the equation $p = pP$ with $p\mathbf{1} = 1$, called the *invariant probability measure* of P.

Another important class of Markov processes are *time-homogeneous Markov jump processes* with continuous time parameter, right-continuous paths, and a countable state space I. If $X(t)$ is such a process, then

$$\mathbf{P}\{X(t+s)=j \mid X(u); u \leq t\} = \mathbf{P}\{X(t+s)=j \mid X(t)\}$$
$$= \mathbf{P}\{X(s)=j \mid X(0)=X(t)\},\ j \in I,\ s,t \geq 0.$$

The function $t \mapsto p_{ij}(t) = \mathbf{P}\{X(t)=j \mid X(0)=i\}$ is called the *transition function* of $X(t)$; $P(t) = (p_{ij}; i,j \in I)$ is the TPM which satisfies the forward Kolmogorov differential equation $P'(t) = P(t)A$, $t \geq 0$. Here A is formally $P'(0+)$, and is called the *infinitesimal matrix* which characterizes the associated process $X = X(t)$. In special cases of X, like "birth-and-death" processes, the adequacy of the infinitesimal matrix A to its respective processes is subject to standard arguments.

The transitions of X occur at some random moments of time $T_0, T_1, \ldots$. Over each time interval $[T_n, T_{n+1})$, X has a constant value $X(T_n)$. The length of $[T_n, T_{n+1})$ is exponentially distributed with parameter indexed by $X(T_n)$. The embedded process $\{X(T_n);\ n = 0,1,\ldots\}$ is a homogeneous Markov chain, called the *minimal embedded Markov chain*, and the transition from state $X(T_n)$ to state $X(T_{n+1})$ is governed by the TPM of the jump chain.

Some classical birth-death queues. The following examples of Markov process methods were successfully applied by Erlang to simple queueing models inspiring many researchers to study more general Markov processes.

1. Erlang's system $M/M/1/\infty$. In this model, it is assumed that the input is an orderly Poisson process with intensity λ, and that a single server processes customers (in order of their arrivals) with service times being independent and exponentially distributed with common parameter μ. The queueing process, $\{Q(t)\}$, which gives the number of customers in the system at time t, is a particular case of the general birth-and-death process whose transition probabilities, $p_{in}(t) = \mathbf{P}\{Q(t)=n \mid Q(0)=i\}$, satisfy the forward Kolmogorov differential equations:

$$\left.\begin{aligned} \frac{dp_{i0}(t)}{dt} &= -\lambda p_{i0}(t) + \mu p_{i1}(t), & n &= 0 \\ \frac{dp_{in}(t)}{dt} &= -(\lambda+\mu)p_{in}(t) + \lambda p_{i,n-1}(t) + \mu p_{i,n+1}(t), & n &\geq 1. \end{aligned}\right\} \tag{1.1}$$

Under the equilibrium condition imposed on the *traffic intensity*, $\rho = \lambda/\mu < 1$, the steady state distribution, $\lim_{t\to\infty} p_{in}(t) = p_n$, exists and is independent of the initial state. In equilibrium, (1.1) reduces to a system of linear equations, which is easy to solve. Furthermore,

$$p_n = (1-\rho)\rho^n,\ n = 0,1,\ldots,$$

has a geometric distribution.

2. Erlang's loss system $M/M/m/0$. In addition to the above hypotheses for the arrival and service processes, we assume that the servicing facility consists of m parallel channels, and that there is no waiting room. This system models telephone traffic where incoming calls are lost (blocked) when all lines are busy. The queueing process, $\{Q(t)\}$ which tracks the number of calls in the system at time t, is again Markov, and in particular, birth-and-death. The equilibrium of such a system exists unconditionally, and the steady state probabilities, $\lim_{t\to\infty} p_{in}(t) = p_n$, $n = 0,1,\ldots,m$, satisfy the famous Erlang *loss formula*:

$$p_n = \frac{\rho^n}{n!}\left(\sum_{j=0}^{m}\frac{\rho^j}{j!}\right)^{-1}, \; n = 0,1,...,m; \; \rho = \lambda/\mu. \qquad (1.2)$$

Interestingly, formula (1.2) also holds true for the case of generally distributed service times, i.e. for the system $M/G/m/0$. The first intuitive proof of formula (1.2) for this general case was given by F. Pollaczek in 1932. B. Sevastyanov gave a rigorous proof in 1957.

Renewal and regenerative processes. Renewal processes characterize a particular class of *point processes.* A point process is formed by an increasing sequence $\tau = \{\tau_n; n = 0,1,...\}$ of random points scattered over the positive real axis. With $\tau_0 = 0$ and *inter-renewal* times $t_{n+1} = \tau_{n+1} - \tau_n$ $(n = 0,1,...)$ being independent and identically distributed (iid), τ is a *renewal process* and τ_n is the nth *renewal point.* Consequently, τ is a sequence of partial sums of iid nonnegative valued random variables. Renewal processes describe input flows in the $GI/...$-type queues. When inter-renewal times t_n $(n = 1,2,...)$ are exponentially distributed (with parameter λ), the renewal process reduces to the Poisson (point) process (with rate λ). The latter represents the arrival stream of customers to an $M/...$-type queue. (For more references see Çinlar [**IV**,13].)

Given τ, N is said to be the *associated counting process* if, for any subset B of $\mathbb{R}_+$, $N(B)$ is the number of renewal points in B. Obviously, for $B = [0,t]$, $N_t = N([0,t])$ is a monotone nondecreasing right-continuous piecewise constant function of t with jumps at τ_n of unit magnitude. If τ is a Poisson point process (with rate λ), the associated counting process N is the "Poisson counting process." For an interval B of length b, $N(B)$ has the Poisson distribution with parameter λb.

The mean $\mathbf{E}[N_t]$ defines a monotone nondecreasing function of t denoted $R(t)$ and called the *renewal function.* Let inter-renewal times, t_n, be arbitrarily distributed with common mean $\frac{1}{\lambda}$. In 1954, W. Smith [**V**,85] established one of the key theorems for stochastic processes, known as the Key Renewal Theorem, which states that for a monotone nondecreasing integrable function g,

$$\lim_{t\to\infty} R*g(t) = \lambda \int_0^\infty g(u)du,$$

where "$*$" stands for the convolution operator. This result plays a central role in deriving steady state distributions of regenerative and semi-regenerative processes.

In queueing applications, it is not uncommon to encounter sequences $\zeta = \{\zeta_n\}$ of partial sums composed of iid random variables $\xi = \{\xi_1,\xi_2,...\}$ valued in $\mathbb{R}$ (instead of $\mathbb{R}_+$). Such a generalization of ζ is referred to as a *recurrent process.* Consider the counting process N associated with a renewal process τ and generalize N as follows. The value n of N at τ_n is replaced by ζ_n, and at all points $t \in [\tau_n,\tau_{n+1})$ will assume constant value ζ_n. The resultant counting-like process, $Z_t = Z([0,t]) = \xi_1 + ... + \xi_{N_t}$, is called a *compound renewal process.* In particular, if N is a Poisson counting process, Z is called a *compound Poisson process.* For positive discrete-valued ξ, Z can model a *batch arrival process* in a queueing system of type $BG/...$; in particular, in $BM/...$ is a queue with a bulk Poisson arrival process.

The renewal process is basic to the class of *regenerative processes* which constitutes an important extension. Consider a stochastic process $X = \{X(t)\}$ that evolves on the positive real axis and takes on values in $\mathbb{R}$. Let τ be a renewal proc-

ooo such that at each instant τ_n, X regenerates; that is, the futures $\{X(t); t \geq \tau_n\}$, $n = 1,2,...$, are the probabilistic replicas of X. The renewal process τ must be "compatible" with X in the following way. For each n, events $\{\tau_n \leq x\}$, $x \geq 0$, belong to the τ_n-history $\{X(u); u \leq \tau_n\}$ of X. One can say that τ_n shows some "relevance" to X. Any random variable with this property is called a *stopping* or *Markov time*. Now, a stochastic process X is called *regenerative* if there is at least one renewal process τ with each τ_n being a stopping time relative to X and at which X regenerates. A typical example of a regenerative process is a jump Markov process with respect to the sequence τ of visits to a particular state. Another notable example is the queueing process $\{Q(t)\}$ in the $GI/G/m$ queue (with $Q(0) = 0$) with respect to the sequence τ of busy period onsets.

Combining a so-called "renewal equation" with the Key Renewal Theorem, one arrives at the following fundamental result for regenerative processes. Let X be a regenerative process with respect to the renewal process τ. Suppose that inter-renewal times have common mean $\frac{1}{\lambda}$. Let $K(t,A) = \mathbf{P}\{X(t) \in A, \tau_1 > t\}$ denote the probability of $X(t)$ to assume a value in a set A during the first inter-renewal time. Then,

$$\lim_{t \to \infty} \mathbf{P}\{X(t) \in A\} = \lambda \int_0^\infty K(u,A)du. \tag{1.3}$$

The Pollaczek-Khintchine formula. One of the most elegant and widely used formulas in queueing, due to Pollaczek and Khintchine, dates back to the thirties. It concerns the distribution of the actual waiting time process, $\{W_n\}$, in the $M/G/1/\infty$ queueing system. In particular, under the equilibrium condition, $\rho = \lambda/\mu < 1$ (where λ is the intensity of the Poisson arrival process and $1/\mu$ is the mean service time), the stationary distribution

$$V_a(x) = \lim_{n \to \infty} \mathbf{P}\{W_n \leq x\} \tag{1.4}$$

exists, and its Laplace-Stieltjes transform (LST), $\Omega(\theta)$, satisfies the formula

$$\Omega(\theta) = \frac{1-\rho}{1-\lambda(1-\beta(\theta))/\theta}, \; Re(\theta) \geq 0, \tag{1.5}$$

where $\beta(\theta)$ is the LST of the service time distribution. Formula (1.5) was obtained by Pollaczek [V,65,66] in 1930, and independently, by Khintchine [V,41] in 1932.

Kendall's embedded processes. During the period 1951-1953, David Kendall introduced in two of his papers [V,38,39], the method of *embedded Markov chains*, even now, still one of the most popular techniques in applied probability, particularly in queueing. We show how his method applies to the queueing process $\{Q(t)\}$ in the $M/G/1/\infty$ system. We define all trajectories of $\{Q(t)\}$ to be right-continuous, and introduce the sequence $\{Q_n = Q(T_n)\}$, where T_n is the time of the nth service completion. The exponentiality of interarrival times of customers is essential for $\{Q_n\}$ to be a time-homogeneous Markov chain whose TPM, P, can be expressed in an explicit form. Under the above mentioned equilibrium condition, $\rho < 1$, the chain is ergodic and its stationary probability vector, $p = (p_0, p_1, ...) > \mathbf{0}$ (vector), is the unique left (stochastic) fixed point of the matrix P, i.e. $p = pP$ with $p_0 + p_1 + ... = 1$. The former equation is equivalent to the functional equation

$$P(u) = \sum_{j \geq 0} P_j(u) p_j, \tag{1.6}$$

where $P(u)$ is the probability generating function (PGF) of p and $P_j(u)$ is the

PGF of the jth row of matrix P. Evaluating $P_j(u)$ and solving (1.6) yields the celebrated Kendall's formula

$$P(u) = p_0 \frac{(1-u)\beta(\lambda - \lambda u)}{\beta(\lambda - \lambda u) - u}, \tag{1.7}$$

where $p_0 = 1 - \rho$ is obtained from the boundary condition, $P(1) = 1$.

The Method of Supplementary Variables. The exceptional utility of the embedded processes method is due to the extensive development of semi-regenerative analysis. Semi-regenerative analysis enables one to "interpolate" values of embedded processes in the values of the original (continuous-time parameter) processes such as $\{Q(t)\}$, thereby making embedded processes a very powerful approach. While the methodology of semi-regenerative processes was introduced in the sixties and evolved during the seventies, Kendall, in 1951, [**V**,38] proposed an alternative method for finding the distribution of the original process. Kendall's idea was fully explored by D. Cox in 1955 [**V**,15] when it was dubbed the method of *supplementary variables.*

The idea in [**V**,15] was to augment the state $Q(t)$ with additional information (i.e. *supplement* it with a *variable*) to a Markov process. The supplementary variable was $U(t)$, the time elapsed from the beginning of service, so that $(Q(t), U(t))$ becomes a two-dimensional homogeneous Markov process. Using Laplace and z-transforms, the corresponding system of Kolmogorov partial differential equations is replaced by a functional equation. An advantage of this method lies in its potential to yield the transient distribution of $Q(t)$ mode in an analytically tractable form. (See, for example, Hokstad [**I**,5] and Keilson and Kooharian [**V**,35].)

Remarkably, Kendall's formula (1.7) also holds true for the original continuous-time parameter process $Q(t)$, but only for the basic $M/G/1/\infty$ system. Any modification of the system, such as one with a bulk arrival process, $BM/G/1/\infty$, splits the results for the embedded process and its continuous-time counterpart.

Takács Process. The term *Takács process* is synonymous with the *virtual waiting time process* in a FIFO (single-server) $GI/G/1$ queue. The "virtual waiting time process" was originally introduced by Takács himself [**V**,92] early in 1955. This process is defined as follows. Consider a $GI/G/1$ queueing model in which customers arrive singly at times $\tau_0 = 0$, $\tau_1, \tau_2, \ldots$, (and are processed by a single server), so that $\{\tau_n\}$ forms a renewal process. Let σ_n be the service time of the nth customer with common distribution $B(x)$ for all n, and suppose that all σ_n are mutually independent as well as $\{\tau_n\}$. Recall that the *actual waiting time*, W_n, of the nth customer is defined as the total service time of all customers in the system ahead of the nth customer (including the one currently in service) at time τ_n. The Takács process, $\{W(t)\}$, is the time needed to serve all customers present in the system at time t. Clearly, $W_n = W(\tau_n -)$.

Let $V = V(t,x) = \mathbf{P}\{W(t) \le x\}$. If the arrival process is Poisson with parameter λ (i.e. the system is $M/G/1$), then the Takács process is Markovian and V satisfies Kolmogorov's equation

$$\left(\frac{\partial}{\partial t} - \frac{\partial}{\partial x}\right) V(t,x) = -\lambda\left[V(t,x) - B * V(t,\cdot)(x)\right], \tag{1.8}$$

with the prescribed initial condition $V(0,x) = \mathbf{P}\{W(0) \le x\}$. This is known as the

"Takács integro differential equation." Here $*$ denotes the convolution operator with respect to the second variable in V. This boundary value problem has a unique solution V given explicitly by Takács in [**I**,9,10], [**II**,161] and [**IV**,66]. That solution, which depends on t (and x), is called, in the theory of differential equations, a *transient solution* as opposed to the *steady state* solution when the process reaches equilibrium. The necessary and sufficient condition for the existence of equilibrium for V is the same as for the queueing process: $\rho = \lambda/\mu < 1$. In 1955, Takács [**V**,92] showed that for the $M/G/1$ system,

$$V_a(x) = V(x) = \lim_{t\to\infty} \mathbf{P}\{W(t) \le x\}, \tag{1.9}$$

where $V_a(x)$ is defined in (1.4). For the general single-server systems, $GI/G/1$, Takács showed that, given $\rho < 1$, the steady state distribution $V(x)$ of the Takács process exists and satisfies the integral equation

$$V(x) = 1 - \frac{1}{\rho} + \frac{1}{\rho} V_a * \widetilde{B}(x), \tag{1.10}$$

where $\widetilde{B}(x) = \mu \int_0^x [1 - B(y)]dy$ and V_a satisfies Lindley's integral equation [**V**,52].

The transient distribution $V(t,x)$ of the Takács process for $GI/G/1$ was obtained by Takács in his widely cited work on fluctuations of partial sums of sequences of random variables (see, for instance, [**V**,103,104]). The solution method utilizes operator recurrence equations.

Semi-Markov and semi-regenerative processes. Semi-Markov processes (SMP) were introduced in 1954 independently by Paul Lévy [**V**,51], Walter Smith [**V**,83] and Lajos Takács [**V**,91]. The contemporary theory of SMP's was developed in the sixties and early seventies and their results and popularity are primarily due to Ronald Pyke [**V**,73] and Erhan Çinlar [**V**,10-12]. SMP's generalize Markov jump processes with countably many states by allowing holding times to be generally distributed. More precisely, given that a Markov process $X = \{X(t)\}$ is in state i, then the transition to state j occurs with probability q_{ij}, and the process holding time in state i is exponentially distributed with parameter λ_i. For an SMP, X, the holding time in state i has a general distribution indexed by i.

The theory of SMP's is essentially based on *Markov renewal theory* which is concerned with *Markov renewal processes* (MRP's). Roughly speaking, an MRP is a component of an SMP, being a chain $\{X_n = X(T_n), T_n;\ n = 0,1,...\}$ embedded in $X(t)$ such that the marginal process $\{X_n\}$ is a Markov chain. The point process $\{T_n\}$ is such that for each n, $T_{n+1} - T_n$ has a general distribution which depends on X_n only.

Formally,

$$\begin{aligned} &\mathbf{P}\{X_{n+1} = j, T_{n+1} - T_n \le t \mid X_0,...,X_n = i; T_0,...,T_n\} \\ &= \mathbf{P}\{X_{n+1} = j, T_{n+1} - T_n \le t \mid X_n = i\} = H_{ij}(t),\ i,j \in I, \end{aligned} \tag{1.11}$$

where I is a countable state space. The matrix $(H_{ij}(t))$ is called the *semi-Markov kernel* over I.

Markov renewal theory was developed in a number of papers dating back to the late fifties and early sixties, of which the 1961 paper by Ronald Pyke [**V**,73] is most widely cited. In contemporary books on stochastic processes, the notion of an SMP is based on the MRP.

The new theory gave rise to generalized queueing systems (in the mid sixties) with arrivals as an SMP by Erhan Çinlar, or with semi-Markov service by Marcel Neuts. In 1969, Çinlar wrote a pivotal survey article [**V**,10] summarizing the main results in Markov renewal theory up to that time. Not only did Markov renewal theory provide an alternative analysis of queues, but it also enabled one to analyze new queueing systems and led to the establishment of characteristics left behind from earlier studies. Especially fruitful was the merging of SMP's and regenerative processes that gave rise to *semi-regenerative processes* (SRP). These were updated by Çinlar in [**IV**,13] and [**V**,11] in 1975.

Recall that a stochastic process $X = \{X(t)\}$ is *regenerative* if there is an increasing sequence of stopping times $T = \{T_n; n = 0,1,2,...\}$ on the time axis (each tied to the history of the process $\{X(u)\colon u \le T_n\}$) forming a renewal process and such that at each T_n, X regenerates itself. That is, relative to T_n, the future process $\{X(u)\colon u \ge T_n\}$ is a probabilistic replica of itself. Conditional regeneration at T_n means that, relative to T_n, the future of X is a probabilistic replica of a version of X with initial value $X(0)$ coinciding with $X_n = X(T_n)$. In addition, $(X,T) = \{X_n, T_n; n = 0,1,...\}$ is an MRP. In other words, X is an SRP if there is an embedded MRP (X,T) such that T is a point process of stopping times at which X regenerates conditionally.

A bridge between regenerative and semi-regenerative processes was recognized. This enabled one to apply the Key Renewal Theorem to SRP's, leading to fundamental results in the theory of semi-regenerative processes and to their numerous applications to the analysis of queueing processes. Specifically, it became possible to find the limiting distribution,

$$\pi = (\pi_k = \lim_{t \to \infty} \mathbf{P}\{X(t) = k \mid X(0) = i\}; k = 0,1,...), \tag{1.12}$$

of X through the limiting distribution $p = (p_0, p_1, ...)$ of the embedded Markov chain $\{X_n\}$, and the "integrated semi-regenerative kernel" $H = \{h_{jk}\}$, defined as

$$h_{jk} = \int_0^\infty \mathbf{P}\{X(t) = k, T_1 > t \mid X(0) = j\}\, dt. \tag{1.13}$$

If $h_j(u)$ is the PGF of the jth row of the matrix H, then the PGF $\Pi(u)$ of π satisfies the formula

$$\Pi(u) = \frac{1}{p\beta} \sum_{j \ge 0} h_j(u) p_j, \tag{1.14}$$

where $\beta = (\beta_0, \beta_1, ...)$, $\beta_k = \mathbf{E}[T_1 \mid X(0) = k]$, and $p\beta$ is scalar product of p and β. For more details, see Dshalalow [**I**,4]. This approach, used in the seventies and eighties (*cf.* Dshalalow [**I**,3] and Schäl [**I**,7]), rejuvenated Kendall's embedded Markov chain analysis of queueing processes in various single-server and many-server queues.

The Theory of Fluctuations. Fluctuation theory dates back to the 1903 work of Lundberg [**V**,55] who studied it in the context of insurance risk. Further research was pursued in the thirties and forties by Cramér [**V**,16] and Täcklind [**V**,111]. Applications to queueing theory were inaugurated in the classical work of D. Lindley [**V**,52] and F. Pollaczek [**V**,67] in 1952 and F. Spitzer [**I**,8] in 1956, inspired by the 1951 work of Kendall [**V**,38]. F. Pollaczek used fluctuation theory as the first step in his analysis of queueing systems with $m \ge 1$ servers. He summarized his results in his monographs [**II**,137] in 1957 and [**II**, 138] in 1961. In a series of papers [**I**,11-15] and [**V**,103,104,105,108], Takács made the most fundamental contributions to fluctuation theory. These were based on the method of Banach algebras, developed by him in the seventies and applied to queueing, waiting time, and busy period processes in a broad class of single- and multi-server queueing systems including those with general service and semi-Markov arrivals. Takács' theory of fluctuations remains one of the most powerful methods in queueing theory. See also Dshalalow and Syski [**V**,28] for additional references.

We demonstrate the notion of fluctuations via an example concerning the actual waiting time process in the $GI/G/1/\infty/FIFO$ queueing system. Suppose that customers arrivals occur at $\tau_0, \tau_1, \tau_2, \ldots$, and form a regular renewal process. Assume that the inter-arrival time distributions are given by $\alpha(\theta) = \mathbf{E}[e^{-\theta(\tau_n - \tau_{n-1})}]$ with mean $\mathbf{E}[\tau_n - \tau_{n-1}] = \frac{1}{\lambda}$, $n = 1,2,\ldots$. Let σ_n be the service time of the nth customer, distributed according to $\beta(\theta) = \mathbf{E}[e^{-\theta\sigma_n}]$ and with mean $\frac{1}{\mu}$. Suppose W_n is the actual waiting time of the nth customer, and that $U_n(\theta) = \mathbf{E}[e^{-\theta W_n}]$, $n = 0,1,\ldots$, ($U_0(\theta)$ is supposed to be a known initial value). We seek the generating function $U(\theta,z) = \sum_{n=0}^{\infty} U_n(\theta) z^n$, for $Re(\theta) \geq 0$ and $|z| < 1$.

It is easy to show that W_n satisfies the recursive relation

$$W_{n+1} = (W_n + \xi_n)^+, n = 0,1,\ldots, \tag{1.15}$$

where

$$\xi_n = \sigma_n - (\tau_{n+1} - \tau_n) \tag{1.16}$$

and $f^+ = max\{0,f\}$. The corresponding relation for the LST can be expressed with the aid of an operator $\mathbf{T}$ defined as follows. (We adopt the formalism and notation used by Takács [**V**,103,104].)

Let $\mathbf{R}_1$ be the space of all complex-valued functions $\Phi(\theta)$ representable as

$$\Phi(\theta) = \mathbf{E}[\zeta e^{-\theta\eta}], \tag{1.17}$$

with the domain on the imaginary axis, where ζ and η are complex-valued (with finite mean) and real-valued random variables, respectively. $\mathbf{R}_1$ is a linear normed space over the field $\mathbb{C}$ (the complex plane) of scalars with the norm $\|\Phi\| = \inf_{\zeta}\{\mathbf{E}[|\zeta|]\}$. Takács showed that $\mathbf{R}_1$ is a commutative Banach algebra. The operator $\mathbf{T}$ is a linear bounded operator from $\mathbf{R}_1$ to itself, defined as

$$\mathbf{T}\Phi(\theta) = (\mathbf{T}\Phi)(\theta) = \mathbf{T}(\Phi(\theta)) = \Phi^+(\theta) = \mathbf{E}[\zeta e^{-\theta\eta^+}],\ Re(\theta) \geq 0, \tag{1.18}$$

and originally studied by Baxter [**I**,2] and Rota [**I**,6].

The LST's of (1.15,1.16) can be expressed as

$$U_{n+1}(\theta) = \mathbf{T}U_n\psi(\theta),\ n = 0,1,\ldots, \tag{1.19}$$

where $\mathbf{T}$ is applied to a Φ in (1.17) with $\zeta = 1$ and $\eta = W_{n+1}$, and $\psi(\theta)$ stands for $\alpha(-\theta)\beta(\theta)$. $\mathbf{T}$ has a fixed point at $U_0(\theta)$ which can be adjoined to (1.19) to form a boundary value problem.

Takács proved that $U(\theta,z) \in \mathbf{R}_1$ and

$$U(\theta,z) = exp\{-\mathbf{T}\Phi(\theta)\}\mathbf{T}\big(U_0(\theta)exp\{-\mathbf{T}^*\Phi(\theta)\}\big), \tag{1.20}$$

where

$$\mathbf{T}^*\Phi = \Phi - \mathbf{T}\Phi, \tag{1.21}$$

$$\Phi(\theta) = log[1 - z\psi(\theta)], \tag{1.22}$$

and

$$\psi(\theta) = \alpha(-\theta)\beta(\theta), \tag{1.23}$$

and he further obtained a simple integral representation of $\mathbf{T}\Phi$. In addition, he derived the following factorization: for $|z| \, \|\psi\| < 1$,

$$1 - z\psi(\theta) = \psi^{+}(\theta,z)\psi^{-}(\theta,z). \tag{1.24}$$

Such a factorization always exists, where the functions ψ^{+} and ψ^{-} satisfy some regularity conditions and are unique up to a factor dependent on z. Once found, ψ^{+} and ψ^{-} are related by

$$\mathbf{T}log[1 - z\psi(\theta)] = log\psi^{+}(\theta,z) + log\psi^{-}(\theta,z). \tag{1.25}$$

For $U_0(\theta) = 1$, i.e. $W_0 = 0$, formula (1.20) reduces to the famous Pollaczek formula

$$U(\theta,z) = exp\{-\mathbf{T}log[1 - z\psi(\theta)]\} \tag{1.26}$$

first revealed in 1952 [V,67]. In 1956 Spitzer [I,8] gave an alternative representation of (1.26):

$$U(\theta,z) = exp \sum_{n \geq 1} \frac{z^n}{n}\mathbf{T}\{\psi(\theta)\}^n. \tag{1.27}$$

Using similar techniques, one can derive distributions of busy period and Takács processes. In particular, the Takács process $\{W(t)\}$ satisfies the relation

$$W(t) = [W_n + \sigma_n - (t - \tau_n)]^{+}, \; t \in (\tau_n, \tau_{n+1}), \; n = 0,1,\dots . \tag{1.28}$$

The distribution of $W(t)$ is sought via iterated LST

$$\widetilde{U}(\theta,q) = \int_0^\infty \mathbf{E}[e^{-\theta W(t)}]e^{-qt}dt, \tag{1.29}$$

which can be expressed in terms of the right-hand side of formula (1.20) where $\psi(\theta)$, defined by (1.23), should be replaced by

$$\psi(\theta,q) = \alpha(q - \theta)\beta(\theta) \tag{1.30}$$

to get $U(\theta,q,z)$. This results in the relation

$$(\theta - q)\widetilde{U}(\theta,q) = \{[1 - \beta(\theta)]U(\theta,q,1) - U_0(\theta)\} - \frac{\theta}{q}\{[1 - \beta(q)]U(q,q,1) - U_0(q)\},$$
$$Re(\theta) \geq 0, \; Re(q) > 0. \tag{1.31}$$

in terms of $U(\theta,q,z)$.

In his fundamental work, not only did Takács generalize Pollaczek's results and his successors', but he has also weakened sufficient conditions of their existence and made the requisite proofs rigorous. For example, Pollaczek assumed that the LST $\psi(\theta)$ is analytic at $\theta = 0$, which, in particular, implies that all moments exist; Takács was able to drop this assumption. Whereas Pollaczek did not prove the uniqueness of the solutions of his integral equations, Takács used Banach algebras to overcome this difficulty. In addition, Takács' formulas are very elegant, easy to use and allow further generalizations.

In order to treat more general queues, such as $SM/G/1$, and to extend Pollaczek's and Spitzer's results, Takács introduced a noncommutative Banach algebra $\mathbf{R}_2$, of matrix functions

$$\Phi(\theta) = [\Phi_{ij}(\theta)]_{i,j \in I}, \; Re(\theta) = 0, \tag{1.32}$$

where $\Phi_{ij}(\theta) \in \mathbf{R}_1$ and I is a countable set. The associated algebraic operations are from the usual matrix algebra, and the norm is

$$\|\Phi\| = \sup_{i \in I} \sum_{j \in I} \|\Phi_{ij}\|_{R_1}. \tag{1.33}$$

The operators $\mathbf{T}$ and $\mathbf{T}^*$ are extended from $\mathbf{R}_1$ (defined in (1.18) and (1.21)) to $\mathbf{R}_2$, i.e., are applied to each element of $\Phi(\theta)$. Then, for any $\Phi(\theta) \in \mathbf{R}_2$ with $|z| \, \|\Phi\| < 1$, there exists a canonical factorization of the Wiener-Hopf type:

$$\mathbf{I} - z\,\Phi(\theta) = \Phi^+(\theta, z)\,\Phi^-(\theta, z), \tag{1.34}$$

where $\mathbf{I} = \{\delta_{ij}\}$, δ_{ij} being the Kronecker function.

Following Takács [V,103,104], we will demonstrate a similar method, applied to the general $SM/G/1$ model, in which both the interarrival and service times of the nth customer depend on its "type" as follows. Let $\{c_n; n = 0,1,...\}$ be a time-homogeneous Markov chain with a countable state space I. Suppose that the sequence $\{(\sigma_n, \tau_{n+1} - \tau_n, c_n)\}$ forms a Markov renewal process, i.e.,

$$\mathbf{P}\{\sigma_n \le x, \tau_{n+1} - \tau_n \le y, c_{n+1} = j \mid \sigma_0,...,\sigma_n, \tau_0,...,\tau_n, c_0,...,c_{n-1}, c_n = i\}$$
$$= \mathbf{P}\{\sigma_n \le x, \tau_{n+1} - \tau_n \le y, c_{n+1} = j \mid c_n = i\} = H_{ij}(x,y). \tag{1.35}$$

Then $H(x,y) = (H_{ij}(x,y); i,j \in I)$ is the semi-Markov kernel and

$$\Phi(\theta,\vartheta) = \int_0^\infty \int_0^\infty e^{-\theta x - \vartheta y} H(dx,dy),\ Re(\theta) \ge 0,\ Re(\vartheta) \ge 0, \tag{1.36}$$

is its LST. For $Re(\vartheta) > 0$, $Re(w) \ge 0$, $|z| \le 1$, and $Re(\theta) = 0$, there exists a canonical factorization

$$\mathbf{I} - z\Phi(\theta + w, \vartheta - \theta) = \Phi^+(\theta + w, \vartheta, z)\Phi^-(\theta + w, \vartheta, z), \tag{1.37}$$

similar to (1.34) with existing matrix inverses of Φ^+ and Φ^-. We seek a distribution of the actual waiting time process $\{W_n\}$, defined by relation (1.15). Let

$$U_{ij}(\theta, n) = \mathbf{E}[e^{-\theta W_n} I_{\{c_n = j\}} \mid c_0 = i], \tag{1.38}$$

where I_A is the indicator function of a set A, and let

$$U_n(\theta) = [U_{ij}(\theta, n)]_{i,j \in I},\ \text{for } Re(\theta) \ge 0,\ n = 0,1,... . \tag{1.39}$$

By (1.15), $U_{ij}(\theta,n)$ satisfies the recurrence formula

$$U_{ij}(\theta, n+1) = \sum_{k \in I} \mathbf{T} U_{ik}(\theta,n)\Phi_{kj}(\theta, -\theta), \tag{1.40}$$

which in the matrix form can be compactly written as

$$U_{n+1}(\theta) = \mathbf{T} U_n \Phi(\theta, -\theta),\ n = 0,1,... . \tag{1.41}$$

Clearly, $\Phi(\theta, -\theta), U_0(\theta) \in \mathbf{R}_2$, $\|\Phi(\theta, -\theta)\| \le 1$, and $\mathbf{T}U_0(\theta) = U_0(\theta)$. Thus, (1.40) and (1.41) yield

$$U(\theta,z) = \mathbf{T}\{U_0(\theta)[\Phi^-(\theta,0,z)]^{-1}\}[\Phi^+(\theta,0,z)]^{-1},\ Re(\theta) \ge 0,\ |z| < 1, \tag{1.42}$$

where $U(\theta,z) = \sum_{n=0}^\infty U_n(\theta) z^n$, and $[\Phi^-(\theta,0,z)]^{-1}$ and $[\Phi^+(\theta,0,z)]^{-1}$ are matrix inverses which can be derived from (1.37).

Similarly, one can find the distribution of the Takács process, busy period process, and queueing process. For more details, see Takács [V,103,104].

Formula $L = \lambda W$. This formula, originally stated by P. Morse [II,120] in 1958, carries one of the most fundamental laws in queueing. According to this law, the mean queue length, L, equals the product of the arrival intensity λ and the average sojourn time, W spent by each customer in the system (i.e., waiting plus service time). This result is valid under very general assumptions on queueing, service and busy period disciplines; it also applies to stochastic systems other than queues.

Morse used this result for some particular queueing models, but conjectured that it should hold true for more general models. The first rigorous proof of $L = \lambda W$ was given by J.D.C Little [**V**,53] in 1961, and, for this reason, $L = \lambda W$ is widely referred to as *Little's formula*. However, even Little's version of $L = \lambda W$ was quite restrictive and its present generality is due to his many successors of which the most notable is Stidham [**V**,88,89]. A comprehensive survey on $L = \lambda W$ and its extensions can be found in Whitt [**V**,112] dated 1991.

1.4 CHAPTERS 2 THROUGH 19

The remainder of this book, Chapters 2 through 19, brings the reader into contact with the most significant methodology in the theory of queues. It is rather difficult to classify all up-to-date methods since they can often be assigned to more than one category. Nevertheless, for each of the following eighteen chapters, one particular category prevailed, thereby leading to the book's natural division into three parts: (I) *Stochastic Methods*, (II) *Analytic Methods*, and (III) *Approximation, Estimates, and Simulation*. A large variety of queueing models are used as vehicles to illustrate a broad range of methods. However, the book does not have the conventional structure of most monographs and textbooks on queueing which are largely organized by queueing models.

Each presentation of a method, or methods, is followed by a discussion of significant open problems and future research directions. The authors hope that this will make it easier for readers to focus their efforts and interest in the respective areas of probability and analysis. Each chapter also contains a large, up-to-date bibliography on pertinent subjects.

ACKNOWLEDGEMENT. I am deeply grateful to Benjamin Melamed who thoroughly read the final draft of this chapter and made numerous improvements throughout. I am very thankful to Jay Yellen, for his careful proofreading and valuable suggestions, and to Hideaki Takagi for kindly faxing me his pivotal articles with comprehensive bibliographies on queueing. Thanks are also due to Gary Russell and Lotfi Tadj for proofreading the manuscript in an earlier version and bibliographic search.

BIBLIOGRAPHY

I. Chapter 1 Special References

[1] Abolnikov, L. and Dukhovny, A., Markov chains with transition delta-matrix: ergodicity conditions, invariant probability measures and applications, *J. Appl. Math. Stoch. Anal.*, **4**, No. 4 (1991), 335-355.

[2] Baxter G., On operator identity, *Pacific J. Math.*, **8** (1958), 649-663.

[3] Dshalalow, J.H., On the multiserver queue with finite waiting room and controlled input, *Adv. Appl. Prob.*, **17** (1985), 408-423.

[4] Dshalalow, J.H., First excess level analysis of random processes in a class of stochastic servicing systems with global control, *Stoch. Anal. Appl.*, **12**:1

(1994), 75-101.

[5] Hokstad, P., A supplementary variable techniques applied to the $M/G/1$ queue, *Scand. J. Stat.*, **2** (1975), 95-98.

[6] Rota, G.-C., Baxter algebras and combinatorial identities. I-II, *Bull. Amer. Math. Soc.*, **75** (1969), 325-364.

[7] Schäl, M., The analysis of queues with state-dependent parameters by Markov renewal processes, *Adv. Appl. Prob.*, **3** (1971), 155-175.

[8] Spitzer, F., A combinatorial lemma and its application to probability theory, *Trans. Amer. Math. Soc.*, **82** (1956), 323-339.

[9] Takács, L., The time dependence of a single-server queue with Poisson input and general service times, *Ann. Math. Stat.*, **33** (1962), 1340-1348.

[10] Takács, L., The distribution of the virtual waiting time for a single-server queue with Poisson input and general service times, *Oper. Res.*, **11** (1963), 261-264.

[11] Takács, L., On a linear transformation in the theory of probability, *Acta Sci. Math.*, **33**, No. 1-2 (1972), 15-24.

[12] Takács, L., On some recurrence equations in a Banach algebra, *Acta Sci. Math. (Szeged)*, **38** (1976), 399-416.

[13] Takács, L., A Banach space of matrix functions and its application in the theory of queues, *Sankhyā*, Ser. A, **38** (1976), 201-211.

[14] Takács, L., On the ordered partial sums of real random variables, *J. Appl. Prob.*, **14** (1977), 75-88.

[15] Takács, L., An identity for ordered partial sums, *J. Comb. Theory*, Ser. A, **23** (1977), 364-365.

II. The Bibliography of Books on Queueing Theory

[1] Agrawal, S.C., *Metamodeling: A Study of Approximations in Queueing Models*, MIT Press, Cambridge, MA 1985.

[2] Allen, A.O., *Probability, Statistics, and Queueing Theory, with Computer Science Applications*, 2nd ed., Academic Press, New York 1990.

[3] Andrejev, V.D. and Kudyavtsev, B.M., *Markov Queueing Systems*, Moscow Institute of Control, 1982 (in Russian).

[4] Anisimov, V.V., Zakusilo, O.K., and Donchenko, V.S., *Elements of Queueing Theory and Asymptotic System Analysis*, Vishcha Shkola, Kiev 1987 (in Russian).

[5] Asmussen, S., *Applied Probability and Queues*, John Wiley, Chichester 1987.

[6] Bacelli, F. and Brémaud, P., *Mathematical Theory of Queues*, Springer-Verlag, Berlin 1990.

[7] Bacelli, F. and Brémaud, P., *Palm Probabilities and Stationary Queues*, Lecture Notes in Statistics, **41**, Springer-Verlag, Berlin 1991.

[8] Bacelli, F. and Brémaud, P., *Elements of Queueing Theory: Palm-Martingale Calculus and Stochastic Recurrence*, Springer-Verlag, New York 1994.

[9] Bagchi, T.P. and Templeton, J.G.C., *Numerical Methods in Markov Chains and Bulk Queues*, Springer-Verlag, Berlin 1972.

[10] Bártfai, P. and Tomkó, J. (eds.), *Point Processes and Queueing Problems*, (Colloq. Math. János Bolyai, **24**, Kesztheley, 1978), North-Holland, Amsterdam 1981.

[11] Basharin, G.P., Bocharov, P.P., and Kogan, Y.A., *Queueing Analysis for*

Computer Networks: Theory and Computational Methods, Nauka, Moscow 1989 (in Russian).

[12] Basharin, G.P., Charkievich, A.D., and Shneps, M.A., *Queueing Theory in Teletraffic*, Nauka, Moscow 1968 (in Russian).

[13] Beckman, P., *Introduction to Elementary Queueing Theory and Telephone Traffic*, Golem Press, Boulder 1968.

[14] Beneš, V.A., *General Stochastic Processes in the Theory of Queues*, Addison-Wesley, Reading, MA 1963.

[15] Bhat, U.N., *A Study of the Queueing Systems* M/G/1 *and* GI/M/1, Springer-Verlag, Berlin 1968.

[16] Bhat, U.N. and Basawa, I.W. (ed.), *Queueing and Related Models*, Clarendon Press, Oxford 1992.

[17] Blanc, J.P.C., *Application of the Theory of Boundary Value Problems in the Analysis of a Queueing Model with Paired Services*, Mathematics Centrum, Amsterdam 1982.

[18] Bocharov, P.P., *Single Server Servicing Systems with Finite Waiting Room*, Friendship University Publishers, Moscow 1985 (in Russian).

[19] Borovkov, A.A., *Stochastic Processes in Queueing Theory*, Springer-Verlag, Berlin 1976.

[20] Borovkov, A.A., *Asymptotic Methods in Queueing Theory*, John Wiley, New York 1984.

[21] Boxma, O.J. and Syski, R. (eds.), *Queueing Theory and its Applications - Liber Amicorum for J.W. Cohen*, Elsevier/North-Holland, Amsterdam 1989.

[22] Brémaud, P., *Point Processes and Queues: Martingale Dynamics*, Springer-Verlag, New York 1981.

[23] Brodi, S.M. and Pogosyan, I.A., *Embedded Stochastic Processes in Queueing Theory*, Naukova Dumka, Kiev 1973 (in Russian).

[24] Bronshtein, O.I. and Dukhovny, I.M., *Priority Queues in Information Networks*, Nauka, Moscow 1976 (in Russian).

[25] Bruell, S.C. and Balbo, G., *Computational Algorithms for Closed Queueing Networks*, North-Holland/Elsevier, Amsterdam 1980.

[26] Bunday, B.D., *Basic Queueing Theory*, Edward Arnold, London 1986.

[27] Burikov, A.D., Malinkovski, Y.V., and Matalytskyi, M.A., *Queueing Theory*, Grodno State University, 1984 (in Russian).

[28] Cao, X.-R., *Realization Probabilities: The Dynamics of Queueing Systems*, Lecture Notes, **194**, Springer-Verlag, New York 1994.

[29] Carmichael, D.G., *Engineering Queues in Construction and Mining*, Ellis Horwood, Chichester 1987.

[30] Chaudhry, M.L., and Templeton, J.G.C., *A First Course in Bulk Queues*, John Wiley, New York 1983.

[31] Cheprasov, V.P., *Elements of Queueing Theory*, Kazan Aviation Institute, 1985 (in Russian).

[32] Chernetskyi, V.I. (ed.), *Stochastic Modeling of Queueing Systems and Networks*, Petrozavodsk State University, 1988 (in Russian).

[33] Chernov, V.P., *Queueing Theory*, Leningrad Financial Economic Institute, 1977 (in Russian).

[34] Chernov, V.P., *Markov Serving Processes*, Leningrad Financial Economic Institute, 1981 (in Russian).

[35] Clarke, A.B. (ed.), *Mathematical Models in Queueing Theory*, Springer-Verlag, Berlin 1974.

[36] Cohen, J.W., *On Regenerative Processes in Queueing Theory*, Lecture Notes, **121**, Springer-Verlag, New York 1976.

[37] Cohen, J.W., *The Single Server Queue*, 2nd ed., North-Holland, Amsterdam 1982.

[38] Cohen, J.W. and Boxma, O.J., *Boundary Value Problems in Queueing System Analysis*, North-Holland, Amsterdam 1983.

[39] Conolly, B., *Queueing Systems*, Ellis Horwood, Chichester 1975.

[40] Conway, A.E. and Georganas, N.D., *Queueing Networks – Exact Computational Algorithms*, MIT Press, Cambridge 1989.

[41] Cooper, R.B., *Introduction to Queueing Theory*, CEEPress, The George Washington University, Washington, D.C. 1990.

[42] Courtois, P.J., *Decomposability: Queueing and Computer System Applications*, Academic Press, New York 1977.

[43] Cox, D.R. and Smith, W.L., *Queues*, Methuen, London 1961.

[44] Cruon, R. (ed.), *Queueing Theory, Recent Developments and Applications*, Elsevier, New York 1967.

[45] Daigle, J.N., *Queueing Theory for Telecommunications*, Addison-Wesley, Reading, MA, 1991.

[46] Danielyan, E.A., *One-Channel Priority Queueing Systems*, Moscow State University, 1969 (in Russian).

[47] Danielyan, E.A., *Priority Problems for Queueing Systems with One Server*, Moscow State University, 1971 (in Russian).

[48] Denyiseva, O.M., *Queueing Systems with Limited Waiting Times*, Radio Svyaz, Moscow 1986 (in Russian).

[49] Descloux, A., *Delay Tables for Finite- and Infinite-Source Systems*, McGraw-Hill, New York 1962.

[50] Disney, R.L. and Kessler, P.C., *Traffic Processes in Queueing Networks: A Markov Renewal Approach*, John Hopkins University Press, Baltimore 1987.

[51] Dobrushin, R.L., Kelbert, M.Ya., Rybko, A.N. and Sukhov, Yu.M., *Qualitative Methods of Queueing Network Theory*, Inst. Problem Peredachi Inform. Acad. Nauk SSSR, Moscow 1986 (in Russian).

[52] Fdida, S. and Pujolle, G., *Modèles de Systèmes et de Réseaux, Tome 2: Files d'Attente*, Eyrolles, Paris 1989.

[53] Fedosejev, Y.N., *Methods of Queueing Systems Analysis*, Moscow Institute of Engineering and Physics, 1982 (in Russian).

[54] Ferschl, F., *Zufallsabhängige Wirtschaftsprozesse. Grundlagen und Anwendungen der Theorie der Wartesysteme* (Ger.), Vienna 1964.

[55] Franken, P., König, D., Arndt, U. and Schmidt, V., *Queues and Point Processes*, Akademie-Verlag, Berlin 1981.

[56] Gani, J. (ed.), *The Craft of Probabilistic Modeling, A Collection of Personal Accounts*, Springer-Verlag, New York 1986.

[57] Gelenbe, E. and Pujolle, G., *Introduction to Queueing Networks*, John Wiley, Chichester 1987.

[58] Ghosal, A., *Some Aspects of Queueing and Storage Systems*, Springer-Verlag, Berlin 1970.

[59] Giffin, W.C., *Queueing: Basic Theory and Applications*, Grid, Columbus 1978.

[60] Gnedenko, B.V. and Kalashnikov, V.V. (eds.), *Queueing Theory. Proceedings of All-Union Seminar*, Institute of System Research, Moscow 1981 (in Russian).

[61] Gnedenko, B.V. and König, D., *Handbuch der Bedienungstheorie: Grundlagen und Methoden (Handbook of Queueing Theory: Foundations and Methods)* (Ger.), Vol. **1**, Akademie-Verlag, Berlin 1983.

[62] Gnedenko, B.V. and König, D., *Handbuch der Bedienungstheorie: Formeln und andere Ergebnisse (Handbook of Queueing Theory: Formulas and Other Results)* (Ger.), Vol **2**, Akademie-Verlag, Berlin 1984..

[63] Gnedenko, B.V. and Kovalenko, I.N., *Introduction to Queueing Theory*, 2nd ed., Revised and Supplemented, Birkhäuser, Boston 1989.

[64] Gnedenko, B.V., Danielyan, E.A., Dimitrov, B.N., Klimov, G.P., and Matvejev, V.F., *Priority Queueing Systems*, Moscow State University, 1973 (in Russian).

[65] Göbel, F., *Queueing Models Involving Buffers*, Math. Centre Tracts, **60**, Mathematisch Centrum, Amsterdam 1976.

[66] Gorncy, L., *Queueing Theory: A Problem Solving Approach*, Petrocelli Books, New York 1981.

[67] Gross, D. and Harris, C., *Fundamentals of Queueing Theory*, 2nd Edition, John Wiley, New York 1985.

[68] Hillier, F.S. and Yu, O.S., *Queueing Tables and Graphs*, Elsevier/North-Holland, Amsterdam 1981.

[69] Homma, T., *Theory of Queues*, Rikogaku-sha, Tokyo 1966 (in Japanese).

[70] Iglehart, D.L. and Shedler, G.S., *Regenerative Simulation of Response Times in Networks of Queues*, Springer-Verlag, New York 1976.

[71] Ivanov, G.A., *A Queue Length of Priority Queueing Systems in Nonstationary Mode*, Moscow State University, 1976 (in Russian).

[72] Ivchencko, G.I., Kashtanov, V.A., and Kovalenko, I.N., *Queueing Theory*, Vyshtshaya Shkola, Moscow 1982 (in Russian).

[73] Ivnitskyi, V.A. and Kasumov, A.B., *Queueing Systems with Nonreliable Server and Dependent Parameters*, Elm, Baku 1986 (in Russian).

[74] Jaiswal, N.K., *Priority Queues*, Academic Press, New York 1968.

[75] Kalashnikov, V.V., *Mathematical Methods in Queueing Theory*, Kluwer Academic Publishers, Dordrecht 1994.

[76] Kalashnikov, V.V. and Rachev, S., *Mathematical Methods for Construction of Queueing Models*, Wadsworth & Brooks/Cole, Pacific Grove, CA 1990.

[77] Kalinina, V.N. and Malychin, V.I., *Computer Solving of Queueing Theory Problems*, Moscow Institute of Control, 1981 (in Russian).

[78] Karpelevich, F.I., *Elements of Markov Chains and Queueing Theory and Their Application to Transport Problems, Part 1 and Part 2*, Moscow Institute of Railway Engineering, 1981 (in Russian).

[79] Karpelevich, F.I. and Kreinin, A., *Heavy Traffic Limits for Multiphase Queues*, American Mathematical Society, 1994.

[80] Kashyap, B.R.K. and Chaudhry, M.L., *An Introduction to Queueing Theory*, Aarkay, Calcutta, 1988.

[81] Kaufmann, A. and Cruon, R., *Les Phénomènes d'Attente*, Dunod, Paris 1961.

[82] Khintchin, A.Y., *Mathematical Methods in the Theory of Queueing*, Charles Griffin, London 1960.

[83] Khintchin, A.Y., *Papers on Mathematical Queueing Theory*, Fizmatgiz, Moscow 1963 (in Russian).

[84] Kingman, J.F.C., *On the Algebra of Queues*, Methuen, London 1966.

[85] Kleinrock, L., *Queueing Systems, Volume I: Theory*, John Wiley, New York

1975.
[86] Kleinrock, L., *Queueing Systems, Volume* II: *Computer Applications*, John Wiley, New York 1976.
[87] Kleinrock, L. and Gail, R., *Solutions Manual for Queueing Systems, Volume I: Theory*, Technology Transfer Institute, Santa Monica, CA 1982.
[88] Kleinrock, L. and Gail, R., *Solutions Manual for Queueing Systems, Volume II: Computer Applications*, Technology Transfer Institute, Santa Monica, CA 1986.
[89] Klimov, G.P, *Stochastic Queueing Systems*, Nauka, Moscow 1966 (in Russian).
[90] Klimov, G.P., *Bedienungsprozesse* (*Queueing Processes*) (Ger.), Birkhäuser Verlag, Basel and Stuttgart 1979.
[91] Klimov, G.P. and Mishkoy, G.K., *Priority Queueing Systems with Orientation*, Moscow State University, 1979 (in Russian).
[92] Kocharyan, T.V. and Kudryavtsev, B.M., *Queueing Theory*, Moscow Institute of Control, 1984 (in Russian).
[93] König, D. and Stoyan, D., *Methoden der Bedienungstheorie* (*Methods of Queueing Theory*) (Ger.), Vieweg, Braunschweig 1976.
[94] König, D., Matthes, K., and Nawrotzki, K., *Verallgemeinerungen der Erlangschen und Engsetschen Formeln*, Academie-Verlag, Berlin 1976.
[95] König, D., Rykov, V., and Stoyan, D., *Queueing Models* (*Markov Models and Markovization Methods*), Moscow Institute of Petrochemical and Gas Industry, 1980 (in Russian).
[96] Konyuchovskyi, V.V., *Stochastic Processes and Certain Problems in Queueing Theory*, Moscow State University, 1968 (in Russian).
[97] Kosten, L., *Stochastic Theory of Service Systems*, Pergamon Press, Oxford University Press, Oxford 1973.
[98] Kozniewska, I. and Wlodarczyk, M.W., *Models of Renewal, Reliability and Queueing Systems*, PWN, Warszawa 1978 (in Polish).
[99] Krampe, H., Kubat, J., and Runge, W., *Queueing Systems*, Verlag Wirtschaft 1974 (in German).
[100] Kuehn, P., *Tables on Delay Systems*, Institute of Switching and Data Technics, University of Stuttgart 1976.
[101] Kunizawa, K. and Homma, T. (eds.), *Encyclopedia of Queueing Theory and Its Applications*, Hirokawa-Shoten, Tokyo 1971 (in Japanese).
[102] Lavowska, E.D., Zahorjan, J., Graham, G.S., and Sevcik, K.C., *Quantitative System Performance: Computer System Analysis Using Queueing Network Models*, Prentice-Hall, Englewood Cliffs, NJ 1984.
[103] Lee, A.M., *Applied Queueing Theory*, MacMillan, London 1966.
[104] LeGall, P., *Les Systèmes avec ou sans Attente et les Processus Stochastiques* (*Systems with or without Waiting and Stochastic Processes*) (Fr.), Dunod, Paris 1962.
[105] Lifshits, A.L. and Malts, E.A., *Statistical Models of Queueing Systems*, Sovetskoje Radio, Moscow 1978 (in Russian).
[106] Lipsky, L., *Queueing Theory: A Linear Algebraic Approach*, Macmillan Publishing Company, New York 1992.
[107] Livshits, B.S. and Fidlin, L.V., *Queueing Systems with Finite Number of Sources*, Svyaz, Moscow 1968 (in Russian).
[108] Lukin, A.I., *Queueing Systems. Analysis of Queueing Systems in Military Practice*, Voenizdat, Moscow 1980 (in Russian).

[109] Makino, T., *Applications of Queues*, Morikita Shuppan, Tokyo 1969 (in Japanese).

[110] Matvejev, V.F. and Ushakov, V.G., *Queueing Systems*, Moscow State University 1984 (in Russian).

[111] Medhi, J., *Recent Developments in Bulk Queueing Models*, Wiley Eastern, New Delhi 1984.

[112] Medhi, J., *Stochastic Models in Queueing Theory*, Academic Press, Boston 1991.

[113] Meyer, K.H.F., *Queueing Systems with Variable Number of Channels*, Lecture Notes in Economics and Mathematical Systems, **61**, Springer-Verlag, Berlin 1971 (in German).

[114] Meyer, C.D. and Plemmeons, R.J. (eds.), *Linear Algebra, Markov Chains, and Queueing Models*, Springer-Verlag, New York 1993.

[115] Mihoc, G.H., *Introducere in Teoria Asteptarii*, Editura Technica, Bucuresti, Romania 1967 (in Romanian).

[116] Mihoc, G.H., Ciucu, G., and Muja, A., *Modele Matematice ale Asteptarii* (*Mathematical Queueing Models*), Editura Academiei Republicii Socialiste Romania, Bucuresti, Romania 1973 (in Romanian).

[117] Mishkoy, G.P., *State Probabilities of Priority Queues in Nonstationary Case*, Shtiintsa, Kishinev 1979 (in Russian).

[118] Miyawaki, K., Nagaoka, T., and Mouri, E., *Queueing Theory and Its Applications*, Nikkan Kogyo Shimbun, Tokyo 1961 (in Japanese).

[119] Morimura, H. and Ohmae, Y., *Applied Queueing Theory*, Nikka Giren, Tokyo 1975 (in Japanese).

[120] Morse, P.M., *Queues, Inventories and Maintenance*, John Wiley, New York 1958.

[121] Murdoch, J., *Queueing Theory: Worked Examples and Problems*, The MacMillan Free Press Ltd., London 1978.

[122] Nazarov, L.V., *Queueing Systems with Spatially Distributed Customers, Part 1 and Part 2*, Moscow State University, 1984 (in Russian).

[123] Nelson, R., *Probability, Stochastic Processes, and Queueing Theory*, Springer-Verlag, New York 1995.

[124] Neuts, M.F., *Matrix-Geometric Solutions in Stochastic Models: An Algorithmic Approach*, The Johns Hopkins University Press, Baltimore 1981.

[125] Neuts, M.F., *Structured Stochastic Matrices of* M/G/1 *Type and their Applications*, Marcel Dekker, New York 1989.

[126] Newell, G.F., *Approximate Stochastic Behavior of n-Server Service Systems with Large n*, Springer-Verlag, Berlin 1973.

[127] Newell, G.F., *Approximate Behavior of Tandem Queues*, Springer-Verlag, Berlin 1979.

[128] Newell, G.F., *Applications of Queueing Theory*, Chapman and Hall, London 1982.

[129] Newell, G.F., *The* $M/M/\infty$ *Service System with Ranked Servers in Heavy Traffic*, Springer-Verlag, New York 1984.

[130] Nishida, T., *Queueing Theory and Applications*, Asakura Shoten, Tokyo 1971 (in Japanese).

[131] Page, E., *Queueing Theory in OR*, Crane Russek & Co., New York 1972.

[132] Panico, J.A., *Queueing Theory*, Prentice-Hall, Englewood Cliffs, NJ 1969.

[133] Papadopoulus, H.T., Heavey, C., and Browne, J., *Queueing Theory in Manufacturing System Analysis and Design*, Chapman and Hall, London 1993.

[134] Peck, L.G. and Hazelwood, R.N., *Finite Queueing Tables*, John Wiley, New York 1958.

[135] Perros, H., *Queueing Networks with Blocking*, Oxford University Press, Oxford 1994.

[136] Perros, H.G. and Altiok, T. (eds.), *Queueing Networks with Blocking*, Elsevier/North-Holland, Amsterdam 1989.

[137] Pollaczek, F., *Problèmes Stochastiques posés par le Phénomène de Formation d'une Queue d'Attente à un Guichet et par des Phénomènes Apparentés (Stochastic Problems due to the Formation Phenomenon of a Waiting Line at a Pay-Desk and to Related Phenomena)* (Fr.), Gauthier-Villars, Paris 1957.

[138] Pollaczek, F., *Théorie Analytique des Problèmes Stochastiques Relatifs à un Groupe de Lignes Téléphoniques avec Dispositif d'Attente (Analytic Theory of Stochastic Problems Related to a Group of Telephone Lines with Waiting Device)* (Fr.), Gauthier-Villars, Paris 1961.

[139] Prabhu, N.U., *Queues and Inventories: A Study of Their Basic Stochastic Processes*, John Wiley, New York 1965.

[140] Prabhu, N.U., *Stochastic Storage Processes: Queues, Insurance Risk and Dams*, Springer-Verlag, New York 1980.

[141] Riordan, J., *Stochastic Service Systems*, John Wiley, New York 1962.

[142] Robertazzi, T., *Computer Networks and Systems: Queueing Theory and Performance Evaluation*, Springer-Verlag, New York 1990.

[143] Rubinstein, R.Y., *Monte Carlo Optimization, Simulation and Sensitivity of Queueing Networks*, John Wiley, New York 1986.

[144] Ruiz-Palá, E., Ávila-Beloso, C., and Hines, W.W., *Waiting Line Models: An Introduction to their Theory and Application*, Reinhold, New York 1967.

[145] Saaty, T.L., *Elements of Queueing Theory*, Dover, New York 1983.

[146] Schassberger, R., *Warteschlangen (Queues)* (Ger.), Springer-Verlag, Vienna 1973.

[147] Schmidt, G., *Über die Stabilität des einfachen Bedienungskanals*, Lecture Notes in Economics and Mathematical Systems **97**, Springer-Verlag, Berlin 1974.

[148] Schwartz, B., *Queueing and Waiting: Studies in the Social Organization of Access and Delay*, University of Chicago Press, Chicago 1975.

[149] Seaman, P.H., *Analysis of Some Queueing Models in Real-Time Systems*, IBM Manual, GF20-000701, 1971.

[150] Seelen, L.P., Tijms, H.C., and Van Hoorn, M.H., *Tables for Multiserver Queues*, Elsevier/North-Holland, Amsterdam 1985.

[151] Shedler, G.S., *Regeneration and Networks of Queues*, Springer-Verlag, New York 1987.

[152] Shelegiya, R.S., *Some Models of Queueing Theory and Operations Research*, Tbilisi State University, 1983 (in Russian).

[153] Skitovich, V.P., *Elements of Queueing Theory*, Leningrad University Publishers 1976 (in Russian).

[154] Solomon, S., *Simulation of Waiting-Line Systems*, Prentice-Hall, Englewood Cliffs, NJ 1983.

[155] Srivastava, H.M. and Kashyap, B.R.K., *Special Functions in Queueing Theory and Related Stochastic Processes*, Academic Press, New York 1983.

[156] Stepanov, S.N., *Numerical Methods for Systems with Repeated Calls*, Nauka, Moscow 1983 (in Russian).

[157] Stoyan, D., *Comparison Methods for Queues and Other Stochastic Models*, English Translation Edited by D.J. Daley, John Wiley, New York 1983.

[158] Suzuki, T., *Table of the Erlang Loss Formula*, Telefonaktiebolaget L.M. Ericsson, Stockholm 1954.

[159] Suzuki, T., *Queues*, Shokabo, Tokyo 1972 (in Japanese).

[160] Szkitovich, V.P., *Elements of Queueing Theory*, Leningrad State University, 1976 (in Russian).

[161] Takács, L., *Introduction to the Theory of Queues*, Oxford University Press, New York 1962.

[162] Takagi, H., *Queueing Theory, Vol. I*, IBM Manual Japan, Ltd., Tokyo 1977 (in Japanese).

[163] Takagi, H., *Queueing Theory, Vol. II*, IBM Manual Japan, Ltd., Tokyo 1977 (in Japanese).

[164] Takagi, H., *Queueing Theory, Vol. III*, IBM Manual Japan, Ltd., Tokyo 1978 (in Japanese).

[165] Takagi, H., *Queueing Analysis. Vol. 3: Discrete-Time Systems*, North-Holland Amsterdam 1993.

[166] Teghem, J.L-T. and Lambotte, J.P., *Modèls d'Attente* M/G/1 *et* GI/M/1 *à Arrivées et Services en Groupes* (*Waiting Models* M/G/1 *and* GI/M/1 *with Bulk Arrival and Service*) (Fr.), Springer-Verlag, Berlin 1969.

[167] Tichonenko, O.M., *Queueing Theory Models in Information Processing Systems*, Znanie, Moscow 1990 (in Russian).

[168] Tilt, B., *Solutions Manual for Robert B. Cooper's Introduction to Queueing Theory*, 2nd ed., North-Holland, New York 1981.

[169] Trivedi, K.S., *Probability and Statistics with Reliability, Queueing, and Computer Science Applications*, Prentice-Hall, Englewood Cliffs, NJ 1982.

[170] Tsaregradskyi, I.P., *Mathematics and Practice* (*Applied Problems of Queueing Theory*), Znanie, Moscow 1978 (in Russian).

[171] van den Berg, J.L, *Sojourn Times in Feedback and Processor Sharing Queues*, Centrum voor Wiskunde en Informatica, Amsterdam 1993.

[172] van Doorn, E., *Stochastic Monotonicity and Queueing Applications of Birth-Death Processes*, Lecture Notes in Statistics **4**, Springer-Verlag, New York 1981.

[173] van Hoorn, M.H., *Algorithms and Approximations for Queueing Systems*, CWI Tracts, **8**, Centre for Mathematics and Computer Science, Amsterdam 1984.

[174] Volkovinskyi, M.I. and Kabalevskyi, A.N., *Analysis of Priority Queueing with Warm-up*, Energoizdat, Moscow 1981 (in Russian).

[175] Walrand, J., *An Introduction to Queueing Networks*, Prentice-Hall, Englewood Cliffs, NJ 1988.

[176] White, J.A., Schmidt, J.W., and Bennett, G.K., *Analysis of Queueing Systems*, Academic Press, New York 1975.

[177] Wolff, R.W., *Stochastic Modeling and the Theory of Queues*, Prentice-Hall, Englewood Cliffs, NJ 1989.

[178] Yashkov, S.F., *Queueing Analysis for Computers*, Radio E Svayaz', Moscow 1989 (in Russian).

[179] Zacharov, G.P. and Varakosin, N.P., *Calculation of the Number of Channels for Queueing Systems*, Svyaz, Moscow 1967 (in Russian).

[180] Zimmermann, G.O. and Störmer, H., *Wartezeiten in Nachrichtenvertmittlungen mit Speichern*, R. Oldenbourg, München 1961.

[181] Zitek, F., *Elements of Queueing Systems Theory*, Academia, Prague 1960 (in Czech).

III. Books on Performance Evaluation of Computer Systems, Communication Networks and Teletraffic Engineering.

[1] Abramson, N. (ed.), *Multiple Access Communications*, IEEE Press, New York 1992.

[2] Abramson, N. and Kuo, F. (eds.), *Computer-Communication Networks*, Prentice-Hall, Englewood Cliffs, NJ 1973.

[3] Abu El Ata, N. (ed.), *Modeling Techniques and Tools for Performance Analysis '85*, Elsevier/North-Holland, Amsterdam 1986.

[4] Agrawal, A., *Analysis of Cache Performance for Operating Systems and Multiprogramming*, Kluwer Academic, Boston 1989.

[5] Agrawala, A.K. and Tripathi, S.K. (eds.), *Performance '83*, North-Holland, Amsterdam 1983.

[6] Ahuja, V., *Design and Analysis of Computer Communication Networks*, McGraw-Hill, New York 1982.

[7] Ajmone Marsan, M., Balbo, G., and Conte, G., *Performance Models of Multiprocessor Systems*, MIT Press, Cambridge, MA 1986.

[8] Akimaru, H. and Cooper, R.B., *Teletraffic Engineering*, Ohm-sha, Tokyo 1985 (in Japanese).

[9] Akimaru, H. and Kawashima, K., *Information Network Traffic: Basics and Applications*, Ohm-sha, Tokyo 1990 (in Japanese).

[10] Akiyama, M. (ed.), *Teletraffic Issues in an Advanced Information Society: ITC-II*, Elsevier/North-Holland, Amsterdam 1985.

[11] Akiyama, M., Kawashima, K., and Kimura, G., *System Design of LAN*, Ohm-sha, Tokyo 1989 (in Japanese).

[12] Alyanah, I.N., *Modeling of Computer Systems*, Mashinostroyenie, Leningrad 1988 (in Russian).

[13] Arató, M. and Farga, L. (eds.), *Mathematical Models in Computer Systems*, Collets, United Kingdom 1981.

[14] Arató, M. and Varga, L. (eds.), *Mathematical Models in Computer Systems*, Académia Press, Budapest, Hungary 1982.

[15] Arató, M., Butrimenko, A., and Gelenbe, E. (eds.), *Performance of Computer Systems*, Elsevier/North-Holland, Amsterdam 1979.

[16] Arató, M., Kótai, I., and Varga, L. (eds.), *Topics in the Theoretical Bases and Applications of Computer Science*, Académia Press, Budapest, Hungary 1985.

[17] Artamonov, G.T. and Brehov, O.M., *Analytical Probability Models of Computer Performance*, Energiya, Moscow 1978 (in Russian).

[18] Artamonov, G.T. and Brehov, O.M., *Analytic Stochastic Models for Computer Systems*, Energiya, Moscow 1987 (in Russian).

[19] Ashton, W.D., *The Theory of Road Traffic Flow*, Methuen, London 1966.

[20] Aven, O.I. and Kogan, Y.A., *Computer Resource Allocation: Algorithms and Models*, Energiya, Moscow 1978 (in Russian).

[21] Aven, O.I., Coffman, E.G., Jr., and Kogan, Y.A., *Stochastic Analysis of Computer Storage*, Reidel, Dordrecht, The Netherlands 1987.

[22] Aven, O.I., Gurin, N.N., and Kogan, Y.A., *Performance Evaluation and Optimization of Computer Systems*, Nauka, Moscow 1982 (in Russian).

[23] Bacelli, F. and Fayolle, G. (eds.), *Modeling and Performance Evaluation Methodology*, Lecture Notes in Control and Information Sciences, **60**, Springer-Verlag, Berlin 1983.

[24] Balbo, G. and Serazzi, G. (eds.), *Computer Performance Evaluation: Modeling Techniques and Tools*, Elsevier/North-Holland, Amsterdam 1992.

[25] Barnes, M.F., *Measurement and Modeling Methods for Computer System Performance*, Inout Two-Nine, Surrey, United Kingdom1979.

[26] Bear, D., *Principles of Telecommunication Traffic Engineering*, P. Peregrinus, London 1976.

[27] Beilner, H. (ed.), *Messung. Modellierung und Bewertung von Rechensystemen*, Springer-Verlag, Berlin 1985.

[28] Beilner, H. and Gelenbe, E. (eds.), *Modeling and Performance Evaluation of Computer Systems*, North-Holland, Amsterdam 1977.

[29] Beilner, H. and Gelenbe, E. (eds.), *Measuring, Modeling and Evaluating Computer Systems*, North-Holland, Amsterdam 1977.

[30] Beizer, B., *Micro-Analysis of Computer System Performance*, Van Nostrand Reinhold, New York 1978.

[31] Beneš, V.A., *Mathematical Theory of Connecting Networks and Telephone Traffic*, Academic Press, New York 1965.

[32] Benwell, N. (ed.), *Benchmarking: Computer Evaluation and Measurement*, Hemisphere, Washington, DC 1975.

[33] Berkeley, G.S., *Traffic and Trunking Principles in Automatic Telephony*, 2nd ed., E. Benn, London 1949.

[34] Bertsekas, D. and Gallager, R.G., *Data Networks*, 2nd ed., Prentice-Hall, Englewood Cliffs, NJ 1992.

[35] Blanc, R.P. and Cotton, I.W. (eds.), *Computer Networking*, IEEE Press, New York 1976.

[36] Boesch, F.T. (ed.), *Large Scale Networks: Theory and Design*, IEEE Press, New York 1976.

[37] Boguslavsky, L.B., *Traffic Control in Computer Networks*, Energoatomizdat, Moscow 1984 (in Russian).

[38] Bolch, G., *Leistungsbewertung von Rechensystemen mittels analytischer Warteschlangenmodelle*, B.G. Teubner, Stuttgart 1989.

[39] Bolch, G., and Akyildiz, I.F., *Analyse von Rechensystemen: Analytische Methoden zur Leistungsbewertung und Leistungsvorhersage*, B.G. Teubner, Stuttgart 1982.

[40] Bonatti, M. (ed.), *Teletraffic Science for New Cost-Effective Systems, Networks and Services: ITC-12*, Elsevier/North-Holland, Amsterdam 1989.

[41] Bonder, V.A., Rondi, N.E., and Yurikov, E.P., *Optimization of Terminal Stochastic Systems*, Mashinostroyenie, Moscow 1987 (in Russian).

[42] Boxma, O.J., Cohen, J.W., and Tijms, H.C. (eds.), *Teletraffic Analysis and Computer Performance Evaluation*, Elsevier/North-Holland, Amsterdam 1986.

[43] Bronshtein, O.I. and Duchovny, I.M., *Priority Service Models for Information and Computer Systems*, Nauka, Moscow 1976 (in Russian).

[44] Bruneel, H. and Kim, B.G., *Discrete-Time Models for Communication Systems Including ATM*, Kluwer, Boston 1993.

[45] Bux, W. and Rudin, H. (eds.), *Performance of Computer-Communication Systems*, Elsevier/North-Holland, Amsterdam 1984.

[46] Buzacott, J.A. and Shanthikumar, J.G., *Stochastic Models of Manufactur-*

ing Systems, Prentice Hall, Englewood Cliffs, NJ 1993.

[47] Cabanel, J.P., Pujolle, G., and Danthine, A. (eds.), *Local Communication Systems: LAN and PBX*, Elsevier/North-Holland, Amsterdam 1987.

[48] Cady, J. and Howarth, B., *Computer System Performance Management and Capacity Planning*, Prentice-Hall, Englewood Cliffs, NJ 1990.

[49] Cellary, W. Gelenbe, E., and Morzy, T., *Concurrency Control in Distributed Database Systems*, Elsevier/North-Holland, Amsterdam 1988.

[50] Chandy, K.M. and Reiser, M. (eds.), *Computer Performance*, North-Holland, Amsterdam 1977.

[51] Chandy, K.M. and Yeh, R.T. (eds.), *Current Trends in Programming Methodology, Vol. III Software Modeling*, Prentice-Hall, Englewood Cliffs, NJ 1978.

[52] Chou, W. (ed.), *Computer Communications, Volume* 1, Prentice-Hall, Englewood Cliffs, NJ 1983.

[53] Chou, W. (ed.), *Computer Communications, Volume* 2, Prentice-Hall, Englewood Cliffs, NJ 1985.

[54] Chu, W.W. (ed.), *Advances in Computer Communications and Networking*, Artech House, Dedham, MA 1979.

[55] Coffman, E.G., Jr. (ed.), *Computer and Job Shop Scheduling Theory*, John Wiley, New York 1976.

[56] Coffman, E.G., Jr. and Denning, P.J., *Operating System Theory*, Prentice-Hall, Englewood Cliffs, NJ 1973.

[57] Courtois, P.J. and Latouche, G. (eds). *Performance '87*, Elsevier/North-Holland, Amsterdam 1988.

[58] Cragon, H.G., *Branch Strategy Taxonomy and Performance Models*, IEEE Computer Science Press, Los Alamitos, CA 1991.

[59] Cravis, H., *Communications Network Analysis*, Lexington Books, D.C. Heath and Company, Lexington, MA 1981.

[60] Dan, A., *Performance Analysis of Data Sharing Environments*, The MIT Press, Cambridge, MA 1991.

[61] Danthine, A. and Spaniol, O. (eds.), *High Speed Local Area Networks II*, Elsevier/North-Holland, Amsterdam 1990.

[62] Decina, M. and Roveri, A., *Code e Traffico nelle reti di Comunicazioni*, La Goliardica Editrice, Roma 1976 (in Italian).

[63] Delhaye, J.L., and Gelenbe, E. (eds.), *High Performance Computing*, Elsevier/North-Holland, Amsterdam 1989.

[64] de Moraes, L.F.M., de Souza e Silva, E., and Soares, L.F.G. (eds.), *Data Communication Systems and Their Performance*, Elsevier/North-Holland, Amsterdam 1988.

[65] Dempster, M.A.H., Lenstra, J.K., and Rinnooy Kan, A.H.G. (eds.), *Deterministic and Stochastic Scheduling*, D. Reidel, Dordrecht, Holland 1981.

[66] Demurjian, S.A., Hsiao, D.K., and Marshall, R.G., *Design Analysis and Performance Evaluation Methodologies for Database Computers*, Prentice-Hall, Englewood Cliffs, NJ 1987.

[67] de Prycker, M., *Asynchronous Transfer Mode: Solution for Broadband ISDN*, Ellis Horwood Ltd., Chichester, England 1991.

[68] Disney, R.L. and Ott, T.J. (eds.), *Applied Probability and Computer Science: The Interface*, Volumes **1** and **2**, Birkhäuser, Boston 1982.

[69] Doll, D.R., *Data Communication: Facilities, Networks and System Design*, John Wiley, New York 1980.

[70] Dowdy, L. and Lowery, C., *PS to Operating Systems: For ACM SIGMETRICS and the Computer Measurement Group (CMG)*, Prentice-Hall, Englewood Cliffs, NJ 1993.

[71] Drummond, M.E., *Evaluation and Measurement Techniques for Digital Computer Systems*, Prentice-Hall, Englewood Cliffs, NJ 1973.

[72] Elldin, A., *Automatic Telephone Exchanges Based on the Link Connection Principle*, L.M. Ericsson, Stockholm 1969.

[73] Elldin, A. and Lind G., *Elementary Telephone Traffic Theory*, L.M. Ericsson, Stockholm 1971.

[74] Everling, W., *Exercises in Computer Systems Analysis*, Lecture Notes in Computer Science, **35**, Springer-Verlag, Berlin 1972.

[75] Fdida, S. and Pujolle, G. (eds.), *Modeling Techniques and Performance Evaluation*, Elsevier/North-Holland, Amsterdam 1987.

[76] Fdida, S. and Pujolle, G., *Modèles de Systèmes et de Réseaux, Tome 1: Performance*, Eyrolles, Paris 1989.

[77] Ferrari, D., *Computer Systems Performance Evaluation*, Prentice-Hall, Englewood Cliffs 1978.

[78] Ferrari, D. (ed.), *Performance of Computer Installations*, Elsevier/North-Holland, Amsterdam 1978.

[79] Ferrari, D. and Spandoni, M. (eds.), *Experimental Computer Performance Evaluation*, Elsevier/North-Holland, Amsterdam 1981.

[80] Ferrari, D., Serazzi, G., and Zeigner, A., *Measurement and Tuning of Computer Systems*, Prentice-Hall, Englewood Cliffs, NJ 1983.

[81] Filipiak, J., *Modeling and Control of Dynamic Flows in Communication Networks*, Springer-Verlag, Berlin 1988.

[82] Filipiak, J., *Real Time Network Management*, Elsevier/North-Holland, Amsterdam 1991.

[83] Filipiak, J. (ed.), *Telecommunication Services for Developing Economies*, Elsevier/North-Holland, Amsterdam 1991.

[84] Flood, J.E. (ed.), *Telecommunication Networks*, Peter Peregrinus Ltd., England.

[85] Fortier, P.J. and Desrochers, G.R., *Modeling and Analysis of Local Area Networks*, CRC Press, Boca Raton, FL 1990.

[86] Fox, J. (ed.), *Proceedings of the Symposium on Computer-Communications Networks and Teletraffic*, Microwave Research Institute Symposia Series, Vol. **22**, Polytechnic Press of the Polytechnic Institute of Brooklyn, New York 1972.

[87] Frankel, T., *Tables for Traffic Management and Design: Book I – Trunking*, Lee's ABC of the Telephone, Geneva, IL 1976.

[88] Franta, W.R. and Chlamtac, I., *Local Networks*, D.C Heath and Co., Lexington, MA 1981.

[89] Freiberger, W., Grenander, U., Margolin, B.H., and Tsao, R.F. (eds.), *Statistical Computer Performance Evaluation*, Academic Press, New York 1972.

[90] Fujiki, M. and Gambe, E., *Theory of Teletraffics*, Maruzen, Tokyo 1980 (in Japanese).

[91] Fuller, S.H., *Analysis of Drum and Disk Storage Units*, Lecture Notes in Computer Science **31**, Springer-Verlag, Berlin 1975.

[92] Gambe, E., *Telephone Switch Traffic*, Denki Tsushin Kyokai, 1966 (in Japanese).

[93] Gambe, E., *Revised Teletraffic Theory and its Applications*, Denshi Tsushin

Gakkai, 1970 (in Japanese).
[94] Gelenbe, E. (ed.), *Performance '84: Models of Computer System Performance*, Elsevier/North-Holland, Amsterdam 1985.
[95] Gelenbe, E. (ed.), *High Performance Computer Systems*, Elsevier/North-Holland, Amsterdam 1988.
[96] Gelenbe, E., *Multiprocessor Performance*, John Wiley, New York 1989.
[97] Gelenbe, E. and Mahl, R. (eds.), *Computer Architectures and Networks Modeling and Evaluation*, North-Holland, Amsterdam 1974.
[98] Gelenbe, E. and Mitrani, I., *Analysis and Synthesis of Computer Systems*, Academic Press, London 1980.
[99] Gerlough, D.L. and Capelle, D.G., *Introduction to Traffic Flow Theory*, Highway Research Board, Washington 1964.
[100] Girard, A., *Routing and Dimensioning in Circuit-Switched Networks*, Addison-Wesley, Reading, MA 1990.
[101] Gopinath, B. (ed.), *Computer Communications*, Proc. Symposia in Applied Mathematics, Vol. **31**, American Mathematical Society, Providence 1984.
[102] Gordon, G.D. and Morgan, W.L., *Communication Satellites: Technology and Performance*, John Wiley, New York 1993.
[103] Gray, J. (ed.), *The Benchmark Handbook: Database and Transaction Processing Systems*, Morgan Kaufmann Pub., 1991.
[104] Haight, F.A., *Mathematical Theories of Traffic Flows*, Academic Press, New York 1963.
[105] Hammond, J.L. and O'Reilly, P.J.P., *Performance Analysis of Local Computer Networks*, Addison-Wesley, Reading, MA 1986.
[106] Haring, G. and Kotsis, G. (eds.), *Performance Measurement and Visualization of Parallel Systems*, Elsevier/North-Holland, Amsterdam 1993.
[107] Harrison, P.G. and Patel, N.M., *Performance Modeling of Communication Networks and Computer Architectures*, Addison-Wesley, Wokingham, England 1993.
[108] Hasegawa, T., Takagi, H., and Takahashi, Y. (eds.), *Computer Networking and Performance Evaluation*, Elsevier/North-Holland, Amsterdam 1986.
[109] Hasegawa, T., Takagi, H., and Takahashi, Y. (eds.), *Performance of Distributed and Parallel Systems*, Elsevier/North-Holland, Amsterdam 1989.
[110] Hasegawa, T., Takagi, H., and Takahashi, Y. (eds.), *Performance of Distributed Systems and Integrated Communication Networks*, Elsevier/North-Holland, Amsterdam 1992.
[111] Hayes, J.F., *Modeling and Analysis of Computer Communications Networks*, Plenum Press, New York 1984.
[112] Hébuterne, G., *Traffic Flow in Switching Systems*, Artech House, Boston 1987.
[113] Hellerman, H. and Conroy, T.F., *Computer System Performance*, McGraw-Hill, New York 1975.
[114] Herzog, U. and Paterok, M. (eds.), *Messung, Modellierung und Bewertung von Rechensystemen*, Springer-Verlag, Berlin 1987.
[115] Highleyman, W.H., *Performance Analysis of Transaction Processing Systems*, Prentice-Hall, Englewood Cliffs, NJ 1989.
[116] Hillston, J., King, P.J.B., and Pooley, R.J. (eds.), *Computer and Telecommunications Performance Engineering*, Springer-Verlag, London 1992.
[117] Hofri, M., *Probabilistic Analysis of Algorithms*, Springer-Verlag, New York 1987.

[118] Iazeolla, G., Courtois, P.J., and Boxma, O.J. (eds.), *Computer Performance and Reliability*, Elsevier/North-Holland, Amsterdam 1988.

[119] Iazeolla, G., Courtois, P.J., and Hordijk, A. (eds.), *Mathematical Computer Performance and Reliability*, Elsevier/North-Holland, Amsterdam 1984.

[120] Inose, H., *An Introduction to Digital Integrated Communication Systems*, P. Peregrinus, Ltd., Stevenage, United Kingdom 1979.

[121] Ionin, G.L., *Teletraffic Theory*, Riga Polytechnic Institute, 1975 (in Russian).

[122] Ionin, G.L. and Sedol, Y.Y., *Statistical Models of Teletraffic Systems*, Radio Svyaz, Moscow 1982 (in Russian).

[123] Jain, R., *The Art of Computer Systems Performance Analysis: Techniques for Experimental Design, Measurement, Simulation and Modeling*, John Wiley, New York 1991.

[124] Kant, K., *Introduction to Computer System Performance Evaluation*, McGraw-Hill, New York 1992.

[125] Kawasaki, H. and Gambe, E., *Introduction to Traffic Theory*, Kyoritsu Shuppan, Tokyo 1959 (in Japanese).

[126] Keiser, G.E., *Local Area Networks*, McGraw-Hill, New York 1989.

[127] Keller, T.C., *CICS: Capacity Planning and Performance Management*, McGraw-Hill, New York 1993.

[128] Kelly, F.P., *Reversibility and Stochastic Networks*, John Wiley, New York 1979.

[129] Kershenbaum, A., *Telecommunications Network Design Algorithms*, McGraw-Hill, Inc., New York 1993.

[130] Kharkevich, A.D. (ed.), *Methods of Teletraffic Theory Development*, Nauka, Moscow 1979 (in Russian).

[131] Kharkevich, A.D. and Garmash, V.A. (eds.), *Information Networks and Their Structures*, Nauka, Moscow 1976 (in Russian).

[132] Kharkevich, A.D. and Garmash, V.A. (eds.), *Information Networks and Their Analysis*, Nauka, Moscow 1978 (in Russian).

[133] Kharkevich, A.D. and Garmash, V.A. (eds.), *Methods and Structures of Teletraffic Systems*, Nauka, Moscow 1979 (in Russian).

[134] Kharkevich, A.D. and Garmash, V.A. (eds.), *Models of Information Networks and Communication Systems*, Nauka, Moscow 1982 (in Russian).

[135] Kharkevich, A.D. and Garmash, V.A. (eds.), *Teletraffic Theory Methods in Decentralized Control Systems*, Nauka, Moscow 1986 (in Russian).

[136] Kharkevich, A.D. and Garmash, V.A. (eds.), *Models and Research in Informatics Systems*, Nauka, Moscow 1988 (in Russian).

[137] King, P., *Computer and Communication System Performance Modeling*, Prentice-Hall, Englewood Cliffs, NJ 1990.

[138] King, P.J.B., Mitrani, I., and Pooley, R.J. (eds.), *Performance '90*, Elsevier/North-Holland, Amsterdam 1990.

[139] Kleinrock, L., *Communication Nets: Stochastic Message Flow and Delay*, Dover, New York 1972.

[140] Klimov, G.P., Lyakhy, A.K., and Matveyev, V.F., *Mathematical Models of Time-Sharing Systems*, Shtiintsa, Kishinyev 1983 (in Russian).

[141] Kobayashi, H., *Modeling and Analysis – An Introduction to System Performance Evaluation and Methodology*, Addison-Wesley, Reading, MA 1978.

[142] Kojima, T., *Study of Teletraffic Calls*, Kagaku Shinko Sha, Tokyo 1949 (in Japanese).

[143] Kolence, K.W., *The Meaning of Computer Measurement: An Introduction to Software Physics*, Institute for Software Engineering, Inc., Palo Alto, CA 1976.

[144] Kolence, K.W., *Introduction to Software Physics*, McGraw-Hill, New York 1985.

[145] Kolesnikov, A.V. and Petuchov, O.A., *Systems Modeling*, Northwestern Polytechnic Institute, Leningrad 1981 (in Russian).

[146] Korolyuk, V.S., *Stochastic Models of Systems*, Kiev, "Lybid" 1993 (in Ukrainian).

[147] Kühn, P. and Schulz, K.M. (eds.), *Messung, Modellierung und Bewertung von Rechensystemen*, Springer-Verlag, Berlin 1983.

[148] Kulkarni, V.G., *Modeling and Analysis of Stochastic Systems*, Chapman and Hall, New York 1995.

[149] Kümmerle, K., Tobagi, F.A., and Limb, J.O. (eds.), *Advances in Local Area Networks*, IEEE Press, New York 1987.

[150] Kuo, F.F. (ed.), *Protocol and Techniques for Data Communication Networks*, Prentice-Hall, Englewood Cliffs, NJ 1981.

[151] Kylstra, F.J. (ed.), *Performance '81*, Elsevier/North-Holland, Amsterdam 1981.

[152] Lavenberg, S.S. (ed.), *Computer Performance Modeling Handbook*, Academic Press, New York 1983.

[153] Leung, C.H.C., *Quantitative Analysis of Computer Systems*, John Wiley, Chichester 1988.

[154] Livshits, B.S., *Calculation of Teletraffic and Losses*, Institute of Teletraffic, Leningrad 1960 (in Russian).

[155] Livshits, B.S., Fidlin, Y.V., and Kharkevich, A.D., *Theory of Telephone and Telegraph Traffic*, Svyaz', Moscow 1971 (in Russian).

[156] Livshits, B.S., Pshenichnikov, A.P., and Kharkevich, A.D., *Teletraffic Theory*, Svyaz', Moscow 1979 (in Russian).

[157] Longo, G. (ed.), *Multi-User Communication Systems*, Springer-Verlag, Wien 1981.

[158] Louchard, G. and Latouche, G. (eds.), *Probability Theory and Computer Science*, Academic Press, New York 1983.

[159] Loukides, M., *System Performance Tuning*, Addison-Wesley, Reading, MA 1991.

[160] Lucantoni, D.M., *An Algorithmic Analysis of a Communication Model with Retransmission of Flawed Messages*, Pitman Advanced Pub. Program, Boston 1983.

[161] MacNair, E.A. and Sauer, C.H., *Elements of Practical Performance Modeling*, Prentice-Hall, Englewood Cliffs, NJ 1985.

[162] Martin, J., *Design of Real-Time Computer Systems*, Prentice-Hall, Englewood Cliffs, NJ 1967.

[163] Martin, J.L. (ed.), *Performance Evaluation of Supercomputers*, North-Holland, Amsterdam 1988.

[164] Masuda, T. and Kameda, H., *Performance Analysis of Operating Systems*, Information Processing Society of Japan, Tokyo 1982 (in Japanese).

[165] Mayama, M., *Traffic Theory and Practice*, Hifumi Shobo, Tokyo 1964 (in Japanese).

[166] McKerrow, P., *Performance Measurement of Computer Systems*, Addison-Wesley, Sydney 1988.

[167] Merrill, H.W., *Merrill's Guide to Performance Evaluation*, SAS Institute Inc., Cary, NC 1980.
[168] Mertens, B. (ed.), *Messung, Modellierung und Bewertung von Rechensystemen*, Springer-Verlag, Berlin 1981.
[169] Mikami, T., Kino, I., and Yoshizawa, Y., *Practice of Computer System Performance Analysis*, Information Processing Society of Japan, Tokyo 1982 (in Japanese).
[170] Mina, R.R., *Introduction to Teletraffic Engineering*, Telephony Publishing Corporation, Chicago 1974.
[171] Mitrani, I., *Modeling of Computer and Communication Systems*, Cambridge University Press, Cambridge, Great Britain 1987.
[172] Molloy, M.K., *Fundamentals of Performance Modeling*, Macmillan, New York 1989.
[173] Morris, M.F. and Roth, P.F., *Computer Performance Evaluation: Tools and Techniques for Effective Analysis*, Van Nostrand Reinhold, New York 1982.
[174] Odoni, A.R., Bianco, L., and Szergö (eds.), *Flow Control of Congested Networks*, Springer-Verlag, Berlin 1987.
[175] Palm, C., *Intensity Variations in Telephone Traffic*, Elsevier/North-Holland, Amsterdam 1988.
[176] Perros, H. (ed.), *High-Speed Communication Networks*, Plenum Press, New York 1992.
[177] Perros, H., Pujolle, G., and Takahashi, Y. (eds.), *Modeling and Performance Evaluation of ATM Technology*, Elsevier/North-Holland, Amsterdam 1993.
[178] Pickholtz, R.L. (ed.), *Local Area and Multiple Access Networks*, Computer Science Press, Rockville, MD 1983.
[179] Pooley, R. and Hillston, J. (eds.), *Computer Performance Evaluation. Modeling Techniques and Tools*, Edinburgh University Press, 1993.
[180] Potier, D. (ed.), *Modeling Techniques and Tools for Performance Analysis*, Elsevier/North-Holland, Amsterdam 1985.
[181] Potthoff, G., *Verkehrsströmungslehre*, Transpress, Berlin 1962.
[182] Pratt, C.W., *A Course in Teletraffic Engineering*, Telecom Australia, Melbourne 1967.
[183] Puigjaner, R. and Potier, D. (eds.), *Modeling Techniques and Tools for Computer Performance Evaluation*, Plenum Press, New York 1989.
[184] Pujolle, G. (ed.), *Performance of Data Communication Systems and Their Applications*, Elsevier/North-Holland, Amsterdam 1981.
[185] Pujolle, G. (ed.), *High-Capacity Local and Metropolitan Area Networks: Architecture and Performance Issues*, Springer-Verlag, Berlin 1991.
[186] Pujolle, G. and Puigjaner, R. (eds.), *Data Communication Systems and Their Performance*, Elsevier/North-Holland, Amsterdam 1990.
[187] Pujolle, G. and Puigjaner, R. (eds.), *Local Communication Systems: LAN and PBX, II*, Elsevier/North-Holland, Amsterdam 1991.
[188] Raghaven, S.V. (ed.), *Local Area Networks*, Elsevier/North-Holland, Amsterdam 1990.
[189] Robertazzi, T. (ed.), *Performance Evaluation of High Speed Switching Fabrics and Networks: ATM, Broadband ISDN and MAN Technology*, IEEE Press, New York 1993
[190] Roberts, J.W. (ed.), *Performance Evaluation and Design of Multiservice Networks*, Final Report of COST 224, Commission of the European Commu-

nities, Directorate-General, Telecommunications, Information Industries and Innovation, Luxemburg 1991.

[191] Rom, R. and Sidi, M., *Multiple Access Protocols: Performance and Analysis*, Springer-Verlag, New York 1990.

[192] Rosen, S., *Lectures on the Measurement and Evaluation of the Performance of Computer Systems*, SIAM, Philadelphia, PA 1976.

[193] Rubas, J. (ed.), *A Course in Teletraffic Engineering*, Telecom Australia, Melbourne 1978.

[194] Ruda, M., and Szirtes, L., *Digital Modeling for Stochastic Systems*, Central Statistical Office, International Computer Education and Information Center, Budapest, Hungary 1979 (in Hungarian).

[195] Ruschitzka, M. (ed.), *Computer Syystems: Performance and Simulation*, Elsevier/North-Holland, Amsterdam 1986.

[196] Sauer, C.H. and Chandy, K.M., *Computer Systems Performance Modeling*, Prentice-Hall, Englewood Cliffs, NJ 1981.

[197] Sauer, C.H. and MacNair, E.A., *Simulation of Computer Communication Systems*, Prentice-Hall, Englewood Cliffs, NJ 1983.

[198] Scherr, A.L., *An Analysis of Time-Shared Computer Systems*, The MIT Press, Cambridge, MA 1967.

[199] Schwartz, M., *Computer-Communication Network Design and Analysis*, Prentice-Hall, Englewood Cliffs, NJ 1977.

[200] Schwartz, M., *Telecommunication Networks: Protocols, Modeling and Analysis*, Addison-Wesley, Reading, MA 1987.

[201] Serrazi, G., *Workload Characterization of Computer Systems and Computer Networks*, Elsevier/North-Holland, Amsterdam 1986.

[202] Sharma, R.L., de Souza, P.J.T., and Inglé, A.D., *Network Systems. Modeling, Analysis and Design*, Van Nostrand Reinhold, New York 1982.

[203] Shneps, M.A., *Numerical Methods in Teletraffic Theory*, Svyaz', Moscow 1974 (in Russian).

[204] Shneps, M.A., *Information Distribution Systems. Methods of Calculation*, Svyaz', Moscow 1979 (in Russian).

[205] Siemens, *Telephone Traffic Theory Tables and Charts*, Siemens Aktiengesellschaft, Berlin 1970.

[206] Sigman, K., *Stationary Marked Point Processes*, Chapman and Hall, London 1994.

[207] Signaevskii, V.A. and Kogan, Y.A., *Analytical Method for Performance Evaluation of High Speed Computers*, Nauka, Moscow 1991 (in Russian).

[208] Smith, C.U., *Performance Engineering of Software Systems*, Addison-Wesley, Reading, MA 1990.

[209] Smith, W.L. and Wilkinson, W.E. (eds.), *Proc. Symp. on Congestion Theory*, University of North Carolina Press, Chapel Hill 1965.

[210] Spaniol, O. and Danthine, A. (eds.), *High Speed Local Area Networks*, 1987.

[211] Spaniol, O. and Danthine, A. (eds.), *High Speed Local Area Networks III*, Elsevier/North-Holland, Amsterdam 1991.

[212] Spirn, J.R., *Program Behavior: Models and Measurements*, North-Holland, New York 1977.

[213] Spragins, J. and Hammond, J., *Telecommunication Networks: Protocols and Design*, Addison-Wesley Pub., Reading, MA 1991.

[214] Stallings, W., *Local Networks: An Introduction*, Macmillan, New York

1984.

[215] Stallings, W., *Local and Metropolitan Area Networks*, 4th ed., Macmillan, New York 1993.

[216] Stiege, G. and Lie, J.S. (eds.), *Messung, Modellierung und Bewertung von Rechensystemen und Netzen*, Springer-Verlag, Berlin 1989.

[217] Störmer, H.E., Behlendorff, E., Bininda, N., Bretschneider, G., Hoffman, E., and Suhlandt, H., *Teletraffic Theory*, Oldenbourg, München 1966, (in German).

[218] Stuck, B.W. and Arthurs, E., *A Computer and Communications Network Performance Analysis Primer*, Prentice-Hall, Englewood Cliffs, NJ 1985.

[219] Svobodova, L., *Computer Performance Measurement and Evaluation Methods: Analysis and Applications*, Elsevier, New York 1976.

[220] Syski, R., *Introduction to Congestion Theory in Telephone Systems*, 2nd ed., Elsevier/North-Holland, Amsterdam 1986.

[221] Takagi, H. (ed.), *Stochastic Analysis of Computer and Communication Systems*, Elsevier/North-Holland, Amsterdam 1990.

[222] Tannenbaum, A., *Computer Networks*, Prentice-Hall, Englewood Cliffs, NJ 1981.

[223] Tasaka, S., *Performance Analysis of Multiple Access Protocols*, The MIT Press, Cambridge, MA 1986.

[224] Tay, Y.C., *Locking Performance in Centralized Databases*, Academic Press, Boston 1987.

[225] Tropper, C., *Local Computer Network Technologies*, Academic Press, New York 1981.

[226] van Dyke, N.M. Gerrand, P., Henderson, W.T., Warfield, R.E., and Addie, R.G. (eds.), *Traffic Theories for New Telecommunications Services*, Elsevier/North-Holland, Amsterdam 1990.

[227] Verma, P.K., *Performance Estimation of Computer Communication Networks: A Structured Approach*, Computer Sci. Press, Rockville, MD 1989.

[228] Viniotis, Y. and Onvural, R.O. (eds.), *Asynchronous Transfer Mode Networks*, Plenum Press, 1993.

[229] Viswanadham, N. and Narahari, Y., *Performance Modeling of Automated Manufacturing Systems*, Prentice-Hall, Englewood Cliffs, NJ 1992.

[230] Walrand, J., *Communication Networks: A First Course*, CRC Press, Boca Raton, FL 1990.

[231] Woodward, M.E., *Discrete-Time Performance Modeling of Packet Switching Networks*, Pentech Press, United Kingdom 1992.

[232] Yao, D. (ed.), *Stochastic Modeling and Analysis of Manufacturing Systems*, Springer-Verlag, New York 1994.

[233] Yemini, Y. (ed.), *Current Advances in Distributed Computing and Communications*, Computer Science Press, Rockville, MD 1987.

[234] Zaychenko, Y. and Gonta, Y.V., *Structural Optimization of Computer Networks*, Technica, Kiev 1986 (in Russian).

IV. Books on Other Topics Related to Queueing

[1] Albin, S.L. and Harris, C.M. (eds.), *Statistical and Computational Issues in Probability Modeling*, *Annals of Oper. Res.*: book edition, Vol. **8** and **9**, J.C. Baltzer A.G., Basel, Switzerland 1987.

[2] Asmussen, S., *Ruin Probabilities*, World Scientific Publishing, Singapore

1995/96.

[3] Baker, K.R., *Introduction to Sequencing and Scheduling*, John Wiley, New York 1966.

[4] Banks, J. and Carson, J.S., *Discrete-Event System Simulation*, Prentice-Hall, Englewood Cliffs, NJ 1984.

[5] Barlow, R.E. and Proschan, F., *Statistical Theory of Reliability and Life Testing*, Holt, Rinehart and Winston, New York 1975.

[6] Baronne, P., Frigessi, A., and Piccioni, M. (eds.), *Stochastic Models, Statistical Methods and Algorithms in Image Analysis*, Springer-Verlag 1992.

[7] Berger, M.A., *An Introduction to Probability and Stochastic Processes*, Springer-Verlag, 1993.

[8] Berman, S.M., *Sojourns and Extremes of Stochastic Processes*, Wadsworth, Pacific Grove, CA 1992.

[9] Bolshakov, I.A. and Rakoshitss, V.S., *Applied Theory of Stochastic Flows*, Soveskoje Radio, Moscow 1978 (in Russian).

[10] Brandt, A., Franken, P. and Lisek, B., *Stationary Stochastic Models*, John Wiley, Chichester 1990.

[11] Brockmeyer, E., Halstrøm, H.L., and Jensen A., *The Life and Works of A.K. Erlang*, The Copenhagen Telephone Company, Copenhagen 1948.

[12] Chung, K.L., *Markov Chains with Stationary Transition Probabilities*, Springer-Verlag, Berlin 1967.

[13] Çinlar, E., *Introduction to Stochastic Processes*, Prentice-Hall, Englewood Cliffs, NJ 1975.

[14] Çinlar, E., Sharpe, M.J., and Burdy, K., *Seminar on Stochastic Processes*, Progress in Probability, Vol. **33**, Birkhäuser, 1992.

[15] Conway, R.W., Maxwell, W.L., and Miller L.W., *The Theory of Scheduling*, Addison-Wesley, Reading, MA 1967.

[16] Cox, D.R., *Renewal Theory*, Methuen, London 1962.

[17] Crane, M.A. and Lemoine, A.J., *An Introduction to the Regenerative Method for Simulation Analysis*, Springer-Verlag, New York 1977.

[18] Daley, D.J. and Vere-Jones, D., *An Introduction to the Theory of Point Processes*, Springer-Verlag, New York 1988.

[19] Dshalalow, J.H. (ed.), *Special Jubilee Issue in Honor of Lajos Takács*, North Atlantic, *JAMSA*, Berlin 1994.

[20] Fry, T.C., *Probability and its Engineering Uses*, Van Nostrand, New York 1928.

[21] Galambos, J. and Gani, J. (eds.), *Studies in Applied Probability*, Applied Probability Trust, *J. Appl. Prob.*, Sheffield 1994.

[22] Giffin, W.C., *Transform Techniques for Probability Modeling*, Academic Press, New York 1975.

[23] Gnedenko, B.V., Belyaev, Yu.K. and Solovyev, A.D., *Mathematical Methods in Reliability Theory*, Academic Press, New York 1969.

[24] Goodman, R., *Introduction to Stochastic Models*, Benjamin/Cummings, Menlo Park, CA 1988.

[25] Graham, A., *Kronecker Products and Matrix Calculus with Applications*, Ellis Horwood, Chichester 1981.

[26] Grandell, J., *Doubly Stochastic Poisson Processes*, Lecture Notes in Mathematics, **529**, Springer-Verlag, Berlin 1976.

[27] Harrison, M.J., *Brownian Motion and Stochastic Flow Systems*, John Wiley, New York 1985.

[28] Hennequin, P.L. (ed.), *Ecole d'été de Probabilités de Saint-Flour XXI – 1991*, Lecture Notes in Mathematics, Vol. **1541**, Springer-Verlag, 1993.

[29] Heyman, D.P. and Sobel, M.J., *Stochastic Models in Operations Research*, Vol. **1**: *Stochastic Processes and Operating Characteristics*, McGraw-Hill, New York 1982.

[30] Heyman, D.P. and Sobel, M.J., *Stochastic Models in Operations Research*, Vol. **2**: *Stochastic Optimization*, McGraw-Hill, New York 1984.

[31] Heyman, D.P. and Sobel, M.J. (ed.), *Stochastic Models* [*Handbooks in OR & MS*, Vol. **2**], North-Holland, Amsterdam 1990.

[32] Jensen, A., *A Distribution Model, Applicable to Economics*, Munksgaard, Copenhagen 1954.

[33] Kalashnikov, V.V., *Topics on Regenerative Processes*, CRC Press, Boca Raton, FL 1994.

[34] Kalashnikov, V.V. and Zolotarev, V.M., *Stability Problems for Stochastic Models*, Springer-Verlag, 1993.

[35] Kallenberg, O., *Random Measures*, 3rd edition, Akademie-Verlag and Academic Press, New York 1983.

[36] Keilson, J., *Green's Function Methods in Probability Theory*, Charles Griffin, London 1965.

[37] Keilson, J., *Markov Chain Models – Rarity and Exponentiality*, Springer-Verlag, New York 1979.

[38] Kingman, J.F.C., *Subadditive Processes*, Ecole d'Eté de Probabilité de Saint-Flour, Springer-Verlag, New York 1976.

[39] Kingman, J.F.C., *Poisson Processes*, Clarendon Press, Oxford 1993.

[40] König, D. and Schmidt, V., *Zufällige Punktprocesse* (*Random Point Processes*) (Ger.), B.G. Teubner, Stuttgart 1992.

[41] Korolyuk, V.S. and Turbin, A.F., *Semi-Markov Processes and Their Applications*, Naukova Dumka, Kiev 1976 (in Russian).

[42] Korolyuk, V.S. and Turbin, A.E., *Mathematical Foundations of the State Lumping of Large Systems*, Kluwer, Dordrecht, Boston, London 1993.

[43] Kovalenko, I.N., *Rare Events Analysis in Estimation of Systems Efficiency and Reliability*, Sov. Radio, Moscow 1980 (in Russian).

[44] Kovalenko, I.N. and Kuznetsov, N.Yu., *Methods of High Reliable Systems Account*, Radio i Svyaz, Moscow 1988 (in Russian).

[45] Kovalenko, I.N. and Nakonechny, A.N., *Approximate Analysis and Optimization of the Reliability*, Naukova Dumka Publ., Kiev 1989 (in Russian).

[46] Kotz, S. and Johnson, N.L. (eds.), *Encyclopedia of Statistical Sciences*, John Wiley, New York 1985.

[47] Law, A.M. and Kelton, W.D., *Simulation Modeling and Analysis*, McGraw-Hill, New York 1982.

[48] Lawler, G.F., *Intersection of Random Walks*, Birkhäuser, 1991.

[49] Lindvall, T., *Lectures on the Coupling Method*, John Wiley, New York 1992.

[50] Matthes, K., Kerstan, J. and Mecke, J., *Infinitely Divisible Point Processes*, John Wiley, Chichester 1978.

[51] Mitrani, I., *Simulation Techniques for Discrete Event Systems*, Cambridge University Press, Cambridge 1982.

[52] Neuts, M.F. (ed.), *Algorithmic Methods in Probability*, Studies in Management Science, 7, North-Holland, Amsterdam 1977.

[53] Neuts, M.F., *Algorithmic Probability: A Collection of Problems*, Chapman and Hall, New York, 1995, (in press).

[54] Parzen, E., *Stochastic Processes*, Holden-Day, San Francisco 1967.

[55] Reiss, R.-D., *A Course on Point Processes*, Springer-Verlag, New York 1993.

[56] Resnick, S.I., *Adventures in Stochastic Processes*, Birkhäuser, Boston 1992.

[57] Ripley, B.D., *Stochastic Simulation*, John Wiley, New York 1987.

[58] Rolski, T., *Stationary Random Processes Associated with Point Processes*, Lecture Notes in Statistics **5**, Springer-Verlag, New York 1981.

[59] Ross, S.M., *Introduction to Probability Models*, Academic Press, Orlando, FL 1993.

[60] Rubinstein, R.Y., *Simulation and Monte Carlo Method*, John Wiley, New York 1981.

[61] Rubinstein, R.Y. and Shapiro, A., *Discrete Event Systems: Sensitivity Analysis and Stochastic Optimization by the Score Function Method*, John Wiley, New York 1993.

[62] Shaked, M. and Shanthikumar, J., *Stochastic Orders and their Applications*, Academic Press, Orlando, FL 1994.

[63] Sigman, K., *Stationary Marked Point Processes*, Chapman and Hall, London 1994.

[64] Stewart, W.J., *Introduction to Numerical Solution of Markov Chains*, Princeton University Press, Princeton, NJ 1995.

[65] Stoyan, D., Kendall, W.S. and Mecke, J., *Stochastic Geometry and its Applications*, John Wiley, Chichester 1991.

[66] Takács, L., *Combinatorial Methods in the Theory of Stochastic Processes*, John Wiley, New York 1967.

[67] Takagi, H., *Analysis of Polling Systems*, MIT Press, Cambridge, MA 1986.

[68] Taylor, H.M. and Karlin, S., *An Introduction to Stochastic Modeling*, Rev. Ed., Academic Press, Boston 1994.

[69] Tijms, H.C., *Stochastic Modeling and Analysis: A Computational Approach*, John Wiley, New York 1986.

[70] Tijms, H.C., *Stochastic Models: An Algorithmic Approach*, John Wiley, New York 1994.

[71] Todorovic, P., *An Introduction to Stochastic Processes and their Applications*, Springer-Verlag, New York 1992.

[72] Tuckwell, C.T., *Stochastic Processes in the Neurosciences*, SIAM Philadelphia, PA 1989.

[73] Whittle, P., *Systems in Stochastic Equilibrium*, John Wiley, New York 1986.

[74] Zikun, W. and Xiangqun, Y., *Birth and Death Processes and Markov Chains*, Springer-Verlag, Berlin 1992.

V. Surveys and Papers of Historical Interest

[1] Abate, J. and Whitt, W., The Fourier series method for inverting transforms of probability distributions, *Queueing Sys.*, **10** (1992), 5-88.

[2] Asmussen, S., Busy period analysis, rare events and transient behavior in fluid flow models, *J. Appl. Math. Stoch. Anal.*, **7**:3 (1994), 269-300.

[3] Bachelier, L., Théorie de la spéculation, *Ann. Sci. Éc. Norm. Sup.*, Paris, series 3, **17** (1900), 21.

[4] Belyaev, Yu.K., Random flows and renewal theory (supplement - survey), in: *Renewal Theory* (by Cox, D.R. and Smith, W.L.) (Rus.), Sovietskoye Radio, Moskow 1967.

[5] Bhat, U.N. and Rao, S.S., Statistical analysis of queueing systems, *Queueing Sys.*, **2** (1987), 217-147.

[6] Bingham, N.H., Fluctuation theory in continuous time, *Adv. Appl. Prob.*, **7** (1975), 705-766.

[7] Bingham, N. H., The work of Lajos Takács on probability theory, in: Galambos, J. and Gani, J. (eds.), *Studies in Applied Probability*, Applied Probability Trust, Sheffield, Papers in honour of Lajos Takács, *J. Appl. Prob.*, Special Volume, **31A** (1994), 29-39.

[8] Borel, E., Sur l'emploi du théoréme de Bernoulli pour faciliter le calcul d'une infinité de coefficients. Application au probléme de l'attente á un guichet (On application of Bernoulli's theorem to facilitate the computation of an infinity of coefficients. Application to the problem of waiting at a pay-desk) , *Comptes Rendus Acad. Sci. Paris*, **214** (1942), 452.

[9] Brémaud, P., Kannurpatti, R. and Mazumdar, R., Event and time averages: A review and some generalizations, *Adv. Appl. Prob.* **24** (1992), 377-411.

[10] Çinlar, E., Markov renewal theory, *Adv. Appl. Prob.*, **1** (1969), 123-187.

[11] Çinlar, E., Markov renewal theory: a survey, *Manag. Sci.*, **7** (1975), 727-752.

[12] Çinlar, E., On semi-Markov processes on arbitrary spaces, *Proc. Camb. Phil. Soc.*, **66** (1969), 381-392.

[13] Coffman, E.G. and Hofri, M., Queueing models of secondary storage devices, *Queueing Sys.*, **2** (1986), 129-168.

[14] Coffman, E.G. and Hofri, M., Queueing models of secondary storage devices, in: *Stochastic Analysis of Computer and Communication Systems*, (ed.: Takagi, H.), Elsevier Sci. Publishers B.V. (North-Holland) 1990, 549-588.

[15] Cox, D.R., The analysis of non-Markovian stochastic processes by the inclusion of supplementary variables, *Proc. Camb. Phil. Soc.*, **51** (1955), 433-441.

[16] Cramér, H., On the mathematical theory of risk, in: *Försäkringaktiebolaget Skandia* (1855-1930) *Jubilee*, Stockholm, **II** (1930), 7-84.

[17] Daley, D.J., Queueing output processes, *Adv. Appl. Prob.*, **8** (1976), 395-415.

[18] Daley, D.J. and Rolski, T., A light traffic approximation for a single-server queue, *Math. Oper. Res.*, **9** (1984), 624-628.

[19] Daley, D.J. and Rolski, T., Light traffic approximations in queues, *Math. Oper. Res.*, **16** (1991), 57-71.

[20] Daley, D.J. and Rolski, T., Light traffic approximations in many-server queues, *Adv. Appl. Prob.*, **24** (1992), 202-218.

[21] Daley, D.J. and Rolski, T., Light traffic approximations in general stationary single-server queues, *Stoch. Proc. Appl.*, **49** (1994), 141-158.

[22] Disney, R.L. and König, D., Queueing networks: a survey of their random processes, *SIAM Review*, **27** (1985), 335-403.

[23] Doeblin, J., Sur deux problémes de Kolmogoroff concernant les chaines dénombrable, *Bull. Soc. Math. Fr.*, **66** (1938), 210-220.

[24] Doob, J.L., Topics in the theory of Markov chains, *Trans. Amer. Math.*

Soc., **52** (1942), 37-64.

[25] Doob, J.L., Markov chains - denumerable case, *Trans. Amer. Math. Soc.*, **58** (1945), 455-473.

[26] Doshi, B.T., Queueing systems with vacations – A survey, *Queueing Sys.*, 1:1 (1990), 29-66,.

[27] Doshi, B., Single-server queues with vacations, in: *Stochastic Analysis of Computer and Communication Systems*, (ed.: Takagi, H.), Elsevier Sci. Publishers B.V. (North-Holland) 1990, 217-265.

[28] Dshalalow, J.H. and Syski R., Lajos Takács and his work, *J. Appl. Math. Stoch. Anal.* (Special Jubillee Issue in Honor of Lajos Takács), **7**:3 (1994), 211-238.

[29] Dobrushin, R.L. and Pechersky, E.A., Large deviations for tandem queueing systems, *J. Appl. Math. Stoch. Anal.*, **7**:3 (1994), 301-330.

[30] Feller, W., On the integro-differential equations of purely discontinuous Markoff processes, *Trans. Amer. Math. Soc.*, **48** (1940), 488-575.

[31] Fischer, W. and Meier-Hellstern, K., The Markov-Modulated Poisson Process (MMPP) cookbook, *Perform. Evaluation*, **18** (1992), 149-171.

[32] Gaver, D.P., Imbedded Markov chain analysis of a waiting-line process in continuous time, *Ann. Math. Stat.*, **30**:3 (1959), 698-720.

[33] Iglehart, D., Weak convergence in queueing theory, *Adv. Appl. Prob.*, **5** (1973), 570-594.

[34] Jagerman, D.L. and Melamed, B., On Markovian traffic with applications to TES processes, *J. Appl. Math. Stoch. Anal.*, **7**:3 (1994), 373-396.

[35] Keilson, J. and Kooharian, A., On time dependent queueing processes, *Ann. Math. Stat.*, **31** (1960), 104-112.

[36] Keller, J.B., Time-dependent queues, *SIAM Rev.* **24** (1982), 401-412.

[37] Kendall, D.G., On the generalized "birth-and-death" process, *Ann. Math. Stat.*, **19** (1948), 1-15.

[38] Kendall, D.G., Some problems in the theory of queues, *J. Roy. Stat. Soc.*, **B13** (1951), 151-185.

[39] Kendall, D.G., Stochastic processes occurring in the theory of queues and their analysis by the method of the imbedded Markov chain, *Ann. Math. Stat.*, **24** (1953), 338-354.

[40] Khintchine, A.Y., Mathematical theory of a stationary queue, (Rus.) *Mat. Sbornik*, **39**:4 (1932), 73-84.

[41] Khintchine, A.Y., Mathematisches über die Erwartung vor einem öffentlichen Schalter, *Mat. Sbornik*, **39** (1932), 73-84.

[42] Kingman, J.F.C., On the algebra of queues, *J. Appl. Prob.*, **3**, (1966), 285-326.

[43] Kolmogorov, A.N., Über die analytischen Methoden in der Wahrscheinlichkeitsrechnung, *Math. Ann.*, **104** (1931), 415-458.

[44] Kolmogorov, A.N., Sur le probléme d'attente (On the problem of waiting), *Rec. Math.*, **38** (1931), 101-106.

[45] Kolmogorov, A.N., Anfangsgründe der Theorie der Markoffschen Ketten mit unendlich vielen möglichen Zuständen, *Rec. Math.*, Moscow, **1**:3 (1936), 1-16.

[46] Korolyuk, V.S., Brodi, S.M., and Turbin, A.F., Semi-Markov processes and their applications (Rus.), in: *VINITI Itogi Nauki i Techniki*, Moscow 1974, 47-98.

[47] Kovalenko, I.N., Rare events in queueing systems - A survey, *Queueing Sys.*

16 (1994), 1-49.

[48] Lemoine, A.J., Networks of Queues – A survey of equilibrium analysis, *Mgt. Sci.*, **24**:4 (1977), 464-481.

[49] Lemoine, A.J., Networks of Queues – A survey of weak convergence results, *Mgt. Sci.*, **24**:11 (1977), 1175-1193.

[50] Lévy, P., Systèmes markoviens et stationnaires. Cas dénombrable, *Ann. Sci. Ecole Norm. Sup.*, **68** (1951), 327-381.

[51] Lévy, P., Systèmes semi-Markoviens à au plus une infinité dénombrable d'états possibles, *Proceedings of the International Congress of Mathematicians, Amsterdam*, **2**, North-Holland, Amsterdam (1954), 294-295.

[52] Lindley, D.V., The theory of queues with a single server, *Proc. Camb. Phil. Soc.*, **48** (1952), 277-289.

[53] Little, J.D.C., A proof for the queueing formula: $L = \lambda W$, *Oper. Res.*, **9** (1961), 383-387.

[54] Lucantoni, D.M., Meier-Hellstern, K.S. and Neuts, M.F., A single-server queue with server vacations and a class of nonrenewal arrival processes, *Adv. Appl. Prob.*, **22** (1990), 676-705.

[55] Lundberg, F., Approximerad framställning av sannolikhetsfunktionen, *Aterförsäkring av kollektivrisker*, Almkvist and Wiksell, Uppsala 1903.

[56] Malyshev, V.A., Wiener-Hopf equations and their applications to the theory of probability (Rus.), in: *VINITI Itogi Nauki i Techniki*, Moscow 1976, 5-36.

[57] Melamed, B., An overview of TES processes and modeling methodology, in: *Performance Evaluation of Computer and Communications Systems* (L. Donatiello and R. Nelson, eds.), 359-393, Lecture Notes, Springer-Verlag 1993.

[58] Melamed, B. and Hill, J.R., A survey of TES modeling applications, *SIMULATION* (to appear).

[59] Melamed, B. and Whitt, W., On arrivals that see time averages, *Oper. Res.*, **38**:1 (1990), 156-172.

[60] Melamed, B. and Whitt, W., On arrivals that see time averages: a martingale approach, *J. Appl. Prob.*, **27** (1990), 376-384.

[61] Miyazawa, M., Rate conservation laws: A survey, *Queueing Sys.* **15** (1994), 1-58.

[62] Neuts, M.F., A versatile Markovian point process, *J. Appl. Prob.*, **16** (1979), 764-779.

[63] Palm, C., Analysis of the Erlang traffic formulae for busy-signal arrangements, *Erricsson Technics*, **6**:4 (1938), 39-58.

[64] Palm, C., Intensitätsschwankungen im Fernsprechsverkehr. Untersuchungen über die Darstellung auf Fernsprechverkehrsprobleme anwendbarer stochastischer Prozesse, *Ericsson Technics*, **1**:44 (1943), 1-189.

[65] Pollaczek, F., Über eine Aufgabe der Wahrscheinlichkeitstheorie, Teil I, *Math. Z.*, **32** (1930), 64-100.

[66] Pollaczek, F., Über eine Aufgabe der Wahrscheinlichkeitstheorie, Teil II, *Math. Z.*, **32** (1930), 729-750.

[67] Pollaczek, F., Sur la répartition des périodes d'occupation ininterrompue d'un guichet, *C.R. Acad. Sci. Paris* **234** (1952), 2042-2044.

[68] Pollaczek, F., Généralisation de la théorie probabiliste des systémes téléphoniques sans dispositif d'attente, *Compt. Rend. Acad. Sci. Paris*, **236** (1953), 1469-1470.

[69] Pollaczek, F., Concerning an analytic method for the treatment of queueing

problems, In: *Proc. Symposium on Congestion Theory*, Smith, W. L., Willkinson, W. E. (eds.), pages 1-42, Univ. North Carolina Press, 1965.

[70] Prabhu, N.U., A bibliography of books and survey papers on queueing systems: Theory and applications, *Queueing Sys.*, **2**:4 (1987), 393-398.

[71] Prabhu, N.U. and Zhu, Y., Markov-modulated queueing systems, *Queueing Sys.* **5** (1989), 215-245.

[72] Prabhu, N.U., Markov renewal and Markov-additive processes - A review and some new results, In: *Proc. of KAIST Math. Workshop* **6**, ed. by B.D. Choi and J.W. Yim, Korea Adv. Inst. Sci. and Tech., Taejon 1991, 57-94.

[73] Pyke, R., Markov renewal processes: definitions and preliminary properties, *Ann. Math. Stat.*, **32** (1961), 1231-1242.

[74] Ramalhoto, M.F., Amaral, J.A., and Cochito, M.T., A survey of J. Little's formula, *Internat. Stat. Rev*, **51** (1983), 255-278.

[75] Ramaswami, V., Nonlinear matrix equations in applied probability: Solution techniques and open problems, *SIAM Review* **30** (1988), 256-263.

[76] Ramaswami, V., From the matrix-geometric to the matrix-exponential, *Queueing Sys.* **6** (1990), 229-260.

[77] Resnick, S.I., Point process, regular variation and weak convergence, *Adv. Appl. Prob.*, **18** (1986), 66-138.

[78] Reynolds, J.F., The covariance structure of queues and related processes - a survey of recent work, *Adv. Appl. Prob.*, **7** (1975), 383-415.

[79] Schreiber, F. and Le Gall, P., In memorial of Félix Pollaczek, (1892-1981), in: Special Issue on Teletraffic Theory and Engineering in Memory of Félix Pollaczek, *Archiv für Elektronik und Übertragungstechnik (AEÜ)*, **47**:5/6 (1993), 275-281.

[80] Sengupta, B., The semi-Markovian queue: Theory and applications, *Stoch. Mod.* **6** (1990), 383-413.

[81] Sevastyanov, B.A., Renewal theory (Rus.), in: *VINITI Itogi Nauki i Techniki*, Moscow 1974, 99-128.

[82] Shaked, M. and Shanthikumar, G.J., Regular, sample path and strong stochastic convexity: A review, *Stochastic Orders and Decision Under Risk*, IMS Lecture Notes-Monograph Series (1991), 320-333.

[83] Smith, W.L., Regenerative stochastic processes, *Proceedings of the International Congress of Mathematicians, Amsterdam*, **2**, North-Holland, Amsterdam (1954), 304-305.

[84] Smith, W., Regenerative stochastic processes, *Proc. Roy. Soc. Edinb.*, **A64** (1954), 9-48.

[85] Smith, W., Renewal theory and its ramifications, *J. Roy. Stat. Soc.*, **B20**:2 (1958), 243-302.

[86] Solovyev, A.D., Analytic methods of the computation and estimation of the reliability, in: *Voprosy Matematicheskoj Teroii Nadezhnosti*, ed. by B.V. Gnedenko, Radio i Svyaz, Moscow 1983, 9-112 (in Russian).

[87] Stidham, S., Regenerative processes in the theory of queues, with applications to the alternating-priority queue, *Adv. Appl. Prob.*, **4** (1972), 542-577.

[88] Stidham, S., Jr., $L = \lambda W$: a discounted analogue and a new proof, *Oper. Res.*, **20** (1972), 1115-1126.

[89] Stidham, S., Jr., A last word on $L = \lambda W$, *Oper. Res.*, **22** (1974), 417-421.

[90] Syski, R., Pollaczek's method in Queueing Theory, in: Special Issue on Teletraffic Theory and Engineering in Memory of Félix Pollaczek, *Archiv für Elektronik und Übertragungstechnik (AEÜ)*, **47**, No. 5/6 (1993), 282-299.

[91] Takács, L., Some investigations concerning recurrent stochastic processes of a certain type. (Hungarian. Russian and English summaries.), *Magyar Tud. Akadémia Alk. Mat. Int. Közl.* **3** (1954) 115-128.

[92] Takács, L., Investigation of waiting time problems by reduction to Markov processes, *Acta Math. Acad. Sci. Hungaricae*, **6**, (1955) 101-129.

[93] Takács, L., On a generalization of the renewal theory, *Magyar Tud. Akadémia Mat. Kut. Int. Közl.* **2**, (1957) 91-103.

[94] Takács, L., On a probability problem in the theory of counters, *Ann. Math. Stat.*, **29** (1958), 1257-1263.

[95] Takács, L., On a sojourn time problem, *Theory Prob. Appl.*, **3**, (1958), 58-65.

[96] Takács, L., On a combined waiting time and loss problem concerning telephone traffic, *Ann. Univ. Sci. Budapest. R. Eötvös. Sect. Math.*, **1**, (1958) 73-82.

[97] Takács, L., A telefon-forgalom elméletének néhány valószinüségszámitási kérdéséröl (On some probability problems in teletraffic) (Hung.), *Magyar Tud. Akadémia Mat. Fiz. Oszt. Közl.*, **8**, (1958) 151-210.

[98] Takács, L., Transient behavior of single-server queueing processes with Erlang input, *Trans. Amer. Math. Soc.* **100**, (1961) 1-28.

[99] Takács, L., The transient behavior of a single-server queueing process with a Poisson input, *Proc. Fourth Berkeley Symp. Math. Stat. and Prob., Univ. of California Press*, **2** (1961), 535-567.

[100] Takács, L., Sojourn time problems, *Ann. Prob.*, **2**, No. 3 (1974), 420-431.

[101] Takács, L., A storage process with semi-Markov input, *Adv. Appl. Prob.*, **7**, No. 4 (1975), 830-844.

[102] Takács, L., Combinatorial and analytic methods in the theory of queues, *Adv. Appl. Prob.*, **7** (1975), 607-635.

[103] Takács, L., On fluctuation problems in the theory of queues, *Adv. Appl. Prob.*, **8**, No. 3 (1976) 548-583.

[104] Takács, L., On fluctuations of sums of random variables, *Adv. Math*, **2** (1978), 45-93.

[105] Takács, L., Fluctuation problems for Bernoulli trials, *SIAM Review*, **21**, (1979) 222-228.

[106] Takács, L., Queues, random graphs and branching processes, *J. Appl. Math. Simul.* **1** (1988), 223-243.

[107] Takács, L., Ballots, queues and random graphs, *J. Appl. Prob.* **26** (1989), 103-112.

[108] Takács, L., Pollaczek's results in fluctuation theory, *Archiv für Elektronik und Übertragungstechnik*, **47** (1993), 322-325.

[109] Takagi, H., B., Queueing analysis of polling models: An update, in: *Stochastic Analysis of Computer and Communication Systems*, (ed.: Takagi, H.), Elsevier Sci. Publishers B.V. (North-Holland) 1990, 267-318.

[110] Takagi, H.B. and Boguslavsky, L.B., A supplementary bibliography of books on queueing analysis and performance evaluation, *Queueing Sys.*, **8**: 3, (1991), 313-322.

[111] Täcklind, S., Sur le risque de ruine dans des jeux inéquitables, *Skand. Akuariet.*, **25** (1942), 1-42.

[112] Whitt, W., A review of $L = \lambda W$ and extensions, *Queueing Sys.*, **9** (1991), 235-286.

[113] Whitt, W., Approximating a point process by a renewal process, I: two

basic methods, *Oper. Res.*, **30**.1 (1982), 125-147.
[114] Wilkinson, R.I., The beginning of switching theory in the United States, *Electrical Engineering* (published by AIEE), (1956).
[115] Yashkov, S.F., Processor-sharing queues: some progress in analysis, *Queueing Sys.*, **2** (1987), 1-17.
[116] Zhang, C.-H., A nonlinear renewal theory, *Ann. Prob.*, **16**:2 (1988), 793-824.

VI. Special Issues

[1] Albin, S.L. and Harris, C.M. (eds.), Statistical and computational issues in probability modeling, *Ann. Oper. Res.*, Vol. **8** and **9**, 1987.
[2] *Performance Evaluation Review*, Vol. **1**, (1972). A publication of the ACM SIGMETRICS (special interest group on measurement and evaluation of computer system performance).
[3] *Computer Performance*, Vol. **1**, (1980) – Vol. **5**, (1984). Butterworth Scientific Ltd., London.
[4] *Performance Evaluation*, Vol. I, (1981) – . Editor-in-chief: H. Kobayashi (1981 – 1986), M. Reiser (1986 – 1990), W. Bux (1991 –). Elsevier/North-Holland, Amsterdam.

Part I

Stochastic Methods

Chapter 2

Queueing methods in the theory of random graphs

Lajos Takács

ABSTRACT In this article several limit theorems are proved for the fluctuations of the queue size during the initial busy period of a queueing process with one server. These theorems are used to find the solutions of various problems connected with the heights and widths of random-rooted trees.

CONTENTS

2.1	Introduction	45
2.2	Combinatorial theorems	46
2.3	A symmetric random walk	47
2.4	A Bernoulli excursion	48
2.5	A Brownian excursion	61
2.6	Single-server queues	63
2.7	Random-rooted trees	67
2.8	Limit distributions for random-rooted trees	72
2.9	Open problems	75
	Bibliography	75

2.1 INTRODUCTION

There is an intrinsic relationship between queueing processes and random graphs. Most of the results of this article are built on this relationship. We shall study the stochastic behavior of the fluctuations of the queue size during the initial busy period of a single-server queueing process and make use of the results obtained for the solutions of various problems connected with the heights and widths of random-rooted trees.

We consider a queueing process with one server. It is supposed that initially, when the server starts working, i customers ($i = 1, 2, \ldots$) are already waiting for service. The server serves these i customers and all the new customers in order of arrival as long as they keep coming. Let $i_0 = i$ and denote by $i_1, i_2, \ldots$ the number of arrivals during the first, second, ... service times respectively. If there are no more customers to serve, the initial busy period ends. The initial busy period consists of n services if and only if

$$i_1 + i_2 + \cdots + i_n = n - i \tag{2.1}$$

and

$$i_1 + i_2 + \cdots + i_r > r - i \text{ for } i \leq r < n. \tag{2.2}$$

0-8493-8074-x/95/$0.00+$.50

If the initial busy period consists of n services, let us associate the following graph with the queueing process considered: The graph has vertex set $(1,2,\ldots,n)$ and vertices r and s where $1 \le r < s \le n$ are joined by an edge if and only if the sth customer arrives during the service time of the rth customer. Evidently, the graph is a forest with vertex set $(1,2,\ldots,n)$. The forest consists of i rooted trees whose roots are vertices $1,2,\ldots,i$. Different queueing processes yield different forests. If $i=1$, the forest reduces to a rooted tree with vertex set $(1,2,\ldots,n)$, vertex 1 being the root of the tree.

If in the queueing process, the numbers of arrivals during the successive service times are random variables, which we shall denote by $\nu_1,\nu_2,\ldots$, and if the initial busy period consists of n services, then the corresponding graph is a random forest having n vertices and consisting of i rooted trees. We assume that $\{\nu_r\}$ is a sequence of independent and identically distributed discrete random variables with distribution

$$\mathbf{P}\{\nu_r = j\} = p_j \tag{2.3}$$

for $j=0,1,2,\ldots$. We obtain various models of random forests by choosing the distribution $\{p_j\}$ in a suitable way.

In what follows, we shall use some combinatorial theorems which are the generalizations of the classical ballot theorem of J. Bertrand [5]. We shall express the various limit distributions as the distributions of some functionals defined on the Brownian excursion $\{\eta^+(t), 0 \le t \le 1\}$. After studying the stochastic behavior of the initial busy period for various queueing processes, we derive some limit theorems for the heights and widths of random-rooted trees.

2.2 COMBINATORIAL THEOREMS

Let $\nu_1,\nu_2,\ldots,\nu_r,\ldots$ be independent discrete random variables which take on nonnegative integers only. Write $N_r = \nu_1+\nu_2+\cdots+\nu_r$ for $r \ge 1$ and $N_0 = 0$.

Theorem 2.1 *We have*

$$\mathbf{P}\{N_r < r \text{ for } 1 \le r \le n \text{ and } N_n = n-i\} = \tfrac{i}{n}\mathbf{P}\{N_n = n-i\} \tag{2.4}$$

for $0 \le i \le n$ *and* $n = 1,2,\ldots$.

Proof. This theorem is a generalization of the classical ballot theorem of J. Bertrand [5]. We shall prove (2.4) by mathematical induction. If $n=1$, then (2.4) is evidently true. Let us suppose that

$$\mathbf{P}\{N_r < r \text{ for } 1 \le r < n \text{ and } N_{n-1} = n-1-j\} = \frac{j}{n-1}\mathbf{P}\{N_{n-1} = n-1-j\} \tag{2.5}$$

for $0 \le j < n$ and $n \ge 2$. If $0 < i < n$ and $n \ge 2$, then by the induction hypothesis

$$\begin{aligned}\mathbf{P}\{N_r < r &\text{ for } 1 \le r \le n \text{ and } N_n = n-i\}\\ &= \sum_{j=i-1}^{n-1} \frac{j}{n-1}\mathbf{P}\{N_{n-1} = n-1-j, N_n = n-i\}\\ &= \mathbf{P}\{N_n = n-i\} - \frac{1}{n-1}\mathbf{E}[N_{n-1}\delta(N_n = n-i)]\end{aligned} \tag{2.6}$$

$$= \mathbf{P}\{N_n = n-i\} - \tfrac{n-i}{n}\mathbf{P}\{N_n = n-i\}.$$

Here $\delta(A)$ denotes the indicator variable of an event A, that is, $\delta(A)=1$ if A occurs and $\delta(A)=0$ if A does not occur. If $i=0$ or $i=n$, then (2.4) is trivially true. Consequently, (2.4) is true for all $0 \le i \le n$ and $n \ge 1$. See L. Takács [43], [44]. □

Define

$$\rho(k) = inf\{r : r - N_r = k, r \ge 0\} \tag{2.7}$$

for $k = 0, 1, 2, \ldots$. If $r - N_r < k$ for all $r \ge 0$, then $\rho(k) = \infty$.

Theorem 2.2 *We have*

$$\mathbf{P}\{\rho(k) = n\} = \tfrac{k}{n}\mathbf{P}\{n - N_n = k\} \tag{2.8}$$

for $n \ge 1$ and $k \ge 0$.

Proof. If $k > n$, then both sides of (2.8) are 0. If $0 \le k \le n$, and $n \ge 1$, then by Theorem 2.1,

$$\begin{aligned}\mathbf{P}\{\rho(k)=n\} &= \mathbf{P}\{r - N_r < k \text{ for } 0 \le r < n \text{ and } N_n = n-k\}\\ &= \mathbf{P}\{N_n - N_r < n - r \text{ for } 0 \le r < n \text{ and } N_n = n-k\} \qquad (2.9)\\ &= \mathbf{P}\{N_i < i \text{ for } 1 \le i \le n \text{ and } N_n = n-k\} = \tfrac{k}{n}\mathbf{P}\{N_n = n-k\}. \quad \square\end{aligned}$$

Theorem 2.3 *Let $f(k_1, k_2, \ldots, k_n)$ be a symmetric function of the variables $k_1, k_2, \ldots, k_n$ where $k_i = 0, 1, 2, \ldots$. Then,*

$$\sum_{\substack{k_1+k_2+\cdots+k_n=k\\ k_1+\cdots+k_r<r \text{ for } 1\le r\le n}} f(k_1, k_2, \ldots, k_n) = \frac{(n-k)}{n} \sum_{k_1+k_2+\cdots+k_n=k} f(k_1, k_2, \ldots, k_n) \tag{2.10}$$

for $0 \le k \le n$.

Proof. We can prove (2.10) by mathematical induction on n if we take into consideration that in (2.10), k_n may take on the values $0, 1, \ldots, k$ where $k \le n$. As an alternative, we can prove (2.10) by the repeated applications of Theorem 2.1. We note that (2.10) still remains valid if we assume only that the function $f(k_1, k_2, \ldots, k_n)$ is invariant under the n cyclic permutations of $(k_1, k_2, \ldots, k_n)$. □

If, in particular,

$$f(k_1, k_2, \ldots, k_n) = g(k_1)g(k_2)\ldots g(k_n), \tag{2.11}$$

then Theorem 2.3 is applicable and in (2.10)

$$\sum_{k_1+k_2+\cdots+k_n=k} f(k_1, k_2, \ldots, k_n) = \textit{Coeff. of } x^k \text{ in } \left\{\sum_{i=o}^{\infty} g(i)x^i\right\}^n. \tag{2.12}$$

2.3 A SYMMETRIC RANDOM WALK

Let $\xi_1, \xi_2, \ldots, \xi_n, \ldots$ be a sequence of independent random variables for which

$$\mathbf{P}\{\xi_n = 1\} = \mathbf{P}\{\xi_n = -1\} = 1/2 \tag{2.13}$$

if $n = 1, 2, \ldots$. Let $\eta_n = \xi_1 + \xi_2 + \cdots + \xi_n$ for $n \ge 1$ and $\eta_0 = 0$. Then $\{\eta_n, n \ge 0\}$

describes a symmetric random walk. We have

$$\mathbf{P}\{\eta_n = 2j - n\} = \binom{n}{j}\frac{1}{2^n} \tag{2.14}$$

for $j = 0, 1, \ldots, n$ and by the central limit theorem

$$\lim_{n\to\infty} \mathbf{P}\{\frac{\eta_n}{\sqrt{n}} \leq x\} = \Phi(x) \tag{2.15}$$

where

$$\Phi(x) = \frac{1}{\sqrt{2\pi}}\int_{-\infty}^{x} e^{-u^2/2}du \tag{2.16}$$

is the normal distribution function.

If we define

$$\rho(k) = inf\{n: \eta_n = k \text{ and } n \geq 0\} \tag{2.17}$$

for $k = 1, 2, \ldots,$ then by (2.8)

$$\mathbf{P}\{\rho(k) = k + 2j\} = \frac{k}{k+2j}\binom{k+2j}{j}\frac{1}{2^{k+2j}} \tag{2.18}$$

for $k = 1, 2, \ldots$ and $j = 0, 1, 2, \ldots$. We have also

$$\mathbf{P}\{\rho(k) \leq n\} = \mathbf{P}\{\eta_n \geq k\} + \mathbf{P}\{\eta_n < -k\} \tag{2.19}$$

for $k \geq 1$ and $n \geq 0$.

If in (2.18), $k = [x\sqrt{2j}]$ where $x > 0$, then we obtain that

$$\lim_{j\to\infty} j\mathbf{P}\{\rho(k) = k + 2j\} = x\varphi(x), \tag{2.20}$$

where

$$\varphi(x) = \frac{1}{\sqrt{2\pi}}e^{-x^2/2} \tag{2.21}$$

is the normal density function.

We note that

$$\sum_{j=0}^{n} \mathbf{P}\{\rho(k) = j\}\mathbf{P}\{\rho(\ell) = n - j\} = \mathbf{P}\{\rho(k+\ell) = n\} \tag{2.22}$$

is valid for any $k = 1, 2, \ldots, \ell = 1, 2, \ldots$ and $n = 1, 2, \ldots$. Furthermore, we have the following identity

$$\binom{2j-1}{j} - \binom{2j-1}{j+m} = 2^{2j}\sum_{i=1}^{m} \mathbf{P}\{\rho(2i) = 2j\} \tag{2.23}$$

for $1 \leq m \leq j$.

2.4 A BERNOULLI EXCURSION

Let us consider a set of sequences each consisting of $2n$ elements such that n elements are equal to $+1$, n elements are equal to -1, and the sum of the first i elements is ≥ 0 for every $i = 1, 2, \ldots, 2n$. The number of such sequences is given by the nth Catalan number,

$$C_n = \binom{2n}{n}\frac{1}{n+1}. \tag{2.24}$$

We have $C_0 = 1$, $C_1 = 1$, $C_2 = 2$, $C_3 = 5$, $C_4 = 14$, $C_5 = 42$, ... Let us choose a sequence at random, assuming that all the possible C_n sequences are equally probable. Denote by η_i^+ $(i = 1, 2, \ldots, 2n)$ the sum of the first i elements in a random sequence and set $\eta_0^+ = 0$. The sequence $\{\eta_0^+, \eta_1^+, \ldots, \eta_{2n}^+\}$ describes a random walk which is usually called a Bernoulli excursion.

The distribution of the maximum. Let us define a random variable δ_n^+ by the following equation

$$\delta_n^+ = max(\eta_0^+, \eta_1^+, \ldots, \eta_{2n}^+). \tag{2.25}$$

If we write

$$\mathbf{P}\{\delta_n^+ \leq k\} = T_n(k)/C_n \tag{2.26}$$

for $k \geq 1$ and $n \geq 1$, then we have

$$T_n(k) = \sum_{j=0}^{n-1} T_j(k-1)T_{n-j-1}(k) \tag{2.27}$$

for $k \geq 1$ and $n \geq 1$ where $T_0(k) = 1$ for $k \geq 0$ and $T_n(0) = 0$ for $n \geq 1$. See Table 1 for $0 \leq n \leq 8$. In proving (2.27), we take into consideration that the event $\delta_n^+ \leq k$ can occur in such a way that the smallest positive integer r for which $\eta_r^+ = 0$ is $r = 2j + 2$ where $j = 0, 1, \ldots, n-1$.

By the reflection principle, we can prove that

$$T_n(k-1) = \sum_j \left[\binom{2n}{n + j(k+1)} - \binom{2n}{n - 1 + j(k+1)}\right] \tag{2.28}$$

for $k \geq 1$ and $n \geq 1$. Also

$$T_n(k-1) = \frac{2^{2n+1}}{k+1}\sum_{r=1}^{k}(cos\frac{r\pi}{k+1})^{2n}(sin\frac{r\pi}{k+1})^2 \tag{2.29}$$

for $k \geq 1$ and $n \geq 1$. See L. Takács [42], [45].

Table 1: $T_n(k)$

n/k	0	1	2	3	4	5	6	7	8	9
0	1	1	1	1	1	1	1	1	1	1
1	0	1	1	1	1	1	1	1	1	1
2	0	1	2	2	2	2	2	2	2	2
3	0	1	4	5	5	5	5	5	5	5
4	0	1	8	13	14	14	14	14	14	14
5	0	1	16	34	41	42	42	42	42	42
6	0	1	32	89	122	131	132	132	132	132
7	0	1	64	233	365	417	428	429	429	429
8	0	1	128	610	1094	1341	1416	1429	1430	1430

The generating function

$$U_k(w) = \sum_{n=0}^{\infty} T_n(k)w^n, \tag{2.30}$$

is convergent for $|w| < 1/4$ and by (2.27) we obtain that

$$U_k(w) = 1 + wU_{k-1}(w)U_k(w), \tag{2.31}$$

that is,

$$U_k(w) = \frac{1}{1 - wU_{k-1}(w)} \tag{2.32}$$

for $k = 1, 2, \ldots$ where $U_0(w) = 1$. By (2.32) we obtain that

$$U_k(w) = P_k(w)/P_{k+1}(w) \tag{2.33}$$

for $k = 0, 1, \ldots$ where

$$P_k(w) = \sum_{j=0}^{[k/2]} (-1)^j \binom{k-j}{j} w^j. \tag{2.34}$$

Theorem 2.4 *We have*

$$\lim_{n\to\infty} \mathbf{P}\{\frac{\delta_n^+}{\sqrt{2n}} \le x\} = F(x) \tag{2.35}$$

where

$$F(x) = \sum_{j=-\infty}^{\infty} (1 - 4j^2x^2)e^{-2j^2x^2} = \frac{\sqrt{2}\pi^{5/2}}{x^3} \sum_{j=0}^{\infty} j^2 e^{-j^2\pi^2/(2x^2)} \tag{2.36}$$

for $x > 0$ and $F(x) = 0$ for $x \le 0$.

Proof. If in (2.28) and (2.29) we put $k = [x\sqrt{2n}]$ where $x > 0$ and let $n \to \infty$, we get (2.36). □

The moments

$$m_r = \int_0^\infty x^r dF(x) \tag{2.37}$$

exist for $r \ge 0$. We have $m_0 = 1$, $m_1 = \sqrt{\pi/2}$, $m_2 = \pi^2/6$, $m_3 = 3\sqrt{\pi}\zeta(3)/\sqrt{8}$, $m_4 = \pi^4/30$. For $r > 1$,

$$m_r = 2(r-1)\Gamma(\tfrac{r}{2} + 1)\zeta(r)/2^{r/2} \tag{2.38}$$

where

$$\zeta(r) = \sum_{n=1}^{\infty} \frac{1}{n^r} \tag{2.39}$$

is the Riemann zeta function.

Theorem 2.5 *We have*

$$\lim_{n\to\infty} \mathbf{E}\left[(\frac{\delta_n^+}{\sqrt{2n}})^r \right] = m_r \tag{2.40}$$

for $r = 0, 1, \ldots$ where m_r is defined by (2.37).

Proof. By formula (2.28) we obtain that

$$\mathbf{E}[(\delta_n^+)^r] = \sum_{j=1}^{r} (-1)^{j-1}\binom{r}{j}(2^j - 1)a_{r-j}(n+1) + (-1)^r \tag{2.41}$$

if $r \ge 1$, where

$$a_r(n) = \frac{2}{\binom{2n}{n}} \sum_{k=1}^{n} (\frac{2k^2}{n} - 1)\binom{2n}{n+k} \sum_{d \mid k} d^r. \tag{2.42}$$

Hence we can derive an asymptotic formula for the moments of δ_n^+ as $n\to\infty$. We can also derive (2.40) for $r=1$ by using the results of N.G. de Bruijn, D.E. Knuth and S.O. Rice [13], and for $r\geq 1$ by using the results of R. Kemp [23]. □

The distribution of the area. Let us define a random variable ω_n by the following equations

$$2n\omega_n = \sum_{i=0}^{2n} \eta_i^+ = 2n\int_0^1 \eta_{[2nt]}^+ dt \tag{2.43}$$

for $n=1,2,\dots$ and set $\omega_0=0$.

The distribution of $2n\omega_n$ is determined by the generating function

$$\varphi_n(z) = C_n \sum_{j=0}^{\binom{n}{2}} \mathbf{P}\{2n\omega_n = n+2j\}z^j \tag{2.44}$$

which can be obtained by the recurrence formula

$$\varphi_n(z) = \sum_{i=1}^{n} \varphi_{i-1}(z)\varphi_{n-i}(z)z^{i-1} \tag{2.45}$$

for $n=1,2,\dots$ and $\varphi_0(z)=1$. In proving (2.45) we take into consideration that the smallest positive integer i for which $\eta_{2i}^+=0$ may be $i=1,2,\dots,n$. See L. Takács [50], [56].

Theorem 2.6 *If $r=0,1,2,\dots$, then*

$$\lim_{n\to\infty} \mathbf{E}[(\omega_n/\sqrt{2n})^r] = M_r \tag{2.46}$$

exists and

$$M_r = K_r \frac{4\sqrt{\pi}\, r!}{\Gamma\left(\frac{3r-1}{2}\right)2^{r/2}} \tag{2.47}$$

where $K_0=-1/2$, $K_1=1/8$ and

$$K_r = \frac{(3r-4)}{4}K_{r-1} + \sum_{j=1}^{r-1} K_j K_{r-j} \tag{2.48}$$

for $r=2,3,\dots$.

Proof. If

$$F(z,w) = w\sum_{n=0}^{\infty} \varphi_n(z)(zw)^n \tag{2.49}$$

for $|z|\leq 1$ and $|w|\leq 1/4$, then by (2.45) we have

$$F(z,w) = w + F(z,w)F(z,zw). \tag{2.50}$$

If $|w|<1/4$, then

$$B_r(w) = \frac{1}{r!}\left(\frac{\partial^r F(z,w)}{\partial z^r}\right)_{z=1} = w\sum_{n=0}^{\infty} C_n \mathbf{E}\left[\binom{(2n\omega_n+n)/2}{r}\right]w^n \tag{2.51}$$

is convergent and by (2.50) we obtain that

$$B_0(w) = w\sum_{n=0}^{\infty} C_n w^n = (1-\sqrt{1-4w})/2, \tag{2.52}$$

and

$$B_r(w) = \sum_{j=0}^{r} B_{r-j}(w)\sum_{i=0}^{j} B_{j-i}^{(i)}(w)w^i/i! \tag{2.53}$$

for $r \geq 1$. By (2.53) we can determine $B_r(w)$ step by step for $r = 1, 2, \ldots$. If we write $R = 1 - 4w$, then

$$B_0(w) = (1 - \sqrt{R})/2, \tag{2.54}$$

and

$$B_1(w) = w(R^{-1} - R^{-1/2})/2. \tag{2.55}$$

By (2.53) the moments $\mathbf{E}[(2n\omega_n)^r]$ can be calculated explicitly for every $r = 1, 2, \ldots$. By (2.53) we can draw the conclusion that $B_r(w)/(wR^{1/2})$ is a polynomial of degree $3r$ in $R^{-1/2}$ if $r \geq 1$. Consequently,

$$B_r(w) = K_r R^{-(3r-1)/2} + \cdots \tag{2.56}$$

where K_r is a constant. In (2.56) the neglected term is 1/2 if $r = 0$. If $r \geq 1$, the neglected terms are constant multiples of $R^{-j/2}$ for $-3 \leq j \leq 3r - 2$. In (2.56), $K_0 = -1/2$, and $K_1 = 1/8$.

If in (2.53) we retain only those terms which contribute to the determination of K_r in (2.56), we obtain that

$$R^{1/2} B_r(w) = \sum_{j=1}^{r-1} B_j(w) B_{r-j}(w) + B_0(w) B_{r-1}^{(1)}(w) w + \cdots \tag{2.57}$$

for $r \geq 2$. Here $w = (1 - R)/4$, and if we form the coefficient of $R^{-(3r-2)/2}$ on both sides of (2.57), we obtain that

$$K_r = \frac{(3r-4)}{4} K_{r-1} + \sum_{j=1}^{r-1} K_j K_{r-j} \tag{2.58}$$

for $r = 2, 3, \ldots$. This proves (2.48).

To find explicit formulas for $\mathbf{E}[(2n\omega_n)^r]$, we need the following expansion: If a is a positive real number, then

$$R^{-a} = (1 - 4w)^{-a} = \sum_{n=0}^{\infty} \binom{-a}{n} (-4)^n w^n \tag{2.59}$$

for $|w| < 1/4$. We can find the asymptotic behavior of $\mathbf{E}[(2n\omega_n)^r]$ as $n \to \infty$ if we use that

$$\binom{-a}{n}(-4)^n = \binom{a+n-1}{n} 4^n \sim \frac{4^n n^{a-1}}{\Gamma(a)} \tag{2.60}$$

as $n \to \infty$.

By (2.51), (2.56) and (2.60)

$$C_n \mathbf{E}\left[\binom{(2n\omega_n + n)/2}{r}\right] \sim K_r \frac{4^{n+1} n^{3(r-1)/2}}{\Gamma\left(\frac{3r-1}{2}\right)} \tag{2.61}$$

for $r = 0, 1, \ldots$ as $n \to \infty$. Because

$$C_n n \sqrt{n\pi} \sim 4^n \tag{2.62}$$

it follows from (2.61) that

$$\mathbf{E}[(n\omega_n)^r] \sim K_r \frac{4\sqrt{\pi} r!}{\Gamma\left(\frac{3r-1}{2}\right)} n^{3r/2} \tag{2.63}$$

for $r = 0, 1, 2, \ldots$ as $n \to \infty$. By (2.63)

$$\lim_{n \to \infty} \mathbf{E}\left[\left(\frac{\omega_n}{\sqrt{2n}}\right)^r\right] = K_r \frac{4\sqrt{\pi} r!}{\Gamma\left(\frac{3r-1}{2}\right) 2^{r/2}} \tag{2.64}$$

for $r = 0, 1, 2, \ldots$. This completes the proof of Theorem 2.6. Table 2 contains K_r and M_r for $r \le 10$. □

Alternatively, we can calculate $K_r (r = 1, 2, \ldots)$ by the following recurrence formula

$$K_r = \frac{(6r+1)}{2(6r-1)} \alpha_r - \sum_{j=1}^{r} \alpha_j K_{r-j} \tag{2.65}$$

for $r \ge 1$ where

$$\alpha_j = \frac{\Gamma(3j + \frac{1}{2})}{\Gamma(j + \frac{1}{2})(36)^j j!} = \frac{(2j+1)(2j+3)\cdots(6j-1)}{(144)^j j!} \tag{2.66}$$

for $j \ge 1$ and $\alpha_0 = 1$. Clearly,

$$\alpha_j = \frac{3}{4}\left(j - 1 + \frac{5}{36j}\right)\alpha_{j-1} \tag{2.67}$$

for $j \ge 1$.

Table 2: The moments M_r

r	K_r	M_r
0	$-\frac{1}{2}$	1
1	$\frac{1}{8}$	$\sqrt{\frac{\pi}{8}}$
2	$\frac{5}{64}$	$\frac{5}{12}$
3	$\frac{15}{128}$	$\frac{15}{32}\sqrt{\frac{\pi}{8}}$
4	$\frac{1105}{4096}$	$\frac{221}{1008}$
5	$\frac{1695}{2048}$	$\frac{565}{2048}\sqrt{\frac{\pi}{8}}$
6	$\frac{414125}{131072}$	$\frac{82825}{576576}$
7	$\frac{59025}{4096}$	$\frac{19675}{98304}\sqrt{\frac{\pi}{8}}$
8	$\frac{1282031525}{16777216}$	$\frac{256406305}{2234808576}$
9	$\frac{242183775}{524288}$	$\frac{16145585}{92274688}\sqrt{\frac{\pi}{8}}$
10	$\frac{1683480621875}{536870912}$	$\frac{304702375}{2790982656}$

To prove (2.65) let us introduce the formal generating function

$$K(z) = \sum_{r=0}^{\infty} (-1)^r K_r z^r. \tag{2.68}$$

By (2.48) we obtain that

$$3z^2K'(z) - zK(z) = 4[K(z)]^2 - 1. \tag{2.69}$$

If

$$K(z) = Ai'(z^{-2/3})z^{1/3}/[2Ai(z^{-2/3})] \tag{2.70}$$

where

$$Ai(z) = \frac{1}{\pi}\int_0^\infty cos\left(\frac{t^3}{3} + tz\right)dt \tag{2.71}$$

is the Airy function, then $K(z)$ satisfies (2.69) and has the following asymptotic series

$$K(z) \sim \sum_{r=0}^{\infty} (-1)^r K_r z^r \tag{2.72}$$

whenever $Re(z) \geq 0$ and $z \to 0$. By comparing the coefficients of similar powers of z on both sides of the equation

$$2Ai(z^{-2/3})K(z) = Ai'(z^{-2/3})z^{1/3}, \tag{2.73}$$

we obtain (2.65).

By (2.48)

$$K_r \geq \frac{3(r-1)}{4}K_{r-1} \tag{2.74}$$

for $r \geq 2$, and $K_1 = 1/8$. Accordingly, $(4/3)^r K_r/(r-1)!$ $(r = 2, 3, \ldots)$ is an increasing sequence. By (2.65),

$$K_r \leq [6r/(6r-1)]\alpha_r \tag{2.75}$$

for $r \geq 1$, and by (2.66),

$$\lim_{r \to \infty}\left(\frac{4}{3}\right)^r \frac{\alpha_r}{(r-1)!} = \frac{1}{2\pi}. \tag{2.76}$$

Thus

$$\lim_{r \to \infty}\left(\frac{4}{3}\right)^r \frac{K_r}{(r-1)!} = \gamma \tag{2.77}$$

exists and $0 < \gamma \leq 1/(2\pi)$. We can prove that $\gamma = 1/(2\pi)$ and, therefore, by (2.47) we have

$$M_r \sim \frac{6r}{\sqrt{2}}\left(\frac{r}{12e}\right)^{r/2} \tag{2.78}$$

as $r \to \infty$.

For the asymptotic distribution of ω_n as $n \to \infty$, we have the following result.

Theorem 2.7 *There exists a distribution function $W(x)$ such that*

$$\lim_{n \to \infty} \mathbf{P}\left\{\frac{\omega_n}{\sqrt{2n}} \leq x\right\} = W(x) \tag{2.79}$$

in every continuity point of $W(x)$. We have $W(x) = 0$ for $x \leq 0$ and $W(x)$ is uniquely determined by its moments

$$\int_0^\infty x^r dW(x) = M_r \tag{2.80}$$

for $r = 0, 1, 2, \ldots$ where M_r is defined by (2.47) and (2.48).

Proof. It follows from (2.77) that

$$\sum_{r=1}^{\infty} M_r^{-1/r} = \infty. \tag{2.81}$$

By (2.46), the sequence $\{M_r\}$ is a moment sequence. Since the condition (2.81) is satisfied, we can conclude from a theorem of T. Carleman [7, p. 81] that there exists one and only one distribution function $W(x)$ such that $W(0) = 0$ and (2.80) holds for $r = 0, 1, 2, \ldots$. By the moment convergence theorem of M. Fréchet and J. Shohat [15], (2.80) implies (2.79). This completes the proof of the theorem. □

Theorem 2.8 *If $x > 0$, we have*

$$W(x) = \frac{\sqrt{6}}{x} \sum_{k=1}^{\infty} e^{-v_k} v_k^{2/3} U(1/6, 4/3, v_k) \tag{2.82}$$

and

$$W'(x) = \frac{2\sqrt{6}}{x^2} \sum_{k=1}^{\infty} e^{-v_k} v_k^{2/3} U(-5/6, 4/3, v_k) \tag{2.83}$$

where $U(a, b, x)$ is the confluent hypergeometric function,

$$v_k = 2a_k^3/(27x^2), \tag{2.84}$$

and $z = -a_k (k = 1, 2, \ldots)$ are the zeros of the Airy function $Ai(z)$ arranged so that $0 < a_1 < a_2 < \ldots < a_k < \ldots$.

Proof. Let

$$\psi(s) = \int_0^{\infty} e^{-sx} dW(x) \tag{2.85}$$

be the Laplace-Stieltjes transform of $W(x)$. We have

$$\psi(s) = \sum_{r=0}^{\infty} (-1)^r M_r s^r / r! \tag{2.86}$$

and the series (2.86) is convergent on the whole complex plane. This follows from (2.78).

Since

$$\left| e^{-z} - \sum_{r=0}^{m-1} \frac{(-1)^r}{r!} z^r \right| = \left| \int_0^z e^{-u} \frac{u^{m-1}}{(m-1)!} du \right| \leq \frac{|z|^m}{m!} \tag{2.87}$$

if $Re(z) \geq 0$ and $m = 1, 2, \ldots$, we obtain that

$$\left| \int_0^{\infty} e^{-sz} [\psi(\sqrt{2} s^{3/2}) - 1] s^{-3/2} ds - 4\sqrt{\pi} \sum_{r=1}^{m-1} (-1)^r K_r z^{-(3r-1)/2} \right|$$

$$\leq 4\sqrt{\pi} K_m |z|^{-(3m-1)/2} \tag{2.88}$$

if $Re(z) \geq 0$. By comparing (2.72) and (2.88), we can draw the conclusion that

$$\int_0^{\infty} e^{-sz} [\psi(\sqrt{2} s^{3/2}) - 1] s^{-3/2} ds = 2\sqrt{\pi} \left[\frac{Ai'(z)}{Ai(z)} + z^{1/2} \right] \tag{2.89}$$

for $Re(z) \geq 0$. In (2.89), the right-hand side is asymptotically equal to $-\sqrt{\pi}/(2z^2)$ if $|\arg z| < \pi$ and $|z| \to \infty$. By inverting (2.89), we obtain that, if

$s > 0$, then

$$\psi(s) = \frac{s\sqrt{2\pi}}{2\pi i} \int_C \frac{Ai'(z)}{Ai(z)} e^{zs^{2/3}2^{-1/3}} dz \tag{2.90}$$

where the path of integration C starts at $-\infty$ on the real axis, encircles the origin in the counterclockwise direction, and returns to the starting point. In (2.90) we used that

$$\frac{2\sqrt{\pi}}{2\pi i} \int_C e^{sz} z^{1/2} dz = -s^{-3/2}. \tag{2.91}$$

By analytical continuation, (2.90) is valid for $Re(s) > 0$ too. The function $Ai(z)$ has zeros only on the negative real axis, namely, $z = -a_k (k = 1, 2, \ldots)$ where $0 < a_1 < a_2 < \ldots < a_k < \ldots$. By the theorem of residues, we obtain that

$$\psi(s) = s\sqrt{2\pi} \sum_{j=1}^{\infty} e^{-a_j s^{2/3} 2^{-1/3}} \tag{2.92}$$

for $Re(s) > 0$. From (2.92), we obtain (2.82) and (2.83) by inversion. Tables 3 and 4 contain $W(x)$ and $W'(x)$ for $0 < x \leq 1.5$. □

Table 3: The function $W(x)$

x	$W(x)$	x	$W(x)$
0.05	0.00000000	0.80	0.86427925
0.10	0.00000000	0.85	0.91153523
0.15	0.00000000	0.90	0.94430335
0.20	0.00000000	0.95	0.96610624
0.25	0.00001007	1.00	0.98005322
0.30	0.00071659	1.05	0.98864280
0.35	0.00858774	1.10	0.99374158
0.40	0.04027493	1.15	0.99666130
0.45	0.11029731	1.20	0.99827530
0.50	0.21745116	1.25	0.99913710
0.55	0.34719272	1.30	0.99958179
0.60	0.48159861	1.35	0.99980363
0.65	0.60641841	1.40	0.99991065
0.70	0.71328366	1.45	0.99996060
0.75	0.79906372	1.50	0.99998317

For the definitions of the confluent hypergeometric function and the Airy function, we refer to L.J. Slater [40], J.C.P. Miller [34], and M. Abramowitz and I.A. Stegun [1]. The first 50 zeros of $Ai(z)$ and $Ai'(z)$ can be found in J.C.P. Miller [34, p. 43] for 8 decimals. See also M. Abramowitz and I.A. Stegun [1, p. 478].

The distribution of the local time. For the Bernoulli excursion $\{\eta_0^+, \eta_1^+, \ldots, \eta_{2n}^+\}$ let us define $\tau_n^+(m)$ $(m = 1, 2, \ldots, 2n)$ as the number of subscripts $r = 1, 2, \ldots, 2n$ for which $\eta_{r-1}^+ = m - 1$ and $\eta_r^+ = m$. If $\eta_{r-1}^+ = m - 1$ and $\eta_r^+ = m$, then we say that a transition $m - 1 \rightarrow m$ occurs at the rth step, and $\tau_n^+(m)$ is the number of transitions $m-1 \to m$ among the $2n$ steps.

Table 4: The function $W'(x)$

x	$W'(x)$	x	$W'(x)$
0.05	0.00000000	0.80	1.11391248
0.10	0.00000000	0.85	0.78816801
0.15	0.00000000	0.90	0.53444938
0.20	0.00000071	0.95	0.34809189
0.25	0.00113999	1.00	0.21811909
0.30	0.04549111	1.05	0.13165761
0.35	0.33036908	1.10	0.07662545
0.40	0.99108575	1.15	0.04303415
0.45	1.80425970	1.20	0.02333689
0.50	2.42954788	1.25	0.01222620
0.55	2.69798891	1.30	0.00619087
0.60	2.63013465	1.35	0.00303102
0.65	2.33507134	1.40	0.00143530
0.70	1.92967395	1.45	0.00065756
0.75	1.50379914	1.50	0.00029152

We have

$$\mathbf{P}\{\tau_n^+(m) = j\} = \sum_{r=j}^{n} (-1)^{r-j}\binom{r}{j} \mathbf{E}\left[\binom{\tau_n^+(m)}{r}\right] \tag{2.93}$$

for $j = 0, 1, \ldots, n$. By (2.93) the distribution of $\tau_n^+(m)$ is determined if we know the binomial moments of $\tau_n^+(m)$.

We shall consider the symmetric random walk defined in Section 2.3 and make use of the formula (2.18) which determines the distribution of $\rho(m)$, the first passage time through m.

Theorem 2.9 *We have*

$$C_n \mathbf{E}[\tau_n^+(m)] = 2^{2n+1}\mathbf{P}\{\rho(2m+1) = 2n+1\} \tag{2.94}$$

and

$$C_n \mathbf{E}\left[\binom{\tau_n^+(m)}{r}\right]$$

$$= 2^{2n+1}\sum_{i_1=1}^{m} \cdots \sum_{i_{r-1}=1}^{m} \mathbf{P}\{\rho(2m+2i_1+\cdots+2i_{r-1}+1) = 2n+1\} \tag{2.95}$$

for $1 \le m \le n$ *and* $r \ge 2$ *where* $\mathbf{P}\{\rho(k) = j\}$ *is determined by* (2.18).

Proof. Denote by A_r $(r = 1, 2, \ldots, n)$ the event that, in the Bernoulli excursion, a transition $m-1 \to m$ takes place at the rth step. Then

$$\mathbf{E}[\tau_n^+(m)] = \sum_{0 \le i \le (2n-m)/2} \mathbf{P}\{A_{m+2i}\}. \tag{2.96}$$

Hence we obtain that

$$C_n \mathbf{E}[\tau_n^+(m)] = 2^{2n+1} \sum_{0 \le i \le (2n-m)/2} \mathbf{P}\{\rho(m) = m + 2i\} \cdot$$

$$\mathbf{P}\{\rho(m+1) = 2n - m - 2i + 1\}. \tag{2.97}$$

By using (2.22), we can express (2.97) in the form of (2.94).

In a similar way, we obtain that

$$\mathbf{E}\left[\binom{\tau_n^+(m)}{r}\right] = \sum_{0 \le j_1 < j_2 < \ldots < j_r \le (2n-m)/2} \mathbf{P}\{A_{m+2j_1} A_{m+2j_2} \cdots A_{m+2j_r}\}. \tag{2.98}$$

Hence we get for $r \ge 2$

$$C_n \mathbf{E}\left[\binom{\tau_n^+(m)}{r}\right] = 2^{2n+1} \sum_{\substack{m+2s+2j \le n \\ s \ge 0,\, j \ge 1}} \mathbf{P}\{\rho(m) = m + 2s\} \cdot$$

$$\mathbf{P}\{\rho(m+1) = 2n - m - 2s - 2j + 1\} \cdot \tag{2.99}$$

$$\cdot \sum_{i_1=1}^{m} \cdots \sum_{i_{r-1}=1}^{m} \mathbf{P}\{\rho(2i_1 + \cdots + 2i_{r-1}) = 2j\}.$$

Here we used (2.23). If we apply (2.22) repeatedly in (2.99), we get (2.95). □

By (2.93), the binomial moments (2.94) and (2.95) determine the distribution of $\tau_n^+(m)$.

Now let us consider the asymptotic behavior of the moments of $\tau_n^+(m)$ whenever $m = [\alpha\sqrt{2n}]$, $\alpha > 0$ and $n \to \infty$.

For $\alpha > 0$ and $r = 0, 1, 2, \ldots$ define

$$I_r(\alpha) = \int_1^\infty e^{-\alpha^2 x^2/2}(x-1)^r dx. \tag{2.100}$$

We have

$$I_0(\alpha) = \frac{\sqrt{2\pi}}{\alpha}[1 - \Phi(\alpha)] \tag{2.101}$$

where $\Phi(\alpha)$ is defined by (2.16),

$$I_1(\alpha) = \frac{1}{\alpha^2} e^{-\alpha^2/2} - \frac{\sqrt{2\pi}}{\alpha}[1 - \Phi(\alpha)] \tag{2.102}$$

and

$$I_{r+1}(\alpha) = \frac{r}{\alpha^2} I_{r-1}(\alpha) - I_r(\alpha) \tag{2.103}$$

for $r = 1, 2, \ldots$.

Theorem 2.10 *If $\alpha > 0$, then,*

$$\lim_{n \to \infty} \mathbf{E}\left[\left(\frac{2\tau_n^+([\alpha\sqrt{2n}])}{\sqrt{2n}}\right)^r\right] = \mu_r(\alpha) \tag{2.104}$$

exists for $r = 0, 1, 2, \ldots$ and we have $\mu_0(\alpha) = 1$,

$$\mu_1(\alpha) = 4\alpha e^{-2\alpha^2} \tag{2.105}$$

and

$$\mu_r(\alpha) = 2^{r+1}\alpha^r r(r-1)\sum_{k=1}^{r}(-1)^{k-1}\binom{r-1}{k-1}k^r[I_{r-1}(2k\alpha)+I_{r-2}(2k\alpha)] \tag{2.106}$$

for $r \geq 2$.

Proof. If $r = 1$, then by (2.94) and (2.20) we obtain (2.105). If in (2.95) we put $i_s = mx_s$, $m = [\alpha\sqrt{2n}]$, $\alpha > 0$ and $\Delta i_s = m\Delta x_s = 1$, then we obtain that

$$\mathbf{E}\left[\binom{\tau_n^+([\alpha\sqrt{2n}])}{r}\right] \sim \frac{n^{r/2}}{r!2^{r/2}}\mu_r(\alpha) \tag{2.107}$$

for $r \geq 2$ as $n \to \infty$ where

$$\mu_r(\alpha) = 2^r\alpha^r r!\int_0^1 \cdots \int_0^1 a(x_1 + x_2 + \cdots + x_{r-1}, \alpha)dx_1dx_2\ldots dx_{r-1} \tag{2.108}$$

and

$$a(x,\alpha) = 2(1+x)e^{-2\alpha^2(1+x)^2}. \tag{2.109}$$

By (2.108),

$$\mu_r(\alpha) = 2^r\alpha^r r!\int_0^{r-1} a(x,\alpha)f_{r-1}(x)dx \tag{2.110}$$

for $r \geq 2$, where

$$f_{r-1}(x) = \sum_{j=0}^{[x]}(-1)^j\binom{r-1}{j}\frac{(x-j)^{r-2}}{(r-2)!} \tag{2.111}$$

for $r \geq 2$ and $x \geq 0$. If $x > r-1$, then $f_{r-1}(x) = 0$. □

In particular, by (2.106) we obtain that

$$\mu_2(\alpha) = 4[e^{-2\alpha^2} - e^{-8\alpha^2}] \tag{2.112}$$

and

$$\mu_3(\alpha) = 12\sqrt{2\pi}[2\Phi(4\alpha) - \Phi(2\alpha) - \Phi(6\alpha)]. \tag{2.113}$$

Let us denote by $H_n(x)$ the nth Hermite polynomial defined by

$$H_n(x) = n!\sum_{j=0}^{[n/2]}\frac{(-1)^j x^{n-2j}}{2^j j!(n-2j)!}. \tag{2.114}$$

We have $H_0(x) = 1$, $H_1(x) = x$ and

$$H_n(x) = xH_{n-1}(x) - (n-1)H_{n-2}(x) \tag{2.115}$$

for $n \geq 2$.

Theorem 2.11 *If* $\alpha > 0$, *then there exists a distribution function* $G_\alpha(x)$ *such that*

$$\lim_{n\to\infty}\mathbf{P}\left\{\frac{2\tau_n^+([\alpha\sqrt{2n}])}{\sqrt{2n}} \leq x\right\} = G_\alpha(x) \tag{2.116}$$

for $x > 0$ *where*

$$G_\alpha(x) = 1 - 2\sum_{k=1}^{\infty}\sum_{j=0}^{k-1}\binom{k-1}{j}\frac{(-1)^j x^j}{j!}e^{-(x+2k\alpha)^2/2}H_{j+2}(x+2k\alpha) \tag{2.117}$$

for $x > 0$,

$$G_\alpha(0) = F(\alpha) \tag{2.118}$$

defined by (2.36) and $G_\alpha(x) = 0$ for $x < 0$. If $x > 0$, then

$$\frac{dG_\alpha(x)}{dx} = 2\sum_{k=1}^{\infty}\sum_{j=1}^{k}\binom{k}{j}\frac{(-1)^{j-1}x^{j-1}}{(j-1)!}e^{-(x+2k\alpha)^2/2}H_{j+2}(x+2k\alpha). \tag{2.119}$$

Proof. By (2.110) it follows that

$$\mu_r(\alpha)/r! < (2\alpha)^r/\alpha \tag{2.120}$$

for $r \geq 2$. Accordingly, there exists one and only one distribution function $G_\alpha(x)$ such that $G_\alpha(x) = 0$ for $x < 0$ and

$$\int_{-0}^{\infty} x^r dG_\alpha(x) = \mu_r(\alpha) \tag{2.121}$$

for $r \geq 0$. By the moment convergence theorem of M. Fréchet and J. Shohat [15], it follows from (2.104) that (2.116) holds in every continuity point of $G_\alpha(x)$. If $|s| < 1/(2\alpha)$, then the Laplace-Stieltjes transform

$$\Psi_\alpha(s) = \int_{-0}^{\infty} e^{-sx}dG_\alpha(x) \tag{2.122}$$

can be expressed as

$$\Psi_\alpha(s) = \sum_{r=0}^{\infty}(-1)^r\mu_r(\alpha)s^r/r!. \tag{2.123}$$

For a fixed α, define

$$a(x) = a(x,\alpha) = 2(1+x)e^{-2\alpha^2(1+x)^2} \tag{2.124}$$

and

$$b(x) = a'(x-1) = 2e^{-2\alpha^2x^2}[1-4\alpha^2x^2]. \tag{2.125}$$

In (2.123), $\mu_0(\alpha) = 1$, $\mu_1(\alpha) = 2\alpha a(0)$ and

$$\mu_r(\alpha) = (2\alpha)^r r!\int_0^\infty a(x)f_{r-1}(x)dx \tag{2.126}$$

for $r \geq 2$. If $r \geq 2$, then,

$$\int_0^\infty a(x)f_{r-1}(x)dx = \sum_{k=1}^{r}\frac{(-1)^{k-1}(r-1)}{(k-1)!(r-k)!}\int_k^\infty a(y-1)(y-k)^{r-2}dy \tag{2.127}$$

and

$$(r-1)\int_k^\infty a(y-1)(y-k)^{r-2}dy = -\int_k^\infty b(y)(y-k)^{r-1}dy. \tag{2.128}$$

By (2.126), (2.127) and (2.128) we obtain that

$$\Psi_\alpha(s) = 1 + \sum_{k=1}^{\infty}\frac{(2\alpha s)^k}{(k-1)!}\int_k^\infty e^{-2\alpha s(y-k)}(y-k)^{k-1}b(y)dy. \tag{2.129}$$

By integrating by parts, we obtain that in (2.129)

$$
\begin{aligned}
&(2\alpha s)^k \int_k^\infty e^{-2\alpha s(y-k)}(y-k)^{k-1}b(y)dy \\
&= (k-1)!b(k) + \int_k^\infty e^{-2\alpha s(y-k)} \frac{d^k(y-k)^{k-1}b(y)}{dy^k}dy.
\end{aligned}
\tag{2.130}
$$

Thus

$$
\Psi_\alpha(s) = 1 + \sum_{k=1}^\infty b(k) + \sum_{k=1}^\infty \frac{1}{(k-1)!} \int_k^\infty e^{-2\alpha s(y-k)} \frac{d^k(y-k)^{k-1}b(y)}{dy^k}dy. \tag{2.131}
$$

If in (2.131) we substitute $x = 2\alpha(y-k)$, then

$$
\Psi_\alpha(s) = 1 + \sum_{k=1}^\infty b(k) + \sum_{k=1}^\infty \frac{1}{(k-1)!} \int_0^\infty e^{-sx} \frac{d^k x^{k-1}b(k+x/(2\alpha))}{dx^k}dx \tag{2.132}
$$

for $|s| < 1/(2\alpha)$. From (2.132) it follows that

$$
G_\alpha(0) = 1 + \sum_{k=1}^\infty b(k) = 1 + 2\sum_{j=1}^\infty (1-4j^2\alpha^2)e^{-2j^2\alpha^2}. \tag{2.133}
$$

This proves (2.118). Furthermore, we have

$$
G_\alpha(x) = 1 + \sum_{k=1}^\infty \frac{1}{(k-1)!} \frac{d^{k-1}x^{k-1}b(k+x/(2\alpha))}{dx^{k-1}} \tag{2.134}
$$

for $x \geq 0$, and

$$
\frac{dG_\alpha(x)}{dx} = \sum_{k=1}^\infty \frac{1}{(k-1)!} \frac{d^k x^{k-1}b(k+x/(2\alpha))}{dx^k} \tag{2.135}
$$

for $x > 0$. Hence (2.117) and (2.119) follow. □

2.5 A BROWNIAN EXCURSION

If $n\to\infty$, then the finite-dimensional distributions of the process $\{\eta^+_{[2nt]}/\sqrt{2n},\ 0 \leq t \leq 1\}$ converge to the corresponding finite-dimensional distributions of a process $\{\eta^+(t),\ 0 \leq t \leq 1\}$ which is called a Brownian excursion process. The Brownian excursion process $\{\eta^+(t),\ 0 \leq t \leq 1\}$ is a Markov process for which $\mathbf{P}\{\eta^+(0) = 0\} = \mathbf{P}\{\eta^+(1) = 0\} = 1$ and $\mathbf{P}\{\eta^+(t) \geq 0\} = 1$ for $0 \leq t \leq 1$. If $0 \leq t \leq 1$, then $\eta^+(t)$ has a density function $f(t,x)$. Obviously, $f(t,x) = 0$ for $x \leq 0$. If $0 < t < 1$ and $x > 0$, then

$$
f(t,x) = \frac{2x^2}{\sqrt{2\pi t^3(1-t)^3}} e^{-x^2/(2t(1-t))}. \tag{2.136}
$$

If $0 < t < u < 1$, then the random variables $\eta^+(t)$ and $\eta^+(u)$ have a joint density function $f(t,x;u,y)$. We have $f(t,x;u,y) = 0$ if $x \leq 0$ or $y \leq 0$. If $0 < t < u < 1$ and $x > 0$, $y > 0$, then

$f(t,x;u,y)$

$$= \frac{\sqrt{8\pi xy}}{\sqrt{t^3(u-t)(1-u)^3}} \varphi\Big(\frac{x}{\sqrt{t}}\Big)\varphi\Big(\frac{y}{\sqrt{1-u}}\Big)\Big[\varphi\Big(\frac{x-y}{\sqrt{u-t}}\Big) - \varphi\Big(\frac{x+y}{\sqrt{u-t}}\Big)\Big] \quad (2.137)$$

where $\varphi(x)$ is the normal density function defined by (2.21). Since $\{\eta^+(t),\ 0 \leq t \leq 1\}$ is a Markov process, the density functions $f(t,x)$ and $f(t,x;u,y)$ completely determine the finite-dimensional distributions of the process $\{\eta^+(t),\ 0 \leq t \leq 1\}$. For the properties of the Brownian excursion process, we refer to P. Lévy [31], [32], K. Itô and H.P. McKean, Jr. [19], K.L. Chung [10] and L. Takács [50].

By the results of Section 2.3, we can prove (2.136) and (2.137). By the reflection principle, we obtain that for $0 \leq k \leq n$

$$C_n \mathbf{P}\{\eta^+_{k+2i} = k\} = 2^{2n+2}\mathbf{P}\{\rho(k+1) = 2i+k+1\}\cdot$$
$$\cdot\mathbf{P}\{\rho(k+1) = 2n-2i+1-k\} \quad (2.138)$$

and

$$C_n \mathbf{P}\{\eta^+_{k+2i} = k \text{ and } \eta^+_{\ell+2j} = \ell\} = 2^{2n+2}\mathbf{P}\{\rho(k+1) = k+2i+1\}\cdot$$
$$\mathbf{P}\{\rho(k+1) = 2n-2j-\ell+1\}\cdot$$
$$\left[\binom{\ell-k+2j-2i}{j-i} - \binom{\ell-k+2j-2i}{j-i+\ell+1}\right]\frac{1}{2^{\ell-k+2j-2i}}. \quad (2.139)$$

If in (2.138), $k = [x\sqrt{2n}]$, $x > 0$, and $k+2i = [2nt]$, $0 < t < 1$, then

$$f(t,x) = \lim_{n\to\infty}\frac{\sqrt{2n}}{2}\mathbf{P}\{\eta^+_{k+2i} = k\}. \quad (2.140)$$

This proves (2.136). If in (2.139), $k = [x\sqrt{2n}]$, $\ell = [y\sqrt{2n}]$, $k+2i = [2nt]$ and $\ell + 2j = [2nu]$ where $x > 0$, $y > 0$ and $0 < t < u < 1$, then

$$f(t,x;u,y) = \lim_{n\to\infty}\frac{n}{2}\,\mathbf{P}\,\{\eta^+_{k+2i} = k \text{ and } \eta^+_{\ell+2j} = \ell\}. \quad (2.141)$$

This proves (2.137).

For the Brownian excursion $\{\eta^+(t),\ 0 \leq t \leq 1\}$ let us define ω^+ as the area

$$\omega^+ = \int_0^1 \eta^+(t)dt, \quad (2.142)$$

and $\tau^+(\alpha)$ for $\alpha \geq 0$ as the local time at level α, that is,

$$\tau^+(\alpha) = \lim_{\epsilon\to 0}\frac{1}{\epsilon}\text{ measure }\{t: \alpha \leq \eta^+(t) < \alpha+\epsilon,\ \ 0 \leq t \leq 1\}. \quad (2.143)$$

By the results of the previous section we can conclude that

$$\mathbf{P}\{\sup_{0\leq t\leq 1} \eta^+(t) \leq x\} = F(x), \quad (2.144)$$

where $F(x)$ is given by (2.36),

$$\mathbf{P}\{\omega^+ \leq x\} = W(x), \quad (2.145)$$

where $W(x)$ is given by (2.82), and

$$\mathbf{P}\{\tau^+(\alpha) \le x\} = G_\alpha(x) \tag{2.146}$$

where $G_\alpha(x)$ is given by (2.117). We note that $G_\alpha(0) = F(\alpha)$ for $\alpha > 0$. In 1952, B.V. Gnedenko and Yu. P. Studnev [17] determined $F(x)$ in the context of order statistics. See also L. Takács [42] and D.P. Kennedy [26]. The distribution function $W(x)$ has been determined by D.A. Darling [11], [12] and G. Louchard [33]. Furthermore, we have

$$\mathbf{P}\{\sup_{\alpha \ge 0} \tau^+(\alpha) \le 2x\} = F(x) \tag{2.147}$$

where $F(x)$ is given by (2.36). For a direct proof of this result, see Th. Jeulin [20, p. 264].

We have also

$$\int_0^\infty x^r dF(x) = m_r, \tag{2.148}$$

$$\int_0^\infty x^r dW(x) = M_r \tag{2.149}$$

and

$$\int_0^\infty x^r dG_\alpha(x) = \mu_r(\alpha) \tag{2.150}$$

for $r = 0, 1, 2, \ldots$ where m_r, M_r and $\mu_r(\alpha)$ are given by (2.37), (2.47), and (2.106), respectively.

2.6 SINGLE-SERVER QUEUES

Let us suppose that, in the time interval $(0, \infty)$, customers arrive at random at a counter and are served singly by one server in order of arrival. It is assumed that the server starts working at time $t = 0$ and, at the time i $(i = 1, 2, \ldots)$, customers are already waiting for service. The initial i customers are numbered $1, 2, \ldots, i$, and the customers arriving subsequently are numbered $i+1, i+2, \ldots$ in the order of their arrivals. Denote by ν_r the number of customers arriving during the service time of the rth customer. This queueing model will be characterized by the initial queue size i and the sequence of random variables $\nu_1, \nu_2, \ldots, \nu_r, \ldots$. Throughout this article, we use the abbreviation $N_r = \nu_1 + \nu_2 + \cdots + \nu_r$ for $r = 1, 2, \ldots$ and $N_0 = 0$.

Denote by ζ_r $(r = 1, 2, \ldots)$ the number of customers in the system immediately after the rth service ends, and write $\zeta_0 = i$. We have

$$\zeta_r = [\zeta_{r-1} - 1]^+ + \nu_r \tag{2.151}$$

for $r \ge 1$ where $[x]^+ = x$ if $x \ge 0$ and $[x]^+ = 0$ if $x < 0$.

Following D.G. Kendall [24], we say that the initial i customers in the queue form the 0th generation. The customers (if any) arriving during the total service time of the initial i customers form the first generation. Generally, the customers (if any) arriving during the total service time of the customers in the $(r-1)$th generation form the rth generation for $r = 1, 2, \ldots$. Denote by $\xi(r)$ $(r = 0, 1, 2, \ldots)$ the number of customers in the rth generation. If $\xi(r) = 0$ for some $r \geq 1$, then $\xi(r+1) = \xi(r+2) = \ldots = 0$. Define

$$\theta(i) = max\{r: \xi(r) > 0\}. \tag{2.152}$$

If $\xi(r) > 0$ for all $r \geq 0$, then $\theta(i) = \infty$.

The time of the server consists of alternating busy periods and idle periods. Denote by $\gamma(i)$ the number of customers served in the initial busy period in the case where the initial queue size is i $(i = 1, 2, \ldots)$. Obviously,

$$\mathbf{P}\{\gamma(i) = n\} = \mathbf{P}\{N_r > r - i \text{ for } i \leq r < n \text{ and } N_n = n - i\} \tag{2.153}$$

if $1 \leq i \leq n$.

In what follows, we assume that $\nu_1, \nu_2, \ldots, \nu_r, \ldots$ is a sequence of independent and identically distributed random variables for which

$$\mathbf{P}\{\nu_r = j\} = p_j \tag{2.154}$$

if $j = 0, 1, 2, \ldots$ where $p_j \geq 0$, and

$$\sum_{j=0}^{\infty} p_j = 1. \tag{2.155}$$

Define

$$d = gcd\{j: p_j > 0\}, \tag{2.156}$$

and

$$a = \sum_{j=0}^{\infty} j p_j. \tag{2.157}$$

If $a < \infty$, define $\sigma \geq 0$ by

$$\sigma^2 = \sum_{j=0}^{\infty} (j-a)^2 p_j. \tag{2.158}$$

If $\nu_1, \nu_2, \ldots, \nu_r, \ldots$ are independent and identically distributed random variables, then in (2.153) we can replace $\nu_1, \nu_2, \ldots, \nu_n$ by $\nu_n, \nu_{n-1}, \ldots, \nu_1$ respectively, without changing the probability. Thus we obtain

$$\mathbf{P}\{\gamma(i) = n\} = \mathbf{P}\{N_r < r \text{ for } 1 \leq r \leq n \text{ and } N_n = n - i\} \tag{2.159}$$

for $1 \leq i \leq n$. By Theorem 2.2 we have

$$\mathbf{P}\{\gamma(i) = n\} = \frac{i}{n}\mathbf{P}\{N_n = n - i\} \tag{2.160}$$

for $1 \leq i \leq n$. Each possible value of $\gamma(i)$ has the form $n = sd + i$ where $s = 0, 1, 2, \ldots$ and actually, $\mathbf{P}\{\gamma(i) = n\} > 0$ if $n = sd + i$ and s is sufficiently large.

The process $\{\zeta_r, r \geq 0\}$. Under the assumption that $\nu_1, \nu_2, \ldots, \nu_r, \ldots$ are independent and identically distributed random variables, the sequence $\{\zeta_r, r \geq 0\}$ is a Markov chain.

Theorem 2.12 *If $i \geq 1$, $n = sd + i$ $(s = 0, 1, 2, \ldots)$ and we assume that in* (2.157) $a = 1$ *and in* (2.158) $0 < \sigma < \infty$, *then*

$$\{\zeta_{[nt]}/(\sigma\sqrt{n}),\ 0 \leq t \leq 1 \mid \gamma(i) = n\} \Rightarrow \{\eta^+(t),\ 0 \leq t \leq 1\}, \tag{2.161}$$

that is the stochastic process on the left-hand side, given the condition $\gamma(i) = n$, converges weakly to the Brownian excursion process $\{\eta^+(t),\ 0 \leq t \leq 1\}$ as $n \to \infty$.

Proof. If $\gamma(i) = n$, then $\zeta_r = i + N_r - r > 0$ for $0 \leq r < n$ and $\zeta_n = i + N_n$

$-n = 0$ and the finite dimensional distributions of the process

$$\{\zeta_{[nt]}/(\sigma\sqrt{n}),\ 0 \le t \le 1 \mid \gamma(i) = n\} \tag{2.162}$$

converge to the corresponding finite dimensional distributions of the Brownian excursion $\{\eta^+(t),\ 0 \le t \le 1\}$ as $n \to \infty$. By a theorem of W.D. Kaigh [21], [22], we have also weak convergence. Previously, B. Belkin [3], [4] considered a variant of (2.161) in which the condition is given by $\gamma(i) > n$. See also D.L. Iglehart [18] and E. Bolthausen [6]. □

A necessary and sufficient condition for weak convergence is that

$$\lim_{h\to 0} \limsup_{n\to\infty} \mathbf{P}\left\{ \max_{\substack{|k-j| \le nh \\ 0 \le j \le k \le n}} |\zeta_k - \zeta_j| > \epsilon\sigma\sqrt{n} \,\middle|\, \gamma(i) = n \right\} = 0 \tag{2.163}$$

for $\epsilon > 0$. Here $n = sd + i$ $(s = 0, 1, 2, \ldots)$ and $h > 0$. See I.I. Gikhman and A.V. Skorokhod [16, pp. 449-450].

Theorem 2.13 *If* $a = 1$, $0 < \sigma < \infty$ *and* $n = sd + i$ $(s = 0, 1, 2, \ldots)$, *then*

$$\lim_{n\to\infty} \mathbf{P}\{max(\zeta_0, \zeta_1, \ldots, \zeta_n) \le x\sigma\sqrt{n} \mid \gamma(i) = n\} = F(x) \tag{2.164}$$

where $F(x)$ *is defined by* (2.36).

Proof. Since the supremum is a continuous functional on the Brownian excursion process, (2.164) immediately follows from (2.161). □

The limit theorem (2.164) suggests that if $0 < \sigma < \infty$ and

$$\sum_{j=0}^{\infty} j^r p_j < \infty, \tag{2.165}$$

for $r \ge 1$, then

$$\lim_{n\to\infty} \mathbf{E}\left[\left(\frac{max(\zeta_0, \zeta_1, \ldots, \zeta_n)}{\sigma\sqrt{n}}\right)^r \,\middle|\, \gamma(i) = n\right] = m_r \tag{2.166}$$

for $r \ge 1$ where m_r is defined by (2.37). This statement is indeed true in two particular cases. If $p_j = (1/2)^{j+1}$ for $j = 0, 1, 2, \ldots$, then the proof for $r = 1$ follows from the results of N.G. de Bruijn, D.E. Knuth, and S.O. Rice [13], and for $r > 1$, from the results of R. Kemp [23]. If $p_0 = p_2 = 1/2$, then the proof follows from the results of Ph. Flajolet and A. Odlyzko [14].

Theorem 2.14 *If* $a = 1$, $0 < \sigma < \infty$ *and* $n = sd + i$ $(s = 0, 1, 2, \ldots)$, *then*

$$\lim_{n\to\infty} \mathbf{P}\{\zeta_1 + \zeta_2 + \cdots + \zeta_n \le x\sigma n^{3/2} \mid \gamma(i) = n\} = W(x) \tag{2.167}$$

where $W(x)$ *is defined by* (2.82).

Proof. Since the integral is a continuous functional on the Brownian excursion process, (2.167) immediately follows from (2.161). □

The process $\{\xi(r),\ r \ge 0\}$. Under the assumption that $\nu_1, \nu_2, \ldots, \nu_r, \ldots$ are independent and identically distributed random variables, the sequence $\{\xi(r), r \ge 0\}$ is a branching process. We can imagine that, in a population, initially we have i $(i = 1, 2, \ldots)$ progenitors and, in each generation, each individual reproduces independently of the others and has probability p_j $(j = 0, 1, 2, \ldots)$ of giving rise to j descendants in the following generation. Then $\xi(r)$ can be interpreted as the number of individuals in the rth generation $(r = 0, 1, 2, \ldots)$. Obviously,

$$\gamma(i) = \sum_{r \ge 0} \xi(r), \tag{2.168}$$

that is, $\gamma(i)$ is the total number of individuals (total progeny) in the branching process if $\xi(0)=i$. Possibly, $\gamma(i)=\infty$.

Evidently, $\{\xi(r), 0 \le r \le \theta(i)\}$ is a subsequence of $\{\zeta_r, 0 \le r \le \gamma(i)\}$, and $\xi(r) = \zeta_s$ if $s=\xi(0)+\xi(1)+\cdots+\xi(r-1)$ for $r \ge 1$. This implies that, if $\gamma(i)=n$, then

$$0 \le \max_{0 \le r \le n} \zeta_r - \max_{0 \le r \le \theta(i)} \xi(r) \le \max_{\substack{|k-j| \le \max \xi(r)(r \ge 0) \\ 0 \le j < k \le n}} |\zeta_k - \zeta_j|. \quad (2.169)$$

If $r \ge 1$, if $a=1$, if $0<\sigma<\infty$, if (2.165) holds, and if $n = sd+i$ $(s=0,1,2,\ldots)$, then for any $\epsilon>0$ we have

$$\left| \mathbf{E}\left[\binom{\xi(m)}{r} \,\Big|\, \gamma(i)=n\right] - \frac{1}{r!}\left(\frac{\sigma\sqrt{n}}{2}\right)^r \mu_r\left(\frac{m\sigma}{2\sqrt{n}}\right) \right| < \epsilon n^{3/2} m^{r-3} \quad (2.170)$$

for sufficiently large m and n where $\mu_r(\alpha)$ is defined by (2.104). For the proof of (2.170), see L. Takács [52]. We note that

$$\lim_{n\to\infty} \sum_{m \ge 0} \mu_r\left(\frac{m\sigma}{2\sqrt{n}}\right)\frac{\sigma}{2\sqrt{n}} = \int_0^\infty \mu_r(\alpha)d\alpha = 2^{(r-1)/2}\Gamma\left(\frac{r+1}{2}\right) \quad (2.171)$$

for $r \ge 1$.

Theorem 2.15 *If $a=1$, $0<\sigma<\infty$ and $n=sd+i (s=0,1,2,\ldots)$, then*

$$\lim_{n\to\infty} \mathbf{P}\{\xi(r) \le x\sigma\sqrt{n} \text{ for } 0 \le r \le \theta(i) \mid \gamma(i)=n\} = F(x) \quad (2.172)$$

where $F(x)$ is defined by (2.36).

Proof. By (2.169)

$$\mathbf{P}\left\{\max_{0\le i \le n} \zeta_i - \max_{m \ge 0} \xi(m) > \epsilon\sigma\sqrt{n} \mid \gamma(i)=n\right\} \quad (2.173)$$

$$\le \mathbf{P}\{\max_{m\ge 0}\xi(m) > nh \mid \gamma(i)=n\} + \mathbf{P}\left\{ \max_{\substack{|j-k| \le nh \\ 0 \le j<k \le n}} |\zeta_k - \zeta_j| > \epsilon\sigma\sqrt{n} \,\Big|\, \gamma(i)=n\right\}$$

for any $\epsilon>0$ and $h>0$. Furthermore, if $r=2$, by (2.170)

$$\mathbf{P}\left\{\max_{m\ge 0}\xi(m) > nh \mid \gamma(i)=n\right\} \le \sum_{m \ge 0} \mathbf{P}\{\xi(m) > nh \mid \gamma(i)=n\}$$

$$\le \frac{1}{n^2h^2}\sum_{m\ge 0}\mathbf{E}\left[[\xi(m)]^2 \mid \gamma(i)=n\right] \sim \frac{1}{n^2h^2}\left(\frac{\sigma\sqrt{n}}{2}\right)^2 \sum_{m \ge 0}\mu_2\left(\frac{m\sigma}{2\sqrt{n}}\right) \quad (2.174)$$

$$\sim \frac{1}{n^2h^2}\left(\frac{\sigma\sqrt{n}}{2}\right)^2\left(\sum_{\alpha\ge 0}\mu_2(\alpha)\Delta\alpha\right)\frac{2\sqrt{n}}{\sigma} \sim \frac{\Gamma(\frac{3}{2})\sigma}{2^{1/2}h^2n^{1/2}}$$

as $n\to\infty$. Here, we made the substitution $\alpha = m\sigma/(2\sqrt{n})$. Then $\Delta\alpha = \sigma\Delta m/(2\sqrt{n})$ and $\Delta m = 1$.

By (2.174),

$$\lim_{n\to\infty}\mathbf{P}\left\{\max_{m \ge 0}\xi(m) > nh \mid \gamma(i)=n\right\} = 0 \quad (2.175)$$

for $h>0$. By (2.163) and (2.175),

$$\lim_{n\to\infty}\mathbf{P}\left\{\max_{0\le i\le n}\zeta_i - \max_{m\ge 0}\xi(m) > \epsilon\sigma\sqrt{n} \mid \gamma(i)=n\right\} = 0 \quad (2.176)$$

for any $\epsilon > 0$. Consequently, by (2.164) and (2.176)

$$\lim_{n\to\infty} \mathbf{P}\left\{\max_{m\geq 0} \xi(m) \leq x\sigma\sqrt{n} \mid \gamma(i) = n\right\}$$

$$= \lim_{n\to\infty} \mathbf{P}\left\{\max_{0\leq i\leq n} \zeta_i \leq x\sigma\sqrt{n} \mid \gamma(i) = n\right\} = F(x). \tag{2.177}$$

This proves (2.172). □

Theorem 2.16 *If $a = 1$, if $0 < \sigma < \infty$, if (2.165) holds for $r \geq 2$ and $n = sd + i$ $(s = 0, 1, 2, \ldots)$, then*

$$\lim_{n\to\infty} \mathbf{P}\left\{\sum_{0\leq r\leq \theta(i)} r\xi(r) \leq x\sigma n^{3/2} \mid \gamma(i) = n\right\} = W(x) \tag{2.178}$$

where $W(x)$ is defined by (2.82).

Proof. For the proof of (2.178), see L. Takács [52]. □

Theorem 2.17 *If $a = 1$, $0 < \sigma < \infty$, and $n = sd + i$ $(s = 0, 1, 2, \ldots)$, then*

$$\lim_{n\to\infty} \mathbf{P}\left\{\frac{2\xi([2\alpha\sqrt{n}/\sigma])}{\sigma\sqrt{n}} \leq x \mid \gamma(i) = n\right\} = G_\alpha(x) \tag{2.179}$$

where $G_\alpha(x)$ is given by (2.117).

Proof. For the proof of (2.179), see L. Takács [52]. □

Theorem 2.18 *If $a = 1$, $0 < \sigma < \infty$, and $n = sd + i$ $(s = 0, 1, 2, \ldots)$, then*

$$\lim_{n\to\infty} \mathbf{P}\left\{\frac{\sigma\theta(i)}{2\sqrt{n}} \leq x \mid \gamma(i) = n\right\} = F(x) \tag{2.180}$$

where $F(x)$ is given by (2.36).

Proof. For the proof of (2.180) we refer to V.F. Kolchin [28], [29, p. 126]. □

Finally, we note that it is plausible that

$$\{2\xi([2\alpha\sqrt{n}/\sigma])/(\sigma\sqrt{n}), \alpha \geq 0 \mid \gamma(i) = n\} \Rightarrow \{\tau^+(\alpha), \alpha \geq 0\}, \tag{2.181}$$

that is, if $n\to\infty$, the stochastic process on the left-hand side, given the condition $\gamma(i) = n$, converges weakly to the stochastic process on the right-hand side.

2.7 RANDOM-ROOTED TREES

A tree is a connected undirected graph which has no cycles, loops or multiple edges. The *root* of a tree is a vertex distinguished from the other vertices. The *height* of a vertex in a rooted tree is the distance from the vertex to the root, that is, the number of edges in the path from the vertex to the root. The *total height* of a rooted tree is the sum of the heights of its vertices. The *width* of a rooted tree is the maximal number of vertices at the same distance from the root.

We shall consider various sets of rooted trees with n vertices. We choose a tree at random in a given set and define $\tau_n(m)$ as the number of vertices at distance m $(m = 0, 1, 2, \ldots)$ from the root. We define

$$\mu_n = \max\{m: \tau_n(m) > 0\} \tag{2.182}$$

as the height of the tree,

$$\delta_n = \max\{\tau_n(m): m \geq 0\} \tag{2.183}$$

as the width of the tree, and

$$\tau_n = \sum_{m \geq 0} m\tau_n(m) \tag{2.184}$$

as the total height of the tree.

Our aim is to find, for various sets of trees, the asymptotic distributions of the random variables μ_n, δ_n, τ_n and $\tau_n(m)$ if $m \to \infty$ and $n \to \infty$. See L. Takács [46]-[59].

Unlabeled trees. Define S_n as the set of all the rooted, oriented (plane) trees with n unlabeled vertices. There are $|S_4| = 5$ different trees with four vertices as shown in Figure 2. The roots of the trees are depicted in black. The branches are ordered. By interchanging the locations of two different branches, relative to the root, we obtain two different trees.

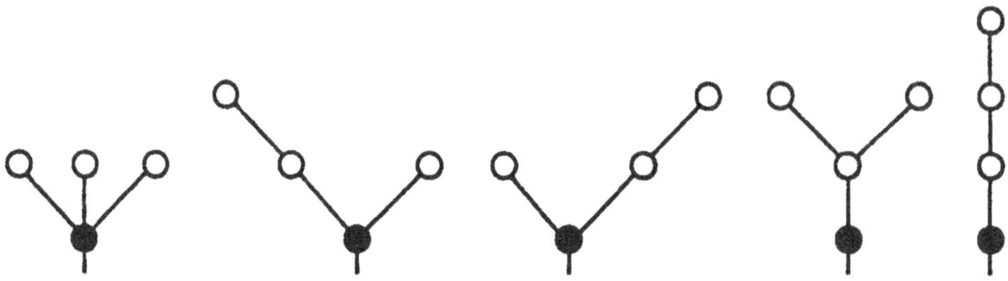

Figure 1: Rooted oriented trees with four unlabeled vertices

As we shall see, the number of trees in S_n is

$$|S_n| = \binom{2n-2}{n-1}\frac{1}{n} \tag{2.185}$$

if $n \geq 1$.

In the set S_n, there are no restrictions on the degrees of the vertices of the trees. However, in many applications, we encounter rooted trees in which the degrees of the vertices are subject to certain constraints. For example, in computer science, trivalent rooted trees play an important role. In a trivalent rooted tree, every vertex has degree 3 except the root which has degree 2 and the end-vertices which have degree 2.

To study various sets of rooted trees in the case where the degrees of the vertices satisfy some constraints, let us define R as a fixed set of nonnegative integers which always contains 0. Denote by $S_n(R)$ the subset of S_n which contains all the trees in S_n in which the degree of the root $\in R$ and, if j is the degree of any other vertex of a tree, then $j-1 \in R$.

Theorem 2.19 *The number of trees in* $S_n(R)$ *is*

$$|S_n(R)| = \frac{1}{n} \text{ Coeff. of } x^{n-1} \text{ in} \left(\sum_{i \in R} x^i\right)^n. \tag{2.186}$$

Proof. We defined S_n as the set of all the rooted, oriented (plane) trees with n unlabeled vertices. Equivalently, we can define S_n as the set of sequences $(i_1, i_2, \ldots, i_n)$ where $i_1, i_2, \ldots, i_n$ are nonnegative integers which satisfy the conditions

$$i_1 + i_2 + \cdots + i_n = n - 1 \tag{2.187}$$

and

$$i_1 + i_2 + \cdots + i_r \geq r \text{ for } 1 \leq r \leq n-1. \tag{2.188}$$

If R is a fixed set of non-negative integers, then let us denote by $S_n(R)$ the set of sequences $(i_1, i_2, \ldots, i_n) \in S_n$ in which $i_r \in R$ for all $r = 1, 2, \ldots, n$.

With every sequence $(i_1, i_2, \ldots, i_n)$ in S_n we associate a rooted tree. The tree has vertex set $(1, 2, \ldots, n)$, and vertex 1 is designated as the root of the tree. Two vertices, r and s $(1 \leq r < s \leq n)$, are joined by an edge if, and only if,

$$i_0 + i_1 + \cdots + i_{r-1} < s \leq i_0 + i_1 + \cdots + i_r \tag{2.189}$$

where $i_0 = 1$. In the tree $(i_1, i_2, \ldots, i_n)$, the root has degree i_1, and vertex r $(1 < r \leq n)$ has degree $i_r + 1$. If $(i_1, i_2, \ldots, i_n) \in S_n(R)$, then the degrees of the vertices are subject to the constrains $i_r \in R$ for $r = 1, 2, \ldots, n$.

Actually, in the above definition of $S_n(R)$, we assign labels to the vertices of the trees, but, if we are not interested, the labels can be disregarded, and each tree can be visualized as an unlabeled tree. We can interpret $|S_n(R)|$ as the number of unlabeled distinct rooted trees in the case where the degrees of the vertices satisfy the constraints imposed by R. The above representation of trees stems from the theory of queues as indicated in the Introduction.

Accordingly, $|S_n(R)|$ is the number of sequences $(i_1, i_2, \ldots, i_n)$ of non-negative integers which satisfy the conditions (2.187), (2.188) and $i_r \in R$ for $r = 1, 2, \ldots, n$. If we apply Theorem 2.3 to the case where $k = n-1$, $k_r = i_{n+1-r}$ for $1 \leq r \leq n$, $g(i) = 1$ for $i \in R$, and $g(i) = 0$ for $i \notin R$, then we obtain (2.186). □

If $R = \{0, 1, 2, \ldots\}$, then $S_n(R) = S_n$ and by (2.186)

$$|S_n| = \frac{1}{n} \text{ Coeff. of } x^{n-1} \text{ in } (1-x)^{-n}. \tag{2.190}$$

Hence (2.185) follows. In a slightly different interpretation, formula (2.185) was proved by A. Cayley [8] in 1859. For the history of the various proofs of (2.185), see L. Takács [50].

Labeled trees. Denote by S_n^* the set of all rooted trees with n labeled vertices. See Figure 2 for S_4^*. There are $|S_4^*| = 64$ rooted trees with four labeled vertices 1, 2, 3, 4. In Figure 2, the roots of the trees are depicted in black, and only one tree is displayed in each of four subsets of S_n^*, (a), (b), (c), (d). The sets (a), (b), (c), (d) contain 24, 12, 24, 4 trees, respectively.

As we shall see, the number of trees in S_n^* is

$$|S_n^*| = n^{n-1} \tag{2.191}$$

if $n \geq 1$.

In the set S_n^*, there are no restrictions on the degrees of the vertices of the trees. To study various subsets of S_n^* in the case where the degrees of the vertices satisfy some constraints, let us define R as a fixed set of nonnegative integers which always contains 0. Denote by $S_n^*(R)$ the subset of S_n^* which contains all the trees in S_n^* in which the degree of the root $\in R$ and, if j is the degree of any other vertex of a tree, then $j-1 \in R$.

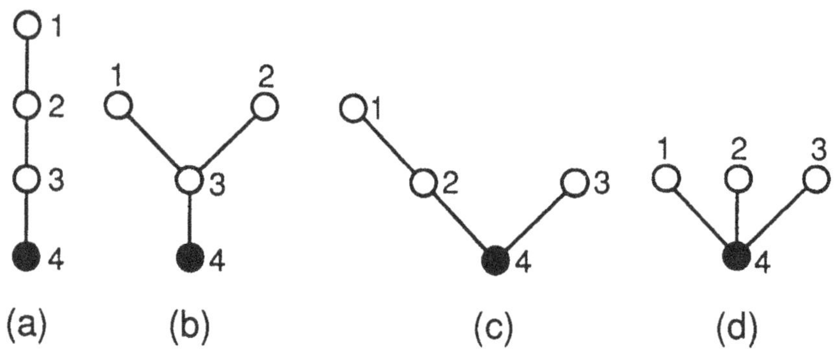

Figure 2: Rooted trees with four labeled vertices

Theorem 2.20 *The number of trees in* $S_n^*(R)$ *is*

$$|\, S_n^*(R) \,| = (n-1)! \text{ Coeff. of } x^{n-1} \text{ in } \Big(\sum_{i \in R} \frac{x^i}{i!}\Big)^n. \tag{2.192}$$

Proof. Let us choose a tree $(i_1, i_2, \ldots, i_n)$ in $S_n(R)$. We can label the vertices of $(i_1, i_2, \ldots, i_n)$ in

$$\frac{n!}{i_1! i_2! \ldots i_n!} \tag{2.193}$$

different ways. If we perform all the possible labelings on all the trees in $S_n(R)$, we get a set of labeled, distinct trees, which we shall denote by $S_n^*(R)$. If $R = \{0, 1, 2, \ldots\}$, we write $S_n^*(R) = S_n^*$. These definitions of S_n^* and $S_n^*(R)$ are in agreement with the earlier ones.

We have

$$|\, S_n^*(R) \,| = \sum_{(i_1, i_2, \ldots, i_n) \in S_n(R)} \frac{n!}{i_1! i_2! \ldots i_n!}. \tag{2.194}$$

If we apply Theorem 2.3 to the case where $k = n - 1$, $k_r = i_{n+1-r}$ for $1 \le r \le n$, $g(i) = 1/i!$ for $i \in R$ and $g(i) = 0$ for $i \notin R$, then we obtain (2.192). □

If $R = \{0, 1, 2, \ldots\}$, then $S_n^*(R) = S_n^*$ and by (2.192)

$$|\, S_n^* \,| = (n-1)! \text{ Coeff. of } x^{n-1} \text{ in } e^{nx}. \tag{2.195}$$

Hence (2.191) follows. Formula (2.191) was discovered by A. Cayley [9] in 1889. For the history of the various proofs of (2.191), see L. Takács [48], [49].

A general model for random trees. The models for random trees discussed above can be considered as particular cases of a more general model defined here. Let us consider S_n, the set of sequences $(i_1, i_2, \ldots, i_n)$ satisfying the conditions (2.187) and (2.188). Let $\{p_j\}$ be a probability distribution defined on the set of nonnegative integers. For $\{p_j\}$, we use the notations (2.156), (2.157), and (2.158). We note that, if $n = sd + 1$ and s is sufficiently large positive integer, then S_n is not empty.

Let us choose a tree at random in S_n, assuming that the probability of a tree represented by $(i_1, i_2, \dots, i_n)$ is

$$p(i_1, i_2, \dots, i_n) = a_n^{-1} p_{i_1} p_{i_2} \dots p_{i_n} \tag{2.196}$$

where

$$a_n = \sum_{(i_1, i_2, \dots, i_n) \in S_n} p_{i_1} p_{i_2} \dots p_{i_n} = \frac{1}{n} \sum_{i_1 + i_2 + \dots + i_n = n-1} p_{i_1} p_{i_2} \dots p_{i_n}. \tag{2.197}$$

If $a_n = 0$, then (2.196) should be interpreted as 0.

If we consider the queueing process defined in Section 6 in the case where the initial queue size $i = 1$ and if $\mathbf{P}\{\gamma(1) = n\} > 0$, we have

$$\mathbf{P}\{\nu_1 = i_1, \nu_2 = i_2, \dots, \nu_n = i_n \mid \gamma(1) = n\} = p(i_1, i_2, \dots, i_n). \tag{2.198}$$

By choosing the distribution $\{p_j\}$ in a suitable way, we obtain each model of the unlabeled and the labeled random-rooted trees discussed above.

If

$$p_j = qp^j \tag{2.199}$$

for $j = 0, 1, 2, \dots$ where $p > 0$, $q > 0$ and $p + q = 1$, then $a = p/q$, $\sigma^2 = p/q^2$ and

$$p(i_1, i_2, \dots, i_n) = \frac{1}{C_{n-1}} \text{ for } (i_1, i_2, \dots, i_n) \in S_n. \tag{2.200}$$

This corresponds to the case where we choose a tree at random in S_n, assuming that all the $|S_n| = C_{n-1}$ trees are equally probable.

Let

$$p_j = p^j / (\sum_{i \in R} p^i) \tag{2.201}$$

for $j \in R$ and $p_j = 0$ for $j \notin R$. In this case,

$$p(i_1, i_2, \dots, i_n) = \frac{1}{|S_n(R)|} \text{ if } (i_1, i_2, \dots, i_n) \in S_n(R). \tag{2.202}$$

This example can be interpreted in the following way. We consider $S_n(R)$ the set of oriented (plane) rooted trees with n unlabeled vertices whenever the degrees of vertices are subject to the constraints imposed by R. We choose a tree at random in $S_n(R)$, assuming that all the possible choices are equally probable.

Let

$$p_j = e^{-\lambda} \frac{\lambda^j}{j!} \text{ for } j = 0, 1, 2, \dots \tag{2.203}$$

where $\lambda > 0$. Then $a = \sigma^2 = \lambda$, and

$$p(i_1, i_2, \dots, i_n) = \frac{n!}{i_1! i_2! \dots i_n!} \frac{1}{n^{n-1}} \text{ for } (i_1, i_2, \dots, i_n) \in S_n. \tag{2.204}$$

In this case, the procedure is the following: We choose a tree at random in the set S_n^* of rooted trees with n labeled vertices, assuming that all the possible choices are equally probable.

Let

$$p_j = (\lambda^j / j!) / (\sum_{i \in R} \lambda^i / i!) \tag{2.205}$$

for $j \in R$ and $p_j = 0$ for $j \notin R$. In this case,

$$p(i_1, i_2, \dots, i_n) = \frac{n!}{i_1! i_2! \dots i_n!} \frac{1}{|S_n^*(R)|} \text{ if } (i_1, i_2, \dots, i_n) \in S_n(R). \tag{2.206}$$

This example can be interpreted in the following way. We consider $S_n^*(R)$ the set of rooted trees with n labeled vertices whenever the degrees of the trees are subject to the constraints imposed by the set R. We choose a tree at random in $S_n^*(R)$, assuming that all the possible choices are equally probable.

If we use the notations introduced in Section 6 for queueing processes, we can write that for the random trees we have

$$\mathbf{P}\{\mu_n = k\} = \mathbf{P}\{\theta(1) = k \mid \gamma(1) = n\}, \tag{2.207}$$

$$\mathbf{P}\{\delta_n = k\} = \mathbf{P}\Big\{\max_{r \geq 0} \xi(r) = k \mid \gamma(1) = n\Big\}, \tag{2.208}$$

$$\mathbf{P}\{\tau_n = k\} = \mathbf{P}\Big\{\sum_{r \geq 0} r\xi(r) = k \mid \gamma(1) = n\Big\}, \tag{2.209}$$

and

$$\mathbf{P}\{\tau_n^+(m) = k\} = \mathbf{P}\{\xi(m) = k \mid \gamma(1) = n\}. \tag{2.210}$$

In proving various limit theorems for the random trees considered, we assume that $a = 1$, that is, $\{\xi(r), r \geq 0\}$ is a critical branching process. If $a = 1$, then $\mathbf{P}\{\gamma(1) < \infty\} = 1$. If we want to apply these limit theorems to the four examples considered in this section, we should choose the parameters p and λ in such a way that the condition $a = 1$ is satisfied.

2.8 LIMIT DISTRIBUTIONS FOR RANDOM-ROOTED TREES

We consider the general model of random-rooted trees defined in Section 2.7. We choose a tree at random in the set S_n. It is assumed that a tree, represented by the sequence $(i_1, i_2, \ldots, i_n)$, has probability $p(i_1, i_2, \ldots, i_n)$ defined by (2.196) and (2.197). Throughout this section, we use the notations (2.156), (2.157) and (2.158) relative to the distribution $\{p_j\}$ and assume that $a = 1$ and $0 < \sigma < \infty$. In some of the theorems, we assume also that (2.165) is satisfied for $r \geq 2$.

The limit distribution of μ_n.

Theorem 2.21 *If $a = 1$, $0 < \sigma < \infty$, and $n = sd + 1$ $(s = 0, 1, 2, \ldots)$, then*

$$\lim_{n \to \infty} \mathbf{P}\Big\{\frac{\sigma\mu_n}{2\sqrt{n}} \leq x\Big\} = F(x) \tag{2.211}$$

where $F(x)$ is given by (2.36).

Proof. We obtain (2.211) by (2.180) and (2.207). □

In the particular case where $p_j = (1/2)^{j+1}$ for $j = 0, 1, 2, \ldots$, I.V. Konovaltsev and E.P. Lipatov [30] proved (2.211). See also N.G. de Bruijn, D.E. Knuth, and S.O. Rice [13] and L. Takács [46]. In this particular case,

$$\mathbf{P}\{\mu_{n+1} < k\} = T_n(k-1)/C_n \tag{2.212}$$

for $k \geq 1$ and $n \geq 1$ where $T_n(k-1)$ is given by (2.28), and Theorem 2.21 is a consequence of Theorem 2.4. In the case where $p_j = e^{-1}/j!$ for $j = 0, 1, 2, \ldots$, A. Rényi and G. Szekeres [38] proved (2.211). See also V.E. Stepanov [41] and V.F. Kolchin [27]. In the general case, Theorem 2.21 was proved by V.E. Kolchin [28], [29, p. 126].

By Theorem 2.21, we expect that, if (2.165) holds for $r \geq 2$, then

$$\lim_{n \to \infty} \mathbf{E}\Big[\Big(\frac{\sigma\mu_n}{2\sqrt{n}}\Big)^r\Big] = m_r \tag{2.213}$$

for $r \geq 0$ where m_r is defined by (2.37). This is indeed proved in some particular cases. If $p_j = (1/2)^{j+1}$ for $j = 0, 1, 2, \ldots$, then N.G. de Bruijn, D.E. Knuth, and S.O. Rice [13] proved (2.213) for $r = 1$ and R. Kemp [23] for $r \geq 1$. If $p_0 = p_2 = 1/2$, Ph. Flajolet and A. Odlyzko [14] proved (2.213) for $r \geq 1$. For $p_j = e^{-1}/j!$ $(j \geq 0)$, formula (2.213) has not been proved yet. A Rényi and G. Szekeres [38] state that (2.213) is true for $r = 1$, but provide no proof.

The limit distribution of δ_n. A direct approach to find the distribution and the asymptotic distribution of δ_n as $n \to \infty$ is not feasible. Fortunately, Theorem 2.15 for the queueing processes makes it possible to determine the asymptotic distribution of δ_n as $n \to \infty$. We have the following result.

Theorem 2.22 *If $a = 1$, $0 < \sigma < \infty$ and $n = sd + 1$ $(s = 0, 1, 2, \ldots)$, then*

$$\lim_{n \to \infty} \mathbf{P}\left\{\frac{\delta_n}{\sigma\sqrt{n}} \leq x\right\} = F(x) \tag{2.214}$$

where $F(x)$ is given by (2.36).

Proof. If, in the queueing process, $\gamma(1) = n$, then

$$\delta_n = max\{\xi(r), r \geq 0\}, \tag{2.215}$$

and obviously,

$$\delta_n \leq \max_{0 \leq r \leq n} \zeta_r \leq 2\delta_n. \tag{2.216}$$

If $\gamma(1) = n$, then by (2.176)

$$\frac{1}{\sqrt{n}}\left(\max_{0 \leq r \leq n} \zeta_r - \delta_n\right) \to 0 \tag{2.217}$$

in probability as $n \to \infty$. Thus if $\gamma(1) = n$, the random variables δ_n and $max_{0 \leq r \leq n} \zeta_r$ have the same asymptotic distribution, and (2.214) follows from (2.172). We note that D. Aldous [2, p. 47] has conjectured that (2.214) is true. □

By (2.214) it is plausible that if (2.165) holds for $r \geq 2$, then

$$\lim_{n \to \infty} \mathbf{E}\left[\left(\frac{\delta_n}{\sigma\sqrt{n}}\right)^r\right] = m_r \tag{2.218}$$

where m_r is defined by (2.37). By (2.166) and (2.216), this is true in two particular cases, namely, when $p_j = (1/2)^{j+1}$ for $j = 0, 1, 2, \ldots$ and when $p_0 = p_2 = 1/2$. A.M. Oldlyzko and H.S. Wilf [35] proved that, if $p_j = e^{-1}/j!$ for $j \geq 0$, then

$$\mathbf{E}[\delta_n] = O(\sqrt{n \log n}) \tag{2.219}$$

as $n \to \infty$.

The limit distribution of τ_n. By Theorem 2.16 we have the following result.

Theorem 2.23 *If $a = 1$, $0 < \sigma < \infty$, if* (2.165) holds for $r \geq 2$ and $n = sd + 1$ $(s = 0, 1, 2, \ldots)$, then the limit,

$$\lim_{n \to \infty} \mathbf{E}\left[\left(\frac{\sigma\tau_n}{2n^{3/2}}\right)^r\right] = M_r \tag{2.220}$$

exists for $r \geq 0$, and M_r is given by (2.47). *Furthermore, we have*

$$\lim_{n \to \infty} \mathbf{P}\left\{\frac{\sigma\tau_n}{2n^{3/2}} \leq x\right\} = W(x) \tag{2.221}$$

where $W(x)$ is given by (2.82).

Proof. Let us introduce the generating functions

$$f(z) = \sum_{j=0}^{\infty} p_j z^j, \tag{2.222}$$

and

$$\Psi(z,w) = \sum_{n=1}^{\infty} \mathbf{P}\{\gamma(1) = n\}\mathbf{E}[z^{\tau_n}]w^n \tag{2.223}$$

defined for $|z| \leq 1$ and $|w| \leq 1$. If we take into consideration that in the queueing process the number of arrivals during the first service time may be $j = 0, 1, 2, \ldots$, we obtain that

$$\Psi(z,w) = wf(\Psi(z, zw)). \tag{2.224}$$

By calculating the successive derivatives of (2.224) at $z = 1$, we can prove that (2.220) holds for $r = 0, 1, 2, \ldots$. For details, see L. Takács [52]. The moments uniquely determine the limit distribution function $W(x)$, and (2.221) follows from (2.220) and from Theorem 2.8. □

The expectation of τ_n has been determined by J. Riordan and N.J.A. Sloan [39] for random-rooted trees with n labeled vertices, and Ju. M. Voloshin [60] for random-rooted trees with n unlabeled vertices.

The limit distribution of $\tau_n(m)$.

Theorem 2.24 *If $a = 1$, $0 < \sigma < \infty$, if (2.165) holds for $r \geq 2$, and if $n = sd + 1$ $(s = 0, 1, 2, \ldots)$, then*

$$\lim_{n\to\infty} \mathbf{E}\left[\left(\frac{2\tau_n([2\alpha\sqrt{n}/\sigma])}{\sigma\sqrt{n}}\right)^r\right] = \mu_r(\alpha) \tag{2.225}$$

exists for $r \geq 0$ and $\alpha > 0$ and $\mu_r(\alpha)$ is defined by (2.104). Furthermore, we have

$$\lim_{n\to\infty} \mathbf{P}\left\{\left(\frac{2\tau_n([2\alpha\sqrt{n}/\sigma])^r}{\sigma\sqrt{n}} \leq x\right\} = G_\alpha(x) \tag{2.226}$$

for $x > 0$ where $G_\alpha(x)$ is given by (2.117).

Proof. Let us introduce the generating function

$$\Phi_m(z,w) = \sum_{n=1}^{\infty} \mathbf{P}\{\gamma(1) = n\}\mathbf{E}[z^{\tau_n(m)}]w^n \tag{2.227}$$

for $|z| \leq 1$ and $|w| \leq 1$. If we take into consideration that, in the queueing process, the number of customers arriving during the first service time may be $j = 0, 1, 2, \ldots$, we obtain that

$$\Phi_m(z,w) = wf(\Phi_{m-1}(z,w)) \tag{2.228}$$

for $m = 1, 2, \ldots$ where

$$\Phi_0(z,w) = z\sum_{n=1}^{\infty} \frac{w^n}{n} \mathbf{P}\{N_n = n-1\}. \tag{2.229}$$

By forming the successive derivatives of (2.228) at $z = 1$, we can prove that (2.225) is true for all $r \geq 0$. For details, see L. Takács [52]. The moments uniquely determine the limit distribution function $G_\alpha(x)$, and (2.226) follows from (2.225) and from Theorem 2.11. □

For random-rooted trees with n labeled vertices, the asymptotic distribution of $\tau_n(m)$ was found by V.E. Stepanov [41]. See also L. Takács [51]. In the context of branching processes and in a different form, the limit theorem (2.226) was found by D.P. Kennedy [25]. By his results, we can conclude that

$$G_\alpha(x) - G_\alpha(0) = \int\int_{\substack{0<u<x/(2\alpha)\\ 0<v<1(4\alpha^2)}} e^{-\alpha^2u^2/(2(1-4\alpha^2v))}(1-4\alpha^2v)^{-3/2}uf(u,v)dudv \tag{2.230}$$

for $x > 0$ and

$$\int_0^\infty\int_0^\infty e^{-su-wv}f(u,v)dudv = \left[\frac{sinh(\sqrt{2w})}{\sqrt{2w}} + s\left(\frac{sinh(\sqrt{w/2})}{\sqrt{w/2}}\right)^2\right]^{-1} \tag{2.231}$$

for $Re(s) > 0$ and $Re(w) \geq 0$.

In the particular case where $s = 0$, formula (2.231) was found by A.G. Pakes [36], [37]. If we suppose that $p_j = (1/2)^{j+1}$ for $j = 0, 1, 2, \ldots$, then we can determine the limit (2.226) by direct calculations, and show that (2.230) is indeed an alternative form of (2.117).

2.9 OPEN PROBLEMS

Let us consider a queueing process with one server. It is supposed, that initially, when the server starts working, one customer is already waiting for service. The server serves this customer and all the new customers in order of arrival as long as they keep coming. The interarrival times and the service times are assumed to be independent sequences of independent and identically distributed positive random variables with expectation 1. If there are no more customers to serve, the initial busy period ends. If the initial busy period consists of n services, let us associate the following graph with the queueing process considered: The graph has vertex set $(1, 2, \ldots, n)$ and vertices r and s where $1 \leq r < s \leq n$ are joined by an edge if, and only if, the sth customer arrives during the service time of the rth customer; the graph is a random tree with vertex set $(1, 2, \ldots, n)$; determine the asymptotic distributions of the height, of the width and of the total height of the random tree.

BIBLIOGRAPHY

[1] Abramowitz, M. and Stegun, I.A., *Handbook of Mathematical Functions,* Dover, New York, 1965.

[2] Aldous, D., The continuum random tree II: An overview, In: *Stochastic Analysis*, Edited by M.T. Barlow and N.H. Bingham, London Mathematical Society Lecture Notes Series No. **167** (1991), Cambridge University Press, 23-70.

[3] Belkin, B., A limit theorem for conditional recurrent random walk attracted to a stable law, *Ann. Math. Stat.* **41** (1970), 146-163.

[4] Belkin, B., An invariance principle for conditional recurrent random walk attracted to a stable law, *Z. Wahrsch. Verw. Geb.* **21** (1972), 45-64.

[5] Bertrand, J., Solution d'un problème, *Comptes Rendus Acad. Sci. Paris* **105** (1887), 369.

[6] Bolthausen, E., On a functional central limit theorem for random walks

conditioned to stay positive, *Ann. Prob.* **4** (1976), 480-485.

[7] Carleman, T., *Les Functions Quasi-Analytiques*, Gauthier-Villars, Paris 1926.

[8] Cayley, A., On the analytic forms called trees, Second part, *Philosophical Magazine* **18** (1859), 374-378. [*The Collected Mathematical Papers of Arthur Cayley*, **IV**, Cambridge University Press, 1891, 112-115.]

[9] Cayley, A., A theorem on trees, *Quart. J. Pure Appl. Math.* **23** (1889), 376-378. [*The Collected Mathematical Papers of Arthur Cayley*, **XIII**, Cambridge University Press, 1897, 26-28.]

[10] Chung, K.L., Excursions in Brownian motion, *Arkiv för Matematik* **14** (1976), 157-179.

[11] Darling, D.A., On the supremum of a certain Gaussian process, *Ann. Prob.* **11** (1983), 803-806.

[12] Darling, D.A., On the asymptotic distribution of Watson's statistic, *Ann. Stat.* **11** (1983), 1263-1266.

[13] de Bruijn, N.G., Knuth, D.E. and Rice, S.O., The average height of planted plane trees, *Graph Theory and Computing*, Edited by: R.C. Read, Academic Press, New York (1972), 15-22.

[14] Flajolet, Ph. and Odlyzko, A., The average height of binary trees and other simple trees, *J. Comp. Sys. Sci.* **25** (1982), 171-213.

[15] Fréchet, M. and Shohat, J., A proof of the generalized second-limit theorem in the theory of probability, *Trans. AMS* **33** (1931), 533-543.

[16] Gikhman, I.I. and Skorokhod, A.V., *Introduction to the Theory of Random Processes*, W.B. Saunders, Philadelphia 1969.

[17] Gnedenko, B.V. and Studnev, Yu. P., Comparison of the effectiveness of several methods of testing homogeneity of statistical material, (Ukrainian), *Dopov. Akad. Nauk Ukr. RSR* **5** (1952), 359-363.

[18] Iglehart, D.L., Functional central limit theorems for random walks conditioned to stay positive, *Ann. Prob.* **2** (1974), 608-619.

[19] Itô, K. and McKean, Jr., H.P., *Diffusion Processes and their Sample Paths*, Springer-Verlag, Berlin 1965.

[20] Jeulin, Th., Application de la théorie du grossissement à l'étude des temps locaux Browniens, In: *Grossissements de filtrations: exemples et applications*, Edited by: Th. Jeulin and M. Yor, Lecture Notes in Mathematics **1118**, Springer-Verlag, New York (1985), 197-304.

[21] Kaigh, W.D., A conditional local limit theorem for recurrent random walk, *Ann. Prob.* **3** (1975), 883-888.

[22] Kaigh, W.D., An invariance principle for random walk conditioned by a late return to zero, *Ann. Prob.* **4** (1976), 115-121.

[23] Kemp, R., On the average stack size of regularly distributed binary trees, In: *Automata, Languages and Programming*, Sixth Colloquium, Graz, Austria, 1979, Edited by: H.A. Maurer, Lecture Notes in Computer Science **71**, Springer-Verlag, Berlin (1979), 340-355.

[24] Kendall, D.G., Some problems in the theory of queues, *J. Royal Stat. Soc.* Ser. B **13** (1951), 151-185.

[25] Kennedy, D.P., The Galton-Watson process conditioned on total progeny, *J. Appl. Prob.* **12** (1975), 800-806.

[26] Kennedy, D.P., The distribution of the maximum Brownian excursion, *J. Appl. Prob.* **13** (1976), 371-376.

[27] Kolchin, V.F., Branching processes, random trees, and a generalized scheme

of arrangements or particles, *Math. Notes* **2** (1977), 386-394.
[28] Kolchin, V.F., Moment of degeneration of a branching process and height of a random tree, *Math. Notes* **24** (1978), 954-961.
[29] Kolchin, V.F., *Random Mappings*, Optimization Software, Inc., New York 1986.
[30] Konovaltsev, I.V., and Lipatov, E.P., Some properties of plane rooted trees, *Cybernetics* **6** (1970), 660-667.
[31] Lévy, P., Sur certains processus stochastiques homogènes, *Compositio Mathematica* **7** (1940), 283-339.
[32] Lévy, P., *Processus Stochastiques et Mouvement Brownien*, Second Edition, Gauthier-Villars, Paris 1965.
[33] Louchard, G., The Brownian excursion area: a numerical analysis, *Comput. Math. Appl.* **10** (1984), 413-417. [Erratum: Ibid **A 12** (1986), 375.]
[34] Miller, J.C.P., *The Airy Integral*, Cambridge University Press 1946.
[35] Odlyzko, A.M. and Wilf, H.S., Bandwidths and profiles of trees, *J. Combin. Theory*, Series B **42** (1987), 348-370.
[36] Pakes, A.G., Some limit theorems for the total progeny of a branching process, *J. Appl. Prob.* **3** (1971), 176-192.
[37] Pakes, A.G., Further results on the critical Galton-Watson process with immigration, *J. Austral. Math. Soc.* **13** (1972), 277-290.
[38] Rényi, A. and Szekeres, G., On the height of trees, *J. Austral. Math. Soc.* **7** (1967), 497-507.
[39] Riordan, J. and Sloane, N.J.A., The enumeration of rooted trees by total height, *J. Austral. Math. Soc.* **10** (1969), 278-282.
[40] Slater, L.J., *Confluent Hypergeometric Functions*, Cambridge University Press 1950.
[41] Stepanov, V.E., On the distribution of the number of vertices in strata of a random tree, *Theory Prob. Appl.* **14** (1969), 65-78.
[42] Takács, L., Remarks on random walk problems, *Publ. Math. Inst. Hung. Acad. Sci.* **2** (1957), 175-182.
[43] Takács, L., The probability law of the busy period for two types of queueing processes, *Oper. Res.* **9** (1961), 402-407.
[44] Takács, L., A generalization of the ballot problem and its application in the theory of queues, *J. Amer. Stat. Assoc.* **57** (1962), 327-337.
[45] Takács, L., Reflection principle, In: *Encyclopedia of Statistical Sciences* **7**, Edited by: S. Kotz and N.L. Johnson, John Wiley, New York (1986), 670-673.
[46] Takács, L., Queues, random graphs and branching processes, *J. Appl. Math. Simul.* **1** (1988), 223-243.
[47] Takács, L., Ballots, queues and random graphs, *J. Appl. Prob.* **26** (1989), 103-112.
[48] Takács, L., On Cayley's formula for counting forests, *J. Combin. Theory* Ser. A **53** (1990), 321-323.
[49] Takács, L., Counting forests, *Discr. Math.* **84** (1990), 323-326.
[50] Takács, L., A Bernoulli excursion and its various applications, *Adv. Appl. Prob.* **23** (1991), 557-585.
[51] Takács, L., On the distribution of the number of vertices in layers of random trees, *J. Appl. Math. Stoch. Anal.* **4** (1991), 175-186.
[52] Takács, L., Conditional limit theorems for branching processes, *J. Appl. Math. Stoch. Anal.* **4** (1991), 262-292.

[53] Takács, L., On the heights and widths of random rooted trees, *International Conference on Random Mappings, Partitions, and Permutations*, Los Angeles, January 3-6, 1992. Abstract: *Adv. Appl. Prob.* **24** (1992), 771.

[54] Takács, L., Limit theorems for random trees, *Proc. National Acad. Sci. USA* **89** (1992), 5011-5014.

[55] Takács, L., On the total heights of random rooted trees, *J. Appl. Prob.* **29** (1992), 543-556.

[56] Takács, L., Random walk processes and their various applications, *Probability Theory and Applications, Essays to the Memory of József Mogyoródi*, Edited by: J. Galambos and I. Kátai, Kluwer Academic Publishers, Dordrecht (1992), 1-32.

[57] Takács, L., Limit distributions for queues and random rooted trees, *J. Appl. Math. Stoch. Anal.* **6** (1993), 189-216.

[58] Takács, L., Enumeration of rooted trees and forests, *Math. Scientist* **18** (1993), 1-10.

[59] Takács, L., The asymptotic distribution of the total heights of random rooted trees, *Acta Sci. Math.* (Szeged) **57** (1993), 613-625.

[60] Voloshin, Ju. M., Enumeration of the terms of the object domains according to the depth of embedding, *Soviet Math. Doklady* **15** (1974), 1777-1782.

Chapter 3

Stationary distributions via first passage times

Søren Asmussen

ABSTRACT This chapter gives a survey of how to express the stationary distribution of a queueing system in terms of first passage time (ruin) probabilities for an associated process and how to solve the corresponding ruin problem. Special attention is given to Siegmund duality (with extensions to Markov-modulated models and stochastic recursions), finite buffer problems, fluid ATM models and Markov-modulated M/G/1 queues. The solution of the ruin problem is in some cases based upon some stopping time identities generalizing the classical work of Wald.

CONTENTS

3.1 Introduction 79
3.2 Random walks and single server queues 81
3.3 Siegmund duality and some of its extensions 82
3.4 Stopping time identities related to Wald's complex exponential martingales 90
3.5 Open problems and further research directions 100
Acknowledgements 101
Bibliography 101

3.1 INTRODUCTION

The duality between stationary distributions and first passage probabilities is a recurrent theme in the theories of queueing, risk and storage. The typical framework is that of two stochastic processes, one a queueing or storage process $\{V_t\}$ and the other process $\{R_t\}$ evolving as a random walk or risk process. The connection is

$$\mathbf{P}(V_T \geq x \mid V_0 = 0) = \mathbf{P}(\tau(x) \leq T), \quad \mathbf{P}(V \geq x) = \mathbf{P}(\tau(x) < \infty), \; x > 0, \tag{3.1}$$

where

$$\tau(x) = inf\{t > 0 : R_t \leq 0 \mid R_0 = x\} \tag{3.2}$$

is the time of ruin, $\mathbf{P}(\tau(x) \leq T)$, resp. $\mathbf{P}(\tau(x) < \infty)$ are the ruin probabilities with finite, resp. infinite horizon, and V denotes the limit in distribution of V_t as $t \to \infty$ (assuming the limit exists and is independent of initial conditions). Presumably the most celebrated classical example of this structure is Lindley's [19] random walk representation of the GI/G/1 queue where V_n is the actual waiting time of customer n and $\{R_n\}$ is a random walk with arbitrary initial value $R_0 = u$ and increments distributed as the difference between an interarrival and a service time. Another important connection exists between compound risk processes and M/G/1 queues; see e.g. Prabhu [24] for an early reference. Additional examples will be met later in the chapter.

0-8493-8074-x/95/$0.00+$.50

Examples where the translation from stationary distributions to first passage time (ruin) probabilities is fruitful certainly exist (say an insurance risk result has initially been developed in a queueing setting). However, as we see it, it is typically easier to say something about first passage time probabilities than stationary distributions. Duality relations of the form (3.1) appear, therefore, primarily useful for studying stationary distributions, and the questions we address here are: How can one translate problems on stationary distributions into ruin problems? Once this has been done, what is the state of the art concerning the evaluation of the ruin probabilities? In studying these issues, we stress computability, motivating Markovian assumptions (including phase-type structure of service times, etc.), and the primary goal is computing the stationary distributions themselves, whereas moments, Laplace transforms etc. are of secondary concern. Simulation methods are logically incorporated into this framework but have been exposed elsewhere ([11] in this book.) Also, we do not look into the role of duality in stability problems (existence of a steady state and convergence to it); see [7], [8] for some relevant references in this direction.

The chapter is a mixture of review material and new results and is organized as follows. We start in Section 3.2 with a review of the classical duality between single server queues and ruin probabilities/maxima for random walks. Section 3.3 deals with general versions of duality relations like (3.1). We first review a general framework developed by Siegmund [28] in the setting of stochastically monotone Markov processes, and proceed to some extensions, monotone stochastic recursions following Asmussen and Sigman [8] and a Markov-modulated version of issues dealt with in [28] (presented here for the first time).

In Section 3.4, we study the problem of how to compute first passage probabilities of the form

$$q^{+}(a,b)=\mathbf{P}(S_{\alpha[-b,a)}\geq a),\ q^{-}(a,b)=\mathbf{P}(S_{\alpha[-b,a)}<b),$$
$$q(a)=\mathbf{P}(\alpha(a)<\infty), \tag{3.3}$$

where

$$\alpha[-b,a)=inf\{n=0,1,\ldots:S_n\notin[-b,a)\},\ \alpha(a)=inf\{n=0,1,\ldots:S_n\geq a\}, \tag{3.4}$$

restricting attention to models with some random walk structure, possibly in a Markov-modulated setting, so that $\{S_t\}$ is a Markov additive process (for some other models, like Example 3.5, there is also a well-developed theory for the ruin probabilities, cf. Asmussen [6], but we do not discuss this here). First passage probabilities of the form (3.3) turn out to determine the stationary distribution of many queueing systems, including some with a finite buffer size. The starting point is a martingale method developed by Wald [31] for simple random walks; the extensions we present are, to a large extent, new. The examples include random walks between two barriers, fluid models with finite or infinite buffers and Markov-modulated M/G/1 queues. In particular, for the Markov-modulated M/PH/1 queue with phase-type service times, we suggest an analysis of the model via an associated fluid model. This provides a new algorithm for the steady-state workload distribution which uses nothing more than linear algebra and avoids nonlinear matrix iteration schemes like those of Asmussen [4] or Sengupta [27].

3.2 RANDOM WALKS AND SINGLE SERVER QUEUES

The reflected version $\{V_n\}$ of a random walk $\{S_n\}$ is defined by the Lindley recursion

$$V_{n+1} = (V_n + X_n)^+; \tag{3.5}$$

here $S_n = X_0 + \ldots + X_{n-1}$ with $X_0, X_1, \ldots$ i.i.d., say with common distribution F. A particularly important example corresponds to the GI/GI/1 queue where V_n is the waiting time of the nth customer, U_n his service time, T_n the time between the arrivals of customers n and $n+1$, and $X_n = U_n - T_n$; see [3], Ch. III.7 for the (easy) details. In the GI/GI/1 setting, the standard stability condition $\rho = \mathbf{E}[U]/\mathbf{E}[T] < 1$ corresponds to F having mean less than 0, and this will be assumed throughout also for the more general case. Then $M = max(0, S_1, S_2, \ldots) < \infty$ a.s.

Obviously, $\{V_n\}$ is a Markov process, and we have:

Theorem 3.1. *The distribution of M is stationary for $\{V_n\}$.*

Proof. The statement means that $\mathbf{P}(V_1 \geq x \mid V_0 = M^*) = \mathbf{P}(M \geq x)$ for all $x > 0$ where M^* is independent of $X_0, X_1, \ldots$ and distributed as M. Write $M = max(0, S_1^*, S_2^*, \ldots)$ where $S_n^* = X_0^* + \ldots + X_{n-1}^*$ and the X_k^* are i.i.d., have common c.d.f. F, and are independent of the X_k. Then, for $x > 0$,

$$\begin{aligned}\mathbf{P}(V_1 \geq x \mid V_0 = M^*) &= \mathbf{P}((V_0 + X_0)^+ \geq x \mid V_0 = M^*)\\ &= \mathbf{P}(M^* + X_0 \geq x) = \mathbf{P}(M^{**} \geq x),\end{aligned}$$

where $M^{**} = max(0, X_0, X_0 + X_0^*, X_0 + X_0^* + X_1^*, \ldots)$. Since M^{**} obviously has the same distribution as M, the proof is complete. □

To conform with the general duality set-up, we first write the event $\{M \geq x\}$ as $\{\tau(x) < \infty\}$ where $\tau(x) = inf\{n: S_n \geq x\}$ (this step requires that the maximum is attained which follows from $S_n \to -\infty$ a.s.). To get a proper dual process, let next

$$R_n = (R_0 - X_{n \wedge \tau(x)})^+.$$

In words, as long as the values are non-negative, $\{R_n\}$ evolves as the random walk with the increments reversed in sign and starting from an arbitrary initial value. If $(-\infty, 0]$ is hit, the value of $\{R_n\}$ is instantaneously reset to 0, and the process remains there forever. It follows that

$$\mathbf{P}(\tau(x) < \infty) = \mathbf{P}(M \geq x) = \mathbf{P}(V \geq x),$$

where V has the stationary distribution of $\{V_n\}$, and $\tau(x)$ is as in (3.2). In this example, the representation in Theorem 3.1 is obviously the more elegant one, but, in more general situations, the dual process $\{R_n\}$ and its ruin probabilities need to be invoked.

As discussed in the Introduction, the key topic of the present chapter is generalizations of this example. Already at this stage, there are some fairly easy ones:

1. We can drop the i.i.d. assumption and require only that the sequence is (strictly) stationary, for simplicity ergodic with $\mathbf{E}[X_n] < 0$. Instead of considering stationary distributions for the Markov chain $\{V_n\}$, we need to consider stationary versions of $\{V_n\}$. To construct such a version, we may assume that the sequence $\{X_n\}_{n=0,\pm 1,\pm 2}$ is double infinite and take

$$V_n = max(0, X_{n-1}, X_{n-1} + X_{n-2}, \ldots) \tag{3.6}$$

(in the proof of Theorem 3.1, this corresponds to taking $X_n^* = X_{-1-n}$).

Then it is easily seen that $\{V_n\}$ is stationary and satisfies the Lindley recursion (3.5). This observation is the starting point for Loynes [21], a reference of substantial impact on later work (a reference relevant to the present chapter is [8]).

2. As a (stationary) input sequence, consider $\{(J_n, X_n)\}$, where we think of $\{J_n\}$ as a background (environmental) process modulating the distribution of the random walk increments $\{X_n\}$. The duality argument just presented then leads to stationarity of the sequence $\{(J_n, X_n, V_n)\}$, with V_n defined as in (3.6). More precisely, with $V = V_0$, $J = J_0$, we have

$$\mathbf{P}(V \geq x, J \in A) = \mathbf{P}\big(max(0, X_{-1}, X_{-1} + X_{-2}, \ldots) \geq x, J_0 \in A\big).$$

This relation is a natural motivation for the Markov-modulated models to be studied later. Namely, assume for simplicity that $\{J_n\}$ is Markovian with a discrete state space and stationary probabilities $\pi_i = \mathbf{P}(J_n = i)$, and that $\{S_n\}$ is a Markov random walk driven by $\{J_n\}$: the increments $\{X_n\}$ are conditionally independent, given $\{J_n\}$, with the distribution of $\{X_n\}$ being F_{ij} when $J_{n-1} = i$, $J_n = j$. Then we can think of $0, X_{-1}, X_{-1} + X_{-2}, \ldots$ as the Markov random walk driven by the time-reversed Markov chain $\{J_n^*\} = \{J_{-n}\}$: the conditional distribution of $X_n^* = X_{-1-n}$ given $\{J_n^*\}$ is $F_{ij}^* = F_{ji}$ when $J_{n-1}^* = i$, $J_n^* = j$.

3. By a continuous time random walk $\{S_t\}_{t \geq 0}$, we understand a Lévy process, defined as the independent sum of a Brownian component, a drift term, and a jump part specified by its Lévy measure. The reflected version is

$$V_t = S_t - \min_{0 \leq v \leq t} S_v, \tag{3.7}$$

and, if the mean drift of $\{S_t\}$ is negative, $\{V_t\}$ has a stationary distribution given by

$$V \overset{\mathfrak{D}}{=} \max_{0 \leq t < \infty} S_t; \quad \text{i.e.,} \quad \mathbf{P}(V \geq x) = \mathbf{P}(\tau(x) < \infty),$$

where $\tau(x) = inf\{t: S_t \geq x\}$. See e.g., [3], Ch. III.8.

3.3 SIEGMUND DUALITY AND SOME OF ITS EXTENSIONS

Siegmund [28] developed a general framework in which to study dualities similar to the ones considered for random walks in Section 3.2. His main result is the following:

Theorem 3.2. *Let $\{V_t\}$ be a Markov process on $[0,\infty)$, with discrete or continuous time. Then the existence of a Markov process $\{R_t\}$ on $[0,\infty]$ with the property that*

$$\mathbf{P}_x(V_t \geq u) = \mathbf{P}_u(R_t \leq x) \tag{3.8}$$

is equivalent to: (i) $\{V_t\}$ is stochastically monotone; (ii) $\mathbf{P}_x(V_t \geq u)$ is a right-continuous function of x for fixed u and t.

Here stochastical monotonicity is defined in the sense of Daley [14], meaning just that $\mathbf{P}_x(V_t \geq u)$ is nondecreasing in x for fixed u and t, and "existence of a Markov process" means existence of a semigroup satisfying the Chapman-Kolmogorov equations. A proof of Theorem 3.2 is contained in the treatment of the Markov-modulated extension in Section 3.3.2 and therefore omitted here.

The crucial step, when applying Siegmund duality, is identifying the dual process which was done in Section 3.2 in the case of a reflected random walk. Here are some further basic examples:

Example 3.3. [28] Assume that $\{V_n\}$ is a discrete-time Markov chain with state space $\{0,1,2,\ldots\}$ and transition probabilities p_{ij}. Then (3.8) with $t=1$ means that the transition probabilities q_{ij} for $\{R_n\}$ are

$$q_{ij} = \begin{cases} 1 & i=j=0 \\ 0 & i=0, 0<j\leq\infty \\ \sum_{k=i}^{\infty}(p_{jk}-p_{(j-1)k}) & i>0, 0\leq j<\infty \\ 1-\lim_{\ell\to\infty}\sum_{k=i}^{\infty}p_{\ell k} & i>0, j=\infty \\ 1 & i=j=\infty \\ 0 & i=\infty, j<\infty \end{cases}$$

(here $p_{(j-1)k}=0$ for $j=0$).

Example 3.4. Let $\{V_n\}_{n=0,1,\ldots}$ be a random walk between two reflecting barriers 0, $b>0$. That is,

$$V_{n+1} = min\{b, (V_n+X_n)^+\},$$

where $X_0, X_1, \ldots$ are i.i.d. with common distribution F (say). Thus the set (b,∞) is transient and $[0,b]$ is absorbing (of course, the natural state space is $[0,b]$, but we need to have $\mathbf{P}_x$ defined also when $x>b$ for the description of $\{R_n\}_{n=0,1,\ldots}$).

If $y>b$, then

$$\mathbf{P}_y(R_1\leq x) = \mathbf{P}_x(V_1\geq y)\leq \mathbf{P}_x(V_1>b)=0$$

for all x, i.e., $\mathbf{P}_y(R_1=\infty)=1$. If $0\leq y\leq b$, then

$$\mathbf{P}_y(R_1\leq x) = \mathbf{P}_x(V_1\geq y) = \mathbf{P}((x+X_0)^+\geq y)$$

$$= \begin{cases} 1 & y=0 \\ \mathbf{P}(y-X_0\leq x) & 0<y\leq b \end{cases}$$

for all $x\geq 0$. Combining these facts shows that $\{R_n\}$ evolves as a random walk $\{S_n\}$ with increments distributed as $-Y_1$ as long as the values are in $(0,b]$. States 0 and ∞ are absorbing. Starting from a state $y\in(b,\infty)$, the next value of $\{R_n\}$ is ∞. An attempt by the random walk to make a transition to $(-\infty,0]$ makes $\{R_n\}$ instantaneously go to the absorbing state 0; if instead the transition is to $y\in(b,\infty)$, the value y is attained in the first step, but after that all values are taken in the absorbing state ∞.

It follows immediately that $\{R_n\}$ is eventually absorbed by either 0 or ∞. Hence the event, $\{\tau(u)<\infty\}$ of absorption at 0, can be identified with $\{S_{\alpha[u-b,u)}\geq u\}$. In particular, in the steady state

$$\mathbf{P}(V\geq u) = \mathbf{P}(S_{\alpha[u-b,u)}\geq u) = q^+(u-b,u). \tag{3.9}$$

This example is implicitly in Lindley [20] and explicitly in Siegmund [28], but seems not to have been noted in queueing theory at all. The Moran dam is an example of a random walk with two reflecting barriers; here $X_n = Y_n - M$ where Y_n is the input at time n and M the (constant) release per unit time. Borovkov [12] and Stadje [30] have derived representations of the steady-state distribution which look somewhat different from (3.9) but in fact are basically the same. To see this, consider w.l.o.g. the integer-valued case where Theorem 2 of [30] states that, for $u = 0, 1, \ldots, b$,

$$\begin{aligned} \mathbf{P}(V = u) &= \mathbf{P}(S_{\alpha[1-b,1)} \le -b)\mathbf{P}(S_{\alpha[u-b+1,u)} = u) \\ &\quad + \mathbf{P}(S_{\alpha[0,b)} \ge b)\mathbf{P}(S_{\alpha[u-b+1,u)} = u - b). \end{aligned}$$

This follows immediately from (3.9) by subtracting the two easily checked identities

$$\begin{aligned} \mathbf{P}(S_{\alpha[u-b,u)} \ge u) &= \mathbf{P}(S_{\alpha[u-b+1,u)} = u-b)\mathbf{P}(S_{\alpha[0,b)} \ge b) \\ &\quad + \mathbf{P}(S_{\alpha[u-b+1,u)} \ge u) \end{aligned}$$

and
$$\begin{aligned} \mathbf{P}(S_{\alpha[u-b+1,u+1)} \ge u+1) &= \mathbf{P}(S_{\alpha[u-b+1,u)} = u)\mathbf{P}(S_{\alpha[1-b,1)} \ge 1) \\ &\quad + \mathbf{P}(S_{\alpha[u-b+1,u)} \ge u+1). \end{aligned}$$

Example 3.5. Assume that $\{V_t\}$ is a storage process with compound Poisson input $\{A_t\}$ and release rate $r(x)$. When $V_t = x$,

$$V_t = V_0 + A_t - \int_0^t r(V_s)ds$$

(taking $r(0) = 0$ to make state 0 reflecting). Then $\{R_t\}$ is a risk process with premium rate $r(x)$ when $R_t = x$ and the accumulated claims A_t before time t being compound Poisson,

$$R_t = R_0 - A_t + \int_0^t r(R_s)ds$$

when $t < \tau$. See Lindley & Resnick [17] for the case $T = \infty$ in (3.1) and Asmussen & Schock Petersen [10] for the case $T < \infty$.

Example 3.6. Assume that $\{V_t\}$ is a continuous time Markov process with state space $\{0, 1, 2, \ldots\}$ and transition intensities λ_{ij}. For a formal argument for identifying the dual process $\{R_t\}$, consider a discrete skeleton $\{R_{nh}\}$ and its transition probabilities p_{ij}^h. Assuming the existence of transition intensities μ_{ij} for $\{R_t\}$, we have

$$p_{ij}^h \approx \mu_{ij}h, \quad i \ne j, \quad p_{ii}^h \approx 1 + \mu_{ii}h,$$

and evaluating the p_{ij}^h by Example 3.3 leads to

$$\mu_{ij} = \begin{cases} 0 & i = 0 \\ \sum_{k=i}^{\infty} (\lambda_{jk} - \lambda_{(j-1)k}) & i > 0, 0 \le j < \infty \\ 1 - \lim_{\ell \to \infty} \sum_{k=i}^{\infty} \lambda_{\ell k} & i > 0, j = \infty \\ 0 & i = \infty. \end{cases}$$

This example is the main one in [28], but it is notable that the rigorous treatment is nontrivial. Indeed, this turns out to be the case in most continuous time examples.

As preparation for the next example, we note the following version of the duality relation:

Proposition 3.7. *Let f, g be C_1 functions with compact support contained in $(0,\infty)$. Then*

$$\int_0^\infty f'(x)\mathbf{E}_x[g(R_t)]dx = -\int_0^\infty g'(y)\mathbf{E}_y[f(V_t)]dy.$$

Proof. Using integration by parts, we get

$$\begin{aligned}\int_0^\infty f'(x)\mathbf{E}_x[g(R_t)]dx &= \int_0^\infty f'(x)dx\int_0^\infty g'(y)\mathbf{P}_x(R_t > y)dy \\ &= \int_0^\infty f'(x)dx\int_0^\infty g'(y)(1-\mathbf{P}_y(V_t \geq x))dy \\ &= -\int_0^\infty f'(x)ds\int_0^\infty g'(y)\mathbf{P}_y(V_t \geq x)dy \\ &= -\int_0^\infty g'(y)\mathbf{E}_y[f(V_t)]dy.\end{aligned}$$

□

Example 3.8. Assume that $\{V_t\}$ is a diffusion process with differential generator $\mathcal{A}f(y) - \mu(y)f'(y) + \sigma^2(y)/2f''(y)$ (the boundary condition at 0 is arbitrary). Let $\mathfrak{B}$ be the generator of $\{R_t\}$ and take f, resp. g in the domain of $\mathcal{A}$, resp. $\mathfrak{B}$. Then,

$$\mathbf{E}_y[f(V_t)] = f(y) + \mathcal{A}f(y)t + o(t), \mathbf{E}_x[g(R_t)] = g(x) + \mathfrak{B}g(x)t + o(t).$$

Since $\int(f'g + fg') = [fg] = 0$ because of the assumption on the support, Proposition 3.7 yields

$$\begin{aligned}\int_0^\infty f'(x)\mathfrak{B}g(x)dx &= -\int_0^\infty g'(y)\mathcal{A}f(y)dy \\ &\quad -\int_0^\infty g'(y)\Big\{\mu(y)f'(y) + \tfrac{1}{2}\sigma^2(y)f''(y)\Big\}dy \\ = \int_0^\infty &\Big\{-\mu(y)f'(y)g'(y) + \tfrac{1}{2}\Big(\sigma^2(y)g''(y) + 2\sigma^{2'}(y)g'(y)\Big)f'(y)\Big\}dy \\ &= \int_0^\infty f'(y)\Big\{\Big(-\mu(y) + \sigma^{2'}(y)\Big)g'(y) + \tfrac{1}{2}\sigma^2(y)g''(y)\Big\}dy.\end{aligned}$$

The truth of this, for all f under consideration, implies

$$\mathfrak{B}g(y) = \Big\{-\mu(y) + \sigma^{2'}(y)\Big\}g'(y) + \tfrac{1}{2}\sigma^2(y)g''(y). \tag{3.10}$$

As noted in [28], the rigorous treatment of this example is again nontrivial (see, however, Cox & Rössler [13]). Some further heuristic formulas for computing the dual process are given in Aldous [1].

3.3.1 Stochastic recursions in duality

A stochastic recursion on $[0,\infty)$ has the form

$$V_{n+1} = f(V_n, U_n), \quad n = 0,1,2,\ldots, \tag{3.11}$$

where $\{U_n\}$ (the *driving sequence*) is a sequence of random elements taking values in some arbitrary space E, and $f:[0,\infty)\times E \to [0,\infty)$ is a function. We denote by $\{V_n(y)\}$ the sequence generated by $V_0 = y$ and (3.11). The setup incorporates, for example, any Markov chain on $[0,\infty)$ where one can take the U_n as i.i.d. uniforms on $(0,1)$, but also much more general examples where one usually just assumes (as will be done in the following) that the driving sequence is stationary. For example, the reflected Markov random walk example (Remark 2 at the end of Section 3.2) is incorporated by taking $U_n = (J_n, J_{n+1}, W_n)$ where $\{J_n\}$ is a stationary Markov chain, the W_n are i.i.d. uniforms on $(0,1)$ (independent of $\{J_n\}$), and $f(y,i,j,w)$ is the inverse of $F_{ij}(\cdot - y)$.

Duality in this setup was considered by Asmussen & Sigman [8]. They as-sumed $f(x,u)$ to be continuous and nondecreasing in x for each fixed $u \in E$ and defined a dual function $g(x,u)$ as the inverse of $f(x,u)$,

$$g(x,u) = inf\{y: f(y,u) \geq x\}; \tag{3.12}$$

here, $g:[0,\infty]\times E \to [0,\infty]$ so that $x=\infty$ is allowed which is essential for examples like random walks between two reflecting barriers. Then $g(x,u)$ is left-continuous, nondecreasing, strictly increasing on the interval $\{x: 0 < g(x,u) < \infty\}$, and satisfies $g(0,u) = 0$, $g(\infty,u) = \infty$. Furthermore, one can retrieve f in terms of g by

$$f(y,u) = sup\{x: g(x,u) \leq y\}, \tag{3.13}$$

and the fundamental connection is

$$g(x,u) \leq y \Leftrightarrow f(y,u) \geq x. \tag{3.14}$$

The dual process $\{R_n\}$ is defined by the recursion

$$R_{n+1} = g(V_n, U_{-n}), \quad n = 0,1,2,\ldots, \tag{3.15}$$

assuming, without loss of generality, that the stationary driving sequence $\{U_n\}_{0,\pm 1,\pm 2,\ldots}$ has doubly infinite time. We denote by $\{R_n(x)\}$ the sequence generated by $R_0 = x$ and (3.15). The basic duality result then states that, for each fixed $m = 0,1,2,\ldots$ and $x,y \in [0,\infty)$, one can construct a coupling satisfying

$$V_m^{(m)}(y) \geq x \Leftrightarrow R_m(x) \leq y, \tag{3.16}$$

where "coupling" means

$$\left\{V_n^{(m)}(y)\right\}_{n=0,\ldots,m} \stackrel{\mathfrak{D}}{=} \{V_n(y)\}_{n=0,\ldots,m}.$$

In fact, one can define $\left\{V_n^{(m)}(y)\right\}_{n=0,\ldots,m}$ by

$$V_0^{(m)}(y) = y,\ V_1^{(m)}(y) = f\left(V_0^{(m)}(y), U_{-m}\right), \ldots, V_m^{(m)}(y) = f\left(V_{m-1}^{(m)}(y), U_{-1}\right);$$

then (3.16) follows from (3.13) by a trivial induction argument.

Since state 0 is absorbing for $\{R_n\}$, we have $\{R_m(x)=0\}=\{\tau(x)\le m\}$ where $\tau(x)=inf\{n: R_n=0\}=inf\{n: R_n\le 0\}$. Letting $m\to\infty$ in (3.16), letting $y=0$, and taking probabilities, then yields the existence of an r.v. V such that

$$\mathbf{P}(V\ge x)=\lim_{m\to\infty}\mathbf{P}(V_m(0)\ge x)=\mathbf{P}(\tau(x)<\infty) \tag{3.17}$$

for all x (typically, V is also the limit of $V_m(y)$ for all $y\ne 0$ but this requires further consideration in the general stationary setup).

The advantage of the setup is twofold: (1) one can incorporate dependency structures much more complicated than Markov chains on $[0,\infty)$; (2) it is typically much easier to identify the dual process by means of (3.13), (3.15) than to invert transition probabilities as in (3.8) (see [8] for an extensive list of examples). The drawback is that, at the moment, we do not see how to develop a similar theory in continuous time. This motivates the extended Siegmund duality concept that we next develop.

3.3.2 Siegmund duality and Markov modulation

We consider a Markov process $\{(J_t, V_t)\}$ with state space $E\times[0,\infty)$, such that the marginal distribution of $\{J_t\}$ corresponds to an irreducible Markov process with finite state space E; the stationary distribution of $\{J_t\}$ then exists and is denoted by $\pi=(\pi_i)_{i\in E}$.

Theorem 3.9. *Assume that* $\mathbf{P}_{i,x}(V_t\ge y, J_t=j)$ *is a right-continuous nondecreasing function of* $x\in[0,\infty)$ *when* $y<\infty$, $i\in E$, $t>0$ *are fixed. Then the formulas*

$$\mathbf{P}_{j,y}(R_t\le x, J_t^*=i)=\frac{\pi_i}{\pi_j}\mathbf{P}_{i,x}(V_t\ge y, J_t=j), 0\le x<\infty, \tag{3.18}$$

$$\mathbf{P}_{j,y}(R_t=\infty, J_t^*=i)=\frac{\pi_i}{\pi_j}\lim_{x\to\infty}\mathbf{P}_{i,x}(V_t<y, J_t=j), \tag{3.19}$$

and

$$\mathbf{P}_{j,\infty}(R_t=\infty, J_t^*=i)=\frac{\pi_i}{\pi_j}\mathbf{P}_i(J_t=j), \tag{3.20}$$

define a Markov process $\{(J_t^*, R_t)\}$ *with state space* $E\times[0,\infty]$, *such that* $\{J_t^*\}$ *is the time-reversed version of* $\{J_t\}$ *and that* $\mathbf{P}_{j,0}(R_t=0)=1$ *for all* j,t.

Proof. Let $p_t((j,y),(i,[0,x]))$ denote the l.h.s. of (3.18). Then we have to show that $\{p_t((j,y),(i,dx))\}$ is a semigroup of transition operators on $E\times[0,\infty]$.

We first note that for $y<\infty$,

$$\int_{x\in[0,\infty]} p_t((j,y),(i,dx))$$

$$=\frac{\pi_i}{\pi_j}\lim_{x\to\infty}\{\mathbf{P}_{i,x}(V_t\ge y, J_t=j)+\mathbf{P}_{i,x}(V_t<y, J_t=j)\}$$

$$=\frac{\pi_i}{\pi_j}\lim_{x\to\infty}\mathbf{P}_{i,x}(J_t=j)$$

$$=\frac{\pi_i}{\pi_j}\mathbf{P}_i(J_t=j). \tag{3.21}$$

Summing over $i\in E$, the r.h.s. becomes $\mathbf{P}_\pi(J_t=j)/\pi_j=1$, which shows that indeed $p_t((j,y),(\cdot,\cdot))$ is a probability measure on $E\times[0,\infty]$. The case $y=\infty$ is similar, though easier.

Further,

$$p_{t+s}((j,y),(i,[0,x]))$$

$$= \frac{\pi_i}{\pi_j}\mathbf{P}_{i,x}(V_{t+s} > y, J_{t+s} = j)$$

$$= \frac{\pi_i}{\pi_j}\sum_{k\in E}\int_{z\in[0,\infty)} \mathbf{P}_{i,x}(V_t \in dz, J_t = k)\mathbf{P}_{k,z}(V_s \geq y, J_s = j)$$

$$= \frac{\pi_i}{\pi_j}\sum_{k\in E}\int_{z\in[0,\infty)} \mathbf{P}_{i,x}(V_t \in dz, J_t = k)\frac{\pi_j}{\pi_k}\int_{u\in[0,z]} p_s((j,y),(k,du))$$

$$= \sum_{k\in E}\frac{\pi_i}{\pi_k}\int_{u\in[0,\infty)} p_s((j,y),(k,du))\int_{z\in[u,\infty)} \mathbf{P}_{i,x}(V_t \in dz, J_t = k)$$

$$= \sum_{k\in E}\frac{\pi_i}{\pi_k}\int_{u\in[0,\infty)} p_s((j,y),(k,du))\mathbf{P}_{i,x}(V_t \geq u, J_t = k)$$

$$= \sum_{k\in E}\frac{\pi_i}{\pi_k}\int_{u\in[0,\infty)} p_s((j,y),(k,du))\frac{\pi_k}{\pi_i}\mathbf{P}_{k,u}(R_t \leq x, J_t^* = i)$$

$$= \sum_{k\in E}\int_{u\in[0,\infty)} p_s((j,y),(k,du))p_t((k,u),(i,[0,x])),$$

so that the semigroup property holds.

We can now rewrite (3.21) as

$$\mathbf{P}_{j,y}(J_t^* = i) = \frac{\pi_i}{\pi_j}\mathbf{P}_i(J_t = j)$$

which is the same as saying that $\{J_t^*\}$ is the time-reversed version of $\{J_t\}$. Finally, by (3.18),

$$\mathbf{P}_{j,0}(R_t = 0) = \sum_{i\in E}\frac{\pi_i}{\pi_j}\mathbf{P}_{i,0}(J_t = j) = \frac{1}{\pi_j}\mathbf{P}_\pi(J_t = j) = 1. \qquad \square$$

Remark 3.10. Letting $H((i,x),(j,y)) = \frac{1}{\pi_j}I(x \geq y, i = j)$, we can rewrite (3.18) as

$$\mathbf{E}_{i,x}[H((J_t,V_t),(j,y))] = \mathbf{E}_{j,y}[H((i,x),(J_t^*,R_t))],$$

which means that $\{(J_t,V_t)\}$ and $\{(J_t^*,R_t)\}$ are dual w.r.t. H in the sense of Liggett [18], p. 84. Some of the setup in [18] is, however, inconvenient for the present purposes. In particular, the requirement of $\{(J_t^*,R_t)\}$ to be Feller (i.e., $p_t((j,y),(\cdot,\cdot))$ to be weakly continuous in y for fixed t) leads to $\mathbf{P}_{i,x}(V_t = 0) = 0$ for $x > 0$ (cf. Theorem 3.5 of [18]), which is not satisfied in the examples we have in mind.

Define $\tau_j(y) = inf\{t > 0: R_t = 0 \mid J_0^* = j, R_0 = y\}$. Assuming the existence of a steady state, we have:

Corollary 3.11. $\mathrm{P}(V \geq y, J = j) = \pi_j\mathrm{P}(\tau_j(y) < \infty)$.

Proof. For each i,

$$\mathbf{P}_{i,0}(V_t \geq y, J_t = j) = \frac{\pi_j}{\pi_i}\mathbf{P}_{j,y}(J_t^* = i, R_t \leq 0)$$
$$= \frac{\pi_j}{\pi_i}\mathbf{P}_{j,y}(J_t^* = i, \tau_j(y) \leq t). \tag{3.22}$$

However, given $\epsilon > 0$ we can choose $T < \infty$ such that $\mathbf{P}(T < \tau_j(y) < \infty) < \epsilon$. Hence, for $t \geq T$, (3.22) is

$$\frac{\pi_j}{\pi_i}\mathbf{P}_{j,y}(J_t^* = i, \tau_j(y) \leq T) + O(\epsilon) = \frac{\pi_j}{\pi_i}\mathbf{E}_{j,y}[\mathbf{P}(J_t^* = i \mid J_T); \tau_j(y) \leq T] + O(\epsilon)$$
$$\overset{t\to\infty}{\to} \pi_j \mathbf{P}(\tau_j(y) \leq T) + O(\epsilon)$$
$$= \pi_j \mathbf{P}(\tau_j(y) < \infty) + O(\epsilon).$$

Letting first $t \to \infty$ and next $\epsilon \downarrow 0$ in (3.22) completes the proof. □

The following heuristical result describes the structure of a process $\{(J_t, V_t)\}$ satisfying the assumptions of Theorem 3.9.

Proposition 3.12. *Assume, in addition to the conditions of Theorem 3.9, that $\{(J_t, V_t)\}$ is strong Markov with right-continuous paths. Then there exist Markov processes $\{V_t^{(i)}\}$ with state space $[0,\infty)$ and (possibly defective) probability measures $K^{(ij;x)}(j \neq i, x \geq 0)$ such that:*

(*a*) *for each y, $\mathbf{P}_x(V_t^{(i)} \geq y)$ is a nondecreasing right-continuous function of x;*

(*b*) *on intervals where $J_t = i$, $\{V_t\}$ evolves as $\{V_t^{(i)}\}$;*

(*c*) *for $y \leq x$, $K^{(ij;y)}([z,\infty)) \leq K^{(ij;x)}([z,\infty))$ for all $z \geq 0$;*

(*d*) *a jump of $\{J_t\}$ from i to $j \neq i$ at time t is accompanied by a jump according to $K^{(ij;x)}$ where $x = S_t -$.*

Conversely, (a)-(d) are sufficient for $\mathbf{P}_{i,x}(V_t \geq y, J_t = j)$ to be a nondecreasing right-continuous function of x.

Proof. (incomplete sketch) Consider the continuous time case (the discrete time case is similar but much easier). Conditioning upon $J_{t_{k-1}}$, it is easy to see that for $0 = t_0 < t_1 < \ldots < t_k = t$,

$$\mathbf{P}_{i,x}(V_t \geq y, J_{t_\ell} = i, \ell = 0, \ldots, k) \tag{3.23}$$

is a nondecreasing right-continuous function of x. Hence, so is

$$\mathbf{P}_{i,x}(V_t \geq y, J_v = i, 0 \leq v \leq t)$$

(as the limit from above of functions of the form (3.23)) and

$$v_t^{(i)}(x,[y,\infty)) = \mathbf{P}_{i,x}(V_t \geq y \mid J_v = i, 0 \leq v \leq t).$$

Further, $\{v_t^{(i)}\}_{t \geq 0}$ can be checked to define a semigroup. Thus, if $\{V_t^{(i)}\}$ is determined by $\{v_t^{(i)}\}$, (*a*) and (*b*) hold. The kernels K can be defined in terms of Lévy systems, and (*c*) then appears intuitively obvious. However, at present we do not have a rigorous proof. □

Example 3.13. Consider the discrete time case with discrete state space $\{0,1,\ldots\}$ for $\{V_t\}$. Then both $\{(J_t, V_t)\}$ and $\{(J_t^*, R_t)\}$ are specified by the transition probabilities, which we represent in terms of the $E \times E$ matrices $\boldsymbol{V}^{(k\ell)}$, $\boldsymbol{R}^{(k\ell)}$, with elements

$$v_{ij}^{(k\ell)} = \mathbf{P}_{i,k}(V_1 = \ell, J_1 = j) \text{ and } r_{ji}^{(\ell k)} + \mathbf{P}_{j,\ell}(R_1 = k, J_1^* = i).$$

It follows immediately that

$$r_{ji}^{(\ell k)} = \frac{\pi_j}{\pi_i} \sum_{m=\ell}^{\infty} \left(v_{ij}^{(k\ell)} - v_{ij}^{((k-1)\ell)} \right), \tag{3.24}$$

interpreting $v_{ij}^{((-1)\ell)}$ as 0.

Example 3.14. Consider the continuous time case with discrete state space $\{0,1,\ldots\}$ for $\{V_t\}$, and assume that $\{(J_t, V_t)\}_{t \geq 0}$ is the minimal process associated with an intensity matrix, which again we represent in terms of the $E \times E$ blocks $V^{(k\ell)}$ with elements $v_{ij}^{(k\ell)}$. Then formally we obtain the intensity matrix of $\{(J_t^*, R_t)\}_{t \geq 0}$ as

$$r_{ji}^{(\ell k)} = \frac{\pi_j}{\pi_i} \sum_{m=\ell}^{\infty} \left(v_{ij}^{(k\ell)} - v_{ij}^{((k-1)\ell)} \right), \tag{3.25}$$

again interpreting $v_{ij}^{((-1)\ell)}$ as 0.

3.4 STOPPING TIME IDENTITIES RELATED TO WALD'S COMPLEX EXPONENTIAL MARTINGALES

3.4.1 Lattice random walks

Let $\{S_n\} = \{X_0 + \ldots + X_{n-1}\}$ be a random walk on the integer lattice with bounded increments. That is, for some integers $n, m < \infty$, we have

$$f_n > 0,\ f_{-m} > 0,\ \sum_{k=-m}^{n} f_k = 1, \tag{3.26}$$

where $f_k = \mathbf{P}(X_0 = k)$. In connection with sequential analysis, Wald [31] encountered the problem of computing the two-boundary first passage time probabilities,

$$p_k^+(a,b) = \mathbf{P}(S_{\alpha[-b,a)} = a + k), k = 0, \ldots, n-1$$

and $$p_\ell^-(a,b) = \mathbf{P}(S_{\alpha[-b,a)} = -b - \ell), \ell = 1, \ldots, m,$$

and suggested the following martingale approach:

Define $\widehat{f}[z] = \mathbf{E}[z^{X_0}] = \sum_{k=-m}^{n} z^k f_k$. Then $z^m \widehat{f}[z]$ is a polynomial of degree $m+n$, and hence the equation $\widehat{f}[z] = 1$ has $m+n$ roots which we denote by $z_1, \ldots, z_{m+n}$. For each i,

$$\mathbf{E}\left[z_i^{S_{n+1}} \mid X_0, \ldots, X_{n-1}\right] = z_i^{S_n} \mathbf{E}\left[z_i^{X_n} \mid X_0, \ldots, X_{n-1}\right] = z_i^{S_n} \mathbf{E}[z_i^{X_1}] = z_i^{S_n},$$

and hence $\{z_i^{S_n}\}$ is a (complex) martingale (integrability is trivial by (3.26)). Further, $sup_{n \leq \alpha[-b,a)} z_i^{S_n}$ is bounded by a constant, again because of (3.26). Hence, optional stopping is justified and yields

$$1 = \mathbf{E}[z_i^{S_0}] = \mathbf{E}[z_i^{S_{\alpha[-b,a)}}] = z_i^a \sum_{k=0}^{n-1} p_k^+(a,b) z_i^k + z_i^{-b} \sum_{\ell=1}^{m} p_\ell^-(a,b) z_i^{-\ell}. \tag{3.27}$$

Letting i vary from 1 to $m+n$, we have as many equations as unknowns and can compute the $p_k^+(a,b), p_\ell^-(a,b)$ by solving linear equations. In our setting, we then have $q^+(a,b) = \sum_0^{n-1} p_k^+(a,b)$, and thus the stationary distribution of the random walk between two reflecting barriers has been obtained; cf. (3.9).

The argument assumes tacitly that the z_i are all distinct. Some discussion is given below. Note that one root equals 1, in which case (3.27) just reduces to the obvious constraint $\sum_0^{n-1} p_k^+(a,b) + \sum_1^m p_\ell^-(a,b) = 1$.

Now assume that instead we want to compute one-boundary first passage time probabilities

$$p_k(a) = \mathbf{P}(S_{\alpha(a)} = a + k), k = 0,\dots,n-1$$

which determine the stationary distribution of the random walk with reflection at 0 since $q(a) = \sum_0^{n-1} p_k(a)$. The conditions for optional stopping at $\alpha(a)$ are *not* satisfied for several reasons, the simplest being that $\alpha(a) < \infty$ fails. However, consider an z_i with $|Re(z_i)| > 1$. Then, letting $b \to \infty$ in (3.27), we get

$$1 = z_i^a \sum_{k=0}^{n-1} p_k(a) z_i^k.$$

Thus, the $p_k(a)$ are available as solutions of linear equations provided there are m distinct z_i with $|Re(z_i)| > 1$. That this is the typical case is not a priori obvious but follows from Corollary 3.16 below. First we give a result locating the roots relative to the unit circle. In order to state it, note that $\{\xi(a)\}_{a=0,1,\dots}$, where $\xi(a) = S_{\tau(a)} - a$ is the overshoot and a Markov chain with state space $\{0,\dots,n-1\}$, and let $\boldsymbol{P}$ denote its transition matrix. Note that if $\mu = \mathbf{E}[X_1] < 0$, then $\{\xi(a)\}$ has finite lifelength, and hence $\boldsymbol{P}$ is substochastic.

Proposition 3.15. *Assume that $\mu < 0$. Then a complex number z with $|Re(z)| > 1$ is a root of the equation $\hat{f}[z] = 1$ with multiplicity k if, and only if, z^{-1} is an eigenvalue of $\boldsymbol{P}$ with multiplicity k. Thus $d_+ = n$ and $d_- = m$ where d_+, resp. d_- are the sum of the multiplicities of the roots with $|Re(z)| > 1$, resp. $|Re(z)| \le 1$.*

Proof. Assume that z is a root of multiplicity $k = 1,2,\dots$ with $|Re(z)| > 1$. Then $\hat{f}[s] - 1 = (s-z)^k g(s)$ where g does not have a pole at z, and hence $\hat{f}^{(\ell)}[z] = 0$, $\ell = 0,\dots,k-1$. This implies that $\mathbf{E}[X_0(X_0-1)\dots(X_0-\ell+1)z^{X_0}] = 0$ for $\ell = 1,\dots k-1$, and hence

$$\mathbf{E}[z^{X_0}] = 1,\ \mathbf{E}[X_0^\ell z^{X_0}] = 0,\ \ \ell = 1,\dots k-1.$$

Define $M_n(\ell) = S_n^\ell z^{S_n}$. Writing

$$M_{n+1}(\ell) = \sum_{j=0}^{\ell} \binom{\ell}{j} X_n^j z^{X_n} S_n^{\ell-j} z^{S_n}$$

and taking conditional expectation given $X_0,\dots,X_{n-1}$, only the $j=0$ term contributes, and hence $\{M_n(\ell)\}$ is a (complex) martingale. Using optional stopping at $\alpha[-b,a)$ and letting $b \to \infty$ yields

$$\mathbf{E}\Big[z^{\xi(a)}; \alpha(a) < \infty\Big] = z^{-a}, \mathbf{E}\Big[(a+\xi(a))^\ell z^{\xi(a)}; \alpha(a) < \infty\Big] = 0, \ell = 1,\dots k-1.$$

An easy induction argument then gives that $\mathbf{E}[\xi(a)^\ell z^{\xi(a)}; \alpha(a) < \infty] = z^{-a} q_\ell(a)$ where $q_\ell(a)$ is a polynomial of degree ℓ. Taking $\ell = k-1$ then shows that z^{-1} is an eigenvalue of $\boldsymbol{P}$ with multiplicity k. It follows that d_+ is at least the dimension m of $\boldsymbol{P}$. Similarly, $d_- \ge n$, and since $d_+ + d_- = m + n$, it follows that $d_- = m$ and $d_+ = n$. From this, the result easily follows. □

Corollary 3.16. (*a*) *Assume that $\mu < 0$. Then the equation $\hat{f}[z] = 1$ has m dis-*

tinct solutions z_i with $|Re(z_i)| > 1$ if, and only if, all eigenvalues of P are distinct. (b) Assume that $\mu \geq 0$. Then the equation $\widehat{f}[z] = 1$ has m distinct solutions z_i with $|Re(z_i)| \geq 1$ if and only if all eigenvalues of P are distinct.

Note that the form of the solution in terms of roots is similar to Smith's [29] classical algorithm for the Laplace transform of the steady-state GI/G/1 waiting time when the service time distribution has a rational Laplace transform. There has been some confusion in the queueing literature concerning the possibility of multiple roots, which seems now settled by Dukhovny [15].

3.4.2 Markovian fluid models

In this and the following examples, the process $\{(J_t, V_t)\}_{t \geq 0}$ of interest has the structure discussed in Section 3.3.2, with the additional feature that $\{V_t\}$ is the reflected (at one boundary 0 or two boundaries 0, b) version of a Markov additive process $\{S_t\}$ defined on $\{J_t\}$; in the one-boundary case, this means that $\{V_t\}$ and $\{S_t\}$ are connected by (3.7). The representation of the stationary distribution given by Corollary 3.11 is then

Corollary 3.17. *In the finite buffer case ($b < \infty$),*

$$\mathbf{P}(J = j, V \geq y) = \pi_j q_j^{(+)}(y). \tag{3.28}$$

Similarly, in the infinite buffer case ($b = \infty$),

$$\mathbf{P}(J = j, V \geq y) = \pi_j q_j(y), \tag{3.29}$$

where $q_j^{(+)}(y)$, $q_j(y)$ are defined as follows: $\{(J_t^*, S_t^*)\}$ is the time-reversed version of the Markov additive process $\{(J_t, S_t)\}$, and

$$\alpha[-x, y) = inf\{t > 0 : S_t^* \notin [-x, y)\}, \quad \alpha(y) = inf\{t > 0 : S_t^* \geq y)\},$$

$$q_j^{(+)}(y) = \mathbf{P}_j(S^*_{\alpha[y-b,y)} \geq y), \; q_j(y) = \mathbf{P}_j(\alpha(y) < \infty).$$

Thus, for the stationary distributions, we need the $q_j^{(+)}$, $q_j(y)$, and the program for computing them is as follows: Determine a matrix $\boldsymbol{K}[\theta]$ such that the matrix $\widehat{\boldsymbol{F}}_t[\theta]$ with ijth element $\mathbf{E}_i[e^{\theta S_t^*}; J_t^* = j]$ is given by $\widehat{\boldsymbol{F}}_t[\theta] = \exp\{t\boldsymbol{K}[\theta]\}$. For any θ such that $|\boldsymbol{K}[\theta]| = 0$ (i.e., 0 is an eigenvalue of $\boldsymbol{K}[\theta]$),

$$\left\{M^{(\theta)}(t)\right\}_{t \geq 0} = \left\{e^{\theta S_t^*} h^{(\theta)}(J_t^*)\right\}_{t \geq 0} \tag{3.30}$$

is a martingale (Proposition 3.18) and, determining sufficiently many such θ, we get a set of equations similar to (3.27) from which the $q_j^{(+)}(y)$, $q_j(y)$ can be computed.

In the following, $\boldsymbol{e}_i$ denotes the ith (column) unit vector, $\boldsymbol{e}$ the (column) vector with 1 at all entries, and for a given set of constants r_i we write $\boldsymbol{\Delta}_r$ or $\boldsymbol{\Delta}(r)$ for the diagonal matrix with the r_i on the diagonal. E.g., in this notation we have $\boldsymbol{\Lambda}^* = \boldsymbol{\Delta}_\pi^{-1}\boldsymbol{\Lambda}'\boldsymbol{\Delta}_\pi$.

Consider now the special case of a fluid flow model; see e.g. Anick *et al.* [2], Asmussen [5], Rogers [26], and references therein. This means that $\{J_t\}$ is a Markov process with $d < \infty$ states and intensity matrix $\boldsymbol{\Lambda}$, whereas $\{V_t\}$ evolves as a linear motion with drift r_i (say) on intervals where $J_t = i$ and has 0 as a reflecting barrier. The model is currently receiving considerable attention in view of its relevance for the modeling of modern ATM (asynchronous transfer mode) technology. Mathematically, the relevant Markov additive process is

$$S_t = \int_0^t r_{J_v} dv. \tag{3.31}$$

We shall consider both the case of an infinite buffer and a finite buffer size $b < \infty$ where also $b > 0$ is reflecting. Quite often, the finite buffer model is the really interesting case and the infinite buffer model is used only as an approximation, say the probability $\mathbf{P}(V = b)$ for the finite model is approximated by $\mathbf{P}(V \geq b)$ for the infinite one. For the infinite buffer case, the representation (3.29) of the steady-state distribution in terms of first passage times was pointed out in Asmussen [5]. For the finite buffer case, an algorithm for computing the steady-state distribution was recently given by Rogers [26], who used spectral expansions along the lines of [2]; the approach we present below is more probabilistic, using duality and martingales.

Assume for simplicity that $r_i \neq 0$ for all $i \in E$ and write

$$E_+ = \{i \in E : r_i > 0\}, \; E_- = \{i \in E : r_i < 0\}.$$

If $b < \infty$, we extend the state space for $\{(J_t, V_t)\}$ from $E \times [0, b]$ to $E \times [0, \infty)$ by assuming that, from an initial value $(J_0, V_0) = (i, x)$ with $x > 0$, the process is immediately reset to state (i, b). Easy but tedious extensions of the arguments of Example 3.3 then show that the dual process $\{(J_t^*, R_t)\}$ evolves as a fluid model with rates $r_i^* = -r_i$ as long as $R_t \in (0, b)$ or $R_t \in (0, b]$, $J_t^* \in E_-$ (recall that $\{J_t^*\}$ is just the time-reversed version of $\{J_t\}$). States of the form $(i, 0)$ are absorbing in the sense that when $(J_0, V_0) = (i, 0)$, then $R_t = 0$ for all t (whereas, of course, $\{J_t^*\}$ just evolves as usual). From an initial value $(J_0^*, R_0) = (i, b)$ with $i \in E_+$, the process is immediately reset to state (i, ∞). A limiting steady state exists always in the finite buffer case and in the infinite buffer it exists if, and only if, $\mu < 0$ where $\mu = \sum_{i \in E} \pi_i r_i$ and $\boldsymbol{\pi}$ is the stationary distribution of $\{J_t\}$; in the ATM applications, $\mu < 0$ usually also holds in the finite buffer case.

Let $\{(J_t^*, S_t^*)\}$ be the Markov additive process defined as in (3.31) with $\{J_t\}$ replaced by $\{J_t^*\}$, and define

$$q_{ji}^{(+)}(y) = \mathbf{P}_j\left(S_{\alpha[y-b,y)} = y, J^*_{\alpha[y-b,y)} = i\right)$$

and

$$q_{ji}(y) = \mathbf{P}_j\left(\alpha(y) < \infty, J^*_{\alpha(y)} = i\right).$$

Then,

$$q_j^{(+)}(y) = \sum_{i \in E_+} q_{ji}^{(+)}(y), \quad q_j(y) = \sum_{i \in E_+} q_{ji}(y),$$

and hence it is sufficient to determine the $q_{ji}^{(+)}(y), q_{ji}(y)$.

Now consider the problem of obtaining a suitable family of martingales. For the fluid model, it follows by an easy differential equation argument that $\boldsymbol{K}[\theta] = \boldsymbol{\Lambda}^* + \theta \boldsymbol{\Delta}_r$ where $\boldsymbol{K}[\theta]$ is defined as above.

Proposition 3.18. *Let* $-\theta$ *be an eigenvalue of* $\boldsymbol{\Delta}^{-1}\boldsymbol{\Lambda}^*$ *and let* $\boldsymbol{h}^{(\theta)} = (h^{(\theta)}(i))_{i \in E}$ *be the corresponding right eigenvector. Then* $\{M^{(\theta)}(t)\}$ *defined in* (3.30) *is a martingale.*

Proof. The assumptions imply that 0 is an eigenvalue of $\boldsymbol{K}[\theta]$ and $\boldsymbol{h}^{(\theta)}$ is the corresponding right eigenvector. This implies $\widehat{\boldsymbol{F}}_t[\theta]\boldsymbol{h}^{(\theta)} = \boldsymbol{h}^{(\theta)} = (h^{(\theta)}(i))_{i \in E}$, which we can rewrite as $\mathbf{E}_j[e^{\theta S_t^*} h^{(\theta)}(J_t^*)] = h^{(\theta)}(j)$. Hence,

$$\mathbf{E}\Big[M^{(\theta)}(t+s)\,|\,M^{(\theta)}(v){:}\,0\le v\le t\Big] = e^{\theta S_t^*}\mathbf{E}\Big[e^{\theta(S_t^*-S_s^*)}h^{(\theta)}(J_{t+s}^*)\,|\,(S_t^*,J_t^*)\Big]$$

$$= e^{\theta S_t^*}\mathbf{E}_{J_t^*}\Big[e^{\theta S_s^*}h^{(\theta)}(J_s^*)\Big]$$

$$= e^{\theta S_t^*}h^{(\theta)}(J_t^*) = M^{(\theta)}(t). \qquad \square$$

The same argument as used for (3.27) now immediately yields a set of linear equations determining the $q_{ji}^{(+)}(u)$ and thereby the stationary distribution for the finite buffer case:

Corollary 3.19. *Assume that the eigenvalues $-\theta_1,\ldots,-\theta_d$ of $\Delta^{-1}\Lambda^*$ are all distinct and let $\boldsymbol{h}^{(\theta_k)} = (h^{(\theta_k)}(i))_{i\in E}$ be the right eigenvector corresponding to $-\theta_k$. Then,*

$$h_j^{(k)} = \sum_{i\in E_+} h_i^{(k)}e^{\theta_k y}q_{ji}^{(+)}(y) + \sum_{i\in E_-} h_i^{(k)}e^{\theta_k(y-b)}q_{ji}^{(-)}(y), \ j,k=1,\ldots,d.$$

Corollary 3.20. *Assume that $\Delta^{-1}\Lambda^*$ has $d_+ = |E_+|$ distinct eigenvalues $-\theta_1,\ldots,-\theta_{d_+}$ with $Re(\theta_i)>0$ and let $\boldsymbol{h}^{(\theta_k)} = (h^{(\theta_k)}(i))_{i\in E}$ be the right eigen vector corresponding to $-\theta_k$. Then*

$$h_j^{(k)} = \sum_{i\in E_+} h_i^{(k)}e^{\theta_k y}q_{ji}(y), \quad j,k=1,\ldots,d_+.$$

The interpretation of the θ_k, $k=1,\ldots,d_+$, is as the eigenvalues of the intensity matrix $U^{(+)}$ of the (terminating) Markov process $\{m_+(x)\}$ defined by $m_+(x) = J^*_{\alpha(x)}$ (the state of the environment when the Markov additive process is at a relative maximum); see [5].

3.4.3 Markov-modulated reflected Brownian motion

Consider a process $\{(J_t,V_t)\}_{t\ge 0}$, such that again $\{J_t\}$ is a Markov process with $d<\infty$ states and intensity matrix Λ, say, whereas $\{V_t\}$ evolves as Brownian motion with drift r_i and variance constant $\sigma_i^2>0$ (say) on intervals where $J_t=i$ and has 0 as a reflecting barrier. See Asmussen [5] (who also considers mixed models where some $\sigma_i^2=0$), Rogers [26] and references therein.

The model is mathematically closely related to fluid flows but in fact is slightly simpler, mainly because there is no need to split E into E_+, E_-. Thus, we shall be brief with the general theory but treat an example in more detail.

Let $\{(J_t^*,S_t^*)\}$ be the Markov additive process such that $\{S_t^*\}$ evolves as Brownian motion with drift r_i and variance constant $\sigma_i^2>0$ (say) on intervals where $J_t^*=i$. We define $\alpha[-x,y)$, $\alpha(y)$, $q_j^{(+)}(y)$, $q_{ji}^{(+)}(y)$ etc., exactly as before, and the representations (3.28), (3.29) of the stationary distribution remain valid.

The matrix $\boldsymbol{K}[\theta]$ equals $\Lambda^*+\Delta(\theta^2\sigma^2/2+\theta r))$ (the diagonal matrix with $\theta^2\sigma_i^2/2+\theta r_i$ as ith diagonal element). If θ satisfies $\boldsymbol{K}[\theta]=0$, (3.30) defines again a martingale, and again we get a set of linear equations, the solution of which determines the stationary distribution in view of (3.28), (3.29):

Corollary 3.21. *Assume that the roots $\theta_1,\ldots,\theta_{2d}$ of $|\boldsymbol{K}[\theta]|=0$ are all distinct and let $\boldsymbol{h}^{(\theta_k)} = (h^{(\theta)}(i))_{i\in E}$ be the right eigenvector corresponding to θ_k. Then,*

$$h_j^{(k)} = \sum_{i\in E} h_i^{(k)} e^{\theta_k y} q_{ji}^{(+)}(y) + \sum_{i \in E} h_i^{(k)} e^{\theta_k (y-b)} q_{ji}^{(-)}(y), \quad k = 1,\ldots,2d. \tag{3.32}$$

Similarly, if there are d distinct positive roots, say $\theta_1 > 0,\ldots,\theta_d > 0$, , then

$$h_j^{(k)} = \sum_{i\in E} h_i^{(k)\theta_k y} q_{ji}(y), \quad k = 1,\ldots,d. \tag{3.33}$$

Note, incidentally, the following alternative way of computing the $q_{ij}^{(+)}(y)$, $q_{jk}^{(-)}(y)$. First compute $q_{ij}(y)$ and (by sign reversion) $r_{jk}(y)$, the probabilities that downcrossing of level $-y$ in state k occurs starting from $J_0^* = j$, and next solve the set of equations

$$q_{ji}(y) = q_{ji}^{(+)}(y) + \sum_{k\in E} q_{jk}^{(-)}(y) q_{ki}(b) \tag{3.34}$$

$$r_{ji}(y) = q_{ji}^{(-)}(y) + \sum_{k\in E} q_{jk}^{(+)}(y) r_{ki}(b). \tag{3.35}$$

Here (3.34) follows by noting that upcrossing of level y may occur in two ways: either $\{S_t\}$ passes to level y without getting below $y-b$, or it first gets to $y-b$, say in state k, and from there to y (the derivation of (3.35) is similar). Similar remarks apply to fluid models. For further ways of computing two-barrier first passage probabilities, see Asmussen & Perry [9].

Example 3.22. Assume that $E = \{1,2\}$, with $\lambda_{12} = 9$, $\lambda_{21} = 21$, $r_1 = 0$, $r_2 = -8$, $\sigma_1^2 = \sigma_2^2 = 4$, so that

$$\widehat{K}[s] = \begin{pmatrix} -9+2s^2 & 9 \\ 21 & -21+2s^2-8s \end{pmatrix}$$

(any two-state Markov process is time reversible, so $\boldsymbol{\Lambda}^* = \boldsymbol{\Lambda}$). The roots of

$$\det(\widehat{K}[s]) = 4s^4 - 16s^3 - 60s^2 + 72s,$$

or, equivalently, the s such that 0 is an eigenvalue of $\widehat{K}[s]$, are 6, 1, 0, -3, with corresponding eigenvectors

$$\boldsymbol{h}_6 = \begin{pmatrix} 1 \\ -7 \end{pmatrix}, \quad \boldsymbol{h}_1 = \begin{pmatrix} 9 \\ 7 \end{pmatrix}, \boldsymbol{h}_0 = \begin{pmatrix} 1 \\ 1 \end{pmatrix}, \quad \boldsymbol{h}_{-3} = \begin{pmatrix} 1 \\ -1 \end{pmatrix}.$$

Assume first that $b = \infty$. Then Equations (3.33) are

$$\begin{aligned} 9 &= e^x\{9q_{11}(x) + 7q_{12}(x)\}, \\ 1 &= e^{6x}\{q_{11}(x) - 7q_{12}(x)\}, \\ 7 &= e^x\{9q_{21}(x) + 7q_{22}(x)\}, \\ -7 &= e^{6x}\{q_{21}(x) - 7q_{22}(x)\}, \end{aligned}$$

with solution

$$\begin{pmatrix} q_{11}(x) & q_{12}(x) \\ q_{21}(x) & q_{22}(x) \end{pmatrix} = \begin{pmatrix} 1 & 9 \\ -7 & 7 \end{pmatrix} \begin{pmatrix} e^{-6x} & 0 \\ 0 & e^{-x} \end{pmatrix} \begin{pmatrix} 1 & 9 \\ -7 & 7 \end{pmatrix}^{-1}$$

$$= \begin{pmatrix} \frac{9}{10}e^{-x} + \frac{1}{10}e^{-6x} & \frac{9}{70}e^{-x} - \frac{9}{70}e^{-6x} \\ \frac{7}{10}e^{-x} - \frac{7}{10}e^{-6x} & \frac{1}{10}e^{-x} + \frac{9}{10}e^{-6x} \end{pmatrix}$$

Noting that $\pi_1 = 7/10$, $\pi_2 = 3/10$, this gives

$$\mathbf{P}(V \geq x, J = 1) = \pi_1\{q_{11}(x) + q_{12}(x)\} = \frac{18}{25}e^{-x} - \frac{1}{50}e^{-6x}, \tag{3.36}$$

$$\mathbf{P}(V \geq x, J = 2) = \pi_2\{q_{21}(x) + q_{22}(x)\} = \frac{6}{25}e^{-x} + \frac{3}{50}e^{-6x}. \tag{3.37}$$

Consider next a finite buffer size, say $b = 1$. Then Equations (3.32) are ($j = 1$)

$$1 = e^{-3x}\{q_{11}^{+}(x) - q_{12}^{+}(x)\} + e^{3-3x}\{q_{11}^{-}(x) - q_{12}^{-}(x)\},$$

$$1 = q_{11}^{+}(x) + q_{12}^{+}(x) + q_{11}^{-}(x) + q_{12}^{-}(x),$$

$$9 = e^{x}\{9q_{11}^{+}(x) + 7q_{12}^{+}(x)\} + e^{x-1}\{9q_{11}^{-}(x) + 7q_{12}^{-}(x)\},$$

$$1 = e^{6x}\{q_{11}^{+}(x) - 7q_{12}^{+}(x)\} + e^{6x-6}\{q_{11}^{-}(x) - 7q_{12}^{-}(x)\}.$$

Solving for $q_{11}^{+}(x)$, $q_{12}^{+}(x)$, $q_{11}^{-}(x)$, $q_{12}^{-}(x)$, we get

$$\begin{aligned} \mathbf{P}(V \geq x, J = 1) &= \pi_1\{q_{11}^{+}(x) + q_{12}^{+}(x)\} \\ &= \frac{7}{10\Delta}\{c_1^{(-3)}e^{3x} + c_1^{(0)} + c_1^{(1)}e^{-x} + c_1^{(6)}e^{-6x}\}, \end{aligned} \tag{3.38}$$

where

$$\Delta = 35e^9 + 3e^8 + 3e^7 + 3e^6 - 3e^3 - 3e^2 - 3e - 35,$$

$$c_1^{(-3)} = -4(e^5 + e^4 + e^3 + e^2 + 3),$$

$$c_1^{(0)} = -32(e^8 + e^7 + e^6 + e^5 + e^4) - 35(e^3 + e^2 + e + 1),$$

$$c_1^{(1)} = 36(e^9 + e^8 + e^7 + e^6 + e^5 + e^4 + e^3 + e^2 + e),$$

and $$c_1^{(6)} = -(e^9 + e^8 + e^7 + e^6).$$

Similar calculations yield

$$\begin{aligned} \mathbf{P}(V \geq x, J = 2) &= \pi_2\{q_{21}^{+}(x) + q_{22}^{+}(x)\} \\ &= \frac{3}{10\Delta}\{c_2^{(-3)}e^{3x} + c_2^{(0)} + c_2^{(1)}e^{-x} + c_2^{(6)}e^{-6x}\}, \end{aligned} \tag{3.39}$$

where

$$c_2^{(-3)} = 4(e^5 + e^4 + e^3 + e^2 + e),$$

$$c_2^{(0)} = -32(e^8 + e^7 + e^6 + e^5 + e^4) - 35(e^3 + e^2 + e + 1),$$

$$c_2^{(1)} = 28(e^9 + e^8 + e^7 + e^6 + e^5 + e^4 + e^3 + e^2 + e),$$

and $$c_2^{(6)} = 7(e^9 + e^8 + e^7 + e^6).$$

As a check, note that

$$c_1^{(-3)} + c_1^{(0)} + c_1^{(1)} + c_1^{(6)} = c_2^{(-3)} + c_2^{(0)} + c_2^{(1)} + c_2^{(6)} = \Delta$$

so that

$$\mathbf{P}(V \geq 0, J = 1) = 7/10 = \pi_1, \quad \mathbf{P}(V \geq 0, J = 2) = 3/10 = \pi_2$$

as they should.

For the alternative route via (3.34), (3.35), we first get

$$\begin{pmatrix} r_{11}(x) & r_{12}(x) \\ r_{21}(x) & r_{22}(x) \end{pmatrix} = \begin{pmatrix} 1 & 1 \\ 1 & -1 \end{pmatrix} \begin{pmatrix} 1 & 0 \\ 0 & e^{-3x} \end{pmatrix} \begin{pmatrix} 1 & 1 \\ 1 & -1 \end{pmatrix}^{-1}$$
$$= \begin{pmatrix} \frac{1}{2}+\frac{1}{2}e^{-3x} & \frac{1}{2}-\frac{1}{2}e^{-3x} \\ \frac{1}{2}-\frac{1}{2}e^{-3x} & \frac{1}{2}+\frac{1}{2}e^{-3x} \end{pmatrix},$$

and thus equations (3.34), (3.35) become ($j = 1$)

$$\frac{9}{10}e^{-x} + \frac{1}{10}e^{-6x} = q_{11}^{+}(x) + q_{11}^{-}(x)\left\{\frac{9}{10}e^{-1} + \frac{1}{10}e^{-6}\right\} + q_{12}^{-}(x)\left\{\frac{9}{70}e^{-1} - \frac{9}{70}e^{-6}\right\},$$
$$\frac{9}{70}e^{-x} - \frac{9}{70}e^{-6x} = q_{12}^{+}(x) + q_{11}^{-}(x)\left\{\frac{7}{10}e^{-1} + \frac{7}{10}e^{-6}\right\} + q_{22}^{-}(x)\left\{\frac{1}{10}e^{-1} - \frac{9}{10}e^{-6}\right\},$$
$$\frac{1}{2} + \frac{1}{2}e^{-3x} = q_{11}^{-}(x) + q_{11}^{+}(x)\left\{\frac{1}{2} + \frac{1}{2}e^{-3}\right\} + q_{12}^{+}(x)\left\{\frac{1}{2} - \frac{1}{2}e^{-3}\right\}, \quad \text{and}$$
$$\frac{1}{2} - \frac{1}{2}e^{-3x} = q_{12}^{-}(x) + q_{11}^{+}(x)\left\{\frac{1}{2} - \frac{1}{2}e^{-3}\right\} + q_{22}^{+}(x)\left\{\frac{1}{2} + \frac{1}{2}e^{-3}\right\}.$$

3.4.4 Markov-modulated M/G/1 queues

We next consider the workload process $\{W_t\}_{t\geq 0}$ of the Markov-modulated M/G/1 queue. The governing environment is a Markov process $\{I_t\}_{t\geq 0}$ with $p < \infty$ states, such that the arrival rate is β_j and the service time (jump size) distribution is B_j on intervals where $I_t = j$. For simplicity, we consider only the infinite capacity case, but the discussion carries over to a finite capacity model where $\{W_t\}$ is reset to $b < \infty$ as soon as (b,∞) is hit. Also, the extension to Neuts' [22] Markovian arrival process, where jumps of $\{I_t\}$ may trigger an arrival, is straightforward and only notationally different (see Asmussen & Perry [9] for a study of this queueing model).

A general derivation of the stationary distribution was first given by Regterschot and de Smit [25], using classical complex plane methods, whereas Asmussen [4] gave a more probabilistic treatment involving duality ideas (the representation of the stationary distribution given in [4] is the same as (3.29)). See also Sengupta [27]. Both papers allow general B_j when calculating moments like $\mathbf{E}[V^k]$ and Laplace transforms, but in the following we assume that each B_j is phase-type which is essential when considering steady-state probabilities like $\mathbf{P}(V \geq x)$. Thus, B_j is the distribution of the lifetime of a terminating Markov process with state space G_j with q_j elements, initial vector $\boldsymbol{\nu}_j$, intensity matrix (phase generator) $\boldsymbol{T}_j$, and exit vector $\boldsymbol{t}_j = -\boldsymbol{T}_j\boldsymbol{e}$; cf. [23].

The relevant Markov additive process is $S_t^* = \sum_1^{N_t^*} U_i - t$ where $\{N_t^*\}$ is the Markov-modulated Poisson process generated by the time-reverse $\{I_t^*\}$ of $\{I_t\}$, and $U_1, U_2, \ldots$ are the corresponding jump sizes (the β_j and the B_j remain unchanged).

The obvious idea in computing the $q_j(y)$ is now to compute the matrix $\boldsymbol{K}[\theta]$, determine the θ_i such that $\boldsymbol{K}[\theta_i]$ has eigenvalue 0 (i.e., $|\boldsymbol{K}[\theta_i]| = 0$, and work with the $M^{(\theta_i)}(t)$ as before. Here it is easily checked that

$$K[\theta] = \Lambda_I^* + (\beta_i(\widehat{B}_i[s]-1))_{diag} - \theta I,$$

where Λ_I^* is the intensity matrix of $\{I_t^*\}$ and $\widehat{B}_j[\theta] = -\nu_j(T_j+\theta I)^{-1}t_j$ is the m.g.f. of B_j; cf. [23]. However, the last step does not work because, typically, the $M^{(\theta_i)}(t)$ are not integrable, so martingale theory is not available. This problem did not occur in our earlier examples because S_t was bounded there (or at least sufficiently light-tailed in the Brownian case). The solution we suggest is to relate $\{(I_t^*, S_t)\}$ to the Markov additive process $\{(J_t^*, S_t^*)\}$ associated with a certain fluid model (with a slight abuse of notation, we write S_t^* in both cases even if the processes are not the same).

The fluid flow representation $\{(J_t^*, S_t^*)\}$ of $\{(I_t^*, S_t^*)\}$ is illustrated in Fig. 1 for the case $p = q_1 = q_2 = 2$. The two environmental states are denoted $\circ$, $\bullet$, the phase space $G_\circ$ for $B_\circ$ has states DIAMOND, HEART, and the phase space $G_\bullet$ for $B_\bullet$ has states CLUB, SPADE. An arrival of $\{(I_t^*, S_t^*)\}$ in state i can then be represented by an G_i-valued Markov process as on Fig. 1(a). The fluid model on Fig. 1(b) $\{(J_t^*, S_t^*)\}$ is obtained by changing the vertical jumps to segments with slope 1. Thus,

$$E = \{\circ, \text{DIAMOND}, \text{HEART}, \bullet, \text{CLUB}, \text{SPADE}\}$$

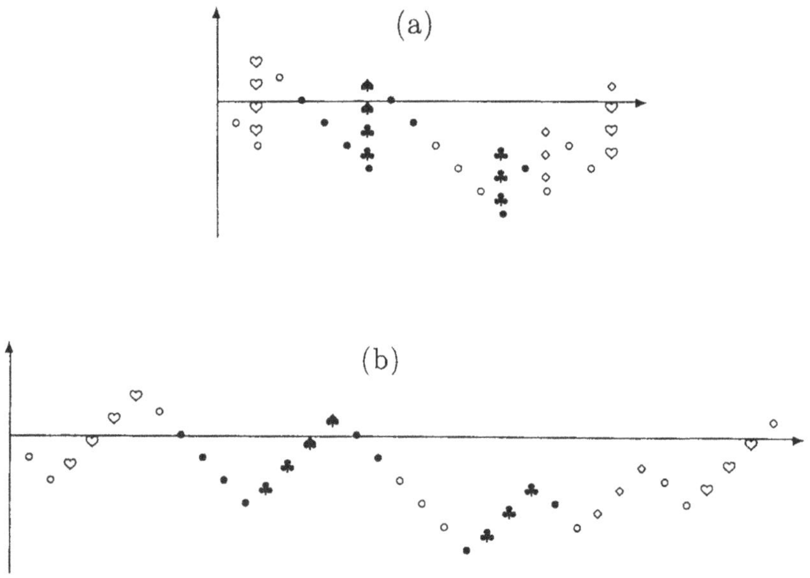

Figure 1

In the general formulation, E is the disjoint union of F and the G_j,

$$E = F \cup \{(i,\alpha): i \in F, \alpha \in G_i\}, \quad r(i) = -1, i \in F, r(i,\alpha) = 1.$$

The intensity matrix for $\{J_t^*\}$ is (taking $p = 3$ for simplicity)

$$
\Lambda^* = \left(\begin{array}{ccc|ccc}
\multicolumn{3}{c|}{\Lambda_I^* - (\beta_i)_{diag}} & \beta_1\nu_1 & 0 & 0 \\
 & & & 0 & \beta_2\nu_2 & 0 \\
 & & & 0 & 0 & \beta_3\nu_3 \\
\hline
t_1 & 0 & 0 & T_1 & 0 & 0 \\
0 & t_2 & 0 & 0 & T_2 & 0 \\
0 & 0 & t_3 & 0 & 0 & T_3
\end{array} \right).
$$

Proposition 3.23. *A complex number θ satisfies $|K[\theta]| = 0$ if, and only if, $-\theta$ is an eigenvalue of*

$$
\Delta_r^{-1}\Lambda^* = \left(\begin{array}{ccc|ccc}
\multicolumn{3}{c|}{(\beta_i)_{diag} - \Lambda_I^*} & -\beta_1\nu_1 & 0 & 0 \\
 & & & 0 & -\beta_2\nu_2 & 0 \\
 & & & 0 & 0 & -\beta_3\nu_3 \\
\hline
t_1 & 0 & 0 & T_1 & 0 & 0 \\
0 & t_2 & 0 & 0 & T_2 & 0 \\
0 & 0 & t_3 & 0 & 0 & T_3
\end{array} \right).
$$

If θ is such a number, consider the vector $\boldsymbol{a}$ satisfying $K[\theta]\boldsymbol{a} = 0$ and the eigenvector $\boldsymbol{b} = (\boldsymbol{c}\ \boldsymbol{d})$ of $\Delta_r^{-1}\Lambda^*$, where $\boldsymbol{c}$, $\boldsymbol{d}$ correspond to the partitioning of $\boldsymbol{b}$ into components indexed by F, resp. $G_1 + \ldots + G_p$. Then (up to a constant)

$$
\boldsymbol{c} = \boldsymbol{a},\ \boldsymbol{d} = -\begin{pmatrix} T_2 + \theta I & 0 & 0 \\ 0 & T_2 + \theta I & 0 \\ 0 & 0 & T_3 + \theta I \end{pmatrix}^{-1} \begin{pmatrix} t_1 & 0 & 0 \\ 0 & t_2 & 0 \\ 0 & 0 & t_3 \end{pmatrix} \boldsymbol{a}.
$$

Proof. Using the well-known determinant identity

$$
\begin{vmatrix} \Sigma_{11} & \Sigma_{12} \\ \Sigma_{21} & \Sigma_{22} \end{vmatrix} = |\Sigma_{22}| \cdot |\Sigma_{11} - \Sigma_{12}\Sigma_{22}^{-1}\Sigma_{21}|,
$$

we get

$$
\left| \begin{array}{ccc|ccc}
\multicolumn{3}{c|}{(\beta_i)_{diag} - \Lambda_I^* + \theta I} & -\beta_1\nu_1 & 0 & 0 \\
 & & & 0 & -\beta_2\nu_2 & 0 \\
 & & & 0 & 0 & -\beta_3\nu_3 \\
\hline
t_1 & 0 & 0 & T_1 + \theta I & 0 & 0 \\
0 & t_2 & 0 & 0 & T_2 + \theta I & 0 \\
0 & 0 & t_3 & 0 & 0 & T_3 + \theta I
\end{array} \right| = 0
$$

$$
\Rightarrow \left| (\beta_i)_{diag} - \Lambda_i^* + \theta I + (\beta_i\nu_i(T_i + \theta I)^{-1}t_i)_{diag} \right| = 0,
$$

which is the same as $|\,\boldsymbol{K}[\theta]\,| = 0$.

For the assertions on the eigenvectors, assume that

$$\left(\Sigma_{11} + \theta I - \Sigma_{12}(\Sigma_{22} + \theta I)^{-1}\Sigma_{21}\right)\boldsymbol{a} = 0,$$

and let $\boldsymbol{d} = -(\Sigma_{22} + \theta I)^{-1}\Sigma_{21}\boldsymbol{a}$, $\boldsymbol{c} = \boldsymbol{a}$. Then

$$\begin{aligned}\Sigma_{21}\boldsymbol{c} + \Sigma_{22}\boldsymbol{d} &= \Sigma_{21}\boldsymbol{a} - (\Sigma_{22} + \theta I - \theta I)(\Sigma_{22} + \theta I)^{-1}\Sigma_{21}\boldsymbol{a} \\ &= \Sigma_{21}\boldsymbol{a} - \Sigma_{21}\boldsymbol{a} - \theta\boldsymbol{d} = -\theta\boldsymbol{d}.\end{aligned}$$

Noting that $\Sigma_{11}\boldsymbol{c} + \Sigma_{12}\boldsymbol{d} = -\theta\boldsymbol{c}$ by definition, it follows that

$$\begin{pmatrix} \Sigma_{11} & \Sigma_{12} \\ \Sigma_{21} & \Sigma_{22} \end{pmatrix}\begin{pmatrix} \boldsymbol{c} \\ \boldsymbol{d} \end{pmatrix} = -\theta\begin{pmatrix} \boldsymbol{c} \\ \boldsymbol{d} \end{pmatrix}. \qquad \square$$

Corollary 3.24. *Assume that equation* $|\,\boldsymbol{K}[\theta]\,| = 0$ *has* $q = q_1 + \ldots + q_p$ *distinct solutions* $\theta_1, \ldots, \theta_q$ *with* $Re(\theta_\nu) > 0$ *and let* $\boldsymbol{b}^{(\nu)} = (\boldsymbol{c}^{(\nu)}\boldsymbol{d}^{(\nu)})$ *be the right eigenvector of* $\Delta_r^{-1}\Lambda^*$ *corresponding to* $-\theta_\nu$, $\nu = 1, \ldots, q$; *cf. Proposition* 3.23. *Then*

$$\mathrm{P}(V \geq x, J = i) = \pi_i \boldsymbol{e}_i'\left(e^{-\theta_1 u}\boldsymbol{c}^{(1)} \ldots e^{-\theta_q u}\boldsymbol{c}^{(q)}\right)\left(\boldsymbol{d}^{(1)} \ldots \boldsymbol{d}^{(q)}\right)^{-1}\boldsymbol{e}.$$

3.5 OPEN PROBLEMS AND FURTHER RESEARCH DIRECTIONS

1. A satisfying rigorous way to compute the Siegmund dual process in continuous time still needs to be developed.
2. A sufficiently general framework allowing the study of stochastic recursions in continuous time still needs to be developed. There is an example in Asmussen and Kella [7] where the recursion is an integral equation, but a general framework should incorporate many other types, e.g., SDE's.
3. It would be highly relevant to extend the theory of [28] and [8] to processes with a more general state space than $[0, \infty)$, even just $[0, \infty)^d$ (an example is the Kiefer-Wolfowitz theory of GI/G/k; see e.g. [3] Ch. XI). However, we are not too optimistic about this possibility. One problem is that the required monotonicity properties fail for many of the processes of interest; also the inverse of the transition probabilities/recursion may not exist or may not be unique.
4. We can give a rigorous proof of Proposition 3.12 under some additional continuity assumptions but not in complete generality.
5. The theory developed in Section 3.3.2 covers some of the important examples of Markov chains/processes of the GI/M/1 or M/G/1 type but not all: it would be desirable to allow boundary conditions as general as in, e.g., Neuts [22] or Hajek [16].
6. It would seem reasonable to conjecture that, in addition to Corollaries 3.19, 3.20, further analogues of the random walk theory (in particular, of Proposition 3.15 and Corollary 3.16) would hold for fluid models, but at present we have no complete proof of this. The problem concerns double roots and may be one of linear algebra more than of probability theory.

ACKNOWLEDGEMENTS

I am indebted to David Aldous for insisting that I should be interested in Siegmund duality and for showing me his notes [1].

BIBLIOGRAPHY

[1] Aldous, D., Brisk applications of Siegmund duality, unpublished notes, University of California, Berkeley (1988).

[2] Anick, D., Mitra, D. and Sondhi, M.M., Stochastic theory of a data-handling system with multiple sources, *Bell System Tech. J.* **61** (1982), 1871-1894.

[3] Asmussen, S., *Applied Probability and Queues*, John Wiley, New York 1987.

[4] Asmussen, S., Ladder heights and the Markov-modulated M/G/1 queue, *Stoch. Proc. Appl.* **37** (1991), 313-326.

[5] Asmussen, S., Stationary distributions for fluid flow models with or without Brownian noise, *Stoch. Mod.* **11**:1 (1995), (to appear).

[6] Asmussen, S., *Ruin Probabilities*, World Scientific Publishing, Co., Singapore 1995/96.

[7] Asmussen, S. and Kella, O., Rate modulation in dams and ruin problems, *J. Appl. Prob.* **32**:4 (1995), (to appear).

[8] Asmussen, S. and Sigman, K., Monotone stochastic recursions and their duals, manuscript, Aalborg University (1994).

[9] Asmussen, S. and Perry, D., On cycle maxima, first passage problems and extreme value theory for queues, *Stoch. Mod.* **8** (1992), 421-458.

[10] Asmussen, S. and Schock Petersen, S., Ruin probabilities expressed in terms of storage processes, *Adv. Appl. Prob.* **20** (1989), 913-916.

[11] Asmussen, S. and Rubinstein, R.Y., Steady-state rare events simulation and its complexity properties. As Chapter 17, in: *Advances in Queueing*, ed. by J.H. Dshalalow, CRC Press, Boca Raton, Florida (1995), 421-454.

[12] Borovkov, A.A., *Asymptotic Methods in Queueing Theory*, John Wiley, New York 1984.

[13] Cox, J.T. and Rösler, U., A duality relation for entrance and exit laws for Markov processes, *Stoch. Proc. Appl.* **16** (1984), 141-156.

[14] Daley, D.J., Stochastically monotone Markov chains, *Z. Wahrscheinlichkeitsth. verw. Geb.* **10** (1968), 305-317.

[15] Dukhovny, A., Multiple roots in some equations of queueing theory, *Stoch. Mod.* **10** (1994), 519-524.

[16] Hajek, B., Birth and death processes on the integers with phases and general boundaries, *J. Appl. Prob.* **19** (1981), 488-499.

[17] Lindley, J.M. and Resnick, S.I., The recurrence classification of risk and storage processes, *Math. Oper. Res.* **3** (1977), 57-66.

[18] Liggett, T.M., *Interacting Particle System*, Springer-Verlag 1985.

[19] Lindley, D., The theory of queues with a single server, *Proc. Cambr. Philos. Soc.* **48** (1952), 277-289.

[20] Lindley, D., Discussion of a paper of C.B. Winsten, *J. Roy. Stat. Soc. Ser.* **B21** (1959), 22-23.

[21] Loynes, R., The stability of a queue with nonindependent interarrival and service times, *Proc. Cambr. Philos. Soc.* **58** (1962), 497-520.

[22] Neuts, M.F., A versatile Markovian point process, *J. Appl. Prob.* **16** (1979), 764-779.

[23] Neuts, M.F., *Matrix-Geometric Solutions in Stochastic Models*, John Hopkins University Press, Baltimore 1981.

[24] Prabhu, N.U., On the ruin problem of collective risk theory, *Ann. Math. Stat.* **32** (1961), 757-764.

[25] Regterschot, G.J.K. and de Smit, J.H.A., The queue M/G/1 with Markov-modulated arrivals and services, *Math. Oper. Res.* **11** (1986), 465-483.

[26] Rogers, L.C.G., Fluid models in queueing theory and Wiener-Hopf factorization of Markov chains, *Ann. Appl. Prob.* **2** (1994), 390-413.

[27] Sengupta, B., The semi-Markov queue: theory and applications, *Stoch. Mod.* **6** (1990), 383-413.

[28] Siegmund, D., The equivalence of absorbing and reflecting barrier problems for stochastically monotone Markov processes, *Ann. Prob.* **4** (1976), 914-924.

[29] Smith, W.L., Distribution of queueing times, *Proc. Cambr. Philos. Soc.* **49** (1953), 449-461.

[30] Stadje, W., A new look at the Moran dam, *J. Appl. Prob.* **30** (1993), 489-495.

[31] Wald, A., *Sequential Analysis*, John Wiley, New York 1947.

Chapter 4

An introduction to spatial queues

Erhan Çinlar

ABSTRACT The aim of this article is to introduce the methodology of random measures into the theory of spatial queues. By the latter, we mean queueing systems where each customer's location in space is taken into account explicitly.

After a few preliminaries, we shall illustrate the uses of random measures on some M/G/∞ type queues and on an M/G/1 queueing system where a single server goes from one customer to the next depending on the configuration of customers present in the space. As we shall point out, a large number of priority queueing systems and networks of queues with different classes of customers are subsumed and extended by these models. On the other hand, the results we shall obtain in the last section (with moving server) are somewhat incomplete and should be regarded as the start of a research program.

CONTENTS

4.1 Preliminaries	103
4.2 Application to M/G/∞ queues	105
4.3 Spatial M/G/∞ queues	107
4.4 Queues with mobile customers	112
4.5 Spatial M/M/1 queues	113
4.6 Open Problems and Final Comments	117
Acknowledgement	118
Bibliography	118

4.1 PRELIMINARIES

Let $(E, \mathcal{E})$ be a measurable space. Given a measure μ on it and a positive measurable function $f: E \mapsto \bar{\mathbb{R}}_+$, we write μf for the integral of f with respect to μ. As usual, when E is a topological space and $\mathcal{E}$ is the Borel σ-algebra on E, we omit $\mathcal{E}$ from the notation and abuse the language somewhat by saying that μ is a measure on E and that f is a Borel function on E.

Let $(\Omega, \mathcal{H}, \mathbf{P})$ be a probability space. The elements ω in Ω are called outcomes; elements of $\mathcal{H}$ are called events. We write $\mathbf{E}$ for the expectation operator corresponding to $\mathbf{P}$.

Random measures. By a *random measure M* on $(E, \mathcal{E})$, simply on E when E is topological and $\mathcal{E}$ is the Borel σ-algebra on E, we mean a transition kernel $M: \Omega \times \mathcal{E} \mapsto \bar{\mathbb{R}}_+$; then,

a) $M(A): \omega \mapsto M(\omega, A)$ is a random variable for each A in $\mathcal{E}$,

b) $M_\omega: A \mapsto M(\omega, A)$ is a measure on $\mathcal{E}$ for each ω in Ω.

Given a positive $\mathcal{E}$-measurable function f, we write Mf for the random variable

0-8493-8074-x/95/\$0.00+\$.50

$$Mf: \omega \mapsto Mf(\omega) = M_\omega f = \int_E M(\omega, dx) f(x). \tag{4.1}$$

By Fubini's theorem, $\mathbf{E}[Mf] = (\mathbf{E}[M])f$, where $\mathbf{E}[M]$ is the measure μ defined by $\mu(A) = \mathbf{E}[M(A)]$, $A \in \mathcal{E}$. Almost all the random measures M we shall encounter will have the following form:

$$M(\omega, A) = \sum_i I(X_i(\omega), A), \tag{4.2}$$

where the sum is over a countable index set of i, each X_i is a random variable taking values in E (and measurable relative to $\mathcal{H}$ and $\mathcal{E}$), and $I(x, A) = 1_A(x) = \delta_x(A)$ is equal to 1 if $x \in A$ and to 0 otherwise. In this case, we say that the X_i *form* the random measure M and that the X_i are the *atoms* of M. If $X_i(\omega)$ and $X_j(\omega)$ differ for all $i \neq j$ for almost every ω, then the points $X_i(\omega)$ are indeed the atoms of the measure M_ω for almost every ω, and M_ω is a counting measure; such M are called random counting measures.

Poisson random measures. These are the most loved random measures. A random measure M on $(E, \mathcal{E})$ is said to be *Poisson with mean* μ if

a) $M(A)$ has the Poisson distribution with mean $\mu(A)$ for each A,

b) $M(A_1), \ldots, M(A_n)$ are independent whenever $A_1, \ldots, A_n$ are disjoint sets in $\mathcal{E}$, $n \geq 2$.

For a Poisson random measure M with mean μ, we indeed have $\mathbf{E}[M] = \mu$ and it is easy to show that

$$\mathbf{E}[Mf] = \mu f, \quad Var(Mf) = \mu(f^2), \quad Cov(Mf, Mg) = \mu(fg) \tag{4.3}$$

$$\mathbf{E}[e^{-Mf}] = \exp_- \int_E \mu(dx)(1 - e^{-f(x)}), \tag{4.4}$$

all these being for arbitrary positive $\mathcal{E}$-measurable functions f and g. The last formula, called the Laplace functional of M, characterizes Poisson random measures: if a random measure M satisfies (4.4) for all positive $\mathcal{E}$-measurable f, then M is a Poisson random measure with mean μ. Note that μ determines, then, the probability law of M, since the left side becomes the joint Laplace transform of $M(A_1)$, $\ldots, M(A_n)$ when $f = a_1 1_{A_1} + \ldots + a_n 1_{A_n}$ with a_i in $\mathbb{R}_+$ and A_i in $\mathcal{E}$. For these facts and much more, we refer to Kallenberg [2] and Kingman [3]. We give next a folk theorem that we will use a number of times.

Theorem 4.1: *Let (X_n) and (Y_n) be sequences of random variables taking values in the measurable spaces $(E, \mathcal{E})$ and $(F, \mathcal{F})$ respectively. Suppose that*

a) *the sequence (X_n) forms a Poisson random measure M on $(E, \mathcal{E})$ with mean μ,*

b) *for each n, the random variable Y_n is conditionally independent of $\{X_j, Y_j\colon j \neq n\}$ given X_n, and there is a transition kernel Q such that*

$$\mathbf{P}\{Y_n \in B \mid X_n = x\} = Q(x, B), \quad x \in E, \ B \in \mathcal{F}.$$

Then, the pairs (X_n, Y_n), $n \geq 1$, form a Poisson random measure N on the product space $(E \times F, \mathcal{E} \otimes \mathcal{F})$ whose mean measure ν is given by

$$\nu(dx, dy) = \mu(dx) Q(x, dy), \quad x \in E, \ y \in F.$$

Remark 4.2. The preceding is a many fold generalization of the following fact: If the sequence (X_n) forms a Poisson random measure M on $(E, \mathcal{E})$ with mean μ, and if h is a measurable transformation from$(E, \mathcal{E})$ into some measurable space $(F, \mathcal{F})$, then the random variables $h \circ X_n$, $n \geq 1$, form a Poisson random measure

N on $(F, \mathcal{F})$ with mean measure $\nu = \mu \circ h^{-1}$. This is immediate from the definition of Poisson random measures (and the fact that $N(\omega, B) = M(\omega, h^{-1}B)$ for every B in $\mathcal{F}$).

4.2 APPLICATION TO M/G/∞ QUEUES

This section is to illustrate the uses of the preceding section in a simple setting. The results we shall obtain are already widespread in the literature, but the streamlined methodology seems less familiar. Consider an infinite server queueing system. Let the customers be labeled somehow. Let T_i be the arrival time (in $\mathbb{R}$) of the customer labeled i, and let Z_i be the length of the corresponding service time. We suppose that the collection (T_i) forms a Poisson random measure on $\mathbb{R}$ with mean measure λ having the form

$$\lambda(dt) = a\,dt, \; t \in \mathbb{R}, \tag{4.5}$$

for some constant $a > 0$ called the arrival rate. Also, suppose that the service lengths Z_i are independent of (T_i) and are independent and identically distributed with the common distribution β on $\mathbb{R}_+$. Then, marking the pairs (T_i, Z_i) yields a point process as in Figure 1 below. According to Theorem 4.1, these pairs form a Poisson random measure N on $\mathbb{R} \times \mathbb{R}_+$ whose mean measure ν has the form

$$\nu(dt, dz) = \lambda(dt)\beta(dz) = a\,dt\,\beta(dz). \tag{4.6}$$

Queue size. Let Q_t be the number of customers in the systems at time t. It is clear that Q_t is equal to the number of atoms (T_i, Z_i) belonging to the wedge $A_t = \{(s,z) \in \mathbb{R} \times \mathbb{R}_+ : z > t - s\}$; this set A_t is the shaded region in the Figure 1 below for $t = t_0$; in other words,

$$Q_t = N(A_t), \quad t \in \mathbb{R}. \tag{4.7}$$

(This simple but important observation should be known more widely; I believe it has first been observed by Prekopa [6].)

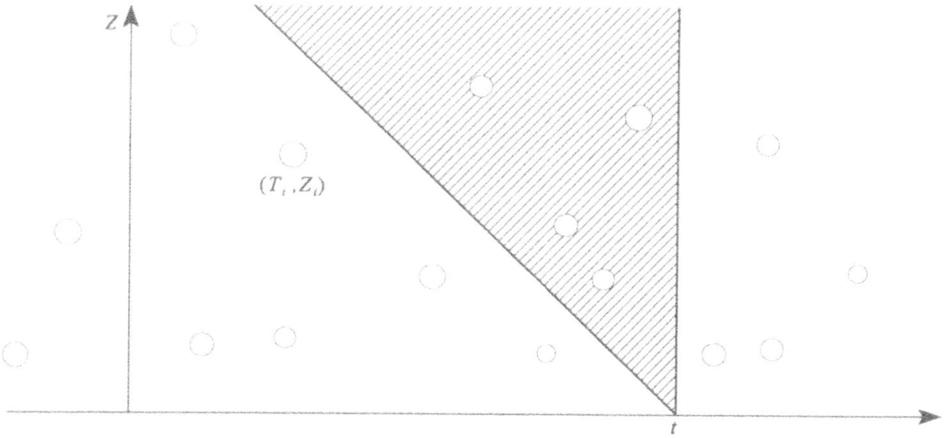

Figure 1.

Since N is a Poisson random measure with mean ν, it is immediate that Q_t has the Poisson distribution with mean

$$\nu(A_t) = \int_{(-\infty,t]} \lambda(ds) \int_{(t-s,\infty)} \beta(dz) = a \int_{\mathbb{R}_+} \beta(dz)z = ab \tag{4.8}$$

independent of t, where b is the expected service time. This result is well-known.

The observation (4.7) can be used to reduce the study of the queue size process to simple problems for the Poisson random measure N. For instance, for fixed times $t < u$,

$$Q_t = N(A_t \backslash A_u) + N(A_t \cap A_u) \text{ and } Q_u = N(A_t \cap A_u) + N(A_u \backslash A_t)$$

where $N(A_t \backslash A_u)$ has the Poisson distribution with mean $a \cdot \mathbf{E}[Z_i \wedge (u-t)]$, $N(A_t \cap A_u)$ has the Poisson distribution with mean $a \cdot \mathbf{E}[(Z_i - (u-t))^+]$, and $N(A_u \backslash A_t)$ has the Poisson distribution with mean $a \cdot \mathbf{E}[Z_i \wedge (u-t)]$ again, all three being independent.

It is possible to extend these results to nonstationary arrivals and arrival time dependent service times with little extra effort. Suppose that the arrival process is a nonstationary Poisson process with an arbitrary mean measure λ on $\mathbb{R}$, and that the service time Z_i has the distribution $\beta(t, dz)$, $z \in \mathbb{R}_+$, if the corresponding arrival time T_i happens to be t. Then Theorem 4.1 applies to show that the pairs (T_i, Z_i) form a Poisson random measure N on $\mathbb{R} \times \mathbb{R}_+$ whose mean measure ν is given by

$$\nu(dt, dz) = \lambda(dt)\beta(t, dz). \tag{4.9}$$

The identity (4.7) is still true, and we conclude that Q_t has the Poisson distribution with mean

$$\nu(A_t) = \int_{(-\infty,t]} \lambda(ds) \int_{(t-s,\infty)} \beta(t, dz). \tag{4.10}$$

Departure process. The departure time of customer i is $D_i = T_i + Z_i = h(T_i, Z_i)$ where $h(t,z) = t + z$. It is now immediate from Remark 4.2 that the departure times D_i form a Poisson random measure on $\mathbb{R}$ whose mean measure is $\nu \circ h^{-1}$. In the stationary case, where ν is given by (4.6), we get

$$\begin{aligned}(\nu \circ h^{-1})f = \nu(f \circ h) &= a \int_{\mathbb{R}} ds \int_{\mathbb{R}_+} \beta(dz) f(s+z) \\ &= a \int_{\mathbb{R}_+} \beta(dz) \int_{\mathbb{R}} ds f(s) = a \int_{\mathbb{R}} ds f(s) = \lambda f\end{aligned}$$

for arbitrary positive Borel functions f on $\mathbb{R}$, which means that

$$\nu \circ h^{-1} = \lambda. \tag{4.11}$$

In other words, in the stationary case, the departure process is Poisson with the same rate as the arrival rate a.

Remaining service times. Fix a time point t in $\mathbb{R}$. For each customer in the system at time t, mark a point r in $\mathbb{R}^0_+ = (0, \infty)$ if r is his remaining service time at that time t. The result is a random measure M_t on $\mathbb{R}^0_+$. Note that M_t is the restriction onto $\mathbb{R}^0_+$ of the random measure on $\mathbb{R}_+$ formed by the points $h(T_i, Z_i)$ where

$$h(s,z) = \begin{cases} s+z-t & \text{if } s < t \quad \text{and} \quad z > t-s, \\ 0 & \text{otherwise.} \end{cases} \tag{4.12}$$

It follows from Remark 4.2 that the latter measure is Poisson on $\mathbb{R}_+$ with mean measure $\nu \circ h^{-1}$, and, therefore, M_t is a Poisson random measure on $\mathbb{R}^0_+$ whose mean measure μ_t is the restriction of $\nu \circ h^{-1}$ onto $\mathbb{R}^0_+$. Now, for positive Borel functions f on $\mathbb{R}^0_+$, in view of (4.12) and (4.6),

$$\mu_t f = \int_{(-\infty,t]} \lambda(ds) \int_{(t-s,\infty)} \beta(dz) f(s+z-t) = a \int_0^\infty dr \beta[r,\infty) f(r).$$

To summarize, M_t is a Poisson random measure with mean measure μ_t given by

$$\mu_t(dr) = a\beta[r,\infty)dr, \quad r > 0. \tag{4.13}$$

In particular, $M_t(\mathbb{R}^0_+) = Q_t$, which is Poisson distributed with mean $\mu_t(\mathbb{R}^0_+) = ab = a\mathbf{E}[Z_i]$ as we have already shown .

4.3 SPATIAL M/G/∞ QUEUES

In the classical M/G/∞ queue reviewed in the preceding section, we spoke of the number of customers in the system without regard to where those customers are in space, or what types those customers were. In this section, we shall take care of such considerations.

We envision a queueing system as follows. The queueing space is some Borel subset D of some Polish space. We shall take $\mathbb{R}$ to be the time set. Each customer is characterized by three random variables: for the customer labeled i, the arrival time is T_i, the spatial location is X_i, and the service time is Z_i. Thus, the customer i arrives at time T_i, picks a site X_i in D, his service starts immediately and lasts Z_i units of time. We shall allow X_i to depend on T_i, and Z_i to depend on both T_i and X_i. We shall let Q_t be the queue's snapshot at time t; that is, $Q_t(A)$ will be the number of customers in the system at time t that are within the Borel subset A of D.

To fix the ideas, the reader may visualize the following. The domain D is the space occupied by some forest. Customers are trees. Arrival of a customer means the sprouting of a plant; it occurs at some time T_i at some site X_i. Service times are the lifetimes of the plants, and the lifetime of a plant depends on birth time T_i and birthplace X_i. The locations of trees that are alive at time t form the random counting measure Q_t.

Of course, D can have various interpretations. For instance, X_i may denote the "type" of customer i, in which case D is the space of types. Or, X_i may be a combination of various factors that somehow influence the service time Z_i. In short, the space D will depend on the application involved.

Arrival-service process. We let L be the random measure on $\mathbb{R} \times D$ formed by the pairs (T_i, X_i). We assume that L is a Poisson random measure with mean λ.

The assumption regarding L is satisfied, in particular, if the arrival times T_i form a Poisson process (possibly nonstationary) and each customer i picks site X_i independent of all others according to the distribution $\alpha(t, dx)$, $x \in D$, if he arrived at $T_i = t$. In this case, further letting $a(t)$ be the arrival rate at time t, we obtain that L is Poisson with mean λ given by

$$\lambda(dt, dx) = a(t)dt\,\alpha(t, dx), \qquad t \in \mathbb{R}, x \in D. \tag{4.14}$$

Regarding the service times, we assume that, for each i, the service time Z_i is conditionally independent of all else given T_i and X_i, and we let

$$\mathbf{P}\{Z_i \in dz \mid T_i = t, X_i = x\} = \beta(t, x, dz). \tag{4.15}$$

Under these assumptions, it follows from Theorem 4.1 that the triplets (T_i, X_i, Z_i) form a Poisson random measure N on $\mathbb{R} \times D \times \mathbb{R}_+$ whose mean ν is given by

$$\nu(dt, dx, dz) = \lambda(dt, dx)\beta(t, x, dz), \qquad t \in \mathbb{R}, x \in D, z \in \mathbb{R}_+. \tag{4.16}$$

The random measure N contains all the information about the queueing system. We shall take up several processes of interest.

Queue size process. For each time t and each Borel subset A of D, let $Q_t(A)$ denote the number of customers in the set A at time t. The random measure Q_t is formed by those X_i corresponding to customers i with $T_i \le t < T_i + Z_i$. In other words, Q_t is a deterministic transformation of N.

Hence, Q_t is a Poisson random measure on D. We compute its mean measure μ_t next. For arbitrary positive Borel functions f on D, we have

$$\begin{aligned} Q_t f &= \sum_i f(X_i) 1_{(-\infty, t]}(T_i) 1_{(t, \infty)}(T_i + Z_i) \\ &= \int_{\mathbb{R} \times D \times \mathbb{R}_+} N(ds, dx, dz) f(x) 1_{(-\infty, t]}(s) 1_{(t, \infty)}(s + z), \end{aligned} \tag{4.17}$$

and hence

$$\begin{aligned} \mu_t f = \mathbf{E}[Q_t f] &= \int \nu(ds, dx, dz) f(x) 1_{(-\infty, t]}(s) 1_{(t, \infty)}(s + z) \\ &= \int_{(-\infty, t] \times D} \lambda(ds, dx) f(x) \int_{(t-s, \infty)} \beta(s, x, dz) \end{aligned} \tag{4.18}$$

Stationary case. Consider the special case where the arrival times T_i form a stationary Poisson process on $\mathbb{R}$ with constant intensity a_0, the arrival site X_i is independent of T_i and has the distribution $\alpha_0(dx)$ on D, and the service time Z_i depends only on X_i and has the distribution $\beta_0(x, dz)$ at site x. Then, (4.14) and (4.15) become

$$\lambda(dt, dx) = a_0 dt \alpha_0(dx), \qquad \beta(t, x, dz) = \beta_0(x, dz). \tag{4.19}$$

Putting (4.19) into (4.18) we see that

$$\mu_t(dx) = a_0 \alpha_0(dx) b(x), \qquad x \in D, \tag{4.20}$$

where $b(x)$ is the expected service time at site x, that is,

$$b(x) = \int_{\mathbb{R}_+} \beta_0(x, dz)z. \tag{4.21}$$

Thus, in this case, sites occupied by customers at time t form a Poisson random measure Q_t whose mean measure μ_t is free of t. Since $Q = \{Q_t, -\infty < t < \infty\}$ is obviously a Markov process, this implies that the process Q is stationary.

Equilibrium. We consider the special case where

$$\lambda(dt, dx) = a(t)dt\, \alpha_0(dx), \quad \beta(t, x, dz) = \beta_0(x, dz); \tag{4.22}$$

that is, the arrival times T_i form a nonstationary Poisson process with intensity $a(t)$ at t, sites are chosen according to the distribution $\alpha_0(dx)$ on D independent of arrival times, and the service time distribution depends on site only. In this case, Q_t is a Poisson random measure on D whose mean μ_t is specified by (put (4.22) into (4.18))

$$\mu_t f = \int_{-\infty}^{t} ds\, a(s) \int_D \alpha_0(dx) f(x) \beta_0(x, (t-s, \infty)). \tag{4.23}$$

Theorem 4.3. *Suppose that the arrival intensity $a(t)$ is bounded by some constant c for all t and that $\lim_{t\to\infty} a(t) = a_0$ exists. Further, suppose that $b(x)$ defined by (4.21) is finite for α_0-almost all x in D. Then, as $t \to \infty$, the random measure Q_t converges in distribution to a Poisson random measure Q with mean μ given by*

$$\mu(dx) = a_0 \alpha_0(dx) b(x), \quad x \in D.$$

Proof. We recall (see Kallenberg [2] for instance) that a sequence of Poisson random measures M_n converges in distribution to a Poisson random measure M if, and only if, the mean measures $\mathbf{E}[M_n]$ converge vaguely to $\mathbf{E}[M]$. Thus, to prove the theorem, it is sufficient to show that, for every positive continuous function f on D with compact support,

$$\lim_{t\to\infty} \mu_t f = \mu f \tag{4.24}$$

where μ_t is specified by (4.23).

Fix such f and consider (4.23). Since the function a is bounded by some constant c, for α_0-almost every x in D,

$$\int_{-\infty}^{0} ds\, a(s) \beta_0(x, (t-s, \infty)) \le c \int_t^{\infty} dr\, \beta_0(x, (r, \infty)) \to 0$$

as $t \to \infty$, since for α_0-almost every x in D we have

$$\int_0^{\infty} dr \beta_0(x, (r, \infty)) = b(x) < \infty.$$

Because $\alpha_0 f < \infty$, it follows that

$$\mu_t f = \int_D \alpha_0(dx) f(x) \int_0^t ds\, a(s) \beta_0(x, (t-s, \infty)) + \epsilon_t \tag{4.25}$$

where $\epsilon_t \to 0$ as $t \to \infty$.

On the other hand, the function $r \to \beta_0(x, (r, \infty))$ is decreasing and Riemann integrable to $b(x)$ for α_0-almost every x. Thus, that function is directly Riemann integrable. Since $a(s) \to a_0$ by assumption as $s \to \infty$, the proof of the key renewal theorem (see Çinlar [1] for instance) applies, and we have, for α_0-almost every x in D,

$$\lim_{t \to \infty} \int_0^t ds\, a(s) \beta_0(x, (t-s, \infty)) = \alpha_0 b(x). \tag{4.26}$$

Putting (4.26) into (4.25), we obtain (4.24) with

$$\mu f = a_0 \int_D \alpha_0(dx) b(x) f(x)$$

thus completing the proof. □

Age and remaining lifetimes. Fix time t. With each customer i who is in the system at time t, we associate his location X_i, his age $U_i(t)$, and his remaining lifetime $V_i(t)$ in the system (so that $U_i(t) + V_i(t) = Z_i$). Let R_t be the random measure on $D \times \mathbb{R}_+ \times \mathbb{R}_+$ formed by those triplets $(X_i, U_i(t), V_i(t))$ corresponding to i with $T_i \leq t < T_i + Z_i$. Thus, for arbitrary f positive and Borel on $D \times \mathbb{R}_+ \times \mathbb{R}_+$

$$R_t f = \sum_i f(X_i, U_i(t), V_i(t)) 1_{(-\infty, t]}(T_i) 1_{(t, \infty)}(T_i + Z_i)$$

$$= \int N(ds, dx, dz) f(x, t-s, s+z-t) 1_{(-\infty, t]}(s) 1_{(t, \infty)}(s+z). \tag{4.27}$$

This shows that $R_t f = N(f \circ h)$ for an appropriate transformation h, that is, $R_t = N \circ h^{-1}$. Hence, since N is Poisson, R_t is a Poisson random measure on $D \times \mathbb{R}_+ \times \mathbb{R}_+$ whose mean measure is given by

$$\mathbf{E}[R_t f] = \int_{(-\infty, t] \times D} \lambda(ds, dx) \int_{(t-s, \infty)} \beta(s, x, dz) f(x, t-s, s+z-t) \tag{4.28}$$

in view of (4.27) and (4.16).

In the stationary case introduced above, using (4.19), we have

$$\mathbf{E}[R_t f] = a_0 \int_D \alpha_0(dx) \int_0^\infty du \int_{\mathbb{R}_+} \beta_0(x, u+dv) f(x, u, v); \tag{4.29}$$

that is, R_t is a Poisson random measure whose mean measure is free of t and is given as

$$a_0 \alpha_0(dx) du \beta_0(x, u+dv), \quad x \in D,\ u \in \mathbb{R}_+, v \in \mathbb{R}_+. \tag{4.30}$$

As a corollary, if we let R_t^- denote the random measure on $D \times \mathbb{R}_+$ formed by the pairs $(X_i, U_i(t))$, we see that R_t^- is again a Poisson random measure with mean measure

$$a_0 \alpha_0(dx) \beta_0(x, (u, \infty)) du, \quad x \in D, u \geq 0, \tag{4.31}$$

since $R_t^-(dx, du) = R_t(dx, du, R_+)$. Similarly, the random measure $\mathbb{R}_t^+$ formed by the pairs $(X_i, V_i(t))$ is a Poisson random measure on $D \times \mathbb{R}_+$ with the same mean measure (4.31).

Workload. Let $W_t(A)$ denote the total workload at time t due to customers situated in the Borel set A of D; that is,

$$W_t(A) = \int_{D\times\mathbb{R}_+} R_t^+(dx, dv) 1_A(x) v, \quad A \subset D. \tag{4.32}$$

Assuming the stationary conditions (4.19), we have shown above that R_t^+ is Poisson with mean measure given by (4.31). It follows from formulas (4.3) applied with $M = R_t^+$ that, for positive Borel f on D,

$$\begin{aligned} \mathbf{E}[W_t f] &= a_0 \int_D \alpha_0(dx) \int_{\mathbb{R}_+} dv \beta_0(x, (v, \infty)) f(x) v \\ &= \tfrac{1}{2} a_0 \int_D \alpha_0(dx) b_2(x) f(x) \end{aligned} \tag{4.33}$$

where

$$b_2(x) = 2 \int_{\mathbb{R}_+} dv \beta_0(x, (v, \infty)) v = \int_{\mathbb{R}_+} \beta_0(x, dz) z^2 \tag{4.34}$$

is the second moment of a service time at location x. More generally, using (4.4), we obtain the Laplace transform of the random variable $W_t f$ (we write $\exp_- x$ for e^{-x})

$$\begin{aligned} \mathbf{E}[e^{-pW_t f}] &= \mathbf{E}[\exp_- \int R_t^+(dx, dv) f(x) p v] \\ &= \exp_- a_0 \int_D \alpha_0(dx) \int_{\mathbb{R}_+} dv \beta_0(x, (v, \infty))(1 - e^{-pvf(x)}). \end{aligned} \tag{4.35}$$

Tombstone process. Finally, we consider the random measure K formed on $D \times \mathbb{R}$ by the pairs $(X_i, T_i + Z_i)$, namely, location X_i and the departure time $T_i + Z_i$ from our space. We do the computations for the stationary case specified by (4.19).

Being a deterministic transformation of N, namely, since $K = N \circ h^{-1}$ with $h(t, x, z) = (x, t + z)$, the random measure K is again Poisson on $D \times \mathbb{R}$, and its mean measure $\mathbf{E}[K]$ is specified, for test functions f on $D \times \mathbb{R}$ that are positive and Borel, by

$$\begin{aligned} \mathbf{E}[Kf] &= \mathbf{E}\left[\int N(dt, dx, dz) f(x, t + z)\right] \\ &= a_0 \int_D \alpha_0(dx) \int_{\mathbb{R}} dt \int_{\mathbb{R}_+} \beta_0(x, dz) f(x, t + z) \\ &= a_0 \int_D \alpha_0(dx) \int_{\mathbb{R}} dt\, f(x, t), \end{aligned} \tag{4.36}$$

which shows that $\mathbf{E}[K]$ is basically the same as the mean measure λ of the arrival process, which was to be expected.

4.4 QUEUES WITH MOBILE CUSTOMERS

This section is devoted to spatial M/G/∞ queues, like the preceding section, but with customers allowed to roam in the space. Mathematically, this amounts to attaching a motion to each customer. Our own interest here is largely methodological and amounts to redoing the basic part of the detailed paper by Massey and Whitt [5]. We refer to their paper for manifold variations and the conceptual richness of the model.

The queueing space is a Borel subset D of some Polish space. Customers are labeled somehow with a countable set of indices. The customer i arrives at time T_i and lands at some random point X_i in D. Once there, he moves in space D according to some stochastic process $\{X_i(t): T_i \leq t < Z_i\}$ and then disappears at time Z_i. The model of the preceding section is the special case where $X_i(t) = X_i$ for all t in $[T_i, Z_i)$, with a slight change in notation: here, the "service-time" is $Z_i - T_i$.

An example of this model arises in mobile telecommunication: T_i is the time and X_i the location of the ith call, $X_i(t)$ is the physical location of the caller at time t, and Z_i is the time at which the call ends. Another example is the model discussed by Lemoine [4]. In that case there is a finite number of nodes in a queueing network; our space D becomes a product space $D = C \times N$, where we say that the customer is at $x = (c, n)$ at time t if the customer belongs to class c and is at node n of the network at that time. Of course, then, the probability law of the motion $X_i(t)$ has to reflect the underlying mechanism for changes of class and node.

Going back to our abstract model, we now introduce the probability assumptions. First, we assume that the pairs (T_i, X_i) form a Poisson random measure on $\mathbb{R} \times D$ with mean measure $\lambda(dt, dx)$. We add an extra point Δ to D, calling Δ the death state, and define

$$Y_i(t) = \begin{cases} X_i(t) & \text{if } T_i \leq t < Z_i, \\ \Delta & \text{if } t \geq Z_i, \end{cases}$$

so that the process $Y_i = \{Y_i(t): t \geq T_i\}$ has state space $\bar{D} = D \cup \{\Delta\}$ with Δ absorbing. For each i, we regard Y_i as a random variable taking values in the function space $E = \bar{D}^{\mathbb{R}}$ with starting time T_i (and the σ-algebra on E being the product σ-algebra generated by the Borel σ-algebra of $\bar{D}$). We assume that the processes Y_i are conditionally independent, given T_i and X_i, and let $\mathbf{P}^{sx}$ denote the common probability law (the same for all customers i) of Y_i, given that $T_i = s$ and $Y_i(T_i) = X_i = x$. Further, we introduce the transition function

$$P_{st}(x, dy) = \mathbf{P}^{sx}\{Y_i(t) \in dy\}, \quad t \geq s, y \in D, \tag{4.37}$$

and note that

$$1 - P_{st}(x, D) = \mathbf{P}^{sx}\{Y_i(t) = \Delta\} = \mathbf{P}^{sx}\{Z_i \leq t\}. \tag{4.38}$$

Main process. Each customer i is characterized by the triplet (T_i, X_i, Y_i) which gives his arrival time, his initial landing point, and his subsequent movement in the space D (and death). We let N denote the random measure on $\mathbb{R} \times D \times E$ formed by those triplets. It follows from Theorem 4.1 that N is a Poisson random measure whose mean ν is given by

$$\nu(ds, dx, dy) = \lambda(ds, dx)\mathbf{P}^{sx}(dy), \quad s \in \mathbb{R}, x \in D, \mathbf{y} \in E; \tag{4.39}$$

recall that E is the collection of all possible paths for customers.

Queueing process. Let Q_t be the random measure formed by the locations $Y_i(t)$ in D of all those customers that are in the system at time t. For arbitrary test functions (positive Borel) f on D,

$$\begin{aligned} Q_t f &= \sum_i f(Y_i(t)) 1_{(-\infty, t]}(T_i) 1_D(Y_i(t)) \\ &= \int N(ds, dx, d\mathbf{y}) f(\mathbf{y}(t)) 1_D(\mathbf{y}(t)) 1_{(-\infty, t]}(s), \end{aligned} \tag{4.40}$$

which shows that Q_t is obtained from N by a deterministic transformation. Hence, by Remark 4.2, Q_t is a Poisson random measure on D whose mean is given by

$$\begin{aligned} \mathbf{E}[Q_t f] &= \int_{(-\infty, t] \times D} \lambda(ds, dx) \int_E \mathbf{P}^{sx}(d\mathbf{y}) 1_D(\mathbf{y}(t)) f(\mathbf{y}(t)) \\ &= \int_{(-\infty, t] \times D} \lambda(ds, dx) \int_D P_{st}(x, dy) f(y) \end{aligned} \tag{4.41}$$

in view of (4.39), (4.38), (4.37).

Stationary case. Similar to considerations outlined in the preceding section, we now assume time-homogeneity by letting

$$\lambda(ds, dx) = a\, ds\, \alpha_0(dx), \quad P_{st}(x, dy) = P^0_{t-s}(x, dy). \tag{4.42}$$

In particular, this makes the motion processes Y_i time-homogeneous. In this case, (4.41) takes the form

$$\mathbf{E}[Q_t f] = a \int_D \alpha_0(dx) \int_D U(x, dy) f(y) \tag{4.43}$$

where U is the potential kernel corresponding to P^0; that is,

$$U(x, B) = \int_0^\infty dt P^0_t(x, B), \quad x \in D, B \subset D. \tag{4.44}$$

Hence, in this case, the law of the Poisson random measure Q_t is the same for all t since its mean (which defines the law) is the same for all t.

Many other processes regarding this system can be studied in particular instances by relating the process of interest to the main process N. In each such application, the work involved is in computations having to do with the probability laws $\mathbf{P}^{sx}$. For several such detailed computations we refer to Lemoine [4] and Massey and Whitt [5].

4.5 SPATIAL M/M/1 QUEUES

In this final section we introduce the harder case of a single-server queue where customers' locations are taken into account explicitly. Here is the rough description.

Demands for service occur over time and space according to a Poisson random measure. A single server moves in space to satisfy the demand. The service duration for a customer situated at x has an exponential distribution with a parameter depending on x. When that service is completed, the customer to be served next is chosen from those waiting at that time according to some rule that depends on the last location x as well as the locations of all other customers then present. To simplify the matter, we assume that the travel time between x and the next service location y can be ignored, so that the service at y starts immediately after that at x is finished. Various, more realistic, extensions are possible.

For visualization purposes, the reader should think of the space as a small town, the server as the only plumber in town, and the customers as houses that need plumbing repairs. Other examples can be drawn from computer networks, machine repair problems in a plant, etc.

As in the earlier sections, we label the customers by indices i. For each customer i, we let T_i denote his arrival time, X_i his location in the space D, and Z_i his service duration. Concerning these, we assume that the pairs (T_i, X_i) form a Poisson random measure N on $\mathbb{R}_+ \times D$ with mean measure having the form

$$\nu(ds, dx) = ds\,\alpha(dx), \quad s \in \mathbb{R}_+, x \in D, \tag{4.45}$$

and that the Z_i are independent of all else given X_i, and

$$\mathbb{P}\{Z_i \in dz \mid X_i = x\} = c(x)e^{-c(x)z}dz, \quad z \geq 0, \tag{4.46}$$

namely, the exponential distribution with mean

$$b(x) = 1/c(x). \tag{4.47}$$

We leave α arbitrary, but the case $\alpha(D) < \infty$ is interesting. Then α has the form

$$\alpha(dx) = a\,\alpha_0(dx), \quad a < \infty, \alpha_0(D) = 1, \tag{4.48}$$

which means that the arrival times T_i form a Poisson process on $\mathbb{R}_+$ with rate a, and each arriving customer picks his location in D according to the probability distribution α_0.

We let Q_t denote the random measure on D formed by the locations of customers that are in the system at time t. Thus, for each t, the random variable Q_t takes values in the space $\mathcal{M} = \mathcal{M}(D)$ of all measures on D. In particular, $Q_t(A)$ is the number of customers in the set A at time t.

We let S_t denote the server's location at time t. If Q_t is not zero, then S_t must belong to the support of Q_t. In other words, the server is with a customer as long as there are customers in the system. If Q_t is zero, then S_t is put equal to Δ, where Δ is some special point outside D. We put $\bar{D} = D \cup \{\Delta\}$.

The service protocol is specified by a function $r: \mathcal{M} \times \bar{D} \mapsto \bar{D}$, called the routing rule, as follows: $r(Q_t, S_t)$ is the location of the customer to be served next given the queueing state Q_t and the server location S_t. In particular, $r(0, \Delta) = \Delta$ and $r(\delta_x, x) = \Delta$, in the latter case, because there is only one customer at x and the server is busy with him. We shall leave r unspecified. We hope that many interesting rules r will be studied separately in the future.

System dynamics. It follows from the assumptions embedded that the queueing process $\{(Q_t, S_t)\colon\ t \geq 0\}$ is a time-homogeneous Markov process with state space $\mathcal{M} \times \bar{D}$. We now describe its time evolution via a coupled system of integral equations.

To this end, we introduce an auxiliary random measure K, called the killing measure, to model the service completions. It is a Poisson random measure on $\mathbb{R}_+ \times \mathbb{R}_+$ with unit intensity and is independent of N. If $K(\omega, \cdot)$ has a point at

(t, z), and if the service rate applicable at time t is greater than z, then this signals the completion of service at that time t. To this end, we extend the definition of the service rate function c onto $\bar{D}$ by setting $c(\Delta) = 0$.

Let f be a positive Borel function on D. Recall that $Q_t f$ is the integral of f with respect to the random measure Q_t and is thus equal to the sum of $f(X_i)$ over all customers i who are in the system at time t. Thus, $Q_t f$ is equal to $Q_{t-} f + f(x)$ if there is an arrival at x at time t, and is equal to $Q_{t-} f - f(S_{t-})$ if there is a departure at time t. The following differential equation expresses this:

$$dQ_t f = \int_D N(dt, dx) f(x) - \int_{z=0}^{\infty} K(dt, dz) 1_{\{c(S_{t-}) > z\}} f(S_{t-}). \tag{4.49}$$

As to the evolution of S_t, we note that the arrivals do not affect S_t unless the queue is empty. Thus, if an arrival occurs at x at time t, and if $Q_{t-} = 0$ then $f(S_{t-}) = f(\Delta) = 0$ and $f(S_t) = f(x)$. Otherwise, arrivals do not affect S_t. If a service is completed at time t, then $f(S_{t-})$ becomes $f(S_t) = f(r(Q_{t-}, S_{t-}))$. We put these observations as a differential equation next:

$$\begin{aligned} df(S_t) &= \int_D N(dt, dx) 1_{\{Q_{t-} = 0\}} f(x) \\ &+ \int_{z=0}^{\infty} K(dt, dz) 1_{\{c(S_{t-}) > z\}} [f(r(Q_{t-}, S_{t-})) - f(S_{t-})]. \end{aligned} \tag{4.50}$$

This completes the specification of system dynamics.

Itô's Lemma and forward equations. Let $\varphi: \mathcal{M} \times \bar{D} \to \mathbb{R}$ be a bounded continuous function (the topology on $\mathcal{M}$ is that corresponding to the vague convergence of measures). We put Itô's equation for $\varphi(Q_t, S_t)$ in the following differential form:

$$\begin{aligned} d\varphi(Q_t, S_t) = & \int_D N(dt, dx) \{[\varphi(Q_{t-} + \delta_x, S_{t-}) - \varphi(Q_{t-}, S_{t-})] 1_{\{Q_{t-} \neq 0\}} \\ & \qquad + [\varphi(\delta_x, x) - \varphi(0, \Delta)] 1_{\{Q_{t-} = 0\}}\} \\ + & \int_{z=0}^{\infty} K(dt, dz) 1_{\{c(S_{t-}) > z\}} [\varphi(Q_{t-} - \delta_{S_{t-}}, r(Q_{t-}, S_{t-})) - \varphi(Q_{t-}, S_{t-})]. \end{aligned} \tag{4.51}$$

This equation is derived from (4.49)-(4.50) by accounting for all possible changes, recalling that δ_x is Dirac measure at x and that $c(\Delta) = 0$ by convention, which makes it impossible to have service completion when the server is at Δ (and thus idle).

Next, we write (4.51) in integral form and take expectations on both sides recalling the rules of stochastic calculus. We obtain

$$\begin{aligned} \mathbf{E}[\varphi(Q_t, S_t)] = & \ \mathbf{E}[\varphi(Q_0, S_0)] \\ & + \int_0^t du \int_D \alpha(dx) \mathbf{E}[(\varphi(Q_{u-} + \delta_x, S_{u-}) - \varphi(Q_{u-}, S_{u-})) 1_{\{Q_u \neq 0\}} \\ & \qquad + (\varphi(\delta_x, x) - \varphi(0, \Delta)) 1_{\{Q_{u-} = 0\}}] \\ & + \int_0^t du \int_0^{\infty} dz \mathbf{E}[1_{\{c(S_{u-}) > z\}} (\varphi(Q_{u-} - \delta_{S_{u-}}, r(Q_{u-}, S_{u-})) - \varphi(Q_{u-}, S_{u-}))]. \end{aligned}$$

Since Q_{u-} and S_{u-} differ from Q_u and S_u, respectively, only for many countable u, the integrals on the right side remain unchanged when we replace Q_{u-} by Q_u and S_{u-} by S_u. This observation and performing the integration over z yield

$$\begin{aligned}\mathbf{E}[\varphi(Q_t,S_t)] &= \mathbf{E}[\varphi(Q_0,S_0)] \\ &+ \int_0^t du \int_D \alpha(dx)\mathbf{E}[(\varphi(Q_u+\delta_x,S_u)-\varphi(Q_u,S_u))1_{\{Q_u\neq 0\}} \\ &\qquad + (\varphi(\delta_x,x)-\varphi(0,\Delta))1_{\{Q_u=0\}}] \\ &+ \int_0^t du\mathbf{E}[c(S_u)(\varphi(Q_u-\delta_{S_u},r(Q_u,S_u))-\varphi(Q_u,S_u))]. \end{aligned} \tag{4.52}$$

Finally, we note that, on the event $\{Q_u=0\}$, we have $\varphi(Q_u,S_u)=\varphi(0,\Delta)$ and $\varphi(Q_u+\delta_x,S_u)=\varphi(\delta_x,\Delta)$. Thus, we may write the integrand in the middle term on the right as

$$\mathbf{E}[\varphi(Q_u+\delta_x,S_u)-\varphi(Q_u,S_u)]+[\varphi(\delta_x,x)-\varphi(\delta_x,\Delta)]\mathbf{P}\{Q_u=0\}.$$

Taking this into account and differentiating on both sides (the right side is definite ly differentiable and therefore so is the left side), we get

$$\begin{aligned}\frac{d}{dt}\mathbf{E}[\varphi(Q_t,S_t)] &= \int_D \alpha(dx)\mathbf{E}[\varphi(Q_t+\delta_x,S_t)-\varphi(Q_t,S_t)] \\ &+ \mathbf{P}\{Q_t=0\}\int_D \alpha(dx)[\varphi(\delta_x,x)-\varphi(\delta_x,\Delta)] \\ &+ \mathbf{E}[c(S_t)(\varphi(Q_t-\delta_{S_t},r(Q_t,S_t))-\varphi(Q_t,S_t))]. \end{aligned} \tag{4.53}$$

This is Kolmogorov's forward equation for the Markov process (Q_t,S_t), $t\geq 0$.

Limiting distribution. Consider the total queue size $Q_t1=Q_t(D)$ at time t. Note that $\{Q_t1: t\geq 0\}$ is the queue size process at an ordinary M/G/1 queue where, assuming the form (4.48) for α, the arrivals are Poisson with rate a and the service times have the density function

$$\int_D \alpha_0(dx)c(x)e^{-c(x)z}, \quad z\geq 0.$$

It is well-known that the M/G/1 queue is positive recurrent if, and only if, the traffic intensity is strictly less than 1; that is, $\{Q_t1: t\geq 0\}$ is positive recurrent if

$$a\cdot\int_D \alpha_0(dx)\frac{1}{c(x)} = \int_D \alpha(dx)b(x) = \alpha b \tag{4.54}$$

is less than 1 (recall the notation (4.47) for $b(x)$ and the integral notation αb). Assume that the traffic intensity αb is less than 1. Since $Q_t1=0$ if, and only if, the measure Q_t is equal to 0, it follows that the state $(0,\Delta)$ in $\mathcal{M}\times\bar{D}$ is a positive recurrent state for the Markov process $\{(Q_t,S_t): t\geq 0\}$. Then, it follows from regeneration theory that (Q_t,S_t) has a limiting distribution. We state this next.

Theorem 4.4. *Suppose that $\alpha b<1$. Then (Q_t,S_t) converges in distribution, as $t\to\infty$, to a pair (Q,S) with values in $\mathcal{M}\times\bar{D}$, and, for arbitrary bounded continuous functions φ on $\mathcal{M}\times\bar{D}$, we have*

$$\int_D \alpha(dx)\mathbf{E}[\varphi(Q+\delta_x,S)-\varphi(Q,S)]$$

$$+\mathbf{P}\{Q=0\}\int_D \alpha(dx)[\alpha(\delta_x,x)-\varphi(\delta_x,\Delta)]$$

$$+\mathbf{E}[c(S)(\varphi(Q-\delta_S,r(Q,S))-\varphi(Q,S))]=0.$$

Proof. That (Q_t,S_t) converges in distribution was justified before starting the theorem. By the definition of convergence in distribution, then, $\mathbf{E}[\varphi(Q_t,S_t)]\to\mathbf{E}[\varphi(Q,S)]$ as $t\to\infty$. Thus, as $t\to\infty$, the left side of (4.53) approaches 0, and we obtain the formula claimed.

Distribution of server's location. Surprisingly, in equilibrium, the server's location does not depend on the serving rule r:

Corollary 4.5. *Suppose $\alpha b<1$ and let S be the limit of S_t in distribution. We have*

$$\mathbf{P}\{S\in dx\}=\alpha(dx)b(x),\quad x\in D,$$

$$\mathbf{P}\{S=\Delta\}=1-\alpha b.$$

Proof. Let f be an arbitrary positive continuous function on D with compact support. Define $\varphi(q,y)=qf$, the integral of f with respect to the measure q in $\mathcal{M}$. Then the formula of Theorem 4.4 reduces to

$$\int \alpha(dx)f(x)+\mathbf{E}[c(S)(-f(S))]=0.$$

Recalling that $c(\Delta)=0$, we may rewrite this as

$$\int_D \mathbf{P}\{S\in dx\}c(x)f(x)=\int_D \alpha(dx)f(x).$$

Since f is arbitrary, this means that

$$\mathbf{P}\{S\in dx\}c(x)=\alpha(dx),\quad x\in D,$$

which is as claimed (since $b(x)=1/c(x)$). Thus,

$$\mathbf{P}\{S\in D\}=\int_D \alpha(dx)b(x)=\alpha b,$$

which completes the proof.

Of course, $\{S=\Delta\}=\{Q=0\}$, and thus

$$\mathbf{P}\{Q=0\}=1-\alpha b \tag{4.55}$$

which is well-known from the theory of M/G/1 queues. □

4.6 OPEN PROBLEMS AND FINAL COMMENTS

1. It is intuitively obvious (though I am unable to prove it), that $r(Q,S)$ in Corollary 4.5 has the same distribution as S.

2. In principle, the formula of Theorem 4.4 specifies the distribution of (Q,S). However, more explicit results are hard to come by. For instance, it would be of great interest to find the mean dispersion of customers in space, namely, an explicit expression for $\mathbf{E}[Qf]$ for arbitrary f. Such computations will require careful choices of φ under specific rules r.

ACKNOWLEDGEMENT

This work was partially supported under ONR Grant No: N00014-92-J-1088 P00005.

BIBLIOGRAPHY

[1] Çinlar, E., *Introduction to Stochastic Processes*, Prentice-Hall, Englewood Cliffs 1975.

[2] Kallenberg, O., *Random Measures*, Akademie Verlag, Berlin 1983.

[3] Kingman, J.F.C., *Poisson Processes*, Clarendon Press, Oxford 1993.

[4] Lemoine, A.J., A stochastic network formulation for complex sequential processes, *Naval Res. Logist. Quart.* **33** (1986), 431-443.

[5] Massey, W.A. and Whitt, W., Networks of infinite server queues with non-stationary Poisson input, *Queueing Sys.* **13** (1993), 183-250.

[6] Prekopa, A., On secondary processes generated by random point distributions of Poisson type, *Ann. Univ. Sci. Budapest de R. Eödvös Nom. Sezdio yMath.* **1** (1958), 153-170.

Chapter 5

Sample-path techniques in queueing theory

Shaler Stidham, Jr. and Muhammad El-Taha[1]

ABSTRACT In this chapter, we use sample-path techniques to derive relations between different performance measures for queues and related stochastic systems. A wide range of topics is reviewed to illustrate the power and versatility of sample-path analysis. We start by reviewing a sample-path version of the renewal-reward theorem ($Y = \lambda X$). It is then shown how $Y = \lambda X$ can be used to give a simple proof of the sample-path rate-conservation law RCL under more general conditions than previously given. We apply $Y = \lambda X$ to obtain simple proofs of relations between continuous-time frequencies of a process defined on a general state space and frequencies at the points of an imbedded point process, including sample-path versions of the stochastic mean-value theorem, the covariance formula, the inverse-rate formula, and necessary and sufficient conditions for $ASTA$, reversed $ASTA$, and conditional $ASTA$. We also show that the RCL implies a sample-path version of the rate-balance formula (equating entrance and exit rates from a given set of states) and give sample-path proofs of relations between forward and backward recurrence-time distributions and their associated equilibrium distributions. Techniques used for deriving relations between (one-dimensional) frequencies are extended to develop a complete sample-path version of the Palm calculus. Strengthened versions of $ASTA$ and $PASTA$ are given, using the theory of martingales. We also review and extend sample-path proofs of $L = \lambda W$ and $H = \lambda G$. Stability of general input-output systems is analyzed, showing that the existence of a limiting frequency distribution implies stability in the sense of equality of input and output rates. We prove our basic stability result by establishing sufficient conditions that are easily verifiable from conditions on primary processes. Processes with non-countable as well as integer state spaces are considered and studied within one unifying framework. We also consider busy-period and idle-period durations and establish their rate stability under the sufficient conditions given previously. Applications to special queueing models and other input-output processes are given. In particular, we give a sample-path proof for rate stability of the workload, queue length, and infinite server queues. Stable versions of Little's formula are also given.

[1]The research of this author was done while he was visiting the Department of Operations Research, University of North Carolina at Chapel Hill (Spring, 1994).

0-8493-8074-x/95/$0.00+$.50

CONTENTS

5.1 Introduction 120
5.2 Background on point processes; $Y = \lambda X$ 122
5.3 Rate-conservation law 124
5.4 Relations between frequencies for a process with an imbedded point process 125
5.5 Systems with countable state space 131
5.6 Sample-path version of the Palm calculus 132
5.7 Filtration of $ASTA$ 136
5.8 Little's formula: $L = \lambda W$ 140
5.9 Generalization of Little's formula: $H = \lambda G$ 147
5.10 Stability conditions for input-output processes 151
5.11 Stability applications 157
5.12 Open problems and new research directions 161
Bibliography 162

5.1 INTRODUCTION

In recent years, sample-path analysis has played an increasingly important role in applied probability in general and in queueing theory in particular. Researchers have come to appreciate the analytical as well as pedagogical advantages of studying a stochastic process in terms of its possible realizations, rather than its distributions.

Examples of the power of pathwise analysis are many and varied. They include, at the fundamental level, the definition and characterization of a stochastic process in terms of the permissible behavior of its realizations, e.g., that they must belong to a certain function space, such as $D[0,\infty)$. This approach lies at the heart of the modern theory of weak convergence of stochastic processes, for example (Billingsley [4]). When combined with coupling arguments, pathwise analysis has led to simple and elegant proofs of fundamental limit theorems for renewal and regenerative processes (see e.g., Lindvall [46, 47], Asmussen [2]) and stationary processes (Loynes [50], Baccelli and Brémaud [3]). In the area of optimal design of queues and other stochastic systems, pathwise analysis and coupling arguments have provided a powerful tool for proving stochastic monotonicity and/or convexity of performance measures such as throughput and waiting times in terms of system parameters such as service rates and number of servers (see e.g., Shanthikumar and Yao [70], Shaked and Shanthikumar [67, 68]). Sample-path arguments have also been used in the analysis of problems in the control of queues to establish that an optimal policy has a particular form. (For a review, see Liu, Nain and Towsley [49].)

In this chapter, we shall focus on the use of sample-path analysis to derive relations between different performance measures for queues and related stochastic systems. Our concern will be with asymptotic pathwise behavior, specifically, with limiting averages and limiting frequencies associated with such measures as queue lengths and workloads. A more precise term for this variety of sample-path analysis might be *asymptotic deterministic analysis* of stochastic systems: "asymptotic" because it deals with long-run behavior and "deterministic" because it focuses on a single fixed realization of a stochastic system. It has been used to derive relations between time averages and customer averages, such as $L = \lambda W$ (Little's formula) and its generalization $H = \lambda G$ (see Stidham [72, 74, 76, 77], Heyman and Stidham [40], Rolski and Stidham [63], Glynn and Whitt [34, 35], Halfin and Whitt [39], Whitt [85], Serfozo [66], El-Taha and Stidham [27], and Dshalalow [17]). Relations between asymptotic state frequencies at arbitrary time points and at times of arrivals to a queueing system have also been derived using sample-path analysis (see Heyman and Stidham [40], Stidham [76], El-Taha [18], Stidham and El-Taha [80], El-Taha and Stidham [23, 24, 25, 26], El-Taha [20], and Stidham [79]). Progress in studying insensitivity using sample path methods has been reported in Stidham [76], Shanthikumar and Sumita [69] and El-Taha and Stidham [23]. Sample-path analysis has also been used to establish stability conditions for input-output processes (see Stidham and El-Taha [81], El-Taha and Stidham [27], Mazumdar et al. [52], and Guillemin and Mazumdar [38]). Some of the techniques and results of sample-path analysis have appeared in the literature in the form of "level crossing analysis" (Brill and Posner [9, 10]), "operational analysis" (Buzen [12, 13], Buzen and Denning [14], and Denning and Buzen [16]) and the "deterministic theory of queues" (Gelenbe [31] and Gelenbe and Finkel [32]).

One of our goals in this chapter is to review each of these types of application of sample-path analysis and show how they all can be derived from a single fundamental theorem: the sample-path version of the renewal-reward theorem ($Y = \lambda X$). This theorem is related to the more powerful, and more familiar, $H = \lambda G$. Many authors (e.g., Heyman and Stidham [40], Sigman [71], Whitt [85, 87], Brémaud [7]) have observed, in both the sample-path and classical stochastic settings, that several well-known relations in queueing theory follow from $H = \lambda G$. One of the themes of the present chapter is that starting with the simpler and more basic formula, $Y = \lambda X$, leads to a more complete understanding of the fundamental nature of the relations in question. Moreover, in some applications, such as relations between asymptotic frequencies and the rate-conservation law RCL, $Y = \lambda X$ by itself suffices. In other cases, such as $L = \lambda W$ and $H = \lambda G$, additional arguments are required.

Throughout this chapter, unless otherwise stated, we make no stochastic assumptions. The quantities and processes defined are deterministic. They may be thought of as representing a fixed sample path of a stochastic system, defined on some probability space.

The chapter is organized as follows. In Section 5.2 we collect some definitions and basic properties of deterministic point processes and present two versions of $Y = \lambda X$. Section 5.3 shows how $Y = \lambda X$ can be used to give a simple proof of the sample-path rate-conservation law RCL under more general conditions than previously given. In Section 5.4, we apply $Y = \lambda X$ to obtain simple proofs of relations between continuous-time frequencies of a process defined on a general state space and frequencies at the points of an imbedded point process, including sample-path versions of the stochastic mean-value theorem, the covariance formula, the inverse-rate formula, and necessary and sufficient conditions for $ASTA$, reversed $ASTA$, and conditional $ASTA$. We also show that the RCL implies a sample-path version of the rate-balance formula (equating entrance and exit rates from a given set of states) and gives sample-path proofs of relations between forward and backward recurrence-time distributions and their associated equilibrium distributions. Section 5.5 specializes some results in Section 5.4 to processes with countable state space. In Section 5.6, we show how the techniques in Section 5.4 for deriving relations between (one-dimensional) frequencies can be extended to develop a complete sample-path version of the Palm calculus. In

Section 5.7, strengthened versions of *ASTA* and *PASTA* are given, using the theory of martingales. (This is the only section where stochastic assumptions are made throughout.) Sections 5.8 and 5.9 review and extend sample-path proofs of $L = \lambda W$ and $H = \lambda G$. In Section 5.10, we analyze the stability of a general input-output system, showing that the existence of a limiting frequency distribution implies stability in the sense of equality of input and output rates. These results can be used to validate the sufficient conditions for the sample-path *RCL*. We prove our basic stability result by establishing sufficient conditions that are easily verifiable from conditions on primary processes. Processes with noncountable, as well as integer state spaces, are considered and studied within one unifying framework. We also consider busy-period and idle-period durations and establish their rate stability under the sufficient conditions given previously. Section 5.11 contains applications to special cases of queueing systems and other input-output processes. In particular, we give a sample-path proof for rate stability of the workload, queue length, and infinite server queues.

5.2 BACKGROUND ON POINT PROCESSES; $Y = \lambda X$

Let $\{T_k, k \geq 1\}$ be a deterministic sequence of time points, with $0 \leq T_k \leq T_{k+1} < \infty$, $k \geq 1$. Define $N(t){:} = \#\{k \geq 1{:}T_k \leq t\}$, $t \geq 0$. We interpret T_k as the time point at which the kth of a sequence of events occurs, such as the kth arrival to a queue, and $N(t)$ as the number of events in $[0, t]$. Note that our definition allows more than one event to occur at the same point, e.g., batch arrivals. That is, the point process $N = \{N(t), t \geq 0\}$ need not be simple. We assume that $T_k \to \infty$ as $k \to \infty$, so that there are only a finite number of events in any finite time interval ($N(t) < \infty$ for all $t \geq 0$). Note that $N(t) \to \infty$ as $t \to \infty$, since $T_k < \infty$ for all $k \geq 1$. Note also that $N(t) = max\{k{:}T_k \leq t\}$ (where $T_0 = 0$), since $\{T_k, k \geq 1\}$ is a nondecreasing sequence.

First, we state the following elementary lemma, which is a sample-path analogue of the elementary renewal theorem. (For a proof see Stidham [72], [74].)

Lemma 5.1. *For any $0 \leq \lambda \leq \infty$, the following are equivalent:*

(a) $N(t)/t \to \lambda$, *as* $t \to \infty$;

(b) $T_n/n \to \lambda^{-1}$, *as* $n \to \infty$.

The following lemmas contain our two versions of a sample-path analogue of the renewal-reward theorem ($Y = \lambda X$).

Lemma 5.2. *Suppose $N(t)/t \to \lambda$, as $t \to \infty$, where $0 \leq \lambda \leq \infty$. Let $\{Y(t), t \geq 0\}$ be a deterministic, nondecreasing and right-continuous real-valued process ($Y(0) = 0$) and define*

$$X_k{:} = Y(T_k) - Y(T_{k-1}), \quad k \geq 1. \tag{5.1}$$

Then

(a) *if $n^{-1}\sum_{k=1}^{n} X_k \to X$, $0 \leq X \leq \infty$, as $n \to \infty$, then $t^{-1}Y(t) \to Y = \lambda X$ as $t \to \infty$, provided that λX is well defined;*

(b) *if $t^{-1}Y(t) \to Y$, $0 \leq Y \leq \infty$, as $t \to \infty$, then $n^{-1}\sum_{k=1}^{n} X_k \to X = \lambda^{-1}Y$ as $n \to \infty$, provided that $\lambda^{-1}Y$ is well defined.*

Lemma 5.3. *Suppose $t^{-1}N(t) \to \lambda$, $0 \leq \lambda \leq \infty$, as $t \to \infty$. Let $\{X_k, k \geq 1\}$ be a deterministic sequence of nonnegative real numbers and define*

$$Y(t){:} = \sum_{k=1}^{N(t)} X_k, \quad t \geq 0. \tag{5.2}$$

Then,

(*a*) *if* $n^{-1}\sum_{k=1}^{n} X_k \to X$, $0 \le X \le \infty$, *as* $n \to \infty$, *then* $t^{-1}Y(t) \to Y = \lambda X$ *as* $t \to \infty$, *provided that* λX *is well defined;*

(*b*) *if* $t^{-1}Y(t) \to Y$, $0 \le Y \le \infty$, *as* $t \to \infty$, *then* $n^{-1}\sum_{k=1}^{n} X_k \to X = \lambda^{-1}Y$ *as* $n \to \infty$, *provided that* $\lambda^{-1}Y$ *is well defined.*

Remark 5.4. As with all the results in this chapter, the results in this section pertain to deterministic processes, or, equivalently, to individual sample paths of stochastic processes. Thus they do not require explicit stochastic assumptions, only the existence of certain limiting averages on the sample path in question. One can apply these results to stochastic models in which the stochastic assumptions imply (via an ergodic theorem or law of large numbers) that the sample-path averages in question exist a.s. For example, suppose that $\{(T_n - T_{n-1}, X_n)\}$ is an ergodic, strictly stationary sequence and that $\{Y(t), t \ge 0\}$ is defined in terms of $\{X_k, k \ge 1\}$ by (5.2). Then the ergodic theorem and Lemma 5.3 imply that $n^{-1}T_n \to \mathbf{E}[T_1] = \lambda^{-1}$ and $n^{-1}\sum_{k=1}^{n} X_k \to \mathbf{E}[X_1]$ *a.s.* as $n \to \infty$, and that $t^{-1}Y(t) \to Y = \mathbf{E}[X_1]/\mathbf{E}[T_1]$ a.s. as $t \to \infty$.

Remark 5.5. Neither the assumption that $Y(t)$ is nondecreasing nor the assumption that X_k is nonnegative is necessary (cf. El-Taha and Stidham [26]). They are imposed here for ease of exposition and because they are satisfied in all the applications of this chapter.

Remark 5.6. If we interpret $Y(t)$ as the cumulative "cost" incurred in $[0, t]$, then both versions of $Y = \lambda X$ make the intuitively plausible statement that *the long-run average cost equals the rate at which points occur times the average cost incurred between two successive points.* The two versions differ as to whether $\{X_k, k \ge 1\}$ is defined in terms of $\{Y(t), t \ge 0\}$ or vice versa. In Lemma 5.2, cost may be accumulated between points, whereas, in Lemma 5.3, all cost is incurred in lump sums at the points of $\{T_k, k \ge 1\}$. On the other hand, in the case of multiple points, Lemma 5.3 allows us to distinguish between the cost contributions X_k of different points T_k that occur at the same time, whereas Lemma 5.2 does not. We shall need both versions for the applications in this chapter. In applications where $Y(t)$ includes both kinds of "costs", one can establish $Y = \lambda X$ by applying Lemmas 5.2 and 5.3 separately to each cost component. We could have stated an omnibus version of $Y = \lambda X$ that would have included each version as a special case (cf. Lemma 3 of Serfozo [66]), but at the cost of a more complicated notation and less facility for handling cases where one or more of the limits is infinite.

Remark 5.7. When the point process $\{N(t), t \ge 0\}$ is simple, Lemma 5.3 follows from Lemma 5.2, since (5.2) implies (5.1) in this case. This is not true in general if $\{N(t), t \ge 0\}$ is not simple, which is why we need two versions of $Y = \lambda X$.

Remark 5.8. The cost interpretation of $Y = \lambda X$ invites a comparison with the queueing formula $L = \lambda W$ and its generalization $H = \lambda G$ (see Sections 5.8 and 5.9 of this chapter), in which λ is the arrival rate of customers, L (or H) is the average cost per unit time, and W (or G) is the average cost per customer. But $Y = \lambda X$ is a simpler and more basic relation, which only applies when the cost X_k associated with the point T_k is incurred entirely between the points T_k and T_{k+1}, which is not the case with $L = \lambda W$ and $H = \lambda G$. As we shall see (in Sections 5.8 and 5.9), sample-path proofs of $L = \lambda W$ and $H = \lambda G$ can be constructed which use one of the versions of $Y = \lambda X$ as a first step, followed by an argument showing that the contribution to the cost associated with the point T_k from outside the interval $[T_k, T_{k+1})$ is asymptotically negligible.

5.3 RATE-CONSERVATION LAW

In this section, we follow El-Taha and Stidham [26] to show how a sample-path version of the *rate-conservation law* (RCL) (Miyazawa [57, 58, 59], Sengupta [64], Brémaud [7], Ferrandiz and Lazar [29], Mazumdar et al. [53], Sigman [71], Whitt [85], [87]) can easily be derived from $Y = \lambda X$ under conditions more general than those given previously in the literature. We follow Sigman [71] in our notation with some minor modifications.

Let $\{x(t), t \geq 0\}$ be a real-valued process, assumed right continuous with left-hand limits. Let $[T_k, k \geq 1\}$ be a point process with counting process $\{N(t), t \geq 0\}$, satisfying the conditions of Section 5.2. In addition, we assume throughout this section that $\{T_k, k \geq 1\}$ is simple: $T_k < T_{k+1}, k \geq 1$. Let $-J_k{:} = x(T_k+) - x(T_k-)$ denote the jump (if any) of $\{x(t), t \geq 0\}$ at T_k. Let $\widetilde{x}(t){:} = x(t) - x(0) + \sum_{k=1}^{N(t)} J_k$, $t \geq 0$, so that

$$x(t) = x(0) + \widetilde{x}(t) - \sum_{k=1}^{N(t)} J_k, \quad t \geq 0. \tag{5.3}$$

Theorem 5.9. *Suppose $t^{-1}N(t) \to \lambda$, $0 \leq \lambda \leq \infty$, and $t^{-1}x(t) \to 0$, as $t \to \infty$. Then*

(*a*) *if $n^{-1}\sum_{k=1}^{n} J_k \to J$ as $n \to \infty$, then $t^{-1}\widetilde{x}(t) \to \widetilde{x} = \lambda J$ as $t \to \infty$, provided that λJ is well defined;*

(*b*) *if $t^{-1}\widetilde{x}(t) \to \widetilde{x}$ as $t \to \infty$, then $n^{-1}\sum_{k=1}^{n} J_k \to J = \lambda^{-1}\widetilde{x}$ as $n \to \infty$, provided that $\lambda^{-1}\widetilde{x}$ is well defined.*

Proof. Let $Y(t){:} = \widetilde{x}(t) + x(0) - x(t) = \sum_{k=1}^{N(t)} J_k$, $t \geq 0$ and apply Lemma 5.3. □

Remark 5.10. Theorem 5.9 contains Theorem 2.1 (as well as Remark 1) of Sigman [71], in which it is assumed that the averages involved are finite, that $\{x(t), t \geq 0\}$ satisfies certain differentiability conditions, and that $\{N(t), t \geq 0\}$ contains the discontinuities of $\{x(t), t \geq 0\}$. If the last condition holds, then $\widetilde{x}(t)$ is the continuous component of $x(t)$. If the differentiability conditions of [71] also hold, then $\widetilde{x}(t)$ is absolutely continuous, that is, $\widetilde{x}(t) = \int_0^t x'(s)ds$, where $x'(t)$ is the right-hand derivative of $x(t)$.

Remark 5.11. It is not difficult to show that $Y = \lambda X$ follows from the RCL, so that the two laws are, in some sense, equivalent. Sigman [71] showed a similar equivalence between RCL and $H = \lambda G$. (See also Whitt [87] and Brémaud [7].) One should not make too much of such equivalences, however, when (as is often the case) the proof that one law follows from another is as difficult as a proof from scratch.

Remark 5.12. In Section 5.10 we shall show that the condition $t^{-1}x(t) \to 0$ is satisfied when the process $\{x(t), t \geq 0\}$ has a proper limiting frequency distribution, provided that $\{x(t), t \geq 0\}$ has bounded variation on finite t-intervals. Although we did not need bounded variation in our proof of the RCL, it is a condition satisfied in most applications where we expect the RCL to hold. Moreover, as Whitt [85] (p. 256) observed, a natural sufficient condition for the differentiability assumptions of Sigman [71] to hold is for $\{x(t), t \geq 0\}$ to have bounded variation on finite t-intervals, with the continuous component being absolutely continuous. The condition that $\{x(t), t \geq 0\}$ have a proper limiting frequency distribution is also a natural one in applications. (Relations between time-average and point-average frequencies are the subject of the next section.)

Remark 5.13. Theorem 5.9 can also be applied to the indicator process $x(t)$: $= 1\{Z(t) \in B\}$ for a process $\{Z(t), t \geq 0\}$ defined on a general state space (see Section 5.4 below). In this case, the condition $t^{-1}x(t) \rightarrow 0$ is trivially satisfied, and we can use the *RCL* to derive rate-balance equations.

5.4 RELATIONS BETWEEN FREQUENCIES FOR A PROCESS WITH AN IMBEDDED POINT PROCESS

In this section, we use $Y = \lambda X$ and the *RCL* to derive various relations between time-average and point-average frequency distributions for processes with imbedded point processes. The material in this section is taken largely from El-Taha and Stidham [26].

Let $Z = \{Z(t), t \geq 0\}$ be a deterministic continuous-time process with state space S. We assume that S is a Polish (complete separable metric) space, with the Borel σ-field $\mathcal{B}(S)$, and that Z is right continuous with left-hand limits. Let $\{T_n, n \geq 1\}$ be an associated deterministic point process with counting process $N = \{N(t), t \geq 0\}$ satisfying the conditions of Section 5.2. (Recall that these conditions do not include a requirement that N be simple.)

The process N may (but does not necessarily) count transition instants of process Z, for example, level crossing or points where upward or downward jumps (e.g., arrivals or departures in a queueing system) occur. In general, transition instants can be points of continuity as well as points of discontinuity of process Z.

For an arbitrary set of states $B \in \mathcal{B}(S)$, define the following limits, when they exist:

$$\lambda: = \lim_{t\rightarrow\infty} \frac{N(t)}{t} = \lim_{n\rightarrow\infty} \frac{n}{t_n}, \tag{5.4}$$

$$p(B): = \lim_{t\rightarrow\infty} t^{-1} \int_0^t 1\{Z(s) \in B\} ds, \tag{5.5}$$

$$X(B): = \lim_{n\rightarrow\infty} n^{-1} \sum_{k=1}^{n} \int_{T_{k-1}}^{T_k} 1\{Z(s) \in B\} ds, \tag{5.6}$$

$$q(B): = \lim_{t\rightarrow\infty} \frac{\int_0^t 1\{Z(s-) \in B\} dN(s)}{\int_0^t 1\{Z(s) \in B\} ds}, \tag{5.7}$$

$$\pi^-(B): = \lim_{n\rightarrow\infty} n^{-1} \sum_{k=1}^{n} 1\{Z(T_k -) \in B\}. \tag{5.8}$$

That is, λ is (as usual) the *rate* of the point process N, $\{p(B), B \in \mathcal{B}(S)\}$ is the *time-average frequency distribution* of the process Z, $q(B)$ is the *conditional rate* of N while $Z(t-)$ is in B, and $\{\pi^-(B), B \in \mathcal{B}(S)\}$ is the *point-average frequency distribution* of $\{Z(t-), t \geq 0\}$. In the "classical" applications to queueing systems, N is the arrival process and $\pi^-(B)$ is the frequency with which arrivals "see" the system in the set of states B. (Recall that $Z(t)$ is right continuous with left limits, so that $Z(T_k -)$ is the state of the system just before the kth "arrival".)

In what follows, we shall make frequent use of the identity

$$\int_0^t 1\{Z(s-)\in B\}dN(s) = \sum_{k=1}^{N(t)} 1\{Z(T_k-)\in B\}. \tag{5.9}$$

The following two theorems are direct applications of $Y = \lambda X$.

Theorem 5.14. *Assume that λ exists, $0 \le \lambda < \infty$. Then $p(B)$ exists if and only if $X(B)$ exists and*

$$p(B) = \lambda X(B), \quad B \in \mathfrak{B}(S). \tag{5.10}$$

Proof. Let $Y(t) = \int_0^t 1\{Z(s)\in B\}ds, t \ge 0, X_k = \int_{T_{k-1}}^{T_k} 1\{Z(s)\in B\}ds,\ k \ge 1.$ That is, $Y(t)$ is the amount of time spent by Z in the set B during $[0,t]$, and X_k is the amount of time spent in B between the $(k-1)^{st}$ and kth points of N. Apply Lemma 5.2. □

The next theorem is a sample-path analogue of the *covariance formula*; see Melamed and Whitt [54]. It gives a relation between the time-average frequency of the set B and the frequency of B "as seen by an arrival".

Theorem 5.15. *Let $B \in \mathfrak{B}(S)$, and assume that the limits λ and $q(B)$ exist, $0 \le \lambda < \infty$, $0 \le q(B) < \infty$. Then $p(B)$ exists if, and only if, $\pi^-(B)$ exists, in which case*

$$p(B)q(B) = \lambda\pi^-(B). \tag{5.11}$$

Proof. Let $Y(t) = \int_0^t 1\{Z(s-)\in B\}dN(s)$, so that $Y(t)$ is the number of points that occur in $[0,t]$ while $Z(s-)\in B$, $t \ge 0$. Let $X_k = 1\{Z(T_k-)\in B\}$, $k \ge 1$ and apply Lemma 5.3, using (5.9) and the fact that $lim_{t\to\infty} t^{-1}Y(t) = p(B)q(B)$, when the limit exists. □

Remark 5.16. Theorems 5.14. and 5.15 play roles similar to those of the Palm inversion formula and the Palm transformation formula in the stochastic theory of marked point processes [30], [3] and stationary processes related to point processes [62]. Stidham [79] has shown how the approach based on $Y = \lambda X$ can be extended to develop a complete sample-path version of the theory of Palm probabilities for processes with imbedded point processes (see Section 5.6).

5.4.1 *ASTA* and related properties

Theorem 5.15 yields necessary and sufficient conditions for the *ASTA* (*A*rrivals *S*ee *T*ime *A*verages) property, as shown in the following theorem. (To simplify the presentation of our results throughout this subsection we shall assume that all relevant limits are well defined.)

Theorem 5.17. (Conditions for *ASTA*) *For all $B \in \mathfrak{B}(S)$,*

$$p(B) = \pi^-(B) \Leftrightarrow q(B) = \lambda.$$

Moreover, the following are equivalent:

(*a*) *$q(B)$ is independent of B, $B \in \mathfrak{B}(S)$;*

(*b*) *$q(B) = \lambda$, for all $B \in \mathfrak{B}(S)$;*

(*c*) *the frequency distributions $\{p(B), B \in \mathfrak{B}(S)\}$ and $\{\pi^-(B), B \in \mathfrak{B}(S)\}$ coincide.*

Proof. Immediate from Theorem 5.15. □

Remark 5.18. This theorem can be used as the starting point for a proof of

PASTA (*P*oisson *A*rrivals *S*ee *T*ime *A*verages). Most proofs of *PASTA* in the literature (see, e.g., Wolff [88]) contain a step in which it is shown that Poisson arrivals together with a lack-of-anticipation assumption (roughly speaking, the future of N odes do not depend on the past or present of Z) imply that $q(B) = \lambda$, for all $B \in \mathfrak{B}(S)$, or the probabilistic equivalent of this statement. El-Taha and Stidham [25] use martingale theory to prove a stronger result under the assumptions of [88]: that the Poisson point process N, restricted to times t when $Z(t) \in B$, is a probabilistic replica of the original point process, that is, it is Poisson with the same rate λ (see Section 5.7).

Reversed processes and *DSTA*. Now we give sample-path versions of the covariance formula and *ASTA*-type results for reversed processes, again using the same framework as for *ASTA*. We begin by defining the following limits, when they exist, for $B \in \mathfrak{B}(S)$:

$$q'(B) := \lim_{t\to\infty} \frac{\int_0^t 1\{Z(s) \in B\} dN(s)}{\int_0^t 1\{Z(s) \in B\} ds}, \tag{5.12}$$

and

$$\pi^+(B) := \lim_{n\to\infty} n^{-1} \sum_{k=1}^{n} 1\{Z(T_k) \in B\}. \tag{5.13}$$

That is, $q'(B)$ is the *conditional rate* of N while the reversed version of process Z is in B, and $\{\pi^+(B), B \in \mathfrak{B}(S)\}$ is the *point-average frequency distribution* of $Z = \{Z(t+), t \geq 0\}$. (Recall that Z is right continuous.) Informally, we refer to $\{\pi^+(B), B \in \mathfrak{B}(S)\}$ as the frequency distribution for Z "just after" a point from N has occurred. Equivalently, it is the frequency distribution of the reversed version of Z "just before" a point from N occurs. In applications to queueing systems, N is often the process that counts departures and hence $\{\pi^+(B), B \in \mathfrak{B}(S)\}$ is the frequency distribution of the state of the system "left behind" by a departure.

By a proof exactly parallel to that used to prove Theorem 5.15, one can derive the following relation between $p(B)$ and $\pi^+(B)$, assuming the relevant limits are well defined;

$$p(B)q'(B) = \lambda\pi^+(B). \tag{5.14}$$

Equation (5.14) immediately gives necessary and sufficient conditions for *DSTA* (*D*epartures *S*ee *T*ime *A*verages).

Theorem 5.19. (Conditions for *DSTA*) *For all* $B \in \mathfrak{B}(S)$,

$$p(B) = \pi^+(B) \Leftrightarrow q'(B) = \lambda.$$

Moreover, the following are equivalent:

(*a*) $q'(B)$ *is independent of* B, $B \in \mathfrak{B}(S)$;

(*b*) $q'(B) = \lambda$, *for all* $B \in \mathfrak{B}(S)$;

(*c*) *the frequency distributions* $\{p(B), B \in \mathfrak{B}(S)\}$ *and* $\{\pi^+(B), B \in \mathfrak{B}(S)\}$ *coincide.*

Theorems 5.17 and 5.19 complement results given in the chapter on the *ASTA* property by Melamed and Yao [56] (Chapter 7 in this book).

In the following subsection, we shall give applications to this and the previous results in this subsection, the case where N counts transitions into a certain subset of states. In particular, we shall show that q and q' satisfy a sample-path version of the well-known relation between conditional intensities in a Markov process and its reversed version.

5.4.2 Inverse-rate formula and transition-rate-balance equations

Another application of Theorems 5.14 and 5.15 yields an *inverse-rate formula*, which can be used to derive several relations between different measures in queueing systems.

Theorem 5.20. (Inverse-rate formula) *Let $B \in \mathfrak{B}(S)$ and assume that the limits λ and $q(B)$ exist, $0 < \lambda < \infty$, $0 < q(B) < \infty$.*

(*a*) *Then $X(B)$ exists if and only if $\pi^-(B)$ exists, in which case*

$$q(B)X(B) = \pi^-(B).$$

(*b*) *Suppose the points of N can only occur while $Z(t) \in B$. Then $X(B)$ exists and*

$$q(B) = X(B)^{-1}.$$

Proof. Theorem 5.14 and Theorem 5.15 together imply that $q(B)X(B) = \pi^-(B)$. Under the assumption that the points of N can only occur while $Z(t) \in B$, we have $\pi^-(B) = 1$, from which part (*b*) follows. □

Remark 5.21. Part (*b*) of Theorem 5.20 can also be proved by a change-of-time argument, that is, restricting Z and N to *B-time* and then applying Lemma 5.1 (cf. Remark 5.18. above and El-Taha and Stidham [25]). It then can be used together with Theorem 5.14. to obtain an alternative proof of Theorem 5.15. A similar approach is taken in Stidham and El-Taha [80] (see also [40], [86]). We believe, however, that the proof of Theorem 5.15 given here is more direct and intuitive.

Remark 5.22. Stidham and El-Taha [80] use part (*b*) of Theorem 5.20 in a sample-path proof of the insensitivity of the frequency distributions in an *LCFS-PR* queue (see Section 5.5).

Until now, we have made no assumptions about the relationship between the definitions of the process Z and the point process N. At one extreme, the points of N could have nothing to do with the evolution of the process Z. At the other extreme, they could be completely defined by that evolution, as is the case, for example, where N counts the transitions of Z between certain sets of states.

We shall now consider this case in further detail. It will be convenient to have a notation that indicates the sets of states involved. Specifically, let $C \in \mathfrak{B}(S)$ and $D \in \mathfrak{B}(S)$ be given, and suppose that T_k $(k \geq 1)$ is the kth time instant at which Z makes a transition from C into D. To avoid pathologies and to make it possible to apply our theory, assume that Z makes transitions from C to D infinitely often in $[0, \infty)$, but, at, most finitely often in every finite t-interval. Then $T_k < \infty$ for all $k \geq 1$ and $T_k \to \infty$ as $t \to \infty$. Moreover, by definition, the point process $\{T_k, k \geq 1\}$ is simple. Let $N = \{N(t), t \geq 0\}$ be the counting process associated with $\{T_k, k \geq 1\}$. Suppose $\lambda(A, A^c)$ (the *exit rate* for A) exists, $0 \leq \lambda(A, A^c) \leq \infty$. Applying Theorem 5.9 and noting that $J_k = 1\{Z(T_k -) \in A\} - 1\{Z(T_k +) \in A\} = 1 - 0 = 1$, we conclude that

$$t^{-1}\tilde{x}(t) \to \lambda(A, A^c), \text{ as } t \to \infty.$$

On the other hand, it follows from the definitions of $x(t)$ and N that $\tilde{x}(t)$ equals the number of transitions of Z into the set A in $[0, t]$. Thus $lim_{t \to \infty} t^{-1}\tilde{x}(t) = \lambda(A^c, A)$, the *entrance rate* for A. Reversing the roles of A and A^c, one has the following theorem (cf. Remark 2 of [71]).

Theorem 5.23. (Rate-balance equation) *For all* $A \in \mathfrak{B}(S)$, $\lambda(A, A^c)$ *is well defined if, and only if,* $\lambda(A^c, A)$ *is well defined, in which case*

$$\lambda(A, A^c) = \lambda(A^c, A).$$

The second application is a sample-path derivation of the relation between the conditional transition rates for Z and for its reversed version. For this application, again, let $A \in \mathfrak{B}(S)$ be given, and let $C = A$ and $D = A^c$, so that N counts the number of transitions of Z out of the set A. Again, we shall write $\lambda(A, A^c)$ for λ, the rate of N, when the latter is well defined by (5.4). For a given set $B \in \mathfrak{B}(S)$, consider the conditional rate $q'(B)$ of N, defined by (5.12). In this case, $q'(B)$ is the conditional transition rate from B to A for the reversed version of Z, and we shall indicate this fact by writing $q'(B, A)$ instead of $q'(B)$. It then follows from equation (5.14) that

$$p(B)q'(B, A) = \lambda(A, A^c)\pi^+(B) = \lambda(A, B), \tag{5.15}$$

assuming the limits are well defined.

Now let $C = B^c$ and $D = B$, so that N counts the number of transitions of Z into the set B. We shall write $\lambda(B^c, B)$ for λ, the rate of N, when the latter is well-defined by (5.4). Consider the conditional rate $q(A)$ of N, defined by (5.12). In this case, $q(A)$ is the conditional transition rate from A to B for the process Z, and we shall indicate this fact by writing $q(A, B)$ instead of $q(A)$. It then follows from Theorem 5.15 that

$$p(A)q(A, B) = \lambda(B^c, B)\pi^-(A) = \lambda(A, B), \tag{5.16}$$

assuming the limits are well-defined. Combining equations (5.15) and (5.16), we obtain

$$p(A)q(A, B) = p(B)q'(B, A), \tag{5.17}$$

which is the sample-path analogue of the relation between the conditional transition rates in a continuous-time Markov chain and its reversed version.

5.4.3 Forward and backward recurrence times

Our next result is a sample-path generalization of a result by Ghahramani [33]. (See also Wolff[89], p. 291-292.) Fix $B \in \mathfrak{B}(S)$, and suppose that T_k $(k \geq 1)$ is the kth time instant at which Z makes a transition out of B. As usual, we assume that Z makes such transitions infinitely often in $[0, \infty)$, but, at most, finitely often in every finite t-interval, so that $T_k < \infty$ for all $k \geq 1$ and $T_k \rightarrow \infty$ as $t \rightarrow \infty$. Let $N = \{N(t), t \geq 0\}$ be the counting process associated with $\{T_k, k \geq 1\}$. Let $\lambda(B, B^c) = lim_{n \rightarrow \infty} t^{-1}N(t)$ when the limit is well-defined. Define

$$X_k(B) := \int_{T_{k-1}}^{T_k} 1\{Z(s) \in B\} ds,$$

$$R(t) := T_{N(t)+1} - t.$$

Thus, $X_k(B)$ is the time spent by Z in the set B between the $k - 1^{st}$ and kth transitions out of set B, and $R(t)$ is the forward recurrence time at t: the time until the next transition out of B after t. Also define

$$X_k(B,x) := \int_{T_{k-1}}^{T_k} 1\{Z(s) \in B, R(s) \le x\}ds, x \in [0,\infty).$$

Now, define the following limits when they exist:

$$X(B) := \lim_{n \to \infty} n^{-1} \sum_{k=1}^{n} X_k(B),$$

$$X(B,x) := \lim_{n \to \infty} n^{-1} \sum_{k=1}^{n} X_k(B,x),$$

and

$$G(x) := \lim_{n \to \infty} n^{-1} \sum_{k=1}^{n} 1\{X_k(B) \le x\},$$

$$G_e(x) := \lim_{t \to \infty} \frac{\int_0^t 1\{Z(s) \in B, R(s) \le x\}ds}{\int_0^t 1\{Z(s) \in B\}ds}.$$

Here, $G(x)$ is interpreted as the asymptotic frequency distribution of the time spent in set B and $G_e(x)$ as the asymptotic frequency distribution of the forward recurrence time, conditioned on Z being in the set B. The next theorem (se El-Taha and Stidham [23, 26]) shows that $G_e(x)$ plays the role of the equilibrium distribution associated with $G(x)$.

Theorem 5.24. *Assume that* $\lambda(B,B^c)$ *exists, and* $0 < \lambda(B,B^c) < \infty$.

(*a*) *If the limits* $X(B)$ *and* $X(B,x)$ *exit, with* $X(B) > 0$, *then* $G_e(x)$ *exists and*

$$G_e(x) = X(B,x)/X(B), x \in [0,\infty). \tag{5.18}$$

(*b*) *If the limit* $G(x)$ *exists for all* $x \in [0,\infty)$ *and the distribution functions,*

$$G_n(x) := n^{-1} \sum_{k=1}^{n} 1\{X_k(B) \le x\},$$

are uniformly integrable, then $G_e(x)$ *exists and*

$$G_e(x) = \frac{\int_0^x (1 - G(y))dy}{\int_0^\infty (1 - G(y))dy}, \quad x \in [0,\infty). \tag{5.19}$$

Remark 5.25. Other forms of the uniform integrability condition are given by El-Taha [19]. Also, Zazanis [91] gives sufficient conditions under which the limiting frequency distribution $G(x)$ exists.

Remark 5.26. If $R(t)$ is replaced by $t - T_{N(t)}$, the backward recurrence time, the theorem remains valid.

Now, we consider how the above theorem applies in a G/G/c queue. A full busy period is defined as the length of time spent in set $A = \{c, c+1, \ldots\}$ between transitions into (out of) set A. Ghahramani [33] proved Theorem 5.24 for a G/G/c queue with c homogeneous servers. Our approach shows that the same result holds in the case of heterogeneous servers as well. Also, Ghaharamani [33] proved Theorem 5.24 in a stationary framework. Our sample-path version reveals that the stationary framework is needed only to guarantee that the limits exist; moreover, our version is valid for noncountable state spaces.

5.5 SYSTEMS WITH COUNTABLE STATE SPACE

In this section, we specialize to discrete-state systems, that is, systems in which S is countable. Let $A \subset S$ be a countable subset of the state space, and use (5.16) with $A = \{i\}$, $B = \{j\}$ and the definitions of $p(i)$ and $q(i,j)$, we obtain $\lambda(A, A^c) = \sum_{i \in A} \sum_{j \in A^c} p(i)q(i,j)$. Using the rate-conservation principle (Theorem 5.23), we obtain the *global-balance conditions*:

$$\sum_{i \in A} \sum_{j \in A^c} p(i)q(i,j) = \sum_{i \in A^c} \sum_{j \in A} p(i)q(i,j), \quad A \in \mathfrak{B}(S). \tag{5.20}$$

For one-dimensional input-output systems, in which $S = \{0,1,\ldots\}$ and $Z(t) = j$ means there are j units in the system, apply the global-balance conditions to the case $A = \{0,1,\ldots,j-1\}$ yields the familiar *equality of upcrossings and downcrossings*:

$$\sum_{i=0}^{j-1} p(i) \sum_{k=j}^{\infty} q(i,k) = \sum_{k=j}^{\infty} p(k) \sum_{i=0}^{j-1} q(k,i), j = 1,2,\ldots. \tag{5.21}$$

For the special case of *left-skip-free transitions*, in which $q(i,j) = 0$, for $j < i-1$, equation (5.21) becomes

$$\sum_{i=0}^{j-1} p(i) \sum_{k=j}^{\infty} q(i,k) = p(j)q(j,j-1), j = 1,2,\ldots. \tag{5.22}$$

Consider a G/G/1 queue. Let $Z(t)$ equal the number of jobs in the system at time t. Let $N(t)$ equal the number of service completions that occur while $Z(s) = i$ (equivalently, the number of transitions from i to $i-1$) in $[0,t]$, and let $A = \{i\}$, for a given i, $i \geq 1$. In this context, we shall find it convenient to write $X(i;i,i-1)$ instead of $X(A)$ and $q(i,i-1)$ instead of $q(A,B)$, noting that $X(i;i,i-1)$ is the average time spent in state i between two successive transitions from i to $i-1$, and $q(i,i-1)$ is the conditional service rate while the system is in state i, $i \geq 1$. Then $q(i,i-1) = [X(i;i,i-1)]^{-1}$. We shall need this result subsequently.

As a particular example of a left-skip-free system, consider an M/G/1 queue with *LCFS-PR* (Last-Come, First-Served, Preemptive Resume) discipline. Let μ denote the reciprocal of the mean service requirement of a job and assume the server works at unit rate, regardless of the state of the system. Under the *LCFS-PR* discipline, a job that arrives when $j-1$ jobs are present begins service immediately and is served at all time instants when there are j jobs present until its service requirement is satisfied. Thus $X(j;j,j-1) = \mu^{-1}$, for all $j, j \geq 1$. But it follows from the inverse-rate formula (see Theorem 5.20 and the example above) that $X(j;j,j-1) = [q(j,j-1)]^{-1}$. Hence, using equation (5.22), *PASTA*, and Theorem 5.17, we have

$$\lambda p(j-1) = \mu p(j), j \geq 1, \tag{5.23}$$

so that $p(j) = (1-\rho)\rho^j$, $j \geq 0$. That is, the state frequency distribution is *insensitive* to the distribution of service requirement in an M/G/1 queue with *LCFS-PR* discipline. Sample-path analysis can be used to show that insensitivity holds in much more general queueing systems with *LCFS-PR* discipline. See Stidham and El-Taha [80] for examples.

5.6 SAMPLE-PATH VERSION OF THE PALM CALCULUS

In Section 5.4, we saw how sample-path analysis can be used to derive relations between time-average and event-average state frequencies for a process with an imbedded point process. These relations, however, were one-dimensional. That is, they were confined to marginal state frequencies, the limiting average fraction of time (or fraction of events) for which the process in question is in a certain state or subset of states.

Stidham [79] showed that the same techniques can also be used to relate the multidimensional frequencies of a process with an imbedded point process. In particular, sample-path analogues of the time-stationary and event-stationary (Palm) probability measures were defined for the processes in question, from which sample-path versions of the Palm transformation and inversion formulas were derived. We now show how this approach can be extended to obtain more general sample-path analogues of results from the Palm calculus. (The material in this section is more advanced than that in the rest of this chapter and can be skipped without loss of continuity).

As in Stidham [79], the framework is that of Baccelli and Brémaud [3], Part I, with which we assume the reader is familiar. Specifically (cf. [3], p. 3) we are given a measurable space $(\Omega, \mathcal{F})$ together with a measurable flow $\{\theta_t\}$, $t \in R$, on $(\Omega, \mathcal{F})$. A locally finite point process N is defined on $(\Omega, \mathcal{F})$ consistent with $\{\theta_t\}$, that is, such that for all $C \in \mathcal{B}(R)$, $\omega \in \Omega$, $t \in R$,

$$N(\theta_t\omega, C) = N(\omega, C + t).$$

The points of N are denoted by $T_n(\omega)$, $n \in Z$, $\omega \in \Omega$, where

$$-\infty \leq \ldots \leq T_{-1}(\omega) \leq T_0(\omega) \leq 0 < T_1(\omega) \leq T_2(\omega) \leq \ldots \leq +\infty.$$

Note that we have not specified a probability measure on $(\Omega, \mathcal{F})$, nor shall we, since our interest is in limiting frequencies and limiting averages associated with a fixed sample point $\omega \in \Omega$.

Let $\omega \in \Omega$ be given, and define the following limiting frequencies, for $A \in \mathcal{F}$, when they are well defined:

$$\mathbf{P}(A;\omega) := \lim_{t\to\infty} t^{-1} \int_0^t (1_A \circ \theta_s)\omega ds, \tag{5.24}$$

$$\mathbf{P}_N(A;\omega) := \lim_{n\to\infty} n^{-1} \sum_{k=1}^{n} (1_A \circ \theta_{T_k(\omega)})\omega. \tag{5.25}$$

As a limiting time average, the measure $\mathbf{P}(\cdot\,;\omega)$ plays a role analogous to that of the stationary probability measure $\mathbf{P}$ in the theory of stationary random-marked point processes (*RMPP*) (Franken et al. [30], Baccelli and Brémaud [3]). As a limiting event average, $\mathbf{P}_N(\cdot\,;\omega)$ corresponds to the Palm probability measure $\mathbf{P}_N$ associated with $\mathbf{P}$. We shall also need sample-path analogues of the expectation operators $\mathbf{E}$ and $\mathbf{E}_N$ associated with $\mathbf{P}$ and $\mathbf{P}_N$, respectively. Like $\mathbf{P}(\cdot\,;\omega)$ and $\mathbf{P}_N(\cdot\,;\omega)$, these operators will be defined as limiting averages associated with the fixed sample point ω. For a nonnegative, measurable function $f:\Omega \rightarrow R$, define

$$\mathbf{E}[f;\omega]: = \lim_{t \rightarrow \infty} t^{-1} \int_0^t (f \circ \theta_s)\omega ds, \tag{5.26}$$

and

$$\mathbf{E}_N[f;\omega]: = \lim_{n \rightarrow \infty} n^{-1} \sum_{k=1}^{n} (f \circ \theta_{T_k(\omega)})\omega, \tag{5.27}$$

when the limits are well defined. With these definitions, we have $\mathbf{P}(A;\omega) = \mathbf{E}[1_A;\omega]$ and $\mathbf{P}_N(A;\omega) = \mathbf{E}_N[1_A;\omega]$.

Also define the following limiting time averages, when they are well defined.

$$\lambda(\omega): = \lim_{t \rightarrow \infty} t^{-1} N(\omega,t) \tag{5.28}$$

$$\lambda(f;\omega): = \lim_{t \rightarrow \infty} t^{-1} \int_0^t (f \circ \theta_s)\omega N(\omega, ds) \tag{5.29}$$

Here, we have abused notation by writing $N(\omega,t)$ for $N(\omega,[0,t])$. In what follows we shall make frequent use of the identity

$$\int_{0+}^{t} (f \circ \theta_s)\omega N(\omega, ds) = \sum_{k=1}^{N(\omega,t)} (f \circ \theta_{T_k(\omega)})\omega. \tag{5.30}$$

Lemma 5.27. *Let $f:\Omega \rightarrow R$ be a nonnegative, measurable function. Suppose $\lambda(\omega)$ is well defined by (5.28), with $0 < \lambda(\omega) < \infty$. Then $\mathbf{E}_N[f;\omega]$ is well-defined by (5.27) if, and only if, $\lambda(f;\omega)$ is well-defined by (5.29), in which case*

$$\mathbf{E}_N[f;\omega] = \lambda(f;\omega)/\lambda(\omega). \tag{5.31}$$

Proof. Let $N(t) = N(\omega,t)$, $t \geq 0$, $X_k = (f \circ \theta_{T_k(\omega)})\omega, k \geq 1, Y(t) = \sum_{k=1}^{N(t)} X_k$, $t \geq 0$, and apply Lemma 5.3, using (5.30). □

Remark 5.28. With $f = 1_A$ for $A \in \mathfrak{F}$, (5.31) reduces to equation (5.10) of Stidham [79]), viz.,

$$\mathbf{P}_N(A;\omega) = \lambda(A;\omega)/\lambda(\omega), \tag{5.32}$$

which is a sample-path version of the Palm transformation. In the stochastic setting of an ergodic stationary marked-point process (cf. Neveu [60], Krickeberg [45], Franken et al. [30], Baccelli and Brémaud [3]), the latter relation can be shown to hold a.s. by virtue of the definition of Palm probability (equation (3.1.1) on p. 9 of [3]) and the ergodic theorem. (See equation (3.1.6) on p. 9 of [3].) The relation (5.31) for an arbitrary nonnegative measurable function f then follows by a standard monotone-class argument, in which f is approximated by simple functions. Lemma 5.27 reveals that neither stationarity nor ergodicity is needed for (5.31) to hold, only the existence of the relevant limiting averages on the sample path in question. We could also have derived (5.31) by a monotone-class argument from (10) in [79], but in the present setting, in which $\mathbf{E}_N[f;\omega]$ is defined as a limiting average rather than as an expectation with respect to the probability measure $\mathbf{P}_N(\cdot;\omega)$, a direct argument such as we have given in Lemma 5.27 is simpler.

The next theorem gives a sample-path version of the Palm inversion formula ([3], pp. 13-14).

Theorem 5.29. *Let $A \in \mathcal{F}$. Suppose $\lambda(\omega)$ is well-defined by (5.28), with $0 < \lambda(\omega) < \infty$. Then $\mathbf{P}(A;\omega)$ is well-defined by (5.24) if, and only if, $\mathbf{E}_N[f;\omega]$ is well defined by (5.27) with $f(\omega) = \int_0^{T_1(\omega)} (1_A \circ \theta_s)\omega ds$, in which case*

$$\mathbf{P}(A;\omega) = \lambda(\omega)\mathbf{E}_N[\int_0^{T_1} (1_A \circ \theta_s)ds;\omega]. \tag{5.33}$$

Proof. Let $Y(t) = \int_0^t (1_A \circ \theta_s)\omega ds$, $N(t) = N(t,\omega)$, $t \geq 0$. Then Lemma 5.2 implies that

$$\mathbf{P}(A;\omega) = \lambda(\omega)[\lim_{n\to\infty} n^{-1}\sum_{k=1}^{n} \int_{T_{k(\omega)}}^{T_{k+1}(\omega)} (1_A \circ \theta_s)\omega ds] \tag{5.34}$$

subject to the limits being well-defined. But

$$\int_{T_k(\omega)}^{T_{k+1}(\omega)} (1_A \circ \theta_s)\omega ds = \int_0^{T_{k+1}(\omega)-T_k(\omega)} (1_A \circ \theta_{s+T_k(\omega)})\omega ds$$

$$= (f \circ \theta_{T_k(\omega)})\omega,$$

where $f(\omega) = \int_0^{T_1(\omega)} (1_A \circ \theta_s)\omega ds$, and we have used the facts that $\theta_s \circ \theta_t = \theta_{s+t}$ and $T_1(\theta_{T_k(\omega)}\omega) = T_{k+1}(\omega) - T_k(\omega)$. The desired result now follows from (5.27) and (5.34). □

Now, let N' be another locally finite point process defined on $(\Omega, \mathcal{F})$ and consistent with $\{\theta_t\}$. The points of N' are denoted by $T'_n(\omega)$, $n \in Z$, $\omega \in \Omega$. Let

$$\lambda'(\omega) := \lim_{t\to\infty} t^{-1}N'(\omega,t), \tag{5.35}$$

provided the limit is well-defined.

Lemma 5.30. *Let $f:\Omega \to R$ be a nonnegative, measurable function. Suppose $\lambda'(\omega)$ is well-defined by (5.35), with $0 < \lambda'(\omega) < \infty$. Then $\lambda(f;\omega)$ is well-defined by (5.29) if, and only if, $\mathbf{E}_{N'}[g;\omega]$ is well-defined by (5.27) with $N = N'$ and $g(\omega) = \int_0^{T'_1(\omega)} (f \circ \theta_s)\omega\ N\ (\omega, ds)$, in which case*

$$\lambda(f;\omega) = \lambda'(\omega)\mathbf{E}_{N'}[\int_0^{T'_1} (f \circ \theta_s)N(ds);\omega]. \tag{5.36}$$

Proof. Let $Y(t) = \int_0^t (f \circ \theta_s)\omega N(\omega, ds)$ and $N(t) = N'(t,\omega)$, $t \geq 0$. Then Lemma 5.2 implies that

$$\lambda(f;\omega) = \lambda'(\omega)[\lim_{n\to\infty} n^{-1}\sum_{k=1}^{n} \int_{T'_k(\omega)}^{T'_{k+1}(\omega)} (f \circ \theta_s)\omega N(\omega, ds)], \tag{5.37}$$

subject to the limits being well-defined. But

$$\int_{T'_k(\omega)}^{T'_{k+1}(\omega)} (f \circ \theta_s)\omega N(\omega, ds) = \int_0^{T'_{k+1}(\omega) - T'_k(\omega)} (f \circ \theta_{s + T'_k(\omega)})\omega N(\omega, ds)$$
$$= (g \circ \theta_{T'_k(\omega)})\omega,$$

where $g(\omega) = \int_0^{T'_1(\omega)} (f \circ \theta_s)\omega N(\omega, ds)$, and we have used the facts that $\theta_s \circ \theta_t = \theta_{s+t}$ and $T'_1(\theta_{T'_k(\omega)}\omega) = T'_{k+1}(\omega) - T'_k(\omega)$. The desired result now follows from (5.27) and (5.37). □

Combining Lemmas 5.27 and 5.30 yields the following sample-path version of Neveu's exchange formula (cf. [3]):

Theorem 5.31. *Let $f: \Omega \to R$ be a nonnegative, measurable function. Suppose $\lambda(\omega)$ is well-defined by* (5.28), *with $0 < \lambda(\omega) < \infty$, and $\lambda'(\omega)$ is well-defined by* (5.35), *with $0 < \lambda'(\omega) < \infty$. Then $\mathbf{E}_N[f;\omega]$ is well-defined by* (5.27) *if, and only if, $\mathbf{E}_{N'}[g;\omega]$ is well-defined by* (5.27) *with $N = N'$ and $g(\omega) = \int_0^{T'_1(\omega)} (f \circ \theta_s)\omega N(\omega, ds)$, in which case*

$$\lambda(\omega)\mathbf{E}_N(f;\omega) = \lambda'(\omega)\mathbf{E}_{N'}[\int_0^{T'_1} (f \circ \theta_s)N(ds);\omega]. \tag{5.38}$$

5.6.1 Application to processes with imbedded point processes

Now, we show how to apply these results to a process with an imbedded point process. This is the context in which the theory of Palm probabilities most often finds application. Here, we shall confine ourselves to the Palm transformation formula and the Palm inversion formula.

Again, assume we are given a measurable space $(\Omega, \mathfrak{F})$ together with a measurable flow $\{\theta_t\}$, $t \in R$, on $(\Omega, \mathfrak{F})$. Let $Z \equiv \{Z(\omega, t), t \in R\}$ be a continuous-time process with state space S, defined on $(\Omega, \mathfrak{F})$. We assume that S is a complete separable metric space, endowed with the Borel σ-field $\mathfrak{B}(S)$. We also assume that Z has right-continuous sample paths with left-hand limits. Let N be a locally finite point process defined on the same measurable space $(\Omega, \mathfrak{F})$. We assume that both Z and N are consistent with $\{\theta_t\}$:

$$Z(\theta_t\omega, s) = Z(\omega, s + t), \quad s \in R;$$
$$N(\theta_t\omega, C) = N(\omega, C + t), \quad C \in \mathfrak{B}(R).$$

We are now in a position to apply these results to the process with an imbedded point process, (Z, N). First, we define limiting averages analogous to those defined in Section 5.6, provided they are well-defined. Let $D_S[0,\infty)$ denote the set of all functions $x: R \to S$ that are right-continuous with left-hand limits and $\mathfrak{D}$ the σ-field generated by the Skorohod topology on $D_S[0,\infty)$ (Ethier and Kurtz [28], p. 117). For $\omega \in \Omega$, $B \in \mathfrak{D}$, define

$$\mathbf{P}\{Z \in B;\omega\} := \lim_{t\to\infty} t^{-1} \int_0^t 1\{Z(\omega, s + \cdot) \in B\}ds, \tag{5.39}$$

$$\mathbf{P}_N\{Z \in B; \omega\}: = \lim_{n \to \infty} n^{-1} \sum_{k=1}^{n} 1\{Z(\omega, T_k(\omega) + \cdot\,) \in B\}, \tag{5.40}$$

$$\lambda_B(\omega): = \lim_{t \to \infty} t^{-1} \int_0^t 1\{Z(\omega, s + \cdot\,) \in B\} N(\omega, ds). \tag{5.41}$$

In words, for a particular $\omega \in \Omega$, $\mathbf{P}\{Z \in B; \omega\}$ is the proportion of time that the shifted processes $Z(\omega, s + \cdot\,)$ fall in the set of paths B; $\mathbf{P}_N\{Z \in B; \omega\}$ is the proportion of points, $T_k(\omega)$, for which the shifted processes $Z(\omega, T_k(\omega) + \cdot\,)$ fall in B; and $\lambda_B(\omega)$ is the rate at which points $T_k(\omega)$ occur for which the shifted processes $Z(\omega, T_k(\omega) + \cdot\,)$ fall in B.

The following corollaries are immediate from Lemma 5.27 and Theorem 5.29, upon setting $A = \{\omega: Z(\omega, \cdot\,) \in B\}$.

Corollary 5.32. *Let $B \in \mathfrak{D}$. Suppose $\lambda(\omega)$ is well-defined by* (5.28), *with $0 < \lambda(\omega) < \infty$. Then $\mathbf{P}_N\{Z \in B; \omega\}$ is well-defined by* (5.40) *if, and only if, $\lambda_B(\omega)$ is well-defined by* (5.41), *in which case*

$$\mathbf{P}_N\{Z \in B; \omega\} = \lambda_B(\omega)/\lambda(\omega). \tag{5.42}$$

Corollary 5.33. *Let $B \in \mathfrak{D}$. Suppose $\lambda(\omega)$ is well-defined by* (5.28), *with $0 < \lambda(\omega) < \infty$. Then $\mathbf{P}\{Z \in B; \omega\}$ is well-defined by* (5.39) *if, and only if, $\mathbf{E}_N[f; \omega]$ is well-defined by* (5.27) *with $f(\omega) = \int_0^{T_1(\omega)} 1\{Z(\omega, s + \cdot\,) \in B\} ds$, in which case*

$$\mathbf{P}\{Z \in B; \omega\} = \lambda(\omega) \mathbf{E}_N[\int_0^{T_1} 1\{Z(s + \cdot\,) \in B\} ds; \omega]. \tag{5.43}$$

5.7 FILTRATION OF *ASTA*

In this section, we address *ASTA*, as well as *PASTA*, in a stochastic framework by taking advantage of the necessary and sufficient conditions for sample-path *ASTA* proved in Section 5.4. Our approach gives an illustration of how sample-path analysis can help provide better understanding of system characteristics in a stochastic environment. (This section contains advanced material and may be skipped without loss of continuity.)

Wolff [88] proved that Poisson arrivals see time averages (*PASTA*) under a lack-of-anticipation assumption (*LAA*), which says that future arrivals occur independently of the present and past behavior of the system. In particular, for a queueing process with an associated arrival point process satisfying *LAA*, if the point process is Poisson, an arriving customer sees the same as a random observer (i.e., time averages are the same as customer averages). More generally, the setting is that of a stochastic process Z (the state of the system) with an imbedded point process N (a sequence of discrete events). In subsequent papers (see [18, 6, 41, 80, 44, 54, 42, 59]) the requirement that the point process be Poisson has been eliminated as a condition for events to see time averages, thus extending *PASTA* to *ASTA* (*A*rrivals *S*ee *T*ime *A*verages). In addition, the *LAA* assumption has been replaced by the weaker Lack-of-Bias assumption (*LBA*) see [54, 55]. Comprehensive reviews are given by Melamed and Whitt [54] and Brémaud et al. [8]. See also the chapter on the *ASTA* property by Melamed and Yao [56].

Rather than extend *PASTA* by weakening the conditions under which it holds, as in [56], we proceed in a different direction, showing, in particular, that the *ASTA* property can be strengthened under Wolff's original assumptions of Poisson arrivals and *LAA*. Specifically, we prove that, if *LAA* holds and the point process satisfies certain conditions (e.g., doubly stochastic Poisson), the B-process restricted to B-time not only has the same rate, but is, in fact, a probabilistic replica of the original point process. In particular, the B-process is Poisson with rate λ for every set B if (and only if) the original point process is Poisson with rate λ. Stidham [73] gave an informal proof of this result for the special case of regenerative processes. Closely related to this problem is conditional *PASTA*; Van Doorn and Regterschot [82] and the Anti-*PASTA* question: If *ASTA* holds, then under what extra conditions is the imbedded (arrival, departure) process of a general process (queueing process) Poisson? Melamed and Whitt [55] show that if *ASTA* holds, and the queueing process is Markovian then, under some regularity conditions, the arrival (departure) process must be Poisson. For other references, see König, Miyazawa and Schmidt [43], Green and Melamed [36], Wolff [90] and Miyazawa and Wolff [59].

Now we formalize the statement of the problem.

5.7.1 Definitions and notation

On a common probability space $(\Omega, \mathfrak{F}, P)$, let $Z \equiv \{Z(t), t \geq 0\}$ be a continuous-time stochastic process with state space S and let $N \equiv \{N(t), t \geq 0\}$ be an associated point process. We assume that S is a complete separable metric space, endowed with the Borel σ-field $\mathfrak{B}(S)$. We also assume that both Z and N have right-continuous sample paths with left-hand limits.

Let T_k represent the kth point from N. For $B \in \mathfrak{B}(S)$, let $M_B \equiv \{M_B(t): t \geq 0\}$ is the point process that counts the number of points that occur while the process $\{Z(t), t \geq 0\}$ is in set B during $[0, t]$, i.e.,

$$M_B(t) = \int_0^t 1\{Z(s-) \in B\} dN(s) \equiv \sum_{k=1}^{N(t)} J(T_k) 1\{Z(T_k -) \in B\} \tag{5.44}$$

where $J(T_k) = N(T_k) - N(T_k -)$ is the size of the jump of N at T_k.

Note that the definition of M_B implies that a point from N at which the process Z jumps from B^c into B is not counted as a point in M_B, while a point from N at which Z jumps from B to B^c is counted as a point in M_B. Now fix $B \in \mathfrak{B}(S)$, and let $U \equiv \{U(t): t \geq 0\}$ be a random-time change defined by

$$U(t) = \int_0^t 1\{Z(s-) \in B\} ds, t \geq 0. \tag{5.45}$$

Conversely, let $V \equiv \{V(u): u \geq 0\}$ be the corresponding original time process, where for a given B-time u, $V(u)$ is obtained from the relation

$$V(u) = inf\{t \geq 0: U(t) \geq u\},\ u \geq 0. \tag{5.46}$$

Our interest is in the process $N_B \equiv \{N_B(u), u \geq 0\}$ obtained from N such that $N_B(u)$ counts the number of points that occur during the first u time units that $\{Z(t), t \geq 0\}$ is in set B, i.e.,

$$N_B(u) = M_B(V(u)), u \geq 0. \tag{5.47}$$

The process N_B and time index u will be referred to informally as the B-process and B-time, respectively, while time t will sometimes be referred to as original time. The two point processes N_B and M_B count the same points using different time clocks.

Remark 5.34. In many applications, the points of N will be a subset of the set of points where Z changes state. (For example, Z could be the queue length process in a G/G/1 queue and N the points at which Z makes a transition from state i to state $i+1$ for some $i \geq 0$ (i.e., an arrival occurs). In this example, our definition of M_B insures that an arrival does not see itself.) In such applications, $\{N(s): 0 \leq s \leq t\}$ is completely determined by $\{Z(s): 0 \leq s \leq t\}$. See Walrand [83], Green and Melamed [36], Wolff [90], and Serfozo [65] for applications to Markov processes that exploit this property. Our more general framework is consistent with that of Wolff [88], Melamed and Whitt [54, 55], Brémaud [6], and König and Schmidt [41] in the sense that it is not necessary that all, or any, of the points of N coincide with changes of state of Z. As an extreme case, N could represent the time points of observations that occur independently of the evolution of Z.

We shall need the following lack-of-anticipation assumption (LAA) introduced by Wolff [88].

Assumption (LAA). For all $t \geq 0$, $\{N(t+h) - N(t); h \geq 0\}$ is independent of $\{Z(\tau): 0 \leq \tau \leq t\}$.

The LAA assumption implies that the number of points from N after time t is independent of the past behavior of the process Z. Melamed and Whitt [55] show that the whole past may be replaced by the present when joint stationarity of Z and N is assumed. We do not assume joint stationarity; thus we must incorporate the past behavior of the process Z in the assumption LAA.

The approach taken here points out a common theme among $ASTA$-type results. First, proofs of $ASTA$-type results typically involve showing that (some version of) the conditional intensity of the point process is the same for all subsets of the state space (a point already observed by [80, 88, 41, 54, 55, 82] among others). Second, with LAA and a Poisson point process (or some variant of it), one can just as easily prove the stronger property that the point process, conditioned on Z being in set B, is a probabilistic replica of the original point process. Third, this stronger property, like $ASTA$ itself, is really intuitive. In fact, the argument via discrete time, given in section 2 of El-Taha and Stidham [25], exploits this intuition rather clearly by using only elementary conditioning arguments.

5.7.2 Main results

In this section it will be shown, using martingale theory, that, in continuous time, the B-process N_B has the same probabilistic structure as the original process N, if LAA holds and N admits a stochastic intensity. In particular, N_B is standard Poisson with intensity λ for all $B \in \mathfrak{B}(S)$ if, and only if, N is Poisson $- \lambda$. As a corollary, we obtain a simple proof of $PASTA$ under weak conditions. The results will be obtained by utilizing the theory of martingales. We exploit the notion that the time instances at which process Z leaves/enters set B, constitute under LAA a sequence of stopping times relative to the σ-field generated by the internal history of processes Z and N. This observation, together with Watanabe characterization of Poisson processes [84] and its extension, given by Brémaud [5], constitute the essence of our approach.

Let $\mathcal{F}_t = \sigma\{N(s), Z(s), 0 \le s \le t\}$ be the internal history of (Z, N), that is, the sub σ-field of $\mathcal{F}$ generated by $\{(N(s), Z(s)), 0 \le s \le t\}$. Assume that $\mathcal{F}_0$ contains all the null events of $\mathcal{F}$. Assume also that the point process N admits an $\mathcal{F}_t$ stochastic intensity $\nu = \{\nu(t); t \ge 0\}$. Note that the process $Z^- \equiv \{Z(t-), t \ge 0\}$ is $\mathcal{F}_t$-predictable, because it is left continuous (Brémaud [5], T5, p. 9).

Now, we prove the following result which is an extension of relation (5.3) and Lemma 3 of Wolff [88].

Lemma 5.35. *Suppose that process M_B is defined as in* (5.44). *Then M_B admits an $\mathcal{F}_t$-intensity $\nu(t)1\{Z(t^-) \in B\}$; and*

(*a*) *for all $t \ge 0$,*

$$\mathbf{E}[M_B(t)] = \mathbf{E}[\int_0^t 1\{Z(s-) \in B\}\nu(s)ds];$$

(*b*) *if $\mathbf{E}[\int_0^t \nu(s)ds] < \infty$ for all $t \ge 0$,*

$$M_B(t) = \int_0^t 1\{Z(s-) \in B\}\nu(s)ds$$

is a mean-zero $\mathcal{F}_t$ martingale.

Proof. The first part of the lemma follows from the definition of stochastic intensity, (see p. 27 of Brémaud [5]). Part (*b*) follows by appealing to T8, p. 27 part (β) of Brémaud [5], where the $Z(t)$ of Brémaud [5] is replaced by $1\{Z(t-) \in B\}$. □

For our next result, we need a B-time version of the intensity ν. To this end, let $\nu_B = \{\nu_B(u); u \ge 0\}$ be the intensity ν observed only during B-time, i.e., when process $\{Z(t-); t \ge 0\}$ is in set B. One can see that ν_B is, in fact, ν restricted to B-time. Similar to (5.47), it is clear that

$$\nu_B(u) = \nu(V(u)).$$

Moreover, let $\mathcal{G}_u = \mathcal{F}_{V(u)}$.

Theorem 5.36. *If $\mathbf{E}[\int_0^t \nu(s)ds] < \infty$ for all $t \ge 0$, then*

$$N_B(u) = \int_0^u \nu_B(s)ds$$

is a mean zero $\mathcal{G}_u$ martingale. Moreover, process N_B admits ν_B as $\mathcal{G}_u$ stochastic intensity.

The next result gives conditions as to when the process N_B has the same probabilistic structure as N.

Corollary 5.37. *Suppose that $\nu \in \mathcal{F}_0$. Then,*

(*a*) *N_B is a homogeneous doubly stochastic Poisson process with $\mathcal{G}_u$-intensity ν for all $B \in \mathcal{B}(S)$ iff N is a homogeneous doubly stochastic Poisson process with $\mathcal{F}_t$-intensity ν. In particular,*

(*b*) *if ν is deterministic, in which case we write, $\lambda = \nu(t)$ for all $t \ge 0$, then the point process N_B is Poisson-λ for every $B \in \mathcal{B}(S)$ iff N is Poisson-λ.*

Proof. Let N be a homogeneous doubly stochastic Poisson process. Then $\nu(t) = \Lambda$, say, where Λ is a nonnegative $\mathcal{F}_0$ random variable (see p. 22 of Brémaud [5]). Thus $\nu_B = \Lambda$, and the proof of (*a*) follows from Theorem 5.36 and the characterization theorem of doubly stochastic Poisson processes, (see T4, p. 25 of Brémaud [5]). Part (*b*) is a special case of (*a*). □

A more elementary version of the above results is given in El-Taha [18].

Corollary 5.37 is also valid if we consider process N to be a point process (not necessarily simple) with stationary independent increments, in which case process N_B will have the same structure as N. This covers compound Poisson processes; see section 2.2 of Daley and Vere-Jones [15]. El-Taha and Stidham [22] give an alternate proof using the discrete-time model of El-Taha and Stidham [25] and limiting arguments that exploit the Skorohod metric and weak convergence theory.

Corollary 5.37 (*b*) also provides an alternate proof for *PASTA*. First define the following sample-path limiting average, when it exists:

$$\bar{\lambda} := \lim_{t \to \infty} N(t)/t.$$

Thus, $\bar{\lambda}$ is the limiting average number of arrivals (points from N) per unit time. *PASTA* now follows from Corollary 5.37 and Theorem 5.17, as the following theorem shows.

Theorem 5.38. (*PASTA*) *Assume that LAA holds for Z and N, that N is Poisson-λ $(0 < \lambda < \infty)$, and that $p(B)$ and $\pi(B)$ exist w.p.1 for each $B \in \mathfrak{B}(S)$. Then the frequency distributions $\{p(B); B \in \mathfrak{B}(S)\}$ and $\{\pi(B); B \in \mathfrak{B}(S)\}$ coincide, w.p.1.*

Proof. Since N is Poisson-λ, the limiting average $\bar{\lambda}$ exists and $\bar{\lambda} = \lambda$ *w.p.*1. It follows from Corollary 5.37 (*b*) that the intensity of N_B equals λ, independent of B, and that N_B is a Poisson process. Thus $q(B)$ exists *w.p.*1 and $q(B) = \lambda$, for each $B \in \mathfrak{B}(S)$. The desired result then follows from Theorem 5.17. □

Because this is a sample-path result, it holds for any (N, Z) for which the limiting frequencies $p(B)$ and $q(B)$, $B \in \mathfrak{B}(S)$, exist *w.p.*1. Thus we have proved Wolff's *PASTA* under conditions essentially equivalent to his. But the intermediate step in our proof, showing that the Poisson-λ process N restricted to any set B is also Poisson-λ, is of independent interest as well as intuitively appealing.

The results in this section may also help in characterizing Poisson flows in queueing systems. Another interesting observation about Corollary 5.37 is that it provides an example of a Poisson process split into different subprocesses with the subprocesses (with appropriate time scaling) inheriting the structure of the original process although the splitting mechanism itself is not memoryless. For example, in an M/G/1 queue, the j-subprocesses $\{N_j(u): u \geq 0\}$, (u measures j-time) are Poisson-λ for all $j = 0, 1, 2, \ldots$ regardless of the fact that the time at which process Z *enters/leaves* state j depends on where it was at the previous transition.

5.8 LITTLE'S FORMULA: $L = \lambda W$

In this section and the next, we review and extend previous research on sample-path proofs of Little's formula ($L = \lambda W$) and its generalization ($H = \lambda G$). In all cases, the starting point is one of the versions of $Y = \lambda X$ (Lemma 5.2 or Lemma 5.3). We begin in this section by giving several sample-path proofs of $L = \lambda W$. The first two are simple and intuitive but require what turns out to be unnecessarily strong assumptions. The third proof requires only a weak asymptotic condition, which is satisfied, in particular, when both λ and W are finite. As we shall see in Section 5.10, this condition is also satisfied under a weak sample-path stability condition. We then present, in the next section, two complementary sample-path approaches to $H = \lambda G$. The first is motivated by Stidham [74] and

Heyman and Stidham [40]. It mimics the third proof of $L = \lambda W$, uses the second version of $Y = \lambda X$, and requires a slightly restrictive regularity condition, albeit one that is satisfied in many applications. The second proof, which uses the first version of $Y = \lambda X$, is more general and is based on Stidham [75, 77]. This proof uses a condition that is necessary and sufficient for $H = \lambda G$ under the assumption that λ and G are both well-defined. We show that this condition is satisfied *w.p.*1 when the bivariate sequence of customer interarrival times and cost functions is strictly stationary.

The material in this section is based in part on Stidham [78]. In the spirit of Little [48] and Stidham [72, 74], we shall consider a general input-output system, fed by a discrete input process of *customers*, each of which spends a certain amount of time in the system, and then departs.

We follow Whitt [85] in the problem setup and notation. The basic data are $\{(A_k, D_k),\ k \geq 1\}$, where $0 \leq A_k \leq A_{k+1} < \infty$, $A_k \leq D_k < \infty$, $k \geq 1$, and A_k and D_k are interpreted as the arrival time and the departure time, respectively, of customer k. We assume that $A_k \to \infty$, as $k \to \infty$, so that there are only a finite number of arrivals in any finite time interval. Let $A(t)\colon = \#\{k\colon A_k \leq t\}$, $D(t)\colon = \#\{k\colon D_k \leq t\}$, $t \geq 0$, so that $A(t)$ and $D(t)$ count the number of arrivals and departures, respectively, in the interval $[0,t]$. Note that, since $A_k < \infty$ for all $k \geq 1$, $A(t) \to \infty$ as $t \to \infty$. Note also that $A(t) = max\{k\colon A_k \leq t\}$, since $\{A_k, k \geq 1\}$ is a nondecreasing sequence. But, in general, we cannot write $D(t) = max\{k\colon D_k \leq t\}$, because $\{D_k,\ k \geq 1\}$ is not necessarily nondecreasing. (It is nondecreasing if the discipline is *first in, first out* (FIFO), that is, if departures occur in the same order as arrivals.) Define

$$L(t)\colon = \#\{k\colon A_k \leq t < D_k\} = A(t) - D(t), t \geq 0, \tag{5.48}$$

$$W_k\colon = D_k - A_k, k \geq 1, \tag{5.49}$$

so that $L(t)$ is the number of customers in the system at time t and W_k is the waiting time in the system of customer k.

If we let $I_k(t)$ denote the indicator of $A_k \leq t < D_k$, then it follows from (5.48) and (5.49) that

$$L(t) = \sum_{k=1}^{\infty} I_k(t)$$

$$W_k = \int_0^{\infty} I_k(t)dt$$

from which we obtain the basic inequality

$$\sum_{k\colon A_k \leq t} W_k \geq \int_0^t L(s)ds \geq \sum_{k\colon D_k \leq t} W_k, t \geq 0. \tag{5.50}$$

All our proofs of $L = \lambda W$ use the basic inequality (5.50) and one of the versions of $Y = \lambda X$ (Lemma 5.2 or Lemma 5.3).

To motivate our first proof of $L = \lambda W$, which is due to Newell [61], we first observe that each of the inequalities in (5.50) is an equality for time points t belonging to idle periods, i.e., such that $L(t) = 0$.

Theorem 5.39. *Suppose* $t^{-1}A(t) \to \lambda$ *as* $t \to \infty$, *where* $0 \leq \lambda \leq \infty$, $n^{-1}\sum_{k=1}^{n}$

$W_k \to W$, as $n \to \infty$, where $0 \leq W \leq \infty$, and $t^{-1} \int_0^t L(s)ds \to L$, as $t \to \infty$, where $0 \leq L \leq \infty$. If $L(t) = 0$ infinitely often as $t \to \infty$, then $L = \lambda W$, provided that λW is well-defined.

Proof. For all t such that $L(t) = 0$, we have

$$t^{-1} \sum_{k: A_k \leq t} W_k = t^{-1} \int_0^t L(s)ds. \tag{5.51}$$

Since the limits λ, W, and L all exist, they must coincide with the respective limits taken through any subsequence tending to ∞. Then taking the limit of each side of equation (5.51) through a sequence of time points t tending to ∞ such that $L(t) = 0$ and using Lemma 5.3 with $T_k = A_k$ and $X_k = W_k$, we conclude that $L = \lambda W$, provided that λW is well-defined. □

Although simple and intuitive, this result is not entirely satisfactory, since it depends on the unnecessary assumption that $L(t) = 0$ infinitely often as $t \to \infty$. For there are queueing systems in which L, λ and W are all well-defined (and finite), but $L(t)$ does not equal 0 infinitely often, or indeed ever. Conversely, the assumption that $L(t) = 0$ infinitely often does not by itself guarantee that the limits L, λ, and W exist (see Stidham [74] for an example).

Thus we seek a proof of $L = \lambda W$ that does not require that $L(t)$ return to zero infinitely often. Our next proof attains this goal but at the cost of assuming a *FIFO* discipline and equality of the departure and arrival rates (but see the corollary for removal of the latter as an explicit assumption). First recall that when the discipline is *FIFO*, customers depart in the same order as they arrive, so that $D_k \leq D_{k+1}$, $k \geq 1$, and $D(t) = max\ \{k: D_k \leq t\}$.

Theorem 5.40. *Suppose $t^{-1}A(t) \to \lambda$ and $t^{-1}D(t) \to \lambda$, as $t \to \infty$, where $0 \leq \lambda \leq \infty$, and $n^{-1}\sum_{k=1}^n W_k \to W$, as $n \to \infty$, where $0 \leq W \leq \infty$. If the discipline is* FIFO, *then $t^{-1} \int_0^t L(s)ds \to L$, as $t \to \infty$ and $L = \lambda W$, provided that λW is well-defined.*

Proof. It follows from the basic inequality (5.50) that

$$t^{-1} \sum_{k: A_k \leq t} W_k \geq t^{-1} \int_0^t L(s)ds \geq t^{-1} \sum_{k: D_k \leq t} W_k. \tag{5.52}$$

If $\lambda = 0$, then the desired result follows immediately from the first inequality in (5.52), the nonnegativity of all quantities involved, and Lemma 5.3, with $T_k = A_k$ and $X_k = W_k$. If $\lambda > 0$, then, since $t^{-1}D(t)$ approaches λ as $t \to \infty$, it follows that $D(t) \to \infty$ as $t \to \infty$. Since the discipline is *FIFO*, $D(t) = max\{k: D_k \leq t\}$. Thus, taking limits of each of the terms in (5.52) and using Lemma 5.3 again, first with $T_k = A_k$ and $X_k = W_k$ and then with $T_k = D_k$ and $X_k = W_k$, it follows that $t^{-1} \int_0^t L(s)ds \to L$ as $t \to \infty$, where $L = \lambda W$, provided that λW is well-defined. □

It turns out that we do not have to assume explicitly that $t^{-1}D(t) \to \lambda$ if $W < \infty$, as we shall show in the corollary below.

Corollary 5.41. *Suppose $t^{-1}A(t) \to \lambda$ as $t \to \infty$, where $0 \leq \lambda \leq \infty$, and $n^{-1}\sum_{k=1}^n W_k \to W$ as $n \to \infty$, where $0 \leq W < \infty$. If the discipline is* FIFO, *then*

$t^{-1}\int_0^t L(s)ds \to L$ *as* $t\to\infty$, *and* $L = \lambda W$, *provided that* λW *is well-defined.*

Proof. From $W < \infty$ it follows that $n^{-1}W_n = n^{-1}(D_n - A_n) \to 0$ as $n\to\infty$. Lemma 5.1 implies that $n^{-1}A_n \to \lambda^{-1}$ and hence $n^{-1}D_n \to \lambda^{-1}$ as $n\to\infty$. Moreover, $D_n < \infty$ for all $n \geq 1$ (since $n^{-1}W_n \to 0$) and $D_n \to \infty$ (since $A_n \to \infty$) as $n\to\infty$. Therefore, applying Lemma 5.1 to $\{D_n, n\geq 1\}$, we conclude that $t^{-1}D(t)\to\lambda$ as $t\to\infty$. The result then follows from Theorem 5.40. □

It is important to recognize where the proof of Theorem 5.40 breaks down when the *FIFO* assumption is not satisfied. Without *FIFO*, it is not generally true that $D(t) = max\{k: D_k \leq t\}$, since departures may not be in the same order as arrivals. Hence we cannot apply Lemma 5.3 with $T_k = D_k$ and $X_k = W_k$. Moreover, it does not follow from (5.50) that $\int_0^t L(s)ds \geq \sum_{k=1}^{D(t)} W_k$, as the following counterexample shows.

Example 5.42. Suppose $A_n = n$, $n \geq 1$, $W_1 = 2$, $W_2 = 7$, $W_n = 2$, $n \geq 3$. Let $t = 5$. Then $D(t) = 2$ (customers 1 and 3 depart in $[0,t]$), but

$$\sum_{k=1}^{D(t)} W_k = W_1 + W_2 = 9 > 8 = \int_0^t L(s)ds > 4 = W_1 + W_3 = \sum_{k:\, D_k \leq t} W_k.$$

This is not a pathological example, since, with a non-*FIFO* discipline, a customer, who departs before a customer who arrived earlier, will necessarily have a shorter waiting time.

While clear from this discussion, the need for the *FIFO* assumption, in order for the inequality $\int_0^t L(s)ds \geq \sum_{k=1}^{D(t)} W_k$ to hold, has been missed by some authors.

Rigorous and correct sample-path proofs of $L = \lambda W$ can be constructed for the non-*FIFO* case, however, if one assumes a weak asymptotic condition, which is satisfied, in particular, if both λ and W are finite, as we shall show presently. We shall need the following lemma, which is of independent interest.

Lemma 5.43. *Suppose* $W_n/A_n \to 0$ *as* $n\to\infty$. *Let* $0 \leq L \leq \infty$. *Then the following are equivalent:*

$$\lim_{t\to\infty} t^{-1}\sum_{k:\, A_k \leq t} W_k = L, \tag{5.53}$$

$$\lim_{t\to\infty} t^{-1}\int_0^t L(s)ds = L, \tag{5.54}$$

and

$$\lim_{t\to\infty} t^{-1}\sum_{k:\, D_k \leq t} W_k = L. \tag{5.55}$$

Proof. Let $\epsilon > 0$ be given. Since $W_n/A_n \to 0$ as $n\to\infty$, there exists an integer N such that $k \geq N$ implies $W_k \leq A_k\epsilon$. Therefore, for all $t \geq 0$,

$$\begin{aligned}\sum_{k:\, D_k \leq t} W_k &= \sum_{k:\, A_k + W_k \leq t} W_k \\ &\geq \sum_{k\geq N:\, A_k(1+\epsilon)\leq t} W_k \\ &\geq \sum_{k:\, A_k(1+\epsilon)\leq t} W_k - \sum_{k\leq N-1} W_k,\end{aligned}$$

which, together with the basic inequality (5.50) , implies

$$\sum_{k:\,A_k \le t} W_k \ \ge \int_0^t L(s)ds \tag{5.56}$$

$$\ge \sum_{k:\,D_k \le t} W_k \tag{5.57}$$

$$\ge \sum_{k:\,A_k(1+\epsilon) \le t} W_k - \sum_{k \le N-1} W_k \tag{5.58}$$

First, we use these inequalities to prove that (5.53) implies (5.54) and (5.55). Suppose (5.53) holds. Then

$$\lim_{t\to\infty} t^{-1} \sum_{k:\,A_k(1+\epsilon) \le t} W_k = (1+\epsilon)^{-1} \lim_{t\to\infty} [t(1+\epsilon)^{-1}]^{-1} \sum_{k:\,A_k \le t(1+\epsilon)^{-1}} W_k$$

$$= (1+\epsilon)^{-1} \lim_{t\to\infty} t^{-1} \sum_{k:\,A_k \le t} W_k$$

$$= (1+\epsilon)^{-1} L.$$

But $\epsilon > 0$ was arbitrary. Hence the desired result follows from this equation and the inequalities (5.56), (5.57), and (5.58), using the fact that $\lim_{t\to\infty} t^{-1} \sum_{k \le N-1} W_k = 0$.

Now we show that (5.54) implies (5.53). (The proof that (5.55) implies (5.53) is similar.) Suppose (5.54) holds. Then (5.56), (5.57), and (5.58) imply that

$$\liminf_{t\to\infty} t^{-1} \sum_{k:\,A_k \le t} W_k \ \ge L$$

$$\ge \limsup_{t\to\infty} t^{-1} \sum_{k:\,A_k(1+\epsilon) \le t} W_k$$

$$= (1+\epsilon)^{-1} \limsup_{t\to\infty} [t(1+\epsilon)^{-1}]^{-1} \sum_{k:\,A_k \le t(1+\epsilon)^{-1}} W_k$$

$$= (1+\epsilon)^{-1} \limsup_{t\to\infty} t^{-1} \sum_{k:\,A_k \le t} W_k$$

where we have used the fact that $\lim_{t\to\infty} t^{-1} \sum_{k \le N-1} W_k = 0$ in the derivation of the second inequality. Since $\epsilon > 0$ was arbitrary, we conclude that these inequalities hold in the limit as $\epsilon \to 0$, so that (5.53) holds.

This completes the proof of Lemma 5.43. □

The following theorem is an immediate consequence of Lemmas 5.3 and 5.43.

Theorem 5.44. *Suppose* $t^{-1}A(t) \to \lambda$ *as* $t \to \infty$, *where* $0 \le \lambda \le \infty$, *and* $W_n/A_n \to 0$ *as* $n \to \infty$. *Then*

(*a*) *if* $n^{-1}\sum_{k=1}^{n} W_k \to W$ *as* $n \to \infty$, *where* $0 \le W \le \infty$, *then* $t^{-1}\int_0^t L(s)ds \to L$ *as* $t \to \infty$, *and* $L = \lambda W$, *provided* λW *is well-defined;*

(*b*) *if* $t^{-1}\int_0^t L(s)ds \to L$ *as* $t \to \infty$, *where* $0 \le L \le \infty$, *then* $n^{-1}\sum_{k=1}^{n} W_k \to W$

as $n\to\infty$, *and* $L=\lambda W$, *provided* $\lambda^{-1}L$ *is well-defined.*

We can use Theorem 5.44 to obtain proofs of $L=\lambda W$ under conditions involving only the existence and finiteness of L, λ, and W. In particular, we do not need to assume that $L(t)=0$ infinitely often or that the discipline is *FIFO*. First, we give an immediate corollary of Theorem 5.44.

Corollary 5.45. *Suppose* $t^{-1}A(t)\to\lambda$ *as* $t\to\infty$, *where* $0\le\lambda<\infty$, *and* $W_n/n\to 0$ *as* $n\to\infty$. *Then*

(*a*) *if* $n^{-1}\sum_{k=1}^{n}W_k\to W$ *as* $n\to\infty$, *where* $0\le W\le\infty$, *then* $t^{-1}\int_0^t L(s)ds\to L$ *as* $t\to\infty$, *and* $L=\lambda W$, *provided* λW *is well-defined.*

(*b*) *if* $t^{-1}\int_0^t L(s)ds\to L$ *as* $t\to\infty$, *where* $0\le L\le\infty$, *then* $n^{-1}\sum_{k=1}^{n}W_k\to W$ *as* $n\to\infty$, *and* $L=\lambda W$, *provided* $\lambda^{-1}L$ *is well-defined.*

Proof. It follows from Lemma 5.1 that $n/A_n\to\lambda<\infty$, and hence $W_n/A_n\to 0$ as $n\to\infty$. The desired result then follows from Theorem 5.44. □

This corollary may be useful in cases where $W=\infty$, but $n^{-1}W_n\to 0$, as is the case, for example, in a stable GI/G/1 queue in which the second moment of the service time is infinite. It also leads immediately to the next corollary, which is the original sample-path version of $L=\lambda W$ contained in Stidham [72, 74].

Corollary 5.46. *Suppose* $t^{-1}A(t)\to\lambda$ *as* $t\to\infty$, *where* $0\le\lambda<\infty$, *and* $n^{-1}\sum_{k=1}^{n}W_k\to W$ *as* $n\to\infty$, *where* $0\le W<\infty$. *Then* $t^{-1}\int_0^t L(s)ds\to L$, *as* $t\to\infty$ *and* $L=\lambda W$.

Proof. Since by hypothesis $W<\infty$, it follows that $n^{-1}W_n\to 0$. The result then follows from Corollary 5.45. □

An alternate approach. The following alternative approach to $L=\lambda W$ is due to Glynn and Whitt [35]. See also Whitt [85] and Serfozo [66].

Define $\widehat{D}_k := max_{1\le j\le k}D_j = max_{1\le j\le k}\{A_j+W_j\}$, $k\ge 1$, that is, $\widehat{D}_k$ is the kth *thorough departure time.* Let $\widehat{D}(t) := \#\{k: \widehat{D}_k\le t\}$, $t\ge 0$; that is, $\widehat{D}(t)$ counts the number of thorough departures in $[0,t]$. Note that $\{\widehat{D}_k, k\ge 1\}$ is a nondecreasing sequence and therefore we can write $\widehat{D}(t)=max\{k: \widehat{D}_k\le t\}$, $t\ge 0$. (This is important for what follows. As we shall see, it allows us to use an argument like that in Theorem 5.40 even in the case of a non-*FIFO* discipline.)

Compare the following theorem with Theorem 5.44, noting that the condition, $lim_{t\to\infty}\widehat{D}(t)/t=\lambda$, replaces the condition, $lim_{n\to\infty}W_n/A_n=0$.

Theorem 5.47. *Suppose* $t^{-1}A(t)\to\lambda$ *and* $t^{-1}\widehat{D}(t)\to\lambda$, *as* $t\to\infty$, *where* $0\le\lambda\le\infty$. *Then*

(*a*) *if* $n^{-1}\sum_{k=1}^{n}W_k\to W$ *as* $n\to\infty$, *where* $0\le W\le\infty$, *then* $t^{-1}\int_0^t L(s)ds\to L$ *as* $t\to\infty$, *and* $L=\lambda W$, *provided* λW *is well -defined;*

(*b*) *if* $t^{-1}\int_0^t L(s)ds\to L$ *as* $t\to\infty$, *where* $0\le L\le\infty$, *then* $n^{-1}\sum_{k=1}^{n}W_k\to W$ *as* $n\to\infty$, *and* $L=\lambda W$, *provided* $\lambda^{-1}L$ *is well-defined.*

Proof. (*a*) Since $\widehat{D}_k\ge D_k$, it follows from the basic inequality (5.50) that

$$\sum_{k: A_k\le t} W_k \ge \int_0^t L(s)ds \ge \sum_{k: \widehat{D}_k\le t} W_k, \quad t\ge 0.$$

Since $\{\widehat{D}_k, k\ge 1\}$ is a nondecreasing sequence, it follows that

$$\frac{A(t)}{t}\frac{1}{A(t)}\sum_{k=1}^{A(t)} W_k \geq \frac{1}{t}\int_0^t L(s)ds \geq \frac{\hat{D}(t)}{t}\frac{1}{\hat{D}(t)}\sum_{k=1}^{\hat{D}(t)} W_k.$$

The desired result then follows upon letting $t \to \infty$.

(b) It follows from the definitions that

$$\int_0^{A_n} L(s)ds \leq \sum_{k=1}^{n} W_k \leq \int_0^{\hat{D}_n} L(s)ds,\ n \geq 1.$$

Hence

$$\frac{A_n}{n}\frac{1}{A_n}\int_0^{A_n} L(s)ds \leq \frac{1}{n}\sum_{k=1}^{n} W_k \leq \frac{\hat{D}_n}{n}\frac{1}{\hat{D}_n}\int_0^{\hat{D}_n} L(s)ds.$$

The desired result then follows upon letting $n \to \infty$ and using Lemma 5.1 applied to the point process $\{\hat{D}(t), t \geq 0\}$. □

Remark 5.48. When $A(t)/t \to \lambda = 0$, it follows from the fact that $\hat{D}(t) \leq A(t)$ that $\hat{D}(t)/t \to \lambda$, so it is not necessary to assume explicitly that the latter condition holds.

The usefulness of this approach depends upon our being able to verify easily that $\hat{D}(t)/t \to \lambda$ in applications. (From the preceding remark, we see that it suffices to consider the case where $\lambda > 0$.) The following lemma (cf. proof of Theorem 2.2 in Whitt [85], Serfozo [66], and Lemma 15) provides some insight into this question. In fact, it shows that, when $A(t)/t \to \lambda$, the condition that $\hat{D}(t)/t \to \lambda$ is essentially equivalent to the condition that $W_n/n \to 0$ (cf. Theorem 5.44 and Corollary 5.45).

Lemma 5.49. *Let* $0 < \lambda \leq \infty$. *The following statements are equivalent:*

(a) $lim_{t\to\infty} A(t)/t = \lambda = lim_{t\to\infty}\hat{D}(t)/t$;

(b) *either* $lim_{t\to\infty} A(t)/t = \lambda$ *or* $lim_{t\to\infty}\hat{D}(t)/t = \lambda$ *and* $lim_{n\to\infty} W_n/n = 0$.

Proof. Suppose (a) holds. Then

$$0 \leq n^{-1}W_n = n^{-1}(D_n - A_n) \leq n^{-1}(\hat{D}_n - A_n) \to \lambda^{-1} - \lambda^{-1} = 0,$$

where we have applied Lemma 5.1 to both $\{A(t), t \geq 0\}$ and $\{\hat{D}(t), t \geq 0\}$. Thus (b) holds.

To show that (b) implies (a), first note that (when the limits exist)

$$\begin{aligned} \lim_{n\to\infty} n^{-1}\hat{D}_n &= \lim_{n\to\infty} n^{-1}\max_{k \leq n}(A_k + W_k) \\ &= \lim_{n\to\infty} n^{-1}(A_n + W_n), \end{aligned}$$

where the second equality follows from the basic property of limits of maxima that, for finite c, and nonnegative sequences $\{a_n\}$, $\{b_n\}$, with b_n nondecreasing, $a_n \to \infty$, and $b_n \to \infty$,

$$b_n^{-1}\max_{k \leq n} a_n \to c \text{ if, and only if, } b_n^{-1}a_n \to c.$$

Suppose that $lim_{n\to\infty} W_n/n = 0$. If $lim_{t\to\infty} A(t)/t = \lambda$, then it follows from Lemma 5.1 applied to $\{A(t), t \geq 0\}$ that $A_n/n \to 1/\lambda$, and hence $\hat{D}_n/n \to 1/\lambda$. Applying Lemma 5.1 to $\{\hat{D}(t), t \geq 0\}$, we conclude that $lim_{t\to\infty}\hat{D}(t)/t = \lambda$. On the other hand, if $lim_{t\to\infty}\hat{D}(t)/t = \lambda$, then it follows from Lemma 5.1 applied to $\{\hat{D}(t), t \geq 0\}$ that $\hat{D}_n/n \to 1/\lambda$ and hence $A_n/n \to 1/\lambda$. Applying Lemma 5.1 to $\{A(t),$

$t \geq 0\}$, we conclude that $lim_{t\to\infty}A(t)/t = \lambda$. Thus we have shown that (*b*) implies (*a*), and the proof of the lemma is complete. □

This lemma shows, in particular, that, when $A(t)/t \to \lambda$, where $0 < \lambda \leq \infty$, then the conditions $\widehat{D}(t)/t \to$ and $W_n/n \to 0$ are equivalent. Thus one has the option of working with whichever of these conditions is easier to verify in a particular application.

Using this lemma, we obtain the following corollary of Theorem 5.47.

Corollary 5.50. *Suppose* $t^{-1}A(t) \to \lambda$ *as* $t \to \infty$, *where* $0 < \lambda \leq \infty$, *and* $W_n/n \to 0$ *as* $n \to \infty$. *Then*

(*a*) *if* $n^{-1}\sum_{k=1}^{n} W_k \to W$ *as* $n \to \infty$, *where* $0 \leq W \leq \infty$, *then* $t^{-1}\int_0^t L(s)ds \to L$ *as* $t \to \infty$, *and* $L = \lambda W$, *provided* λW *is well-defined;*

(*b*) *if* $t^{-1}\int_0^t L(s)ds \to L$ *as* $t \to \infty$, *where* $0 \leq L \leq \infty$, *then* $n^{-1}\sum_{k=1}^{n} W_k \to W$ *as* $n \to \infty$, *and* $L = \lambda W$, *provided* $\lambda^{-1}L$ *is well-defined.*

Note that Corollary 5.50 is nearly identical to Corollary 5.45 (the difference being the range of allowable values for λ), although the routes followed to the two versions were quite different. One can combine Corollaries 5.45 and 5.50 and their proofs to obtain a version valid for all $0 \leq \lambda \leq \infty$.

5.9 GENERALIZATION OF LITTLE'S FORMULA: $H = \lambda G$

Little's formula has an economic interpretation, which sheds light on its generality and also suggests possible extensions. Let us interpret $I_k(t)$ as the cost per unit time associated with customer k at time t, so that customer k incurs a cost of one monetary unit per unit time while in the system (i.e., while $A_k \leq t < D_k$) and zero cost otherwise. Under this assumption, $L(t)$ is the total cost rate at time t and W_k is the total cost incurred by customer k, so that $L = \lambda W$ says simply (and plausibly) that the long-run average cost per unit time equals the arrival rate of customers times the long-run average cost per customer.

The generalizations to $H = \lambda G$ arise naturally if one allows a more general cost-rate function than the indicator $I_k(t)$. To this end, let $f_k(t)$ denote the rate at which customer k incurs cost at time t, $k \geq 1$, $t \geq 0$. Define

$$H(t) := \sum_{k=1}^{\infty} f_k(t), \quad t \geq 0 \tag{5.59}$$

and

$$G_k := \int_0^{\infty} f_k(s)ds, \quad k \geq 1, \tag{5.60}$$

so that $H(t)$ is the total cost rate at time t and G_k is the total cost incurred by customer k.

As we shall see in applications of $H = \lambda G$, the functions $\{f_k(t), t \geq 0\}$ need not be indexed by customers arriving to a queueing system, but rather by an arbitrary point process, $\{T_k, k \geq 1\}$. Our formal setup is as follows (cf. Stidham [75, 77]):

As in Section 5.2, we are given a deterministic sequence of time points, $\{T_k, k \geq 1\}$, with $0 \leq T_k \leq T_{k+1} < \infty$, $k > 1$, and we define $N(t) := max\{k \geq 1: T_k \leq t\}$, $t \geq 0$, so that $N(t)$ is the number of points in $[0,t]$. As in Section 5.2, we assume that $T_k \to \infty$ as $k \to \infty$, so that there are only a finite number of events in any finite time interval ($N(t) < \infty$ for all $t \geq 0$), and we note that $N(t) \to \infty$ as

$t \to \infty$, since $T_k < \infty$ for all $k \geq 1$. Associated with each time point T_k, there is a function $f_k: [0, \infty) \to [0, \infty)$. The bivariate sequence $\{(T_k, f_k(\cdot)), k \geq 1\}$ constitutes the basic data, in terms of which the behavior of the system is described. We assume that $f_k(t)$ is Lebesgue integrable on $t \in [0, \infty)$, for each $k \geq 1$.

With $H(t)$ and G_k defined by (5.59) and (5.60), respectively, define the following limiting averages, when they exist:

$$\lambda: = \lim_{t \to \infty} t^{-1} N(t), \tag{5.61}$$

$$H: = \lim_{t \to \infty} t^{-1} \int_0^t H(s)ds, \tag{5.62}$$

$$G: = \lim_{n \to \infty} n^{-1} \sum_{k=1}^{n} G_k. \tag{5.63}$$

We seek conditions under which $H = \lambda G$. We shall present two approaches.

5.9.1 Approach based on $L = \lambda W$

Following Stidham [74] and Heyman and Stidham [40], suppose that the bivariate sequence $\{(T_k, f_k(\cdot)), k \geq 1\}$ satisfies the following condition:

Condition L. There exists a sequence $\{W_k, k \geq 1\}$ such that

(*a*) $W_k / T_k \to 0$ as $k \to \infty$; and

(*b*) $f_k(t) = 0$ for $t \notin [T_k, T_k + W_k)$.

In economic terms, Condition L says that all the cost associated with the kth point (e.g., the kth customer) is incurred in a finite time interval beginning at the point (e.g., the arrival of the customer), and that the lengths of these intervals cannot grow at the same rate as the points themselves, as $k \to \infty$. This is a stronger-than-necessary condition for $H = \lambda G$ (as we shall see in the next subsection), but it is satisfied in most applications to queueing systems, in which the time points T_k and $T_k + W_k$ correspond to customer arrivals and departures, respectively, and is natural to assume that customers can incur cost only while they are physically present in the system.

Under Condition L we have the following generalization of the basic inequality (5.50) for $L = \lambda W$.

$$\sum_{k: T_k \leq t} G_k \geq \int_0^t H(s)ds \geq \sum_{k: T_k + W_k \leq t} G_k, \quad t \geq 0. \tag{5.64}$$

Using (5.64) and essentially the same arguments as in Section 5.8, we obtain the following generalization of Lemma 5.43.

Lemma 5.51. *Suppose Condition L holds. Let $0 \leq H \leq \infty$. Then the following are equivalent:*

$$\lim_{t \to \infty} t^{-1} \sum_{k: T_k \leq t} G_k = H, \tag{5.65}$$

$$\lim_{t \to \infty} t^{-1} \int_0^t H(s)ds = H, \tag{5.66}$$

and

$$\lim_{t \to \infty} t^{-1} \sum_{k: T_k + W_k \leq t} G_k = H. \tag{5.67}$$

The following theorem (cf. Theorem 5.44 in Section 5.8) is then an immediate consequence of this lemma and Lemma 5.3 (the second version of $Y = \lambda X$).

Theorem 5.52. *Suppose $t^{-1}N(t) \to \lambda$ as $t \to \infty$, where $0 \le \lambda \le \infty$, and Condition L holds. Then*

(a) *if $n^{-1}\sum_{k=1}^{n} G_k \to G$ as $n \to \infty$, where $0 \le G \le \infty$, then $t^{-1}\int_0^t H(s)ds \to H$ as $t \to \infty$,*

(b) *if $t^{-1}\int_0^t H(s)ds \to H$ as $t \to \infty$, where $0 \le H \le \infty$, then $n^{-1}\sum_{k=1}^{n} G_k \to G$ as $n \to \infty$, and $H = \lambda G$, provided $\lambda^{-1}H$ is well-defined.*

5.9.2 Alternate approach

In this section, we shall give an alternate sample-path proof of $H = \lambda G$. Our goal is to find a necessary as well as sufficient condition for $H = \lambda G$, assuming that the averages λ and G exist. This condition (of course) is weaker than Condition L. We shall show that it is implied by a slightly stronger pathwise-uniform-integrability condition, and we argue that the latter is a natural sample-path analogue of the strict stationarity condition required by proofs of $H = \lambda G$ based on the theory of stationary random marked point processes (cf. Franken et al. [30], Baccelli and Brémaud [3]). In contrast to the approach in the previous subsection, here we use the first (rather than the second) version of $Y = \lambda X$.

Applying Lemma 5.2 with $Y(t) = \int_0^t H(s)ds$ and $X_n = \int_{T_{n-1}}^{T_n} H(s)ds$, we im-mediately obtain the following lemma.

Lemma 5.53. *Suppose $t^{-1}N(t) \to \lambda$ as $t \to \infty$, where $0 \le \lambda \le \infty$. Then*

(a) *if $n^{-1}\int_0^{T_n} H(s)ds \to G$ as $n \to \infty$, where $0 \le G \le \infty$, then $t^{-1}\int_0^t H(s)ds \to H$ as $t \to \infty$, and $H = \lambda G$, provided that λG is well-defined;*

(b) *if $t^{-1}\int_0^t H(s)ds \to H$ as $t \to \infty$, where $0 \le H \le \infty$, then $n^{-1}\int_0^{T_n} H(s)ds \to G$ as $n \to \infty$, and $H = \lambda G$, provided that $\lambda^{-1}H$ is well-defined.*

Lemma 5.53 reveals that the key to proving $H = \lambda G$ by this approach is showing that $n^{-1}\int_0^{T_n} H(s)ds$ and $n^{-1}\sum_{k=1}^{n} G_k$ approach the same limit G as $n \to \infty$. Now, it follows from the definitions of $H(t)$ and G_n that

$$\int_0^{T_n} H(s)ds = \sum_{k=1}^{n} G_k + r_n, \tag{5.68}$$

where

$$r_n := \sum_{k=n+1}^{\infty} \int_0^{T_n} f_k(s)ds - \sum_{k=1}^{n} \int_{T_n}^{\infty} f_k(s)ds \tag{5.69}$$

The two terms in (5.69) are: (A) the cost incurred in $[0, T_n]$ by customers who arrive after time T_n and (B) the cost incurred after time T_n by customers $k = 1, \ldots, n$. The following theorem is a consequence of Lemma 5.53 and (5.68).

Theorem 5.54. *Suppose $t^{-1}N(t) \to \lambda$ as $t \to \infty$, where $0 \le \lambda \le \infty$, and*

$n^{-1}\sum_{k=1}^{n} G_k \to G$ *as* $n \to \infty$, *where* $0 \le G \le \infty$, *and* λG *is well-defined. Then the following are equivalent:*

(*a*) $t^{-1}\int_0^t H(s)ds \to H = \lambda G$ *as* $t \to \infty$;

(*b*) $n^{-1} r_n \to 0$ *as* $n \to \infty$ (*Condition R*).

Thus, when the limits λ and G exist, Condition R is both necessary and sufficient for $H = \lambda G$.

Example 5.55. $L = \lambda W$. Let $T_k = A_k$, the arrival time of customer k to a queueing system, and let W_k be the time spent by customer k in the system, $k \ge 1$. Suppose $f_k(t) = 1\{A_k \le t < A_k + W_k\}$, that is, $f_k(t)$ is the indicator of whether customer k is in the system at time t. Then $G_n = W_n$, $H(t) = L(t)$, the number of customers in the system at time t, and $H = \lambda G$ is the sample-path version of Little's formula, $L = \lambda W$. Theorem 5.54 shows that, if both the arrival rate, λ, and the average waiting time, W, exist and λW is well-defined, then the average number of customers in the system, L, is well-defined and $L = \lambda W$ if, and only if, $r_n = o(n)$ as $n \to \infty$ (Condition R). Since $f_k(t) = 0$ for $t < A_k$, the term A vanish es in this case and

$$-r_n = B = \sum_{k=1}^{n} \int_{A_n}^{\infty} f_k(s)ds.$$

It is not difficult to verify that the term B is $o(n)$ as $n \to \infty$ if, and only if, $W_n / A_n \to 0$ as $n \to \infty$, thus providing an independent route to the results in Section 5.8.

5.9.3 Alternative sufficient conditions

Instead of verifying Condition R directly in applications, sometimes it is easier to work with intermediate conditions that imply $r_n = o(n)$. In the special case of $L = \lambda W$, for example, we have seen (Section 5.8) that $W < \infty$ implies $W_n = o(n)$, which is sufficient for $r_n = o(n)$. Here is another example.

Consider the case in which $f_n(t) = 0$ for $t < T_n$. (For example, job n incurs no cost until it arrives.) Then, it is not difficult to show that each of the terms (A) and (B) of r_n is $o(n)$ if the sequence $\{f_k(T_k + \cdot), k \ge 1\}$ is uniformly Lebesgue integrable, that is, if

$$\lim_{a \to \infty} \sup_k \int_{T_k + a}^{\infty} f_k(s)ds = 0.$$

This condition holds *w.pr.*1 in the stationary case, for example (see below).

5.9.4 Stationary stochastic models

Now, we show how to relate our sample-path version of $H = \lambda G$ to the corresponding relation between expectations in the context of a stationary stochastic model (cf. Brumelle [11]). Suppose that the bivariate sequence $\{(T_n - T_{n-1}, G_n), n \ge 1\}$ $(T_0 = 0)$ is strictly stationary. Then the ergodic theorem implies that, with probability one, $G = \mathbf{E}[G_1 \mid \mathfrak{I}]$ and $\lambda^{-1} = \mathbf{E}[T_1 \mid \mathfrak{I}]$, where $\mathfrak{I}$ is the invariant σ-field of $\{(T_n - T_{n-1} < G_n), n \ge 1\}$. If in addition $\{(T_n - T_{n-1}, G_n), n \ge 1\}$ is ergodic and Condition R holds *w.pr.*1, then

$$H = \mathrm{E}[G_1]/\mathrm{E}[T_1] \ w.pr.1.$$

Thus, in this approach, we see that Condition R is all that is needed to conclude that the basic relation between time averages and customer averages, $H(\omega) = \lambda(\omega)G(\omega)$, holds for a particular sample point ω, while the stationarity assumption is used only to establish that the limiting averages, λ and G, exist and coincide with the expectations of the corresponding stationary random variables, with probability one.

In our discussion of the special case of Little's formula, we saw that Condition R holds, and hence $H = \lambda G$ (that is, $L = \lambda W$), on each sample path for which λ and G are well-defined and finite, provided only that $f_k(t) = 1\{A_k \leq t < A_k + W_k\}$. This specification of $f_k(t)$ implies that it is not possible for a customer to leave the system and return later to continue his sojourn. (A customer who did so would have to be relabeled as a new arrival and begin a new sojourn each time he returns to the system.) In this case, the finiteness of λ and G is enough to imply that Condition R holds. This implication does not hold, however, when customers are allowed to leave and re-enter, as demonstrated by a counterexample in Heyman and Stidham [40].

Thus, for general relations between time averages and customer averages to hold, it is clear that Condition R cannot be removed as an assumption without strengthening the stationarity assumption. Our next result concerns such a strengthening and its consequences. For notational convenience, define $\tau_n := T_n - T_{n-1}, k_n(\cdot) := f_n(T_n + \cdot)$, $n \geq 1$.

Theorem 5.56. *Suppose* $\{(\tau_n, k_n(\cdot)), n \geq 1\}$ *is strictly stationary with* $\mathrm{E}[\tau_1 \mid \mathcal{I}] > 0$ *and* $\mathrm{E}[G_1 \mid \mathcal{I}] < \infty$, *a.s., where* $\mathcal{I}$ *is the invariant σ-field of* $\{(\tau_n, k_n(\cdot)), n \geq 1\}$. *Then Condition R holds w.pr.1 and hence*

$$H = \lambda G = \mathrm{E}[G_1 \mid \mathcal{I}]/\mathrm{E}[\tau_1 \mid \mathcal{I}], \quad w.pr.1.$$

Proof. It suffices to show that each of the terms, (A) and (B), in (5.69) is $o(n)$ as $n \to \infty$ $w.pr.1$ under the strict-stationarity condition. This can be done by showing that the sequence $\{f_k(T_k + \cdot), k \geq 1\}$ is uniformly Lebesgue integrable (cf. previous subsection). For details, see Stidham [75, 77]. □

This theorem shows that the strict-stationarity condition is stronger than our necessary and sufficient sample-path Condition R.

5.10 STABILITY CONDITIONS FOR INPUT-OUTPUT PROCESSES

The issue of stability for nonstationary processes, using a sample-path framework, has recently been considered by several authors. El-Taha and Stidham [23, 24, 26] provide a sample-path characterization of (rate) stability and establish connections between rate stability and other measures of interest, such as tightness and the finiteness of the limiting average number of customers in a queueing system. Mazumdar, Guillemin, Badrinath, and Kannurpatti [52], study rate stability in the context of the workload process in a G/G/1 queue using sample-path arguments. Stidham and El-Taha [81] consider an input-output process with a single output stream and establish rate stability conditions using only sample-path information available from primary processes. Guillemin and Mazumdar [38] provide a pathwise proof for rate stability of the workload process in a multiserver queue with *FCFS* discipline. See, also, Mazumdar et al. [51] and Altman, Foss, Riehl, and

Stidham [1].

In this and subsequent sections, we investigate sample-path conditions for stability in general input-output processes. A process is said to be *rate stable* if its evolution is $o(t)$ as $t \to \infty$ (see El-Taha and Stidham [23, 24, 26], Stidham and El-Taha [81], Mazumdar et al. [52], Guillemin and Mazumdar [38], and Altman, Foss, Riehl, and Stidham [1]). We establish conditions for rate stability that can be verified from information on input (primary) processes in a deterministic framework that makes it possible to characterize the sample-path behavior of non-stationary stochastic processes. We also discuss the relation between rate stability and other forms of stability such as average stability and tightness of the family of empirical distributions associated with the process under investigation.

We shall work with the framework of Section 5.4, that is, our object of study will be a deterministic one-dimensional continuous-time process $Z = \{Z(t), t \geq 0\}$ with state space $S = R$. In this section, we shall also assume that $\{Z(t), t \geq 0\}$ is an *input-output* process, in which $Z(t) \geq 0$ represents quantity (e.g., number of customers or workload) in a system.

5.10.1 Definitions and characterization

In the derivation of the *RCL* in Section 5.3, the condition $t^{-1}x(t) \to 0$ played a crucial role as a necessary and sufficient condition for rate conservation to hold. It could be regarded as a kind of sample-path *stability* condition. In the present context, we shall, also, find that the condition $t^{-1}Z(t) \to 0$ acts as a stability condition.

Suppose then that $S = R^+$ with the Borel σ-field $\mathfrak{B}(R^+)$ of subsets generated by the open sets. As usual, we assume that $\{Z(t), t \geq 0\}$ is right continuous with left-hand limits. Suppose also that

$$Z(t) = Z(0) + A(t) - D(t), \; t \geq 0, \tag{5.70}$$

where $A(t)(D(t))$ is the cumulative input (output) to the system in $[0, t]$ and $\{A(t), \; t \geq 0\}$ and $\{D(t), t \geq 0\}$ are both nondecreasing, right-continuous processes. Thus $Z(t)$ has bounded variation on finite t-intervals.

Definition 5.57. An input-output process $\{Z(t), t \geq 0\}$ is said to be *rate stable* if

$$\lim_{t \to \infty} t^{-1}Z(t) = 0. \tag{5.71}$$

The following lemma is immediate from (5.70).

Lemma 5.58. *Suppose* $t^{-1}A(t) \to \alpha < \infty$ *as* $t \to \infty$. *Then the input-output process* $\{Z(t), t \geq 0\}$ *is stable if and only if* $t^{-1}D(t) \to \alpha$ *as* $t \to \infty$.

Thus, an input-output process is rate stable if the long-run input and output rates exist and are finite and equal. Rate stability embodies at least one intuitive notion of stability, which corresponds to the behavior exhibited by queues and other stochastic systems in which the long-run input and output rates are in balance. We have used the term *rate stable* in order to emphasize the distinction between our definition and the more familiar definitions in the stochastic literature, e.g., the existence of a limiting (or limiting average) probability distribution (cf. e.g. Rolski [62]). Another time-averaging concept of sample-path stability is given in the following definition.

Definition 5.59. An input-output process $\{Z(t), t \geq 0\}$ is said to be *average stable* if

$$\lim_{t\to\infty} t^{-1}\int_0^t Z(s)ds = Z < \infty. \quad (5.72)$$

Average stability appears to be stronger than rate stability. However, when it comes to pathwise behavior of an input-output process, the existence and finiteness of the asymptotic input rate and average stability are both needed to guarantee rate stability as the following theorem shows.

Theorem 5.60. *Consider an input-output process defined by* (5.70) *that is average stable. Suppose that there exists a constant* α, $0 < \alpha < \infty$ *such that*

$$t^{-1}A(t)\to\alpha, \text{ as } t\to\infty. \quad (5.73)$$

Then the input-output process $\{Z(t), t \geq 0\}$ *is rate stable.*

Theorem 5.60 is a slightly stronger version than Lemma 2.4 proved in El-Taha and Stidham [24].

It is well-known that if $n^{-1}\sum_{k=1}^{n} x_k \to x$ as $n\to\infty$, $0 < x < \infty$, then $n^{-1}x_n \to 0$ as $n\to\infty$ (cf, e.g., Section 5.8). In the continuous-variable case additional conditions are needed for a similar result to hold as indicated in Theorem 5.60.

Example 5.61. We give a counterexample to show that average stability is not, by itself, sufficient for rate stability to hold. Let

$$Z(t) = \begin{cases} k, & k - 1/k \leq t < k, \quad k = 1,2,\ldots \\ 0, & \text{otherwise.} \end{cases}$$

Now,

$\dfrac{Z(k-1/k)}{k-1/k} = \dfrac{k}{k-1/k} \to 1$ as $k\to\infty$. (In fact, $0 = \liminf Z(t)/t < \limsup Z(t)/t = 1$.)

Therefore, $Z(t)/t \not\to 0$ as $t\to\infty$. Now,

$$t - 1 \leq \int_0^t Z(s)ds \leq t$$

for all $t > 0$. Divide by t, then take limits as $t\to\infty$ to obtain $Z = 1$. This shows that both (5.73) and (5.72) are needed for rate stability to hold.

The following theorem gives another sufficient condition for rate stability. (See El-Taha and Stidham [26].)

Theorem 5.62. *Suppose* $t^{-1}A(t)\to\alpha$ *as* $t\to\infty$, *where* $0 < \alpha < \infty$. *Suppose also that*

$$\lim_{z\to\infty} \limsup_{t\to\infty} t^{-1}\int_0^t 1\{Z(s) > z\}ds = 0. \quad (5.74)$$

Then the input-output process $\{Z(t), t \geq 0\}$ *is rate stable.*

Remark 5.63. Condition (5.74) is equivalent to an assertion of *tightness* of the family of distribution functions,

$$F_t(z) := t^{-1}\int_0^t 1\{Z(s) \leq z\}ds, t \geq 0.$$

See, e.g., Billingsley [4].

The following corollary applies this result to the case where there is a proper limiting frequency distribution for $Z(t)$. The proof is an immediate consequence of Prohorov's theorem ([4], p. 37).

Corollary 5.64. *Suppose* $t^{-1}A(t) \rightarrow \alpha$ *as* $t \rightarrow \infty$*, where* $0 < \alpha < \infty$*. Suppose also that* $F_t(z) \rightarrow F(z)$ *as* $t \rightarrow \infty$*,* $z \geq 0$*, where* $F(\infty) = 1$*. (That is, F is a proper probability distribution.) Then the input-output process* $\{Z(t), t \geq 0\}$ *is rate stable.*

Remark 5.65. Theorem 5.62 and Corollary 5.64 can be applied to the process $\{x(t), t \geq 0\}$, considered in the derivation of the *RCL* in Section 5.3, to obtain sufficient conditions for $t^{-1}x(t) \rightarrow 0$, provided that $\{x(t), t \geq 0\}$ has bounded variation on finite t-intervals, as we noted in Remark 5.12.

Remark 5.66. For some special cases of input-output processes, such as the workload in a G/G/1 queue, it is possible to demonstrate rate stability under elementary conditions on *primary* quantities, such as the input rate and the output rate (while in sufficiently large states). (See subsection 5.10.2 below, and Stidham and El-Taha [81] and the references cited therein.) By contrast, Theorem 5.62 and Corollary 5.64 give sufficient conditions for rate stability that are expressed in terms of *secondary* quantities, namely, the occupation-time process associated with $\{Z(t), t \geq 0\}$. In fact, the main contribution of Theorem 5.62 and Corollary 5.64 lies, not in establishing easily verified sufficient condition for rate stability, but rather in showing that rate stability is a *necessary* condition for tightness of the family of distributions $\{F_t(\cdot), t \geq 0\}$ (condition (5.74)) and the convergence of $F_t(\cdot)$ to a proper distribution $F(\cdot)$.

5.10.2 Rate stability condition

In this subsection, we focus on rate stability. In particular, we consider the *input-output* process defined by (5.70) and show that this process is rate stable if the exogenous input rate and the conditional output rate, while the process is in sufficiently large states, are both well-defined, and the latter is greater than the former. These rate stability conditions can be checked from parameters of primary processes and thus can be verified *a priori*. We give a preliminary result that is of independent interest.

Lemma 5.67. *Consider the input-output process* $\{Z(t), t \geq 0\}$ *defined by* (5.70). *Let* α *and* δ*, be nonnegative constants. Suppose*

(*a*) *the input process satisfies*

$$\limsup_{t \rightarrow \infty} t^{-1}A(t) \leq \alpha, \text{ and} \tag{5.75}$$

(*b*) *the output process satisfies*

$$\liminf_{t \rightarrow \infty} \frac{\int_0^t 1\{Z(s) > z_0\} dD(s)}{\int_0^t 1\{Z(s) > z_0\} ds} \geq \delta, \tag{5.76}$$

where $0 < \alpha < \delta < \infty$.

Then, the event $\{Z(t) \leq z_0\}$ *occurs infinitely often as* $t \rightarrow \infty$*. That is, for every* $T_0 \geq 0$*, there exists a* $t \geq T_0$ *such that* $Z(t) \leq z_0$*.*

Remark 5.68. The above result is of independent interest. It gives sufficient conditions for the existence of construction points (points that start busy periods) for the $\{Z(t), t \geq 0\}$ process (see Baccelli and Brémaud [3], pp. 38-46). For example, let $Z(t)$ represent the queue length at time t in a G/G/1 queue. Then $\{Z(t) = 0\}$ represents the event that the server is idle. Let $\{T_n, n \geq 1\}$ be a sequence of arrival instants such that $Z(T_n) > 0$ for all $n \geq 1$. Then, the sequence

$\{b_k, k \geq 1\}$ of construction points is defined by $b_0 = 0$ and $b_k: = min\{T_n : T_n > b_{k-1}, Z(T_n -) = 0\}$, for $k = 1, \ldots$.

The following theorem gives sufficient conditions for rate stability of the input-output process $\{Z(t), t \geq 0\}$ (see Stidham and El-Taha [81]).

Theorem 5.69. *Consider the input-output process* $\{Z(t), t \geq 0\}$ *defined by* (5.70). *Suppose*

$$\lim_{t \to \infty} t^{-1} A(t) = \alpha, \tag{5.77}$$

and there exists a real number $z_0 \geq 0$ *such that*

$$\lim_{t \to \infty} \frac{\int_0^t 1\{Z(s) > z_0\} dD(s)}{\int_0^t 1\{Z(s) > z_0\} ds} = \delta, \tag{5.78}$$

where $0 < \alpha < \delta$. *Then the process* $\{Z(t), t \geq 0\}$ *defined by* (5.70) *is rate-stable.*

Remark 5.70. Since $Z(t)$ has bounded variation on finite t-intervals, $1\{Z(t) > z_0\}$ is a measurable function of t, and hence, the Lebesgue-Stieltjes integral $\int_0^t 1\{Z(s) > z_0\} dD(s)$ is well-defined for each $t \geq 0$.

We point out that δ may be interpreted as the long-run average amount of work that can be processed per unit time by the server, while busy.

In general, we allow for the possibility that some output can occur while $Z(t) \leq z_0$. In many applications (e.g., the G/G/1 queue) this is not the case: either the server is busy and producing output at rate δ or it is idle and producing no output. In such cases, the following corollary of Theorem 5.69 provides additional results for the special case $z_0 = 0$. (It actually makes a slightly weaker assumption.)

Corollary 5.71. *Suppose the conditions* (5.77) *and* (5.78) *of Theorem* 5.69 *are satisfied with* $z_0 = 0$, *and* $0 < \alpha < \delta$. *Suppose also that*

$$\lim_{t \to \infty} t^{-1} \int_0^t 1\{Z(s) = 0\} dD(s) = 0. \tag{5.79}$$

Then $p(0): = lim_{t \to \infty} t^{-1} \int_0^t 1\{Z(s) = 0\} ds$ *is well-defined and*

$$p(0) = 1 - \rho, \tag{5.80}$$

where $\rho: = \alpha / \delta$.

Proof. In (5.70), divide by t and take limits as $t \to \infty$, using Theorem 5.69, to obtain

$$0 = \alpha - \delta \lim_{t \to \infty} t^{-1} \int_0^t 1\{Z(s) > 0\} ds, \tag{5.81}$$

from which the desired result follows, upon noting that

$$p(0) = 1 - lim_{t \to \infty} t^{-1} \int_0^t 1\{Z(s) > 0\} ds. \qquad \square$$

Remark 5.72. Corollary 5.71 provides a pure sample-path proof of (5.80), making assumptions only on "primary" quantities, thus generalizing the result for the workload in a G/G/1 queue in Guillemin et al. [37]. Note that the limit on the r.h.s. of (5.81) is well defined, by the assumptions of the corollary. El-Taha and Stidham [27] extend Theorem 5.69 and Corollary 5.71 to multiserver input-output processes.

Remark 5.73. If the limit in (5.79) exists but does not equal zero, then one can obtain an extension of the above corollary. Specifically, suppose that

$$\lim_{t\to\infty} \frac{\int_0^t 1\{Z(s)=0\}dD(s)}{\int_0^t 1\{Z(s)=0\}ds} = \beta. \tag{5.82}$$

Then $\beta < \delta$ and

$$\lim_{t\to\infty} t^{-1}\int_0^t 1\{Z(s)=0\}ds = \frac{\delta-\alpha}{\delta-\beta}. \tag{5.83}$$

This formula allows for output to occur when $Z(t)=0$ (e.g., a server is idle).

5.10.3 Busy period fluctuations

In this section, we show that, under the conditions for rate stability of $\{Z(t), t\geq 0\}$, the sequence of durations of busy cycles is also rate stable. Consider a single-server system. Then, Lemma 5.67 shows the existence of infinitely many cycles; in other words, it shows the existence of an infinite sequence $\{T_n\}$ such that $T_n \to \infty$ as $n\to\infty$ and $Z(T_n)=0$. Our next result shows that in single-server queues busy cycles are rate stable.

Let $b_0=0$ and, for $n=1,2,\ldots$, define

$$b_n = inf\{t > b_{n-1}: Z(t-)=0, Z(t)>0\},$$

$$e_n = inf\{t > b_n: Z(t-)>0, Z(t)=0\},$$

and

$$B_n = e_n - b_n.$$

Now, let

$$I_n: = b_{n+1} - e_n,$$

$$C_n: = B_n + I_n,$$

and

$$\tilde{A}_n: = A(b_{n+1}) - A(b_n).$$

We interpret b_n and e_n as the beginning and the end, respectively, of the nth busy period. We also interpret B_n, I_n, C_n, and $\tilde{A}_n$, respectively, as the length of the nth busy period, idle period, and busy cycle, and the input during the nth busy cycle. Under the conditions of Lemma 5.67, $b_n\to\infty$ as $n\to\infty$.

Theorem 5.74. *Consider the input-output process* $\{Z(t), t\geq 0\}$ *defined by* (5.70). *Suppose that*

$$\lim_{t\to\infty} t^{-1}\int_0^t 1\{Z(s)=0\}dD(s) = 0. \tag{5.84}$$

Suppose also that the input process satisfies

$$\lim_{t\to\infty} t^{-1}A(t) = \alpha, \tag{5.85}$$

and the output process satisfies

$$\lim_{t\to\infty} \frac{\int_0^t 1\{Z(s) > 0\}dD(s)}{\int_0^t 1\{Z(s) > 0\}ds} = \delta, \tag{5.86}$$

where $0 \leq \alpha < \delta$. *Then,*

(*a*) $C_n/b_n \to 0$, *as* $n \to \infty$, *and*

(*b*) $\tilde{A}_n/b_n \to 0$, *as* $n \to \infty$.

It follows from Theorem 5.74 that busy-period and idle-period durations are rate stable.

Remark 5.75. Theorem 5.74 remains valid if we replace the hypothesis that the conditions of Corollary 5.71 are satisfied by the condition that the limit

$$p(0) := \lim_{t\to\infty} t^{-1} \int_0^t 1\{Z(s) = 0\}ds$$

is well-defined and $0 < p(0) \leq 1$

Theorem 5.74 and its extension to multiserver queues are given in El-Taha and Stidham [27].

5.11 STABILITY APPLICATIONS

In order to be useful in applications to stochastic models, the sufficient conditions for rate stability given in Theorem 5.69 should be easily verified *a priori*. That is, it should be possible to calculate the rates α and δ and check whether $\alpha < \delta$, based on primary quantities associated with the definition of the model, such as the arrival processes, number of servers, and service requirements in the system model. In this section, we give several applications in which this is the case.

5.11.1 Rate stability conditions for G/G/1 queues

The G/G/1 queue is defined by the sequence $\{(A_n, S_n), n \geq 1\}$, where A_n is the time of nth arrival, and S_n is the server requirement of the nth arrival. (Customers need not be served in order of arrival, but a server is never idle when customers are waiting.) It is also assumed that the server works at nonnegative rate $\delta < \infty$. Let $N(t) = max\{n: A_n \leq t\}$ denote the number of customers that arrive in $[0,t]$. In all applications, we assume that servers are kept busy whenever possible. We also assume a work-conserving queue discipline.

The workload of a G/G/1/∞ queue. In this special case we have $\{Z(t), t \geq 0\} \equiv \{W(t), t \geq 0\}$, and $A(t) = \sum_{k=1}^{N(t)} S_k$. Here $\{W(t), t \geq 0\}$ is the workload process. The output process is given by $D(t) = \int_0^t \delta 1\{Z(s) > 0\}ds$, for all $t \geq 0$. Thus we have

$$W(t) = \sum_{k=1}^{N(t)} S_k - \int_0^t \delta 1\{Z(s) > 0\}ds. \tag{5.87}$$

Note that under these conditions the limit in (5.78) exists and is equal to 1. Thus we obtain the following result.

Theorem 5.76. *Consider the workload process* $\{W(t), t = 0\}$ *in the* G/G/1 *queue described by* (5.87). *Suppose*

(a) $lim_{t\to\infty} t^{-1}N(t) = \lambda$, *where* $0 \le \lambda < \infty$;
(b) $lim_{n\to\infty} n^{-1}\sum_{k=1}^{n} S_k = S$, *where* $0 \le S < \infty$,
with $\alpha := \lambda S < \delta$. *Then the workload process* $\{W(t), t \ge 0\}$ *is rate stable.*

Remark 5.77. Consider the case where $\{Z(t), t \ge 0\}$ is the workload process in a G/G/1-*FCFS* queue. Under the above conditions, we have

$$\frac{W_k}{T_k} = \frac{Z(T_k -)}{T_k} \to 0,$$

and hence, $(W_k + S_k)/T_k \to 0$ as $k \to \infty$, where W_k and T_k are, respectively, the waiting time in queue and the arrival time of customer k. Thus, it follows from the results in Heyman and Stidham [40] that

$$\lim_{t\to\infty} t^{-1} \int_0^t Z(s)ds = \lambda[\lim_{n\to\infty} n^{-1} \sum_{k=1}^{n} (W_k S_k + S_k^2/2)],$$

if either or these limits is well-defined. Thus we see that, in order for the limiting average $lim_{t\to\infty} t^{-1} \int_0^t Z(s)ds$ to be finite, it is necessary that the empirical second moment of the work requirement, i.e.,

$$\lim_{n\to\infty} n^{-1} \sum_{k=1}^{n} S_k^2,$$

be finite. A similar observation has been made by Mazumdar et al. [52], under the additional assumption that $Z(t)^2 = o(t)$ as $t \to \infty$.

Remark 5.78. Note that batch arrivals and/or bulk service are allowed, provided the batch and/or bulk size is bounded. Theorem 5.76 extends results in [81], [38], and [52]. Guillemin and Mazumdar [38] provide an alternate pathwise proof of Theorem 5.76 in the special case when the servers are homogeneous and the discipline is *FCFS*.

Suppose that the hypotheses of Theorem 5.76 hold. Then $\delta = 1$, $\alpha = \lambda S < 1$, thus (5.80) reduces to $p(0) = 1 - \rho$, where $\rho := \lambda S$. This gives a set of conditions under which $0 < p(0) \le 1$. Consequently, it follows from Remark 5.75, following Theorem 5.74, that the busy period durations and the inputs during busy periods are rate stable for this queueing system.

Number of customers in a G/G/1/∞ queue. In this special case, $\{Z(t), t \ge 0\} \equiv \{L(t), t \ge 0\}$ is the queue-length process. That is, $L(t)$ is the number of customers in the system (including customers in service). Customers need not be served in order of arrival, but a server is never idle when customers are waiting. Let $A(t) \equiv N(t)$ denote the number of customers that arrive and $D(t) = \sum_{i=1}^{c} \int_0^t 1\{Z(s) > 0\}dD(s)$ the number that depart in $[0,t]$. Then $Z(t) = L(t)$ defined by (5.70) is the number of customers in the system at time t.

Theorem 5.79. *Consider the* G/G/1 *queue described in this subsection. Suppose*

$$\lim_{n\to\infty} n^{-1} A_n = \lambda^{-1}, \quad 0 \le \lambda < \infty,$$

$$\lim_{n\to\infty} n^{-1} \sum_{k=1}^{n} S_k = S, \quad 0 \le S < \infty,$$

and that $\lambda S < \delta$. *Then the queue-length process* $\{L(t), t \ge 0\}$ *is rate stable.*

Proof. The proof follows by appealing to Theorem 5.69, Theorem 5.74 and Theorem 5.76. □

Batch arrivals and/or bulk service are permitted in this model by setting the appropriate A_n and S_n equal to zero.

5.11.2 Infinite server queues: G/G/∞/∞

In this special case, we consider rate stability of infinite server queues. In this case, sufficient conditions for rate stability can be verified from conditions on primary quantities, that is, the arrival process and service requirements in the system model, by appealing to Theorem 5.60. We assume that servers are kept busy whenever possible. We also assume a work-conserving queue discipline. Now, define

$$L(t) = \sum_{k=1}^{\infty} 1\{A_k \le t < A_k + S_k\}, \quad t > 0, \tag{5.88}$$

and

$$W(t) = \sum_{k=1}^{N(t)} S_k - \int_0^t L(s)ds, \tag{5.89}$$

where $\{L(t), t \ge 0\}$ and $\{W(t), t \ge 0\}$ are the queue-length and workload processes, respectively. Note that $S_k = \int_0^{\infty} 1\{T_k \le s < T_k + S_k\}ds$. Thus we obtain the inequality (see Section 5.8)

$$\sum_{k: T_k \le t} S_k \ge \int_0^t L(s)ds \ge \sum_{k: T_k + S_k \le t} S_k. \tag{5.90}$$

The following result given in El-Taha [21] gives a natural set of primary conditions for the rate stability of both the workload and the queue-length processes.

Theorem 5.80. *Consider the infinite server queue described by* (A_k, S_k). *Suppose* $n^{-1}A_n \to \lambda^{-1}$ *as* $n \to \infty$, $0 < \lambda < \infty$; *and* $n^{-1}\sum_{k=1}^{n} S_k \to S$ *as* $n \to \infty$, $0 < S < \infty$. *Then,*

(*a*) *the queue-length process* $\{L(t), t \ge 0\}$ *is rate stable, and*

(*b*) *the workload process* $\{W(t), t \ge 0\}$ *is rate stable.*

Proof. It follows from Lemma 5.43 and the inequality (5.90) that $\{L(t), t \ge 0\}$ is average stable; therefore, it is rate stable by Theorem 5.60. Part (*b*) follows if we divide (5.89) by t and take limits as $t \to \infty$. □

5.11.3 More examples

In this section we give several additional models in which sufficient conditions for rate stability can be verified *a priori*.

A reservoir with state-dependent release rule. Let $A(t)$ be an arbitrary nondecreasing right-continuous function, and suppose $D(t) = \int_0^t \delta(A(s))ds$, where $\delta(z)$ is a nonnegative Lebesgue-measurable function of $z \in [0, \infty)$, with $\delta(z) = \delta$ for all $z > z_0 \ge 0$. Here, we interpret $A(t)$ as the amount of water entering the reservoir in $[0, t]$ and $\delta(z)$ as the state-dependent release rate, so that $Z(t)$ defined by (5.70) is the amount of water in the reservoir at time t. Then, condition (5.78) of Theorem 5.69 holds. If in addition, condition (5.77) holds, with $0 < \alpha < \delta$, then Theorem 5.69 implies that $\{Z(t), t \ge 0\}$ is rate-stable.

As in the G/G/1 workload model, if $A(t) = \sum_{k=1}^{N(t)} S_k$, where $t^{-1}N(t) \to \lambda$ as $t \to \infty$, and $n^{-1}\sum_{k=1}^{n} S_k \to S$ as $n \to \infty$, then $t^{-1}A(t) \to \alpha = \lambda S$ as $t \to \infty$, provided λS is well-defined. But Theorem 5.69 allows much more general input processes. All that is required is that the condition (5.77) hold, with $\alpha < \delta$. For example, the input process $\{A(t), t \geq 0\}$, could be a Levy process (a limit of compound Poisson processes, with an infinite number of jumps in every finite time interval) or it could have both pure-jump and continuous components.

Vacation model with threshold. Consider a G/G/1 queue in which the server is turned off, when the workload reaches zero, and remains off until the workload reaches the threshold level z_0, at which point the servers is turned on and left on until the workload again reaches zero. While on, the server completes work at a constant rate δ. As in our first example, we have $A(t) = \sum_{k=1}^{N(t)} S_k$, where $N(t)$ is the number of customer arrivals in $[0, t]$, and S_k is the work requirement of customer k. If $t^{-1}A(t) \to \alpha < \delta$, then Theorem 5.69 implies that the system is rate-stable.

Fluid model with time-varying flow rates. Suppose $A(t) = \int_0^t \alpha(X(s))ds$, and $D(t) = \int_0^t 1\{Z(s) > 0\}\delta(X(s))ds$, $t \geq 0$, where $X(t)$ is a nonnegative, measurable real-valued function of $t \geq 0$, and both $\alpha(x)$ and $\delta(x)$ are nonnegative, measurable functions of $x \geq 0$. Thus, input and output occur at flow rates that depend on the *environment*, which is modeled by the auxiliary process, $\{X(t), t \geq 0\}$. In stochastic fluid models, $\{X(t), t \geq 0\}$ is typically a finite-state, continuous-time Markov chain, with stationary distribution $\{\pi(j), j = 1, 2, \ldots, m\}$. In this case, we have (*a.s.*)

$$\lim_{t \to \infty} t^{-1} \int_0^t \alpha(X(s))ds = \sum_{j=1}^{m} \pi(j)\alpha(j),$$

$$\lim_{t \to \infty} t^{-1} \int_0^t \delta(X(s))ds = \sum_{j=1}^{m} \pi(j)\delta(j),$$

and Theorem 5.69 can be used to show that $\{Z(t), t \geq 0\}$ is rate-stable if

$$\alpha = \sum_{j=1}^{m} \pi(j)\alpha(j) < \sum_{j=1}^{m} \pi(j)\delta(j) < \delta.$$

5.11.4 Little's formula for stable queues

Recall from Section 5.8 (Lemma 5.43) that the condition that $W_n/A_n \to 0$ as $n \to \infty$ is necessary and sufficient for $L = \lambda W$, when λ and W are both well-defined. This condition cannot be verified directly (without additional assumptions) from input data such as interarrival and service times. In this section, however, we show that the *rate stability* conditions established in Section 5.10 are sufficient to verify that the above condition for Little's formula holds.

The single-server case. The input data for the G/G/1 queue consists of the sequence $\{(A_n, S_n), n \geq 1\}$, where A_n is the arrival instant and S_n the work requirement of customer n, $n \geq 1$. In the special case where $\{Z(t), t \geq 0\}$ is the workload process in the G/G/1 queue, we have $A(t) = \sum_{k=1}^{N(t)} S_k$ and $D(t) = \int_0^t \delta 1\{Z(s) > 0\}ds$, for all $t \geq 0$. Here $\{N(t), \ t \geq 0\}$ is the point process which counts the number of customer arrivals: $N(t) = max\{n: A_n \leq t\}$, $t \geq 0$.

Our goal is to prove that $L = \lambda W$ under minimal stability conditions on the input process, $\{A(t), t \geq 0\}$.

Theorem 5.81. *Suppose* $t^{-1}N(t) \to \lambda$ *as* $t \to \infty$, *and* $n^{-1}\sum_{k=1}^{n} S_k \to S$ *as* $n \to \infty$, *where* $0 < \alpha := \lambda S < \delta$. *Then* $W_n / A_n \to 0$ *as* $n \to \infty$ *and hence* (a) *and* (b) *of Theorem* 5.44 *hold.*

Proof. Let $b(n) := \sum_{k=1}^{\infty} b_k 1\{b_k \leq A_n < e_k\}$ and $B(n) := \sum_{k=1}^{\infty} (e_k - b_k) 1\{b_k \leq A_n < e_k\}$. That is, $b(n)$ and $B(n)$ are the beginning and the duration, respectively, of the busy period that corresponds to the nth arrival. It follows from Theorem 5.74 that

$$A_n^{-1} W_n \leq A_n^{-1} B(n) \leq b(n)^{-1} B(n) \to 0 \text{ as } n \to \infty.$$

The desired result then follows by appealing to Theorem 5.44. □

The multiserver case. In the special case of G/G/c queue, we have $A(t) = \sum_{k=1}^{N(t)} S_k$ for all $t \geq 0$. Here again, $\{N(t), t \geq 0\}$ is the point process which counts the number of customer arrivals, and $\{S_k, k \geq 1\}$ is the sequence of work requirements of the arriving customers.

Unlike the single-server case, in multiserver queues, we need the additional assumption that the queue discipline is work-conserving and nonpreemptive. The following result is proved in El-Taha and Stidham [27].

Corollary 5.82. *Assume the queue discipline is work-conserving and nonpreemptive. Suppose also the conditions of Theorem* 5.81 *hold. Then* $W_n / A_n \to 0$ *as* $n \to \infty$ *and hence,* (a) *and* (b) *of Theorem* 5.44 *hold.*

The above corollary is valid under rather weak conditions, for example, servers can be heterogeneous as well as homogeneous. Furthermore, although a customer cannot be preempted once in service, it may be switched from a fast to a slow server or vice versa.

We note also that, in the stable versions of $L = \lambda W$, we did not require the assumption that the sequence of departure times be finite.

5.12 OPEN PROBLEMS AND NEW RESEARCH DIRECTIONS

Fundamental relations. In this chapter, we address fundamental relations between a process and an imbedded point process. Those relations have been shown to hold in great generality under the assumption that relevant limits exist. The proofs were basic and elementary. Our approach reveals that stochastic assumptions, such as stationarity, are needed only to prove the existence of limits and/or to construct stationary versions of the processes in question. One question of interest is to introduce more such relations in this deterministic framework. Another interesting issue is to investigate the minimal stochastic assumptions necessary to insure the existence of such processes. The section on martingale *ASTA* is a model of what we have in mind.

Stability. In the section on stability we discussed, in some depth, *rate stability* of input-output processes and its consequences. We briefly introduced the notion of *average stability*. Moreover, one can treat the notion of pathwise tightness as a form of stability. The later two notions of stability were briefly discussed. The results presented in this chapter and references therein represent a first step in that direction. Investigating those types of pathwise-stability conditions and other forms of stability is a promising area of research that still need to be explored.

Applying those concepts in the study of communication systems, also, is a promising field of research. A step in that direction has been the work of Altman et al. [1].

One specific problem is to investigate under what conditions can one prove a converse for Theorem 5.62.

Networks of queues. A question of interest is how can we specialize and apply the fundamental relations, discussed above, in the case of queueing networks. Furthermore, in the case of stability, we focused on input-output processes with emphasis on single and multichannel queues. The question here is how to extend those results to cover open queueing networks. Since *rate stability* is "weaker" than classical stability, which requires the existence of "stationary distributions", we expect *rate stability* to hold in open queueing networks, without even Markovian routing, under conditions that mimic and extend those discussed in the stability section.

Another problem of interest is to investigate sample-path stable versions of Little's formula for networks of queues.

Insensitivity. Only a few pathwise insensitivity results have appeared in the literature so far. One such result is the insensitivity of the *LCFS-PR* queue. Pathwise results for round robin discipline and variants of the infinite server and the Loss models are under investigation. Using sample-path analysis, how much more can we say about this issue?

Communication networks. A basic characteristic of communication networks is that they reach steady state rather fast. Asymptotic analysis becomes essential to understand the behavior of models representing such systems. Sample-path analysis which is well suited to study limiting behavior, when combined with other approaches, can provide a powerful tool to study the performance of these models.

BIBLIOGRAPHY

[1] Altman, E., Foss, S.G., Riehl, E.R., and Stidham Jr., S., Performance bounds and pathwise stability for generalized vacation and polling systems, Technical Report UNC/OR TR93-8, Dept. of Oper. Res., Univ. of North Carolina at Chapel Hill 1993.

[2] Asmussen, S., *Applied Probability and Queues*, John Wiley, New York 1987.

[3] Baccelli, F. and Brémaud, *Palm Probabilities and Stationary Queues*, Lecture Notes in Statistics **41**, Springer-Verlag, New York 1987.

[4] Billingsley, P., *Weak Convergence*, John Wiley, New York 1968.

[5] Brémaud, P., *Point Processes and Queues: Martingale Dynamics*, Springer-Verlag, New York 1981.

[6] Brémaud, P., Characteristics of queueing systems observed at events and the connection between stochastic intensity and Palm probability, *Queueing Sys.* **5** (1989), 99-112.

[7] Brémaud, P., An elementary proof of Sengupta's invariance relation and a remark on Miyazawa's conservation principle, *J. Appl. Prob.* **28** (1991), 950-954.

[8] Brémaud, P., Kannurpatti, R. and Mazumdar, R., Event and time averages: A review, *J. Appl. Prob.* **24** (1992), 377-411.

[9] Brill, P.H. and Posner, M.J.M., Level crossing in point process applied to queues: Single server case, *Oper. Res.* **25** (1977), 662-674.

[10] Brill, P.H. and Posner, M.J.M., The system point method in exponential queues: A level crossing approach, *Math. Oper. Res.* **6** (1981), 31-49.

[11] Brumelle, S., On the relation between customer and time averages in queues, *J. Appl. Prob.* **8** (1971), 508-520.

[12] Buzen, J.P., Fundamental operational laws of computer system performance, *Acta Informatica* **7** (1976), 167-182.

[13] Buzen, J.P., Operational analysis: An alternative to stochastic modeling, In: *Performance of Computer Installations*, ed. by D. Ferrari (1978), 175-194.

[14] Buzen, J.P. and Denning, P.J., Measuring and calculating queue length distributions, *IEEE Trans. Computer* **C-18** (1980), 33-44.

[15] Daley, D. and Vere-Jones, *An Introduction to the Theory of Point Processes*, Springer-Verlag, Berlin, New York 1988.

[16] Denning, P.J. and Buzen, J.P., The operational analysis of queueing network models, *Computing Surveys* **10** (1978), 225-261.

[17] Dshalalow, J., On applications of Little's formula, *J. Appl. Math. Stoch. Anal.* **6**:3 (1993), 271-276.

[18] El-Taha, M., *Sample-Path Analysis of Queueing Systems: New Results*, Ph.D. thesis, School of Engineering, Graduate Program in O.R., NCSU, Raleigh 1986.

[19] El-Taha, M., Sample-path relations between time averages and asymptotic distributions, preprint, Dept. of Math and Stats., USM (1989).

[20] El-Taha, M., On conditional *ASTA*: A sample-path approach, *Stoch. Mod.* **8** (1992), 157-177.

[21] El-Taha, M., A note on sample-path stability for infinite-server input-output processes, preprint (1994).

[22] El-Taha, M. and Stidham Jr., S., An extension of *ASTA*, Technical Report, UNC/OR TR90/6, Dept. of Oper. Res., Univ. of North Carolina at Chapel Hill (1990).

[23] El-Taha, M. and Stidham Jr., S., Sample-path analysis of stochastic discrete-event systems, In: *Proc. of the 30th IEEE CDC Meeting* (1991), 1145-1150.

[24] El-Taha, M. and Stidham Jr., S., Deterministic analysis of queueing systems with heterogeneous servers, *Theoretical Computer Science* **106** (1992), 243-264.

[25] El-Taha, M. and Stidham Jr., S., A filtered *ASTA* property, *Queueing Sys.* **11** (1992), 211-222.

[26] El-Taha, M. and Stidham Jr., S., Sample-path analysis of stochastic discrete-event systems, *Discrete Event Dynamic Systems* **3** (1993), 325-346.

[27] El-Taha, M. and Stidham Jr., S., Sample-path conditions for multiserver input-output processes, *J. Appl. Math. Stoch. Anal.* **7** (1994), 437-456.

[28] Ethier, S. and Kurtz, T., *Markov Processes, Characterization and Convergence*, John Wiley, New York 1986.

[29] Ferrandiz, J. and Lazar, A., Rate conservation for stationary processes, *J. Appl. Prob.* **28** (1991), 146-158.

[30] Franken, P., König, D., Arndt, U. and Schmidt, V., *Queues and Point Processes*, John Wiley 1982.

[31] Gelenbe, E., Stationary deterministic flow in discrete systems I, *Theor. Comp. Sci.* **23** (1983), 107-128.

[32] Gelenbe, E. and Finkel, D., Stationary deterministic flows II: The single ser-

ver queue, *Theor. Comp. Sci.* **52** (1987), 269-280.

[33] Ghahramani, S., On remaining full busy periods of GI/GI/c queues and their relation to stationary point processes, *J. Appl. Prob.* **27** (1990), 232-236.

[34] Glynn, P. and Whitt, W., A central limit theorem version of $L = \lambda W$, *Queueing Sys.* **2** (1986), 191-215.

[35] Glynn, P. and Whitt, W., Extensions of the queueing relations $L = \lambda W$ and $H = \lambda G$, *Oper. Res.* **37** (1989), 634-644.

[36] Green, L. and Melamed, B., An Anti-*PASTA* result for Markovian systems, *Oper. Res.* **38** (1990), 173-175.

[37] Guillemin, F., Badrinath, V. and Mazumdar, R., Les techniques trajectorielles appliquées aux files d'attente nonstationnaires, preprint (1991).

[38] Guillemin, F. and Mazumdar, R., On pathwise behavior of multiserver queues, *Queueing Sys.*, **15** (1994), 279-288.

[39] Halfin, S. and Whitt, W., An extremal property of the *FIFO* discipline via an ordinal version of $L = \lambda W$, *Stoch. Mod.* **5** (1989), 515-529.

[40] Heyman, D.P. and Stidham Jr., S., The relation between customer and time averages in queues, *Oper. Res.* **28** (1980), 983-994.

[41] König, D. and Schmidt, V., *EPSTA*: The coincidence of time-stationary and customer-stationary distributions, *Queueing Sys.* **5** (1989), 247-264.

[42] König, D. and Schmidt, V., Extended and conditional versions of the *PASTA* property, *Adv. Appl. Prob.* **22** (1990), 510-512.

[43] König, D., Miyazawa, D. and Schmidt, V., On the identification of Poisson arrivals in queues with coinciding time-stationary and customer-stationary state distributions, *J. Appl. Prob.* **20** (1983), 860-871.

[44] König, D., Schmidt, V. and Van Doorn, E.V., On the *PASTA* property and a further relationship between customer and time averages in stationary queueing systems, *Stoch. Mod.* **5** (1989), 261-272.

[45] Krickeberg, K., Processus pontuels en statistique, In: *Ecole d'été de Probabilités de Saint Flour X*-1980, ed. by P.L. Hennequin, Lecture Notes in Math **929** (1982), 206-313.

[46] Lindvall, T., A probabilistic proof of Blackwell's renewal theorem, *Ann. Prob.* **5** (1977), 482-485.

[47] Lindvall, T., On coupling of continuous time renewal process, *J. Appl. Prob.* **19** (1982), 82-89.

[48] Little, J.D.C., A proof for the queueing formula: $L = \lambda W$, *Oper. Res.* **9** (1961), 383-387.

[49] Liu, Z., Nain, P. and Towsley, D., Sample path methods in the control of queues, submitted for publication (1994).

[50] Loynes, R.M., The stability of queues with nonindependent interarrival and service times, *Proc. Camb. Phil. Soc.* **58** (1962), 497-520.

[51] Mazumdar, R., Badrinath, V., Guillemin, F. and Rosenberg, C., Pathwise rate conservation and queueing applications, preprint 1991.

[52] Mazumdar, R., Guillemin, F., Badrinath, V. and Kannurpatti, R., On pathwise behavior of queues, *Oper. Res. Letters* **12** (1992), 263-270.

[53] Mazumdar, R., Kannurpatti, R. and Rosenberg, C., On a rate conservation law for nonstationary processes, *J. Appl. Prob.* **28** (1991), 762-770.

[54] Melamed, B. and Whitt, W., On arrivals that see time averages, *Oper. Res.* **38** (1990), 156-172.

[55] Melamed, B. and Whitt, W., On arrivals that see time averages: A martin-

gale approach, *J. Appl. Prob.* **27** (1990), 376-384.

[56] Melamed, B. and Yao, D., The *ASTA* property, as Chapter 7 in this book, 195-224.

[57] Miyazawa, M., The derivation of invariance relations in complex queueing systems with stationary inputs, *Adv. Appl. Prob.* **15** (1983), 874-885.

[58] Miyazawa, M., The intensity conservation law for queues with randomly changed service rate, *J. Appl. Prob.* **22** (1985), 408-418.

[59] Miyazawa, M. and Wolff, R.W., Further results on *ASTA* for general stationary processes and related problems, *J. Appl. Prob.* **28** (1990), 729-804.

[60] Neveu, J., Processus ponctuels, In: *Ecole d'été de Probabilités de Saint Flour VI*-1976, ed by: P.L. Hennequin, Lecture Notes in Math. **598**, Springer-Verlag, Heidelberg (1977), 249-447.

[61] Newell, G., *Applications to Queueing Theory*, Chapman and Hall, London 1971.

[62] Rolski, T., *Stationary Random Processes Associated with Point Processes*, Lecture Notes in Statistics, Springer-Verlag, New York 1981.

[63] Rolski, T. and Stidham Jr., S., Continuous versions of the formulas $L = \lambda W$ and $H = \lambda G$, *Oper. Res. Letters* **2** (1983), 211-215.

[64] Sengupta, B., An invariance relation for the G/G/1 queue, *Adv. Appl. Prob.* **21** (1989), 956-957.

[65] Serfozo, R., Poisson functionals of Markov processes and queueing networks, *Adv. Appl. Prob.* **21** (1989), 595-611.

[66] Serfozo, R., More about Little's laws for waiting times and utility processes, School of Ind. and Sys. Eng., Georgia Institute of Technology 1993.

[67] Shaked, M. and Shanthikumar, G.J., Stochastic convexity and its applications, *Adv. Appl. Prob.* **20** (1988), 427-446.

[68] Shaked, M. and Shanthikumar, G.J., Regular, sample path and strong stochastic convexity: A review, *Stochastic Orders and Decision Under Risk,* IMS Lecture Notes-Monograph Series (1991), 320-333.

[69] Shanthikumar, G.J. and Sumita, U., On G/G/1 queues with *LIFO-P* service discipline, *J. Oper. Res. Soc. Japan* **29** (1986), 220-231.

[70] Shanthikumar, G.J. and Yao, D.D., Strong stochastic convexity: Closure properties and applications, *J. Appl. Prob.* **28** (1991), 131-145.

[71] Sigman, K., A note on a sample-path conservation law and its relation with $H = \lambda G$, *Adv. Appl. Prob.* **23** (1991), 662-665.

[72] Stidham Jr., S., $L = \lambda W$: A discounted analogue and a new proof, *Oper. Res.* **20** (1972), 708-732.

[73] Stidham Jr., S., Regenerative processes in the theory of queues with applications to the alternating-priority queue, *Adv. Appl. Prob.* **4** (1972), 542-577.

[74] Stidham Jr., S., A last word on $L = \lambda W$, *Oper. Res.* **22** (1974), 417-421.

[75] Stidham Jr., S., On the relation between time averages and customer averages in stationary random-marked processes, Technical report 79-1, Dept. of Ind. Eng., NCSU, Raleigh 1979.

[76] Stidham Jr., S., Sample-path analysis of queues, In: *Appl. Prob. Comp. Sci.: The Interface*, ed. by R. Disney and T. Ott (1982), 41-70.

[77] Stidham, Jr., S., On the relation between time averages and customer averages in queues, In: *Variational Methods and Stochastic Analysis*, ed. by H.J. Kimn and D.M. Chung, Proc. Workshop in Pure Math **9** (1990), 243-278.

[78] Stidham Jr., S., A comparison of sample-path proofs of $L = \lambda W$, preprint

(1991).

[79] Stidham Jr., S., A note on a sample-path approach to Palm probabilities, *J. Appl. Prob.* **31** (1994), (to appear).

[80] Stidham Jr., S and El-Taha, M., Sample-path analysis of processes with imbedded point processes, *Queueing Sys.* **5** (1989), 131-165.

[81] Stidham Jr., S. and El-Taha, M., A note on a sample-path stability conditions for input-output processes, *Oper. Res. Letters* **14** (1993), 1-7.

[82] Van Doorn, E.A. and Regterschot, G.J.K., Conditional *PASTA*, *Oper. Res. Letters* **7**:5 (1988), 229-232.

[83] Walrand, J., *An Introduction to Queueing Networks*, Prentice-Hall, NJ 1988.

[84] Watanabe, S., On discontinuous additive functionals and Levy measures of a Markov process, *Jpn. J. Math.* **34** (1964), 53-70.

[85] Whitt, W., A review of $L = \lambda W$ and extensions, *Queueing Sys.*, **9** (1991), 235-268.

[86] Whitt, W., Correction note on $L = \lambda W$, *Queueing Sys.* **15** (1992), 431-432.

[87] Whitt, W., $H = \lambda G$ and Palm transformation, *Adv. Appl. Prob.* **24** (1992), 755-758.

[88] Wolff, R., Poisson arrivals see time averages, *Oper. Res.* **30** (1982), 223-231.

[89] Wolff, R., *Stochastic Modeling and the Theory of Queues*, Prentice-Hall, New Jersey 1989.

[90] Wolff, R., A note on *PASTA* and anti-*PASTA* for continuous-time Markov chains, *Oper. Res.* **38** (1990), 176-177.

[91] Zazanis, M., Sample-path analysis of level crossing for the workload process, *Queueing Sys.*, **11** (1992), 419-428.

Chapter 6

Markov-additive processes of arrivals

António Pacheco[1] and N.U. Prabhu

ABSTRACT A Markov-additive process (MAP) of arrivals is a process $(\mathbf{X}, J)$ on the state space $\mathbb{N}^r \times E$ such that the increments in $\mathbf{X}$ correspond to arrivals. A typical example is that of different classes of arrivals at a queueing system. In this chapter, we investigate the lack of memory property, interarrival times and moments of the number of counts. We then consider transformations of the process that preserve the Markov-additive property, such as linear transformations, patching of independent processes, and linear combinations. Random time transformations are also investigated. Finally, we consider secondary recordings that generate new arrival processes from the original process. These include, in particular, marking, coloring, and thinning. For Markov-Bernoulli recording, the secondary process, in each case, turns out to be an MAP of arrivals.

CONTENTS

6.1	Introduction	167
6.2	Some properties of MAPs	173
6.3	MAPs of arrivals	175
6.4	Some properties of MAPs of arrivals	178
6.5	Transformations of MAPs of arrivals	183
6.6	Markov-Bernoulli recording of MAPs of arrivals	188
6.7	Brief discussion of some directions for future research	193
	Bibliography	193

6.1 INTRODUCTION

Markov-additive processes (MAPs) are a class of Markov processes which have important applications to queueing models. An MAP $(\mathbf{X}, J)$ is a Markov process whose transition probability measure is translation invariant in the *additive component* $\mathbf{X}$. In this chapter, we treat the case where the *Markov component* J has discrete (finite or countable) state space E.

We consider the vector space $\mathbb{R}^r$ with the usual componentwise addition of vectors and multiplication of a vector by a scalar and denote the points $\mathbf{x} = (x_1, x_2, \ldots, x_r)$ of $\mathbb{R}^r$, $r \geq 1$, by using boldface. We will let the index set $\mathcal{T}$ be either $\mathbb{R}_+ = [0, +\infty)$ or $\mathbb{N} = \{0, 1, 2, \ldots\}$.

Definition 6.1. **(Markov-Additive Process (MAP))** A process $(\mathbf{X}, J) = \{(\mathbf{X}, J)(t),\ t \in \mathcal{T}\}$ on the state space $\mathbb{R}^r \times E$ is an *MAP* if

---

[1]Research supported by Grant BD/645/90-RM from JNICT.

0-8493-8074-x/95/$0.00+$.50

(*i*) $(\mathbf{X}, J)$ is a Markov process;

(*ii*) for $s, t \in \mathcal{T}$, the conditional distribution of $[\mathbf{X}(s+t)-\mathbf{X}(s), J(s+t)]$, given $[\mathbf{X}(s), J(s)]$, depends only on $J(s)$.

We augment E with a compactification point Δ and let $E_\Delta = E \cup \{\Delta\}$. We denote $\zeta = inf\{t \in \mathcal{T}: J(t) = \Delta\}$ and assume that $J(t) = \Delta$ and $\mathbf{X}(t) - \mathbf{X}(\zeta) = \mathbf{0}$ almost surely (a.s.) for $t \geq \zeta$. We consider only MAPs $(\mathbf{X}, J)$ having a transition probability measure in the sense of Blumenthal ([3], Definition I.(2.1)). From Definition 6.1 it follows that the transition probability measure of an MAP is given by

$$\mathbf{P}\{\mathbf{X}(s+t) \in A, J(s+t) = k \mid \mathbf{X}(s) = \mathbf{x}, J(s) = j\}$$
$$= \mathbf{P}(\mathbf{X}(s+t) - \mathbf{X}(s) \in A - \mathbf{x}, J(s+t) = k \mid J(s) = j\} \tag{6.1}$$

for $s, t \in \mathcal{T}$, $j, k \in E_\Delta$, $\mathbf{x} \in \mathbb{R}^r$ and $A \in \mathcal{R}^r$, where $\mathcal{R}^r$ is the σ-algebra of Borel subsets of $\mathbb{R}^r$. Thus, an MAP is a Markov process whose transition probability measure is translation invariant in the additive component. Since $(\mathbf{X}, J)$ is Markov, it follows easily from (6.1) that J is Markov and that $\mathbf{X}$ has conditionally independent increments. Given the states of J, i.e., for $0 \leq t_1 \leq \ldots \leq t_n (n \geq 2)$, the increments

$$\mathbf{X}(t_1) - \mathbf{X}(0), \mathbf{X}(t_2) - \mathbf{X}(t_1), \ldots, \mathbf{X}(t_n) - \mathbf{X}(t_{n-1})$$

are conditionally independent given $J(0)$, $J(t_1), \ldots, J(t_n)$. Since, in general, $\mathbf{X}$ is non-Markovian, we may, therefore, call J the Markov component and $\mathbf{X}$ the additive component of the MAP $(\mathbf{X}, J)$.

In the trivial case where J has only one state, $\mathbf{X}$ is a Markov process with additive increments. Thus, MAPs are extensions of processes with additive increments, but, in general, the additive component of an MAP does not have additive increments.

Definition 6.1 is in the spirit of Çinlar [5], who considers a more general state space for J. A second definition of MAPs is given by Çinlar [6], who starts with the Markov component J and defines $\mathbf{X}$ as a process having properties relative to J that imply, in particular, property (*ii*) of Definition 6.1. In discrete time, this amounts to viewing the additive component $\mathbf{X}$ as sums of random variables defined on a Markov chain, rather than formulating $(\mathbf{X}, J)$ as a *Markov Random Walk* (MRW). In this approach, $\mathbf{X}$ bears a causal relationship with J, which is assumed to be more or less known at the beginning. This is the case in many applications where J represents an extraneous factor such as the environment (e.g., in some Markov-modulated queueing systems). However, in other applications, the phenomenon studied gives rise in a natural fashion to $\mathbf{X}$ and J jointly, and it is important to study the evolution of $(\mathbf{X}, J)$ as a Markov process. In such situations, Definition 6.1 is a natural one to use. The basic references to MAPs are still Çinlar [5, 6]. For more recent work on MAPs, see Prabhu [19].

In the case $\mathcal{T} = \mathbb{N}$, an MAP is said to be an MRW. Due to the discrete time feature of MRWs, their study has evolved in part independently of general MAPs. If a continuous time MAP is embedded at an appropriately selected sequence of times, the resulting process will turn out to be an MRW, which may be used to study properties of the original MAP (see e.g., Pacheco and Prabhu [18]). We recall that a Markov process $(\mathbf{S}^*, J^*) = \{(\mathbf{S}_n^*, J_n^*), n \in \mathbb{N}\}$ on $\mathbb{R}^r \times E$ is an MRW if its transition probability measure has the property

$$P\{\mathbf{S}_{m+n}^* \in A, J_{m+n}^* = k \mid \mathbf{S}_m^* = \mathbf{x}, J_m^* = j\}$$

$$= P\{\mathbf{S}^*_{m+n} - \mathbf{S}^*_m \in A - \mathbf{x}, J^*_{m+n} = k \mid J^*_m = j\} \tag{6.2}$$

for $m, n \in \mathbb{N}$, $j, k \in E_\Delta$, $\mathbf{x} \in \mathbb{R}^r$ and $A \in \mathfrak{R}^r$. When the additive component $\mathbf{S}^*$ of an MRW $(\mathbf{S}^*, J^*)$ takes values on $\mathbb{R}^r_+$, $(\mathbf{S}^*, J^*)$ is said to be a *Markov renewal process* MRP. MRPs are thus discrete versions of MAPs with the additive component taking values in $\mathbb{R}^r_+$ $(r \geq 1)$, and they are called *Markov subordinators.* A survey of MRWs and some new results have recently been given in Prabhu, Tang and Zhu [21]. The single major reference to MRPs is Çinlar [4]; for a recent review of the literature on MRPs, see Prabhu [19]. Markov subordinators are reviewed in Prabhu and Zhu [22], where applications to queueing systems are considered.

A particularly important class of Markov subordinators is the class of MAPs $(\mathbf{X}, J)$ with the additive component taking values in the set of nonnegative (multi-dimensional) integers. The increments of $\mathbf{X}$ may correspond to events. The standard example is that of different classes of arrivals at a queueing system. Without loss of generality (w.l.o.g.), we use the term *arrivals* to denote the events studied. We then say that $(\mathbf{X}, J)$ is an *MAP of arrivals*, and call $\mathbf{X}$ (the additive component of the MAP) the *arrival component.* Thus for $i = 1, 2, \ldots, r$,

$$X_i(t) = \text{total number of class } i \text{ arrivals in } (0, t].$$

In the discrete time case, MAPs of arrivals are MRPs; some special cases of these have been considered in the applied literature (see Lucantoni [11] and Neuts [15] for references). We consider only continuous time-homogeneous MAPs of arrivals; the general theory of these processes follows from Neveu's [16] construction of Markov subordinators.

Since an MAP of arrivals $(\mathbf{X}, J)$ is a Markov subordinator on $\mathbb{N}^r \times E$, it is a Markov chain. Moreover, since $(\mathbf{X}, J)$ is time-homogeneous, it is characterized by its transition rates, which are translation invariant in the arrival component $\mathbf{X}$. Thus, it suffices to give for $j, k \in E$ and $\mathbf{m}, \mathbf{n} \in \mathbb{N}^r$ the transition rate from $(\mathbf{m}, j)$ to $(\mathbf{m} + \mathbf{n}, k)$, which we denote simply by $\lambda_{jk}(\mathbf{n})$ (since the rate does not depend on $\mathbf{m}$). Note that, whenever the Markov component is in state j, the following three types of transitions in $(\mathbf{X}, J)$ may occur with respective rates:

(*i*) arrivals without change of state in J occur at rate $\lambda_{jj}(\mathbf{n})$, $\mathbf{n} > \mathbf{0}$;
(*ii*) changes of state in J without arrivals occur at rate $\lambda_{jk}(\mathbf{0})$, $k \neq j$;
(*iii*) arrivals with change of state in J occur at rate $\lambda_{jk}(\mathbf{n})$, $k \neq j$, $\mathbf{n} > \mathbf{0}$.

We denote $\Lambda_{\mathbf{n}} = (\lambda_{jk}(\mathbf{n}))$, $\mathbf{n} \in \mathbb{N}^r$, and say that $(\mathbf{X}, J)$ is a *simple MAP of arrivals* if $\Lambda_{(n_1, \ldots, n_r)} = 0$ if $n_l > 1$ for some l. In the case when $\mathbf{X}(= X)$ has state space $\mathbb{N}$, we say that (X, J) is a *univariate MAP of arrivals.*

The identification of the meaning of each of the above transitions is very important for applications. In particular, arrivals are clearly identified by transitions of type (*i*) and (*iii*), so that we may talk about arrival epochs and interarrival times and define complex operations like thinning of MAPs of arrivals. Moreover, the parametrization of these processes is simple and, by being Markov chains, they have "nice" structural properties. These are computationally important since it is a simple task to simulate Markov chains, although some problems may arise when transition rates do not have any special structure; this is not usually the case in practice.

There has not been a systematic study of MAPs of arrivals, and the literature has been focused on univariate MAPs of arrivals with finite Markov component J (but most of the time without a connection to the theory on MAPs). The following is a brief account of the history and terminology for this class of processes which has no aim of completeness due to the vast literature on these processes.

The emphasis is on the evolution of the structure of the processes, and, in the following, the reference is to univariate MAPs of arrivals with finite Markov component, investigated in the applied literature.

The simplest process corresponds to the case where the Markov component has only one state. Then X is a continuous time Markov process with stationary and independent increments on $\mathbb{N}$, and hence it is a compound Poisson process. In the terminology for MAPs of arrivals, the process is characterized by the rates $\Lambda_n = \lambda_n$ of arrivals of batches of size n ($n \geq 1$), and its generating function is

$$G(z;t) = \exp\left\{-t\sum_{n\geq 1}\lambda_n(1-z^n)\right\}. \tag{6.3}$$

The compound Poisson process allows only transitions of type (i) (arrivals without change of state in the Markov component) which are the only ones admissible anyway. If we leave this simple one-state case, and get into genuine examples of MAPs of arrivals, transitions of types (ii) or (iii) have to be allowed. To our knowledge, the first defined (simple) MAP of arrivals, with transitions of types (i) and (ii), was the *Markov-modulated Poisson process* (*MMPP*), which is a Cox process with intensity rate modulated by a finite Markov chain. This process was first used in queueing models by P. Naor and U. Yechialli, followed by M.F. Neuts; a recent compilation of results and relevant references on the MMPP is given by Fischer and Meier-Hellstern [8]. The MMPP was defined as an MAP of arrivals by Prabhu [19].

We may also consider the MMPP where batches of arrivals are allowed. This process may be constructed from m independent compound Poisson processes $X_1, X_2, \ldots, X_m$ by observing the process X_j whenever the Markov component J is in state j (this follows from Neveu's [16] characterization of Markov subordinators with a finite Markov component). This setting suggests immediately that the process has important properties, such as lack of memory, being closed under Bernoulli thinning and under superpositioning of independent processes, similar to those of the compound Poisson process. The MMPP is an MAP of arrivals with rate matrices Λ_n ($n \geq 1$) being diagonal matrices. Its generating function is a natural matrix extension of the generating function of the compound Poisson process, namely,

$$G(z;t) = \exp\left\{t\left[Q - \sum_{n\geq 1}\Lambda_n(1-z^n)\right]\right\} \tag{6.4}$$

where Q is the generator matrix of the Markov component J. In fact, all univariate MAPs of arrivals have this generating function, with the Λ_n not necessarily diagonal. As with all Cox processes, MMPPs are more *bursty* than the Poisson process in the sense that the variance of the numbers of counts is greater than its mean (see Kingman [9]), whereas for the Poisson process these quantities are equal.

Allowing only transitions of types (i) and (ii) amounts to stating that only one of the components X, J changes state at a time; this is consistent with viewing X having a causal relationship with J. Type (iii) transitions are sometimes called *Markov-modulated arrivals* in the literature.

Transitions of types (i)-(iii) were allowed for the first time by Rudemo [24] who considered a simple process. Thus, in addition to arrivals as in the MMPP, there may be arrivals at transition epochs of the Markov chain J. Explicitly, at the time a transition from state j to state k occurs in J, an arrival occurs if

$(j,k) \in A$ and no arrival occurs otherwise, where $A \subseteq \{(i,l) \in E^2: i \neq l\}$. This process is thus a simple MAP of arrivals for which Λ_1 is not necessarily diagonal. The restrictive feature of the process is that

$$\lambda_{jk}(0)\lambda_{jk}(1) = 0 \qquad (j,k \in E), \tag{6.5}$$

which may be inconvenient in modelling.

The restrictive condition (6.5) is not present in the *phase-type* (PH) renewal process introduced by Neuts (see Neuts [14]) as a generalization of the Poisson process, containing modifications of the Poisson process such as the Erlangian and hyperexponential arrival processes. Here the interarrival times have a phase-type distribution, i.e., they are identified as times until absorption in a finite state Markov chain with one absorbing state. The process may be used to model sources that are less bursty than the Poisson sources.

The first defined univariate MAP of arrivals, that achieves the full generality possible when the Markov component has a finite state space, was the versatile Markovian arrival process of Neuts, the so-called N-process. The definition of this process is a constructive one (see Neuts [14]), and in, the original formulation, special care was taken to include arrivals of the same type as in both the MMPP and the PH renewal process. In addition to the transitions of types (*i*)-(*iii*) (in the set of transient states of a finite Markov chain), the process allows for phase-type arrivals with one absorbing state. As follows easily from Proposition 6.8, the inclusion of phase-type arrivals does not add any additional generality to the class of N-processes, and, in fact, makes the model overparametrized.

Lucantoni, Meier-Hellstern, and Neuts [10] defined a second (simple) Markovian arrival process, which was generalized to allow for batch arrivals by Lucantoni, who named it batch Markovian arrival process (BMAP). Although the classes of BMAPs and N-processes are the same, the BMAP has a simpler definition, given in Remark 6.7 below, than the N-process. For a history of the BMAP and its applications, with special emphasis on matrix-geometric methods and a very extensive list of references, see Lucantoni [11]. The simple BMAP has been referred to as MAP; this is unfortunate since this process is just a particular case of a Markov additive process, and MAP is also the standard acronym for Markov additive processes (first used by Çinlar [5,6]).

Extending Neuts' idea of phase-type arrivals with one absorbing state, Machihara [12] considered a case of simple phase-type arrivals with more than one absorbing state (see Example 6.6), and defined a process which is closely related to the phase-type Markov renewal process we discuss in Example 6.12. This process was extended to the case of batch arrivals by Yamada and Machihara [27].

The *point processes of arrivals* (sequence of arrival epochs along with the corresponding batch sizes) associated with the classes of univariate MAPs of arrivals with finite Markov component, BMAPs, and Yamada and Machihara's arrival processes are the same (see Proposition 6.13).

MAPs of arrivals arise in important applications, as components of more complex systems, especially in queueing and data communication models. They have been used to model overflow from trunk groups, superpositioning of packeted voice streams, and input to ATM (Asynchronous Transfer Mode) networks, which will be used in high-speed communication networks (see e.g., [8,11,12,14,15,26,27] for references). They have also been used to establish theoretical queueing results and to investigate constructions on arrival streams (see e.g., [11,15] for references). Queues with Markov additive input have, also, been the subject of much study (see e.g., [8,10,11,12,14,20,22,26,27] and references cited therein). The output from queues with Markov additive input has also been considered; in general, this process is not an MAP of arrivals. In particular, Olivier and Walrand [17] have shown that the output from an MMPP/M/1 queue is not an MAP of arrivals unless the input is Poisson. Takine, Sengupta, and Hasegawa [26] study the extent to which the output process *conforms* to the input process generated by an MAP of arrivals for a high-speed communications network with the *leaky bucket* for *regulatory access mechanism* (or traffic shaping mechanism). The class of stationary univariate MAPs of arrivals with a finite Markov component was shown to be dense in the family of all stationary point processes on $\mathbb{R}_+$ by Asmussen and Koole [1] (here an MAP is said to be stationary if its Markov component is stationary). This fact and the considerable tractability of MAPs of arrivals make the class of MAPs of arrivals very important for applications. For more details on applications of MAPs of arrivals, we refer the reader to the papers cited, and in particular, to [11,15,22].

Our analysis of MAPs of arrivals is based on what we can say from the Markovian nature of these processes, rather than by looking at them as point processes. The use of the Markovian framework makes it possible to show the role the Markov component has for MAPs of arrivals and is well suited to answer questions on these processes. In Section 6.2, we study some properties of MAPs, with emphasis on transformations that preserve the Markov additive property. MAPs of arrivals are formally introduced in Section 6.3 along with some examples, and some of their elementary properties are studied in Section 6.4; these include lack of memory, interarrival times, moments of the number of counts, and a law of large numbers. In Section 6.5, we consider transformations of MAPs of arrivals, both deterministic and random. In Section 6.6 we define a type of secondary recording which includes important special cases of marking and thinning. We consider, in particular, Markov-Bernoulli recording of MAPs of arrivals, for which the secondary process turns out to be an MAP of arrivals. Finally, in Section 6.7, we briefly outline some of the research directions on MAPs of arrivals we find most likely to be followed.

In the rest of this section, we introduce some further notations to be used in the chapter. For a stochastic process Y defined on a probability space $(\Omega, \mathcal{F}, \mathcal{P})$, we let $\{\mathcal{F}_t^Y\}$ be the filtration generated by Y, so that $\mathcal{F}_t^Y = \sigma(Y(s), 0 \leq s \leq t)$ for $t \in \mathcal{T}$, where for a family $\mathcal{G}$ of random variables $\sigma(\mathcal{G})$ is the σ-field generated by $\mathcal{G}$. As usual, for a transformation $T: \Omega_1 \rightarrow \Omega_2$, we denote $T^{-1}(A) = \{\omega_1 \in \Omega_1: T(\omega_1) \in A\}$ for $A \subseteq \Omega_2$.

We denote by $\mathbf{e}$ a vector with unit elements, with the dimension being clear from the context. Suppose $A = (a_{jk})$ and $B = (b_{jk})$ are matrices of the same order. We define the *Schur* or *entrywise multiplication* of A and B by $A \bullet B = (a_{jk} b_{jk})$. If for countable sets $F_1, \ldots, F_4$, we have $A = (a_{jk})_{F_1 \times F_2}$ and $B = (b_{il})_{F_3 \times F_4}$, then we define the *Kronecker sum* of A and B by

$$A \oplus B = (c_{(j,i)(k,l)})_{(F_1 \times F_3) \times (F_2 \times F_4)} = (a_{jk} \delta_{il} + \delta_{jk} b_{il}).$$

For a countable set F, we let $B(F)$ be the space of bounded real functions of F, $l_\infty(F)$ be the Banach space of real sequences $a = (a_j)_{j \in F}$ with the norm $\| a \| = \supset_{j \in F} | a_j |$, and $B(l_\infty(F))$ be the space of bounded linear operators on $l_\infty(F)$, an element W of which may be identified by a matrix $W = (w_{jk})_{j,k \in F}$ with norm

$$\| W \| = \supset_{j \in F} \sum_{k \in F} | w_{jk} |.$$

We consider in $\mathbb{R}^r$ partial order relations $\leq$, $<$, $\geq$ and $>$ such that

$$\mathbf{x} \leq \mathbf{y} \Leftrightarrow x_i \leq y_i,\ i = 1,2,\ldots,r; \qquad \mathbf{x} < \mathbf{y} \Leftrightarrow \mathbf{x} \leq \mathbf{y} \text{ and } \mathbf{x} \neq \mathbf{y};$$

and $\mathbf{x} \geq \mathbf{y} \Leftrightarrow \mathbf{y} \leq \mathbf{x}$, $\mathbf{x} > \mathbf{y} \Leftrightarrow \mathbf{y} < \mathbf{x}$.

6.2 SOME PROPERTIES OF MAPS

In this section, we investigate some general properties of MAPs. We assume w.l.o.g. that, if $(\mathbf{X}, J)$ is an MAP, then $\mathbf{X}(0) = \mathbf{0}$ a.s. For clarity, we consider only time-homogeneous MAPs. Then, the distribution of $(\mathbf{X}, J)$ is characterized by its transition probability measure,

$$P_{jk}(A;t) = \mathbf{P}\{\mathbf{X}(t) \in A, J(t) = k \mid J(0) = j\}; \tag{6.6}$$

if

$$P_{jk}(A;t) \geq 0, \quad P_{jk}(A;0) = \delta_{jk} 1_{\{\mathbf{0} \in A\}}, \quad \sum_{k \in E_\Delta} P_{jk}(\mathbb{R}^r;t) \leq 1. \tag{6.7}$$

We let $P(A;t) = (P_{jk}(A;t))$, and when A is a singleton $\{\mathbf{x}\}$, we write $P(\mathbf{x};t)$ instead of $P(\{\mathbf{x}\};t)$. From (6.6) the transition probabilities of J are found to be $\pi_{jk}(t)$, where

$$\Pi(t) = (\pi_{jk}(t)) = (\mathbf{P}\{J(t) = k \mid J(0) = j\}) = (P_{jk}(\mathbb{R}^r;t)), \tag{6.8}$$

with

$$\pi_{jk}(t) \geq 0, \quad \pi_{jk}(0) = \delta_{jk}, \quad \sum_{k \in E_\Delta} \pi_{jk}(t) \leq 1, \tag{6.9}$$

$$\pi_{jk}(t+s) = \sum_{l \in E_\Delta} \pi_{jl}(t)\pi_{lk}(s). \tag{6.10}$$

We say that $(\mathbf{X}, J)$ is a *strong* MAP if for any stopping time T we have

$$\mathbf{P}\{\mathbf{X}(T+t) - \mathbf{X}(T) \in A,\ J(T+t) = k \mid \mathcal{F}_T^{(\mathbf{X},J)}\} = P_{J(T)k}(A;t) \tag{6.11}$$

for $t \in \mathcal{T}$, $k \in E_\Delta$, and $A \in \mathcal{R}^r$. The following result presents important properties of deterministic transformations of MAPs.

Theorem 6.2. *Let $A \in \mathcal{R}^r$ and $B \in \mathcal{R}^s$.*

(a) (***Linear transformations of MAPs***) *Suppose that $(\mathbf{X}, J)$ is an MAP on $\mathbb{R}^r \times E$ with transition probability measure $P^{\mathbf{X}}$, and $T: \mathbb{R}^r \to \mathbb{R}^s$ is a linear transformation. If $\mathbf{Y} = T(\mathbf{X})$, then $(\mathbf{Y}, J)$ is an MAP on $\mathbb{R}^s \times E$ with transition probability measure $P^{\mathbf{Y}}$ such that*

$$P_{jk}^{\mathbf{Y}}(B;T) = P_{jk}^{\mathbf{X}}(T^{-1}(B);t). \tag{6.12}$$

(b) (***Patching together independent MAPs***) *Suppose that $(\mathbf{X}, J_1)$ and $(\mathbf{Y}, J_2)$ are MAPs on $\mathbb{R}^r \times E_1$ and $\mathbb{R}^s \times E_2$ with transition probability measures $P^{\mathbf{X}}$ and $P^{\mathbf{Y}}$, respectively. If $(\mathbf{X}, J_1)$ and $(\mathbf{Y}, J_2)$ are independent, then $((\mathbf{X}, \mathbf{Y}), (J_1, J_2))$ is an MAP on $\mathbb{R}^{r+s} \times E_1 \times E_2$ with transition probability measure $P^{(\mathbf{X},\mathbf{Y})}$ such that*

$$P_{(j_1,j_2)(k_1,k_2)}^{(\mathbf{X},\mathbf{Y})}(A \times B;t) = P_{j_1k_1}^{\mathbf{X}}(A;t)P_{j_2k_2}^{\mathbf{Y}}(B;t). \tag{6.13}$$

Proof. (a) Blumenthal ([2], Theorem II (1.2)) states sufficient conditions under which transformations of Markov processes are themselves Markov. These conditions do not apply necessarily to our case, but we may prove the desired result more directly as follows.

Suppose $0 = t_0 < t_1 < \ldots < t_n < t$, $\mathbf{y}_p \in \mathbb{R}^s (0 \leq p \leq n)$, $B \in \mathcal{R}^s$ and let

$$A_p = \{(\mathbf{Y},J)(t_p) = (\mathbf{y}_p, j_p)\} \quad (0 \le p \le n).$$

From the fact that T is a measurable transformation we have

$$A_p = \{(\mathbf{X},J)(t_p) \in T^{-1}(\{\mathbf{y}_p\}) \times \{j_p\}\} \in \sigma((\mathbf{X},J)(t_p)) \quad (0 \le p \le n),$$

and, similarly, $A_t = \{(\mathbf{X},J)(t) \in T^{-1}(B) \times \{k\}\}$ belongs to $\sigma((\mathbf{X},J)(t))$. Since $(\mathbf{X},J)$ is an MAP, these facts imply that

$$\begin{aligned}
\mathbf{P}\{\mathbf{Y}(t) \in B, J(t) = k \mid \bigcap_{0 \le p \le n} A_p\} &= \mathbf{P}\{A_t \mid \bigcap_{0 \le p \le n} A_p\} = P\{A_t \mid A_n\} \\
&= P\{\mathbf{Y}(t) \in B, J(t) = k \mid \mathbf{Y}(t_n) = \mathbf{y}_n, J(t_n) = j_n\} \\
&= P\{\mathbf{Y}(t) - \mathbf{Y}(t_n) \in B - \mathbf{y}_n, J(t) = k \mid A_n\} \\
&= P\{\mathbf{X}(t) - \mathbf{X}(t_n) \in T^{-1}(B - \mathbf{y}_n), J(t) = k \mid J(t_n) = j_n\} \\
&= P\{\mathbf{X}(t - t_n) \in T^{-1}(B - \mathbf{y}_n), J(t - t_n) = k \mid J(0) = j_n\} \\
&= P\{\mathbf{Y}(t - t_n) \in B - \mathbf{y}_n, J(t - t_n) = k \mid J(0) = j_n\}.
\end{aligned}$$

This shows that $(\mathbf{Y},J)$ is an MAP with the given transition probability measure.

(*b*) Since $(\mathbf{X},J_1)$ and $(\mathbf{Y},J_2)$ are independent time-homogeneous Markov processes, it is clear that $(\mathbf{X},\mathbf{Y},J_1,J_2)$ is a time-homogeneous Markov process. Property (*ii*) of MAPs for $((\mathbf{X},\mathbf{Y}),(J_1,J_2))$ follows, using arguments similar to the ones needed to prove that $(\mathbf{X},\mathbf{Y},J_1,J_2)$ is Markov, and its proof is omitted. □

We note that in Theorem 6.2, if the original processes are strong MAPs, so are the transformed processes. In addition, if the original processes are Markov subordinators, so are the resulting processes obtained from patching together independent processes or by taking linear transformations with nonnegative coefficients. Similar remarks apply to Corollary 6.3.

Corollary 6.3. *Suppose* $\alpha, \beta \in \mathbb{R}$, $A \in \mathcal{R}^r$ *and* $B \in \mathcal{R}^s$.

(*a*) (***Marginals of MAPs***) *If* $(\mathbf{X},J)$ *is an MAP on* $\mathbb{R}^r \times E$ *with transition probability measure* $P^{\mathbf{X}}$ *and* $\mathbf{Z} = (X_{i_1}, X_{i_2}, \dots, X_{i_s})$ *with* $1 \le i_1 < \dots < i_s \le r$, *then* $(\mathbf{Z},J)$ *is an MAP on* $\mathbb{R}^s \times E$. *If w.l.o.g.* $\mathbf{Z} = (X_1, X_2, \dots, X_s)$, *then* $(\mathbf{Z},J)$ *has transition probability measure*

$$P_{jk}(B) = P^{\mathbf{X}}_{jk}(B \times \mathbb{R}^{r-s}). \tag{6.14}$$

(*b*) (***Linear combinations of dependent MAPs***) *If* $\mathbf{X} = (X_1, \dots, X_r)$, $\mathbf{Y} = (Y_1, \dots, Y_r)$ *and* $((\mathbf{X},\mathbf{Y}),J)$ *is an MAP on* $\mathbb{R}^{2r} \times E$ *with transition probabi lity measure* $P^{(\mathbf{X},\ \mathbf{Y})}$, *then the process* $(\alpha\mathbf{X} + \beta\mathbf{Y}, J)$ *is an MAP on* $\mathbb{R}^r \times E$ *with transition probability measure*

$$P_{jk}(A;t) = P^{(\mathbf{X},\mathbf{Y})}_{jk}(\{\mathbf{z} = (\mathbf{x},\mathbf{y}) \ \ (\mathbf{x},\mathbf{y} \in \mathbb{R}^r) : \alpha\mathbf{x} + \beta\mathbf{y} \in A\}; t). \tag{6.15}$$

(*c*) (***Linear combinations of independent MAPs***) *If* $(\mathbf{X},J_1)$ *and* $(\mathbf{Y},J_2)$ *are independent MAPs on* $\mathbb{R}^r \times E_1$ *and* $\mathbb{R}^r \times E_2$ *with transition probability measures* $P^{\mathbf{X}}$ *and* $P^{\mathbf{Y}}$, *respectively, then* $(\alpha\mathbf{X} + \beta\mathbf{Y}, (J_1,J_2))$ *is an MAP on* $\mathbb{R}^r \times E_1 \times E_2$ *with transition probability measure*

$$P_{(j_1,j_2)(k_1,k_2)}(A;t) = \int_{\{\mathbf{z} = (\mathbf{x},\mathbf{y})(\mathbf{x},\mathbf{y} \in \mathbb{R}^r) : \alpha\mathbf{x} + \beta\mathbf{y} \in A\}} P^{\mathbf{X}}_{j_1k_1}(d\mathbf{x};t) P^{\mathbf{Y}}_{j_2k_2}(d\mathbf{y};t). \tag{6.16}$$

Proof. Statements (*a*) and (*b*) follow from Theorem 6.2 (*a*) since marginals and linear combinations are special cases of linear transformations. To prove statement (c), we first patch together $(\mathbf{X}, J_1)$ and $(\mathbf{Y}, J_2)$ to get in their product probability space the process $((\mathbf{X}, \mathbf{Y}), (J_1, J_2))$, which by Theorem 6.2 (b) is an MAP with its probability transition measure given by (6.13). Using (b), it then follows that $(\alpha\mathbf{X} + \beta\mathbf{Y}, (J_1, J_2))$ is an MAP of arrivals with transition probability measure as given. □

If $(\mathbf{X}, J)$ is an MAP on $\mathbb{R}^r \times E$ and $0 = t_0 \leq t_1 \leq t_2 \leq \ldots$ is a deterministic sequence, then the process $(S^*, J^*) = \{(S_n^*, J_n^*) = (\mathbf{X}(t_n), J(t_n)), n \in \mathbb{N}\}$ is an MRW on $\mathbb{R}^r \times E$ (which is time-homogeneous in case $t_n = nh (n \in \mathbb{N})$ for some $h > 0$). In order for the embedded process to be an MRW, the embedded sequence does not need to be deterministic, and more interesting embeddings may be obtained by using a random sequence of embedding times.

As for MAPs, we will consider the time-homogeneous case of MRWs, in which the second probability in (6.2) does not depend on m. Since $(\mathbf{S}^*, J^*)$ is an MAP, it suffices to define the one-step transition probability measure $V_{jk}(A) = \mathbf{P}\{\mathbf{S}_1^* \in A; J_1^* = k \mid J_0^* = j\}$, for $j, k \in E_\Delta$ and $A \in \mathcal{R}^r$, along with the initial distribution of J^*.

Theorem 6.4. *Suppose* $(\mathbf{X}, J)$ *is an MAP on* $\mathbb{R}^r \times E$, *and* $T_n^* (n \geq 0)$ *are stopping times such that* $0 = T_0^* \leq T_1^* \leq \ldots < \infty$ *a.s. Denote* $J_n^* = J(T_n^*)$, *and* $\mathbf{S}_n^* = \mathbf{X}(T_n^*)$, *for* $n \in \mathbb{N}$. *If* $(\mathbf{X}, J)$ *is a strong MAP, then* $(T^*, \mathbf{S}^*, J^*) = \{(T_n^*, \mathbf{S}_n^*, J_n^*), n \in \mathbb{N}\}$ *is an MRW on* $\mathbb{R}_+ \times \mathbb{R}^r \times E$. *Moreover, in case the conditional distribution of* $(T_{n+1}^* - T_n^*, S_{n+1}^* - S_n^*, J_{n+1}^*)$ *given* J_n^* *does not depend on* n, *then* $(T^*, \mathbf{S}^*, J^*)$ *is a homogeneous MRW with one-step transition probability measure*

$$V_{jk}(A \times B) = \mathbf{P}\{T_1^* \in A; \mathbf{X}(T_1^*) \in B; J(T_1^*) = k \mid J(0) = j\}$$

for $j, k \in E_\Delta, A \in \mathcal{R}_+$ *and* $B \in \mathcal{R}^r$.

In applications, the most common use of Theorem 6.4 is for the case where the embedding points are the successive transition epochs in J.

6.3 MAPS OF ARRIVALS

In this and the following sections, we consider time-homogeneous MAPs of arrivals. Transitions of an MAP of arrivals $(\mathbf{X}, J)$ on $\mathbb{N}^r \times E$ may be identified by the states of J immediately before (j) and after (k); the transition occurs along with the corresponding observed increment in $\mathbf{X}$: $\mathbf{n}$. Thus we characterize each transition by one element of the set

$$S_r(E) = \{(j, k, \mathbf{n}) \in E^2 \times \mathbb{N}^r : (\mathbf{n}, k) \neq (\mathbf{0}, j)\}. \tag{6.17}$$

For a set $C \subseteq S_r(E)$ and transition rates $\lambda_{jk}(\mathbf{n})$, we let $\lambda(C) = \{\lambda_{jk}(\mathbf{n}) : (j, k, \mathbf{n}) \in C\}$. The process $(\mathbf{X}, J)$ is parametrized by the set of transition rates $\lambda(S_r(E))$. In general, some of the rates will be zero, thus it suffices to give the positive rates $\lambda(S_r^+(E))$ where

$$S_r^+(E) = \{(j, k, \mathbf{n}) \in S_r(E) : \lambda_{jk}(\mathbf{n}) > 0\}. \tag{6.18}$$

In the following, we always assume that $\lambda_{jk}(\mathbf{n}) = 0$ for $(j, k, \mathbf{n}) \notin S_r(E)$. For the transition probability measure of $(\mathbf{X}, J)$ we have

$$P(\mathbf{X}(h) = \mathbf{n}, J(h) = k \mid J(0) = j) = \begin{cases} \lambda_{jk}(\mathbf{n})h + o(h) & (\mathbf{n}, k) \neq (\mathbf{0}, j), \\ 1 - \gamma_j h + o(h) & (\mathbf{n}, k) = (\mathbf{0}, j), \end{cases} \tag{6.19}$$

with

$$\gamma_j = \sum_{\{(k,\mathbf{n}):\,(j,k,\mathbf{n})\,\in\, S_r^+(E)\}} \lambda_{jk}(\mathbf{n}) < \infty. \tag{6.20}$$

A process with (6.20) is said to be *stable*. The infinitesimal generator of $(\mathbf{X}, J)$ is given by

$$\mathcal{A}f(\mathbf{m}, j) = \sum_{\{(k,\mathbf{n}):\,(j,k,\mathbf{n})\,\in\, S_r^+(E)\}} \lambda_{jk}(\mathbf{n})[f(\mathbf{m}+\mathbf{n}, k) - f(\mathbf{m}, j)], \tag{6.21}$$

with f being a bounded real function on $\mathbb{N}^r \times E$. This shows that the process is determined by matrices $\Lambda_{\mathbf{n}} = (\lambda_{jk}(\mathbf{n}))$, with $\mathbf{n} \in \mathbb{N}^r$. Let

$$\Gamma = (\gamma_j \delta_{jk}), \qquad \Lambda = (\lambda_{jk}) = \sum_{\mathbf{n} > \mathbf{0}} \Lambda_{\mathbf{n}}, \quad \Sigma = (\sigma_{jk}) = \sum_{\mathbf{n} \in \mathbb{N}^r} \Lambda_{\mathbf{n}}, \tag{6.22}$$

and

$$Q = (q_{jk}) = \Sigma - \Gamma, \quad Q^{\mathbf{0}} = (q^{\mathbf{0}}_{jk}) = \Lambda_{\mathbf{0}} - \Gamma. \tag{6.23}$$

By inspection of its entries, it is easy to see that $\Lambda(\Sigma)$ is the matrix of transition rates in J associated with arrivals (either arrivals or non-arrivals), Q is the generator matrix of the Markov component J and $Q^{\mathbf{0}}$ is the generator matrix associated with nonarrival transitions in $(\mathbf{X}, J)$. The process $(\mathbf{X}, J)$ is thus characterized by $(Q, \{\Lambda_{\mathbf{n}}\}_{\mathbf{n} > \mathbf{0}})$, and we say that $(\mathbf{X}, J)$ is an MAP of arrivals with $(Q, \{\Lambda_{\mathbf{n}}\}_{\mathbf{n} > \mathbf{0}})$-*source*. This term is a natural extension of the (Q, Λ)-source term commonly used in the literature on MMPPs, where $\Lambda = \Lambda_1$.

In applications, it makes sense to assume that J is nonexplosive, and, for a stable MAP of arrivals $(\mathbf{X}, J)$, this implies that $(\mathbf{X}, J)$ is also nonexplosive, so that the number of arrivals in finite time intervals is a.s. finite. Thus in the rest of the chapter, we consider nonexplosive MAPs of arrivals, which, in particular, are strong Markov. For simplicity, we impose a slightly stronger condition for $(\mathbf{X}, J)$, namely

$$\gamma = \sum_{j \in E} \gamma_j = \sum_{j \in E} \sum_{\{(k,\mathbf{n}):\,(j,k,\mathbf{n})\,\in\, S_r^+(E)\}} \lambda_{jk}(\mathbf{n}) < \infty. \tag{6.24}$$

However, we note that (6.24) is not needed for some of the results we give and that (6.24) holds trivially when E is finite. For $s, t \geq 0$, $j, k \in E$, $\mathbf{n} \in \mathbb{N}^r$ the transition probability measure of $(\mathbf{X}, J)$ is such that

$$P_{jk}(\mathbf{n}; t) = \mathbf{P}\{\mathbf{X}(t) = \mathbf{n},\ J(t) = k \mid J(0) = j\}; \tag{6.25}$$

if

$$P_{jk}(\mathbf{n}; t) \geq 0, \quad P_{jk}(\mathbf{n}; 0) = \delta_{(\mathbf{0}, j)(\mathbf{n}, k)}, \quad \sum_{k \in E} P_{jk}(\mathbb{N}^r; t) = 1. \tag{6.26}$$

The Chapman-Kolmogorov equations are

$$P_{jk}(\mathbf{n}; t+s) = \sum_{l \in E} \sum_{\mathbf{0} \leq \mathbf{m} \leq \mathbf{n}} P_{jl}(\mathbf{m}; t) P_{lk}(\mathbf{n} - \mathbf{m}; s), \tag{6.27}$$

and obviously $\Pi(t) = (\pi_{jk}(t)) = (P_{jk}(\mathbb{N}^r; t))$. From (6.19), we have for $h > 0$ and $j \neq k$

$$\pi_{jk}(h) = P_{jk}(\mathbb{N}^r; h) = \sum_{\mathbf{n} \in \mathbb{N}^r} [\lambda_{jk}(\mathbf{n})h + o(h)] = \sigma_{jk}h + o(h)$$

which shows that Q is the generator matrix of the Markov component J of an MAP of arrivals with $(Q, \{\Lambda_{\mathbf{n}}\}_{\mathbf{n} > \mathbf{0}})$-source. In the rest of the section, we give some examples of MAPs or arrivals.

Example 6.5. (*Arrivals, departures and overflow from a Markovian network*) Suppose we have a Markovian network with r nodes and let $J_i(t)$ $(1 \leq i \leq r)$ be the number of units at node i at time t and $J = (J_1, J_2, \ldots, J_r)$; suppose also that nodes $j_1, j_2, \ldots, j_s$ $(s \le r)$ have finite capacity while the rest have infinite capacity. Let Y_i be the number of external units which entered node i, Z_l be the number of units which left the system from node l, and W_p the overflow at node j_p by time t, and denote $\mathbf{X} = (\mathbf{Y}, \mathbf{Z}, \mathbf{W})$ with

$$\mathbf{Y} = (Y_1, Y_2, \ldots, Y_r), \quad \mathbf{Z} = (Z_1, Z_2, \ldots, Z_r), \qquad \mathbf{W} = (W_1, W_2, \ldots, W_s).$$

The process $(\mathbf{X}, J)$ is an MAP of arrivals on a subset of $\mathbb{N}^{3r+s}$. This holds even with batch input and state dependent input, output or routing rates. We could also include arrival components counting the number of units going from one set of nodes to another.

The fact that $(\mathbf{X}, J)$ is an MAP holds in particular for the queueing networks with dependent nodes and concurrent movements studied by Serfozo [25]. Networks are systems with inherent dependencies, which may be either outside the control of the manager of the system or introduced by the manager. Dependencies may be introduced to avoid congestion, balance the workload at nodes, increase the throughput, etc.; thus the study of queueing networks with dependencies is very important for applications.

Example 6.6. (*Compound phase-type* (*CPH*) *arrival processes*) Consider a continuous time Markov chain J^* on $\{1, 2, \ldots, m, m+1, \ldots, m+r\}$ with stable and conservative infinitesimal generator matrix

$$Q^* = \begin{bmatrix} Q^0 & \Lambda^* \\ 0 & 0 \end{bmatrix} \tag{6.28}$$

with Q^0 being an $m \times m$ matrix, so that states $m+1, \ldots, m+r$ are absorbing. After absorption into state $m+l$, the Markov chain is instantaneously restarted (independently of previous restartings) into transient state k with probability α_{lk}; moreover, if absorption is from state j, then associate with it an arrival of a batch of size $n > 0$ (independently of the size of other batches) with probability $p_{jl}(n)$.

This model was used by Machihara [12] to model service interruptions in a queueing system, where interruptions are initiated at absorption epochs. Machihara viewed these interruption epochs as the arrival epochs of phase-type Markov renewal customers. If the service interruptions are due to physical failure with the absorption state indicating the type of failure, then it becomes important to record the type of failure (absorption state) which occurs at each failure time (absorption epoch). Accordingly, we let $X_l(t)$ $(1 \le l \le r)$ be the number of arrivals associated with adsorptions into state $m+l$ in $(0, t]$, and let J be the Markov chain on states $E = \{1, 2, \ldots, m\}$ obtained by carrying out the described instantaneous restartings after absorptions and requiring the sample functions of the resulting process to be right-continuous. If we let $\alpha = (\alpha_{lk})$ and

$$P_n = (p_{jl}(n)), \quad \Psi_n = (\psi_{jl}(n)) = \Lambda^* \bullet P_n (n \ge 1), \quad Q = Q^0 + \sum_{n \ge 1} \Psi_n \alpha, \tag{6.29}$$

then it is easy to see that J has a stable and conservative infinitesimal generator matrix Q. Moreover, if we let $\mathbf{X} = (X_1, X_2, \ldots, X_r)$ then $(\mathbf{X}, J)$ is an MAP of arrivals on $\mathbb{N}^r \times E$ with a $(Q, \{\Lambda_{\mathbf{n}}\}_{\mathbf{n} > \mathbf{0}})$-source where, for $\mathbf{n} = (n_1, n_2, \ldots, n_r) > \mathbf{0}$,

$$\lambda_{jk}(\mathbf{n}) = \begin{cases} \psi_{jl}(n_l)\alpha_{lk} & \sum_{p=1}^{r} n_p = n_l, \\ 0 & \text{otherwise.} \end{cases}$$

Note that arrivals occur only in one of the coordinates of the arrival component at each time. Similarly, if we let $X(t)$ be the total number of arrivals in $(0,t]$, then the process (X,J) is a univariate MAP of arrivals with $(Q,\{\Psi_n\alpha\}_{n>0})$-source. We call (X,J) a *CPH arrival process* with representation $(Q^0,\Lambda^*,\{P_n\}_{n>0},\alpha)$.

For control of the system it is important to know which types of failures occur most frequently in order to minimize the loss due to service interruptions, for which we need to consider the process $(\mathbf{X},J)$ instead of the CPH arrival process (X,J). This shows that multivariate arrival components may be needed for an appropriate study of some systems. The use of (X,J) is justified only if different failures produce similar effects and have approximately equal costs.

Remark 6.7. If (Y,J) is an MAP of arrivals on $\mathbb{N}\times\{1,2,\ldots,m\}$ with $(Q,\{\Lambda_n\}_{n>0})$-source, and we use the lexicographic ordering of the states of (Y,J), then (Y,J) has an infinitesimal generator matrix with upper triangular block structure

$$\begin{bmatrix} Q^0 & \Lambda_1 & \Lambda_2 & \Lambda_3 & \cdots \\ 0 & Q^0 & \Lambda_1 & \Lambda_2 & \cdots \\ 0 & 0 & Q^0 & \Lambda_1 & \cdots \\ \cdots & \cdots & \cdots & \cdots & \end{bmatrix}. \tag{6.30}$$

A BMAP with representation $\{D_k, k\geq 0\}$ has generator matrix of the form (6.30) with $D_0=Q^0$ and $D_n=\Lambda_n(n\geq 1)$. For BMAPs, it is sometimes assumed that Q^0 is invertible and Q is irreducible, but these conditions are not essential.

Proposition 6.8. *The class of CPH arrival processes is equivalent to the class of univariate MAPs of arrivals with a finite Markov component (i.e., the class of BMAPs).*

Proof. A CPH arrival process (X,J) with representation $(Q^0,\Lambda^*,\ \{P_n\}_{n>0},\alpha)$ is a univariate MAP of arrivals with $(Q,\{(\Lambda^*\bullet P_n)\alpha\}_{n>0})$-source, as shown in Example 6.6, and has a finite Markov component.

Conversely, suppose we are given a MAP of arrivals (Y,J) on $\mathbb{N}\times\{1,2,\ldots,m\}$ with $(Q,\{\Lambda_n\}_{n>0})$-source. For $n>0$, we can obtain matrices $P_n=(p_{jk}(n))$ such that $\Lambda_n=\Lambda\bullet P_n$, with $\{p_{jk}(n), n>0\}$ being a probability function. The process (Y,J) is then indistinguishable from a CPH arrival process with representation $(Q^0,\ \Lambda,\{P_n\}_{n>0},I)$. □

Proposition 6.8 shows that the same thing may be accomplished either with phase-type arrivals or with arrivals of types (i)-(iii). If, in practice, the process studied suggests the consideration of both phase-type arrivals and arrivals of types (i)-(iii), then the individualization of these two classes of arrivals may be done easily using MAPs of arrivals with a multivariate arrival component.

6.4 SOME PROPERTIES OF MAPS OF ARRIVALS

For MAPs of arrivals, we now investigate the partial lack of memory property, interarrival times, moments of the number of counts, and a strong law of large numbers.

We consider first the partial lack of memory property. Since the arrival component $\mathbf{X}$ has non-decreasing sample functions (i.e., $\mathbf{X}(t)\geq\mathbf{X}(s)$ a.s. for $0\leq s\leq t$), the epoch of first arrival becomes

$$T^\circ = inf\{t{:}\,\mathbf{X}(t)>\mathbf{0}\}. \tag{6.31}$$

Theorem 6.9. *If* $(\mathbf{X}, J)$ *is an MAP of arrivals on* $\mathbb{N}^r \times E$ *with* $(Q, \{\Lambda_{\mathbf{n}}\}_{\mathbf{n}>\mathbf{0}})$-*source, then:*

(*a*) $T^{\circ} > t \Leftrightarrow \mathbf{X}(t) = \mathbf{0}$ *a.s.*

(*b*) T° *is a stopping time.*

(*c*) (***Partial lack of memory property***) *For* $t \in \mathbb{R}_+$ *and* $j, k \in E$ *denote*

$$U^{\mathbf{0}}_{jk}(t) = P\{T^{\circ} > t, J(t) = k \mid J(0) = j\} \tag{6.32}$$

and $U^{\mathbf{0}}(t) = (U^{\mathbf{0}}_{jk}(t))$. *Then, the family of matrices* $\{U^{\mathbf{0}}(t), t \in \mathbb{R}_+\}$ *forms a semigroup, i.e.*

$$U^{\mathbf{0}}(t+s) = U^{\mathbf{0}}(t)U^{\mathbf{0}}(s) \quad (t, s \in \mathbb{R}_+). \tag{6.33}$$

(*d*) *Moreover,* $U^{\mathbf{0}}(t) = e^{tQ^{\mathbf{0}}}$, *so that* (6.33) *reads*

$$e^{(t+s)Q^{\mathbf{0}}} = e^{tQ^{\mathbf{0}}}e^{sQ^{\mathbf{0}}} \quad (t, s \in \mathbb{R}_+). \tag{6.34}$$

Proof. Statement (*a*) follows easily from the fact that $(\mathbf{X}, J)$ is a Markov subordinator, and statement (*b*) is an immediate consequence of (*a*).

(*c*) From (*a*), $U^{\mathbf{0}}_{jk}(u) = P_{jk}(\mathbf{0}; u)$ for $j, k \in E$ and $u \in \mathbb{R}_+$. From the Chapman-Kolmogorov equations (6.27), we find that the semigroup property (6.33) holds.

(*d*) From (6.32) and (6.19), using (*a*), we have, for $j, k \in E$,

$$U^{\mathbf{0}}_{jk}(h) = \begin{cases} \lambda_{jk}(\mathbf{0})h + o(h) & j \neq k, \\ 1 - \gamma_j h + o(h) & j = k, \end{cases} \tag{6.35}$$

so that $\{U^{\mathbf{0}}(t), t \geq 0\}$ is a continuous semigroup with infinitesimal generator matrix $Q^{\mathbf{0}}$. Moreover, $U^{\mathbf{0}}(0) = I$, $[U^{\mathbf{0}}(t)]' = U^{\mathbf{0}}(t)Q^{\mathbf{0}}$, and $\| Q^{\mathbf{0}} \| = \| \Lambda_{\mathbf{0}} - \Gamma \| \leq \| \Lambda_{\mathbf{0}} \| + \| \Gamma \| \leq 2\gamma < \infty$, which implies that $U^{\mathbf{0}}(t) = e^{tQ^{\mathbf{0}}}$, in view of Lemma 6.10. □

Lemma 6.10. *If* $R \in B(l_\infty(E))$, *the matrix differential equation* $Q'(t) = Q(t)R$ *with the condition* $Q(0) = I$ *has* $Q(t) = e^{tR}$ *as its unique solution in* $B(l_\infty(E))$.

The partial lack of memory property has been investigated for the MMPP by Prabhu [19] as a natural extension of the well known lack of memory property of the Poisson process. From (6.32), we see that $Q^{\mathbf{0}}$ is the generator matrix of the transitions in J not associated with arrivals in $(\mathbf{X}, J)$. We note that $Q^{\mathbf{0}}$ was used as a building block of the BMAP, the CPH arrival process, and Yamada and Machihara's arrival process (which is discussed in Example 6.12).

We define the successive arrival epochs of the MAP of arrivals $(\mathbf{X}, J)$ as

$$T^{\circ}_p = inf\{t: \mathbf{X}(t) > \mathbf{X}(T^{\circ}_{p-1})\} \quad (p \geq 1), \tag{6.36}$$

where $T^{\circ}_0 = 0$, so that $T^{\circ}_1 = T^{\circ}$. Owing to the presence of the Markov process J, we expect the interarrival times $T^{\circ}_p - T^{\circ}_{p-1}$ to be Markov-dependent.

Theorem 6.11. (***Interarrival times***) *Denote* $\mathbf{X}^{\circ}_p = \mathbf{X}(T^{\circ}_p)$ *and* $J^{\circ}_p = J(T^{\circ}_p)$. *Then the process* $\{T^{\circ}_p, \mathbf{X}^{\circ}_p, J^{\circ}_p), p \geq 0\}$ *is an MRP whose one-step transition pro-bability density is given by*

$$V(t, \mathbf{n}) = (v_{jk}(t, \mathbf{n})) = e^{tQ^{\mathbf{0}}}\Lambda_{\mathbf{n}}. \tag{6.37}$$

Proof. Since $(\mathbf{X}, J)$ is nonexplosive it is a strong Markov process. Now since

$$T^{\circ}_{p+1} - T^{\circ}_p = inf\{t - T^{\circ}_p: \mathbf{X}(t) > \mathbf{X}(T^{\circ}_p)\}$$

and

$$\mathbf{X}^{\circ}_{p+1} - \mathbf{X}^{\circ}_p = \mathbf{X}(T^{\circ}_{p+1}) - \mathbf{X}(T^{\circ}_p),$$

we see that, given J°_0, $(T^{\circ}_1, \mathbf{X}^{\circ}_1, J^{\circ}_1), \ldots, (T^{\circ}_p, \mathbf{X}^{\circ}_p, J^{\circ}_p)$, the distribution of

$(T^\circ_{p+1}-T^\circ_p,\mathbf{X}^\circ_{p+1}-\mathbf{X}^\circ_p,J^\circ_{p+1})$ depends only on J°_p and is the same as that of $(T^\circ_1,\mathbf{X}^\circ_1,J^\circ_1)$, given J°_0. This implies that $\{(T^\circ_p,\mathbf{X}^\circ_p,J^\circ_p),p\geq 0\}$ is an MRP. Moreover, since the probability of two or more transitions in $(\mathbf{X},J)$ in a time interval of length h is of order $\circ(h)$, we have, for the one-step transition probability density of the MRP,

$$\begin{aligned}(\mathbf{P}\{T^\circ_{p+1}-T^\circ_p &\in (t,t+dt],\mathbf{X}^\circ_{p+1}-\mathbf{X}^\circ_p=\mathbf{n},J^\circ_{p+1}=k\mid J^\circ_p=j\})\\ &=(\sum_{l\in E}P_{jl}(\mathbf{0};t)P_{lk}(\mathbf{n};dt))+\circ(dt)=U^{\mathbf{0}}(t)\Lambda_{\mathbf{n}}dt+\circ(dt),\end{aligned}$$

which leads to (6.37), in view of Theorem 6.9 (d). □

Interarrival times have received much consideration in the applied literature. Some authors have used the MRP $\{(T^\circ_p,\mathbf{X}^\circ_p,J^\circ_p)\}$ as defined in Theorem 6.11 to characterize the arrival process of special cases of MAPs of arrivals (e.g., Neuts [15]). From Theorem 6.11, we can define a semi-Markov process $(\mathbf{X}^*,J^*)$ by letting $(\mathbf{X}^*,J^*)(t)=(\mathbf{X}^\circ_p,J^\circ_p)$, for $T^\circ_p\leq t<T^\circ_{p+1}$, but this process gives only partial information on the Markov component of the MAP of arrivals $(\mathbf{X},J)$.

Example 6.12. *(Compound phase-type Markov renewal process (CPH-MRP)).* Consider a CPH arrival process (X,J) with representation $(Q^0,\Lambda^*,\{P_n\}_{n>0},\alpha)$ (see Example 6.6). We let $\{T^\circ_p\}$ be the epochs of increments in X and define $J^\circ_p=J(T^\circ_p)$, $X^\circ_p=X(T^\circ_p)$. Then by Theorem 6.11, the process $\{(T^\circ_p,X^\circ_p,J^\circ_p),\ p\in\mathbb{N}\}$ is an MRP with transition probability density

$$V(t,n)=(v_{jk}(t,n))=e^{tQ^0}(\Lambda^*\bullet P_n)\alpha.$$

We shall call this process *CPH-MRP* with representation $(Q^0,\Lambda^*,\{P_n\}_{n>0},\alpha)$. Its associated point process of arrivals is $\{(T^\circ_p,X^\circ_p-X^\circ_{p-1}),\ p\geq 1\}$; this process has information about the arrival epochs and the batch sizes of arrivals but not of the Markov component of the MRP.

Closely related to this is the process defined for simple arrivals by Machihara [12] and for the general case by Yamada and Machihara [27]; in fact, in the context of Example 6.6, the components T°_p, X°_p of the two processes are the same, while the Markov component is J^*, where J^*_p represents the state the pth absorption moves into. This MRP $\{(T^\circ_p,X^\circ_p,J^*_p)\}$ has transition probability density

$$V^*(t,n)=\alpha e^{tQ^0}(\Lambda^*\bullet P_n),$$

and the authors state that it has the representation $(\alpha,Q^0,\Lambda^*,\{P_n\}_{n>0})$. Yamada and Machihara assume that Q^0 is invertible, but this condition is not essential. We note that, unless the initial distribution π° of J° and π^* of J^* are chosen so that $\pi^\circ=\pi^*\alpha$, the point processes of arrivals associated with $\{(T^\circ_p,X^\circ_p,J^\circ_p)\}$ and $\{(T^\circ_p,X^\circ_p,J^*_p)\}$ need not be stochastically equivalent.

Our definition of CPH-MRPs does not coincide with the definitions in [12, 15, 27]. In Neuts [15], PH-MRPs are MRPs with interarrival times with a phase-type distribution.

Proposition 6.13. *The classes of point process of arrivals associated with CPH-MRPs, Yamada and Machihara's processes, univariate MAPs of arrivals with a finite Markov component, and BMAPs are equal.*

Proof. From Proposition 6.8 it is clear that the classes of point processes of arrivals associated with CPH-MRPs, univariate MAPs of arrivals with finite Markov component, and BMAPs are equal. We now show that the class of point processes of arrivals associated with Yamada and Machihara's processes and univariate MAPs of arrivals with a finite Markov component are the same. Suppose we are given a Yamada and Machihara's process with representation $(\alpha, Q^0, \Lambda, \{P_n\}_{n>0})$ and initial distribution π^* for the Markov component. From Examples 6.6 and 6.12, it is easy to see that this process has the same associated point process of arrivals as a univariate MAP of arrivals (X, J) with $(Q^0 + \sum_{n \geq 1}(\Lambda^* \bullet P_n)\alpha, \{(\Lambda^* \bullet P_n)\alpha\}_{n>0})$-source and initial distribution $\pi^*\alpha$ for J. Conversely, a univariate MAP of arrivals with a finite Markov component, $(Q, \{\Lambda_n\}_{n>0})$-source and initial distribution π for J has the same associated point process of arrivals as a Yamada and Machihara's process with representation $(I, Q^0, \Lambda, \{P_n\}_{n>0})$ and initial distribution π for the Markov component, where P_n is chosen so that $\Lambda_n = \Lambda \bullet P_n$. □

For $\mathbf{n} \in \mathbb{N}^r$ and $\mathbf{z} \in \mathbb{R}^r_+$ such that $\mathbf{0} \leq \mathbf{z} \leq \mathbf{e}$, and with $\mathbf{z^n} = \prod_{i=1}^r z_i^{n_i}$, we denote

$$\Phi(\mathbf{z}) = \sum_{\mathbf{n} \in \mathbb{N}^r} \Lambda_{\mathbf{n}} \mathbf{z}^{\mathbf{n}}, \tag{6.38}$$

and

$$G(\mathbf{z}; t) = (G_{jk}(\mathbf{z}, t)) = \left(\sum_{\mathbf{n} \in \mathbb{N}^r} P_{jk}(\mathbf{n}; t) \mathbf{z}^{\mathbf{n}} \right). \tag{6.39}$$

By considering the process over the time intervals $(0, t]$ and $(t, t + dt]$, we are led to the following matrix differential equation

$$\frac{d}{dt} G(\mathbf{z}; t) = G(\mathbf{z}; t)[\Phi(\mathbf{z}) - \Gamma]. \tag{6.40}$$

Using Lemma 6.10, this implies the following result.

Theorem 6.14. *The generating function matrix of an MAP of arrivals with $(Q, \{\Lambda_{\mathbf{n}}\}_{\mathbf{n}>0})$-source is given by*

$$G(\mathbf{z}; t) = \exp\{t[\Phi(\mathbf{z}) - \Gamma]\} = \exp\left\{ t\left[Q - \sum_{\mathbf{n}>\mathbf{0}} \Lambda_{\mathbf{n}}(1 - \mathbf{z}^{\mathbf{n}}) \right] \right\} \tag{6.41}$$

and $\prod(t) = e^{tQ}$.

Next, we give results for the moments of the number of counts of MAPs of arrivals. These have been considered for some particular cases of MAPs of arrivals by different authors; see in particular [8, 13, 14, 19]. For the derivations, we use the generating function matrix of MAPs, as given in Theorem 6.14. For a linear combination with nonnegative integer coefficients $Y = \alpha \mathbf{X}$ of $\mathbf{X}$ and $p \in \mathbb{N}$, we let

$$\mathbb{E}_p^Y(t) = (\mathbf{E}[Y^p(t); J(t) = k \mid J(0) = j]) = \left(\sum_{\mathbf{n} \in \mathbb{N}^r} (\alpha \mathbf{n})^p P_{jk}(\mathbf{n}; t) \right), \tag{6.42}$$

$$\bar{\Sigma}_p^Y = \sum_{\mathbf{n} \in \mathbb{N}^r} (\alpha \mathbf{n})^p \Lambda_{\mathbf{n}}. \tag{6.43}$$

Theorem 6.15. *For $1 \leq i \leq r$, if $\| \bar{\Sigma}_1^{X_i} \| < \infty$, then*

$$\mathbb{E}_1^{X_i}(t) = \int_0^t \Pi(s) \bar{\Sigma}_1^{X_i} \Pi(t - s) ds, \tag{6.44}$$

whereas if $\| \bar{\Sigma}_2^{X_i} \| < \infty$, then

$$\mathbb{E}_2^{X_i}(t) = \int_0^t \left[2\mathbb{E}_1^{X_i}(s) \bar{\Sigma}_1^{X_i} + \Pi(s) \bar{\Sigma}_2^{X_i} \right] \Pi(t - s) ds. \tag{6.45}$$

Proof. Since $G(\mathbf{e}; t) = \Pi(t) = e^{tQ}$ and $[\frac{d}{dz_i} G(\mathbf{z}; t)]_{\mathbf{z}=\mathbf{e}} = \mathbb{E}_1^{X_i}(t)$, using (6.40), we find that $\mathbb{E}_1^{X_i}(t)$ satisfies the differential equation

$$\frac{d}{dt}\mathbb{E}_1^{X_i}(t) - \mathbb{E}_1^{X_i}(t)Q = e^{tQ}\bar{\Sigma}_1^{X_i}. \qquad (6.46)$$

Postmultiplying both members of (6.46) by e^{-tQ}, we get $\frac{d}{dt}[\mathbb{E}_1^{X_i}(t)e^{-tQ}] = e^{tQ}\bar{\Sigma}_1^{X_i}e^{-tQ}$. In case $\| \bar{\Sigma}_1^{X_i} \| < \infty$, this implies (6.44), by Lemma 6.10 and the facts that $\mathbb{E}_1^{X_i}(0) = (0)$ and $\Pi(t) = e^{tQ}$, $\forall t \geq 0$.

Since $[\frac{d^2}{dz_i^2}G(\mathbf{z};t)]_{\mathbf{z}=\mathbf{e}} = \mathbb{E}_2^{X_i}(t) - \mathbb{E}_1^{X_i}(t)$, again using (6.40), we find that $\mathbb{E}_2^{X_i}(t) - \mathbb{E}_1^{X_i}(t)$ satisfies the differential equation

$$\frac{d}{dt}[\mathbb{E}_2^{X_i}(t) - \mathbb{E}_1^{X_i}(t)] - [\mathbb{E}_2^{X_i}(t) - \mathbb{E}_1^{X_i}(t)]Q = 2\mathbb{E}_1^{X_i}(t)\bar{\Sigma}_1^{X_i} + e^{tQ}[\bar{\Sigma}_2^{X_i} - \bar{\Sigma}_1^{X_i}].$$

In view of (6.46), it follows that $\frac{d}{dt}\mathbb{E}_2^{X_i}(t) - \mathbb{E}_2^{X_i}(t)Q = 2\mathbb{E}_1^{X_i}(t)\bar{\Sigma}_1^{X_i} + e^{tQ}\bar{\Sigma}_2^{X_i}$. Proceeding as before with (6.46), and since $\mathbb{E}_2^{X_i}(0) = (0)$, we obtain (6.45) in case $\| \bar{\Sigma}_2^{X_i} \| < \infty$. □

If J is irreducible with stationary distribution $\pi = (\pi_j)$, we call the version of $(\mathbf{X}, J)$, for which $J(0)$ has distribution π, the *stationary version* of the MAP of arrivals $(\mathbf{X}, J)$. We note that $\pi\Pi(t) = \pi$, $\forall t \geq 0$, and $\Pi(t) \to \mathbf{e}\pi$, as $t \to \infty$.

Corollary 6.16. *Suppose J is irreducible with stationary distribution $\pi = (\pi_j)$. For the stationary version of $(\mathbf{X}, J)$ and with $1 \leq i \leq r$, the following statements hold.*

(a) *If $\| \bar{\Sigma}_1^{X_i} \| < \infty$, then*

$$(\mathbf{E}[X_i(t); J(t) = k]) = t\pi\bar{\Sigma}_1^{X_i}\mathbf{e}\pi + \pi\bar{\Sigma}_1^{X_i}(I - e^{tQ})(\mathbf{e}\pi - Q)^{-1}, \qquad (6.47)$$

and, in particular,

$$\mathbf{E}[X_i(t)] = t\pi\bar{\Sigma}_1^{X_i}\mathbf{e}. \qquad (6.48)$$

(b) *If $\| \bar{\Sigma}_2^{X_i} \| < \infty$, then*

$$\mathbf{E}[X_i^2(t)] = \left[t\pi\bar{\Sigma}_1^{X_i}\mathbf{e}\right]^2 + t\pi\bar{\Sigma}_2^{X_i}\mathbf{e} + 2t\pi\bar{\Sigma}_1^{X_i}C\bar{\Sigma}_1^{X_i}\mathbf{e}$$

$$- 2\pi\bar{\Sigma}_1^{X_i}(I - e^{tQ})(\mathbf{e}\pi - Q)^{-2}\bar{\Sigma}_1^{X_i}\mathbf{e} \qquad (6.49)$$

with $C = (I - \mathbf{e}\pi)(\mathbf{e}\pi - Q)^{-1}$ and, in particular,

$$\frac{Var(X_i(t))}{t} \xrightarrow{\ \ } \pi\left[\bar{\Sigma}_2^{X_i} + 2\bar{\Sigma}_1^{X_i}C\bar{\Sigma}_1^{X_i}\right]\mathbf{e}. \qquad (6.50)$$

(c) *If $\| \bar{\Sigma}_2^{X_i} \| < \infty$ $(1 \leq i \leq r)$, then for $1 \leq i, l \leq r$,*

$$\frac{Cov(X_i(t), X_l(t))}{t} \xrightarrow{\ \ } \pi\left[\sum_{\mathbf{n} \in \mathbb{N}^r} n_i n_l \Lambda_{\mathbf{n}} + \bar{\Sigma}_1^{X_i}C\bar{\Sigma}_1^{X_l} + \bar{\Sigma}_1^{X_l}C\bar{\Sigma}_1^{X_i}\right]\mathbf{e}. \qquad (6.51)$$

Proof. (a) and (b) The following identities can be proved easily:

$$\int_0^t (e^{sQ} - \mathbf{e}\pi)ds = (I - e^{tQ})(\mathbf{e}\pi - Q)^{-1} \qquad (6.52)$$

$$\int_0^t \int_0^s e^{uQ} du\, ds = \tfrac{1}{2}\mathbf{e}\pi t^2 + \left(tI - \int_0^t e^{sQ}ds\right)(\mathbf{e}\pi - Q)^{-1}. \qquad (6.53)$$

The statements follow by using these identities in Theorem 6.15.

(c) Proceeding as in Theorem 6.15 with $z_i = z_l = z$ and taking derivatives with respect to z, we would conclude that

$$\mathbb{E}_2^{X_i + X_l}(t) = \int_0^t [2\mathbb{E}_1^{X_i + X_l}(s)\bar{E}_1^{X_i + X_l} + \Pi(s)\bar{\Sigma}_2^{X_i + X_l}]\Pi(t-s)ds.$$

By the arguments used to prove (a) and (b), we may then conclude that

$$\frac{Var(X_i(t) + X_l(t))}{t} \to \pi \ [\bar{\Sigma}_2^{X_i + X_l} + 2(\bar{\Sigma}_1^{X_i} + \bar{\Sigma}_1^{X_l})C(\bar{\Sigma}_1^{X_i} + \bar{\Sigma}_1^{X_l})]\mathbf{e}.$$

The statement follows easily from this and (6.50), by using the fact that

$$Cov(X_i(t), X_l(t)) = \tfrac{1}{2}[Var(X_i + X_l(t)) - Var(X_i(t)) - Var(X_l(t))]. \qquad \square$$

In the case where J has a stationary distribution $\pi = (\pi_j)$, we let

$$\lambda = (\lambda_1, \ldots, \lambda_r), \qquad \lambda_i = \pi \bar{\Sigma}_1^{X_i}\mathbf{e} \quad (1 \leq i \leq r). \tag{6.54}$$

Theorem 6.17. (***Law of Large Numbers***) *Suppose J is irreducible and has stationary distribution $\pi, \lambda_i < \infty$, $\forall i$. Then, for all initial distributions,*

$$\frac{\mathbf{X}(t)}{t} \to \lambda \ \ a.s. \tag{6.55}$$

Proof. Consider the sequence $\{T_p\}_{p \geq 0}$ of successive transition epochs in $(\mathbf{X}, J)$ with $T_0 = 0$. We assume w.l.o.g. that $T_1 < \infty$ a.s., since otherwise $\mathbf{X}(t) \equiv \mathbf{0}$ a.s., $\Lambda_{\mathbf{n}} = 0$ ($\mathbf{n} \in \mathbb{N}^r$) and everything is satisfied trivially. If for fixed i $(1 \leq i \leq r)$ we define the MRP $\{T_p, Y_p, J_p\}$, with $J_p = J(T_p)$ and $Y_p = X_i(T_p)$ $(p \geq 0)$, then

$$(E_j(Y_1)) = (\mathbf{E}[Y_1 \mid J_0 = j]) = \left(\frac{1}{\gamma_j}\sum_{k \in E} \ \sum_{\mathbf{n} \in \mathbb{N}^r} n_i \lambda_{jk}(\mathbf{n})\right) = \Gamma^{-1}\bar{\Sigma}_1^{X_i}\mathbf{e}$$

and $(E_j(T_1)) = (\mathbf{E}[T_1 \mid J_0 = j]) = (1/\gamma_j) = \Gamma^{-1}\mathbf{e}$. Moreover, $\{J_p\}$ is an irreducible discrete time Markov chain with stationary distribution $\pi^* = \beta^{-1}\pi\Gamma$, where $\beta = \Sigma \pi_k \gamma_k$. This implies, by Theorem 12 in Prabhu, Tang and Zhu [21], that a.s.

$$\frac{Y_n}{n} \to \sum_{j \in E} \pi_j^* E_j(Y_1) = \beta^{-1}\pi\bar{\Sigma}_1^{X_i}\mathbf{e} \ \text{ and } \ \frac{T_n}{n} \to \sum_{j \in E} \pi_j^* E_j(T_1) = \beta^{-1} \tag{6.56}$$

for any initial distribution. For $t \in \mathbb{R}_+$ we let $n(t) = max\{p \in \mathbb{N} : T_p \leq t\}$; by (6.24) $n(t) < \infty$ a.s. Moreover, $n(t) \uparrow \infty$ a.s. as $t \uparrow \infty$. Now, since

$$\frac{Y_{n(t)}}{n(t)}\frac{n(t)}{n(t)+1}\frac{n(t)+1}{T_{n(t)+1}} = \frac{Y_{n(t)}}{T_{n(t)+1}} \leq \frac{X_i(t)}{t} \leq \frac{Y_{n(t)+1}}{T_{n(t)}} = \frac{Y_{n(t)+1}}{n(t)+1}\frac{n(t)+1}{n(t)}\frac{n(t)}{T_{n(t)}}$$

using (6.56), $\frac{X_i(t)}{t} \to \pi\bar{\Sigma}_1^{X_i}\mathbf{e}$ a.s., and (6.55) follows. $\square$

We note that, for the stationary version of $(\mathbf{X}, J)$, (6.55) may be obtained from (6.48) and (6.50) by using Chebyshev's inequality, as in the proof of the law of large numbers for the Poisson process in Kingman [9].

6.5 TRANSFORMATIONS OF MAPS OF ARRIVALS

In this section, we consider transformations of MAPs of arrivals. We study first deterministic transformations and later random transformations.

Considering transformations of MAPs, and for $\mathbf{Z} = D\mathbf{X}$ with $D_{p \times r}$ being a matrix of constants with values in $\mathbb{N}$, it is useful to define for $\mathbf{m} > \mathbf{0}$

$$\Lambda_{\mathbf{m}}^{\mathbf{Z}} = \sum_{\{\mathbf{n} \in \mathbb{N}^r : D\mathbf{n} = \mathbf{m}\}} \Lambda_{\mathbf{n}}. \tag{6.57}$$

Theorem 6.18. (*a*) (***Linear transformations of MAPs of arrivals***) *If* $(\mathbf{X}, J)$ *is an MAP of arrivals on* $\mathbb{N}^r \times E$ *with* $(Q, \{\Lambda_{\mathbf{n}}\}_{\mathbf{n} > \mathbf{0}})$*-source and* $\mathbf{Z} = D\mathbf{X}$ *where* $D_{p \times r}$ *is a matrix of constants with values in* $\mathbb{N}$, *then* $(\mathbf{Z}, J)$ *is an MAP of arrivals on* $\mathbb{N}^p \times E$ *with* $(Q, \{\Lambda_{\mathbf{m}}^{\mathbf{Z}}\}_{\mathbf{m} > \mathbf{0}})$*-source.*

(*b*) (***Patching together independent MAPs of arrivals***) *Suppose* $(\mathbf{X}, J_1)$ *and* $(\mathbf{Y}, J_2)$ *are MAPs of arrivals on* $\mathbb{N}^r \times E_1$ *and* $\mathbb{N}^s \times E_2$ *with sources* $(Q^1, \{\Lambda_{\mathbf{n}}^1\}_{\mathbf{n} > \mathbf{0}})$ *and* $(Q^2, \{\Lambda_{\mathbf{m}}^2\}_{\mathbf{m} > \mathbf{0}})$, *respectively. If* $(\mathbf{X}, J_1)$ *and* $(\mathbf{Y}, J_2)$ *are independent, then* $((\mathbf{X}, \mathbf{Y}), (J_1, J_2))$ *is an MAP of arrivals on* $\mathbb{N}^{r+s} \times E_1 \times E_2$ *with* $(Q^1 \oplus Q^2, \{\Lambda_{(\mathbf{n},\mathbf{m})}\}_{(\mathbf{n},\mathbf{m}) > \mathbf{0}})$ *-source, where for* $(\mathbf{n}, \mathbf{m}) > \mathbf{0}$

$$\Lambda_{(\mathbf{n},\mathbf{m})} = [\Lambda_{\mathbf{n}}^1 1_{\{\mathbf{m} = \mathbf{0}\}}] \oplus [1_{\{\mathbf{n} = \mathbf{0}\}} \Lambda_{\mathbf{m}}^2]. \tag{6.58}$$

Proof. (*a*) The fact that $(\mathbf{Z}, J)$ is an MAP follows from Theorem 6.2 (*a*), and, since D has entries with values in $\mathbb{N}$, it follows that $\mathbf{Z}$ takes values in $\mathbb{N}^p$. Thus $(\mathbf{Z}, J)$ is an MAP of arrivals on $\mathbb{N}^p \times E$; moreover, using (6.12) and (6.19), we have for $(j, k, \mathbf{m}) \in S_p(E)$

$$\begin{aligned} \mathbf{P}(\mathbf{Z}(h) = \mathbf{m}, J(h) = k \mid J(0) = j) &= \sum_{\{\mathbf{n}: D\mathbf{n} = \mathbf{m}\}} \mathbf{P}(\mathbf{X}(h) = \mathbf{n}, J(h) = k \mid J(0) = j) \\ &= h \sum_{\{\mathbf{n}: D\mathbf{n} = \mathbf{m}\}} \lambda_{jk}(\mathbf{n}) + o(h). \end{aligned}$$

This implies that $(\mathbf{Z}, J)$ has $(Q, \{\Lambda_{\mathbf{m}}^{\mathbf{Z}}\}_{\mathbf{m} > \mathbf{0}})$-source.

(*b*) Using Theorem 6.2 (*b*), it follows easily that $((\mathbf{X}, \mathbf{Y}), (J_1, J_2))$ is an MAP of arrivals on $\mathbb{N}^{r+s} \times E_1 \times E_2$. From the independence of J_1 and J_2, it is well known (and easy to check) that (J_1, J_2) has generator matrix $Q^1 \oplus Q^2$. We let P^1 and P^2 be the transition probability measures of $(\mathbf{X}, J_1)$ and $(\mathbf{Y}, J_2)$, respectively. For $((j_1, j_2), (k_1, k_2), (\mathbf{n}, \mathbf{m})) \in S_{r+s}(E_1 \times E_2)$, the transition probability measure of $((\mathbf{X}, \mathbf{Y}), (J_1, J_2))$ is such that

$$\begin{aligned} &(P_{(j_1, j_2)(k_1, k_2)}((\mathbf{n}, \mathbf{m}); h)) \\ &\quad = (\lambda_{j_1 k_1}^1(\mathbf{n}) 1_{\{\mathbf{m} = \mathbf{0}\}} \delta_{j_2 k_2} + \delta_{j_1 k_1} l_{\{\mathbf{n} = \mathbf{0}\}} \lambda_{j_2 k_2}^2(\mathbf{m}))h + o(h) \\ &\quad = ([\Lambda_{\mathbf{n}}^1 1_{\{\mathbf{m} = \mathbf{0}\}}] \oplus [1_{\{\mathbf{n} = \mathbf{0}\}} \Lambda_{\mathbf{m}}^2])h + o(h), \end{aligned}$$

which proves the statement. □

Some consequences of Theorem 6.18 are stated in Corollary 6.19, which is stated without proof. The statements in Corollary 6.19 may be proved in a way similar to that used to prove the consequences of Theorem 6.2 which constitute Corollary 6.3.

Corollary 6.19. *Assume that* $\alpha, \beta \in \mathbb{N}$.

(*a*) (***Marginals of MAPs of arrivals***) *If* $(\mathbf{X}, J)$ *is an MAP of arrivals on* $\mathbb{N}^r \times E$ *with* $(Q, \{\Lambda_{\mathbf{n}}\}_{\mathbf{n} > \mathbf{0}})$*-source and* $\mathbf{Z} = (X_{i_1}, \ldots, X_{i_p})$ *with* $1 \leq i_1 < \ldots < i_p \leq r$, *then* $(\mathbf{Z}, J)$ *is an MAP of arrivals on* $\mathbb{N}^p \times E$ *with* $(Q, \{\Lambda_{\mathbf{m}}^{\mathbf{Z}}\}_{\mathbf{m} > \mathbf{0}})$*-source, where for* $\mathbf{m} > \mathbf{0}$

$$\Lambda_{\mathbf{m}}^{\mathbf{Z}} = \sum_{\{\mathbf{n}: (n_{i_1}, n_{i_2}, \ldots, n_{i_p}) = \mathbf{m}\}} \Lambda_{\mathbf{n}}. \tag{6.59}$$

(*b*) (***Linear combination of dependent MAPs of arrivals***) *If* $\mathbf{X} = (X_1, \ldots, X_r)$, $\mathbf{Y} = (Y_1, \ldots, Y_r)$ *and* $((\mathbf{X}, \mathbf{Y}), J)$ *is an MAP of arrivals of* $\mathbb{N}^{2r} \times E$ *with* $(Q, \{\Lambda_{(\mathbf{n},\mathbf{m})}\}_{(\mathbf{n},\mathbf{m}) > \mathbf{0}})$*-source, then the process* $(\mathbf{Z}, J) = (\alpha\mathbf{X} + \beta\mathbf{Y}, J)$ *is an MAP of arrivals on* $\mathbb{N}^r \times E$ *with* $(Q, \{\Lambda_{\mathbf{a}}^{\mathbf{Z}}\}_{\mathbf{a} > \mathbf{0}})$*-source, where for* $\mathbf{a} > \mathbf{0}$

$$\Lambda_{\mathbf{a}}^{\mathbf{Z}} = \sum_{\{\mathbf{n},\mathbf{m} \in \mathbb{N}^r : \alpha\mathbf{n} + \beta\mathbf{m} = \mathbf{a}\}} \Lambda_{(\mathbf{n},\mathbf{m})}. \tag{6.60}$$

(*c*) (***Linear combinations of independent MAPs of arrivals***) *Suppose* $(\mathbf{X}, J_1)$ *and* $(\mathbf{Y}, J_2)$ *are independent MAPs of arrivals on* $\mathbb{N}^r \times E_1$ *and* $\mathbb{N}^r \times E_2$ *with sources* $(Q^1, \{\Lambda_{\mathbf{n}}^1\}_{\mathbf{n} > \mathbf{0}})$ *and* $(Q^2, \{\Lambda_{\mathbf{m}}^2\}_{\mathbf{m} > \mathbf{0}})$, *respectively. If* $\mathbf{Z} = \alpha\mathbf{X} + \beta\mathbf{Y}$ *and* $J = (J_1, J_2)$, *then* $(\mathbf{Z}, J)$ *is an MAP of arrivals on* $\mathbb{N}^r \times E_1 \times E_2$ *with* $(Q^1 \oplus Q^2, \{\Lambda_{\mathbf{a}}^{\mathbf{Z}}\}_{\mathbf{a} > \mathbf{0}})$*-source. Here,*

$$\Lambda_{\mathbf{a}}^{\mathbf{Z}} = \sum_{\{\mathbf{n},\mathbf{m} \in \mathbb{N}^r : \alpha\mathbf{n} + \beta\mathbf{m} = \mathbf{a}\}} \Lambda_{(\mathbf{n},\mathbf{m})} \tag{6.61}$$

for $\mathbf{a} > \mathbf{0}$, *where* $\Lambda_{(\mathbf{n},\mathbf{m})}$ *is given by* (6.58).

Theorem 6.18 and Corollary 6.19 show that the class of MAPs of arrivals is closed under important transformations. Moreover, the transition rates of the transformed processes are easily obtained. We note that Corollary 6.19 (*b*) and (*c*) could have been stated for linear combinations with a finite number of terms. Corollary 6.19 (*c*) leads to the following result which has been used to establish certain asymptotic results and to study the effect of multiplexing bursty traffic streams in an ATM network (see [8, 11] for references).

Corollary 6.20. (***Finite sums of independent MAPs of arrivals***) *If* $(\mathbf{X}_1, J_1)$, $\ldots$, $(\mathbf{X}_K, J_K)$ *are independent MAPs of arrivals on* $\mathbb{N}^r \times E_1, \ldots, \mathbb{N}^r \times E_K$ *with sources* $(Q^1, \{\Lambda_{\mathbf{n}}^1\}_{\mathbf{n} > \mathbf{0}}), \ldots, (Q^K, \{\Lambda_{\mathbf{n}}^K\}_{\mathbf{n} > \mathbf{0}})$, *then* $(\mathbf{X}_1 + \mathbf{X}_2 + \ldots + \mathbf{X}_K, (J_1, J_2, \ldots, J_K))$ *is an MAP of arrivals on* $\mathbb{N}^r \times E_1 \times E_2 \times \ldots E_K$ *with source*

$$(Q^1 \oplus Q^2 \oplus \ldots \oplus Q^K, \{\Lambda_{\mathbf{n}}^1 \oplus \Lambda_{\mathbf{n}}^2 \oplus \ldots \oplus \Lambda_{\mathbf{n}}^K\}_{\mathbf{n} > \mathbf{0}}). \tag{6.62}$$

Proof. It suffices to prove the statement for $K = 2$. Using Corollary 6.19 (*c*) we conclude that $(\mathbf{X}_1 + \mathbf{X}_2, (J_1, J_2))$ is an MAP of arrivals on $\mathbb{N}^r \times E_1 \times E_2$ with $(Q_1 \oplus Q_2, \{\Lambda_{\mathbf{n}}^*\}_{\mathbf{n} > \mathbf{0}})$ where for $\mathbf{n} > \mathbf{0}$

$$\Lambda_{\mathbf{n}}^* = \Lambda_{\mathbf{n}}^1 \oplus 0 + 0 \oplus \Lambda_{\mathbf{n}}^2 = \Lambda_{\mathbf{n}}^1 \oplus \Lambda_{\mathbf{n}}^2. \qquad \square$$

Corollary 6.20 shows that the class of MAPs of arrivals is closed under finite superpositions of independent processes, which is a generalization of the similar result for Poisson processes and for MMPPs. The proof for the MMPP in Neuts [14] is based essentially on properties of the Poisson process (this is because the MMPP process may be viewed as constructed from a series of independent Poisson processes such that the *i*th Poisson process is observed only when the Markov component is in state *i*, as remarked in Section 6.1). For more general MAPs of arrivals, it is better to base the corresponding proof more directly on the Markov property (as in proof given above), which is also the basic property of the Poisson process leading to the additive property.

In computational terms, the rapid (geometric) increase in the dimensionality of the state space of the Markov component of $(\mathbf{X}_1 + \mathbf{X}_2 + \ldots + \mathbf{X}_K, (J_1, J_2, \ldots, J_K))$, as the number K of added independent MAPs of arrivals increases, limits the utility of Corollary 6.20. We show in Example 6.21 that this unpleasant situation may be avoided in the case where the processes added have identical parameters.

Example 6.21. (*Superposition of independent and identical MAPs of arrivals*) Suppose that the processes $(\mathbf{X}_i, J_i)$ $(1 \le i \le K)$ are independent MAPs of arrivals on $\mathbb{N}^r \times \{0,1,\ldots,m\}$ with common rate matrices $\Lambda_{\mathbf{n}}$. If we define

$$J_p^*(t) = \#\{1 \le i \le K: J_i(t) = p\} \quad (1 \le p \le m),$$

then $(J_1^*, J_2^*, \ldots, J_m^*)$ is a Markov process on the state space

$$E = \{(i_1, i_2, \ldots, i_m) \in \mathbb{N}^m: 0 \le i_1 + i_2 + \ldots + i_m \le K\}.$$

Moreover, it may be seen that the process $(\mathbf{X}_1 + \mathbf{X}_2 + \ldots + \mathbf{X}_K, (J_1^*, J_2^*, \ldots, J_m^*))$ is an MAP of arrivals on $\mathbb{N}^r \times E$ whose transitions rates $\lambda_{\mathbf{ij}}^*(\mathbf{n})$ are such that for $(\mathbf{i},\mathbf{j},\mathbf{n}) \in S_r(E)$

$$\lambda_{\mathbf{ij}}^*(\mathbf{n}) = \begin{cases} 0 & \#d(\mathbf{i},\mathbf{j}) > 2, \\ i_l \lambda_{lp}(\mathbf{n}) & i_l - j_l = j_p - i_p = 1, d(\mathbf{i},\mathbf{j}) = \{l, p\}, \\ \sum_{l=0}^m i_l \lambda_{ll}(\mathbf{n}) & \mathbf{i} = \mathbf{j}, \end{cases}$$

where

$$i_0 = K - \sum_{l=1}^m i_l, \quad j_0 = K - \sum_{l=1}^m j_l, \quad d(\mathbf{i},\mathbf{j}) = \{l: 0 \le l \le m, i_l \ne j_l\}.$$

Particular cases of this example have been considered by some authors (see e.g., Fischer and Meier-Hellstern [8] and Neuts [14] for the superpositioning of two-state MMPPs, which are also called *switched Poisson processes*).

In Example 6.21, the reduction in the number of states from$(J_1, \ldots, J_K)$ to $(J_1^*, \ldots, J_m^*)$ increases with the number of added processes K and is especially significant when K is large compared with m. In the special case where the individual MAPs of arrivals have a two-state Markov component $(m = 1)$, the state space is reduced from 2^K states to $K + 1$ states. This result is important in applications where MAPs of arrivals with a two-state Markov component have been used to model the input from bursty sources in communications systems (see [8, 11] for references).

Theorem 6.22. (***Random time transformations of MAPs of arrivals***) *Suppose that $(\mathbf{X}, J)$ is an MAP of arrivals on $\mathbb{N}^r \times E$ with $(Q, \{\Lambda_{\mathbf{n}}\}_{\mathbf{n} > \mathbf{0}})$-source, f is a non-negative function on E, $A = (a_{jk}) = (f(j)\delta_{jk})$, and*

$$A(t) = \int_0^t f(J(s))ds. \tag{6.63}$$

(*a*) *With $(\mathbf{X}^A(t), J^A(t)) = (\mathbf{X}(A(t)), J(A(t)))$, the process $(\mathbf{X}^A, J^A)$ is an MAP of arrivals on $\mathbb{N}^r \times E$ with $(AQ, \{A\Lambda_{\mathbf{n}}\}_{\mathbf{n} > \mathbf{0}})$-source.*

(*b*) *Suppose $\inf_{j \in E} f(j) > 0$, and for $t \in \mathbb{R}_+$ let*

$$B(t) = \inf\{x > 0: A(x) \ge t\}. \tag{6.64}$$

With $(\mathbf{X}^B(t), J^B(t)) = (\mathbf{X}(B(t)), J(B(t)))$, the process $(\mathbf{X}^B, J^B)$ is an MAP of arrivals on $\mathbb{N}^r \times E$ with $(A^{-1}Q, \{A^{-1}\Lambda_{\mathbf{n}}\}_{\mathbf{n} > \mathbf{0}})$-source.

Proof. (*a*) Since $A(t) \in \mathcal{F}_t^J$ and (A, J) is a time-homogeneous Markov subordinator, properties (*i*) and (*ii*) of MAPs for $(\mathbf{X}^A, J^A)$ follow directly from the corresponding properties for the MAP $(\mathbf{X}, J)$, and moreover, $(\mathbf{X}^A, J^A)$ is time-homogeneous. Thus, $(\mathbf{X}^A, J^A)$ is a time-homogeneous MAP with the same state space as $(\mathbf{X}, J)$. We denote by P^A the transition probability measure of $(\mathbf{X}^A, J^A)$. For

$(j,k,\mathbf{n}) \in S_r(E)$, we have

$$P^A_{jk}(\mathbf{n};h) = \int_0^{f(j)h} e^{-\gamma_j u}\lambda_{jk}(\mathbf{n})e^{-\gamma_k f(k)(h-u)}du + o(h) = f(j)\lambda_{jk}(\mathbf{n})h + o(h),$$

from which we conclude that $(\mathbf{X}^A, J^A)$ has $(AQ, \{A\Lambda_{\mathbf{n}}\}_{\mathbf{n}>\mathbf{0}})$-source.

(*b*) First note that, since $B(t) \le x \Leftrightarrow A(x) \ge t$, it follows that $B(t)$ is a stopping time for J. Using the fact that $(\mathbf{X}, J)$ is a strong MAP and (A, J) is a time-homogeneous Markov subordinator, this implies that properties (*i*) and (*ii*) of MAPs for $(\mathbf{X}^B, J^B)$ follow directly from the corresponding properties for $(\mathbf{X}, J)$, and that $(\mathbf{X}^B, J^B)$ is a time-homogeneous process. Thus $(\mathbf{X}^B, J^B)$ is a time-homogeneous MAP with the same state space as $(\mathbf{X}, J)$. We denote by P^B the transition probability measure of $(\mathbf{X}^B, J^B)$. For $(j,k,\mathbf{n}) \in S_r(E)$, we have

$$\begin{aligned} P^B_{jk}(\mathbf{n};h) &= \int_0^{[f(j)]^{-1}h} e^{-\gamma_j u}\lambda_{jk}(\mathbf{n})e^{-\gamma_k \frac{h-f(j)u}{f(k)}}du + o(h) \\ &= [f(j)]^{-1}\lambda_{jk}(\mathbf{n})h + o(h), \end{aligned}$$

from which we conclude that $(\mathbf{X}^B, J^B)$ has $(A^{-1}Q, \{A^{-1}\Lambda_{\mathbf{n}}\}_{\mathbf{n}>\mathbf{0}})$-source. □

The functional A used in Theorem 6.22 is a continuous additive functional of the Markov component of (X, J) (for the definition, see e.g., [3, 6, 20]). Functionals of the same type as A in (6.63) have been considered in the applied literature (e.g., Neuts [15], Pacheco and Prabhu [18, 20]); they may be used, in particular, to model the service in queueing systems with variable service rate. We note that the Markov additive property would have been preserved in Theorem 6.22 with $(\mathbf{X}, J)$ being a strong MAP and A being an additive functional of J, not necessarily of the form (6.63).

Corollary 6.23. *Suppose $(\mathbf{X}, J_1)$ is an MAP of arrivals on $\mathbb{N}^r \times E_1$ with $(Q^1, \{\Lambda^1_{\mathbf{n}}\}_{\mathbf{n}>\mathbf{0}})$-source, J_2 is a nonexplosive Markov chain on E_2 with generator matrix Q^2, and $(\mathbf{X}, J_1)$ and J_2 are independent. We let $J = (J_1, J_2)$ and assume that for a nonnegative function f on E_2*

$$A(t) = \int_0^t f(J_2(s))ds.$$

(*a*) *With $(\mathbf{X}^A(t), J^A(t)) = (\mathbf{X}(A(t)), J(A(t)))$, the process $(\mathbf{X}^A, J^A)$ is an MAP of arrivals on $\mathbb{N}^r \times E_1 \times E_2$ with $(AQ, \{A\Lambda_{\mathbf{n}}\}_{\mathbf{n}>\mathbf{0}})$-source, where*

$$Q = Q^1 \oplus Q^2, \qquad A = (f(j_2)\delta_{(j_1,j_2)(k_1,k_2)}), \quad \Lambda_{\mathbf{n}} = \Lambda^1_{\mathbf{n}} \oplus 0.$$

(*b*) *Suppose $\inf_{j_2 \in E_2} f(j_2) > 0$, and let $(\mathbf{X}^B(t), J^B(t)) = (\mathbf{X}(B(t)), J(B(t)))$, with $B(t)$ as in (6.64). Then the process $(\mathbf{X}^B, J^B)$ is an MAP of arrivals on $\mathbb{N}^r \times E_1 \times E_2$ with source*

$$(A^{-1}Q, \{A^{-1}\Lambda_{\mathbf{n}}\}_{\mathbf{n}>\mathbf{0}}).$$

Proof. If we define $O(t) = \mathbf{0} \in \mathbb{N}^r$, it is easy to see that (O, J_2) is an MAP of arrivals on $\mathbb{N}^r \times E_2$ with $(Q^2, \{(0)\}_{\mathbf{n}>\mathbf{0}})$-source, and is independent of $(\mathbf{X}, J_1)$. Using Theorem 6.18 (*b*), it then follows that $(\mathbf{X}, J)$ is an MAP of arrivals on $\mathbb{N}^r \times E$ with source

$$(Q, \{\Lambda_{\mathbf{n}}\}_{\mathbf{n}>\mathbf{0}}) = (Q^1 \oplus Q^2, \{\Lambda^1_{\mathbf{n}} \oplus (0)\}_{\mathbf{n}>\mathbf{0}}).$$

The statements now follow by using Theorem 6.22. □

Neuts [15] considered a random time transformation of particular MRPs associated with the simple BMAP, which is related to the transformations considered in Corollary 6.23; his approach is not applicable to multivariate arrival processes.

If in Corollary 6.23, $\mathbf{X}(=X)$ is a Poisson process with rate λ, $J_2 = J$ is a Markov chain on $\{1,2,\ldots,m\}$ with generator matrix Q, and X and J are independent, then (X^A, J^A) is an MMPP with $(AQ, \lambda A)$-source and (X^B, J^B) is an MMPP with $(A^{-1}Q, \lambda^{-1}A)$-source. Thus the transformations described in Theorem 6.22 transform Poisson processes into MMPPs. It is also easy to see that MMPPs are also transformed into MMPPs, and the same is true for the transformations in Theorem 6.18 and Corollary 6.19; this shows that MMPPs have many closure properties.

6.6 MARKOV-BERNOULLI RECORDING OF MAPS OF ARRIVALS

Secondary recording of an arrival process is a mechanism that, from an *original* arrival process, generates a *secondary* arrival process. A classical example of secondary recording is the Bernoulli thinning which records each arrival in the original process, independently of all others, with a given probability p or removes it with probability $1-p$. For MAPs of arrivals $(\mathbf{X}, J)$, we consider secondary recordings that leave J unaffected.

A simple example for which the probability of an arrival being recorded varies with time is the case where there is a recording station (or control process), which is *on* and *off* from time to time, so that arrivals in the original process are recorded in the secondary process during periods in which the station is operational (*on*). The control process may be internal or external to the original process and may be of variable complexity (e.g., rules for access of customer arrivals into a queueing network may be simple or complicated and may depend on the state of the network at arrival epochs or not).

Another example is the one in which the original process counts arrivals of batches of customers into a queueing system, while the secondary process counts the number of individual customers, which is more important than the original arrival process in case service is offered to customers one at a time (this is a way in which the compound Poisson process may be obtained from the simple Poisson process). Here, the secondary process usually counts more arrivals than the original process, whereas in the previous examples the secondary process always records fewer arrivals than the original process (which corresponds to thinning).

Secondary recording is related to what is called *marking* of the original process (see e.g., Kingman [9]). Suppose that each arrival in the original process is given a *mark* from a space of marks M, independently of the marks given to other arrivals (e.g., for Bernoulli thinning with probability p each point is marked "recorded" with probability p or "removed" with probability $1-p$). If we consider the process that accounts only for arrivals which have marks on a subset C of M, it may be viewed as a secondary recording of the original process. A natural mark associated with cells moving in a network is the pair of origin and destination nodes of the cell. If the original process accounts for traffic generated in the network, then the secondary process may represent the traffic generated between a given pair of nodes or between two sets of nodes. The marks may represent priorities. These are commonly used in modelling access regulators to communications network systems. A simple example of an access regulator is the one in which cells (arrivals) which are judged in violation of the "contracts" between the network and the user are marked and may be dropped from service under specific situations of congestion in the network; other cells are not marked and are carried through the network (see e.g., Elwalid and Mitra [7]). The secondary processes of arrivals of marked and unmarked cells are of obvious interest.

Consider a switch which receives inputs from a number of sources, and suppose that the original process accounts for inputs from those sources. If we are interested in studying one of the sources, in particular, we should observe a secondary process which accounts only for input from this source; this corresponds to a marginal case of the original process. In case all sources generate the same type of input, what is relevant to study the performance of the system is the total input arriving at the switch; this is a secondary process which corresponds to the sum of the coordinates of the original arrival process. This shows that some of the transformations of arrival processes considered in Section 6.5 may also be viewed as special cases of secondary recording of an arrival process.

Suppose that $(\mathbf{X}, J)$ is an MAP of arrivals on $\mathbb{N}^r \times E$ with $(Q, \{\Lambda_{\mathbf{n}}\}_{\mathbf{n} > \mathbf{0}})$-source. We consider a recording mechanism that independently records with probability $r_{jk}(\mathbf{n}, \mathbf{m})$ an arrival in $\mathbf{X}$ of a batch of size $\mathbf{n}$ associated with a transition from j to k in J as an arrival of size $\mathbf{m}$ in the secondary process (with $\mathbf{m} \in \mathbb{N}^s$ for some $s \geq 1$). The operation is identified by the set R of recording probabilities

$$R = \{R_{(\mathbf{n},\mathbf{m})} = (r_{jk}(\mathbf{n},\mathbf{m})): \quad \mathbf{n} \in \mathbb{N}^r, \mathbf{m} \in \mathbb{N}^s\} \tag{6.65}$$

where $r_{jk}(\mathbf{n}, \cdot)$ is a probability function on $\mathbb{N}^s$, and $r_{jk}(\mathbf{0}, \mathbf{m}) = \delta_{\mathbf{m}\mathbf{0}}$. We call this recording *Markov-Bernoulli recording with probabilities in* R and denote the resulting secondary process as $(\mathbf{X}^R, J)$. Thus, $(\mathbf{X}^R, J)$ is a process on $\mathbb{N}^s \times E$ which is nondecreasing in $\mathbf{X}^R$, and which increases only when $\mathbf{X}$ increases, i.e.,

$$\mathbf{X}^R(t) = \mathbf{X}^R(T_p^\circ) \quad (T_p^\circ \leq t \leq T_{p+1}^\circ), \tag{6.66}$$

with $\{T_p^\circ\}$ given by (6.36). Moreover, for $\mathbf{n} > \mathbf{0}$

$$r_{jk}(\mathbf{n},\mathbf{m}) = \mathbf{P}\{\mathbf{X}^R(T_{p+1}^\circ) - \mathbf{X}^R(T_p^\circ) = \mathbf{m} \mid \ A_{jk}(\mathbf{n})\} \tag{6.67}$$

with $A_{jk}(\mathbf{n}) = \{\mathbf{X}(T_{p+1}^\circ) - \mathbf{X}(T_p^\circ) = \mathbf{n}, J(T_{p+1}^\circ -) = j, J(T_{p+1}^\circ) = k\}$.

Theorem 6.24. (***Markov-Bernoulli recording***) *Suppose that* $(\mathbf{X}, J)$ *is an MAP of arrivals on* $\mathbb{N}^r \times E$ *with* $(Q, \{\Lambda_n\}_{n>0})$*-source and* $R = \{R_{(\mathbf{n},\mathbf{m})}, \mathbf{n} \in \mathbb{N}^r, \mathbf{m} \in \mathbb{N}^s\}$ *is a set of recording probabilities. Then:*

(*a*) *The process* $(\mathbf{X}, \mathbf{X}^R, J)$ *is an MAP of arrivals on* $\mathbb{N}^{r+s} \times E$ *with source*

$$(Q, \{\Lambda_{\mathbf{n}} \bullet R_{(\mathbf{n},\mathbf{m})}\}_{(\mathbf{n},\mathbf{m}) > \mathbf{0}}). \tag{6.68}$$

(*b*) *The process* $(\mathbf{X}^R, J)$ *is an MAP of arrivals on* $\mathbb{N}^s \times E$ *with source*

$$(Q, \{\Lambda_{\mathbf{m}}^R\}_{\mathbf{m} > \mathbf{0}}) = \left(Q, \left\{\sum_{\mathbf{n} > \mathbf{0}} \Lambda_{\mathbf{n}} \bullet R_{(\mathbf{n},\mathbf{m})}\right\}_{\mathbf{m} > \mathbf{0}}\right). \tag{6.69}$$

Proof. (*a*) From (6.66) and (6.67), it is easy to see that $(\mathbf{X}, \mathbf{X}^R, J)$ is an MAP of arrivals on $\mathbb{N}^{r+s} \times E$. If $P^{\mathbf{X}}$ is the probability transition measure of $(\mathbf{X}, J)$, the transition probability measure $\mathbf{P}$ of $(\mathbf{X}, \mathbf{X}^R, J)$ is such that for $j, k \in E$ and $(\mathbf{n}, \mathbf{m}) > \mathbf{0}$

$$P_{jk}((\mathbf{n},\mathbf{m}); h) = P_{jk}^{\mathbf{X}}(\mathbf{n}; h) r_{jk}(\mathbf{n},\mathbf{m}) + o(h) \tag{6.70}$$

since the probability of two or more transitions in $(\mathbf{X}, J)$ in time h is of order $o(h)$. In any case, since $r_{jk}(\mathbf{0},\mathbf{m}) = 0$ for $\mathbf{m} > \mathbf{0}$, (6.70) gives for $(\mathbf{n},\mathbf{m}) > \mathbf{0}$

$$P_{jk}((\mathbf{n},\mathbf{m});h) = \lambda_{jk}(\mathbf{n})r_{jk}(\mathbf{n},\mathbf{m})h + o(h).$$

This implies that, in fact, $(\mathbf{X}, \mathbf{X}^R, J)$ has $(Q, \{\Lambda_{\mathbf{n}} \bullet R_{(\mathbf{n},\mathbf{m})}\}_{(\mathbf{n},\mathbf{m}) > \mathbf{0}})$-source.

(*b*) The statement follows from (*a*), using Corollary 6.19 (*a*). □

We now give two examples of Markov-Bernoulli recording. In Example 6.25, we view the overflow process from a state dependent M/M/1/K system as as special case of secondary recording of the arrival process of customers to the system, and, in Example 6.26, we give a more elaborated example of secondary recording.

Example 6.25. (*Overflow from a state dependent* M/M/1/K *system*) We consider a Markov-modulated M/M/1/K system with batch arrivals with (independent) size distributions $\{p_n\}_{n > 0}$. When there is an arrival of a batch with n customers and only $m < n$ positions are available, only m customers from the batch enter the system. Assume that the service rate is μ_j and the arrival rate of batches is α_j, whenever the number of customers in the system is j. We let $J(t)$ be the number of customers in the system at time t and $X(t)$ be the number of customer arrivals in $(0, t]$.

The process (X, J) is an MAP of arrivals on $\mathbb{N} \times \{0, 1, \ldots, K\}$ with rates

$$\lambda_{jk}(n) = \begin{cases} \alpha_j p_n & k = min(j+n, K), \\ \mu_j & k = j-1, n = 0, \\ 0 & \text{otherwise.} \end{cases}$$

Its source is $(Q, \{\Lambda_n\}_{n > 0})$, where Q is obtained from the matrices $\{\Lambda_n\}_{n > 0}$. If we define $X^R(t)$ as the overflow from the system in $(0, t]$, then it is readily seen that (X^R, J) is a Markov-Bernoulli recording of (X, J) with recording probabilities $r_{jk}(n, m) = 1_{\{m = n - (k-j)\}}$ for $n, m > 0$. Thus, (X^R, J) is an MAP of arrivals on $\mathbb{N} \times \{0, 1, \ldots, K\}$ with $(Q, \{\Lambda_m^R\}_{m > 0})$-source, where $\lambda_{jk}^R(m) = \alpha_j 1_{\{k = K\}} p_{m + (K-j)}$ for $m > 0$.

Example 6.26. Suppose $\mathbf{X} = (X_1, \ldots, X_r)$, $\mathbf{Y} = (Y_1, \ldots, Y_s)$ and $(\mathbf{X}, \mathbf{Y}, J_2)$ is an MAP of arrivals on $\mathbb{N}^{r+s} \times E_2$ with $(Q, \{\Lambda_{(\mathbf{n},\mathbf{m})}\}_{(\mathbf{n},\mathbf{m}) > \mathbf{0}})$-source. We are interested in keeping only the arrivals in $\mathbf{X}$ which are preceded by an arrival in $\mathbf{Y}$ without simultaneous arrival in $\mathbf{X}$. Denote by $\mathbf{X}^R$ the arrival counting process which we obtain by this operation. For simplicity, we assume that E_2 is finite and $\Lambda_{(\mathbf{n},\mathbf{m})} = 0$ if $\mathbf{n} > \mathbf{0}$ and $\mathbf{m} > \mathbf{0}$.

We let $\{T_p^{\mathbf{X}}\}_{p \geq 0}$ and $\{T_p^{\mathbf{Y}}\}_{p \geq 0}$ be the successive arrival epochs in $\mathbf{X}$ and $\mathbf{Y}$, respectively; denote $J = (J_1, J_2)$ with

$$J_1(t) = \begin{cases} 1 & max\{T_p^{\mathbf{Y}} : T_p^{\mathbf{Y}} \leq t\} > max\{T_p^{\mathbf{X}} : T_p^{\mathbf{X}} \leq t\}, \\ 0 & \text{otherwise.} \end{cases}$$

It can be checked in a routine fashion that $((\mathbf{X}, \mathbf{Y}), J)$ is an MAP of arrivals on $\mathbb{N}^{r+s} \times \{0, 1\} \times E_2$ with $(Q^*, \{\Lambda^*_{(\mathbf{n},\mathbf{m})}\}_{(\mathbf{n},\mathbf{m}) > \mathbf{0}})$-source (with the states of (J_1, J_2) ordered in lexicographic order), where $\Lambda^*_{(\mathbf{n},\mathbf{m})} = 0$ if $\mathbf{n} > \mathbf{0}$ and $\mathbf{m} > \mathbf{0}$, and

$$Q^* = \begin{bmatrix} Q - \sum_{\mathbf{m} > \mathbf{0}} \Lambda_{(\mathbf{0},\mathbf{m})} & \sum_{\mathbf{m} > \mathbf{0}} \Lambda_{(\mathbf{0},\mathbf{m})} \\ \sum_{\mathbf{n} > \mathbf{0}} \Lambda_{(\mathbf{n},\mathbf{0})} & Q - \sum_{\mathbf{n} > \mathbf{0}} \Lambda_{(\mathbf{n},\mathbf{0})} \end{bmatrix}, \quad \Lambda^*_{(\mathbf{n},\mathbf{m})} = \begin{bmatrix} \Lambda_{(\mathbf{n},\mathbf{0})} & \Lambda_{(\mathbf{0},\mathbf{m})} \\ \Lambda_{(\mathbf{n},\mathbf{0})} & \Lambda_{(\mathbf{0},\mathbf{m})} \end{bmatrix},$$

if either $\mathbf{n}=\mathbf{0}$ or $\mathbf{m}=\mathbf{0}$. It is easy to see that $(\mathbf{X}^R, J)$ is a Markov-Bernoulli recording of $((\mathbf{X},\mathbf{Y}),J)$ with recording probabilities $r_{(p,j)(q,k)}((\mathbf{n},\mathbf{m}),\mathbf{l}) = 1_{\{p=1,\mathbf{n}=\mathbf{l}\}}$, for $(\mathbf{n},\mathbf{m})>\mathbf{0}$. Thus, $(\mathbf{X}^R, J)$ is an MAP of arrivals on $\mathbb{N}^r \times \{0,1\} \times E_2$ with $(Q^*, \{\Lambda_{\mathbf{n}}^R\}_{\mathbf{n}>\mathbf{0}})$-source, where

$$\Lambda_{\mathbf{n}}^R = \begin{bmatrix} 0 & 0 \\ \Lambda_{(\mathbf{n},\mathbf{0})} & 0 \end{bmatrix}.$$

A particular case of this example with $r=1$ and Y being a Poisson process was considered briefly by Neuts [15].

We now consider the special case of *Markov-Bernoulli marking* for which the space of marks M is discrete. Specifically, each arrival of a batch of size $\mathbf{n}$ associated with a transition from state j to state k in J is given a mark $m \in M$ with (*marking*) probability $c_{jk}(\mathbf{n},m)$, independently of the marks given to other arrivals. If we assume w.l.o.g. that $M=\{0,1,\ldots,K\} \subseteq \mathbb{N}$, Markov-Bernoulli marking becomes a special case of Markov-Bernoulli recording for which the *marked process* $(\mathbf{X},\mathbf{X}^R,J)$ is such that $(\mathbf{X}^R,J)$ is a Markov-Bernoulli recording of $(\mathbf{X},J)$ with recording probabilities

$$R_{(\mathbf{n},m)} = (r_{jk}(\mathbf{n},m)) = (1_{\{m\in M\}} c_{jk}(\mathbf{n},m))$$

for $\mathbf{n}>\mathbf{0}$ and $m\in\mathbb{N}$.

Theorem 6.27. *With the conditions described, we have the following.*

(*a*) (***Markov-Bernoulli marking***) *The marked process* $(\mathbf{X},\mathbf{X}^R,J)$ *is an MAP of arrivals on* $\mathbb{N}^{r+1}\times E$ *with source*

$$\{Q, \{\Lambda_{\mathbf{n}} \bullet R_{(\mathbf{n},m)}\}_{(\mathbf{n},m)>\mathbf{0}}). \tag{6.71}$$

(*b*) (***Markov-Bernoulli coloring***) *Suppose that the set of marks is finite, i.e.,* $K<\infty$. *We identify mark* $m\in M$ *as color* m, *and let* $\mathbf{X}^{(m)}$ *be the counting process of arrivals of batches colored* m, *for* $0\le m\le K$. *The process* $(\mathbf{X}^{(0)},\ldots,\mathbf{X}^{(K)},J)$ *is an MAP of arrivals on* $\mathbb{N}^{(K+1)r}\times E$ *with* $(Q,\{\Lambda_{\mathbf{a}}^*\}_{\mathbf{a}>\mathbf{0}})$*-source. Here, for* $\mathbf{a}=(\mathbf{a}^{(0)},\ldots,\mathbf{a}^{(K)})>\mathbf{0}$ *with* $\mathbf{a}^{(m)}\in\mathbb{N}^r (0\le m\le K)$,

$$\Lambda_{\mathbf{a}}^* = \left(\sum_{m=0}^{K} 1_{\{A_{\mathbf{a}}=\{\mathbf{a}^{(m)}\}\}} \Lambda_{\mathbf{a}^{(m)}} \bullet R_{(\mathbf{a}^{(m)},m)}\right), \tag{6.72}$$

where $A_{\mathbf{a}} = \{\mathbf{a}^{(l)}: 0\le l\le K, \mathbf{a}^{(l)}>\mathbf{0}\}$.

Proof. Statement (*a*) is a consequence of Theorem 6.24 (*a*). Statement (*b*) follows from (*a*), Theorem 6.24 (*b*), and the fact that $(\mathbf{X}^{(0)},\ldots,\mathbf{X}^{(K)},J)$ is a Markov-Bernoulli recording of $(\mathbf{X},\mathbf{X}^R,J)$ with recording probability matrices

$$R^*_{((\mathbf{n},m)(\mathbf{n}^{(0)},\ldots,\mathbf{n}^{(K)}))} = (1_{\{\mathbf{n}^{(m)}=\mathbf{n},\mathbf{n}^{(l)}=\mathbf{0}\ (0\le l\le K, l\ne m)\}})$$

for $0\le m\le K$ and $\mathbf{n},\mathbf{n}^{(0)},\ldots,\mathbf{n}^{(K)}\in\mathbb{N}^r$, with $(\mathbf{n},m)>\mathbf{0}$. □

Theorem 6.27 (*b*) is an extension of the Coloring Theorem for Poisson processes (see Kingman([9], Section 5.1)). An important consequence of Theorem 6.27 is the following result.

Corollary 6.28. *Suppose* $(\mathbf{X},J)$ *is an MAP of arrivals on* $\mathbb{N}^r\times E$ *with* $(Q,\{\Lambda_{\mathbf{n}}\}_{\mathbf{n}>\mathbf{0}})$*-source, and* $\{B_m\}_{0\le m\le K}$ *is a finite partition of* $\mathbb{N}^r-\{\mathbf{0}\}$. *If for* $0\le m\le K$, *we let* $\mathbf{X}^{(m)}$ *be the counting process of arrivals with batch sizes in* B_m, *then* $(\mathbf{X}^{(0)},\ldots,\mathbf{X}^{(K)},J)$ *is an MAP of arrivals on* $\mathbb{N}^{(K+1)r}\times E$ *with* $(Q,\{\Lambda_{\mathbf{a}}^*\}_{\mathbf{a}>\mathbf{0}})$*-source, where for* $\mathbf{a}=(\mathbf{a}^{(0)},\ldots,\mathbf{a}^{(K)})>\mathbf{0}$, *with* $\mathbf{a}^{(m)}\in\mathbb{N}^r (0\le m\le K)$ *and*

$$\Lambda^*_{\mathbf{a}} = \begin{cases} \Lambda_{\mathbf{a}^{(m)}} & A_{\mathbf{a}} = \{\mathbf{a}^{(m)}\} \subseteq B_m, \\ (0) & otherwise, \end{cases} \tag{6.73}$$

where $A_{\mathbf{a}} = \{\mathbf{a}^{(l)}: 0 \leq l \leq K, \mathbf{a}^{(l)} > \mathbf{0}\}$.

Proof. The statement follows from Theorem 6.27 (b) by considering a Markov-Bernoulli marking of $(\mathbf{X}, J)$ with marking probabilities $c_{jk}(\mathbf{n}, m) = 1_{\{\mathbf{n} \in B_m\}}$. □

Note that the processes $(\mathbf{X}^{(p)}, J)$ and $(\mathbf{X}^{(q)}, j)$ with $p \neq q$ have no common arrival epochs a.s. When J has more than one state, $\mathbf{X}^{(p)}$ and $\mathbf{X}^{(q)}$ $(0 \leq p, q \leq K)$ are not independent, except perhaps in very special cases. In case J has only one state and $r = 1$, X is a compound Poisson process and Corollary 6.28 implies that $X^{(m)}$ $(0 \leq m \leq K)$ are independent compound Poisson processes.

When the primary and secondary processes are defined on the same state space and the recording probabilities are such that $R_{(\mathbf{n},\mathbf{m})} = 0$ for $\mathbf{m} > \mathbf{n}$, we say that the associated secondary recording is a *thinning*, and the secondary process is a *thinned process.* If $\mathbf{X}$ and $\mathbf{X}^R$ are the counting processes of arrivals in the original and the thinned processes, we may also consider the process $\mathbf{X}^L = \mathbf{X} - \mathbf{X}^R$ which counts lost arrivals. From Theorem 6.24, we know that the *loss* process $(\mathbf{X}^L, J)$ is also an MAP of arrivals, since we may interchange recorded arrivals with nonrecorded arrivals. However, for Markov-Bernoulli thinning we are able to give sharper results.

Theorem 6.29. (***Markov-Bernoulli thinning***) *Suppose that* $(\mathbf{X}, J)$ *is an MAP of arrivals on* $\mathbb{N}^r \times E$ *with* $(Q, \{\Lambda_{\mathbf{n}}\}_{\mathbf{n} > 0})$*-source and* $R = \{R_{(\mathbf{n},\mathbf{m})}\}$ *is a set of thinning probabilities. Then the process* $(\mathbf{X}^R, \mathbf{X}^L, J)$ *is an MAP of arrivals on* $\mathbb{N}^{2r} \times E$ *with source*

$$Q, \{\Lambda_{\mathbf{n}+\mathbf{m}} \bullet R_{(\mathbf{n}+\mathbf{m},\mathbf{n})}\}_{(\mathbf{n},\mathbf{m}) > 0}). \tag{6.74}$$

Proof. Using arguments similar to the ones in the proof of Theorem 6.24 (a), we may conclude that the process $(\mathbf{X}, \mathbf{X}^R, \mathbf{X}^L, J)$ is an MAP of arrivals on $\mathbb{N}^{3r} \times E$ with source

$$\left(Q, \left\{[\Lambda_{\mathbf{n}} \bullet R_{(\mathbf{n},\mathbf{m})}]1_{\{\mathbf{p} = \mathbf{n} - \mathbf{m}\}}\right\}_{(\mathbf{n},\mathbf{m},\mathbf{p}) > \mathbf{0}}\right). \tag{6.75}$$

Using Corollary 6.19 (a), the statement follows. □

Example 6.30. Suppose that X is a Poisson process with rate λ, J is a stable finite Markov chain with generator matrix Q, and X and J are independent. We view J as the state of a recording station and assume that each arrival in J is recorded, independently of all other arrivals, with probability p_j or nonrecorded with probability $1 - p_j$, whenever the station is in state j. We let $X^R(t)$ be the number of recorded arrivals in $(0, t]$.

Using Theorem 6.18 (b), Theorem 6.29, and Corollary 6.19 (a) we may conclude that (X^R, J) is an MMPP with $(Q, (\lambda p_j \delta_{jk}))$-source. In the special case where the probabilities p_j are either 1 or 0, the station is *on* and *off* from time to time, with the distribution of the on and off periods being Markov dependent.

6.7 BRIEF DISCUSSION OF SOME DIRECTIONS FOR FUTURE RESEARCH

With applications to queueing systems in view, we have considered only MAPs with the additive component $\mathbf{X}$ having the state space $\mathbb{N}^r$. However, many of the results established in the chapter have obvious extensions to the case where **X** has state space $\mathbb{R}^r$. A systematic development of the necessary theory for this general case will be very useful in other applications.

Our analysis is based on the Markovian nature of an MAP of arrivals. However, it is possible to characterize this process as a point process driven by a Markov chain. We believe that some of the transformations considered in Sections 6.5 and 6.6 of the chapter can then be formulated within the framework of this point process. In particular, these transformations could be unified under the title of "state-dependent partitioning" of MAPs of arrivals, as was done by Resnick [23] for transformations of the Poisson process.

A currently active research area of applications of MAPs of arrivals is the area of communications systems. A problem that arises here concerns the output process generated by an MAP of arrivals under some regulatory access mechanism into a high-speed communications network. This problem is important since one of the critical decisions to be made for the complete implementation of the network is to choose the access regulator mechanism to be used (see e.g., Elwalid and Mitra [7] and Takine et al. [26]). Applications of this type are likely to stimulate further research on MAPs of arrivals.

Essential for the use of MAPs of arrivals in practice is the existence of good methods and efficient numerical algorithms to fit these processes. Thus, there is an important need for the development of such methods and algorithms for general MAPs of arrivals. There has been some work on this problem (see Lucantoni [11] and Fischer and Meier-Hellstern [8], and references therein), but this work has been confined to the case where the Markov component has a small number of states.

With implications for applications is the study of limit theorems for MAPs of arrivals and for systems involving MAPs of arrivals, as e.g., queueing networks with input process being an MAP of arrivals.

BIBLIOGRAPHY

[1] Asmussen, S. and Koole, G., Marked point processes as limits of Markovian arrival streams, *J. Appl. Prob.* **30** (1993), 365-372.

[2] Blumenthal, R.M., *Excursions of Markov Processes*, Birkhäuser, Boston 1992.

[3] Blumenthal, R.M. and Getoor, R.K., *Markov Processes and Potential Theory*, Academic Press, New York 1968.

[4] Çinlar, E., Markov renewal theory, *Adv. Appl. Prob.* **1** (1969), 123-187.

[5] Çinlar, E., Markov additive processes. I, *Z. Wahrscheinlichkeitsth. verw. Geb.* **24** (1972), 85-93.

[6] Çinlar, E., Markov additive processes. II, *Z. Wahrscheinlichkeitsth. verw. Geb.* **24** (1972), 95-121.

[7] Elwalid, A.I. and Mitra, D., Analysis and design of rate-based congestion control of high speed networks, I: Stochastic fluid models, access regulation, *Queueing Sys.* **9** (1991), 29-64.

[8] Fischer, W. and Meier-Hellstern, K., The Markov-Modulated Poisson Process (MMPP) cookbook, *Perform. Evaluation* **18** (1992), 149-171.

[9] Kingman, J.F.C., *Poisson Processes*, Oxford University Press, New York 1993.

[10] Lucantoni, D.M., Meier-Hellstern, K.S. and Neuts, M.F., A single-server

queue with server vacations and a class of non-renewal arrival processes, *Adv. Appl. Prob.* **22** (1990), 676-705.

[11] Lucantoni, D.M., The BMAP/G/1 queue: A tutorial, In: *Models and Techniques for Perform. Eval. of Computer and Commun. Systems*, ed. by L. Donatiello and R. Nelson, Springer-Verlag 1993, 330-358.

[12] Machihara, F., Completion time of service unit interrupted by PH-Markov renewal customers and its applications, In: *Teletraffic Science for New Cost-Effective Systems, Networks and Services*, Proc. 12th Intern. Teletraffic Congress (ITC-12), Torino, Italy (1988), North-Holland, Amsterdam 1989, 1508-1514.

[13] Narayana, S. and Neuts, M.F., The first two moment matrices of the counts for the Markovian arrival process, *Stoch. Mod.* **8** (1992), 459-477.

[14] Neuts, M.F., *Structured Stochastic Matrices of* M/G/1 *Type and their Applications*, Marcel Dekker, New York 1989.

[15] Neuts, M.F., Models based on the Markovian arrival process, *IEICE Trans. Commun.* **E75-B** (1992), 1255-1265.

[16] Neveu, J., Une généralisation des processus à accroissements positifs indépendants, *Abh. Math. Sem. Univ. Hamburg* **25** (1961), 36-61.

[17] Olivier, C. and Walrand, J., On the existence of finite dimensional filters for Markov modulated traffic, *J. Appl. Prob.*, (to appear).

[18] Pacheco, A. and Prabhu, N.U., A Markovian storage model, Techn. Report No. **1079**, School of Operations Research and Industrial Engineering, Cornell University, Ithaca, New York 1993.

[19] Prabhu, N.U., Markov renewal and Markov-additive processes - A review and some new results, In: *Proc. of KAIST Math. Workshop* **6**, ed. by B.D. Choi and J.W. Yim, Korea Adv. Inst. of Science and Tech., Taejon 1991, 57-94.

[20] Prabhu, N.U. and Pacheco, A., A storage model for data communication systems, *Queueing Sys.* (1995), (to appear).

[21] Prabhu, N.U., Tang, L.C. and Zhu, Y., Some new results for the Markov random walk, *J. Math. Phys. Sci.* **25** (1991), 635-663.

[22] Prabhu, N.U. and Zhu, Y., Markov-modulated queueing systems, *Queueing Sys.* **5** (1989), 215-245.

[23] Resnick, S.I., *Adventures in Stochastic Processes*, Birkhäuser, Boston 1992.

[24] Rudemo, M., Point processes generated by transitions of Markov chains, *Adv. Appl. Prob.* **5** (1973), 262-286.

[25] Serfozo, R.F., Queueing networks with dependent nodes and concurrent movements, *Queueing Sys.* **13** (1993), 143-182.

[26] Takine, T., Sengupta, B. and Hasegawa, T., A conformance measure for traffic shaping in high-speed networks with an application to the leaky bucket, (to appear).

[27] Yamada, H. and Machihara, F., Performance analysis of a statistical multiplexer with control on input and/or service processes, *Perform. Evaluation* **14** (1992), 21-41.

Chapter 7
The ASTA property

Benjamin Melamed and David D. Yao[1]

ABSTRACT *ASTA* (*A*rrivals *S*ee *T*ime *A*verages) is concerned with properties of stochastic systems where "event" averages sampled over certain sequences of time epochs are equal to time averages. We present a detailed review of three approaches to *ASTA:* (*i*) the elementary approach that treats event averages as stochastic Riemann-Stieltjes integrals; (*ii*) the martingale approach, which exploits properties of the compensators and intensities of point processes, the Doob-Meyer decomposition, and the martingale strong law of large numbers; and (*iii*) the Palm calculus approach that focuses on the stationary setting. We also illustrate the applications of *ASTA* in queueing networks. In particular, we demonstrate that for Markovian queues, a key *ASTA* condition, the *lack of bias assumption* (LBA), is in fact equivalent to quasi-reversibility, and that LBA is preserved when quasi-reversible queues are connected into a network.

CONTENTS

7.1 Introduction 195
7.2 ASTA setting and problem definition 197
7.3 Synopsis of *ASTA* results 199
7.4 The Riemann-Stieltjes sum approach to *ASTA* 207
7.5 The martingale approach to *ASTA* 209
7.6 Queueing network applications 215
7.7 Summary and open problems 222
Acknowledgements 223
Bibliography 223

7.1 INTRODUCTION

The term *ASTA* (*A*rrivals *S*ee *T*ime *A*verages) connotes the equality of statistics in a time series to corresponding statistics in an embedded time series, the embedding times being typically random. While *ASTA* can be mathematically formulated in a fairly general setting, the name, *ASTA*, reflects its historical origins in queueing theory [7, 26, 31].

The *ASTA* property in queueing context can be informally described as follows. Consider a stable queueing system and the arrival stream of discrete jobs offered to it (the underlying time parameter may be continuous or discrete, but batch arrivals are precluded for simplicity). Suppose that the state process of the system (typically queue length) has equilibrium statistics equaling the corresponding long-term time averages over sample paths; such time averages will be referred to as *time-observed* statistics. Let now each arriving job observe the state of the

[1]Supported in part by NSF Grant MSS-92-16490.
0-8493-8074-x/95/$0.00+$.50

queueing system, but with the arriving job excluded from the observed statistic; this is sometimes referred to as the "state of the system just prior to arrival", a notion to be made precise in the sequel. System state observations embedded at job arrival times will be referred to as *job-observed* statistics. In particular, *job averages* are obtained by averaging the sequence of job-observed statistics. In this setting, the *ASTA* property is said to hold for the state process, if the time-observed and corresponding job-observed statistics coincide asymptotically, as the time interval and number of jobs, respectively, increase to infinity. Two related issues are the existence of these quantities and whether the existence of one implies the existence of the other. Notice that the need to exclude the arriving job from the job-observed state stems from the simple observation that, otherwise, *ASTA* is a priori precluded, in most cases, due to the biasing effect of the arrivals; for example, arriving jobs which are allowed "to see themselves", can never observe an empty system.

Evidently, the setting of the informal discussion above can be generalized at several levels. First, job-observed statistics may be embedded at departure times, i.e., jobs observe the "state of the system just after departure", or in a queueing network context, jobs may observe the state at "traffic times", i.e., at time instants when jobs are routed from one node to another. Second, the embedding times need not be related to the physical movement of jobs; embedding can be effected at any sequence of random time points. However, for *ASTA* to have a chance of holding, it is generally necessary to define the embedding state, so as to remove any biasing effect introduced by the embedding sequence. And third, the state process and the random embedding times can be abstracted to fairly general stochastic processes, entirely divorced from a queueing setting; any pair consisting of a real-valued stochastic process and a point process over a common probability space will do. In this general setting, the queueing terminology is modified: time-observed statistics become *time averages*, and job-observed statistics become *event averages*.

At its most general, the *ASTA* setting and associated *ASTA* problem (to be described below) may be profitably viewed simply as a *sampling problem*. To fix the ideas, we couch the discussion in engineering-like terminology. Imagine some random system (e.g., a physical device or network) whose stochastic state evolves in time (continuous or discrete), with the proviso that its state or a function thereof may be observed (sampled) only at a subset of discrete time points. Suppose that a set of state functions has been selected, and a class of sampling procedures, obeying this proviso, has been devised. Usually, no other restrictions are imposed on the sampling mechanism. Admissible sampling procedures are allowed to be random or deterministic, and they may or may not interact with the state of the system under observation; the term "interaction" will be used in informal discussion in lieu of stochastic dependence. Next, form the event averages of interest for each admissible sampling procedure, and compare each of them to the corresponding time averages (note that the latter is just another sampling procedure, except that it permits continual observation of the system without time gaps and, therefore, is excluded from the admissible class). Informally, the *ASTA* problem seeks to answer the following question: For which state functions and admissible sampling procedures will the limiting event averages coincide with the corresponding limiting time averages? A more refined *ASTA* problem would seek information of the *estimation bias* (the difference between the event average and the time average): Do the event averages overestimate or underestimate the corresponding time averages, and if so, by how much?

To illustrate that the theoretical *ASTA* problem is of practical interest, consider a computer operating system which collects some useful statistics on its own operation (e.g., CPU utilization, execution times of jobs, etc.). For reasons of efficiency, the system is permitted to collect those statistics only at prescribed time points, and statistics collection is combined with other housekeeping tasks; otherwise, the attendant overhead may well degrade system performance. System designers would then like to know whether the event averages are good approximations of the "true" statistics and whether the former are hopelessly biased with respect to the latter.

A qualitative first-cut analysis of the *ASTA* problem can be given in terms of the interaction (dependence) between the sampling procedure and the observed state. Clearly, if the sampling process is stochastically independent of the observed state process, then no bias will ensure. This case is mathematically uninteresting and may not be feasible in practice (see, e.g., the example above). Conversely, if the sampling procedure "interacts strongly" with the observed state, then biasing effects should be fully expected. This is, in fact, what happens in most queueing systems, where the sampling procedure observes the state at job arrival times. Unlike "outside observers" which are nonintrusive, "job observers" interact with the state of the queue, since each ends up joining the queue. This example makes it clear that the aforementioned requirement that "jobs may not see themselves" is designed to lessen that interaction, which would otherwise preclude *ASTA* in most cases. Thus, the *ASTA* problem can be informally viewed as seeking to delineate the "upper limits" of dependence between the sampling procedure and the observed state, within which the *ASTA* condition holds true.

Remarkably, a host of *ASTA* conditions has been identified. The most celebrated instance of these is the *PASTA* Theorem, which is due to R. Wolff [31] (see also [15]). Furthermore, the general *ASTA* problem (lack of bias), as well as its refined variant (direction of bias), both have a surprisingly simple resolution (see Melamed and Whitt [20]). Ignoring secondary technical details, the bias can be written in terms of the sampling rate and the pointwise covariance between the observed state and the conditional intensity of the counting process which models the sampling procedure (the point process at whose arrival points the state is sampled); the sign and magnitude of the bias are proportional to the covariance (sign and magnitude) and inversely proportional to the sampling rate. A necessary and sufficient condition for *ASTA* to hold is zero covariance (correlation).

The rest of this chapter is organized as follows. Section 7.2 sets up the *ASTA* problem formally and provides related definitions. Section 7.3 presents a synopsis of *ASTA* results. Section 7.4 sketches *ASTA*-related proofs, using the stochastic Riemann-Stieltjes sum approach, whose merits are its intuitiveness and technical simplicity. Section 7.5 generalizes the setting and the results, using a martingale approach. Section 7.6 presents *ASTA* examples for queueing systems, mainly queueing networks. Finally, Section 7.7 presents a brief summary of the chapter and some open problems.

7.2 *ASTA* SETTING AND PROBLEM DEFINITION

The mathematical setting of *ASTA* will be presented next. Let $(\Omega, \mathcal{F}, P)$ be the underlying probability space over which all stochastic processes will be defined. Let

$X \equiv \{X(t): t \geq 0\}$ be a stochastic process taking values in a complete separable metric space, E, with the associated σ-algebra, $\mathcal{E}$, generated by the open sets. Let $N = \{N(t): t \geq 0\}$ be a simple counting process, which is nonnull and nonexplosive, i.e., $0 < \mathrm{E}[N(t)] < \infty$, $t > 0$. The jump times of N form a stochastic sequence $T = \{T_n: n \geq 0\}$, where $T_n = \inf\{t > 0: N(t) \geq n\}$ for $n \geq 1$, and $T_0 = 0$ by convention (0 may not be an arrival point). The process X is referred to as a *state process* and represents the stochastic evolution of the system of interest. The counting process N is referred to as the *arrival process*, and its jump times serve as embedding times of the state process. More specifically, the *embedded (state) process* is the stochastic sequence $X^{(N)} \equiv \{X^{(N)}(n) = X(T_n): n \geq 0\}$.

For technical reasons, it is customary to work with transformed state processes of the form $U \equiv \{U(t): t \geq 0\}$, where $U(t) = f(X(t))$ and f is a bounded real-valued measurable function on $\mathcal{E}$. The function f is designed to extract the relevant state information from the state description of X. For example, in queueing settings, f typically yields the node or network state, excluding the observing job. For arrival streams, this is often attained by taking the left continuous version, $U(t) = f(X(t-))$, of the original state, while for departure streams, the right continuous version, $U(t) = f(X(t+))$, of the original state is taken instead; for traffic streams within a queueing network, f extracts the left continuous state of the origination node and the right continuous state of the destination node. In particular, f can be an indicator function, so that the corresponding $U(t)$ can extract detailed stochastic information from X, whose limiting time averages are the state probabilities of X. Both X and U will be referred to as state processes (the ambiguity to be resolved by context), and, similarly, for the embedded state processes $X^{(N)}$ and $U^{(N)}$. We mention that the distinction between the state processes, X and U, as well as their respective embedded ones, is retained for modeling purposes.

The *ASTA* problem has several related variants, all of which involve various notions of stochastic convergence. Ultimately, these variants imply the *ASTA* property in terms of convergence in distribution (weak convergence, denoted $\Rightarrow$); see Billingsley [3]. Equality in distribution is denoted by $\stackrel{d}{=}$. This fundamental notion of *ASTA* is defined next.

Definition 7.1. (The *ASTA* problem) Suppose that

$$U(t) \underset{t \to \infty}{\Rightarrow} U(\infty) \text{ and } U^{(N)}(n) \underset{n \to \infty}{\Rightarrow} U^{(N)}(\infty). \tag{7.1}$$

The *ASTA problem* is concerned with finding sufficient and/or necessary conditions for

$$U(\infty) \stackrel{d}{=} U_{\infty}^{(N)}. \tag{7.2}$$

In this case, we say that *ASTA holds* for the pair (U, N).

Other variants of the *ASTA* problem are designed to facilitate mathematical manipulation by imposing additional structure which implies (7.2) in Definition 7.1. The two main variants are described below.

The first variant, referred to as the *sample path framework*, assumes that $U(\infty)$ and $U^{(N)}(\infty)$ correspond to the unique stationary distributions of U and $U^{(N)}$, respectively. Consider the two stochastic processes, $V \equiv \{V(t): t \geq 0\}$ and $\{W(t):\ t \geq 0\}$, given by

$$V(t) \equiv \frac{1}{t}\int_0^t U(s)ds, \quad t \geq 0 \tag{7.3}$$

and
$$W(t) \equiv \frac{1}{N(t)} \sum_{n=1}^{N(t)} U^{(N)}(n) = \frac{\int_0^t U(s)dN(s)}{N(t)}, \quad t \geq 0, \tag{7.4}$$

where $V(0) = 0$ in (7.3), and in (7.4), it is understood that $W(t) = 0$, whenever $N(t) = 0$. Observe that Equation (7.3) is a time average, whereas Equation (7.4) is the corresponding N-embedded event average, where the sum admits a simple stochastic Riemann-Stieltjes integral representation. The *ASTA* condition for the pair (U, N) then assumes the form

$$\lim_{t \to \infty} [W(t) - V(t)] = 0, \ w.p.1, \tag{7.5}$$

where the existence of the limits is part of the problem.

The second variant, referred to as the *expectations framework*, was first introduced in [20]. It is formulated in terms of the following expected values (instead of random variables),

$$\bar{V}(t) \equiv \tfrac{1}{t}\mathbf{E}\left[\int_0^t U(s)ds\right], \quad t \geq 0 \tag{7.6}$$

and
$$\bar{W}(t) \equiv \frac{1}{\mathbf{E}[N(t)]} \mathbf{E}\left[\sum_{n=1}^{N(t)} U^{(N)}(n)\right] = \frac{\mathbf{E}\left[\int_0^t U(s)dN(s)\right]}{\mathbf{E}[N(t)]}, \quad t \geq 0, \tag{7.7}$$

where $\bar{V}(0) = 0$ in (7.6), and in (7.7), it is understood that $\bar{W}(t) = 0$, whenever $\mathbf{E}[N(t)] = 0$. The *ASTA* condition for the pair (U, N) is then stated, in terms of ordinary limits, as

$$\lim_{t \to \infty} \bar{V}(t) = \bar{V}(\infty) = \bar{W}(\infty) = \lim_{t \to \infty} \bar{W}(t), \tag{7.8}$$

where the existence of the limits is part of the problem. This case is of interest when we have the weak convergences,

$$V(t) \underset{t \to \infty}{\Rightarrow} V(\infty) = \bar{V}(\infty) = \lim_{t \to \infty} \bar{V}(t), \tag{7.9}$$
$$W(t) \underset{t \to \infty}{\Rightarrow} W(\infty) = \bar{W}(\infty) = \lim_{t \to \infty} \bar{W}(t), \tag{7.10}$$

which coincide with the previous formulation. Equations (7.9)-(7.10) hold in a regenerative framework or in stationary ergodic framework (see Remark 7.24). These are usually sufficiently general for most practical purposes.

7.3 SYNOPSIS OF *ASTA* RESULTS

This section presents a synopsis of *ASTA* results and attendant definitions. Generally, the discussion will proceed from strong sufficient conditions to weaker ones, culminating with necessary and sufficient ones. As usual, U is assumed bounded throughout.

7.3.1 *LAA*, *PASTA*, and anti-*PASTA*

We start with *LAA* and *PASTA*, the two pioneering *ASTA*-related properties, later refined and generalized beyond their original setting.

Definition 7.2. (*LAA*) Let U be left continuous and let N be right continuous. The pair (U, N) is said to satisfy the *Lack of Anticipation Assumption*

(LAA), if for all $t \geq 0$, $\{N(t+s) - N(t): s \geq 0\}$ is independent of $\{U(s)\colon 0 \leq s \leq t\}$.

Intuitively, LAA restricts the iteration between U and N by requiring that, at any time point, future arrivals and the past state history be independent; thus, future arrivals cannot be "anticipated" by past states. LAA is the sufficient condition for $PASTA$ in the following celebrated theorem, cast in the sample-path framework.

Theorem 7.3. (*PASTA* Theorem) *Let N be a Poisson Process, and suppose that LAA holds for the pair (U, N). Then, $V(t) \underset{t\to\infty}{\to} V(\infty)$ w.p.1 iff $W(t) \underset{t\to\infty}{\to} W(\infty)$ w.p.1, in which case $V(\infty) = W(\infty)$ w.p.1.*

An early heuristic variant of LAA and $PASTA$ appears in Strauch [26]. A complete proof, using martingale techniques, is due to R. Wolff [31], which partly draws upon Whitt [30]. A discrete-time analog of Theorem 7.3, in which intervals between points have a geometric distribution, was first derived in Halfin [13]. For more recent work in the discrete-time setting, see Makowski et al. [18].

The $PASTA$ Theorem is often invoked in queueing context, where Poisson arrivals are frequently used to model arrival processes. LAA then holds due to the independence of future Poisson increments of the σ-algebra generated by the state process to date. Consequently, the $ASTA$ Theorem implies that the equilibrium ("outside-observer") queue length distribution coincides with its embedded counterpart (at arrival time points). In particular, for queues with exponential servers, this fact simplifies the computation of equilibrium distributions for job waiting times (time spent in queue only) and job sojourn times (time spent in queue and in service). To see that, note that both the waiting time and sojourn time distributions are completely determined by the queue length *ahead* of the arriving job, that is, by the state "seen" by that job; more specifically, given $U^{(N)}(n) = k$, the waiting and sojourn time distributions are convolutions of k and $k+1$ service times, respectively. The unconditional distributions are then obtained with respect to the known "outside-observer" queue length distribution.

The converse of the $PASTA$ property can be formulated as follows.

Definition 7.4. (The anti-*PASTA* problem) Given that $ASTA$ holds for a pair, (U, N), find conditions under which N is a Poisson process.

The term, anti-$PASTA$, was coined by Green and Melamed [12]. It is an obvious choice for the converse of $PASTA$, in view of the inadvertent "culinary" nature of the latter. Some anti-$PASTA$ results will be given later in a Markovian setting.

7.3.2 *ASTA* under the *WLAA* condition

As its name suggests, the *W*eak *L*ack of *A*nticipation *A*ssumption ($WLAA$) is a weaker version of LAA, i.e., it strengthens the interaction between U and N without destroying the $ASTA$ property (see [20]).

Definition 7.5. (*WLAA*) Let U be left continuous and let N be right continuous. The pair (U, N) is said to satisfy the *W*eak *L*ack of *A*nticipation *A*ssumption ($WLAA$), if for each $t \geq 0$ there exists $s_0 > 0$, such that $U(t)$ and $N(t+s) - N(t)$ are uncorrelated for all $0 \leq s \leq s_0$.

The definition of $WLAA$ weakens that of LAA in three ways. First, the entire past state history, $\{U(s): 0 \leq s \leq t\}$, is replaced by the present state, $U(t)$, alone; second, the entire future history of arrivals, $\{N(t+s) - N(t): s \geq 0\}$, is replaced by finite-horizon future increments, $\{N(t+s) - N(t): 0 \leq s \leq s_0\}$; and third, stochastic independence is replaced by lack of correlation. Thus, LAA

implies *WLAA*, but not vice versa. The *WLAA* condition is closely related to an independence condition introduced by König and Schmidt [15]; see Theorem 1.6.6 of Franken et al. [9].

A sufficient condition for *ASTA* in terms of *WLAA* is cast in the expectation framework as follows.

Theorem 7.6. (*ASTA* under *WLAA*) *Suppose that the pair (U, N) satisfies WLAA. Assume further that, at least, one of the two following conditions holds:*

1. $\mathbf{E}[U(t)]$ *is independent of* $t \geq 0$,
2. *for each* $t \geq 0$ *there exists* $s_0 > 0$, *such that* $\mathbf{E}[N(t+s) - N(t)] = \lambda s$ *for* $0 \leq s \leq s_0$.

Then, $\bar{V}(t) = \bar{W}(t)$ *for all* $t > s_0$, *so that* $\bar{V}(t) \underset{t\to\infty}{\to} \bar{V}(\infty)$ *iff* $\bar{W}(t) \underset{t\to\infty}{\to} \bar{W}(\infty)$, *in which case* $\bar{V}(\infty) = \bar{W}(\infty)$.

Clearly, if U is stationary and N has stationary increments, then Conditions 1 and 2 in Theorem 7.6 will be satisfied. However, there is no requirement that N be a Poisson Process, though *WLAA* imposes a considerable restriction on N. The proof of Theorem 7.6, utilizing the Riemann-Stieltjes Sum approach in the expectations framework, will be sketched in Section 7.4.

7.3.3 *ASTA* under the *LBA* condition

The *L*ack of *B*ias *A*ssumption (*LBA*) generalizes both *LAA* and *WLAA* essentially to the utmost generality. Ignoring some minor technical details, *LBA* provides a fairly general necessary and sufficient condition for *ASTA*, thereby delineating the strongest interaction possible between U and N, which still preserves the *ASTA* property. The discussion of this section is based on [20].

The *LBA* condition does not directly involve the counting process, N. Rather, *LBA* requires the existence of a *conditional stochastic intensity* process, $\eta_U \equiv \{\eta_U(t): t > 0\}$, of N given U. The general definition of the stochastic intensity concept requires highly technical conditions under martingale approach and is deferred until Section 7.5. In this section, however, we present a more restrictive definition, but one that has the advantage of being intuitive.

Definition 7.7. (Conditional stochastic intensity) Let U be left continuous and let N be right continuous. The conditional intensity, $\eta_U(t)$, of N at t given $U(t)$, is the random variable,

$$\eta_U(t) = \lim_{s\to 0} \frac{\mathbf{E}[N(t+s) - N(t) \mid U(t)]}{s}, \quad t \geq 0, \tag{7.11}$$

provided the limit exists.

One motivation for calling η_U an intensity process is suggested by Equation (7.11). A more transparent motivation is offered by the fact that, if the interchange of the limit and expectation operation is admissible in (7.11), then

$$\begin{aligned} \mathbf{E}[\eta_U(t)] &= \mathbf{E}\left[\lim_{s\to 0} \frac{\mathbf{E}[N(t+s) - N(t) \mid U(t)]}{s}\right] \\ &= \lim_{s\to 0} \mathbf{E}\left[\frac{\mathbf{E}[N(t+s) - N(t) \mid U(t)]}{s}\right] = \lambda(t), \end{aligned} \tag{7.12}$$

where $\lambda(t) = (d/dt)\mathbf{E}[N(t)]$ is the rate (or intensity) of $N(t)$. It is useful to think of $\eta_U(t)$ as a conditional hazard, that is, as the (random) rate of a new arrival (in the counting process, N) at time t, given that the state process is $U(t)$. Thus, η_U encapsulates the "decision" of whether or not to sample U at time t, having observed its value, $U(t)$. The interaction between U and N should, thus, be intuitively clear: for *ASTA* to hold, the sampling decision should not depend "too much" on the state; otherwise, biased statistics would ensue, that is, $\bar{W}(\infty) - \bar{V}(\infty) \neq 0$. It will turn out that lack of bias is ensured by the *LBA* property, as defined below.

Definition 7.8. (*LBA*) Let U be left continuous and let N be right continuous. The pair (U, N) is said to satisfy the *L*ack of *B*ias *A*ssumption (*LBA*), if U and η_U are pointwise uncorrelated, i.e., $\text{Cov}[U(t), \eta_U(t)] = 0$, for all $t \geq 0$.

Formally, *LBA* weakens *WLAA* by substituting a pointwise condition for an interval-based condition. This fact, together with Definition 7.7, render *LBA* a local condition in the sense that the interaction between U and N (through η_U) need be known only in an infinitesimal forward interval and, then, only through covariances which are a measure of *linear* dependence. The *LBA* definition and terminology are motivated, in retrospect, by the following lemma.

Lemma 7.9. (The Bias formula) *Let U be left continuous and let N be right continuous. Suppose that*

1. *U and N are jointly stationary in the sense that*

$$(U(t), N(t+s) - N(t)) \stackrel{d}{=} (U(t+h), N(t+h+s) - N(t+h)),\ t, s, h > 0$$

2. *η_U is well-defined and obeys Equation* (7.12), *so that*

$$\mathbf{E}[\eta_U(t)] \equiv \lambda, \quad t \geq 0, \tag{7.13}$$

where $\lambda = \lim_{s \to 0} \frac{N(s)}{s}$.

Then, both $\bar{V}(t)$ and $\bar{W}(t)$ are independent of t and the bias, $\bar{W}(t) - \bar{V}(t)$, obeys the relation

$$\bar{W}(t) - \bar{V}(t) = \frac{\text{Cov}[U(0), \eta_U(0)]}{\lambda}, \quad t \geq 0. \tag{7.14}$$

The Bias formula (7.14) states that the *ASTA* condition can be attained either by decreasing the magnitude of the covariance in the numerator or by increasing the sampling rate in the denominator. In the first case, the magnitude of the covariance, being proportional to the bias, can be viewed as a direct measure of the interaction between U and N, where the sign of the covariance determines the direction of the bias. Thus, a positive (negative) covariance means that the event average overestimates (underestimates) its time average counterpart. In the second case, sampling at a higher rate decreases the bias inverse-proportionately. *ASTA* holds when the bias vanishes, i.e., when either the covariance vanishes or the rate is infinite. The latter case is not feasible, but, intuitively, sampling U at an infinite rate is informationally equivalent to U itself. Overall, the covariance in the numerator of (7.14) plays the most fundamental role in the *ASTA* problem, motivating the *LBA* condition. This is explicitly summarized in the following theorem.

Theorem 7.10. (*ASTA* under *LBA*) *Assume that the conditions of Lemma 7.9 hold for the pair (U, N). Then, $\bar{W}(\infty) = \bar{V}(\infty)$ iff LBA holds for (U, N).*

While the *LBA* condition is given in terms of covariances, it can be shown that it assumes the form of stochastic independence, provided it holds for a sufficiently rich class of functions f. Specifically, consider the class $\mathcal{U}$ consisting of all stochastic processes of the form $U(t) = f(X(t))$ over all bounded continuous real-valued functions, f. Alternatively, $\mathcal{U}$ may be taken as the class of all indicator functions of events in terms of X.

Theorem 7.11. (*ASTA* under lack of dependence) *Assume that the* *conditions of Lemma 7.9 hold for each pair (U,N), where $U \in \mathcal{U}$ and N is a fixed counting process. Then, the following are equivalent:*

1. *ASTA holds for (X,N) in the sense that $X(\infty) \stackrel{d}{=} X^{(N)}(\infty)$,*
2. *LBA holds for each pair (U,N), for all $U \in \mathcal{U}$,*
3. *$\eta_X(t) = \mathrm{E}[\eta_X(t)] \equiv \lambda$ w.p.1 for all $t \geq 0$, and*
4. *$\eta_X(t)$ is independent of $X(t)$, for all $t \geq 0$.*

Thus, *ASTA* is equivalent here to the fourth condition above, which is stated in terms of pointwise independence (and therefore is nominally stronger than pointwise lack of correlation); however, being a local (pointwise) condition, it is still weaker than *WLAA*.

We mention that, excluding *PASTA*, both *WLAA* and *LBA* are not easy to verify, except when X or U are Markovian. In that case, the conditional intensity process can be defined rather simply in terms of the Markovian transition intensities (rates) of the corresponding infinitesimal generator. Finally, we mention that *WLAA* and *LBA* are not vacuous conditions. One can exhibit non-Poisson counting processes of interest for which *ASTA* holds. These will be demonstrated in Section 7.6.

7.3.4 Conditional *ASTA*

The term "Conditional *ASTA*" refers to a generalization of *ASTA* to conditional statistics; see Van Doorn and Regterschot [27]. The setting requires a modification and augmentation of the old one, but nothing essentially new is required.

The Conditional *ASTA* setting assumes that X and N are as before, but these are augmented as follows. In addition to X, let $Z \equiv \{Z(t): t \geq 0\}$ be another process on the underlying probability space, $(\Omega, \mathcal{F}, P)$, having the same sample path properties as X. Further, in addition to f, let g and h be bounded measurable real-valued functions, where g is defined on the state space of Z and h is defined on the cross product of the state spaces of X and Z. Finally, in addition to U, define two other transformed stochastic processes, $G \equiv \{G(t) = g(A(t)): t \geq 0\}$ and $H \equiv \{H(t) = h(X(t), Z(t)): t \geq 0\}$.

Definition 7.12. (The Conditional *ASTA* problem) Suppose that

$$\frac{H(t)}{G(t)} \underset{t\to\infty}{\Rightarrow} \frac{H(\infty)}{G(\infty)} \text{ and } \frac{H^{(N)}(n)}{G^{(N)}(n)} \underset{n\to\infty}{\Rightarrow} \frac{H^{(N)}(\infty)}{G^{(N)}(\infty)}. \tag{7.15}$$

The *Conditional ASTA* problem is concerned with finding sufficient and/or necessary conditions for the equality

$$\frac{H(\infty)}{G(\infty)} \stackrel{d}{=} \frac{H^{(N)}(\infty)}{G^{(N)}(\infty)}. \tag{7.16}$$

In this case, we say that *Conditional ASTA holds* for the triple (G, H, N).

Calling (7.16) Conditional *ASTA* is justified by the observation that the particular choices, $h = 1_{A\times B}$ and $g = 1_B$, where A and B are measurable state sets of X and Z, respectively, correspond to conditional probabilities of the form

$$\mathrm{P}\{X(t) \in A \mid Z(t) \in B\},\ \mathrm{P}\{X^{(N)}(n) \in A \mid Z^{(N)}(n) \in B\},$$

whose equality in the limit (as $t\to\infty$) is the counterpart of the weak convergence condition (7.2) in ordinary *ASTA* setting.

The two variants of the *ASTA* problem can be extended to their conditional counterparts as well. In analogy to $V(t)$ in (7.3) and $W(t)$ in (7.4), the sample path framework for the Conditional $ASTA$ problem considers the conditional time average process, V^*, and conditional event average process, W^*, given by

$$V^*(t) = \frac{\int_0^t H(s)ds}{\int_0^t G(s)ds}, \quad t > 0 \tag{7.17}$$

and

$$W^*(t) = \frac{\int_0^t H(s)dN(s)}{\int_0^t G(s)dN(s)}, \quad t > 0. \tag{7.18}$$

Similarly, in analogy to $\bar{V}(t)$ in (7.6) and $\bar{W}(t)$ in (7.7), the expectations framework for the Conditional $ASTA$ problem considers the corresponding sequences of expectation ratios.

$$\bar{V}^*(t) = \frac{\mathbf{E}[\int_0^t H(s)ds]}{\mathbf{E}[\int_0^t G(s)ds]}, \quad t > 0 \tag{7.19}$$

and

$$\bar{W}^*(t) = \frac{\mathbf{E}[\int_0^t H(s)dN(s)]}{\mathbf{E}[\int_0^t G(s)dN(s)]}, \quad t > 0. \tag{7.20}$$

It should be clear that the entire theory presented, up to this point, will go through, by applying the various conditions, with minor modifications, to the pairs (H, N) and (G, N), separately. In fact, the new process (X, Z) can be viewed as being of the same species as the old process X. Moreover, $H(t) = h(X(t), Z(t))$ and $G(t) = g(Z(t))$ are just two candidate bounded continuous real-valued functions of $(X(t), Z(t))$ of the same species as $U(t)$. Thus, the Conditional $ASTA$ problem can be treated via the tools of the ordinary $ASTA$ problem, and, for this reason, we proceed here no further. The omitted details may be found in Van Doorn and Regterschot [27] and in Melamed and Whitt [20].

7.3.5 *ASTA* under time reversal

The $PASTA$ Theorem in Wolff [31] was stated for left continuous state processes, U, and right continuous counting processes, N. In queueing context, the semantics of these assumptions correspond to *jobs in an arrival process observing the state of the queue just before joining it.* There is, however, a conceptual duality between these semantics and those corresponding to *jobs in a departure process observing the state of the queue just after departing from it.* The duality is obtained by time reversal, namely, by considering a stochastic process $Y \equiv \{Y(t): -\infty < t < \infty\}$ and its time-reversed version $\tilde{Y} \equiv \{\tilde{Y}(t) = Y(-t): -\infty < t < \infty\}$.

Note carefully that "time reversal" is a concept distinct from "time reversibility"; any process can be reversed in time, but only certain processes are time-reversible, namely, their probability law coincides with that of their time-reversed version. In queueing context, time reversal can be visualized as jobs moving through a queue backward in time. Thus, arrivals in Y become departures in $\tilde{Y}$ and vice versa. Moreover, reversing time in a left continuous U results in a right continuous $\widetilde{U}$, while reversing time in a right continuous $\widetilde{N}$ results in a left continuous N. However, the arrival time points, T_n, remain unchanged under time reversal, though their enumeration changes per force. More importantly, all limiting time averages, $\bar{V}(\infty)$, and event averages, $\bar{W}(\infty)$, along each sample path remain unchanged under time reversal, provided only that the integrals (sums) considered are invariant under the order of integration (summation). Thus, the foregoing *ASTA* results can be applied to "new" traffic processes in the time-reversed system and "old" state processes in the original setting, yielding dual *ASTA* results. Such results typically extend *ASTA* from arrival-embedded to departure-embedded statistics and will be referred to as *Reverse ASTA*.

To illustrate this idea, we restate some *Reverse ASTA* conditions in terms of *ASTA* conditions for pairs (U, N) in ordinary time; these generally require the replacement of forward time intervals with backward time intervals. (Refer to Serfozo [23] for related conditions and results.)

Definition 7.13. (***Reverse ASTA* conditions**) Let U be right continuous and let N be left continuous.

The *R*everse *L*ack of *A*nticipation *A*ssumption (*Reverse LAA*) holds for (U, N), if, for all $t \geq 0$, $\{N(t) - N(t-s): 0 \leq s \leq t\}$ is independent of $\{U(s): s \geq 0\}$.

The *R*everse *W*eak *L*ack of *A*nticipation *A*ssumption (*Reverse WLAA*) holds for (U, N), if for each $t \geq 0$ there exists $s_0 > 0$, such that $U(t)$ and $N(t-s, t)$ are uncorrelated for all $0 \leq s \leq s_0$ with $t - s \geq 0$.

The *R*everse *L*ack of *B*ias *A*ssumption (*Reverse LBA*) holds for (U, N), if $U(t)$ and $\widetilde{\eta}_U(t)$ are pointwise uncorrelated for all $t \geq 0$, where $\widetilde{\eta}_U(t) = \lim_{s \to 0} \frac{\mathbf{E}[N(t-s, t) \mid U(t)]}{s}$ is the reverse conditional intensity of N at t given $U(t)$.

The results of the previous *ASTA* theorems, will then hold under the following modifications.

Theorem 7.14. (***ASTA* under time reversal**) *Let U be right continuous and N left continuous. Then*

1. *The results of Theorem* 7.3 *hold, provided LAA in Definition* 7.2 *is replaced by Reverse LAA in Definition* 7.13.
2. *The results of Theorem* 7.6 *hold, provided WLAA in Definition* 7.5 *is replaced by Reverse WLAA in Definition* 7.13, *and Condition* 2 *in Theorem* 7.6 *is replaced by the condition:*
   (2′) *For each $t \geq 0$ there exists positive s_0, such that $\mathbf{E}[N(t) - N(t-s)] = \lambda s$, for $0 \leq s \leq s_0$ with $t - s \geq 0$.*
3. *The results of Theorem* 7.10 *hold, provided LBA in Definition* 7.8 *is replaced by Reverse LBA in Definition* 7.13, *η_U in Definition* 7.7 *is replaced by $\widetilde{\eta}_U$ in Definition* 7.13, *and finally, Condition* 1 *in Lemma* 7.9 *is replaced by the condition:*
   (1′) *U and N are jointly stationary in the sense that*

$$(U(t), N(t) - N(t-s)) \stackrel{d}{=} (U(t+h), N(t+h) - N(t+h-s)), \quad t, s, h > 0.$$

The time-reversed version of Theorem 7.11 is similar.

7.3.6 *ASTA* in Markovian setting with transition counts

When U is a Markovian state process and N is a counting process of transitions in U, a sweeping simplification of the *ASTA* problem takes place. In this specialized setting, the *ASTA*, *PASTA* and anti-*PASTA* properties merge and become essentially equivalent. Many useful queueing models fall within the scope of this simplification; see Melamed [19].

The setting of this case is as follows. Let U be a right continuous Markov process. For simplicity, we assume that the state space, E of U, is countable, though the results go through for the uncountable case as well (see [19]) Assume further that the infinitesimal generator matrix, Q of U, with elements $q(i,j)$, has a bounded diagonal and that U has a unique long-run distribution vector ξ. Let $\Theta \subset E \times E$ be a so-called *traffic set* of U. Define a sequence of random times, T_n, recursively by

$$T_n = \begin{cases} 0, & n = 0 \\ inf\{t > T_{n-1}: (U(t-), U(t+)) \in \Theta\}, & n > 0 \end{cases}$$

and take N to be the right continuous counting process of the random times T_n, above. Let ψ^+ and ψ^- be the long-run distributions of $\{U(T_n+): n = 0, 1, \ldots\}$ and $\{U(T_n-): n = 0, 1, \ldots\}$, respectively. It can be shown (see [19]) that

$$\psi^+(k) = \frac{\sum_{j:(j,k) \in \Theta} \xi(j) q(j,k)}{\sum_{(i,j) \in \Theta} \xi(i) q(i,j)} \tag{7.21}$$

and

$$\psi^-(k) = \frac{\xi(k) \sum_{j:(k,j) \in \Theta} q(k,j)}{\sum_{(i,j) \in \Theta} \xi(i) q(i,j)}. \tag{7.22}$$

When U has an uncountable state space, then additional technical measure-theoretical assumptions are made, such as imposing Harris recurrence on U. However, Equations (7.21) and (7.22) remain structurally unchanged, with the sums replaced by corresponding integrals. See [19] for details.

Recall that the $ASTA$ problem seeks conditions under which $\xi = \psi^+$ for the right continuous version of U, or $\xi = \psi^-$ for the left continuous version of U. The key conditions that bring about this state of affairs are pointwise independence conditions of U and N, which may be viewed as weaker versions of LAA; refer to Section 7.5.3 for precise details. Intuitively, the Markov property obviates the requirement that future arrivals be independent of the entire past history of a Markovian state process, U. In fact, independence of the present (Markovian) state of U will do. The Markovian analogue of Theorem 7.3 is stated below.

Theorem 7.15. **(*ASTA* theorem for Markovian state processes)** *Let U be a regular Markov jump process. Then $ASTA$ holds for the pair (U,N) iff either of the following holds:*

1. *U is left continuous and the pair (U,N) is forward-pointwise independent, i.e, for all $t > 0$, $U(t)$ and $\{N(t+s) - N(s): s \geq 0\}$ are independent.*
2. *U is right continuous and the pair (U,N) is backward-pointwise independent, i.e., for all $t > 0$, $U(t)$ and $\{N(t) - N(t-s): 0 \leq s \leq t\}$ are independent. In fact, the latter condition is equivalent to pointwise independence of the pair (U,N), i.e., for all $t > 0$, $U(t)$ and $N(t)$ are independent.*

In each case above, $PASTA$ and anti-$PASTA$ are equivalent to $ASTA$.

A proof for the second case in Theorem 7.15 appears in Melamed [19]. This case is a consequence of more general results on traffic-embedded invariant distributions in Markovian jump processes with right continuous sample paths. The corresponding version for the left continuous case can be obtained via time-reversal arguments (see e.g., Green and Melamed [12]). Recall that time reversal does not affect ξ, but merely interchanges left continuity and right continuity, as well as forward-pointwise and backward-pointwise independence.

Theorem 7.15 is a variant of the Arrival Theorem class of *ASTA* results. Applications of the Arrival Theorem to queueing networks may be found in Section 7.6.

7.4 THE RIEMANN-STIELTJES SUM APPROACH TO *ASTA*

The Riemann-Stieltjes Sum approach to *ASTA* uses a stochastic integral representation for the embedded sums used in event averages. Specifically, since the counting process N is nondecreasing, its realizations induce measures of the form $dN(\omega, s) = N(\omega, s+) - N(\omega, s-)$ on the nonnegative half-line for almost every sample point, ω. Notice that $dN(s) = 1$ whenever $s = T_k$ for some $k > 0$, but $dN(s) = 0$, otherwise. Omitting the ω argument for typographical clarity, we can consequently write $w.p.1$

$$\sum_{k=1}^{N(t)} U_k^{(N)} = \sum_{k-1}^{N(t)} U(T_k) = \int_0^t U(s)dN(s)$$

$$= \lim_{n\to\infty} \sum_{k=0}^{n-1} U(kt/n)[N((k+1)t/n) - N(kt/n)], \tag{7.23}$$

where the last sum is the Riemann-Stieltjes sum of the preceding Riemann-Stieltjes integral. From Apostol [1], Chapter 9, the integral in (7.23) is well-defined by virtue of the $w.p.1$ boundedness and left continuity of U and the $w.p.1$ right continuity of N. We point out that for the Riemann-Stieltjes integral to be well-defined, it suffices for U to have $w.p.1$ no discontinuities from the right (left) where N has its right (left) discontinuities; in fact, limits from the left and right are not actually needed. If, however, U and N have common discontinuities from the left or from the right with positive probability, then there can be difficulties ([1]) This is avoided by the assumption that U is left continuous (right continuous), whereas N is assumed to be right continuous (left continuous). Note also that the Riemann-Stieltjes sum above need not have equal-size subintervals in the partition of $[0, t]$, so long as the width of the largest interval approaches zero in the limit. Moreover, the kth term of the nth sum in (7.23) can be evaluated for U anywhere in the closed interval $[kt/n, (k+1)t/n]$. However, for our purposes, it is important to use the left end point (right end point for time-reversed cases).

Recalling that U is bounded (because f is), and $\mathbf{E}[N(t)] < \infty$, $t > 0$, we have $w.p.1$,

$$\sum_{k=0}^{n-1} U(kt/n)[N((k+1)t/n) - N(kt/n)] \le [\sup_{0 \le s \le t} |U(s)|]N(t) < \infty. \tag{7.24}$$

Applying the Lebesgue Dominated Convergence Theorem to interchange the expectation and limit operations in (7.24), we get

$$\mathbf{E}[\sum_{k=1}^{N(t)} U(T_k)] = \lim_{n\to\infty} \sum_{k=0}^{n-1} \mathbf{E}[U(kt/n)\{N((k+1)t/n) - N(kt/n)\}]. \tag{7.25}$$

We are now in a position to illustrate the Riemann-Stieltjes Sum approach to the $ASTA$ problem. The proof of two $ASTA$ theorems will be now sketched in the expectations framework.

Proof of Theorem 7.6. By (7.25) and $WLAA$ (Definition 7.5),

$$\mathbf{E}[\sum_{k=1}^{N(t)} U(T_k)] = \lim_{n\to\infty} \sum_{k=0}^{n-1} \mathbf{E}[U(kt/n)]\mathbf{E}\,[N((k+1)t/n) - N(kt/n)].$$

Suppose that Condition 1 in Theorem 7.6 holds, so that $\mathbf{E}[U(t)] = \mathbf{E}[U(0)]$ for all t. Then,

$$\begin{aligned}\mathbf{E}[\sum_{k=1}^{N(t)} U(T_k)] &= \mathbf{E}[U(0)] \lim_{n\to\infty} \sum_{k=0}^{n-1} \mathbf{E}[N((k+1)t/n) - N(kt/n)] \\ &= \mathbf{E}[U(0)]\mathbf{E}[N(t)].\end{aligned}$$

Hence,

$$\bar{W}(t) = \mathbf{E}[U(0)] = \frac{1}{t}\int_0^t \mathbf{E}[U(s)]ds = \bar{V}(t), \quad t > 0,$$

and $ASTA$ has been established.

Suppose that Condition 2 in Theorem 7.6 holds, so that $\mathbf{E}[N(t+s) - N(t)] = \lambda s$ for $t \geq 0$ and $0 \leq s \leq s_0$. Then

$$\mathbf{E}[\sum_{k=1}^{N(t)} U(T_k)] = \lim_{n\to\infty} \frac{\mathbf{E}[N(t/n)]}{t/n} \sum_{k=0}^{n-1} \frac{t}{n}\,\mathbf{E}[U(kt/n)] = \lambda \int_0^t \mathbf{E}[U(s)]ds.$$

Since $\lambda = \mathbf{E}[N(t)/t]$, it follows that $\bar{W}(t) = \bar{V}(t)$ for $t \geq s_0$, whence $ASTA$ follows again. □

Proof of Theorem 7.10. First, assuming that f is continuous, we have

$$\begin{aligned}&\mathbf{E}\left[\frac{1}{t}\int_0^t U(s)dN(s)\right] \\ &= \mathbf{E}\left[\lim_{n\to\infty}\frac{1}{t}\sum_{k=0}^{n-1} U(kt/n)\{N((k+1)t/n) - N(kt/n)\}\right] \text{ by (7.23)} \\ &= \lim_{n\to\infty}\frac{1}{t}\sum_{k=0}^{n-1} \mathbf{E}[U(kt/n)\{N((k+1)t/n) - N(kt/n)\}] \text{ by (7.25)} \\ &= \lim_{n\to\infty}\frac{n}{t}\mathbf{E}[U(0)N(t/n)] \text{ by Condition 1 in Lemma 7.9} \\ &= \lim_{n\to\infty}\frac{n}{t}\mathbf{E}\left[U(0)\mathbf{E}[N(t/n)\mid U(0)]\right] \text{ by the calculus of conditioning} \\ &= \mathbf{E}\left[U(0)\lim_{n\to\infty}\frac{n}{t}\mathbf{E}[N(t/n)\mid U(0)]\right] \text{ by (7.11) and (7.12)} \\ &= \mathbf{E}[U(0)\eta_U(0)] \text{ by (7.11),}\end{aligned}$$

where in the second to last step, the convergence and uniform integrability in (7.11) and (7.12) remain unaltered on multiplication by the bounded random variable $U(0)$. Hence, from (7.7),

$$\bar{W}(t) = \frac{t}{\mathbf{E}[N(t)]}\,\mathbf{E}[U(t)\eta_U(t)].$$

But by (7.6) and the assumed stationarity of U,

$$\bar{V}(t) = \mathbf{E}[U(t)].$$

The result follows from the above for U with continuous f, in view of the relation

$$\mathbf{E}[N(t)] = \lambda t = \mathbf{E}[\eta_U(t)]t.$$

The extension to all measurable f, for which the expectations are well-defined, is obtained by taking limits; see Billingsley [3], pp. 7-9. □

We conclude this section by pointing out that the Riemann-Stieltjes Sum approach is highly intuitive. In particular, it makes it easy to identify the *WLAA* and *LBA* as useful *ASTA* conditions. To see that, observe that both of these *ASTA* conditions are based on lack of correlation, or equivalently, on the algebraic equality of expectations of products to the corresponding products of expectations.

7.5 THE MARTINGALE APPROACH TO *ASTA*

This section employs the martingale approach (refer to [2, 5, 6, 18, 21], among others) to treat the *ASTA* Problem in the sample path framework. The treatment requires a more elaborate setting, but the results are technically more general then the Riemann-Stieltjes Sum approach. An alternative sample path approach is due to Stidham and El-Taha (see [8], and [25] as Chapter 5 in this book).

7.5.1 The martingale approach setting

As before, we start with a general state process, $X \equiv \{X(t): t \geq 0\}$, and a counting process, $N \equiv \{N(t): t \geq 0\}$, over a common probability space, $(\Omega, \mathfrak{F}, P)$. Both X and N are assumed to be right continuous with left limits.

Let $\{\mathfrak{F}_t\}_{t \geq 0}$ be the filtration (an increasing family of σ-fields),

$$\mathfrak{F}_t = \sigma\{(X(s), N(s)): s \leq t\},$$

generated by the pair, (X, N). Henceforth, all martingales and stochastic intensities, as well as all measurability and predictability properties, are defined with respect to this filtration, unless otherwise specified.

As in Section 7.2, let $U(t) = f(X(t-))$ be bounded and left continuous. The left continuity is required as a technical condition to ensure predictability, namely, $U(t) \in \mathfrak{F}_{t-}$ for all t, where $\mathfrak{F}_{t-} = \cup_{s<t}\mathfrak{F}_s$. If $M \equiv \{M(t): t \geq 0\}$ is a martingale, then the boundedness and predictability of U imply that the stochastic integral,

$$(U \circ M)(t) \equiv \int_0^t U(s) dM(s), \tag{7.26}$$

is also martingale, a fact that will be used later, in particular, in proving Theorem 7.18 below.

The (unique) Doob-Meyer decomposition of N yields

$$N(t) = M(t) + A(t), \tag{7.27}$$

where $M \equiv \{M(t): t \geq 0\}$ is a martingale, and $A \equiv \{A(t): t \geq 0\}$ is a predictable, increasing process, with $A(0) = 0$. The process A is known as the *compensator* (or "dual predictable projection") of the counting process N. The martingale, M, is assumed to be square integrable.

Finally, N is assumed to be a *simple* counting process, i.e., of a *unit* jump size. Let

$$\Delta N(t) \equiv N(t) - N(t-), \text{ and } \Delta A(t) \equiv A(t) - A(t-).$$

Then, based on (7.27), and by virtue of the martingale property of M and the predictability of A, we have

$$\mathbf{E}[\Delta N(t) \mid \mathcal{F}_{t-}] = \mathbf{E}[\Delta A(t) \mid \mathcal{F}_{t-}] = \Delta A(t),$$

implying $\Delta A(t) \leq 1$ for all t, since $\Delta N(t) \leq 1$. Consequently, the *quadratic variation* process, $\langle M \rangle$, satisfies, for all $t \geq 0$,

$$\langle M \rangle(t) = \int_0^t [1 - \Delta A(s)] dA(s) \leq A(t). \tag{7.28}$$

Note that when A is continuous, i.e., $\Delta A(t) = 0$ for all t, we have

$$\langle M \rangle(t) = A(t), \quad \forall t. \tag{7.29}$$

We shall need the following SLLN (strong law of large numbers) for the martingale M.

Lemma 7.16. (Special case of Liptser [17])
(*i*) *If* $\langle M \rangle(\infty) < \infty$, *then* $M(\infty) < \infty$;
(*ii*) *If* $\langle M \rangle(\infty) = \infty$, *then* $\lim_{t \to \infty} M(t)/\langle M \rangle(t) = 0$.

Another mild regularity condition is $N(\infty) = \infty$.

Lemma 7.17. $N(\infty) = \infty$ *if, and only if,* $A(\infty) = \infty$. *Furthermore, either condition implies*

$$\lim_{t \to \infty} M(t)/A(t) = 0,$$

or equivalently,

$$\lim_{t \to \infty} N(t)/A(t) = 1.$$

Proof. (*i*) Suppose $N(\infty) = \infty$. If $A(\infty) < \infty$, then $\langle M \rangle(\infty) < \infty$ from (7.28), implying $M(\infty) < \infty$ (Lemma 7.16). Hence, $N(\infty) < \infty$ follows from (7.27), a contradiction.

(*ii*) Suppose $A(\infty) = \infty$. If $\langle M \rangle(\infty) = \infty$, then $M(t)/\langle M \rangle(t) \to 0$, from Part (*ii*) in Lemma 7.16, and hence, (cf. (7.28)),

$$|M(t)| / A(t) \leq |M(t)| / \langle M \rangle(t) \to 0.$$

If, however, $\langle M \rangle(\infty) < \infty$, then $M(\infty) < \infty$ (Lemma 7.16) and hence, $M(t)/A(t) \to 0$. Thus, in either case, $M(t)/A(t) \to 0$, which is equivalent to $N(t)/A(t) \to 1$, following the decomposition in (7.27). Hence, we conclude that $N(\infty) = \infty$. □

Next, assume that N admits a stochastic intensity, $\nu \equiv \{\nu(t): t \geq 0\}$ of the form

$$\nu(t) = \lim_{h \to 0} \frac{1}{h} \mathbf{E}[N(t+h) - N(t) \mid \mathcal{F}_t]. \tag{7.30}$$

Substituting (7.27) into (7.30), and again making use of the martingale property of M and the predictability of A, we have

$$\begin{aligned}
\nu(t) &= \lim_{h \to 0} \frac{1}{h} \mathbf{E}[M(t+h) - M(t) + A(t+h) - A(t) \mid \mathcal{F}_t] \\
&= \lim_{h \to 0} \frac{1}{h} \mathbf{E}[A(t+h) - A(t) \mid \mathcal{F}_t] \\
&= \lim_{h \to 0} \frac{1}{h}[A(t+h) - A(t)].
\end{aligned}$$

That is,

$$\nu(t) = \frac{d}{dt}A(t), \text{ or } A(t) = \int_0^t \nu(s)ds.$$

Hence, $A(t)$ is absolutely continuous, and in particular, (7.29) holds. Note that ν is nonnegative, and predictable, i.e., $\nu(t) \in \mathcal{F}_{t-}$, for all t. Assume that for each t, $\mathbf{E}[\nu(t)] < \infty$, and define

$$r(t) \equiv \mathbf{E}[\nu(t)], \text{ and } R(t) \equiv \int_0^t r(s)ds.$$

Applying Fubini's theorem, we get $R(t) = \mathbf{E}[A(t)]$, for all t.

Another well-known fact (e.g., Brémaud [4]) is that for any bounded, predictable process $\{Y(t): t \geq 0\}$, one has

$$\mathbf{E}[\int_0^t Y(s)dN(s)] = \mathbf{E}[\int_0^t Y(s)\nu(s)ds], \ \forall t. \tag{7.31}$$

This is so, because, from (7.27), the left-hand side of (7.31) can be written as

$$\mathbf{E}[\int_0^t Y(s)dN(s)] = \mathbf{E}[\int_0^t Y(s)dM(s)] + \mathbf{E}[\int_0^t Y(s)dA(s)].$$

The first term on the right-hand side above equals 0, since $\int_0^t Y(s)dM(s)$ is a martingale (as discussed earlier), while the second term equals the right-hand side of (7.31).

For the rest of this section, we always assume that N is a simple counting process, with $N(\infty) = \infty$, that admits the decomposition in (7.27) and the stochastic intensity in (7.30).

7.5.2 The general case

This section establishes a version of *ASTA* in the sample path framework. Define

$$V(t) = \frac{\int_0^t r(s)U(s)ds}{R(t)} \quad \text{and} \quad W(t) = \frac{\int_0^t U(s)dN(s)}{N(t)}, \tag{7.32}$$

and note that, while $W(t)$ in (7.32) remains the same as in (7.4), $V(t)$ in (7.32) is slightly more general than the definition in (7.3): here it is a *weighted* time average, with weight $r(s)$ for each $s \in (0, t]$. The goal is, again, to prove *ASTA* in the sense of

$$\lim_{t\to\infty} [W(t) - V(t)] = 0. \tag{7.33}$$

Unless otherwise specified, all limits involving stochastic processes are, henceforth, in the almost sure sense.

Letting

$$V^{(A)}(t) = \frac{\int_0^t U(s)dA(s)}{A(t)},$$

we first prove an intermediate, key result.

Theorem 7.18.

$$\lim_{t\to\infty} [W(t) - V^{(A)}(t)] = 0.$$

Proof.

$$W(t) - V^{(A)}(t) = \frac{A(t)}{N(t)} \frac{\int_0^t U(s)[dN(s) - dA(s)]}{A(t)} + \left[\frac{A(t)}{N(t)} - 1\right] \frac{\int_0^t U(s)dA(s)}{A(t)}$$

$$= \frac{A(t)}{N(t)} \frac{\int_0^t U(s)dM(s)}{A(t)} + \left[\frac{A(t)}{N(t)} - 1\right] \frac{\int_0^t U(s)dA(s)}{A(t)}, \tag{7.34}$$

where the second equality makes use of the Doob-Meyer decomposition in (7.27). Note that the second term on the right-hand side above vanishes as $t \to \infty$, since $A(t)/N(t) \to 1$ by Lemma 7.17 and U is bounded. On the other hand, since U is bounded and predictable, it follows from (7.26) that

$$\widetilde{M}(t) \equiv (U \circ M)(t) = \int_0^t U(s)dM(s),$$

is itself a martingale, and, in view of (7.29), the associated quadratic variation process is

$$\langle \widetilde{M} \rangle(t) = (U^2 \circ \langle M \rangle)(t) = \int_0^t U^2(s)dA(s).$$

As in the proof of Part (*ii*) of Lemma 7.17, and with the aid of Lemma 7.16, we may deduce that $\langle \widetilde{M} \rangle(\infty) < \infty$ implies $\widetilde{M}(\infty) < \infty$, and hence, $\widetilde{M}(t)/A(t) \to 0$ (since $A(\infty) = \infty$); whereas, $\langle \widetilde{M} \rangle(\infty) = \infty$ implies $\widetilde{M}(t)/\langle \widetilde{M} \rangle(t) \to 0$. Now, the boundedness of U implies $U^2(t) \leq B$, for some positive constant B; whence, regardless whether $\widetilde{M}(\infty) < \infty$ or $\widetilde{M}(\infty) = \infty$, we have

$$\frac{\widetilde{M}(t)}{A(t)} = \frac{\widetilde{M}(t)}{\langle \widetilde{M} \rangle(t)} \frac{\langle \widetilde{M} \rangle(t)}{A(t)} \leq B \frac{\widetilde{M}(t)}{\langle \widetilde{M} \rangle(t)} \to 0.$$

It then follows that the first term on the right side of (7.34) vanishes, too, as $t \to \infty$. □

A key necessary and sufficient condition for *ASTA* is presented in the following theorem.

Theorem 7.19. *ASTA holds in the sense of* (7.33) *iff* $\lim_{t \to \infty}[V^{(A)}(t) - V(t)] = 0$, *i.e.,*

$$\lim_{t \to \infty} \left[\frac{\int_0^t U(s)dA(s)}{A(t)} - \frac{\int_0^t U(s)r(s)ds}{R(t)} \right] = 0. \tag{7.35}$$

Suppose now that the intensity ν, and hence the compensator, A, is a deterministic function. That is,

$$r(t) \equiv \mathbf{E}[\nu(t)] = \nu(t) \text{ and } R(t) \equiv \mathbf{E}[A(t)] = A(t), \ w.p.1, \ \forall t.$$

Hence, (7.35) is trivially satisfied, since $V^{(A)}(t) = V(t)$ *w.p.*1 for all t. Furthermore, from Watanabe's characterization of Poisson processes (see, e.g., [29]), N is an $\mathcal{F}_t$-Poisson process. In addition, since, $\mathbf{E}[\nu(t)] = \nu(t)$, *w.p.*1, ν is also an intensity with respect to $\{\mathcal{F}_t^N\}$, the filtration generated by N alone. Hence, N is also an $\mathcal{F}_t^N$-Poisson process. This reduces to the case of *PASTA* (cf. Theorem 7.3).

Corollary 7.20. (*PASTA*) *ASTA holds when N has an $\mathcal{F}_t$-intensity ν, satisfying $\nu(t) = \mathbf{E}[\nu(t)]$ w.p.1. That is, ASTA holds when N is a Poisson process (with a possibly time-dependent intensity function).*

7.5.3 The stationary ergodic case

An important setting for *ASTA* is the stationary ergodic case, in which (7.35) is reduced to

$$\lim_{t\to\infty}[V^{(A)}(t) - V(t)] = \frac{\mathbf{E}[U(t)\nu(t)]}{\mathbf{E}[\nu(t)]} - \mathbf{E}[U(t)] = 0. \tag{7.36}$$

Since $U(t) = f(X(t-))$, (7.36) is satisfied iff

$$\mathrm{Cov}[\nu(t), f(X(t-))] = 0, \tag{7.37}$$

and because $\mathbf{E}[U(t)\nu(t)] = \mathbf{E}\big[f(X(t-))\mathbf{E}[\nu(t) \mid X(t-))\big]$, a sufficient condition is

$$\mathbf{E}[\nu(t) \mid X(t-)] = \mathbf{E}[\nu(t)]. \tag{7.38}$$

Corollary 7.21. *Under stationary ergodicity, ASTA holds iff* (7.37) *holds, or if* (7.38) *holds.*

Remark 7.22. Condition (7.37) coincides with *LBA* in Definition 7.8. Condition (7.38) is slightly stronger: it is equivalent to requiring that (7.37) hold for all bounded and measurable function f; as such it is analogous to Condition 2 in Theorem 7.11 Henceforth, we shall mainly use this slightly stronger version, still referring to it as *LBA*.

Remark 7.23. It is easy to verify that *LBA* in (7.38) is implied by the following sufficient condition,

$$\mathbf{E}[\Delta N(t) \mid X(t-)] = \mathbf{E}[\Delta N(t)] \quad \forall t. \tag{7.39}$$

Clearly, (7.39) and hence (7.38) are implied by *LAA* and *WLAA*, and both are satisfied when N is a Poisson process, with respect to $\mathcal{F}_t$. In other words, when N is Poisson (with respect to $\mathcal{F}_t$), stationary or non-stationary, *PASTA* follows. No additional condition is necessary. In particular, *LBA* is implied by the Poisson property. (In Wolff's version of *PASTA* (cf. Theorem 7.3), the condition *LAA* is required because the Poisson process is defined with respect to the filtration $\mathcal{F}_t^N$, instead of the larger filtration $\mathcal{F}_t$, as is the case here.)

Remark 7.24. In the stationary ergodic case, the sample path and expectation frameworks of *ASTA* coincide (in the sense of the (7.41) below), *LBA* being the key condition for both. Similarly to (7.6) and (7.7), denote

$$\bar{V}(t) = \frac{\mathbf{E}[\int_0^t U(s)r(s)ds]}{R(t)}, \quad \bar{W}(t) = \frac{\mathbf{E}[\int_0^t U(s)dN(s)]}{\mathbf{E}[N(t)]}. \tag{7.40}$$

Then in the stationary ergodic case,

$$\bar{V}(\infty) = V(\infty) = \mathbf{E}[U(t)].$$

Also, applying (7.30) and (7.27) to the numerator and the denominator of $\bar{W}(t)$, respectively, one has

$$\bar{W}(t) = \frac{\mathbf{E}[\int_0^t U(s)\nu(s)ds]}{\mathbf{E}[A(t)]},$$

whence,

$$\bar{W}(\infty) = V^{(A)}(\infty) = \frac{\mathbf{E}[U(t)\nu(t)]}{\mathbf{E}[\nu(t)]}.$$

Thus, under *LBA*, in the sense of (7.36),

$$\bar{W}(\infty) = V^{(A)}(\infty) = V(\infty) = \bar{V}(\infty).$$

On the other hand, from Theorem 7.18, we have $W(\infty) = V^{(A)}(\infty)$. Thus, in the stationary ergodic case, *LBA* leads to

$$\bar{W}(\infty) = \bar{V}(\infty) = V(\infty) = W(\infty). \tag{7.41}$$

Consider now Corollary 7.21 in Markovian setting, thereby revisiting Section 7.3.6. Specifically, suppose X is a Markov process (in addition to being stationary and ergodic), and assume that $\mathcal{F}_t = \mathcal{F}_t^X$. The latter implies that $\mathcal{F}_t^N \subset \mathcal{F}_t^X$, which is typically satisfied in queueing applications. For instance, in an M/M/1 queue, with X and N being, respectively, the state process and the arrival process, all the jumps of N correspond to the upward transitions of X. Utilizing the predictability of ν and the Markov property of X, we have

$$\nu(t) = \mathbf{E}[\nu(t) \mid \mathcal{F}_{t-}] = \mathbf{E}[\nu(t) \mid \mathcal{F}_{t-}^X] = \mathbf{E}[\nu(t) \mid X(t-)]. \tag{7.42}$$

Hence, LBA in (7.38) becomes $\nu(t) = \mathbf{E}[\nu(t)]$, implying that N is a Poisson process (with respect to both $\mathcal{F}^X$ and $\mathcal{F}^N$). This is the *anti-PASTA* property in Definition 7.4. Since LBA is generally satisfied when N is a Poisson process (see Remark 7.22), LBA and the Poisson property become equivalent in the Markovian setting.

Corollary 7.25. *Suppose that the state process X is Markovian (in addition to being stationary and ergodic), and that $\mathcal{F}_t^N \subset \mathcal{F}_t^X$ for all t. Then, LBA in the sense of* (7.38) *holds iff N is a Poisson process.*

The relations between $ASTA$, LBA and the Poisson property are summarized below (here only, the symbol '$\Rightarrow$' denotes logical implication rather than weak convergence).

- In general, N Poisson$\Rightarrow LBA$ (Remark 7.22), Poisson$\Rightarrow ASTA$ (Corollary 7.20).
- In the stationary ergodic case, N Poisson$\Rightarrow LBA \Leftrightarrow ASTA$ (Corollaries 7.20 and 7.21).
- In the stationary ergodic Markovian case, N Poisson$\Leftrightarrow LBA \Leftrightarrow ASTA$ (Corollary 7.25).

7.5.4 Connections to Palm calculus

Some interesting connections become apparent in the stationary setting between Corollary 7.21 and the so-called *Palm calculus* (see Brémaud [5], Baccelli and Brémaud [2], and König and Schmidt [15]). The stationary setting means that (X, N) are assumed to be jointly stationary on the probability space, $(\Omega, \mathcal{F}, \mathrm{P})$.

Let $\mathbf{P}^0$ denote the *Palm probability* associated with N. Let $T_0 < T_1 < \ldots$ be the sequence of arrival instants of N, and let N^0 be the *synchronized* version of N, in the sense that $T_0 = 0$ (i.e., the origin is an arrival instant). The *Palm probabi lity (measure)*, $\mathbf{P}^0$, associated with N^0, is defined by

$$\mathbf{P}^0(B) = \mathbf{P}\{N^0 \in B\}.$$

Henceforth, $\mathbf{E}$ and $\mathbf{E}^0$ will denote the expectations with respect to $\mathbf{P}$ and $\mathbf{P}^0$, respectively.

A key result of the Palm calculus is Papangelou's Lemma (see Papangelou [22]).

Lemma 7.26. (Papangelou's Lemma) $\mathbf{P}^0 \ll \mathbf{P}$ *iff N admits an intensity ν with respect to* $(\mathbf{P}, \mathcal{F}_t)$, *in which case one has the Radon-Nikodym derivative*

$$\frac{d\mathbf{P}^0}{d\mathbf{P}} = \frac{\nu}{\mathbf{E}[\nu]},$$

where the time index, t, is omitted from $\nu(t)$, due to the assumed stationarity.

Making use of Lemma 7.16, one has

$$\mathbf{E}^0[f(X)] = \int_{\omega \in \Omega} f(X(\omega))\mathbf{P}^0(d\omega) = \int_{\omega \in \Omega} f(X(\omega))\frac{\mathbf{P}^0(d\omega)}{\mathbf{P}(d\omega)}\mathbf{P}(d\omega)$$

$$= \int_{\omega \in \Omega} f(X(\omega))\frac{\nu}{\mathbf{E}[\nu]}\mathbf{P}(d\omega) = \frac{\mathbf{E}[f(X)\nu]}{\mathbf{E}[\nu]}.$$

In other words, the event average of the Palm version of the process $U \equiv f(X)$ is equal to $\mathbf{E}[f(X)\nu]/\mathbf{E}[\nu]$. Hence, $\mathbf{E}^0[f(X)] = \mathbf{E}[f(X)]$ iff X and ν are uncorrelated, as per (7.37) in the earlier analysis of stationary ergodic case of Section 7.5.3; see Theorem 7.18 and Remark 7.24. The connection between *ASTA* and Papangelou's result in Lemma 7.16 was established in Brémaud [5], which also contains a proof of the latter.

The relation derived above,

$$\mathbf{E}^0[f(X)] = \frac{\mathbf{E}[f(X)\nu]}{\mathbf{E}[\nu]}, \tag{7.43}$$

is commonly known as the *Palm formula.* The converse relation

$$\mathbf{E}[f(X)] = \mathbf{E}[\nu]\mathbf{E}^0\left[\int_0^{T_1} f(X(s-))ds\right], \tag{7.44}$$

is called the *Palm inversion formula.* Taking $f(x) \equiv 1$ in (7.44) implies that $\mathbf{E}^0[T_1] = 1/\mathbf{E}[\nu]$. Thus, for *ASTA* to hold (or equivalently, for $\mathbf{E}^0[f(X)] = \mathbf{E}[f(X)]$ to hold), it suffices to have

$$\mathbf{E}^0\left[\int_0^{T_1} f(X(s-))ds\right] = \mathbf{E}^0[T_1]\mathbf{E}^0[f(X)]. \tag{7.45}$$

ASTA through the Palm inversion formula is due to König and Schmidt [15].

7.6 QUEUEING NETWORK APPLICATIONS

It turns out that Corollaries 7.21 and 7.25 have interesting connections to the so-called *quasi-reversible* queues and networks thereof (Kelly [14]). Most of the properties of quasi-reversible queues and networks, to be discussed below, are known results. However, their connection to *LBA* and *ASTA* appears to be new.

7.6.1 Quasi-reversible queues

Let $\mathbf{X} = \{\mathbf{X}(t): -\infty < t < \infty\}$ be a Markov chain with a countable state space $\mathbf{S}$ and rate matrix $Q = [q(\mathbf{x}, \mathbf{y})]$. Let $\{\pi(\mathbf{x}): \mathbf{x} \in \mathbf{S}\}$ be an invariant (ergodic) distribution of $\mathbf{X}$. The time-reversed version of $\mathbf{X}$ is the process $\widetilde{\mathbf{X}} = \{\widetilde{\mathbf{X}}(t): -\infty < t < \infty\}$ over $\mathbf{S}$, defined by $\widetilde{\mathbf{X}}(t) = \mathbf{X}(-t)$. The notion of a quasi-reversible queue is motivated by the following result, which relates a stationary Markov chain to its time-reversed version (Kelly, [14], p. 28).

Proposition 7.27. (Time reversal in a Markovian process) *The time-reversed version,* $\widetilde{\mathbf{X}}$ *of* $\mathbf{X}$, *is also a stationary Markov chain, governed by a rate matrix,* $\widetilde{Q} = [\widetilde{q}(\mathbf{x}, \mathbf{y})]$, *satisfying*

$$\pi(\mathbf{x})\widetilde{q}(\mathbf{x}, \mathbf{y}) = \pi(\mathbf{y})q(\mathbf{y}, \mathbf{x}), \quad \forall \mathbf{x}, \mathbf{y} \in \mathbf{S}, \quad \mathbf{x} \neq \mathbf{y}. \tag{7.46}$$

Moreover, π is also an invariant distribution of $\widetilde{\mathbf{X}}$.

It is easy to verify that (7.46) implies the relation,

$$\sum_{\mathbf{y} \neq \mathbf{x}} q(\mathbf{x}, \mathbf{y}) = \sum_{\mathbf{y} \neq \mathbf{x}} \widetilde{q}\,(\mathbf{x}, \mathbf{y}) \quad \forall \mathbf{x} \in \mathbf{S}. \tag{7.47}$$

Henceforth, we restrict consideration to queueing processes. Specifically, assume that $\mathbf{X}$ represents a queue with job classes in a countable set C, and denote the arrival rate of class-c jobs by $\lambda^{(c)}$. Further, let $\mathbf{A}^{(c)}$ be the traffic set (see Section 7.3.6), denoting the set of all state pairs $(\mathbf{x}, \mathbf{y})$, such that $\mathbf{y}$ is a state reachable from $\mathbf{x}$ due to a class-c job's arrival. Similarly, $\mathbf{D}^{(c)}$ is the analogous traffic set corresponding to class-c job departures. Clearly, $1[(\mathbf{x}, \mathbf{y}) \in \mathbf{D}^{(c)}] = 1[(\mathbf{y}, \mathbf{x}) \in \mathbf{A}^{(c)}]$, where $1[\cdot]$ is an indicator function of the condition enclosed in brackets.

Definition 7.28. **(Quasi-reversibility of a Markovian queue)** $\mathbf{X}$ is called a *Markovian queue* if, as the state process of the queue, it is a stationary and ergodic Markov chain. A Markovian queue, $\mathbf{X}$, is termed a *quasi-reversible queue*, if

$$\lambda^{(c)} = \sum_{\mathbf{y}} q(\mathbf{x}, \mathbf{y}) 1[(\mathbf{x}, \mathbf{y}) \in \mathbf{A}^{(c)}] = \sum_{\mathbf{y}} \widetilde{q}\,(\mathbf{x}, \mathbf{y}) 1[(\mathbf{x}, \mathbf{y}) \in \mathbf{A}^{(c)}], \quad \mathbf{x} \in \mathbf{S}, \tag{7.48}$$

for all job classes, $c \in C$.

Condition (7.48) requires that the rate, $\lambda^{(c)}$, be a constant *independent* of the state $\mathbf{x}$. Together with the Markov property of $\mathbf{X}$, this implies that the arrival process of each job class is Poisson. Furthermore, the time-reversed version of a quasi-reversible queue is *by definition* just another quasi-reversible queue, with the same Poisson arrivals. Since time-reversed arrivals correspond to departures in the original queue (and vice versa), one can immediately deduce that the departure process is Poisson as well.

Proposition 7.29. *A quasi-reversible queue satisfies the following properties:*

(*i*) *Each arrival process of class-c jobs forms a Poisson process with rate $\lambda^{(c)}$ for all $c \in C$, and these are mutually independent.*

(*ii*) *Each departure process of class-c jobs forms a Poisson process with rate $\lambda^{(c)}$ for all $c \in C$, and these are mutually independent. Moreover, past departures up to time t are independent of the state of the system at time t.*

Thus, a quasi-reversible queue has the *Poisson-in-Poisson-out* property, namely, both input and output flows are Poisson.

7.6.2 Quasi-reversible networks

A quasi-reversible network consists of quasi-reversible queues, connected through a very general routing scheme, and in particular, the network itself possesses the Poisson-in-Poisson-out property.

A quasi-reversible network consists of a node set $M = \{1, \ldots, m\}$, and a fictitious node 0 that denotes both the "source" and "sink" of the network. Jobs belong to classes, drawn from a set C. External jobs arrive at the network (from the source node, 0) as a Poisson process of rate λ_0. Each arrival joins node i as a class-c job with probability $r_{0,i}^{(c)}$. On service completion at node i, a class-c is routed to node j and becomes a class-d job, with probability $r_{i,j}^{(c,d)}$, or leaves the network with probability $r_{i,0}^{(c)} = 1 - \sum_{(j,d) \in M \times C} r_{i,j}^{(c,d)}$, where $r_{0,0}^{(c,d)} = 0$, for all $c, d \in C$.

In the context of queueing networks, we consistently denote mathematical objects associated with node i, by appending the subscript i to them. For example, network states have the form $\mathbf{x} = (\mathbf{x}_i)_{i=1}^{m}$, where $\mathbf{x}_i$ is the state of node i; in a similar vein, $\mathbf{A}_i^{(c)}$ denotes the traffic set of all state pairs, $(\mathbf{x}_i, \mathbf{y}_i)$, such that $\mathbf{y}_i$ is reachable from $\mathbf{x}_i$ due to an arrival of a class-c job at node i from any network node (including network sources) and similarly for $\mathbf{D}_i^{(c)}$. Accordingly, let $\lambda_i^{(c)}$ denote the total equilibrium traffic rate of class-c jobs through node i (including both external arrivals and internal transitions). Then the $\lambda_i^{(c)}$ obey the so-called *traffic equations*,

$$\lambda_i^{(c)} = \lambda_0 r_{0,i}^{(c)} + \sum_{(j,d) \in M \times C} \lambda_j^{(d)} r_{j,i}^{(d,c)} \quad \text{and} \quad (i,c) \in M \times C. \tag{7.49}$$

The service mechanism at each node is immaterial, except that each queue *in isolation* is required to be quasi-reversible. Specifically, the analog of (7.48) is assumed to hold for all $(i,c) \in M \times C$, namely,

$$\lambda_i^{(c)} = \sum_{\mathbf{y}_i \in \mathbf{S}_i} q_i(\mathbf{x}_i, \mathbf{y}_i) 1[(\mathbf{x}_i, \mathbf{y}_i) \in \mathbf{A}_i^{(c)}] = \sum_{\mathbf{y}_i \in \mathbf{S}_i} \tilde{q}_i(\mathbf{x}_i, \mathbf{y}_i) 1[(\mathbf{x}_i, \mathbf{y}_i) \in \mathbf{A}_i^{(c)})]. \tag{7.50}$$

Proposition 7.30. *A quasi-reversible network has a product-form invariant distribution*

$$\pi(\mathbf{x}) = \prod_{i=1}^{m} \pi_i(\mathbf{x}_i),$$

where π_i is the invariant (marginal) distribution of the queue at node i when operated in isolation and fed with a set of independent Poisson arrivals with rates $\lambda_i^{(c)}$, $c \in C$, and the $\lambda_i^{(c)}$ are the solution to the traffic equations in (7.49). *In particular, the departure processes from the network are independent Poisson processes.*

Proposition 7.30 is proved by pairing the network with its time-reversed version, which is itself a network of quasi-reversible queues (each being the time-reversed version of its counterpart in the original network), and, then, analyzing both networks together based on Proposition 7.27.

Note carefully that, when quasi-reversible queues are connected into a quasi-reversible network, total input streams into nodes (consisting of both external arrivals and internal transitions) *may or may not be Poisson processes*, depending on network topology; see, e.g., Walrand [28], Chapter 4. In view of Proposition 7.29, quasi-reversibility may not hold at individual nodes in a quasi-reversible network, though it holds for the network as a whole. Nevertheless, *LBA* and *ASTA* still hold, as will be explained below.

7.6.3 *ASTA*, *LBA* and quasi-reversibility

Corollaries 7.21 and 7.25 have interesting connections to quasi-reversibility. In particular, as we will demonstrate in Theorem 7.31 below, *LBA* and quasi-reversibility are equivalent.

In queueing applications, *LBA* for the arrival stream takes the form

$$\mathbf{E}[\alpha(t) \mid \mathbf{X}(t-)] = \mathbf{E}[\alpha(t)], \tag{7.51}$$

where $\mathbf{X}$ denotes the state process, $\alpha = \{\alpha(t): t \geq 0\}$ denotes the stochastic intensity of the arrival process, $N_{\mathbf{A}}$, in $\mathbf{X}$, and $\mathbf{A} \subset \mathbf{S} \times \mathbf{S}$ is the corresponding traffic set of state transitions. In quasi-reversible queues, the departure process, $N_{\mathbf{D}}$, in $\mathbf{X}$, can be viewed as the arrival process, $\tilde{N}_{\mathbf{A}}$, in the time-reversed version, $\tilde{\mathbf{X}}$ (refer to Section 7.3.5). With $\delta = \{\delta(t): t \geq 0\}$ denoting the stochastic intensity of the departure process, *LBA* for the departure stream takes the analogous form,

$$\mathbf{E}[\delta(t) \mid \mathbf{X}(t+)] = \mathbf{E}[\delta(t)], \tag{7.52}$$

where $\delta(t)$ denotes the stochastic intensity of the departure process, $N_{\mathbf{D}}$, in $\mathbf{X}$, and $\mathbf{D} \subset \mathbf{S} \times \mathbf{S}$ is the corresponding traffic set of state transitions. Note, that we write $\mathbf{X}(t+) = \mathbf{X}(t)$ (even though $\mathbf{X}$ is right continuous) for emphasis and contrast with arrival processes.

Theorem 7.31. *For a stationary Markovian queue,* $\mathbf{X}$*, possessing an ergodic distribution,* π*, the following are equivalent:*

(*i*) $\mathbf{X}$ *is quasi-reversible, i.e.,* (7.50) *holds,*

(*ii*) *LBA holds for* $(\mathbf{X}, N_{\mathbf{A}})$ *and* $(\mathbf{X}, N_{\mathbf{D}})$*, i.e.,* (7.51) *and* (7.52) *are satisfied, and*

(*iii*) *The arrival and the departure processes are Poisson.*

Proof. We verify that (7.51) and (7.52) are equivalent to the two equations in (7.50) (ignoring the indices c and i in the latter). Making use of the relation between a queue and its time-reversed version (cf. Proposition 7.27), we have,

$$\begin{aligned}
\mathbf{E}[\delta(t) \mid \mathbf{X}(t+) = \mathbf{y}] &= \frac{\mathbf{E}[q(\mathbf{X}(t-), \mathbf{y})\mathbf{1}[\mathbf{X}(t+) = \mathbf{y}]]}{\mathbf{E}[\mathbf{1}[\mathbf{X}(t+) = \mathbf{y}]]} \\
&= \sum_{\mathbf{x} \in \mathbf{S}} \frac{\pi(\mathbf{X})q(\mathbf{x}, \mathbf{y})\mathbf{1}[(\mathbf{x}, \mathbf{y}) \in \mathbf{D}]}{\pi(\mathbf{y})} \\
&= \pi(\mathbf{y}) \sum_{\mathbf{x} \in \mathbf{S}} \frac{\tilde{q}\,(\mathbf{y}, \mathbf{x})\mathbf{1}[(\mathbf{y}, \mathbf{x}) \in \mathbf{A}]}{\pi(\mathbf{y})} \\
&= \sum_{\mathbf{x} \in \mathbf{S}} \tilde{q}\,(\mathbf{y}, \mathbf{x})\mathbf{1}[(\mathbf{y}, \mathbf{x}) \in \mathbf{A}], \quad \mathbf{y} \in \mathbf{S}.
\end{aligned} \tag{7.53}$$

On the other hand,

$$\alpha(t) = \sum_{\mathbf{y} \in \mathbf{S}} q(\mathbf{X}(t-), \mathbf{y})\mathbf{1}[(\mathbf{X}(t-), \mathbf{y}) \in \mathbf{A}]. \tag{7.54}$$

Suppose that the queue $\mathbf{X}$ is quasi-reversible. Combining (7.50) and (7.53), we get

$$\mathbf{E}[\delta(t) \mid \mathbf{X}(t+) = \mathbf{y}] = \mathbf{E}[\delta(t)] = \lambda_0, \tag{7.55}$$

whence (7.52) holds. In addition, (7.50) and (7.54) together imply that $\alpha(t) \equiv \lambda_0$, whence (7.51) holds.

For the converse, (7.52) implies (7.55), which, on combining with (7.53), yields one of the two equations in (7.50). Further, (7.51) implies that $\alpha(t)$ is a deterministic function, independent of $\mathbf{X}(t-)$. Due to stationarity, $\alpha(t)$ must be a constant, which in turn must be equal to λ_0, due to equilibrium. Hence, the other equation in (7.50) follows from (7.54).

Thus, (*i*) and (*ii*) are equivalent. The equivalence of (*ii*) and (*iii*) follows Corollary 7.25. □

Theorem 7.31 and Corollary 7.20 ensure that *PASTA* holds in a quasi-reversible queue, so that both arrivals and departures there see time averages. Recall that in quasi-reversible networks, internal Poisson traffic is not guaranteed. However, *LBA* and *ASTA* still hold. To see that, pick some node i and let $\alpha_i^{(c)}(t)$ denote the stochastic intensity of the total arrival process of class-c jobs at node i, including both external arrivals and internal transitions (from *other* nodes). Then,

$$\alpha_i^{(c)}(t) = \lambda_0 r_{0,i}^{(c)} + \sum_{d \neq c,\, j \neq i} \mu_j^{(d)}(\mathbf{X}_j(t-)) r_{j,i}^{(d,c)},$$

where

$$\mu_j^{(d)}(\mathbf{X}_j(t-)) = \sum_{\mathbf{y}_j \in \mathbf{S}_j} q_j(\mathbf{X}_j(t-), \mathbf{y}_j)\mathbf{1}[(\mathbf{X}_j(t-), \mathbf{y}_j) \in \mathbf{D}_j^{(d)}]$$

is the (state-dependent) service rate for class-d jobs at node j, $\mathbf{D}_j^{(d)} \subset \mathbf{S}_j \times \mathbf{S}_j$ is the corresponding traffic set, and $\mathbf{X}_j$ is the state at node j with state space $\mathbf{S}_j$. Since the product-form invariant distribution in Proposition 7.30 implies that $\mathbf{X}_i$ is independent of $\mathbf{X}_j$ for any $j \neq i$, we have

$$\mathbf{E}[\alpha_i^{(c)}(t) \mid \mathbf{X}_i(t-)] = \mathbf{E}[\alpha_i^{(c)}(t)],$$

implying that LBA is satisfied. Consequently, arrivals at node i do see time averages there; in particular, arrivals see the invariant distribution, π_i, provided the arriving job is excluded.

Do arrivals at node i see time averages at the other nodes as well? The answer is affirmative, and the argument proceeds through an examination of the departure process at node i, along with its time-reversed version. Let $\delta_i^{(c)} = \{\delta_i^{(c)}(t)\colon\ t \geq 0\}$ denote the stochastic intensity of the departure process of class-c jobs from node i, excluding those leaving the network. Denoting by $\mathbf{X}_{-i} = (\mathbf{X}_1, \ldots, \mathbf{X}_{i-1}, \mathbf{X}_{i+1}, \ldots, \mathbf{X}_m)$ the partial state process of $\mathbf{X} = (\mathbf{X}_1, \ldots, \mathbf{X}_m)$, with the ith component omitted, we have

$$\delta_i^{(c)}(t) = \mu_i^{(c)}(\mathbf{X}_i(t-))[1 - r_{i,0}^{(c)}],$$

which is independent of $\mathbf{X}_{-i}(t-)$. Hence, departures from node i that are routed to other parts of the network observe time averages at all the other nodes. Note that here $\mathbf{X}_i(t-)$ denotes the state at node i *just before* the departure of the job in motion, i.e., it includes the departing job. Similarly, $\mathbf{X}_{-i}(t-)$ denotes the state of the rest of the network, which *does not* include the job in motion.

The above argument applies to the time-reversed network as well. Hence, we may conclude that *arrivals* at node i, also, see time averages at all the other nodes, *just after* arrival. The following proposition summarizes the foregoing discussion.

Proposition 7.32. *In an open network of quasi-reversible queues (such as the one in Proposition* 7.30), *LBA holds, at each node i,*

- *for the arrival process* $N_{\mathbf{A}_i^{(c)}}$, *with stochastic intensity* $\alpha_i^{(c)}$, *and the left continuous state process,* $\mathbf{X}_i^{(c)}(t-)$,
- *for the arrival process* $N_{\mathbf{A}_i^{(c)}}$, *with stochastic intensity* $\alpha_i^{(c)}$, *and the right continuous state process,* $\mathbf{X}_{-i}^{(c)}(t+)$, *and*
- *for the departure process* $N_{\mathbf{D}_i^{(c)}}$, *with stochastic intensity* $\delta_i^{(c)}$, *and the left continuous state process,* $\mathbf{X}_{-i}^{(c)}(t-)$.

Hence, ASTA holds. In particular, arrivals at node i see time averages at node i, just before *arrival at node i; they see time averages at all other nodes* just after *arrival at node i.*

Remark 7.33. Note that Proposition 7.32 also implies that any job in *transition* sees the ergodic distribution of the whole network, but with itself excluded from the state description. Intuitively, the job in transition may be considered as visiting an extra node, which is not part of the original network, and from there it observes the ergodic distribution of all other nodes, which constitute the original network. It is crucial that the observation be made *just after* arrival at the extra node to ensure that the job in motion does not "see itself". To make this argument rigorous, it only remains to let the service rate at the extra node go to infinity, so that, in the limit, no time is actually spent at the extra node. This argument was pointed out in [20].

As noted in [20] through a network example, it is also possible to use Theorem 7.18, or equivalently, the Palm formula in (7.43), to derive job averages directly. This class of results falls in the domain of the so-called *Arrival Theorem*, to be illustrated in detail in the next two examples.

Example 7.34. *The Arrival Theorem in quasi-reversible networks.* In this example, it will be shown directly (in contrast to the indirect argument in Remark 7.33) that any job in transition sees the ergodic state distribution, *with itself excluded.*

Let $\mathbf{X}$ denote the state of the open network in Proposition 7.32, associated with a job's transition from node i to node j and changing class from c to d. The state does *not* include the job in transition. Let f be a bounded, measurable function on $\mathbf{S}$. Then, conditioning upon the states immediately before and immediate ly after the transition, and making use of the relations in (7.50), we have

$$\begin{aligned}
\mathbf{E}[f(\mathbf{X})\delta_{i,j}^{(c,d)}] &= \sum_{\mathbf{X}_{-(i,j)}} \prod_{k \neq i,j} \pi_k(\mathbf{x}_k) \sum_{\mathbf{x}_i} \sum_{\mathbf{y}_i} \pi_i(\mathbf{y}_i) q_i(\mathbf{y}_i, \mathbf{x}_i) \mathbf{1}[(\mathbf{y}_i, \mathbf{x}_i) \in \mathbf{D}_i^{(c)}] \\
&\quad \times \sum_{\mathbf{x}_j} \pi_j(\mathbf{x}_j) \sum_{\mathbf{y}_j} q_j(\mathbf{x}_j, \mathbf{y}_j) \mathbf{1}[(\mathbf{x}_j, \mathbf{y}_j) \in \mathbf{A}_j^{(d)}] f(\mathbf{X}) \frac{r_{i,j}^{(c,d)}}{\lambda_j^{(d)}} \\
&= r_{i,j}^{(c,d)} \sum_{\mathbf{X}_{-(i,j)}} \prod_{k \neq i,j} \pi_k(\mathbf{x}_k) \sum_{\mathbf{x}_j} \pi_j(\mathbf{x}_j) f(\mathbf{x}) \\
&\quad \times \sum_{\mathbf{x}_i} \pi_i(\mathbf{x}_i) \sum_{\mathbf{y}_i} \tilde{q}_i(\mathbf{x}_i, \mathbf{y}_i) \mathbf{1}[(\mathbf{x}_i, \mathbf{y}_i) \in \mathbf{A}_i^{(c)}] \\
&= \lambda_i^{(c)} r_{i,j}^{(c,d)} \sum_{\mathbf{X}_{-(i,j)}} \prod_{k \neq i,j} \pi_k(\mathbf{x}_k) \sum_{\mathbf{x}_i} \pi_i(\mathbf{x}_i) \sum_{\mathbf{x}_j} \pi_j(\mathbf{x}_j) f(\mathbf{x}) \\
&= \mathbf{E}[\delta_{i,j}^{(c,d)}] \mathbf{E}[f(\mathbf{X})].
\end{aligned}$$

Hence, following the Palm formula (7.43), we have $\mathbf{E}^0[f(\mathbf{X})] = \mathbf{E}[f(\mathbf{X})]$. In particular, letting $f(\mathbf{X}) = \mathbf{1}[\mathbf{X} = \mathbf{x}]$, it follows that a job in transition from node i to node j sees the ergodic network state distribution, but with itself excluded from the state description.

Note that the derivation above does *not* constitute a verification of *LBA*, since the state, $\mathbf{X}$, seen by the job in transition with itself excluded, is neither the state $\mathbf{X}(t-)$ nor the state $\mathbf{X}(t+)$, as required by *LBA*. Note carefully that if the above were a verification of *LBA*, it would imply that the process in question were Poisson, following Corollary 7.25. In particular, this does *not* contradict Theorem 7.15, since the state seen by a job in transition (with itself excluded) does *not* constitute a Markov process.

Example 7.35. *The Arrival Theorem in closed networks.* In this example, the same approach is applied to a *closed* network. This example is similar to Example 4 in [20], where a closed network with multiple servers at each node is analyzed, but the closed network here is more general.

Consider a single-class network, with N jobs circulating among its m nodes. The (state-dependent) service rate at node i is denoted by $\mu_i(n)$, when there are n jobs there, subject to $\mu_i(0) = 0$ and $\mu_i(n) > 0$ for $n > 0$. The routing matrix is $R = [r_{i,j}]$, which is a stochastic matrix (each rows sums to unity). For any vector $\mathbf{z} = (z_1, \ldots, z_m)$, denote $|\mathbf{z}| = \sum_i z_i$. Let $\mathbf{v} = (v_1, \ldots, v_m)$ be a solution to the system of equations

$$v_i = \sum_{j=1}^{m} v_j r_{j,i}, \quad i = 1,\dots,m.$$

Consider the network state process, $\mathbf{X} = (\mathbf{X}_1,\dots,\mathbf{X}_m)$, of job totals at the nodes in equilibrium. Let $\mathbf{Y} = (\mathbf{Y}_1,\dots,\mathbf{Y}_m)$ be a set of *independent* nonnegative integer-va lued random variables, satisfying

$$v_i \mathbf{P}\{\mathbf{Y}_i = n-1\} = \mu_i(n)\mathbf{P}\{\mathbf{Y}_i = n\}, \quad n = 1,2,\dots; \ i = 1,\dots,m. \tag{7.56}$$

It is known that the stationary (invariant) distribution of $\mathbf{X}$ is

$$\mathbf{P}\{\mathbf{X} = \mathbf{x}\} = \prod_{i=1}^{m} \frac{\mathbf{P}\{\mathbf{Y}_i = \mathbf{x}_i\}}{\mathbf{P}\{|\mathbf{Y}| = N\}}, \quad \mathbf{x} \in \mathbf{S}_N, \tag{7.57}$$

where $\mathbf{S}_N = \{\mathbf{x} = (\mathbf{x}_1,\dots,\mathbf{x}_m) \colon \mathbf{x}_i \ge 0, \, i \in M, \, |\mathbf{x}| = N\}$.

From the joint distribution, (7.57), one may compute the marginal distributions,

$$\mathbf{P}\{\mathbf{X}_i = \mathbf{x}_i\} = \mathbf{P}\{\mathbf{Y}_i = \mathbf{x}_i\} \sum_{\mathbf{x}_{-i} \in \mathbf{S}_{N-\mathbf{x}_i}} \frac{\mathbf{P}\{\mathbf{Y}_{-i} = \mathbf{x}_{-i}\}}{\mathbf{P}\{|\mathbf{Y}| = N\}}, \tag{7.58}$$

and verify that

$$\mathbf{E}[\mu_i(\mathbf{X}_i)] = v_i \frac{\mathbf{P}\{|\mathbf{Y}| = N-1\}}{\mathbf{P}\{|\mathbf{Y}| = N\}}. \tag{7.59}$$

Since the $\mathbf{X}_i$ in the closed network are *not* independent, due to the constraint $|\mathbf{X}(t)| = N$, one cannot expect *LBA* to hold. However, as mentioned earlier, one may use Theorem 7.18, or equivalently, the Palm formula in (7.43), to derive event averages, and compare them against their time-average counterparts. To this end, pick a node, say 1, and consider its departure process, with stochastic intensity $\delta_1(t) = \mu_1(\mathbf{X}_1(t-))$. We are interested in the distribution of the state, as seen by a departing job from node 1, but with the departing job excluded from the state description. Now, while $\mathbf{P}\{\mathbf{X}(t-) \in B\} = \mathbf{P}[\mathbf{Y} \in B \mid |\mathbf{Y}| = N]$ (see (7.57)), the state, $\mathbf{X}(t+)$, left behind by the departing job (with itself excluded), follows the *ergodic* distribution, $\mathbf{P}\{\mathbf{Y} - \mathbf{e}_1 \mid |\mathbf{Y}| = N\}$, where $\mathbf{e}_i$ is the unit vector of coordinate i. Making use of (7.58), one has

$$\begin{aligned}
\mathbf{E}[f(\mathbf{X})\mu_1(\mathbf{X}_1)] &= \mathbf{E}[f(\mathbf{Y} - \mathbf{e}_1)\mu_1(\mathbf{Y}_1) \mid |\mathbf{Y}| = N] \\
&= \sum_{x_1 = 1}^{N} \sum_{\mathbf{x}_{-1} \in \mathbf{S}_{N-\mathbf{x}_1}} f(\mathbf{x}_1 - 1, \mathbf{x}_{-1}) \frac{\mathbf{P}\{\mathbf{Y}_1 = \mathbf{x}_1\}\mathbf{P}\{\mathbf{Y}_{-1} = \mathbf{x}_{-1}\}}{\mathbf{P}\{|\mathbf{Y}| = N\}} \\
&= v_1 \sum_{\mathbf{x}_1 = 1}^{N} \sum_{\mathbf{x}_{-1} \in \mathbf{S}_{N-\mathbf{x}_1}} f(\mathbf{x}_1 - 1, \mathbf{x}_{-1}) \frac{\mathbf{P}\{\mathbf{Y}_1 = \mathbf{x}_1 - 1\}\mathbf{P}\{\mathbf{Y}_{-1} = \mathbf{x}_{-1}\}}{\mathbf{P}\{|\mathbf{Y}| = N]} \\
&= v_1 \sum_{\mathbf{x}_1 = 0}^{N-1} \sum_{\mathbf{x}_{-1} \in \mathbf{S}_{N-1-\mathbf{x}_1}} f(\mathbf{x}_1, \mathbf{x}_{-1}) \frac{\mathbf{P}\{\mathbf{Y}_1 = \mathbf{x}_1]\mathbf{P}\{\mathbf{Y}_{-1} = \mathbf{x}_{-1}\}}{\mathbf{P}\{|\mathbf{Y}| = N\}} \\
&= v_1 \frac{\mathbf{P}\{|\mathbf{Y}| = N-1\}}{\mathbf{P}\{|\mathbf{Y}| = N\}} \mathbf{E}[f(\mathbf{Y}_1, \mathbf{Y}_{-1}) \mid |\mathbf{Y}| = N-1] \\
&= \mathbf{E}[\mu_1(\mathbf{X}_1)]\mathbf{E}[f(\mathbf{Y}_1, \mathbf{Y}_{-1}) \mid |\mathbf{Y}| = N-1].
\end{aligned}$$

Thus, the (event) average observed by jobs departing from node 1 is

$$\mathbf{E}^0[f(\mathbf{X})] = \mathbf{E}[f(\mathbf{Y}_1, \mathbf{Y}_{-1}) \mid |\mathbf{Y}| = N-1].$$

In particular, for $f(\mathbf{X}) = 1[\mathbf{X} = \mathbf{x}]$,

$$\mathbf{P}^0\{\mathbf{X}=\mathbf{x}\} = \mathbf{P}\{\mathbf{Y}=\mathbf{x} \mid |\mathbf{Y}| = N-1\}.$$

The above result is the well-known Arrival Theorem in a closed queueing network; see e.g., Kelly [14], Sevcik and Mitrani [24], and Lavenberg and Reiser [16].

Again, we note that the above does not constitute a verification of *LBA*, since the state $\mathbf{X}$ is associated with a job in transition. It is neither $\mathbf{X}(t-)$ nor $\mathbf{X}(t+)$ as required in the verification of *LBA*.

To summarize, we have established that for Markovian queues (under stationary ergodicity), *LBA* is equivalent to quasi-reversibility. While quasi-reversibility is, in general, lost when the queue is connected into a network, *LBA* is preserved. For a quasi-reversible queue in isolation, *LBA* implies Poisson arrivals, and hence, *PASTA*; in a network of quasi-reversible queues, *LBA* implies (or more precisely, is equivalent to) *ASTA*. Hence, it appears that *LBA*, and hence, *ASTA*, are more fundamental properties that characterize quasi-reversible queues.

Finally, we note that in a network of quasi-reversible queues, *LBA* can also be used to identify Poisson flows. Along this line, conditions that are similar to those in Serfozo [23] and in Walrand [28] (pp. 166-167) can be established.

7.7 SUMMARY AND OPEN PROBLEMS

Three approaches to the *ASTA* problem may be discerned in the literature. The first is the elementary approach, which treats event averages as stochastic Riemann-Stieltjes integrals. This is the approach taken in Melamed and Whitt [20] and exemplified here in Section 7.4. The second is the martingale approach, which exploits properties of the martingale decomposition, stochastic integration, martingale SLLN, and the theory of stochastic intensities and compensators, leading to *ASTA* in the most general and direct manner. This approach is taken in Brémaud, Kannurpatti and Mazumdar [6], and in Melamed and Whitt [21], among others. It is also the approach followed in Section 7.5. The third approach focuses on the stationary case and views the problem as relating two expectations with respect to the underlying probability measure, $\mathbf{P}$, and its Palm-measure counterpart, $\mathbf{P}^0$, respectively. This approach is briefly summarized in Section 7.5.4. It is due to König and Schmidt [15], which first presented *ASTA* in the form of the Palm inversion formula [see (7.44)]. The connection between this approach and the martingale approach was brought out in Brémaud [5], through Papangelou's Lemma (Lemma 7.16), which relates the stochastic intensity to the Radon-Nykodym derivative $d\mathbf{P}^0/d\mathbf{P}$. This result reveals that the role played by the counting (or, "arrival") process N, through its intensity, is no more than a change of measure, from P to P^0.

The *LBA* condition, (7.37) and (7.38), is central to *ASTA*, particularly in the stationary ergodic setting and in queueing network applications. The intuition behind *LBA* is that for *ASTA* to hold, the time points ("arrival" epochs) selected for making observations, should not bias the observed state process.

To conclude, we comment on some open problems as candidates for further studies.

- To date, the best-known examples of *ASTA*, *in the absence* of Poisson flows, are all in the area of stationary Markovian queues, and in particular, networks of quasi-reversible queues (also known as "product-form" networks). This is hardly surprising, in view of the equivalence between *LBA* and quasi-reversibility (Theorem 7.31) and the fact that *LBA* is preserved in a network of quasi-reversible queues (Section 7.6.3). On the other hand, it remains an interesting and perhaps challenging open problem to identify *ASTA* results in other nonstationary and non-Markovian settings.

- The martingale approach to *ASTA* in Section 7.5 is based on a martingale strong law of large numbers; refer to Lemma 7.16. It turns out that one can also make use of a martingale central limit theorem to study the *ASTA* problem (Yao [32], Glynn, Melamed and Whitt [11]). This approach leads to results that are useful in constructing confidence intervals for estimators of job and time averages and in performing statistical tests. For details, refer to [11] and Glasserman [10]. Further studies along this line, we believe, should lead to more fruitful applications of *ASTA*.
- *ASTA* was originally motivated, in the queueing context, by the need to relate job (or, customer) based performance (e.g., delay) to system performance (e.g., time-average work load). Hence, the discrete (or, "embedded") observation epochs are naturally chosen as the arrival epochs of jobs or customers. In other applications, however, the choice of such sampling epochs might be far from obvious. A quick example is a fluid network that processes *continuous* flows, instead of discrete jobs. Formulating and studying an analogous *ASTA* problem in this setting should be an interesting undertaking.

ACKNOWLEDGEMENTS

We are grateful to Ward Whitt for his careful reading of the manuscript and his valuable comments. We thank Arif Merchant for reading and commenting on the manuscript.

BIBLIOGRAPHY

[1] Apostol, T.M., *Mathematical Analysis*, Addison-Wesley, New York 1957.
[2] Baccelli, F. and Brémaud, P., *Elements of Queueing Theory: Palm-Martingale Calculus and Stochastic Recurrences*, Springer-Verlag, New York 1994.
[3] Billingsley, P., *Convergence of Probability Measures*, John Wiley, New York 1968.
[4] Brémaud, P., *Point Processes and Queues*, Springer-Verlag, New York 1981.
[5] Brémaud, P., Characteristics of queueing systems observed at events and the connection between stochastic intensity and Palm probability, *Queueing Sys.* **5** (1989), 99-112.
[6] Brémaud, P., Kannurpatti, R. and Mazumdar, R., Event and time averages: A review and some generalizations, *Adv. Appl. Prob.* **24** (1992), 377-411.
[7] Descloux, A., On the validity of a particular subscriber's view, *Proc. Fifth International Traffic Conf.* New York (1967), 309.
[8] El-Taha, M. and Stidham Jr., S., Sample-path stability conditions for multi-server input-output processes, *J. Appl. Math. Stoch. Anal.* 7 (1994), 437-456.

[9] Franken, P., König, D., Arndt, U. and Schmidt, V., *Queues and Point Processes*, Akademie-Verlag, Berlin 1981. (Also Wiley, Chichester 1982).

[10] Glasserman, P., Filtered Monte-Carlo, *Math. Oper. Res.* **18** (1993), 610-634.

[11] Glynn, P., Melamed, B. and Whitt, W., Estimating customer and time averages, *Oper. Res.* **41** (1993), 400-408.

[12] Green, L. and Melamed, B., An anti-*PASTA* result for Markovian systems, *Oper. Res.* **38** (1990), 173-175.

[13] Halfin, S., Batch delays versus customer delays, *Bell Syst. Tech. J.* **62** (1983), 2011-2015.

[14] Kelly, F.P., *Reversibility and Stochastic Networks*, Wiley, New York 1979.

[15] König, D. and Schmidt, V., Imbedded and nonimbedded stationary characteristics of queueing systems with varying service rate and point processes, *J. Appl. Prob.* **17** (1980), 753-767.

[16] Lavenberg, S.S. and Reiser, M., Stationary state probabilities at arrival instants for closed queueing networks with multiple types of customers, *J. Appl. Prob.* **17** (1980), 1048-1061.

[17] Liptser, R.Sh., A strong law of large numbers for local martingales, *Stochastics* **3** (1980), 217-228.

[18] Makowski, A., Melamed, B. and Whitt, W., On averages seen by arrivals in discrete time, *Proc. of the 28th IEEE CDC*, Tampa, Florida 1989.

[19] Melamed, B., On Markov jump processes imbedded at jump epochs and their queueing-theoretic applications, *Math. Oper. Res.* **7** (1982), 111-128.

[20] Melamed, B. and Whitt, W., On arrivals that see time averages, *Oper. Res.* **38** (1990), 156-172.

[21] Melamed, B. and Whitt, W., On arrivals that see time averages: A martingale approach, *J. Appl. Prob.* **27** (1990), 376-384.

[22] Papangelou, F., Integrability of expected increments of point processes and a related change of time, *Am. Math. Soc.* **165** (1972), 483-506.

[23] Serfozo, R.F., Poisson functionals of Markov processes and queueing networks, *Adv. Appl. Prob.* **21** (1989), 595-611.

[24] Sevcik, K.C. and Mitrani, I., The distribution of queueing network states at input and output instants, *J. Assoc. Comput. Mach.* **28** (1981), 358-371.

[25] Stidham Jr., S. and El-Taha, M., Sample-path techniques in queueing theory, *Chapter 5*, in this book, 119-166.

[26] Strauch, R.E., When a queue looks the same to an arriving customer as to an observer, *Mgt. Sci.* **17** (1970), 140-141.

[27] Van Doorn, E.A. and Regterschot, G.J.K., Conditional *PASTA*, *Oper. Res. Letters* **7** (1988), 229-232.

[28] Walrand, J., *An Introduction to Queueing Networks*, Prentice-Hall, New Jersey 1988.

[29] Watanabe, S., On discontinuous additive functionals and Lévy measures of a Markov process, *Japan J. Math.* **34** (1964), 53-70.

[30] Whitt, W., A note on Poisson arrivals seeing time averages, unpublished paper, AT&T Bell Laboratories, Murray Hill, New Jersey 1979.

[31] Wolff, R.W., Poisson arrivals see time averages, *Oper. Res.* **30** (1982), 223-231.

[32] Yao, D.D., On Wolff's *PASTA* martingale, *Oper. Res.* **40** (1992), 352-355.

Chapter 8

Campbell's formula and applications to queueing

Volker Schmidt and Richard F. Serfozo

ABSTRACT Campbell's formula for Palm probabilities is a basic tool for deriving properties of stationary queueing systems and stationary processes in general. This study reviews Campbell's formula and gives new insights into its versatility by establishing several equivalent versions of it. A few of these are known (e.g., the exchange formula and $H = \lambda G$). Also included are applications involving integrals with respect to random product-measures, waiting times in systems, rate conservation laws, sojourn times of processes, travel times in networks, ladder heights in risk processes and virtual delays in queueing systems.

CONTENTS

8.1 Introduction 225
8.2 Campbell's formula 226
8.3 Equivalent versions of Campbell's formula 228
8.4 Applications to stationary queueing processes 231
8.5 An application to risk processes 235
8.6 Application of the exchange formula 237
8.7 Open problems 239
Bibliography 240

8.1 INTRODUCTION

Many features of a queueing system with stationary dynamics are expressed in terms of Palm probability distributions. The classic example is Little's law

$$\mathbf{E}[X(0)] = \mathbf{E}[M((0,1])]\mathbf{E}_M[W_0],$$

which says that the expected number of customers in the system at time 0 equals the expected arrival rate times the expected waiting time for a customer. The first two expectations are with respect to the usual probability measure of the process X, while the last expectation $\mathbf{E}_M$ is with respect to the Palm probability $\mathbf{P}_M$ associated with an arrival occurring at time 0. Another example involving Palm probabilities is

$$\mathbf{P}(X(0) > \ell) = \mathbf{E}[M_\ell((0,1])]\mathbf{E}_{M_\ell}[\text{Sojourn time of } X \text{ above } \ell], \tag{8.1}$$

where M_ℓ is the point process of times at which the process X exceeds ℓ. Such expressions relate probabilities or expectations of functionals of X under $\mathbf{P}$ to probabilities or expectations under Palm probabilities. A key tool for deriving such

0-8493-8074-x/95/$0.00+$.50

relations is Campbell's formula for Palm probabilities. This formula is a special case of Fubini's theorem for interchanging expectation and integral operations. Here the integrals are with respect to random product-measures and one passes from the usual probability to a Palm probability.

Several studies, see e.g., [10], [34], [40], [41], [55], point out that, besides Campbell's formula, there are other general formulas for functionals of random point processes that are suitable for specific applications: the exchange formula for point process ([40], [41]), Swiss Army formula ([10]), the exchange formula for general random measures ([34]), etc. For some of these formulas, their equivalence has been pointed out in [10]. In the present chapter we show, however, that all these formulas are equivalent to Campbell's formula. This is justified by subtle manipulations, with integrals and Palm probabilities similar to the early work of Mecke [33]; see also [31]. Applications of Campbell's formula to stationary queueing systems have a long history, see e.g., the comprehensive survey [53]. Recently, there have been several new applications of this formula to queueing and related stochastic models; see e.g., [2], [4], [5], [6], [7], [16], [26], [27], [28], [29], [37], [48], [50], [56]. We discuss some of these new applications in Sections 8.4 to 8.6 of the present chapter, including a brief review of classical ones like Little laws and conservation laws. In particular, we consider sojourn and travel times in stationary queueing networks, ladder height distributions, virtual delay and attained waiting time in stationary single-server queues. Concerning applications of Campbell's formula in stochastic geometry and, in particular, in stereology, we refer to [53] and to the literature cited in this monograph; see also [25].

Although the results involving Campbell's formula are for expectations of stationary systems, these expectations can also be interpreted as limiting averages of sample path functionals; this follows directly from corresponding ergodic theorems. Furthermore, there are analogous sample-path results for limiting averages for nonstationary systems, see e.g., [48], [51], [52], [54].

The rest of the chapter is organized as follows. In Section 8.2, we define Palm probabilities and present Campbell's formula. The setting covers random measures as well as random point processes on the Euclidean space $\mathbb{R}^d$ or an arbitrary complete separable metric space. Section 8.3 contains equivalent versions of Campbell's formula, and Sections 8.4-8.6 contain examples and applications. We end in Section 8.7 with a discussion of some open problems.

8.2 CAMPBELL'S FORMULA

We will use the following notation. Let $(E, \mathcal{E})$ denote the Euclidean space $\mathbb{R}^d$ with its σ-algebra of Borel sets. All the results herein also hold when E is an arbitrary complete, separable metric space that is a group under the operation $+$. Let $(\Omega, \mathcal{A}, \mathbf{P})$ be an arbitrary probability space, and let $\theta = \{\theta_t : t \in E\}$ be a measurable stochastic process with values in Ω such that $\theta_s(\theta_t(\omega)) = \theta_{s+t}(\omega)$ and $\theta_0(\omega) = \omega$, for $\omega \in \Omega$ and $s, t \in E$. The process θ is called a flow on Ω. We assume that the probability measure $\mathbf{P}$ on Ω is *invariant* under θ:

$$\mathbf{P}(\theta_t \in A) = \mathbf{P}(A), \quad t \in E, \ A \in \mathcal{A}.$$

Consequently, θ is a stationary process (the distribution of the process $\{\theta_{t+u} : t \in E\}$ is independent of u). A stochastic process (or random element or field) $X = \{X(t) : t \in E\}$ with values in some measurable space E' is *θ-compatible* if $X(t)$

$= f(\theta_t)$, $t \in E$, for some function $f:\Omega \to E'$. All functions herein are assumed to be measurable. Each θ-compatible process is clearly stationary since θ is. Accordingly, we call X a *θ-compatible stationary process* (sometimes X is said to be homogeneous or consistent with θ or generated by θ).

Let M be a *random measure* on E, where $M(A)$ denotes the random mass associated with $A \in \mathcal{E}$. That is, M is a measurable map from Ω to the space $\mathcal{M}$ of all measures on E that are finite on compact sets; the σ-algebra used on $\mathcal{M}$ is the smallest one under which the map $\mu \to \mu(A)$ is measurable for each $A \in \mathcal{E}$. Assume that M is a *θ-compatible stationary random measure.* That is, $M(\cdot + t) = \phi(\theta_t)$, $t \in E$, for some $\phi:\Omega \to \mathcal{M}$. The stationarity of θ ensures that M is stationary (the distribution of $M(\cdot + t)$ is independent of t). Then $\mathbf{E}[M(A)] = \lambda_M |A|$, where $|A|$ denotes the Lebesgue measure of the set A and $\lambda_M \geq 0$. The constant λ_M is the *intensity* of M. We assume that λ_M is positive and finite. The random measure M is called a *point process on E* if each of its realizations is an atomic, integer-valued measure on E, and we use the notation

$$M(A) = \sum_n 1(T_n \in A), \quad A \in \mathcal{E},$$

where the T_n's are the point (or unit-mass) locations of M. Here $1(S)$ is the indicator function that is 1 or 0 according as the statement S is true or false. Note also that the (random) atoms of M can be numbered in such a way that the mappings $T_n:\Omega \to \mathbb{R} \cup \{\infty\}$ are measurable for each $n \in \mathbf{Z}$, where $\mathbf{Z} = \{\ldots, -1, 0, 1, \ldots\}$. In case $E = \mathbb{R}$, we assume that the point locations are indexed such that $\ldots \leq T_{-2} \leq T_{-1} < 0 \leq T_0 \leq T_1 \leq T_2 \ldots$ The point process M is *simple* if its atoms have unit mass.

We shall study the stationary random measure M under its Palm probability measure $\mathbf{P}_M$ defined on Ω by

$$\mathbf{P}_M(A) = \frac{1}{\mathbf{E}[M(B)]}\mathbf{E}[M\{t \in B: \theta_t \in A\}]$$

$$= \frac{1}{\lambda_M |B|}\mathbf{E}\left[\int_B 1(\theta_t \in A)M(dt)\right], \quad A \in \mathcal{A}, \tag{8.2}$$

where $B \in \mathcal{E}$ is an arbitrary, fixed set with $0 < |B| < \infty$. This probability is well-defined in that the right side of (8.2) does not depend on the choice of the set B because of the stationarity assumptions above, see e.g., [33]. We let $\mathbf{E}_M$ denote the expectation under $\mathbf{P}_M$. When M is a simple point process, $\mathbf{P}_M$ can be interpreted as the conditional probability under the condition that at the origin $t = 0$ there is a point of M, and an alternate expression of (8.2) is

$$\mathbf{P}_M(A) = \frac{1}{\lambda_M |B|}\mathbf{E}\left[\sum_n 1(\theta_{T_n} \in A, T_n \in B)\right], \quad A \in \mathcal{A}. \tag{8.3}$$

Many probabilities and expectations under $\mathbf{P}$ have natural representations in terms of probabilities and expectations under $\mathbf{P}_M$. They can be obtained by applying Fubini's theorem to expectations of integrals of M, as in the following *Campbell formula* (where dt denotes integration with respect to Lebesgue measure). This is named after Campbell, who initiated such computations in [13] and [14] for point processes of radioactive particle emissions.

Theorem 8.1. (***Campbell's Formula***) *For any* $f: E \times \Omega \to \mathbb{R}_+$,

$$\mathbf{E}\left[\int_E f(t, \theta_t)M(dt)\right] = \lambda_M \mathbf{E}_M\left[\int_E f(t, \theta_0)dt\right]. \tag{8.4}$$

Proof. This can be proved, first, for indicator functions f by the definition of $\mathbf{P}_M$, then, for linear combinations of indicators, and finally, for general functions by monotone convergence. Another somewhat more direct approach is to apply Fubini's theorem as follows. We can write

$$\int_E f(t,\theta_t)M(dt) = \int_{\Omega \times E} f(t,a)\bar{M}(d(a,t)),$$

where $\bar{M}$ is the random measure defined by $\bar{M}(A \times B) = \int_B 1(\theta_t \in A)M(dt)$. Note that the dummy variable $a \in \Omega$ is not the same as the suppressed ω associated with the randomness of θ_t and M. By the definition (8.2) of $\mathbf{P}_M$, we know that $\mathbf{E}[\bar{M}(A \times B)] = \lambda_M \mathbf{P}_M(A)\int_B dt$. Then by Fubini's theorem, the left side of (8.4) equals

$$\mathbf{E}\left[\int_{\Omega \times E} f(t,a)\bar{M}(d(a,t))\right] = \int_{\Omega \times E} f(t,a)\mathbf{E}[\bar{M}(d(a,t))]$$

$$= \lambda_M \int_{\Omega \times E} f(t,a)\mathbf{P}_M(da)dt,$$

which equals the right side of (8.4). □

This result and its many ramifications are discussed in standard references such as [3], [15], [18], [25], [31], [33], [43], [47], [53]; see also the recent monographs [5], [50]. Note that (8.4) holds under very general assumptions on E; see [33] where the case is considered that E is an arbitrary locally compact Abelian group. The next sections describe equivalent versions of Campbell's formula and a few applications. We end this section by noting that Campbell's formula (8.4) allows us to express the probability $\mathbf{P}$ in terms of $\mathbf{P}_M$. This result, along with the definition of $\mathbf{P}_M$, shows that $\mathbf{P}_M$ is uniquely determined by $\mathbf{P}$ and vice versa.

Corollary 8.2. (***Inversion Formula***) *For any $A \in \mathcal{A}$,*

$$P(A) = \lambda_M E_M\left[\int_E 1(\theta_t \in A)g(-t,\theta_t)dt\right], \tag{8.5}$$

where $g: E \times \Omega \to \mathbb{R}_+$ is any function that satisfies $\int_E g(t,\omega)M(dt) = 1$. Hence,

$$P(A) = \lambda_M E_M\left[\int_B 1(\theta_t \in A)dt\right], \tag{8.6}$$

where B is any random set such that $M(B) = 1$. If case M is a point process on $\mathbb{R}$, a convenient choice is $B = (T_n, T_{n+1}]$ for any n.

Proof. Expression (8.5) follows from (8.4) with $f(t,\omega) = g(t,\omega)1(\theta_{-t} \in A)$. And (8.6) follows from (8.5) with $g(t,\omega) = 1(t \in B(\omega))$. □

8.3 EQUIVALENT VERSIONS OF CAMPBELL'S FORMULA

In this section, we present several examples of Campbell-type formulas that are actually equivalent versions of (8.4). For this discussion, we assume, as in the previous section, that M is a θ-compatible stationary random measure on E.

Example 8.3. (*Marked point processes*) Many applications of Campbell's formula (8.4) in the literature are for functionals of marked point processes described as follows. Suppose the stationary random measure M is a simple point process and that $\{\xi_n : n \in \mathbf{Z}\}$ is a sequence of *marks* of M that take values in some measurable space E'. Assume that $\xi_n = h(\theta_{T_n})$, for some $h: \Omega \to E'$. The sequence $[(T_n, \xi_n): n \in \mathbf{Z}\}$ is a *marked point process* that is stationary (i.e., the distribution of $\{(T_n - t, \xi_n): n \in \mathbf{Z}\}$ is independent of t, see for instance [3], [5], [18], [25], [50] for the properties of marked point processes that we use herein). The term mark also applies loosely to any variables associated with the point locations T_n, but we will consider only marks of the form $\xi_n = h(\theta_{T_n})$. Now, consider the process

$$X(t) = \sum_n f(t - T_n, \xi_n), \quad t \in E, \tag{8.7}$$

where $f: E \times E' \to \mathbb{R}$. Note that by the change of variable $u = s - t$,

$$X(t) = \int_E f(t-s, h(\theta_s))M(ds) = \int_E f(-u, h(\theta_u(\theta_t)))\phi(\theta_t)(du), \tag{8.8}$$

where $\xi_n = h(\theta_{T_n})$ and $M(A+t) = \phi(\theta_t)(A)$. This shows that $X(t)$ is a function of θ_t, and so X is a θ-compatible stationary process. Furthermore,

$$\mathbf{E}[X(0)] = \lambda_M \mathbf{E}_M\left[\int_E f(s, \xi_0)ds\right], \tag{8.9}$$

provided this expectation exists. To prove this, it suffices to assume that f is non-negative. Now, (8.9) follows by taking the expectation of (8.8) for $t = 0$ and using (8.4) with $f(s, \omega)$ replaced by $f(-s, h(\omega))$ and changing $-s$ to s.

The next sections contain several applications of the process X, where $E = \mathbb{R}$ and

$$X(t) = \sum_n g(t - T_n, \xi_n)1(T_n + \alpha_n \le t \le T_n + \beta_n), \tag{8.10}$$

and $(\xi_n, \alpha_n, \beta_n)$ are marks of M such that $\beta_n \ge \alpha_n$. In this case,

$$\mathbf{E}[X(0)] = \lambda_M \mathbf{E}_M\left[\int_{\alpha_n}^{\beta_n} g(s, \xi_0)ds\right]. \tag{8.11}$$

Expression (8.9) for $E = \mathbb{R}$ is sometimes referred to as $H = \lambda G$ (for expectations). Note that formula (8.9) with h as the identity function is Campbell's formula (8.4). Consequently, formula (8.9) is equivalent to Campbell's formula for point processes. A similar statement holds for marked random measures which we discuss next.

Example 8.4. (*Marked random measures*) The ideas in the preceding example also extend to marked random measures defined as follows. Suppose M is a θ-compatible stationary random measure on E. Assume that each t in the support of M has associated with it a quantity $\xi_t = h(\theta_t)$, where $h: \Omega \to E'$. We call ξ_t a *mark of* M at the location t. Then, as above, $X(t) = \int_E f(t-s, \xi_s))M(ds)$ is a θ-compatible stationary process whose mean is given by (8.9).

Example 8.5. (*Integrals of product measures*) Suppose that M and M' are θ-compatible stationary random measures on E. Then for any $h: E^2 \times \Omega \to \mathbb{R}_+$,

$$\mathbf{E}\left[\int_{E^2} h(t,s,\theta_t)M'(ds)M(dt)\right] = \lambda_M \mathbf{E}_M\left[\int_{E^2} h(t,s+t,\theta_0)M'(ds)dt\right]. \quad (8.12)$$

This follows from Campbell's formula (8.4) with

$$f(t,\theta_t) = \int_E h(t,s+t,\theta_t)M'(ds+t) = \int_E h(t,s+t,\theta_t)\phi'(\theta_t)(ds).$$

One can view (8.12) as a "conditional" Campbell formula for the bivariate random measure $M(ds \times dt) = M'(ds)M(dt)$, where the right side of (8.12) is like "conditioning" on the M part of $\widetilde{M}$. Note that formula (8.12) with M' as Lebesgue measures and $h(t,s,\omega) = f(t,\omega)1(s \in (0,1])$ is Campbell's formula (8.4). Consequently, (8.12) is equivalent to (8.4). Expression (8.12) also extends to h that may be negative as well as nonnegative and the measure M' may be a signed random measure: $M'(A) = M'_1(A) - M'_2(A)$, where M'_1 and M'_2 are nonnegative random measures. In this case, one applies (8.12) separately to the integrals of the positive and negative parts of h under the measures M'_1 and M'_2, provided that the sum of the expectations is well-defined (possibly infinite).

It may also be of interest to consider the integral

$$\widetilde{M}(B) = \int_{B \times E} h(t,s,\theta_t)M'(ds)M(dt),$$

as a θ-compatible stationary random measure. Then, its intensity, according to (8.12) is

$$\lambda_{\widetilde{M}} = \lambda_M \mathbf{E}_M\left[\int_{B \times E} h(t,s+t,\theta_0)M'(ds)dt\right], \quad (8.13)$$

where $|B| = 1$. This is essentially equations (3.11) and (6.2) in Brémaud [10], called the *Swiss Army formula*. He proves this version of Campbell's formula by using an integration-by-parts representation for the integral $\widetilde{M}(B)$ and establishes (8.13) first for bounded integrals and, then, proceeds to unbounded ones by a monotone convergence argument.

Example 8.6. (*Exchange formula*) Let M and M' be θ-compatible stationary random measures on E. Then for any $f: E \times \Omega \to \mathbb{R}_+$,

$$\lambda_M \mathbf{E}_M\left[\int_E f(t,\theta_t)M'(dt)\right] = \lambda_{M'}\mathbf{E}_{M'}\left[\int_E f(-t,\theta_0)M(dt)\right]. \quad (8.14)$$

This is Neveu's exchange formula [35,36]. To see this, fix $B \in \mathcal{E}$ such that $|B| = 1$. Then applying (8.12) to M and then to M', we have

$$\begin{aligned}
\text{Left side of (8.14)} &= \lambda_M \mathbf{E}_M\left[\int_{E^2} f(t,\theta_t)1(s+t \in B)M'(dt)ds\right] \\
&= \mathbf{E}\left[\int_{E^2} f(t-s,\theta_t)1(s \in B)M'(dt)M(ds)\right] \\
&= \lambda_{M'}\mathbf{E}_{M'}\left[\int_{E^2} f(-s,\theta_0)1(t \in B)M(ds)dt\right] \\
&= \text{Right side of (8.14)}.
\end{aligned}$$

Note that (8.14) is *Neveu's exchange formula* which has been obtained in [40], [41] for point processes; for the case of general random measures see also [34]. Like the previous examples, (8.14) is equivalent to Campbell's formula (8.4). The latter follows from the former when M is Lebesgue measure (in this case, $\mathbf{P}_M = \mathbf{P}$ and, consequently $\mathbf{E}_M = \mathbf{E}$). When $E = \mathbb{R}$ and at least one of the random measures M, M' is a point process on $\mathbb{R}$, applications in queueing of exchange formulas of the form (8.14) have been discussed in [3], [5], [9], [11], [18], [20], [26], and [27].

Example 8.7. (*Integrals and exchange formulas for random kernels*) Let $(E', \mathcal{E}')$ denote a complete, separable metric space and suppose that $K\colon E \times \Omega \times \mathcal{E}' \to \mathbb{R}$ is such that $K(t, \omega, \cdot)$ is a measure on E' for each t, ω. The K is called a *random kernel from E to E'*. The following formula is a variation of (8.12). For any $h\colon E \times E' \times \Omega \to \mathbb{R}_+$,

$$\mathbf{E}\left[\int_E \left(\int_{E'} h(t, s, \theta_t) K(t, \theta_t, ds)\right) M(dt)\right]$$

$$= \lambda_M \mathbf{E}_M\left[\int_E \left(\int_{E'} h(t, s, \theta_0) K(t, \theta_0, ds)\right) dt\right]. \tag{8.15}$$

This follows from Campbell's formula (8.4) with $f(t, \omega) = \int_{E'} h(t, s, \omega) K(t, \omega, ds)$. Furthermore, formula (8.15) is equivalent to (8.4) since the latter follows from (8.15) with $h(t, s, \omega) = f(t, \omega)$ and $K(t, \omega, E') = 1$.

The following is a variation of the exchange formula (8.14). Suppose K and K' are random kernels from E to E that are compatible with respect to θ such that

$$M(dt)K(t, ds) = M'(dt)K'(t, ds)$$

(the omegas in the kernels are now suppressed). Then, similarly to the proof of (8.14), it follows by two applications of (8.15) that

$$\lambda_M \mathbf{E}_M\left[\int_E h(s, \theta_s) K'(0, ds)\right] = \lambda_{M'} \mathbf{E}_{M'}\left[\int_E h(-s, \theta_0) K(0, ds)\right].$$

This is another equivalent version of (8.4).

The versions of Campbell's formula (8.4) in the preceding examples are a portmanteau of equivalent formulas that one could choose to use depending on one's application. Although one should be cautious in viewing all potential applications through the prism of any particular formula, it appears that Campbell's formula (8.4), because of its simplicity, is the easiest to use for new applications.

8.4 APPLICATIONS TO STATIONARY QUEUEING PROCESSES

This section gives several applications of Campbell's formula (8.9) for marked point processes. The notation here is similar to that in Sections 8.2 and 8.3.

Example 8.8. (*Little's law for waiting times*) Consider a queueing system or any abstract input-output system in which discrete units or customers periodically enter and eventually leave the system. For example, the system may be a certain subset of nodes in a stochastic network that processes units as they move among the nodes. Our interest is in the expected waiting time of the units in the system. Suppose units enter the system according to a θ-compatible stationary point process M on $\mathbb{R}$. Let W_n denote the waiting time in the system of the nth unit that arrives at time T_n. Assume that the stationary process θ contains enough information such that the W_n's are marks of M (i.e., W_n is a function of $\{\theta_{T_n+t}: t \in \mathbb{R}\}$, which is θ seen at time T_n). The stationary process

$$X(t) = \sum_n 1(T_n \leq t \leq T_n + W_n) \tag{8.16}$$

records the number of units in the system at time t. There are many examples of such stationary systems including the G/G/m queue and systems that are parts or functions of stationary Markovian or regenerative phenomena; see for instance [48]. Then Campbell's formula (8.6) for marked point processes yields

$$\mathbf{E}[X(0)] = \lambda_M \mathbf{E}_M[W_0]. \tag{8.17}$$

This is *Little's law* for expected waiting times in stationary systems. A typical example for a stationary Jackson network applies to the number of units $X(t)$ in a sector J (set of nodes) of the network at time t and the total sojourn time W_n in sector J of the nth unit to enter J. Another example is the slightly different situation in which $X(t)$ denotes the number of units in sector J waiting in queues "before" getting their services and W_n is the total time the nth unit visiting J waits in queues before getting services during its sojourn in J. These and many other examples follow, without further analysis, simply by defining the process X and times W_n appropriately.

In this regard, the law (8.17) also applies to the queueing system above in which the input process M is no longer simple but may include batch arrivals. Specifically, for fixed $j \leq k$, consider the waiting time of a unit that is the jth one in a batch of size k. The arrival times of these units are given by the point process

$$M_{jk}(A) = \sum_n 1(T_n \in A, T_{n-j-1} < T_{n-j} = \ldots = T_{n+k-j} < T_{n+k-j+1}).$$

This is a θ-compatible stationary point process since M is. Let $T_n(j,k)$ denote the arrival times associated with M_{jk}, and let $W_n(j,k)$ denote the waiting time of the unit that arrives at the time $T_n(j,k)$. Assume that these waiting times are marks of M_{jk}. Now, the number of the jth units in a batch of size k that are in the system at a time is

$$X_{jk}(t) = \sum_n 1(T_n(j,k) \leq t \leq T_n(j,k) + W_n(j,k)). \tag{8.18}$$

Then, similarly to (8.17),

$$\mathbf{E}[X_{jk}(0)] = \lambda_{M_{jk}} \mathbf{E}_{M_{jk}}[W_0(j,k)]. \tag{8.19}$$

Example 8.9. (*Workloads for service systems*) Suppose the queuing system described in the preceding example is work conserving. Then the *workload process* representing the sum of the remaining service times of the units in the system at time t is given by

$$\begin{aligned} Z(t) = \sum_n [&U_n 1(T_n \leq t \leq T_n + W_n) \\ &+ (T_n + W_n + U_n - t) 1(T_n + W_n \leq t \leq T_n + W_n + U_n)], \end{aligned}$$

where W_n is the duration of time the nth unit waits in the queue before its service and U_n is the unit's service time. The first part of the sum is the workload of those units still waiting in the queue at time t and the other part of the sum is the workload of those units that have already entered service. Assume that the process θ contains enough information such that (W_n, U_n) are marks of the arrival process M. Then, applying Campbell's formula (8.9) for marked point processes to the two parts of the sum, we have

$$\mathbf{E}[Z(0)] = \lambda_M \mathbf{E}_M[U_0 W_0 + \int_{W_0}^{W_0+U_0} (W_0 + U_0 - s)ds]$$

$$= \lambda_M \mathbf{E}_M[U_0 W_0 + U_0^2/2].$$

This formula is due to Brumelle [12].

Example 8.10. (*Rate conservation laws for Radon-Nikodym derivatives*) Suppose M and Λ are θ-compatible stationary random measures on E. Assume that Λ is absolutely continuous with respect to M, and so

$$\Lambda(A) = \int_A \frac{d\Lambda}{dM}(t) M(dt).$$

The derivative process $\frac{d\Lambda}{dM}(t)$ is also necessarily θ-compatible and stationary. By Campbell's formula (8.4) with $f(t, \theta_t) = \frac{d\Lambda}{dM}(t)\mathbf{1}(t \in A)$, we have

$$\lambda_\Lambda \mid A \mid \; = \mathbf{E}[\Lambda(A)] = \lambda_M \mid A \mid \mathbf{E}_M\Big[\frac{d\Lambda}{dM}(0)\Big].$$

This yields the *rate conservation law*

$$\lambda_\Lambda = \lambda_M \mathbf{E}_M\Big[\frac{d\Lambda}{dM}(0)\Big]. \tag{8.20}$$

This says that the ratio of rates $\lambda_\Lambda / \lambda_M$ equals the expectation of the derivative of Λ with respect to M. This law is related to those reviewed by Miyazawa [34], which we discuss shortly. Note also that the law (8.20) holds when Λ is a signed random measure and $\mathbf{E}[\Lambda(A)]$ exists (possibly finite).

A variation of the preceding law is as follows. Suppose Λ is a signed random measure and $M = M_1 + \ldots + M_k$ is a partition of M into singular parts M_i. That is, there is a partition $B_1, \ldots, B_k$ of E such that $M_i(B_j) = 0$ w.p.1 when $i \neq j$. Assume that $\Lambda, M_1, \ldots, M_k$ are θ-compatible stationary measures and that Λ is absolutely continuous with respect to each M_i. Thus, we can write

$$\Lambda(A) = \int_A \frac{d\Lambda}{dM}(t) M(dt) = \sum_{i=1}^{k} \int_A \frac{d\Lambda}{dM_i}(t) M_i(dt).$$

Then, applying (8.20), we have

$$\lambda_\Lambda = \sum_{i=1}^{k} \lambda_{M_i} \mathbf{E}_{M_i}\Big[\frac{d\Lambda}{dM_i}(0)\Big]. \tag{8.21}$$

This *rate conservation law for partitions* says that the rate of Λ is a linear combination of the rates λ_{M_i} of the parts M_i of M.

The law (8.21) applies to real-valued stationary processes as follows. Suppose $X = \{X(t): t \in \mathbb{R}\}$ is a θ-compatible stationary real-valued stochastic process that is absolutely continuous with respect to $M = M_1 + \ldots + M_k$. Also, assume X is absolutely continuous with respect to each M_i. Now, $\Lambda(a, b] = X(b) - X(a)$ defines a signed random measure. Then under the assumptions above, we can write

$$X(t) - X(0) = \int_{(0,t]} \frac{dX}{dM}(s) M(sd) = \sum_{i=1}^{k} \int_{(0,t]} \frac{dX}{dM_i}(s) M_i(ds).$$

Thus, by (8.21) and $\mathbf{E}[\Lambda(0,1]] = 0$, we have

$$0 = \sum_{i=1}^{k} \lambda_{M_i} \mathbf{E}_{M_i}\left[\frac{dX}{dM_i}(0)\right] \tag{8.22}$$

provided that the expectations in (8.22) exist. This rate conservation law says that the expected rate of change of X under all of its types of changes is 0. Note that a closely related variant of (8.22) reads

$$\mathbf{E}[X'(0)] = \sum_{i=1}^{k} \lambda_{M_i} \mathbf{E}_{M_i}[X(0-) - X(0)], \tag{8.23}$$

see [35]. An extension of (8.23) to the case $E = \mathbb{R}^d$ has been given in [30].

Example 8.11. (*Sojourns of stationary processes*) Suppose $\{Y_i(t): t \in \mathbb{R}\}$ is a θ-compatible stationary process with state space E_i, for $i = 1,\ldots,m$. For fixed $A_i \in \mathcal{E}_i$, the process

$$X(t) = \sum_{i=1}^{m} 1(Y_i(t) \in A_i) \tag{8.24}$$

records the number of the processes Y_i that are visiting these special sets at time t. The point process

$$M(B) = \sum_{t \in B} 1(X(t) = X(t-) + 1), \tag{8.25}$$

counts the number of the processes that enter the special sets during the time set B. Clearly, X and M are θ-compatible stationary processes. For simplicity, assume that M is simple (at most one process enters its special set at any time) and that $0 < \lambda_M < \infty$. Let γ_n denote the index i on the process Y_i that enters its set A_i at time T_n (where we assume that $Y_{\gamma_n}(T_n -) \notin A_{\gamma_n}, Y_{\gamma_n}(T_n) \in A_{\gamma_n}$). Consider the times

$$W_n = inf\{t > 0: Y_{\gamma_n}(T_n + t) \notin A_{\gamma_n}\},$$

which is the sojourn time of the process Y_{γ_n} in its set A_{γ_n}. In other words, W_n is the nth sojourn time of an arbitrary process in its special set. Clearly γ_n and W_n are marks of M. Note that X is also of the form (8.16), and so Little's law (8.17) in this case is

$$\sum_{i=1}^{m} \mathbf{P}(Y_i(t) \in A_i) = \lambda_M \mathbf{E}_M[W_0]. \tag{8.26}$$

When $m = 1$, this yields the well-known equality

$$\mathbf{P}(Y(t) \in A) = (\text{rate at which } Y \text{ enters } A)\mathbf{E}_M[\text{sojourn time of } Y \text{ in } A].$$

For example, if Y is a stationary Jackson network process whose distribution is known, then this formula is useful for evaluating the expected idle time of a set of nodes (A is the event that Y records 0 units in the set of nodes). Another example we mentioned in (8.1) yields the expected time that the total number of units in a set of nodes exceeds a certain limit ℓ (A denotes this event).

Example 8.12. (*Travel times in stationary networks*) Consider a closed network in which ν units move among the set of nodes $\{1,\ldots,m\}$. Let $Z_i(t)$ denote the node where unit i is located at time t. Assume that $\{(Z_1(t),\ldots,Z_m(t)): t \in \mathbb{R}\}$ is a θ-compatible stationary process. For instance, this may represent a stationary, closed Jackson network (this representation by unit locations contains a little more information than the standard representation by queue lengths).

Our interest is in the expected time it takes an arbitrary unit to travel from one set of nodes J to another set K. To analyze this, first note that unit i at time t is making a traverse from J to K if $Z_i(\cdot + t) \in A_i$, where A_i is the set of all sample paths z of the process Z_i such that

$$\tau_0(z) \equiv sup\{s \leq 0: z(s) \in J, z(u) \notin K, s \leq u \leq 0\} > -\infty, \text{ and}$$

$$\tau_1(z) \equiv inf\{s > 0: z(0) \in K, z(s) \notin J\} < \infty.$$

Then, $\tau_1(z) - \tau_0(z)$ is the travel time between J and K for the path z that is traversing from J to K at time 0. Note that this time excludes the possibility that the path returns to J one or more times before it reaches K (this type of travel time, however, could be analyzed similarly by defining A_i accordingly). Now, the number of units $X(t)$ traveling between J and K at time t is as in (8.24), where $Y_i(t) = Z_i(\cdot + t)$. Also the number of units $M(B)$ that begin a traverse from J to K in the time set B is as in (8.25). The X and M are θ-compatible stationary processes. For simplicity, we assume that, at most, one unit may change its node location at any time, the expected number of changes in any finite time period is finite, and there is a positive probability that a change can take place. Consequently, M is a simple point process with $0 < \lambda_M < \infty$. Then from (8.26), we have

$$\mathbf{E}[\text{number of units traveling from } J \text{ to } K] = \lambda_M \mathbf{E}_M[\text{travel time from } J \text{ to } K].$$

The first two terms are tractable for many network processes including Jackson processes. Furthermore, this formula for expected travel times also applies to an infinite family of travel times defined by stopping times or by functionals of θ; see [28].

Recently, Campbell's formula has, also, been applied in getting higher order approximations for stationary queues under light traffic. This is reported in another chapter of this volume; see [7]. Furthermore, in [37], extensive use of Campbell's formula has been made to derive expected attained sojourn times in insensitive queueing networks.

8.5 AN APPLICATION TO RISK PROCESSES

In Section 8.4, we applied Campbell's formula to functionals of θ-compatible stationary marked point processes. Typically, these functionals $\{X(t)\}$, $\{Y(t)\}$, $\{Z(t)\}$ themselves are θ-compatible stationary processes. Now, we consider a slightly different situation. Namely, Campbell's formula is used to give a short proof of a generalization of a classical formula of W. Feller [17] for the ascending ladder height of a random walk, which is a *nonstationary* functional of the underlying stationary input. This formula is of basic importance for the M/G/1 queue and standard risk processes with compound Poisson input. The following is a related application considered in [2]; see also [38].

Assume that M is a θ-compatible stationary point process representing times T_n at which insurance claims occur at an insurance company. The claims are described by marks (U_n, V_n) of M, where U_n is the monetary value of the claim at time T_n and V_n is the type of claim. Assuming that the insurance premiums accumulate at a unit rate, then, the claims minus the premiums in the time period $[0, t]$ is given by

$$X(t) = \sum_{k=1}^{M((0,t])} U_k - t. \tag{8.27}$$

This process X is called the *risk process* of an insurance stock. Note that X is not stationary.

We will consider the behavior of the risk process when it crosses 0, which represents the "ruin" of an insurance stock. The *time of ruin* (or first ladder epoch of X) is

$$\tau = inf\{t > 0: X(t) > 0\}.$$

The surplus just before the ruin is $Z_- = -X(\tau - 0)$ and the *severity of ruin* (or ladder height) is $Z_+ = X(\tau)$. Note that $\tau = T_\kappa$, where $\kappa = inf\{k: X(T_k) > 0\}$ is the claim number that triggers the ruin. The claim size at the ruin is $U_\kappa = Z_- + Z_+$, and its type is V_κ.

It turns out that the joint distribution of the random vector $(Z_-, Z_+, U_\kappa, V_\kappa)$ can be expressed by characteristics of the Palm distribution $\mathbf{P}_M$, which is the distribution of the system seen from the arrival epoch of a typical claim. Namely, suppose the underlying flow θ is ergodic and $\rho \equiv \lambda \mathbf{E}_M[U_0] \le 1$. Then

$$\mathbf{P}(Z_- \ge a, Z_+ \ge b, V_\kappa \in A, \tau < \infty) = \lambda_M \int_{a+b}^{\infty} \mathbf{P}_M(U_0 \ge u, V_0 \in A)du \tag{8.28}$$

for every $a, b \ge 0$ and $A \in \mathbb{R}$. In particular,

(*a*) $\mathbf{P}(\tau < \infty) = \rho$ and

(*b*) The conditional distribution of (U_κ, V_κ) given $\tau < \infty$ is obtained from the (Palm) distribution of (U_0, V_0) by the change of measure given by the likelihood ratio $U_0/\mathbf{E}_M[U_0]$. That is,

$$\mathbf{E}[g(U_\kappa, V_\kappa); \tau < \infty] = \rho \mathbf{E}_M\left[\frac{U_0}{\mathbf{E}_M[U_0]} g(U_0, V_0)\right] = \lambda_M \mathbf{E}_M[U_0 g(U_0, V_0)]$$

for every nonnegative function g.

(*c*) The conditional distribution of (Z_-, Z_+) given $U_\kappa, V_\kappa, \tau < \infty$ is that of $(U_\kappa S, U_\kappa(1-S))$ where S is uniform on $(0,1)$ and independent of U_κ, $V_\kappa, \tau < \infty$.

The **proof of (8.28)** can be sketched as follows. We use a further equivalent version of Campbell's formula (8.4) which sometimes is called the *inverted-flow variant* of (8.4). Namely, for any $f: E \times \Omega \to \mathbb{R}_+$

$$\mathbf{E}\left[\int_E (t, \theta_0) M(dt)\right] = \lambda_M \mathbf{E}_M\left[\int_E g(t, \theta_{-t}) dt\right]. \tag{8.29}$$

To see that (8.29) holds, it suffices to use the notation $g(t, \theta_0) = f(t, \theta_t)$ in (8.4) which means that $f(t, \theta_0) = g(t, \theta_{-t})$. Next, consider the mapping

$$g(s, \theta_0) = \mathbf{1}\{Z_- > a, Z_+ > b, V_\kappa \in A\} \mathbf{1}\{\tau = s\},$$

where, with a slight abuse of notation, we will also view Z_-, Z_+, τ as functionals of θ_0 in the sense that $Z_- \equiv Z_-(\theta_0), Z_+ \equiv Z_+(\theta_0)$ and $\tau \equiv \tau(\theta_0)$. Then from Campbell's formula (8.2), we have

$$\text{Left side of (8.28)} \quad = \mathbf{E}\left[\sum_{n=-\infty}^{\infty} g(T_n, \theta_0)\right] = \lambda_M \mathbf{E}_M\left[\int_{\mathbb{R}} g(t, \theta_{-t}) dt\right]$$

$$= \lambda_M \int_{\mathbb{R}_+} \mathbf{P}_M(Z_-(\theta_{-t}) \geq a, Z_+(\theta_{-t}) \geq b,$$

$$V_\kappa(\theta_{-t}) \in A, \tau(\theta_{-t}) = t)dt$$

$$= \text{Right side of (8.28)},$$

where the last equality follows by an argument involving a time reversal and an occupation measure; for this and other details of the proof, see [2]. □

Using (8.28), one can show that the stationary virtual-waiting-time distribution in a single-server queue with Markov modulated Poisson arrival process is stochastically larger than the stationary virtual-waiting-time distribution in an (averaged) M/G/1 queue with the same arrival rate, provided that the Markov background chain is stochastically monotone. We also remark that, equivalently, one can state that a stochastically monotone Markov-modulation of the input increases the probabilities of ruin (see [1], for further details).

Moreover, if in (8.27) the decreasing part $D(t) = t$ is replaced by a θ-compatible stochastic process $\{D(t)\}$ with continuous nondecreasing trajectories, under some mild conditions a formula similar to (8.28) can be obtained from the exchange formula (8.14) where, on the left-hand side of (8.28), the basic probability measure $\mathbf{P}$ must be replaced by the Palm measure $\mathbf{P}_D$ induced by $\{D(t)\}$. As a consequence, one obtains a formula for the ascending ladder-height distribution of the stationary virtual-waiting-time process in general single-server queues; see [39].

8.6 APPLICATION OF THE EXCHANGE FORMULA

In this section, we present a further queueing application of the exchange formula (8.14). Consider the virtual-waiting-time process $\{Z(t)\}_{t \in \mathbb{R}}$ for the stable single-server queue as in Examples 8.8 and 8.9. Let $\tilde{Z}(t)$ be *the time already spent in the system* by the customer in service at time t ($\tilde{Z}(t) = 0$ if the queue is empty), i.e., $\tilde{Z}(t) = t - T_n$ on $t \in [T_n + W_n, T_n + W_n + U_n)$. Sometimes, $\tilde{Z}(t)$ is called *attained waiting time*. In [46], it has been shown that

$$Z(0) \stackrel{\mathfrak{D}}{=} \tilde{Z}(0) \tag{8.30}$$

and, in the sequel, a number of alternative proofs of this result have been given (see [9], [45], [49]). In [42], a similar result has been derived concerning the distribution of attained and residual service times in stationary queueing systems.

Note, however, that (8.30) is a simple consequence of the fact that the intervals, during which the processes Z and $\tilde{Z}$ are above a given level $x \geq 0$, say, are identical but shifted to each other by x. In particular, $\{Z(0) > x\} = \{\tilde{Z}(x) > x\}$ for each $x > 0$. Thus, because the process $\tilde{Z}$ is stationary, we have

$$\mathbf{P}(Z(0) > x) = \mathbf{P}(\tilde{Z}(x) > x) = \mathbf{P}(\tilde{Z}(0) > x), \tag{8.31}$$

i.e., (8.30) follows. Moreover, the following extension of (8.30) holds. In connection with this, we consider not only the event $\{Z(0) > x\}$ that the virtual waiting time at time 0 is greater than x, but that the virtual-waiting-time process Z is above the level x in the whole interval $[0, s)$ for an arbitrary fixed $x > 0$. Analogously, we consider the events that the process $\tilde{Z}$ is above x in the interval $(-s, 0]$, and that the process $\tilde{Z}$ is above the level x in the intervals $[0, s)$ and

$(-s,0]$, respectively. For the random variables

$$S_x^+ = inf\{t \geq 0: Z(t) < x\}, S_x^- = inf\{t \geq 0: Z(-t) < x\}$$

and

$$\widetilde{S}_x^+ = inf\{t \geq 0: \widetilde{Z}(t) < x\}, \ \widetilde{S}_x^- = inf\{t \geq 0: \widetilde{Z}(-t) < x\}$$

we get the following result: The random vectors (S_x^+, S_x^-) and $(\widetilde{S}_x^+, \widetilde{S}_x^-)$ are identically distributed. Moreover, for each $s_1, s_2 \geq 0$, we have

$$\mathbf{P}(\widetilde{S}_x^+ > s_1, \widetilde{S}_x^- > s_2) = \mathbf{P}(S_x^+ > s_1, S_x^- > s_2)$$

$$= \lambda_M \int_{s_1+s_2}^{\infty} \mathbf{P}_M(S_x^- < u)du, \tag{8.32}$$

where M denotes the stationary point process of those epochs, when the virtual-waiting-time process Z crosses the level $x > 0$ from the above. Note that from (8.32) we get, in particular, that

$$\mathbf{P}(S_x^+ > 0) = \mathbf{P}(S_x^+ > 0, S_x^- > 0) = \mathbf{P}(\widetilde{S}_x^+ > 0, \widetilde{S}_x^- > 0) = \mathbf{P}(\widetilde{S}_x^- > 0)$$

which is equivalent to (8.31).

Proof of (8.32). By the same argument which lead to (8.31), we get the first equality in (8.32). Next, by using the exchange formula (8.14), we prove the second equality in (8.32). For this purpose, besides the random measure M counting the downcrossings of Z below the level x, we consider a further stationary random measure M' given by $M'(B) = \int_B \mathbf{1}_{\{Z(u) > x\}} du$. Observe that the Palm distribution $\mathbf{P}_{M'}$ is given by

$$\lambda_{M'}\mathbf{P}_{M'}(\omega) = \mathbf{1}(Z(0) > x)(\omega)\mathbf{P}(d\omega). \tag{8.33}$$

Furthermore, for each $s_1, s_2 \geq 0$, let the function $f: \mathbb{R} \times \Omega \to \mathbb{R}_+$ be given by

$$f(u,\omega) = \begin{cases} 1 & \text{if } s_1 + s_2 < u \leq S_x^-(\omega) \\ 0 & \text{otherwise.} \end{cases}$$

Then, by (8.14) and (8.33), we have

$$\mathbf{P}(S_x^+ > s_1, S_x^- > s_2) = \mathbf{E}\left[\int_{\mathbb{R}} f(t,\theta_t)M(dt)\mathbf{1}(Z(0) > x)\right]$$

$$= \lambda_{M'}\mathbf{E}_{M'}\left[\int_{\mathbb{R}} f(t,\theta_t)M(dt)\right] = \lambda_M\mathbf{E}_M\left[\int_{\mathbb{R}} f(-t,\theta_0)M'(dt)\right]$$

$$= \lambda_M\mathbf{E}_M[(S_x^- - (s_1+s_2))^+] = \lambda_M \int_{s_1+s_2}^{\infty} \mathbf{P}_M(S_x^- > u)du.$$

Thus, (8.32) follows. □

The quantity $\mathbf{P}_M(S_x^- > u)$ can be interpreted as probability that a typical period during which Z is uninterruptedly above the level x is greater then u. This is in accordance with the fact that the Palm distribution $\mathbf{P}_M$ is invariant with respect to pointwise shifting, i.e.,

$$\mathbf{P}_M(\theta_{T_2} \in A) = \mathbf{P}_M(A) \text{ for each } A \in \mathcal{A}. \tag{8.34}$$

This invariance property of $\mathbf{P}_M$ immediately follows from the definition of the Palm distribution $\mathbf{P}_M$ given in (8.2). Namely, from (8.2) we get that, putting $B = [0,t)$,

$$\begin{aligned} &| \mathbf{P}_M(A) - \mathbf{P}_M(\theta_{T_2} \in A) | \\ &\leq \frac{1}{\mathbf{E}[M([0,t))]} \mathbf{E}\left[\left| \sum_{j=1}^{M([0,t))} \left(\mathbf{1}(\theta_{T_j} \in A) - \mathbf{1}(\theta_{T_{j+1}} \in A)\right)\right|\right] \\ &\leq \frac{2}{\lambda_M t}. \end{aligned}$$

This gives (8.34) because t can be chosen arbitrarily large.

8.7 OPEN PROBLEMS

We end this chapter with a few comments on open problems related to Campbell's formula.

Nonhomogeneous Palm probabilities. Although the focus of this chapter is on stationary systems, many of the results should have analogues for nonstationary systems. In particular, Palm probabilities for nonstationary processes are defined in terms of Radon-Nikodym derivatives with respect to so-called Campbell measures; see for instance [22] and [26]. This setting also yields a Campbell formula analogous to (8.4). It would be of interest to find out to what extent (and/or under what conditions) the equivalent versions of this Campbell's formula in Section 8.3 also apply to nonstationary systems. In this regard, it would be interesting to identify special classes of nonstationary processes that arise naturally in practical problems.

Sample path formulas for averages in nonstationary systems. The formulas in Sections 8.3 and 8.4 are for expectations of functionals of stationary systems. These expectations have analogues as limiting averages for nonstationary systems. It would be of interest to see how the formulas in Sections 8.3 and 8.4 carry over to these contexts. This might shed more insight into the duality between sample-path analyses and Palm probability results.

Higher moment formulas and L_2 convergence. Campbell's formula and its relatives also apply to higher moments of quantities such as waiting times in queues, but the expressions are not as tractable as those for means. Are these some classes of queueing processes in which the second moments of customer waiting times can be related to moments of the queue length (aside from elementary Markovian systems where we can relate these by generating functions)? Second moments arise naturally when considering L_2 convergence of sums of variables. Why is there no L_2 convergence for sums of waiting times? We can get these for some regenerative systems, and maybe the techniques there would be useful for stationary systems. Additional mixing conditions or coupling assumptions may be in order here.

Ladder heights for diffusions. An interesting problem would be to derive a formula like (8.28) for marked ladder heights in the case when the accumulation of premiums is described by a diffusion process $\{D(t)\}$ with stationary increments (and with positive linear drive). Also, duality relations between risk processes with (randomly) varying premium rates and single-server queues with varying service rates appears to be a promising area for future research.

Exchange formulas. It would be interesting to prove sample path analogues of (8.31) and (8.32) for nonstationary systems. Also of interest would be extensions of these formulas to corresponding characteristics of random fields $\{Z(t): t \in R^d\}$. Here, the concern would be stochastic properties of areas where $Z(t)$ is above a given level and where the areas can have different sizes, shapes, orientations. Also, they can be located in different directions and distances, seen from the origin (which is chosen at random, in the stationary case).

BIBLIOGRAPHY

[1] Asmussen, S., Frey, A., Rolski, T., and Schmidt, V., Does Markov-modulation increase the risk?, preprint, Universities of Aalborg, Ulm and Wroclaw (1994).

[2] Asmussen, S. and Schmidt., V., Ladder height distributions with marks, *Stoch. Proc. Appl.* (1995), (to appear).

[3] Baccelli, F. and Brémaud, P., *Palm Probabilities and Stationary Queues*, Lecture Notes in Statistics **41**, Springer, Berlin 1987.

[4] Baccelli, F. and Brémaud, P., Virtual customers in sensitivity and light traffic analysis via Campbell's formula for point processes, *Adv. Appl. Prob.* **25** (1993), 221-234.

[5] Baccelli, F. and Brémaud, P., *Elements of Queueing Theory*, Springer, Berlin 1995.

[6] Błaszczyszyn, B., Factorial moment expansion for stochastic systems, *Stoch. Proc. Appl.* (1995), (to appear).

[7] Błaszczyszyn, B., Rolski, T. and Schmidt, V., Light-traffic approximations in queues and other stochastic models, as Chapter 15 in this book, 379-408.

[8] Brandt, A., Franken, P. and Lisek, B., *Stationary Stochastic Models*, John Wiley, Chichester 1990.

[9] Brémaud, P., An elementary proof of Sengupta's invariance relation and a remark on Miyazawa's conservation principle, *J. Appl. Prob.* **28** (1991), 950-954.

[10] Brémaud, P., A Swiss Army formula of Palm calculus, *J. Appl. Prob.* **30** (1993), 40-51.

[11] Brémaud, P., Kannurpatti, R. and Mazumdar, R., Event and time averages: A review, *Adv. Appl. Prob.* **24** (1992), 377-411.

[12] Brumelle, S.L., On the relation between customer and time averages in queues, *J. Appl. Prob.* **8** (1971), 508-520.

[13] Campbell, N.R., The study of discontinuous phenomena, *Proc. Cambridge Philos. Soc.* **15** (1909), 117-136.

[14] Campbell, N.R., Discontinuities in light emission, *Proc. Cambridge Philos. Soc.* **15** (1910), 310-328.

[15] Daley, D. and Vere-Jones, D., *An Introduction to the Theory of Point Processes*, Springer-Verlag, New York 1988.

[16] Dshalalow, J.H., On applications of Little's formula, *J. Appl. Math. Stoch. Anal.* **6** (1993), 271-275.

[17] Feller, W., *An Introduction to Probability Theory and its Applications* **2**, (2nd edition), John Wiley, New York 1971.

[18] Franken, P., König, D., Arndt, U. and Schmidt, V., *Queues and Point Processes*, John Wiley, Chichester 1982.

[19] Glynn, P.W. and Whitt, W., Extensions of the queueing relations $L = \lambda W$ and $H = \lambda G$, *Oper. Res.* **37** (1989), 634-644.

[20] Guillemin, F. and Mazumdar, R., Excursions of the workload process in G/GI/1 queues, *Stoch. Proc. Appl.* (1995), (to appear).

[21] Halfin, S. and Whitt, W., An extremal property of the FIFO discipline via an ordinal version of $L = \lambda W$, *Stoch. Mod.* **5** (1989), 515-529.

[22] Kallenberg, O., *Random Measures*, 3rd edition, Akademie-Verlag and Academic Press, New York 1983.

[23] Keilson, J. and Servi, L.D., A distributional form of Little's law and the Fuhrmann-Cooper decomposition, *Oper. Res. Letters* **9** (1988), 239-247.

[24] Keilson, J. and Servi, L.D., A distributional form of Little's law, *Oper. Res. Letters* **7** (1988), 223-227.

[25] König, D. and Schmidt, V., *Random Point Processes*, Teubner, Stuttgart 1992 (in German).

[26] Konstantopoulos, P. and Zazanis, M.A., Sensitivity analysis for stationary and ergodic queues, *Adv. Appl. Prob.* **24** (1992), 738-750.

[27] Konstantopoulos, P. and Zazanis, M.A., Sensitivity analysis for stationary and ergodic queues: Additional results, *Adv. Appl. Prob.* **26** (1994), 556-560.

[28] Kook, K. and Serfozo, R.F., Travel and sojourn times in stochastic networks, *Ann. Appl. Prob.* **3** (1992), 228-252.

[29] Kroese, D.P. and Schmidt, V., Light-traffic analysis for queues with spatially distributed arrivals, *Math. Oper. Res.* (1995), (to appear).

[30] Last, G. and Schassberger, R., A flow conservation law for surface process, preprint, University of Braunschweig (1994).

[31] Matthes, K., Kerstan, J. and Mecke, J., *Infinitely Divisible Point Processes*, John Wiley, Chichester 1978.

[32] McKenna, J., A generalization of Little's law to moments of queue lengths and waiting times in closed product-form queueing networks, *J. Appl. Prob.* **26** (1989), 121-133.

[33] Mecke, J., Stationary random measures on locally compact Abelian groups, *Z. Wahrscheinlichkeitstheorie verw. Geb.* **9** (1967), 36-58 (in German).

[34] Miyazawa, M., Note on generalizations of Mecke's formula and extensions of $H = \lambda G$, *J. Appl. Prob.* (1994), (to appear).

[35] Miyazawa, M., Rate conservation laws: A survey, *Queueing Sys.* **15** (1994), 1-58.

[36] Miyazawa, M., Palm calculus for a process with a stationary random measure and its applications, *Queueing Sys.* **17** (1994), 183-211.

[37] Miyazawa, Schassberger, R. and Schmidt, V., On the structure of insensitive generalized semi-Markov processes with reallocation and with point-processes input, *Adv. Appl. Prob.* **27** (1995), (to appear).

[38] Miyazawa, M. and Schmidt, V., On ladder height distributions of general risk processes, *Ann. Appl. Prob.* **3** (1993), 763-776.

[39] Miyazawa, M. and Schmidt, V., On level crossings of stochastic processes with stationary bounded variations and continuous decreasing components, preprint, Science University of Tokyo and University of Ulm (1994).

[40] Neveu, J., Sur les mesures de Palm de deux processus ponctuels stationaires, *Z. Wahrscheinlichkeitstheorie verw. Geb.* **34** (1976), 199-203.

[41] Neveu, J., Processus ponctuels, In: *Ecole d'été de Probabilités de St. Flour VI*, Lecture Notes Math **598**, Springer, Berlin (1976), 249-447.

[42] O'Donovan, T.M., Distribution of attained and residual service in general queueing systems, *Oper. Res.* **22** (1974), 570-575.

[43] Rolski, T., *Stationary Random Processes Associated with Point Processes*, Lecture Notes in Statistics **5**, Springer-Verlag, New York 1981.

[44] Rolski, T. and Stidham Jr., S., Continuous versions of the queueing formulas $L = \lambda W$ and $H = \lambda G$, *Oper. Res. Letters* **2** (1983), 211-215.

[45] Sakasegawa, H. and Wolff, R., The equality of the virtual delay and attained waiting time distribution, *Adv. Appl. Prob.* **22** (1990), 257-259.

[46] Sengupta, B., An invariance relationship for the $G/G/1/\infty$ queue, *Adv. Appl. Prob.* **21** (1989), 956-957.

[47] Serfozo, R.F., Point processes, In: *Handbooks in Oper. Res. and Management Science* **2**: *Stochastic Models*, ed. by D.P. Heymann, and M.J. Sobel, North Holland, Amsterdam (1990), 1-93.

[48] Serfozo, R.F., Little laws for utility processes and waiting times in queues, *Queueing Sys.* **17** (1994), 137-181.

[49] Sigman, K., A note on a sample-path rate conservation law and its relationship with $H = \lambda G$, *Adv. Appl. Prob.* **23** (1991), 662-665.

[50] Sigman, K., *Stationary Marked Point Processes*, Chapman and Hall, London 1991.

[51] Stidham Jr., S., Sample-path analysis of queues, In: *Applied Prob.-Computer Science: The Interface*, ed. by R.L. Disney and T.J. Ott, **2**, Birkhäuser, Boston (1982), 41-70.

[52] Stidham Jr., S. and El-Taha, M., Sample-path analysis of processes with imbedded point processes, *Queueing Sys.* **5** (1989), 131-165.

[53] Stoyan, D., Kendall, W.S. and Mecke, J., *Stochastic Geometry and its Applications*, John Wiley, Chichester 1991.

[54] Whitt, W., A review of $L = \lambda W$ and extensions, *Queueing Sys.* **9** (1991), 235-268.

[55] Whitt, W., $H = \lambda G$ and the Palm transformation, *Adv. Appl. Prob.* **24** (1992), 755-758.

[56] Wortmann, M.A. and Disney, R.L., On the relationship between stationary and Palm moments of backlog in the G/G/1 priority queue, In: *Queueing and Related Models*, ed. by U.N. Bhat and I.V. Basawa, Clarendon Press, Oxford (1992), 161-174.

Chapter 9

Excess level processes in queueing

Jewgeni H. Dshalalow

ABSTRACT This chapter analyzes the behavior of one- and two-dimensional marked point processes (with dependent marks) about some fixed level. The author obtains a joint transformation of the first excess level, first passage time, and the index of the point process (that labels the first passage time). Obtained results are demonstrated on various queueing systems, such as those with quorum, vacations, N-policy, D-policy, and "vector" customers, specifically N-D-policy.

CONTENTS

9.1 Introduction	243
9.2 Preliminaries	247
9.3 The process with discrete-valued components	249
9.4 The process with continuous and mixed components	253
9.5 Examples of queueing models	254
9.6 Pretermination processes for a special case	256
9.7 Open problems and future research directions	259
Bibliography	260

9.1 INTRODUCTION

Many processes encountered in queueing require the so-called *level-crossing analysis*, i.e., analysis of the characteristics of stochastic processes crossing a fixed level, say L. Some classical problems apply to waiting time processes (see i.e. Cohen [9] and Shanthikumar [32]). Other problems refer to input processes. Typically, an input process, A, is observed at random moments of time, $\tau = \{\tau_1, \tau_2, ...\}$, at which A may increase or remain unchanged. Therefore, A is a monotone nondecreasing jump process relative to τ. [Of course, A may also alter at epochs of time other than τ but no such information is available or it is not significant to the analysis.] One is interested in when (i.e., at what instants, τ_n, $n = 1,2,...$, of time) the process A *hits*, i.e., reaches or exceeds, level L for the first time, and what the *first excess level* (greater than or equal to L) would be. A value of n for which the first excess occurs is called the *index of the first excess* and denoted by ν_1. We call τ_{ν_1} the *first passage time*. Below we will give examples of situations that apply to various queueing systems.

A) QUORUM SYSTEMS. Consider a single-server queue of M/G/1-type. Assume that the server capacity is r and that the server takes exactly r customers for service. If upon a service completion the queue level falls below r, the server idles until the queue again accumulates at least r customers. Only then is the service of the next group of r customers restored. If the input A to the system is bulk, it is

0-8493-8074-x/95/$0.00+$.50

more likely to exceed r at some arrival time than to reach r, as it would when customers arrive singly. To analyze the queueing process in such a system (specifically, the embedded queueing process over the moments of successive completions of service), one needs to know the first passage time, τ_{ν_1}, when the excess occurs and the value of queue, A_{ν_1} (i.e. the *first excess level*), at time τ_{ν_1}. Such a system is denoted by $M^X/G^r/1$ and called an *r-quorum system*. A practical generalization would be a system in which server capacity increases to $R \geq r$, where r is the minimal number of customers the server is permitted to process. Such a system is denoted by $M^X/G^{r,R}/1$ and called an *(r, R)-quorum system*. Many authors treated different processes in a variety of quorum systems. For instance, Jacob and Madhusoodanan [21] considered a variant of the $M/G^{r,R}/1$ queue with single arrivals and vacationing server. Abolnikov and Dshalalow [2] and Abolnikov, Dshalalow and Dukhovny [4] studied an $M^X/G^{r,R}/1$ queue with resting server and bulk arrival in which results of first excess level theory were used.

B) N-POLICY SYSTEMS. A standard N-policy service discipline refers to the class of M/G/1 queues (with single arrivals) where the server initiates a busy period only after the queue reaches level N. An idle period starts when the system becomes empty. Busy periods are usually preceded by "start-up" times, which, in many real-world systems, it is desirable to reduce. Consequently, the N-policy is one such measure. With the exception of works by H.S. Lee and Srinivasan [25] and H.W. Lee, S.S. Lee and Chae [26], the analysis of the queueing process in N-policy models has not been significantly modified in the two and a half decades since Heyman's widely referred to work [20] in 1968. Bulk arrivals in such systems was a desirable upgrade that seemed to be an impossibility without first excess level theory or something similar. The above cited authors [25,26], however, utilized probabilistic arguments to obtain special characteristics for the first excess level applied to the queueing process with continuous time parameter. A meaningful generalization of standard N-policy systems was made by Muh [28,29] with the aid of the first excess level analysis. In his work, the author allows a bulk input flow in combination with quorum and N-policy, in notation $M^X/G_N^{r,R}/1$. Within busy periods, the server takes batches of customers between r and R. An idle period starts, when the queue falls below r (customers), and ends when the system accumulates to at least N customers, which corresponds to the first excess A_{ν_1} of level N. Busy periods are preceded by start-up times. In various modifications in [28,29], N can be greater than or equal to R or less than R. In the latter case, the system is allowed to take any number of customers during start-up periods in excess of A_{ν_1} such that batches to be serviced do not exceed R. The results of [28,29] formally generalize those in [25,26] giving a more elegant analytic representation of a variant of Pollaczek-Khintchine formula. However, unlike [25,26], the author of [28,29] considered the embedded process.

C) QUEUES WITH SERVER VACATIONS. A standard queueing system with server vacations is one when the server goes on vacations each time the system becomes empty. Queues with vacationing servers first appeared in the seventies and quickly became popular because of their practical use. The literature on queueing with vacationing servers included systems with various vacationing policies, such as: single vacations, multiple exhaustive vacations (when the server goes on vacations as long as the system is empty on his returns), and Bernoulli scheduled vacations (when the server goes on vacations with probability p when the queue is not empty). A comprehensive survey of queues with vacations is made by Doshi [10]. Vacation times of the server are generally distributed differently from service durations. A combination of quorum systems and systems with vacationing servers gives rise to an upgrade of standard queues with server vacations. In such systems, a vacation sequence starts whenever the queue falls below level r and terminates when, upon the server's return, at least r (or N, in the case of N-policy) customers are present in the system. Such systems were studied in Dshalalow [11] and Dshalalow and Yellen [18]. (A special case of systems with quorum, vacations and single arrivals was studied in Jacob and Madhusoodanan [21] mentioned in Example A.)

D) SYSTEMS WITH VACATIONS AND BERNOULLI SCHEDULE. A modification of *Bernoulli schedule* is one in which any vacation sequence is terminated by an arbitrary random variable. A classical vacation scheme with Bernoulli discipline was introduced and studied by Keilson and Servi [22]. In their model of type GI/G/1, a single channel goes on vacation when the queue becomes empty. A vacation sequence is continued until at least one customer is present in the system upon completion of a vacation period by the server. If on service completion the queue is not empty, the server goes on vacation with probability p and resumes service with probability $1-p$. Other models of type M/G/1 with Bernoulli schedule were studied by Ramaswami and Servi [31] and by Keilson and Servi [23] with state dependent Bernoulli schedule. Kella [24] suggested a generalized Bernoulli scheme according to which a single server goes on k consecutive vacations with probability p_k if the queue upon his return is empty. All these systems can be further generalized by employing the first excess level theory developed in Dshalalow [14], where the sequence τ can be terminated by an arbitrary random variable ν_2 before A hits level L. The first excess index then becomes $\Delta = \min\{\nu_1,\nu_2\}$, the first passage time is τ_Δ, and the first excess level is A_Δ. The results obtained in [14] can be applied to various situations in queueing systems, when a vacation sequence must be interrupted due to circumstances. (See examples in Section 9.3 and a discussion in Section 9.7.)

E) D-POLICY SYSTEMS. A standard D-policy system refers to the following discipline. A busy period does not begin until the system's cumulative job (i.e. the total amount of all service durations of all customers waiting in the system) hits level D. In this system, as usual, each idle period begins when the system becomes empty. Such systems were studied by Balachandran and Tijms [7] and Li and Niu [27]. A useful generalization of a D-policy system would be one with quorum and batch arrivals. [We need to assume that customers are processed singly since otherwise, it would be hard to predict the cumulative job of the system should the sizes of batches (to be taken for service) vary. In other words, the capacity of the server is 1, and it stops further service if the system becomes empty.] Then the system needs to accumulate enough customers in the system so that their cumulative service time is at least D before the server begins to work (starting perhaps with warm-up). Over a period of time when the server does not process customers he may rest, which is assumed in [7] and [27], or, alternatively, go on vacations. (See Example 9.6 in Section 9.5 and a discussion in Section 9.7.)

F) OTHER STOCHASTIC SYSTEMS. In Abolnikov and Dshalalow [1], the authors considered a compound delayed renewal process (τ,A), where τ is a delayed renewal process marked by another (independent) delayed renewal process A which is to hit a fixed level L at one of the instants of time, $\tau_0, \tau_1, \ldots$. In [14], the process of hitting level L by A could be interrupted by some discrete-valued random variable, $\nu_2 \geq 0$, independent of A and competing with ν_1 for termination of (τ, A). Such a process can be useful in inventories and power stations. For example, suppose that an inventory is characterized by a stream of supplies and demands, and there are two (possibly random) levels, p_u (an upper level) and p_l (a lower level). Suppose also that a specific decision is to be made when either the total inventory level crosses level p_u or the first demand arrives. In this case, the predictable instant when a decision must be made and the predictable inventory level will be τ_Δ and A_Δ, respectively, defined above. In the situation where the inventory decreases due to demand, a decision can be made when either the inventory level crosses p_l from above or a random variable ν_2 (representing the occurrence of the first demand) takes on a specific value, whichever comes first. A similar problem can be set for a power station with the power level oscillating between the two prescribed critical levels. Such a process can be approximated by a "cruder" embedded jump process of essential jumps.

G) PRETERMINATION LEVEL. First excess level problems can also be associated with catastrophes whose occurrences it may be desirable to predict at the instant of time $\tau_{\Delta-1}$ (called *pretermination time*) preceding the first passage time, τ_Δ. The distribution of $A_{\Delta-1}$ (the *pretermination level*) is of main interest. Such information may be necessary in inventories, stochastic control, insurance risk and in stock market models. In the majority of cases it is possible to arrive at simple explicit results. The treatment of this problem is rendered in Section 9.6.

H) SYSTEMS WITH VECTOR ARRIVALS. One of the practical combinations of quorum, N-policy and D-policy systems, may be one when the server will not start processing customers until the queue accumulates to at least N customers or the cumulative job hits D, whichever of the two situations occur first. Another scenario of an N-D-policy system may be one in which customers arrive along with their other (random) components which are discrete- or continuous-valued (a queueing system with "two-dimensional arrivals"). The main component of each arrival may represent the batch size (number of customers in that arrival) and the second component might be another random variable associated with that batch of customers (e.g. customers' baggage, cumulative account balance, investment or debt, amount of fluid, etc.). Here the server also rests or goes on vacations, and then resumes service as soon as the total number of customers is at least N or the associated cumulative value of the second component hits D. Dshalalow in [15] gave examples of such systems and introduced related *first excess level processes*, *first passage times*, and other characteristics.

The first excess level theory belongs to the theory of *Fluctuations of Stochastic Processes* extensively studied in the past [6,8,19,30,33-40]. Lajos Takács seems to have made the most significant contributions to this theory with applications to queueing and other stochastic models. He studied behaviors of stochastic processes more general than renewal processes (such as recurrent and semi-Markov processes). In his well-known works [35-40], Takács extended results of Andersen [6], Baxter [8], Pollaczek [30], Spitzer [33] and others. He studied general fluctuation phenomena by solving relevant recurrent operator equations in terms of operators acting in classes of Banach algebras. The latter was formalized and described by him in [40].

This chapter surveys results on the first excess level theory applied to one- and two-dimensional compound renewal processes. Most of these results are based on papers, Abolnikov and Dshalalow [1], Dshalalow [14], and Dshalalow [15], in which the formulas were obtained in closed analytical and numerically tractable forms.

The problems studied in these papers were motivated by classes of stochastic models listed in Examples A-H in which systems alter their modes as soon as their input processes hit certain specified levels. Some of the results in this chapter may overlap with those of Takács, but they are based on different methods. As mentioned in Example H, if a targeted queueing process has more than one random component, the *hitting time* or *first passage time* of such a multidimensional process will be the first instant of time when one of the components of the process hits its associated level. This problem was introduced and studied in Dshalalow [15]. Formally, a two-dimensional marked process (τ, A, B), whose marks A and B are not assumed to be independent, is to hit $L = (p_1, p_2)$. L is *hit* by (A, B) if $A > p_1$ or $B > p_2$ at one of the instants of time, $\tau_0, \tau_1, \ldots$. If ν_1 and ν_2 are first excess level indices of A and B, respectively, the random variable $\Delta = \min\{\nu_1, \nu_2\}$ is the *termination index*, and the first passage time of (τ, A, B) is τ_Δ.

Observe that, if A and B are independent, B has constant increments ("interarrivals") equal to 1 and p_2 takes over the role of a random variable ν_2 (now arbitrarily nonnegative and discrete-valued), then this can be reduced to the first excess level problem studied in [14], after some transformations, as shown in Section 9.3 of this chapter. Thus, we will treat it as a special case of the two-dimensional first excess level problem in [15].

The main results of this chapter are obtained in Sections 9.3 and 9.4 for the first excess level processes applied to marked renewal processes with vector marks of dependent components. They are essentially based on studies conducted in [15]. For functionals of the termination index, first passage time and first excess level process formulas are obtained in analytically tractable forms and they generalize previous results in Abolnikov and Dshalalow [1] and Dshalalow [14]. Sections 9.3 and 9.4 also discuss all special cases and instruct how to employ them in various queueing models. Section 9.5 discusses how to apply special cases to two rather general classes of queues. Section 9.7 suggests projects of practical models (in the light of the first excess level theory) and related open problems.

9.2. PRELIMINARIES

All random variables and stochastic processes throughout this chapter will be considered on a probability space $\{\Omega, \mathcal{F}, P\}$. Let $Z = \sum_{n \geq 0} (A_n, B_n) \varepsilon_{\tau_n}$ (ε_a is a point mass) be a marked counting delayed renewal process with the following assumptions. $(A, B) = \sum_{n=0}^{\infty} \varepsilon_{(A_n, B_n)}$ on $(\Psi, \mathfrak{B}(\Psi))$, $\Psi \subseteq \mathbb{R}_+$, and $\tau = \sum_{n \geq 0} \varepsilon_{\tau_n}$ on $(\mathbb{R}_+, \mathfrak{B}(\mathbb{R}_+))$ ($\mathfrak{B}$ is the Borel σ-algebra) are delayed renewal processes, such that Z is obtained from τ by position independent marking. The "interarrivals," i.e., random vectors $(X_n = A_n - A_{n-1}, Y_n = B_n - B_{n-1})$, $A_{-1} = B_{-1} = 0$, $n = 0, 1, \ldots$, are valued in $\mathbb{R}_+ \times \mathbb{R}_+$, independent of each other, and for $n \geq 1$, identically distributed. [However, for each n, the components, X_n and Y_n, need not be independent.]

Denote

$$a_0(u, v) = \mathbf{E}[u^{X_0} v^{Y_0}],\ a(u, v) = \mathbf{E}[u^{X_1} v^{Y_1}],\ h_0(\theta) = \mathbf{E}[e^{-\theta \tau_0}],$$

and

$$h(\theta) = \mathbf{E}[e^{-\theta(\tau_1 - \tau_0)}].$$

Let p_1 and p_2 be two nonnegative real numbers representing two levels, one for A and the second for B. We will say that (A,B) *hits level* $L = (p_1,p_2)$, or, equivalently, L is the *first excess level* for (A,B) if, for some n, A_n or B_n exceed at least one of their respective levels, p_1 or p_2, for the first time. The *first passage time* is the instant of time when (A,B) hits L. More formally, if

$$\nu_1 = \nu_1(p_1) = \inf\{k\colon A_k > p_1\} \text{ and } \nu_2 = \nu_2(p_2) = \inf\{k\colon B_k > p_2\}$$

and

$$\Delta = \min\{\nu_1, \nu_2\},$$

which is the *index of the first excess* of L by (A,B) or just the *termination index*, then τ_Δ is the *first passage time* when (A,B) hits L. Consequently, (A_Δ, B_Δ) is the *first excess level.*

Let f and g be integrable functions with respect to relevant Borel measures and weight functions. Define

$$D_p^a(f(p))(z) = (1-z)\sum_{p \geq 0} z^p f(p), \ |z| < 1, \tag{a}$$

$$D_p^b(f(p))(s) = s\int_{p=0}^{\infty} e^{-sp} f(p)dp, \ Re(s) > 0. \tag{b}$$

and call them a- and b-type operators. Let

$$D_{p_1,p_2}(g(p_1,p_2))(x,y) = D_{p_1}\{D_{p_2}(g(p_1,p_2))\}(x,y), \tag{9.1}$$

where on the right-hand side of (9.1) the operators may be of types a or b or mixed. [Here and throughout we drop the superscripts a and b whenever one of them or both are not specified.] Note that due to Fubini's theorem, the operators on the right-hand side of (9.1) are commutative.

Lemma 9.1. *Let I_G be the indicator function of a set G. Then, for a- and b-type operators, it holds true that*

$$D_{p_1}(I_{\{\nu_1 = j\}})(x) = x^{A_{j-1}} - x^{A_j}, \ j = 0,1,\ldots, \tag{9.2}$$

and

$$D_{p_2}(I_{\{\nu_2 = k\}})(y) = y^{B_{k-1}} - y^{B_k}, \ k = 0,1,\ldots, \tag{9.2a}$$

where

$$A_{-1} = B_{-1} = 0.$$

Proof. From the definition of ν_1, we deduce that

$$I_{\{\nu_1 = j\}} = I_{[0,p_1]}(A_{j-1}) - I_{[0,p_1]}(A_j), \ j = 0,1,\ldots .$$

This yields the above assertion for the operators of types a and b. □

Define the operators:

$$\mathfrak{D}_x^k(\cdot) = \lim_{x\to 0}\frac{1}{k!}\frac{\partial^k}{\partial x^k}\frac{1}{1-x}(\cdot) \tag{9.3}$$

and

$$\mathfrak{D}_{x,y}^{j,k}(\cdot) = \lim_{(x,y)\to(0,0)}\frac{1}{j!k!}\frac{\partial^k}{\partial x^k}\frac{\partial^j}{\partial y^j}\frac{1}{(1-x)(1-y)}(\cdot). \tag{9.4}$$

For various special cases throughout the remainder of this chapter, we notice a few elementary properties of the operator $\mathfrak{D}_y^p$:

P1) $\mathfrak{D}_y^p$ is a linear operator with fixed points at every constant function.

P2) For any function g, analytic at zero,

$$\mathfrak{D}_y^p(y^k g(y)) = \begin{cases} 0, & p < k \\ \mathfrak{D}_y^{p-k}(g(y)), & p \geq k \end{cases}$$

P3) With the notation

$$\delta(m,p) = \begin{cases} 1, & m \leq p \\ 0, & m > p \end{cases}$$

and Properties P1 and P2 we derive

$$\mathfrak{D}_y^p\left\{\frac{y^m}{1-\gamma y}\right\} = \delta(m,p)\frac{1-\gamma^{p-m+1}}{1-\gamma}.$$

We shall be interested in an analytic representation of the functional

$$E = \mathbf{E}[\xi^{\Delta} e^{-\theta \tau_{\Delta}} u^{A_{\Delta}} v^{B_{\Delta}}] \tag{9.5}$$

to which we apply the transformation defined in (9.1). In Section 9.3 we will treat E for (A,B) with both components discrete. Section 9.4 will discuss (A,B) with continuous and mixed components.

9.3. THE PROCESS WITH DISCRETE-VALUED COMPONENTS

Denote

$$\mathfrak{F} = \mathfrak{F}(\xi,\theta,u,v;\, x,y) = D_{p_1}^a D_{p_2}^a \{\mathbf{E}[\xi^{\Delta} e^{-\theta\tau_{\Delta}} u^{A_{\Delta}} v^{B_{\Delta}}]\}(x,y),$$

and

$$\mathfrak{F} = \mathfrak{F}_1 + \mathfrak{F}_2 + \mathfrak{F}_3,$$

such that

$$\mathfrak{F}_i = D_{p_1}^a D_{p_2}^a \{\mathbf{E}[\xi^{\Delta} e^{-\theta\tau_{\Delta}} u^{A_{\Delta}} v^{B_{\Delta}} I_{G_i}]\}(x,y),$$

and

$$G_1 = \{\nu_1 < \nu_2\},\ G_2 = \{\nu_1 = \nu_2\},\ G_3 = \{\nu_1 > \nu_2\}.$$

(I_G is the indicator function of a set G.) By Fubini's theorem,

$$\mathfrak{F}_1 = \sum_{k \geq 1} \sum_{j=0}^{k-1} D_{p_1}^a D_{p_2}^a \left\{\mathbf{E}[\xi^j e^{-\theta\tau_j} u^{A_j} v^{B_j} I_{\{\nu_1 = j\}} I_{\{\nu_2 = k\}}]\right\}(x,y).$$

From Lemma 9.1 and by the independence of τ_j from the rest in $\mathbf{E}$, we have

$$\mathfrak{F}_1 = h_0(\theta) \sum_{k \geq 1} \sum_{j=0}^{k-1} \{\xi h(\theta)\}^j \mathbf{E}\left[u^{A_j} v^{B_j} (x^{A_{j-1}} - x^{A_j})(y^{B_{k-1}} - y^{B_k})\right].$$

The last expression can be rewritten as

$$\mathfrak{F}_1 = h_0(\theta) \sum_{k \geq 1} \sum_{j=0}^{k-1} \{\xi h(\theta)\}^j E_1 E_2,$$

where E_1 and E_2 are the two factors extracted from $\mathfrak{F}_1$ by using the independent increments property of (A,B):

$$E_1 = \mathbf{E}\Big[[(ux)^{A_{j-1}}u^{X_j} - (ux)^{A_j}](vy)^{B_j}\Big]$$

and

$$E_2 = [1 - b(y)]\{b(y)\}^{k-1-j}, \text{ with } b(y) = a(1,y).$$

After some standard transformations, E_1 reduces to

$$E_1 = \begin{cases} a_0(u,vy) - a_0(ux,vy), & j = 0 \\ a_0(ux,vy)[a(u,vy) - a(ux,vy)]\{a(ux,vy)\}^{j-1}, & j > 0, \end{cases}$$

and this yields

$$\frac{1}{h_0(\theta)}\mathfrak{F}_1 = a_0(u,vy) - a_0(ux,vy) + \frac{\xi h(\theta)a_0(ux,vy)[a(u,vy) - a(ux,vy)]}{1 - \xi h(\theta)a(ux,vy)}.$$

An analogous expression for $\mathfrak{F}_3$ can be easily deduced from that for $\mathfrak{F}_1$ by interchanging the roles of the variables u and x with v and y:

$$\frac{1}{h_0(\theta)}\mathfrak{F}_3 = a_0(ux,v) - a_0(ux,vy) + \frac{\xi h(\theta)a_0(ux,vy)[a(ux,v) - a(ux,vy)]}{1 - \xi h(\theta)a(ux,vy)}.$$

Now,

$$\mathfrak{F}_2 = \sum_{k \geq 0} D_{p_1} D_{p_2}\Big\{\mathbf{E}\big[\xi^k e^{-\theta\tau_k} u^{A_k} v^{B_k} I_{\{\nu_1 = \nu_2 = k\}}\big]\Big\}(x,y).$$

By Lemma 9.1, $\mathfrak{F}_2$ transforms to

$$\frac{1}{h_0(\theta)}\mathfrak{F}_2 = a_0(u,v) - a_0(ux,v) + a_0(u,vy) - a_0(ux,vy)$$

$$+ \frac{\xi h(\theta)a_0(ux,vy)[a(u,v) - a(ux,v) - a(u,vy) + a(ux,vy)]}{1 - \xi h(\theta)a(ux,vy)}.$$

Performing the summation of $\mathfrak{F}_1$, $\mathfrak{F}_2$ and $\mathfrak{F}_3$ we have

$$\frac{1}{h_0(\theta)}\mathfrak{F} = a_0(u,v) - a_0(ux,vy)\frac{1 - \xi h(\theta)a(u,v)}{1 - \xi h(\theta)a(ux,vy)}. \tag{9.6}$$

The above can be summarized in the following statement.

Theorem 9.2. *The functional*

$$\mathfrak{F} = \mathfrak{F}(\xi,\theta,u,v;\, x,y) = D^a_{p_1} D^a_{p_2}\{\mathbf{E}[\xi^{\Delta} e^{-\theta\tau_{\Delta}} u^{A_{\Delta}} v^{B_{\Delta}}]\}(x,y),$$

of the termination index, first passage time, and first excess level satisfies formula (9.6).

Now, taking into consideration (a) and applying $\mathfrak{D}^{p_1,p_2}_{x,y}$ (introduced in Section 9.2) to (9.6), we can restore the functional E:

$$\frac{1}{h_0(\theta)}\mathbf{E}[\xi^{\Delta} e^{-\theta\tau_{\Delta}} u^{A_{\Delta}} v^{B_{\Delta}}]$$

$$= a_0(u,v) - \big(1 - \xi h(\theta)a(u,v)\big)\mathfrak{D}^{p_1,p_2}_{x,y}\left\{\frac{a_0(ux,vy)}{1 - \xi h(\theta)a(ux,vy)}\right\}. \tag{9.7}$$

The "marginal functional", with respect to the component A, is

$$\frac{1}{h_0(\theta)}\mathbf{E}[\xi^{\Delta}e^{-\theta\tau_{\Delta}}u^{A_{\Delta}}] = a_0(u) - \big(1-\xi h(\theta)a(u)\big)\mathfrak{D}_x^{p_1}\left\{\frac{a_0(ux)}{1-\xi h(\theta)a(ux)}\right\}, \quad (9.8)$$

where $a_0(u) = a_0(u,1)$ and $a(u) = a(u,1)$. This coincides with formula (3.4) in Dshalalow [14], which is almost identical to (9.7), but the present case deals with a two-dimensional renewal process.

Below, we consider a few important special cases and their applications. For this reason, we apply only operator $\mathfrak{D}_y^{p_2}$ to (9.6):

$$\mathfrak{D}_y^{p_2}\frac{1}{h_0(\theta)}\mathfrak{F} = a_0(u,v) - [1-\xi h(\theta)a(u,v)]\mathfrak{D}_y^{p_2}\left\{\frac{a_0(ux,vy)}{1-\xi h(\theta)a(ux,vy)}\right\}. \quad (9.9)$$

SPECIAL CASE 1. In the event that the components of (A,B) are independent, a_0 and a are factorizable:

$$a_0(u,v) = a_0(u)b_0(v) \text{ and } a(u,v) = a(u)b(v).$$

Then, from (9.9) we have that

$$\mathfrak{D}_y^{p_2}\frac{1}{h_0(\theta)}\mathfrak{F} = \frac{1}{h_0(\theta)}D_{p_1}^a\left\{\mathbf{E}[\xi^{\Delta}e^{-\theta\tau_{\Delta}}u^{A_{\Delta}}v^{B_{\Delta}}]\right\}(x)$$

$$= a_0(u)b_0(v) - a_0(ux)\big(1-\xi h(\theta)a(u)b(v)\big)\mathfrak{D}_y^{p_2}\{\Gamma(y)\} \quad (9.10)$$

where

$$\Gamma(y) = \frac{b_0(vy)}{1-\xi h(\theta)a(ux)b(vy)}. \quad (9.11)$$

SPECIAL CASE 2. Let $B_0 = i$ (≥ 0) and $Y_1 = 1$ *a.s.*, i.e., $b_0(z) = z^i$ and $b(z) = z$. This case can describe the situation when the first level needs to be reached or exceeded in no more than in p_2 steps. The i initial steps may be accumulated from the previous cycle. This is sometimes useful for embedded Markov chains.

For this special case, Γ defined in (9.11), will become

$$\Gamma(y) = v^i\frac{y^i}{1-\gamma(x)y},$$

and, by P1-P3, we have

$$\mathfrak{D}_y^{p_2}\Gamma = v^i\delta(i,p_2)\frac{1-\gamma^{p_2-i+1}(x)}{1-\gamma(x)},$$

where $\gamma(x) = \xi h(\theta)a(ux)v$. Substituting this into (9.10), we have

$$\mathfrak{D}_y^{p_2}\frac{1}{h_0(\theta)}\mathfrak{F} = \frac{1}{h_0(\theta)}D_{p_1}^a\left\{\mathbf{E}[\xi^{\Delta}e^{-\theta\tau_{\Delta}}u^{A_{\Delta}}v^{B_{\Delta}}]\right\}(x)$$

$$= a_0(u)v^i - a_0(ux)\big(1-\gamma(1)\big)v^i\delta(i,p_2)\frac{1-\gamma^{p_2-i+1}(x)}{1-\gamma(x)}. \quad (9.12)$$

Noticing that $B_{\Delta} = i+\Delta$, we can cancel v^i in (9.12). Then, since the information on Δ is already included in the functional $\mathbf{E}$, we can set $v = 1$, thereby arriving at

$$\frac{1}{h_0(\theta)}D_{p_1}^a\left\{\mathbf{E}[\xi^{\Delta}e^{-\theta\tau_{\Delta}}u^{A_{\Delta}}]\right\}(x)$$

$$= a_0(u) - a_0(ux)\big(1-\gamma(1)\big)\delta(i,p_2)\frac{1-\gamma^{p_2-i+1}(x)}{1-\gamma(x)}, \quad (9.13)$$

where, now, $\gamma(x) = \xi h(\theta)a(ux)$.

A useful application of the last formula is when p_2 becomes a random variable. A special case of this, $i = 1$, was considered in [14]. It was assumed there that the process A will be terminated when A hits level p_1 or if an (arbitrary discrete-valued) random variable, p_2, takes on some value (thereby giving the number of steps the compound renewal process (A,τ) will take), whichever of the two events occurs first. [Consequently, the random variable A_Δ was called in [14] the *termination process*.] If A is a fragment of the bulk input stream in an M/G/1 type queue and the increments of A are taken over vacation times of the server, then, this situation generalizes the Bernoulli schedule vacation discipline.

Now, if we assume p_2 to be a random variable (with the probability generating function W), we can treat the right-hand side of (9.13) as

$$\frac{1}{h_0(\theta)} D^a_{p_1}\Big\{\mathbf{E}[\xi^{\Delta} e^{-\theta\tau_\Delta} u^{A_\Delta} \mid p_2]\Big\}(x),$$

and, by applying $\mathbf{E}$ to (9.13), we get

$$\frac{1}{h_0(\theta)} D^a_{p_1}\Big\{\mathbf{E}[\xi^{\Delta} e^{-\theta\tau_\Delta} u^{A_\Delta}]\Big\}(x)$$

$$= a_0(u) - a_0(ux)(1-\gamma(1))\frac{\mathbf{T}_i W(1) - \gamma^{-i+1}(x)\mathbf{T}_i W(\gamma(x))}{1-\gamma(x)}, \tag{9.14}$$

where the operator $\mathbf{T}_i g$ gives the i-tail of the expansion of a function g in Taylor series in a vicinity of zero. Specifically, for $i = 1$, (9.14) reduces to

$$\frac{1}{h_0(\theta)} D^a_{p_1}\Big\{\mathbf{E}[\xi^{\Delta} e^{-\theta\tau_\Delta} u^{A_\Delta}]\Big\}(x)$$

$$= a_0(u) - a_0(ux)(1-\gamma(1))\frac{1 - W(\gamma(x))}{1-\gamma(x)}, \tag{9.15}$$

which is identical to formula (2.3a) in [14]. Recall that $\gamma(x) = \xi h(\theta)a(ux)$.

SPECIAL CASE 3. In the condition of Special Case 1, we assume that

$$b_0(z) = b(z) = \frac{z(1-q)}{1-qz},$$

i.e., B is a regular renewal process with successive increments (batches) distributed geometrically with parameter $1-q$. Then, Γ in (9.11) can be transformed into

$$\Gamma(y) = (1-q)v\frac{y}{1-y\gamma(x)},$$

where

$$\gamma(x) = v\{q + \xi h(\theta)a(ux)(1-q)\}. \tag{$*$}$$

Using properties P1-P3 above, we have

$$\mathfrak{D}^{p_2}_y \Gamma = (1-q)v\delta(1,p_2)\frac{1-\gamma^{p_2}(x)}{1-\gamma(x)}, \tag{9.16}$$

which we substitute into (9.10) to get

$$\mathfrak{D}^{p_2}_y \frac{1}{h_0(\theta)} \mathfrak{F} = \frac{1}{h_0(\theta)} D^a_{p_1}\Big\{\mathbf{E}[\xi^{\Delta} e^{-\theta\tau_\Delta} u^{A_\Delta} v^{B_\Delta}]\Big\}(x)$$

$$= \frac{v(1-q)}{1-qv}\left\{a_0(u) - a_0(ux)(1 - v\gamma(1))\delta(1,p_2)\frac{1-\gamma^{p_2}(x)}{1-\gamma(x)}\right\}. \tag{9.17}$$

Again, assuming p_2 to be a random variable, as in Special Case 2, we arrive at

$$\mathfrak{D}_y^{p_2} \frac{1}{h_0(\theta)} \mathfrak{F} = \frac{1}{h_0(\theta)} D_{p_1}^a \left\{ \mathbf{E}[\xi^{\Delta} e^{-\theta\tau_\Delta} u^{A_\Delta} v^{B_\Delta}] \right\}(x)$$

$$= \frac{v(1-q)}{1-qv}\left\{a_0(u) - a_0(ux)(1 - v\gamma(1))\frac{1-W(\gamma(x))}{1-\gamma(x)}\right\}. \tag{9.18}$$

with γ defined by (*).

9.4. THE PROCESS WITH CONTINUOUS AND MIXED COMPONENTS

This refers to the case of b-type operator D, applied to transformation (9.5), and treats a continuous-valued process (A,B). Only formula (9.6) needs to be considered, since the remaining analysis from Section 9.3 also applies to the continuous case.

With the substitutions $u = e^{-\vartheta_1}$ and $v = e^{-\vartheta_2}$ we have that

$$\mathfrak{F} = \mathfrak{F}(\xi,\theta,e^{-\vartheta_1},e^{-\vartheta_2};\, e^{-s_1},e^{-s_2})$$

$$= D_{p_1} D_{p_2}\{\mathbf{E}[\xi^{\Delta} e^{-\theta\tau_\Delta} e^{-\vartheta_1 A_\Delta} e^{-\vartheta_1 B_\Delta}]\}(s_1,s_2),$$

and the formula (similar to (9.6)) will, then, be

$$\frac{1}{h_0(\theta)}\mathfrak{F} = \alpha_0(\vartheta_1,\vartheta_2) - \alpha_0(\vartheta_1+s_1,\vartheta_2+s_2)\frac{1-\xi h(\theta)\alpha(\vartheta_1,\vartheta_2)}{1-\xi h(\theta)\alpha(\vartheta_1+s_1,\vartheta_2+s_2)}, \tag{9.19}$$

where

$$\alpha_0(\vartheta_1,\vartheta_2) = a_0(e^{-\vartheta_1},e^{-\vartheta_2}), \text{ and } \alpha(\vartheta_1,\vartheta_2) = a(e^{-\vartheta_1},e^{-\vartheta_2}). \tag{9.20}$$

Now we assume that A is discrete and B is continuous. A formula analogous to (9.6) and (9.7) will read

$$\frac{1}{h_0(\theta)}\mathfrak{F} = \mathcal{A}_0(u,\vartheta_2) - \mathcal{A}_0(ux,\vartheta_2+s_2)\frac{1-\xi h(\theta)\mathcal{A}(u,\vartheta_2)}{1-\xi h(\theta)\mathcal{A}(ux,\vartheta_2+s_2)}, \tag{9.21}$$

with

$$\mathcal{A}_0(u,\vartheta_2) = a_0(u,e^{-\vartheta_2}) \text{ and } \mathcal{A}(u,\vartheta_2) = a(u,e^{-\vartheta_2}). \tag{9.22}$$

The special case

$$D_{p_2}^b\{\mathbf{E}[u^{A_\Delta} e^{-\vartheta_2 B_\Delta}]\}(s_2)$$

$$= \mathcal{A}_0(u,\vartheta_2) - (1-\mathcal{A}(u,\vartheta_2))\mathfrak{D}_x^{p_1}\left\{\frac{\mathcal{A}_0(ux,\vartheta_2+s_2)}{1-\mathcal{A}(ux,\vartheta_2+s_2)}\right\} \tag{9.23}$$

(first excess level) will be used in the following section.

9.5. EXAMPLES OF QUEUEING MODELS

Example 9.3. An M/G/1 quorum N-policy system with or without server vacations. Consider a class of M/G/1-type queues with bulk Poisson input flow, quorum, group service and multiple vacations. Suppose that batches of customers arrive at the system in accordance with marked Poisson counting process $Z = \sum_{n \geq 1} A_n \varepsilon_{t_n}$ (with position-independent marking), where mark A_n represents the total number of customers that arrive at the system by the time t_n, and X_n is the size of the nth arriving batch. (X_n) are independent and identically distributed random variables with the given probability generating function $\mathbf{E}[u^{X_1}] = \alpha(u)$. The Poisson point process $\{t_n\}$ of arriving batches is stationary with rate λ. A single channel processes customers in batches not exceeding R (its capacity).

Let $\{T_n\}$ and $\{\sigma_n\}$ be the point processes of successive service completions and service durations, respectively. Both are independent of the input process. Denote by $Q(t)$ the queueing process with continuous time parameter t. If $Q_n = Q(T_n+) < r$ (where r is a positive integer not exceeding R), the server rests until the input replenishes the queue to at least N, after which the server resumes its work. Alternatively, he goes on vacations and returns to the system to check if the queue has increased to at least N; if not, he leaves the system again. We assume that the server does not interrupt any single vacation even though the system may hit its level within a vacation period. The service duration of the nth group, σ_n, is distributed according to S_i (an arbitrary PDF) if $Q_n = i$, $i = 0,1,\ldots$. The PDF of any vacation time is V. So the server stops processing new customers whenever the queue level falls below r and resumes service when the queue reaches or exceeds level N.

During a period of time, when the server is unavailable for customers, he may either rest within an idle period or leave on vacations initiating a vacation sequence. The time interval $[T_n, T_{n+1})$ is called the nth *service cycle*. If $Q_n \geq r$, the length of the nth service cycle is equal to σ_n. Otherwise, it is a sum of the length of an idle period and σ_n (in a system with no vacations) or it totals all vacation periods and σ_n (in a system with vacations). Let γ_n denote the total number of customers that arrive within a period of time after T_n during which the server either idles or goes on vacations and before he resumes his service (for convenience call any such period of time *free*), and let $\widetilde{\gamma}_n$ be the total number of customers entering the system during the nth service. Then, the transitions of the embedded process $\{Q_n\}$ satisfy the following obvious recursive relation:

$$Q_1 = \begin{cases} (Q(\tau_\nu) - R)^+ + \widetilde{\gamma}_1, & Q_0 < r \\ (Q_0 - R)^+ + \widetilde{\gamma}_1, & Q_0 \geq r, \end{cases} \tag{9.24}$$

Here, $Q(\tau_\nu)$ is the queue level at the end of the first free period, and τ_ν is the first passage time. In our case, $p_1 = N - 1$. Various options, when $R \leq N$ or $R \geq N$, were considered by Muh [28,29] in his work on N-policy systems without vacations. In one model when $r \leq N \leq R$, considered by Muh, the server might take some customers from the total of W_1 that arrive at the system during his warm-up time in the event $Q(\tau_\nu) < R$. Then, relation (9.24) is modified to

$$Q_1 = \begin{cases} (Q(\tau_\nu) + W_1 - R)^+ + \widetilde{\gamma}_1, & Q_0 < r \\ (Q_0 - R)^+ + \widetilde{\gamma}_1, & Q_0 \geq r. \end{cases}$$

Abolnikov and Dshalalow [2] studied a special case of the general model without vacations and for $N = R$.

For convenience, we will rewrite formula (9.8) as functionals for the first excess level and the index of the first excess:

$$\text{[First excess]} \qquad \mathbf{E}[u^{A_\nu}] = a_0(u) - (1 - a(u))\mathfrak{D}_x^{p_1}\left\{\frac{a_0(ux)}{1 - a(ux)}\right\}, \tag{9.25}$$

$$\text{[Index]} \qquad \mathbf{E}[\xi^\nu] = 1 - (1 - \xi)\mathfrak{D}_x^{p_1}\left\{\frac{a_0(x)}{1 - \xi a(x)}\right\}. \tag{9.26}$$

Due to the above relation for $\{Q_n\}$, we shall be interested in $\mathbf{E}^i[\cdot] = \mathbf{E}[\cdot \mid Q_0 = i]$. Consequently, in the above formulas, $a_0(u)$ will equal u^i. We will also set $p_1 = N - 1$, and, taking into account property P2 in Section 9.2, we get

$$\text{[first excess]}\ E_i(u) = \mathbf{E}^i[u^{A_\nu}] = u^i - u^i(1 - a(u))\mathfrak{D}_x^{N-i-1}\left\{\frac{1}{1 - a(ux)}\right\}, \tag{9.27}$$

$$\text{[Index]} \qquad \mathbf{E}^i[\xi^\nu] = 1 - (1 - \xi)\mathfrak{D}_x^{N-i-1}\left\{\frac{1}{1 - \xi a(x)}\right\}, \tag{9.28}$$

with the mean value of index

$$\mathbf{E}^i[\nu] = \mathfrak{D}_x^{N-i-1}\left\{\frac{1}{1 - a(x)}\right\}. \tag{9.29}$$

Let X be a discrete-valued random variable with

$$f(u) = \mathbf{E}[u^X] = \textstyle\sum_{j \geq 0} P\{X = j\}u^j.$$

Define $\mathcal{H}^R f(u) = \mathbf{E}[u^{(X-R)^+}]$.

Lemma 9.4. *$\mathcal{H}^R$ is a linear operator with fixed points at constant functions; its analytic representation* is:

$$\mathcal{H}^R f(u) = u^{-R} f(u) + \mathfrak{D}_x^{R-1}\{f(x) - u^{-R} f(ux)\}. \tag{9.30}$$

Theorem 9.5. *Let $\beta_i(\theta)$ be the Laplace-Stieltjes transform of the PDF S_i (i.e., the distribution of the service duration of a group given that, by the preceding completion of service, the length was i). Suppose that $S_i = S$ and $\beta_i = \beta$ for $i \geq R$. Then, given $\alpha\lambda s < R$ (α is the mean value of an arriving batch and $s = \int x dS(x)$), Markov chain $\{Q_n\}$ is ergodic and the generating function $P(u)$ of the invariant probability vector $\boldsymbol{P} = (p_0, p_1, \ldots)$ of $\{Q_n\}$ satisfies the following formula:*

$$P(u) = \frac{\sum_{i=0}^{R-1} p_i \beta_i(\lambda - \lambda\alpha(u))\{u^R \mathcal{H}^R E_i(u) - u^i\}}{u^R - \beta(\lambda - \lambda\alpha(u))}, \tag{9.31}$$

where E_i satisfies formula (9.27) with $a(u) = \alpha(u)$ for a system with resting server and $a(u) = V^*(\lambda - \lambda\alpha(u))$ for a system with vacationing server. (V^* is the Laplace-Stieltjes transform of the PDF V of the vacation times.) The probabilities $p_0, p_1, \ldots, p_{R-1}$ satisfy a system of R linear equations. (See Abolnikov and Dshalalow [2] for reference.)

An explicit analytic formula for the generating function of the stationary distribution of the queueing process with continuous time parameter can be obtained by methods similar to those in Abolnikov and Dshalalow [2] and Dshalalow [13]. This formula also contains an expression relative to the functional for the first excess level.

Example 9.6. An M/G/1 N-D-policy quorum system with or without server vacations. Assume that the input stream is as in the above example, but the server takes just one customer at a time. An idle period or a vacations sequence starts when the queue level falls below level r and ends when the queue hits N or the cumulative job in the system hits D, whichever comes first. In terms of the first excess level theory, the joint input to the system is a two-dimensional marked process Z, introduced at the beginning of Section 9.2, with (generally dependent) marks $(A_n = \sum_{k=0}^{n} X_k, B_n = \sum_{k=0}^{n} Y_k)$, where X_k represents the increment of the input process over the interval $(t_{k-1}, t_k]$ (no vacations) or $(\hat{t}_{k-1}, \hat{t}_k]$ (with vacations, where $\hat{t}_k$ is the start of the kth vacation within the first free period, $k = 1,2,...$), and Y_k represents the cumulative job brought by X_k customers over the same period of time. Clearly,

$$Y_n = \sum_{j=A_{n-1}+1}^{A_n} \sigma_j, \; n = 0,1,..., \tag{9.32}$$

where, $A_n = X_0 + ... + X_n$, $A_{-1} = 0$, $X_0 = Q_0 < r$, and σ_i is the ith service duration. The process (A,B) (with $B_n = Y_0 + ... + Y_n$) is terminated when it hits level $L = (N,D)$. To use formula (9.23) for the mixed case, we first derive $\mathcal{A}(u,\vartheta_2)$ by applying standard probability arguments:

$$\mathcal{A}(u,\vartheta_2) = \mathbf{E}[u^{X_1} e^{-\vartheta_2 Y_1}] = \mathbf{E}[u^{X_1} e^{-\vartheta_2 \sum_{j=X_0+1}^{X_0+X_1} \sigma_j}]$$
$$= \mathbf{E}[\mathbf{E}[u^{X_1} e^{-\vartheta_2 \sum_{j=X_0+1}^{X_0+X_1} \sigma_j} \mid X_1]] = a(u\beta(\vartheta_2)). \tag{9.33}$$

Obviously,

$$\mathcal{A}_0(u,\vartheta_2) = \mathbf{E}[u^{X_0} e^{-\vartheta_2 \sum_{j=1}^{X_0} \sigma_j} \mid Q_0 = i] = [u\beta(\vartheta_2)]^i. \tag{9.34}$$

(We let $\sum_{j=1}^{k}(\cdot) = 0$ for $k = 0$.) Now, substituting $\mathcal{A}$ and $\mathcal{A}_0$ into (9.7) and by property P2 of the operator $\mathfrak{D}$, we get

$$D_{p_2}^{b}\{\mathbf{E}[u^{Q_{\tau\Delta}} e^{-\vartheta_2 B_\Delta} \mid Q_0 = i]\}(s_2) = D_{p_2}^{b}\{\mathbf{E}[u^{A_\Delta} e^{-\vartheta_2 B_\Delta} \mid X_0 = A_0 = i]\}(s_2)$$
$$= [u\beta(\vartheta_2)]^i - [u\beta(\vartheta_2 + s_2)]^i[1 - a(u\beta(\vartheta_2))]\mathfrak{D}_x^{N-i-1}\left\{\frac{1}{1 - a(ux\beta(\vartheta_2 + s_2))}\right\}. \tag{9.35}$$

Specifically,

$$D_{p_2}^{b}\{\mathbf{E}[u^{Q_{\tau\Delta}} \mid Q_0 = i]\}(s_2) = D_{p_2}^{b}\{E_i(u)\}(s_2)$$
$$= u^i - [u\beta(s_2)]^i[1 - a(u)]\mathfrak{D}_x^{N-i-1}\left\{\frac{1}{1 - a(ux\beta(s_2))}\right\}. \tag{9.36}$$

In the end, the probability generating function of the stationary distribution of the embedded queueing process can be obtained in a similar manner, as in the previous example of this section, by extracting E_i from (9.36) and substituting it into a formula similar to (9.31).

9.6. PRETERMINATION PROCESSES FOR A SPECIAL CASE

The results of this section are based on Dshalalow [14]. In the light of Special Case

2, the counting measure $Z = \sum_{n \geq 0}(A_n, B_n)\varepsilon_{\tau_n}$ now contains (X_n, Y_n) as increments of the two-dimensional renewal process (A,B) whose components A and B are now independent. The increments Y_n are all equal to 1, and component B is supposed to reach random value ν_2. The random variable ν_2 is supposed to be distributed as $P_{\nu_2} = \sum_{j \geq 0} q_j \varepsilon_j$ with $W(z) = \mathbf{E}[z^{\nu_2}]$. The process (A,B) will be terminated if A hits p_1 or if B takes on value ν_2, whichever of the two events occurs first. As a matter of fact, many excess level processes that are encountered in applications can be associated with catastrophes, which it is worthwhile to predict exactly one step before they occur. We will consider all random factors associated with the termination as Δ, τ_Δ and A_Δ and analyze their preceding values. Define the random variable

$$\Delta^- = \begin{cases} 0, & \Delta = 0 \\ \Delta - 1, & \Delta > 0 \end{cases}$$

and call it the *pretermination index*. Correspondingly, introduce the random variables τ_{Δ^-} and A_{Δ^-}, the *pretermination time* and *pretermination level*, respectively. We set $\Delta^- = \tau_{\Delta^-} = A_{\Delta^-} = 0$ for all $\omega \in \{A_0 > p_1\}$. In other words, all pretermination processes are set to zero when the initial level hits p_1.

Let $\Phi_{p_1}(\theta,\xi,u) = E[e^{-\theta\tau_{\Delta^-}}\xi^{\Delta^-}u^{A_{\Delta^-}}]$ and $\Phi(\theta,\xi,u;x) = D_{p_1}\Phi_{p_1}(\theta,\xi,u)$.

Theorem 9.7. *For the a- and b-type transformation D_{p_1}, functional Φ satisfies the relation:*

$$\Phi(\theta,\xi,u,x) = 1 - a_0(x)(1-q_0)$$
$$+ h_0(\theta)a_0(xu)\left\{[1-a(x)]\,\frac{1 - W\{\xi a(xu)h(\theta)\}}{1-\xi a(xu)h(\theta)} + a(x)\,\frac{W\{\xi a(xu)h(\theta)\} - q_0}{\xi a(xu)h(\theta)}\right\}. \tag{9.37}$$

Proof. Denote $\Phi_{p_1}(\theta,\xi,u) = \Phi^{(1)}_{p_1}(\theta,\xi,u) + \Phi^{(2)}_{p_1}(\theta,\xi,u)$, where

$$\Phi^{(1)}_{p_1}(\theta,\xi,u) = \mathbf{E}[e^{-\theta\tau_{\Delta^-}}\xi^{\Delta^-}u^{A_{\Delta^-}}I_{\{\Delta=0\}}] \quad \text{and}$$

$$\Phi^{(2)}_{p_1}(\theta,\xi,u) = \mathbf{E}[e^{-\theta\tau_{\Delta^-}}\xi^{\Delta^-}u^{A_{\Delta^-}}I_{\{\Delta>0\}}].$$

From $I_{\{\Delta=0\}} = I_{(p_1,\infty)}(A_0) + I_{\{0\}}(\nu_2) - I_{(p_1,\infty)}(A_0)I_{\{0\}}(\nu_2)$, we have that

$$\Phi^{(1)}_{p_1}(\theta,\xi,u) = q_0 + P\{A_0 > p_1\}(1-q_0).$$

Now, we get $\Phi^{(2)}_{p_1}(\theta,\xi,u)$:

$$= \mathbf{E}\left[I_{\{1,2,\ldots\}}(\nu_2)\sum_{j=0}^{\nu_2-1}\mathbf{E}\left[e^{-\theta\tau_j}\xi^j u^{A_j}\left(I_{[0,p_1]}(A_j) - I_{[0,p_1]}(A_{j+1})\right)\right]\right]$$
$$+ h_0(\theta)\mathbf{E}[I_{\{1,2,\ldots\}}(\nu_2)\{\xi h(\theta)\}^{\nu_2-1}$$
$$\cdot \sum_{j>\nu_2}\mathbf{E}[u^{A_{\nu_2-1}}\{I_{[0,p_1]}(A_{j-1}) - I_{[0,p_1]}(A_j)\} \mid \nu_2]].$$

Thus, by Lemma 9.1 and elementary transformations, we have $\Phi^{(2)}(\theta,\xi,u,x)$

$$= h_0(\theta)a_0(xu)\left\{[1-a(x)]\frac{1-W\{\xi a(xu)h(\theta)\}}{1-\xi a(xu)h(\theta)} + a(x)\frac{W\{\xi a(xu)h(\theta)\} - q_0}{\xi a(xu)h(\theta)}\right\}.$$

On the other hand, for the a- or b-type transformation D_{p_1},

$$\Phi^{(1)}(\theta,\xi,u,x) = 1 - a_0(x)(1-q_0),$$

finally leading to formula (9.37). □

Examples

1) For $\nu_2 = \infty$ the renewal process A is to run only until it crosses a specified level. Then we have $q_0 = 0$ and $W(u) \equiv 0$, and the functional Φ reduces to

$$\Phi(\theta,\xi,u,x) = 1 - a_0(x) + h_0(\theta)a_0(xu)\,\frac{1-a(x)}{1-\xi a(xu)h(\theta)}\ .$$

2) Consider the discrete case of the component A under the scenario of Special Case 2. Theorem 9.7 can be reformulated for that case as:

Theorem 9.8. *If the renewal process A is discrete-valued, the joint functional $\Phi_r(\theta,\xi,u)$ of the pretermination time, pretermination index, and pretermination level can be found from the formula:*

$$\Phi_{p_1}(\theta,\xi,u) = q_0 + (1-q_0)P\{A_0 > p_1\}$$

$$+\, h_0(\theta)\mathfrak{D}_x^{p_1}\left(a_0(xu)\left\{[1-a(x)]\,\frac{1-W\{\xi a(xu)h(\theta)\}}{1-\xi a(xu)h(\theta)} + a(x)\,\frac{W\{\xi a(xu)h(\theta)\}-q_0}{\xi a(xu)h(\theta)}\right\}\right). \tag{9.38}$$

If $W(u) = \frac{1-q}{1-uq}$, then $q_0 = W(0) = 1-q$ and thus formula (9.38) reduces to

$$\Phi_{p_1}(\theta,\xi,u) = 1 - qP\{A_0 \le p_1\} + qh_0(\theta)\mathfrak{D}_x^{p_1}\left\{\frac{a_0(xu)[1-qa(x)]}{1-q\xi a(xu)h(\theta)}\right\}. \tag{9.39}$$

Specifically, the generating function of the pretermination level $A_{\Delta}-$ is

$$\Phi_{p_1}(0,1,u) = 1 - qP\{A_0 \le p_1\} + q\mathfrak{D}_x^{p_1}\left\{\frac{a_0(xu)[1-qa(x)]}{1-qa(xu)}\right\}. \tag{9.40}$$

With geometric distribution for X_i, since $a(x) = \frac{(1-a)x}{1-ax}$, the mean pretermination level is

$$\mathbf{E}[A_{\Delta}-] = q\Big\{p_1!a_{0(p_1)}(1) - \textstyle\sum_{i=0}^{p_1-1}(-1)^i\binom{-2}{i}a_{0p_1-1-i}$$

$$+\,\frac{q}{1-b}a_{0(p_1-1)}(1) + \frac{a^{p_1+1}}{1-a}a_{0(p_1-1)}\left(\tfrac{1}{a}\right) - \frac{b^{p_1+1}}{1-b}a_{0(p_1-1)}\left(\tfrac{1}{b}\right)\Big\}, \tag{9.41}$$

where $a_{0(k)}(x) = \sum_{j=0}^{k} a_{0j}x^j$, $b = a + q(1-a)$, and $\binom{\alpha}{\beta} = \frac{\alpha\cdots(\alpha-\beta+1)}{\beta!}$ for some nonnegative integer β and α real.

For $a_0(x) = x^i$ $(i \le p_1)$, we have

$$\Phi_{p_1}(\theta,\xi,u) = 1 - q + qu^ih_0(\theta)\mathfrak{D}_x^{p_1-i}\left\{\frac{1-qa(x)}{1-q\xi a(xu)h(\theta)}\right\}. \tag{9.42}$$

With geometric distribution for X_i, we obtain

$$\Phi_{p_1}(\theta,\xi,u) = 1 - q + qu^i\Big\{u + (1-u)\left(1 - \frac{q}{1-bu}\right) + \frac{q(1-u)}{a-bu}a^{p_1-i+1}$$

$$-\frac{q(1-u)(1-a)}{(1-bu)(a-bu)}(bu)^{p_1-i+1}\Big\}. \tag{9.43}$$

3) In the "continuous case," we have, as in Section 9.4,

$$\Phi^*(\theta,\xi,\vartheta,s)=\Phi(\theta,\xi,e^{-\vartheta},e^{-s})=1-\gamma(s)(1-q_0)$$
$$+h_0(\theta)\gamma(s+\vartheta)\left\{[1-\alpha(s)]\frac{1-W\{\xi\alpha(s+\vartheta)h(\theta)\}}{1-\xi\alpha(s+\vartheta)h(\theta)}+\alpha(s)\frac{W\{\xi\alpha(s+\vartheta)h(\theta)\}-q_0}{\xi\alpha(s+\vartheta)h(\theta)}\right\}. \tag{9.44}$$

Specifically, for $\nu_2=\infty$, $q_0=0$ and $W(u)\equiv 0$, formula (9.44) reduces to

$$\Phi^*(\theta,\xi,\vartheta,s)=\Phi(\theta,\xi,e^{-\vartheta},e^{-s})$$
$$=1-\gamma(s)+h_0(\theta)\gamma(s+\vartheta)\left\{[1-\alpha(s)]\frac{1}{1-\xi\alpha(s+\vartheta)h(\theta)}\right\}. \tag{9.45}$$

9.7. OPEN PROBLEMS AND FUTURE RESEARCH DIRECTIONS

The following projects are related to the first excess level theory and may be of practical interest.

1) It seems plausible to assume that formula (9.6) holds true for the marked point process with n dependent marks $\{X^{(i)},\ i=1,\dots,n\}$. A formula, similar to (9.6) would read:

$$\frac{1}{h_0(\theta)}\mathfrak{F}=a_0(u_1,\dots,u_n)-a_0(u_1x_1,\dots,u_nx_n)\frac{1-\xi h(\theta)a(u_1,\dots,u_n)}{1-\xi h(\theta)a(u_1x_1,\dots,u_nx_n)}. \tag{9.46}$$

However, the same arguments applied to the n-dimensional case may be too awkward.

2) For one-dimensional recurrent marked process Z (with real-valued marks A_n), it is reasonable to assume that an analytically tractable result can be obtained by observing (A) at the instants $\delta_1,\delta_2,\dots,$ of stopping times of $\{\tau_n\}$ at which $A_n > A_{n-1}$. One of the methods developed by Takács [40] can be combined with those in Section 9.3 to study the distribution of the first excess level. This may lead to alternative ways to arrive at results obtained by Takács using the theory of fluctuations.

3) Consider the following generalization of M/G/1-type D-N-policy systems. (We will be using the terminology introduced in Section 9.5.) Assume that the input stream is compound Poisson. If $1\le Q_0<r$, a single server with capacity R goes on not more than ν_2 vacations, where ν_2 is a random variable valued in $\{0,1,\dots\}$. Specifically, if $\nu_2=0$, the server stays in the system and continues processing customers. His vacations sequence can be terminated earlier if, by the end of a vacation segment, the queue hits level N, whichever of the two events comes first. For $Q_0=0$, the server goes on vacations as long as it is necessary for the queue to accumulate to at least N customers or for the server to make ν_3 vacations, whichever comes first. If the sequence of vacations is terminated by ν_3 and the queue is still empty, the whole process repeats, and so on. The

whole system can be made with state dependent service, vacations, and system modulated input. Methods can be applied, developed in Dshalalow [12], combined with the the first excess level theory treated in Special Case 2, formula (9.15). Observe that if ν_2 is a Bernoulli random variable, $r = \infty$, $\nu_3 = 1$ *a.s.*, and $N = 1$, the system overlaps with one (Bernoulli schedules) introduced and studied in Keilson and Servi [22]. [Note, however, that the system studied by Keilson and Servi was GI/G/1, whereas the suggested model is with batch Poisson (perhaps modulated) input.] The above system also generalizes a vacation model studied by Kella [24] (mentioned in the introduction, Example C) which, in turn, is an M/G/1-type queue with a generalized Bernoulli vacation discipline.

4) A practical queueing system of type M/G/1 with batch arrivals, N-D-policy, and vacations would be one initiated in Example 9.6 of Section 9.5. It seems possible to get analytically tractable formulas by using results in Example 9.6 combined with techniques in Dshalalow [13].
5) It is reasonable to study classes of queueing systems with vector arrivals of independent streams of batches of customers. Suppose two flows of batches enter the system, at times $\tau_n^{(s)}$, $s = 1$ (flow 1), and 2 (flow 2), in accordance with two independent Poisson processes. The superposed process would be $\{\tau_n\}$, so that if, for some n, τ_n belongs to $\tau_n^{(1)}$, then $X_n^{(2)} = 0$, and vice versa. Using the results of Section 9.3, one can analyze various scenarios of systems where hitting levels by one of the input processes is of interest.
6) The above cases can be extended to those for G/G/1-type queues.

BIBLIOGRAPHY

[1] Abolnikov, L. and Dshalalow, J.H., A first passage problem and its applications to the analysis of a class of stochastic models, *J. Appl. Math. Stoch. Analysis*, **5**:1 (1992), 83-98.

[2] Abolnikov, L. and Dshalalow, J.H., On a multilevel controlled bulk queueing system $M^X/G^{r,R}/1$, *J. Appl. Math. Stoch. Anal.*, **5**:3 (1992), 237-260.

[3] Abolnikov, L., Dshalalow, J.H. and Dukhovny, A.M., A multilevel controlled bulk queueing system with vacationing server, *Oper. Res. Letters*, **13** (1993), 183-188.

[4] Abolnikov, L., Dshalalow, J.H. and Dukhovny, A.M., First passage processes in queueing system $M^X/G^r/1$ with service delay discipline, *Intern. J. Math. & Math. Sci.*, **17**:3 (1994), 571-586.

[5] Abolnikov, L., Dshalalow, J.H. and Dukhovny, A.M., Stochastic analysis of a controlled bulk queueing system with continuously operating server: continuous time parameter queueing process, *Stat. Prob. Letters*, **16** (1993), 121-128.

[6] Andersen, E.S., On the fluctuations of sums of random variables, II, *Math. Scand.*, **2** (1954), 195-223.

[7] Balachandran, K. and Tijms, H., On the D-policy for the M/G/1 queue, *Mgt. Sci.*, **21** (1975), 1073-1076.

[8] Baxter G., On operator identity, *Pacific J. Math.*, **8** (1958), 649-663.

[9] Cohen, J., On up- and downcrossings, *J. Appl. Prob.*, **14** (1977), 405-410.

[10] Doshi, B., Single-server queues with vacations. In: *Stochastic Analysis of Computer and Communication Systems*, 217-265, edited by H. Takagi, Else-

vier Sci. Publ. B.V. (North-Holland), 1990.

[11] Dshalalow, J.H., On a first passage problem in general queueing systems with multiple vacations, *J. Appl. Math. Stoch. Anal.*, **5**:2 (1992), 177-192.

[12] Dshalalow, J.H., Single-server queues with controlled bulk service, random accumulation level, and modulated input, *Stoch. Anal. Appl.*, **11**:1 (1993), 29-41.

[13] Dshalalow, J.H., First excess level analysis of random processes in a class of stochastic servicing systems with global control, *Stoch. Anal. Appl.*, **12**:1 (1994), 75-101.

[14] Dshalalow, J.H., On termination time processes, in *Studies in Applied Probability*, Edited by J. Galambos and J. Gani, Essays in honor of Lajos Takács, *J. Appl. Prob.*, Special Volume **31A** (1994), 325-336.

[15] Dshalalow, J.H., First excess levels of vector processes, *J. Appl. Math. Stoch. Anal.*, **7**:3 (1994), 457-464.

[16] Dshalalow, J.H. and Tadj, L., A queueing system with random server capacity and multiple control, *Queueing Sys.*, **14** (1993), 369-384.

[17] Dshalalow, J.H. and Tadj, L., On applications of first excess level random processes to queueing systems with random server capacity and capacity dependent service time, *Stoch. Stoch. Reports*, **45** (1993), 45-60.

[18] Dshalalow, J.H. and Yellen, J., Bulk input queues with quorum and multiple vacations, *Math. Probl. Engin.* (to appear).

[19] Dynkin, E., Some limit theorems for sums of independent random variables with infinite mathematical expectations, *Sel. Transl. Math. Statist. Prob.*, **1** (1961), 171-189.

[20] Heyman, D.P., Optimal operating policies for M/G/1 queueing systems, *Oper. Res.* **16** (1968), 362-382.

[21] Jacob, M. and Madhusoodanan, T., Transient solution for a finite capacity $M/G^{a,b}/1$ queueing system with vacations to the server, *Queueing Sys.*, **2** (1987), 381-386.

[22] Keilson, J. and Servi, L., Oscillating random walk models for GI/G/1 vacation systems with Bernoulli schedules, *J. Appl. Prob.*, **23** (1986), 790-802.

[23] Keilson, J. and Servi, L., Blocking probabilities for M/G/1 vacation systems with occupancy level dependent schedules, *Oper. Res.* **37**:1 (1989), 134-140.

[24] Kella, O., Optimal control of the vacation scheme in an M/G/1 queue, *Oper. Res.*, **38**:4 (1990), 724-728.

[25] Lee, H.S. and Srinivasan, M.M., Control policies for the $M^X/G/1$ queueing system, *Mgt. Sci.*, **35**:6 (1989), 708-721.

[26] Lee, H.W., Lee, S.S. and Chae, K.C., Operating characteristics of $M^X/G/1$ queue with N-policy, *Queueing Sys.*, **15** (1994), 387-399.

[27] Li, J. and Niu S., The waiting time distribution for the GI/G/1 queue under *D*-policy, *Prob. Engineer. Inform. Sci.*, **6** (1992), 287-308.

[28] Muh, D.C.R., A bulk queueing system under N-policy with bilevel service delay discipline and start-up time, *J. Appl. Math. Stoch. Anal.*, **6**, No. 4 (1993), 359-384.

[29] Muh, D.C.R., On a Class of N-Policy Multilevel Control Queueing Systems, Doctoral Thesis, Florida Tech, 1994.

[30] Pollaczek, F., Sur la répartition des périodes d'occupation ininterrompue d'un guichet, *C.R. Acad. Sci. Paris*, **234** (1952), 2042-2044.

[31] Ramaswami, R. and Servi, L, The busy period of the M/G/1 vacation

model with a Bernoulli schedule, *Stoch. Mod.*, **4**:3 (1988), 507-521.

[32] Shanthikumar, J.G., Level crossing analysis of priority queues and a conservation identity for vacation models, *Nav. Res. Logist. Quart.*, **36** (1989), 797-806.

[33] Spitzer, F., A combinatorial lemma and its application to probability theory, *Trans. Amer. Math. Soc.*, **82** (1956), 323-339.

[34] Takács, L., *Combinatorial Methods in the Theory of Stochastic Processes*, John Wiley, New-York 1967.

[35] Takács, L., On a linear transformation in the theory of probability, *Acta Sci. Math.*, **33**:1-2 (1972), 15-24.

[36] Takács, L., Sojourn time problems, *Ann. Prob.*, **2**, No. 3 (1974), 420-431.

[37] Takács, L., A storage process with semi-Markov input, *Adv. Appl. Prob.*, **7**: 4 (1975) 830-844.

[38] Takács, L., On some recurrence equations in a Banach algebra, *Acta Sci. Math.*, **38**:3-4 (1976), 399-416.

[39] Takács, L., On fluctuation problems in the theory of queues, *Adv. Appl. Prob.*, **8**:3 (1976) 548-583.

[40] Takács, L., On fluctuations of sums of random variables, *Adv. Math.*, **2** (1978), 45-93.

Part II

Analytic Methods

Chapter 10

Matrix-analytic methods in queueing theory

Marcel F. Neuts[1]

ABSTRACT Broad classes of queueing models that have embedded Markov renewal processes of M/G/1 or GI/M/1-type have been analyzed by matrix-analytic methods. The earlier results of that methodology are discussed, respectively, in the 1989 and 1981 books by Neuts. We survey the developments since the publication of these books, discuss some of the varied applications of the methods, and outline the current research trends. Some open problems are formulated.

CONTENTS

10.1 The matrix-analytic methods 265
10.2 The phase-type distributions 269
10.3 The Markovian arrival process 271
10.4 Matrix-geometric solutions 272
10.5 The M/G/1 structure 277
10.6 Open problems and new research directions 281
Bibliography 284

10.1 THE MATRIX-ANALYTIC METHODS

The M/G/1 and GI/M/1 queues are classical classroom examples of analytically tractable queueing models. Underlying that tractability are well-structured Markov renewal processes, embedded respectively at service completions and at arrival epochs. Many variants of the M/G/1 and GI/M/1 queues have been discussed. These include a variety of threshold and vacation models, special services at the starts of busy periods, limitations on the numbers of customers admitted during services, and service with gating. In most of these, embedded Markov renewal processes (*MRP*s) with substantially the same structure as the basic cases remain present. They account for the detailed, explicit results obtainable for these variants.

Poisson arrivals or exponential service times induce a simple Markovian behavior of the queue during the sojourn intervals of these embedded *MRP*s. It therefore suffices to derive the matrix renewal function of the embedded *MRP* and to relate other time-dependent probability distributions for the queue to it. Steady-state distributions for the stable queue are then obtained by a limiting argument based on the key renewal theorem.

Under the impetus of applications in communications engineering, see e.g., Saito [144], discrete-time analogues of the basic queues have recently generated much interest. The corresponding lack-of-memory property of the Bernoulli process and the geometric distribution yields a similar formal structure of the embed-

[1]This research was supported by NSF Grant Nr. DMI-9306828.
0-8493-8074-x/95/$0.00+$.50

ded *MRPs*. That leads to a completely analogous mathematical analysis.

Similarity in the analytic derivations for related models usually points to a shared underlying *structure*. For the classical queues, including GI/G/1, one unifying structure is that of a general random walk on R^+ with a single boundary state corresponding to the empty queue. That random walk has been studied in great detail. The most salient results rest on the exchangeability of the step sizes, and the analysis is strongly combinatorial. An alternative approach exploits Markovian properties. It is often called the *method of supplementary variables*. In brief, one carries along enough supplementary random variables to retain a Markov process for which, one hopes, the appropriate (usually integral) equations can be solved.

The matrix-analytic methods inherit aspects from both sets of ideas. There is an analogous random walk on a semi-infinite strip of states. Many derivations also use supplementary variables. What is new to these methods is the introduction of models for the service and input processes to which the *elementary* analysis of the M/G/1 and GI/M/1 queues can be adapted and generalized. The results are expressed in explicit matrix formulas that are well-suited for algorithmic implementation.

In the matrix-analytic methods, we use a weakened version of the lack-of-memory property. We work with service time distributions or input processes that allow a Markovian dependence on one of a *finite number of phases*. Qualitatively, that greatly enhances the versatility of the probability distributions and point processes available for modeling practice. The cost of that versatility lies in the matrix formalism needed to analyze the generalized models. The following table lists some classical notions with the corresponding generalizations used in the matrix-analytic methodology:

Elementary	**Generalization**
The exponential distribution	The *PH*-distributions
The Poisson process	Markovian arrival processes (*MAP*)
The geometric distribution	Discrete *PH*-distributions
The Bernoulli process	The discrete *MAPs*
The GI/M/1 queue	Queues of the GI/M/1-type
The M/G/1 queue	Queues of M/G/1-type

In discussing the state of the art on these subjects, the treatment in my books [109] and [112] serves as departure point. Unless otherwise noted, we use the same terminology and notation. Classical material will not be repeated. Even so, this chapter cannot give a systematic, complete treatment of the extensive developments since these books were published. The new results are grouped under broad headings. To convey further insight and to show how everything fits together, we offer occasional comments on them. However, for the proofs and derivations, we must refer to the appropriate articles. With few exceptions, references already listed in [109] or [112] are not repeated. For classical results of queueing theory, no sources are indicated. The still lengthy reference list is meant to inform on the recent developments, not to overwhelm with bibliographical material.

A brief history. The matrix methods for queues found their origin in the 1960s in work of Çinlar and Neuts on the M/SM/1 and SM/M/1 queues and in the finding of Wallace that some specific block-tridiagonal generators of computer performance models have matrix-geometric steady-state probabilities. Other specific models led to a recurring mathematical argument related to the Rouché roots of a determinental equation. These were always related to the spectrum of the transform matrix of a semi-Markov matrix, but for a long time, the emphasis on transform methods hid the underlying probabilistic features. A personal dedication to computationally tractable solutions led me, during the 1970s, to examine the real-time integral equations for the semi-Markov matrices of some basic first passage times. That was the crucial transition to probabilistic arguments and matrix analysis. While that took place simultaneously for the M/G/1 and GI/M/1 structures, an invitation to give the Mathematical Science lectures at the John Hopkins University in 1976 and to write a book based on these, gave early emphasis to the matrix-geometric theorem and its application.

Between 1976 and 1983, I was fortunate to have an excellent group of associates and students at the University of Delaware. S. Chakravarthy, S. Kumar, G. Latouche, D. Lucantoni, and V. Ramaswami pursued matrix-analytic methods and their applications with an interest and passion only matched by my own. Together, we added to the theory of phase-type distributions, the Markovian process and the models of M/G/1 structure. By solving problems of direct applied interest, some of us drew attention to the applicability of these developments which quickly found wide acceptance in the communications and computer performance community.

By 1985, the matrix analysis of rich classes of queueing models was a mature subject. My 1989 book on the M/G/1 structure clearly reflects a research area that was growing vigorously. That growth has since gone unabated. The variety of applications, treated mostly in the engineering literature, is astonishing. On the mathematical side, many further theoretical questions have been answered. The needs of computational implementations for ever larger matrices have led to major new algorithms. This survey deals with the developments since my second book appeared. Clearly, structural and probabilistic analysis has been successful for important classes of stochastic models. Its mode of thinking is now being brought to bear on other structures that hold promise for the study of some queueing networks and of priority queues. The subject is far from exhausted.

Notation. The following notation is used throughout. A *continuous PH-distribution* $F(\cdot)$ is represented by a row vector of probabilities $\boldsymbol{\alpha}$ of dimension m, satisfying $0 < \boldsymbol{\alpha}\mathbf{e} \leq 1$, and a nonsingular matrix T of order m with negative diagonal and nonnegative off-diagonal elements. $\mathbf{e}$ is a column vector of ones. The PH-distribution is then given by

$$F(x) = 1 - \boldsymbol{\alpha}\exp(Tx)\mathbf{e}, \quad \text{for } x \geq 0.$$

The column vector $\boldsymbol{T}^\circ$ is defined by $\boldsymbol{T}^\circ = -T\mathbf{e}$. The matrix $T + \mathbf{T}^\circ\boldsymbol{\alpha}$ may be assumed to be irreducible. $(\boldsymbol{\alpha}, T)$ is then called an *irreducible* representation. The basic properties of PH-distributions are discussed in Chapter 2 of [109] and Chapter 5 of [112].

The *Markovian arrival process* (MAP) is parameterized by a sequence of $m \times m$ matrices $\{D_k\}$. D_0 is nonsingular, has negative diagonal elements and nonnegative off-diagonal elements. The matrices D_k, $k \geq 1$, are nonnegative and $D = \sum_{k=0}^{\infty} D_k$ is an irreducible generator. If $D_k = 0$, for all $k \geq 2$, the MAP has single arrivals. It is then a natural generalization of the Poisson process. To distinguish that case from the general one where group arrivals may occur, some authors use the abbreviation $BMAP$, where B stands for batch. Analogous parameterizations hold for discrete PH-distributions and MAPs. The abbreviation $DBMAP$

is sometimes used to emphasize that a discrete *MAP* may have group arrivals. The same analysis applies to single and group arrivals, so we do not make that distinction. The *versatile Markovian arrival process* in Chapter 5 of [112] is equivalent to the *MAP*, but the notation used there is less elegant. Reviews of the basic properties and modeling applications of *MAP*s are found in Lucantoni [93] and Neuts [119].

Models of GI/M/1-*type* have an embedded *MRP* whose transition probability matrix $Q(\cdot)$ has the canonical form

$$Q(x) = \begin{vmatrix} B_0(x) & A_0(x) & 0 & 0 & 0 & \cdots \\ B_1(x) & A_1(x) & A_0(x) & 0 & 0 & \cdots \\ B_2(x) & A_2(x) & A_1(x) & A_0(x) & 0 & \cdots \\ B_3(x) & A_3(x) & A_2(x) & A_1(x) & A_0(x) & \cdots \\ \cdot & \cdot & \cdot & \cdot & \cdot & \end{vmatrix},$$

where the $A_k(x)$ and $B_k(x)$, $k \geq 0$, $x \geq 0$, are matrices of probability massfunctions satisfying the conditions for $Q(x)$ to be a valid semi-Markov matrix. We shall limit our discussion to the case where all coefficient matrices are $m \times m$ and the matrix $A = \sum_{k=0}^{\infty} A_k(\infty)$ is *irreducible* and stochastic. As we discussed in Chapter 1 of [109], the theory is easily adapted to generators Q or modified to handle cases with more complex boundary behavior. A remark in Section 10.5 of the present chapter addresses the issue of reducibility of the matrix A.

Using terminology that reflects the natural partition of the state space, we refer to the sets of states $\{(i, j)\}$ with a fixed i as *level* **i**. The level 0 is called the boundary level. One of the nice features of the matrix methods is that the distinct behavior of the models at the boundary can be studied separately from that in the homogeneous part consisting of the levels **i** with $i \geq 1$. That is useful in examining alternative models whose behavior differs only at the boundary.

The embedded *MRP* of a *model of* M/G/1-*type* has a transition probability matrix $Q(\cdot)$ of the form

$$Q(x) = \begin{vmatrix} B_0(x) & B_1(x) & B_2(x) & B_3(x) & B_4(x) & \cdots \\ C_0(x) & A_1(x) & A_2(x) & A_3(x) & A_4(x) & \cdots \\ 0 & A_0(x) & A_1(x) & A_2(x) & A_3(x) & \cdots \\ 0 & 0 & A_0(x) & A_1(x) & A_2(x) & \cdots \\ \cdot & \cdot & \cdot & \cdot & \cdot & \end{vmatrix},$$

where $C_0(x)$ and the $A_k(x)$, $B_k(x)$, $k \geq 0$, $x \geq 0$, are matrices of probability massfunctions satisfying the conditions for $Q(x)$ to be a valid semi-Markov matrix. The coefficient matrices $A_k(x)$ are $m \times m$, and we shall only deal with cases where $A = \sum_{k=0}^{\infty} A_k(\infty)$ is an *irreducible* stochastic matrix. The matrix $B_0(x)$ is $m_1 \times m_1$ and $C_0(x)$ and the $B_k(x)$ for $k \geq 1$ are of dimensions $m \times m_1$ and $m_1 \times m$, respectively. The first row and the matrix $C_0(x)$ allow for the treatment of boundary behavior that can be entirely different from that at higher levels. We emphasize that boundary behavior does not necessarily refer to what happens when the queue is empty. For example, in a queue where, below a certain threshold K, there is a different service time distribution, all states needed to describe those shorter queues are included in the boundary level 0.

The general mathematical discussion is also easily translated to handle generators or cases where the massfunctions in $Q(\cdot)$ concentrate on the integers, as in the discrete-time queues. What really matters is the structure of the transition probability matrix, not these detailed features of the coefficient matrices.

In some algorithmic methods, specifically those based on aggregation-disaggregation, steady-state probabilities can be computed by exploiting only the natural partition of the transition probability matrix. These methods do not require spatial homogeneity away from the level 0, and they are also suitable for structured Markov chains with finitely many states. However, in addition to steady-state probabilities, the matrix-analytic methods also provide a thorough mathematical analysis of other distributions and descriptors of the models. Although inspired by a search for computable solutions, they are, in the first place, a unified mathematical treatment of rich, useful classes of probability models.

10.2 THE PHASE-TYPE DISTRIBUTIONS

Because of their versatility and ease of computation, PH-distributions have gained widespread use in probability modeling. New *closure properties* were established; see Bobbio and Trivedi [32], Neuts [120]. We often encounter different probability models requiring formally similar constructions of random variables. An example of such a construction is that of the *augmented service duration* in queues with service interruptions. If the original duration of the task has a PH-distribution and the random mechanism causing the interruptions has special properties, the augmented duration still has a PH-distribution with an explicit representation. As observed in Assaf and Levikson [24], under PH-assumptions, many derived waiting times can be viewed as absorption times in finite Markov chains. Mere preservation of the PH-property holds in great generality. However, for practical use, explicit and computationally manageable representations of PH-distributions are essential.

The characterization and the geometry of PH-distributions. How versatile are certain subclasses of PH-distributions? Which properties guarantee that a probability distribution $F(\cdot)$ on $[0,\infty)$ is of phase-type? What are the smallest numbers of phases or parameters needed to represent a given PH-distribution? Examination of these fundamentally important questions has led to nice results, some proved by highly imaginative methods.

In studying the versatility of mixtures of Erlang distributions, Dehon and Latouche [55] recognized the geometric nature of this problem. Using additional, innovation geometric considerations, O'Cinneide [125] established the following characterization of PH-distributions.

> A probability distribution $F(\cdot)$ on $[0,\infty)$ is PH if and only if: (a) It has a rational Laplace-Stieltjes transform. (b) Apart from a possible mass of 0, it has a continuous, *positive* density on $(0,\infty)$. (c) Its Laplace-Stieltjes transform has a unique pole of maximum real part.

A somewhat different characterization theorem (Thm. 1.2 in [125]) holds for discrete PH-distributions. The geometry of the set of PH-distributions explains why the representations of some distributions "at the boundary" of that set may require many phases. The number of phases needed to represent a PH-distribution also depends on its coefficient of variation. Aldous and Shepp [4] proved that the order of a PH-distribution is at least as large as the inverse of its coefficient of variation minus one. O'Cinneide [124, 126] obtained lower bounds on the order of a PH-distribution based on poles of its Laplace-Stieltjes transform. If the transform of a PH-distribution has only real poles, that distribution has a bidiagonal representation. Further geometric or algebraic aspects of PH-distributions are discussed in Maier [99] and O'Cinneide [124-127].

Geometric considerations are also important in reducing the number of parameters needed to represent PH-distributions with m phases. In an early paper on PH-distributions, Cumani showed that, starting from any representation with an *upper-triangular* matrix T, there is an equivalent representation with a matrix having nonzero elements on the diagonal and the superdiagonal. For a detailed discussion of triangular PH-distributions (TPH), see O'Cinneide [127]. Clearly, a TPH-distribution corresponds to the absorption time distribution in a Markov chain with a cyclefree graph. It is not yet known whether there is a special subclass of representation $(\boldsymbol{\alpha}, T)$ with m phases by which *all* PH-distributions having m phases can be represented. The class PH_m° with representations $(\boldsymbol{\alpha}, T)$, where T is a matrix of the form

$$T = \begin{vmatrix} -\lambda_1 & \lambda_1 & 0 & \cdots & 0 & 0 & 0 \\ 0 & -\lambda_2 & \lambda_2 & \cdots & 0 & 0 & 0 \\ \cdot & \cdot & \cdot & \cdots & \cdot & \cdot & \cdot \\ \cdot & \cdot & \cdot & \cdots & \cdot & \cdot & \cdot \\ 0 & 0 & 0 & \cdots & -\lambda_{m-2} & \lambda_{m-2} & 0 \\ 0 & 0 & 0 & \cdots & 0 & -\lambda_{m-1} & \lambda_{m-1} \\ p_1\lambda_m & p_2\lambda_m & p_3\lambda_m & \cdots & p_{m-2}\lambda_m & p_{m-1}\lambda_m & -\lambda_m \end{vmatrix},$$

and $\boldsymbol{\alpha}$ is a general probability vector, is perceived as a likely candidate for spanning all PH-distributions with m phases.

Such theoretically important questions also have practical significance. In fitting PH-distributions to data or in approximating other useful probability distributions by those of phase-type, we use nonlinear optimization algorithms. These can take advantage of the special structure of the matrices for the class PH_m°. That class is certainly versatile, even should it turn out not to be exhaustive. Any major computational gains from exploiting its special structure are worthwhile.

Laplace-Stieltjes transforms of PH-distributions are rational functions. It is interesting to ask when a given rational function is the transform of a PH-distribution and, if so, how a matrix representation can be computed from it. For special families of rational functions, these issues can be resolved. For that problem and related material, see Chemla [48] and Commault and Chemla [52].

Fitting of and approximation by *PH*-distributions. Approximations by mixtures of Erlang distributions were examined in doctoral dissertations by Mary A. Johnson and Leonhard Schmickler. Results of combining fitting of moments and some other features, such as percentiles, by techniques of nonlinear optimization are reported in Johnson and Taaffe [72-74], Johnson [75, 76], and Schmickler [147]. An efficient implementation of the EM algorithm for fitting

PH-distributions is discussed in Asmussen, Nerman and Olsson [22]. That paper also gives extensive set of references. Software packages and computational experience with them are described in Asmussen, Häggström and Nerman [20], Bobbio and Telek [33], Lang [80], and Schmickler [147]. The phases in the representation of a *PH*-distribution do not, in general, have a physical significance. However, Faddy [59-61] reports on his experience with fitting specially structured *PH*-distributions to medical, pharmacological or sociological data sets. In several instances, practically significant meanings could be ascribed to the phases of the fitted models.

10.3 THE MARKOVIAN ARRIVAL PROCESS

The *MAP* is now recognized as a natural and useful matrix generalization of the Poisson process. Given the phases of the Markov chain with generator D at the endpoints of non-overlapping intervals, the numbers of events in those intervals remain *conditionally independent.* That property accounts for the preservation of features such as the embedded *MRP*s of a variety of queueing models. There is a natural formalism involving matrix generating functions, matrix exponential functions, and Kronecker products and sums by which many calculations, classical for Poisson models, can be extended to their versatile generalizations with *MAP*s. In addition to those in Chapter 5 of Neuts [112], many examples of *MAP*s and of constructions on point processes that lead to *MAP*s are found in Neuts [119].

Before elaborating on all the good properties of *MAP*s, we note that these do not include a useful generalization of the *conditional uniformity property* of the Poisson process. That is the property that, conditional on n events in $(0, t)$, the times of these events are distributed as the order statistics of n independent uniform random variables on that interval.

At the expense of larger matrices, it remains possible to work with *superpositions* of independent *MAP*s and with certain types of random *thinning* operations. However, nothing of the simplicity in the analysis of uniform order statistics and spacings, due to the conditional uniformity property, remains valid for the spacings induced by the events of *MAP*s. To give a specific example, most known formulas for the M/G/1 queue have tractable matrix analogues for MAP/G/1. In contrast, the MAP/G/∞ queue is analytically intractable. The simple analysis of M/G/∞ rests on the conditional uniformity property of Poisson arrivals.

Asmussen and Koole [21] show that, to a given general marked point process N, there is a sequence $\{N_k\}$ of *MAP*s with the property that $N_k \rightarrow N$ in distribution as $k \rightarrow \infty$. That property, though much deeper, is the analogue of the denseness property of *PH*-distributions. Practically, it implies that very general point processes can, in principle, be approximated by appropriate *MAP*s.

The utility of the *MAP* is most evident in queueing theory where they serve as a versatile, yet tractable family of models for "bursty" input processes and for some types of service processes. However, they also remain sufficiently tractable to serve as benchmarks in exploring physical properties and descriptors of general point processes on the real line. For the *MAP*, the standard moment descriptors of point processes are given by explicit matrix formulas, (Narayana and Neuts [105]), and the analysis of some nonconventional constructions remains tractable. For investigations of such constructions, we refer to the treatment of *local Poissonification* in Neuts, Liu and Narayana [119] and to the exploration of the intuitive quality of *burstiness* in Neuts [121]. A novel area of applications for *PH*-distributions and *MAP*s is *risk theory.* For problems in risk theory treated by matrix-analytic methods, we refer to Asmussen and Rolski [17], Asmussen and Bladt [23], and Stanford and Stroiński [151].

Adapting a *MAP* to point process data is much more difficult than fitting a *PH*-distribution to scalar data. Work to date has concentrated on the *MMPP*, where the rate parameters λ_j have a clear significance and the modulating process can be treated as a "hidden Markov chain". What has been accomplished so far is well-treated in Rydén [141, 142].

For use in simulations, it is easy to generate realizations of *MAP*s. The visualization of simulated *MAP*s, and their effect as input streams to queues, is an excellent way for practitioners to appreciate the versatility of the *MAP*. As for *PH*-distributions, it will take time before these point processes are familiar to many. Concrete practical uses should then raise new theoretical questions.

10.4 MATRIX-GEOMETRIC SOLUTIONS

The principal result on *MRP*s of GI/M/1-type is that the invariant probability vector $\mathbf{x}$ of their (irreducible) embedded Markov chain is *matrix-geometric.* If $\mathbf{x}$ is partitioned into vectors $\mathbf{x}_i$, $i \geq 0$ according to the levels i, then

$$\mathbf{x}_i = \mathbf{x}_0 R^i, \quad \text{for } i \geq 0, \tag{10.1}$$

where the matrix R is the minimal nonnegative solution to the equation

$$R = \sum_{k=0}^{\infty} R^k A_k. \tag{10.2}$$

The Markov chain is positive recurrent if, and only if, all eigenvalues of R lie *inside* the unit disk. If the matrix A is irreducible, the necessary and sufficient condition for that to hold is that $\boldsymbol{\pi}\boldsymbol{\beta} > 1$, where $\boldsymbol{\pi}$ is the invariant probability vector of A and $\boldsymbol{\beta} = \sum_{k=1}^{\infty} k A_k \mathbf{e}$. If the spectral radius η of R is less than one, the matrix

$$B[R] = \sum_{k=0}^{\infty} R^k B_k,$$

is irreducible and stochastic. The vector $\mathbf{x}_0$ is proportional to the invariant probability vector of $B[R]$. It is fully determined by solving the equations

$$\mathbf{x}_0 = \mathbf{x}_0 B[R], \quad \text{and} \quad \mathbf{x}_0 (I - R)^{-1} \mathbf{e} = 1. \tag{10.3}$$

The matrix R has a nice probabilistic significance, discussed in detail in Chapter 1 of Neuts [109]. The adaption of that interpretation and of the equations (10.2) and (10.3) to generators of GI/M/1-type is also discussed there. Mathematically elementary, but very useful generalizations to more involved boundary behavior are also given there. Application to models, such as the GI/PH/1 queue and to the rich class of queueing models leading to *QBD*-processes, is discussed in Neuts [109]. Since 1981, the matrix-geometric theorem has found many more applications and has inspired mathematical research which we now review.

The operator-geometric theorem. The portion of the state space of Markov chains of GI/M/1-type consisting of the levels i with $i \geq 1$ is simply a Cartesian product of countably many copies of the set $\{1, \ldots, m\}$. On the set, the Markov chain has a spatially homogeneous behavior that is skipfree to the right. Tweedie

[154] considered a class of discrete-time Markov process whose state space is the Cartesian product of countably many copies of an (essentially) arbitrary set E. For the Markov process with a similar spatially homogeneous, right skipfree transition probability operator, if an invariant probability measure exists, it has an *operator-geometric form.* The operator R has a probabilistic interpretation extending that of the matrix case. It satisfies a nonlinear operator equation. Tweedie's theorem, which is of considerable theoretical interest, shows how strongly that result is related to the structure of the models of GI/M/1-type. It implies, for instance, that several two-dimensional models of notorious analytic difficulty, such as the shortest queue problem, have matrix-geometric stationary distributions for which, however, R is an infinite matrix.

The caudal characteristic curve. For queueing models with a matrix-geometric invariant vector, the Perron-Frobenius eigenvalue η of the matrix R can be a useful descriptor. A graph of η as a function of some parameter θ of the queueing model is called a *caudal characteristic curve.* The notion was introduced in Neuts [111], where, also, several examples with useful interpretations are discussed. For convenience, limiting this discussion to the case where R is irreducible, we note that

$$R^i = \eta^i \mathbf{vu} + o(\eta^i), \qquad \text{as } i \to \infty.$$

$\mathbf{u}$ and $\mathbf{v}$ are the left and right eigenvectors of R corresponding to η and normalized by $\mathbf{ue} = \mathbf{uv} = 1$. That implies that as $i \to \infty$,

$$\mathbf{x}^i = \eta^i(\mathbf{x}_0\mathbf{v})\mathbf{u} + o(\eta^i) = K\eta^i\mathbf{u} + o(\eta^i), \tag{10.4}$$

where $K = \mathbf{x}_0\mathbf{v}$, and, also, that

$$\frac{\mathbf{x}^i}{\mathbf{x}^i\mathbf{e}} = \mathbf{u} + o(1). \tag{10.5}$$

Formulas (10.4) and (10.5) yield probabilistic interpretations for η and $\mathbf{u}$. For large values of i, the eigenvalue η approximates the ratio of the fractions of time spent in levels $\mathbf{i}+1$ and $\mathbf{i}$. For large i, the component u_j of the positive probability vector $\mathbf{u}$ approximates the conditional probability of being in the state (i, j), given that the embedded Markov chain is in the level $\mathbf{i}$.

Caudal characteristic curves, therefore, provide an indication of the persistence of long queues. In addition to the examples in Neuts [111], they have been used in Latouche [81] and Neuts and Rao [114] for the exploration of queueing models. Under special assumptions, η can be computed without knowing the matrix R. For such cases, we refer to Neuts and Takahashi [110] and Neuts [111]. For example, it requires only a modest computation to evaluate caudal characteristic curves for the GI/PH/c queue with heterogeneous servers, a model for which the order of the matrix R is typically so large as to render computation of its steady-state distributions infeasible. For that and for several related models, the computation of η, also, yields the decay parameter ξ of the exponential tail of various steady-state waiting time distributions. The continuous analogue of this asymptotic result is the basis for the "equivalent bandwidth" approach developed for fluid flow models by Elwalid and Mitra [58]; Whitt [155] is a good survey of these developments.

In some applications, such as those treated in Chapter 5 of Neuts [112], the GI/M/1-type model describes two interacting queues, one with a buffer of limited capacity. The second state index j often describes the number of jobs in that finite buffer. $\mathbf{u}$ can then be interpreted as the probability density of the contents of the second buffer when the queue in the first unit is heavily loaded.

Busy periods. The busy period distribution of the GI/PH/1 queue long defied analysis. Generalizing the probabilistic interpretation of R, Ramaswami [129] introduced a matrix $R(x)$ of probability massfunctions to which the busy period distribution is simply related. The important significance of that matrix is further clarified in Ramaswami [137, 138]. For a *MRP* of GI/M/1-type, starting in the state (i, j) of level **i**, $i \geq 0$, $\hat{R}_{jj'}(n, \nu; dx)$ is the elementary probability that the process visits the state $(i+\nu,\ j')$ at the nth step which occurs in $(x, x+dx)$ and avoids the level **i** at all intermediate steps. Writing

$$R^{(\nu)}(z,s) = \sum_{n=1}^{\infty} z^n \int_0^{\infty} e^{-sx} d\hat{R}(n, \nu; x),$$

he shows that $R^{(\nu)}(z,s) = [R^{(1)}(z,s)]^{\nu}$, and that $R = R^{(1)}(1,0)$. The matrix $R(x)$ arising in the treatment of the busy period has the transform $R^{(1)}(1,s)$. Omitting the superscript 1, it is shown that $R(z,s)$ is the minimal solution to the equation

$$R(z,s) = z\sum_{k=0}^{\infty} R^k(z,s)\tilde{A}_k(s), \tag{10.6}$$

which is clearly entirely analogous to (10.7) for the matrix $\tilde{G}(z,s)$. The approach is general, and also serves to obtain equations for the busy periods distributions of other related models. Function analytic and formal solution properties of (10.6) and (10.7) are reviewed in Ramaswami [134], a basic reference for the main mathematical properties of these equations.

Waiting time distributions. By examining a Markov process on the state space $\{1,\ldots,m\} \times [0,\infty)$, Sengupta [148] derived a *matrix-exponential* formula for the steady-state waiting time distributions for a family of models that includes the GI/PH/1 queue. His approach was essential to a result generalizing Takàcs' equation and which is described in Section 10.5. In Ramaswami [138], the relationship between that result and the matrix-geometric theorem is further elucidated. In Sengupta [150], matrix-exponential representations are obtained for waiting times in queues with semi-Markovian arrivals, and it is observed that, there, marginal distributions are also of phase-type.

Matrix-geometric solutions and *PH*-distributions. By a simple, but insightful argument, Sengupta [149] proved the unexpected result that the sequence $\{\mathbf{x}_0 R^i \mathbf{e}\}$, the "marginal" of a matrix-geometric solution is a discrete PH-density, and he constructed its representation. In Sengupta [150] and in Asmussen [18], it was noted that several similar results hold for steady-state waiting time distributions of a variety of queues. While proofs of these theorems are transparent, the results themselves are unexpected. Several waiting times can be viewed as absorption times in *infinite* Markov chains with appropriate initial conditions. What was unexpected, and is therefore striking, is that their distributions are also the probability laws of absorption times in appropriate *finite-state* chains.

Algorithmic methods. As is typical for mathematical results, the matrix-geometric theorem elucidates the nature of the solution to a problem. It clarifies what *needs* to be computed, but it does not provide universally applicable methods to do that feasibly. Major progress has been made in the numerical analysis aspects of matrix-analytic methods, but given the diversity and size of the applications, the quest for more efficient and faster algorithms is bound to continue indefinitely.

The extensive review by Latouche [83] compares all the classical iterative methods for computing R and treats the implementation of parallel algorithms for which matrix methods are ideally suited. That article is an essential reference to those interested in numerical implementations.

Successful implementations of *spectral methods* (the method of Rouché roots) have been reported by several authors, e.g., in Daigle and Lucantoni [53]. Since the Rouché roots are precisely the eigenvalues of R, the method should work well in cases where R can be accurately computed via its spectral representation. For examples with multiple Rouché roots, see Dukhovny [57].

With additional mathematical work, it is sometimes possible to develop algorithms well-suited to the solution of special models. That is the case, for instance, for the GI/PH/1 queue. At first glance, it appears that extensive computation and much storage is required just to find the coefficient matrices A_k. Already in 1984, Lucantoni and Ramaswami [91, 131, 132] developed a clever, special algorithm to compute R and many other features of that and related queues in which the computation of the matrices A_k is avoided entirely. Results of extensive test of that algorithm, also for some multiserver queues, are described in Ramaswami and Latouche [136]. In the same spirit, Latouche and Ramaswami [85] constructed a new algorithm for the steady-state distributions of PH/PH/1. Their algorithm involves computations with matrices of order $m+n$, where m and n are the numbers of phases in the representations of the interarrival and service time distributions. In other algorithms, the matrix R is of dimension mn or else, much setup computation is required.

Quasi-birth-and-death processes. When the transition probability matrix $Q(\cdot)$ is block tridiagonal, the model is both of M/G/1 and GI/M/1 type. Such cases can be transparently analyzed in great detail. Most familiar are the *QBD*-processes, the continuous parameter Markov chains with a generator of that type, which have found many applications. Models leading to specific *QBD*-models are discussed in Daigle and Lucantoni [53], Nelson [107], Nelson and Iyer [106], and Nelson and Squillante [108], but many more have been described in the literature on specific applications.

The special structure of *QBD*s was cleverly exploited by Latouche and Ramaswami [84] to yield a highly efficient algorithm to solve the matrix-quadratic equations

$$R^2A_2 + RA_1 + A_0 = 0, \quad \text{and} \quad A_0G^2 + A_1G + A_2 = 0$$

for R and G. Their algorithm is quadratically convergent. It very substantially reduces the number of iterations, even in cases for which exceedingly slow convergence of standard iterative methods had been reported.

The idea behind the algorithm is elegant and simple. The relationship between the R- and G-matrices of blocks consisting of 2^k and 2^{k+1} levels is analyzed and provides the essential step of their recursive scheme. The actual algorithm does not involve the high order matrices corresponding to those macroblocks, but works with matrices of the order m of the original coefficient matrices. In [86], the same authors survey algorithms for first passage times in *QBD*s and propose a new, efficient algorithm of their own. A phase substitution algorithm is treated in Kao and Lin [77]. The transient behavior of *QBD* processes is discussed in Zhang and Coyle [160].

In *QBD* processes, transitions to higher or lower levels often correspond to arrivals and departures, respectively, in queues. For the simplest case of a stationary M/M/1 queue, the joint transform of the distribution of the numbers

$N_A(t)$ and $N_D(t)$ of arrivals and services completions during an interval of length t has long been known. In Neuts [116], the corresponding joint matrix transform of the numbers of transitions up and down is obtained for an arbitrary stationary *QBD*. That derivation thoroughly exploits the GI/M/1 and M/G/1 structures of *QBD*s. As noted there, challenging algorithmic problems remain in the numerical study of that joint distribution. When it becomes practically useful, the inversion algorithms mentioned in Section 10.5 should make its computation feasible.

Perturbation analysis. As for other models in applied mathematics, one would like to know how descriptors of a queue depend on one or more of its parameters. In exploring how the descriptors change as a parameter θ ranges from, say, θ_0 to θ_1, we need to compute the solutions for successive θ-values as efficiently as possible. A classical approach to such problems is by perturbation analysis.

The importance of these problems was first noted by Latouche [82] who carried out a perturbation analysis for a phase-type queue with small correlations between arrivals. He observed that various technical issues, such as the differentiability of queue descriptors with respect to parameters, are rather difficult. An M/M/1 queue in a Markovian environment (see Chapter 6 of [109]), is examined under perturbations in Chang and Nelson [47]. He [70] shows that, under general conditions, the matrices R or G may be repeatedly differentiated with respect to a parameter, but the expressions for higher derivations are exceedingly involved. Conditions for R or G to be analytic in θ have not yet been established. Perturbation analysis of the matrix-analytic solutions is a challenging subject. Tracking algorithms and related optimal design problems are promising subjects for further research.

State-dependent *QBD*s. Several researchers had informally noted that for *QBD*s with level-dependent matrices $A_0(i)$, $A_1(i)$, and $A_2(i)$, the steady-state probabilities, if they exist, must satisfy $\mathbf{x}_{i+1} = \mathbf{x}_i R(i)$, where $R(i)$ is a matrix with a probabilistic interpretation similar to that of R in the homogeneous case. The matrices $R(i)$ can be recursively computed. With an infinite sequence of equations involved and without an explicit stability condition, the usefulness of these results seemed to be limited. However, that property became the focus of much attention since Ramaswami and Taylor [139, 140] found that it sheds light on a link between matrix-geometric and product-form solutions. Computational procedures based on an extension of the algorithm of Latouche and Ramaswami [84] to *QBD*s with level dependence are discussed in Bright and Taylor [42].

***QBD*s with finite state spaces.** A great variety of finite-capacity queues are related to finite Markov chains with block tridiagonal generators. Because of their interest to applications, such *QBD*s with finite state spaces, also with level dependent coefficient matrices, now have an extensive literature. In some cases, formulas of a matrix-geometric form can be found for the steady-state probabilities; see Albores and Bocharov [3], Bocharov [36], Bocharov and Naumov [34], Gün and Makowski [67], Gün [68], and Hajek [69]. The *folding algorithm* of Li [87] and Ye and Li [156-158] exploits the structure of finite *QBD*s to reduce computation of the invariant probability vector. It reduces the time and space complexity of solving the steady-state equations from $O(N^3K)$ and $O(N^2K)$ to $O(N^3\log_2 K)$ and $O(N^2 \log_2 K)$, respectively. Successful, efficient implementations for very large problems are reported.

10.5 THE M/G/1 STRUCTURE

Most results for the GI/M/1 structure stem from elaborations of the central matrix-geometric theorem. In comparison, the M/G/1 structure is mathematically more challenging and its applications also appear to be more diverse. The material in Neuts [112] has been greatly enriched by developments since 1989.

Fundamental period. The *fundamental period*, discussed in Chapters 2 and 3 of Neuts [112], is crucial to the theory. We recall that the joint distribution of the number of downward transitions (service completions) during and the time length of the fundamental period satisfies a nonlinear matrix integro-difference equation whose transform version is

$$\tilde{G}(z,s) = z\sum_{k=0}^{\infty} \tilde{A}_k(s)\tilde{G}^k(z,s), \tag{10.7}$$

where

$$\tilde{A}_k(s) = \int_0^{\infty} e^{-sx} dA_k(x), \quad \text{for } k \geq 0.$$

For positive recurrence of the embedded *MRP* it is necessary that the minimal nonnegative solution G of the equation

$$G = \sum_{k=0}^{\infty} A_k G^k, \tag{10.8}$$

be stochastic. If the matrix A is irreducible, the necessary and sufficient condition for G to be stochastic is that $\rho = \boldsymbol{\pi\beta} \leq 1$. $\boldsymbol{\pi}$ is again the invariant probability vector of A and $\boldsymbol{\beta} = \sum_{k=1}^{\infty} kA_k\mathbf{e}$. For queues, ρ is the traffic intensity. With additional, straightforward moment assumptions on the boundary matrices $\{B_k(\cdot)\}$, the *MRP* is positive recurrent if and only if $\rho < 1$.

If the invariant probability vector $\mathbf{x}$ of the embedded Markov chain is partitioned into vectors $\mathbf{x}_i$, $i \geq 0$, according to the levels i, then the vector generating function $\mathbf{X}(z) = \sum_{i=1}^{\infty} \mathbf{x}_i z^i$, satisfies

$$\mathbf{X}(z)[zI - A^*(z)] = z\mathbf{x}_0 B(z) - z\mathbf{x}_1 A_0, \tag{10.9}$$

for all z such that $\|z\| \leq 1$. $A^*(z)$ and $B(z)$ are respectively the matrices $\sum_{k=0}^{\infty} A_k z^k$ and $\sum_{k=1}^{\infty} B_k z^k$. The derivation of moments from formula (10.9) is discussed in Chapter 3 of Neuts [112].

By probabilistic arguments, it is shown in Neuts [112] how the vectors $\mathbf{x}_0$ and $\mathbf{x}_i$ can be found, once G has been computed. A stable recursive procedure for computing the vectors $\mathbf{x}_i$, $i \geq 1$, is developed in Ramaswami [135]. Useful simplifications in the solution of the boundary equations are noted in Schellhaas [145].

Before proceeding to the review of new results, we remark on the significance of the reducibility of A (and therefore of all the coefficient matrices A_k).

Reducibility of the matrix A. Cases where A is reducible frequently occur in models in communications engineering. They require the more belabored treatment in Chapter 3 of [112]. Reducibility is linked to the physical behavior of these models. The semi-infinite strip of the states not in level **0** is now a union of disjoint strips. On those strips corresponding to the irreducible subclasses of A, the Markov chain behaves as when A is irreducible. In addition, there may also be a (transient) strip of states from which the chain can move to an "irreducible" strip, but to which it can return only by passing through the level **0**. In well-formulated models, the overall embedded *MRP* is irreducible as the various strips are linked through the boundary level. The analysis of such models essentially pieces together the strips corresponding to the subclasses of A, with the transient strip requiring special treatment.

Although the general approach is clear, it is advisable to treat cases of reducibility with care. Appropriate matrices R or G must be computed for each irreducible strip, and the transient strip (if present) requires special calculations. On the positive side, the orders of the matrices for the various strips is smaller than for a comparable model in which A is irreducible. Examples, among several extant, are found in the discussion of a dynamic priority queue in Fratini [63] and of a flow control model in Liu and Neuts [88].

The analysis of specific classes of queues. One of the benefits of the matrix-analytic methods is that most known results for the M/G/1 queue can be extended in a systematic and analogous manner to the MAP/G/1 and even to the MAP/SM/1 queue. In [94], Lucantoni gives a lucid exposition of that analogy for the MAP/G/1 queue. He shows how, by relating queue length and waiting time distributions to the embedded MRP, natural analogues of nearly all known explicit results for the M/G/1 queue are generalized to MAP/G/1. That analogy was already established in the dissertation of V. Ramaswami, but the newer notation for the MAP serves to make it even more apparent. Equations for the time-dependent behavior of MAP/G/1 are derived in a similar manner in Lucantoni, Choudhury, and Whitt [95]. By using efficient methods for transform inversion algorithms in conjunction with matrix analysis, many practical uses of transient analysis are now well within reach.

For the MAP/SM/1 queue, treated in Lucantoni and Neuts [97], substantially the same analysis can be carried out. The coefficient matrices $A_k(x)$ are given by

$$A_k(x) = \int_0^x P(k;u) \otimes dH(u), \quad \text{for } x \geq 0,\ k \geq 0.$$

The states in level **i**, for $i \geq 0$, are now triples $(i;j,h)$, where j is the phase of the MAP arrival process and h is the type of service initiated after the current service completion. If D is $m \times m$ and the semi-Markov matrix $H(\cdot)$ describing the service process is $n \times n$, each level consists of mn states (written in lexicographic order).

An area of current interest is the extension of some results on priority queues to MAP input. An elegant treatment of such a priority model is found in Takine and Hasegawa [152]. Dynamic priority queues leading to models of M/G/1-type are analyzed in Ramaswami and Lucantoni [130] and Fratini [63].

Two forms of the equation for $\widetilde{G}(z,s)$. While studying the vacation model in [92], we encountered some difficulty in proving, from the form (10.7) of the equation for G, that two matrices are commutative. That step was crucial in verifying that our new results were compatible with Doshi's factorization for the waiting time distribution in GI/G/1 with vacations. Following the approach in Sengupta [148], Neuts [113] derived an alternative integral equation for $\widetilde{G}(z,s)$. The needed commutativity then immediately followed from it. The other equation for $\widetilde{G}(z,s)$, valid for MAP/G/1 (with single arrivals), is

$$\widetilde{G}(z,s) = z \int_0^\infty \exp\{-[sI - D_0 - D_1\widetilde{G}(z,s)]u\} dH(u). \tag{10.10}$$

For the M/G/1 queue, the equations (10.7) and (10.10) are identical. For *general* models of M/G/1, only the form (10.7) can be expected. The integral equation (10.10) is therefore a consequence of *MAP* input. Although for some time it was not clear, the relationship between the two equations was elucidated by a combinatorial argument in Lucantoni and Neuts [96]. Generalized to semi-Markovian services and group arrivals, the corresponding equation for the MAP/SM/1 queue reads

$$\widetilde{G}(z,s) = z\int_0^\infty [I \otimes dH(t)]\exp\{-[sI \otimes I - D_0 \otimes I - \widetilde{G}^*(z,s)]t\}.$$

where $\widetilde{G}^*(z,s) = \sum_{k=1}^{\infty}(D_k \otimes I[\widetilde{G}(z,s)]^k$, in terms of standard notation for that queue.

Factorization theorems. Several steady-state distributions for GI/G/1 queues with vacations and for variants of the M/G/1 queue can be written as convolution products of other distributions. For the principal references, see Neuts [119, p. 59]. A matrix factorization of transform matrices related to a waiting time distribution of a MAP/G/1 queue was obtained in Lucantoni, Meier-Hellstern and Neuts [92]. In [56], Doshi generalized his earlier method of proof for GI/G/1 to show that such matrix factorizations hold in great generality. Of course, tractability of the factors depends on specific structure of each model. Further results for vacation models with *MAP* input are found in Matendo [101-104] and Schellhaas [146].

Duality of the structures of GI/M/1 and M/G/1-type. There is a classical duality between the elementary M/G/1 and GI/M/1 queues. Many of the shared properties of these queues are understood with reference to a random walk on the real line and its reversal. The relationship between the matrix models coefficient matrices $\{A_k(\cdot)\}$ is similarly related to a dual pair of skipfree Markov renewal random walks whose state space is the set of lattice points in the strip $\{(i,j): -\infty < i < \infty, 1 \le j \le m\}$. The consequences of that duality are examined in Asmussen and Ramaswami [16], and Ramaswami [137]. An essential insight is the proper definition of the coefficient matrices $\{A_k^\circ(\cdot)\}$ for the dual of the random walk with the matrices $\{A_k(\cdot)\}$. That construction, first given in Ramaswami [137], is as follows: The matrix $\Delta(\boldsymbol{\pi})$ is the diagonal matrix with the (positive) components of the invariant probability vector $\boldsymbol{\pi}$ as its diagonal elements. The matrices $\{A_k^\circ(\cdot)\}$ are defined by

$$A_k^\circ(x) = \Delta^{-1}(\pi)[A_k(x)]'\Delta(\pi), \quad \text{for } x \ge 0, k \ge 0,$$

where the prime denotes transpose. As is easily verified, this defines a proper sequence of coefficient matrices. The resulting random walks are also reversals of each other and the R and G matrices for both are simply related.

Further properties of the matrix equations for G and R. The nonlinear matrix equations (10.1) and (10.8) for R and G are interesting in themselves. For stable queues, there is one solution, the minimal nonnegative one, with a clear probabilistic significance. Its uniqueness is well-known. However, there are other solutions. For example, in a stable Markov chain of GI/M/1-type, there is at least one solution of (10.1) of spectral radius one. Similarly, for M/G/1-type models, when the minimal solution is substochastic, there always exists a second, stochastic solution. The eigenvalues of all solutions must satisfy the equation

$$\det[zI - A^*(z)] = 0,$$

so one can ask how its roots (not necessarily only those in the closed unit disk) arise in the spectra of the various solutions. What are the properties of the eigenspaces for the eigenvalues of the various solution matrices? A thorough examination of these questions, also leading to the many technically important results on the spectral methods based on Rouché roots, is found in Gail, Hantler and Taylor [64, 65].

Asymptotic behavior of the invariant probability vector. For Markov chains of GI/M/1-type, classical results on the behavior of high powers of nonnegative matrices easily lead to the asymptotic results in Section 10.4. Asymptotic results for the vectors $\mathbf{x}_i$ in M/G/1-type models are noticeably harder. The tail behavior of various steady-state waiting time distributions is closely related to that of the vectors $\mathbf{x}_i$. The articles by Falkenberg [62], Abate, Choudhury and Whitt [2], and Choudhury and Whitt [49] give the state of the art. In addition to algorithms for higher order moments, Choudhury and Lucantoni [51] propose a practical procedure to determine the exponential or geometric decay rates.

For the scalar case, the asymptotic behavior of the invariant probabilities $\{x_i\}$ depends on the nature of the first singularity of the generating function $A^*(z)$. The behavior of $A^*(z)$ at that singularity is translated to that of $\{x_i\}$ by using tauberian theorems. If the first singularity of $A^*(z)$ is a simple pole at some $\sigma > 1$, then the steady-state probabilities have a geometrically decaying tail with parameter $1/\sigma$.

For the matrix case, the asymptotic properties are linked to the Perron-Frobenius eigenvalued $\xi(z)$ of $A^*(z)$. If $\xi(z)$ is analytic in some disk of radius > 1, the first singularity of $A^*(z)$ occurs at the smallest $\sigma > 1$, for which $\sigma = \xi(\sigma)$. The authors of [62] and [2] primarily deal with the case where σ is a simple pole. As they note, that is not the only case, and more complex asymptotic behavior than geometric decay is possible. Related asymptotic methods are used in Baiocchi [25] and Baiocchi and Blefari-Melazzi [26] to obtain useful approximations to the loss probabilities of some finite-capacity queues. Light traffic approximations for multi-server queues with MMPP input are obtained in Blaszczyszyn, Frey and Schmidt [28].

Asymptotic properties of cycle maxima. For queues of M/G/1-type, the time between entrances into the level 0 corresponds to the *busy cycle*. The maximum virtual waiting time V during a busy cycle is a random variable of practical interest. Asmussen and Perry [19] is a thorough study of the (exponential) tail behavior of its distribution. Detailed results, illustrated by numerical examples, are obtained for the MAP/MMPH/1 queue. In that model, customers arrive singly according to a *MAP*. If, at the arrival, there is a phase transition $j \rightarrow j'$, the customer is of the type jj' and has a *PH*-service time distribution whose representations can depend on that type. The authors denote that service process by *MMPH*. Algorithms for the asymptotic parameters are given. As an elementary byproduct, an explicit formula for the distribution of V is obtained for the M/PH/1 queue.

Efficient transform inversion. In a series of papers, J. Abate, G.L. Choudhury, D.M. Lucantoni, and W. Whitt [1,50,51,95], have demonstrated the high degree of efficiency and the practical accuracy of several methods for the numerical inversion of probability generating functions and Laplace-Stieltjes transforms. They have examined specific applications to the time-dependent and the steady-state distributions of MAP/G/1 queues. With the solution to the nonlinear equations for the transform matrices at a rather small number of *complex* values of the transform variables z or s, strikingly accurate inverse transforms can be computed. By the numerical implementation of a steepest descent method, high order moments of various distributions can also be evaluated, see Choudhury and Lucantoni [51]. In some cases, it is possible to obtain estimates of the constants in asymptotic formulas from these.

Quasi-stationary distributions. Informally, these describe the state probabilities of a Markov chain conditioned on the event that the chain has spent a long time in a subset C of its state space. For finite sets C and under some irreducibility assumptions, the quasi-stationary density is the left eigenvector corresponding to the Perron-Frobenius eigenvalue of the principal submatrix defined by the state indices in C. For the structured matrices of interest here, quasi-stationary densities, corresponding to particular infinite sets C, can be characterized. Of particular interest is the set of all levels **i** with $i \geq K > 0$. The corresponding quasi-stationary density then describes the queue length, given that a long excursion above the level **K** is in course. The related results are found in Kijima [79], Makimoto [100], and Kijima and Makimoto [78].

Finite-capacity queues of M/G/1-type. Motivated by various applications, finite-capacity queues with the M/G/1 structure have attracted much attention. While the presence of a second boundary reduces the number of explicitly matrix formulas that can be derived, several efficient recurrence formulas for first passage time distributions and their moments can be derived. Discussions of models with capacity constraints are found in Blondia [29,30], Chakravarthy [43,44,46], and Neuts and Rao [117].

Discrete queues. Discrete queues with either an M/G/1 or a GI/M/1 structure are discussed in Alfa [8,9], Alfa and Chakravarthy [5,7], Alfa, Dolhun and Chakravarthy [6], Alfa and Frigui [10], Alfa and Neuts [11], Daigle and Tang [54], Gail, Hantler, Konheim, and Taylor [66], and Liu and Neuts [89].

10.6 OPEN PROBLEMS AND NEW RESEARCH DIRECTIONS

In writing this review, I was struck by the high level of maturity attained by the matrix-analytic methodology since its early developments in the 1970s. This area has seen an unusually rapid integration of mathematical and applications related results. That accounts, I believe, for the wealth of new problems, algorithmic and software developments, and novel applications that have come about, especially during the past five years. The subject area is growing so vigorously that any current treatment, this review included, is likely to be somewhat outdated by the time of its publication. To inform researchers in this field of the latest findings, to bring persons with kindred interests in contact, and to avoid that good people waste time in duplicated efforts, I have, already for several years, coordinated the electronic Matrix-Analytic Bulletin. Four or five times per year, it goes by e-mail to all persons who express interest and keep me informed of their current whereabouts in Cyberspace.

In this concluding section, I shall outline some broad directions of current research and also list a few specific problems that are important to this subject. Their selection is mine, although I have obviously benefited from conversations with many other specialists. Some questions arose in applications; others address mathematical clarifications. There is little difference between these. In this area, the partition between theory and applications, if it exists at all, is highly permeable. To facilitate exposition, I have grouped related questions under common headings.

Semi-Markov matrices. The relation of the matrix methods to Markov renewal theory is evident. The matrix formalism we use is a calculus of semi-Markov (SM) matrices. A sequence of $m \times m$ nonnegative matrices $\{A_k\}$, whose sum A is stochastic, is a discrete SM-matrix. The coefficient matrices $\{A_k(\cdot)\}$ define a bivariate SM-matrix, with one integral and (usually) a continuous second variable. Their sum $A(\cdot)$ is the usual SM-matrix of Markov renewal theory. A summary of properties of SM-matrices and of the Perron-Frobenius eigenvalue $\delta(\cdot)$ of the Laplace-Stieltjes transform is given in the Appendix to Neuts [112]. The function $\delta(\cdot)$ is important in discussing moments and plays an essential role in asymptotic analysis.

A systematic treatment of semi-Markov matrices, which are natural analogues of scalar probability distributions, would be welcome. Further properties of the function $\delta(\cdot)$, such as any probabilistic significance it may have, should be useful. Specifically, a deepened understanding of the relationship between the nature of the first singularities of the elements of the semi-Markov matrix and that of $\delta(\cdot)$ would add to the current asymptotic results.

It appears that the function $\delta(\cdot)$ will also play a crucial role in the examination of geometric and exponential convergence of MRPs and of the relaxation time of models of the GI/M/1 and M/G/1 types. To date, such problems have not been examined in general and they appear to be technically difficult.

Structure and efficient algorithms. Most progress on algorithms, to date, has come from clever exploitation of special structure, such as that of QBD processes. In recent applications of which I am aware, the matrices A_k, while of high order, have further special structure. In one instance, they have the simple form

$$A_k = \begin{vmatrix} 0 & A_k(1) & 0 & 0 & 0 & \cdots & 0 & 0 \\ 0 & 0 & A_k(2) & 0 & 0 & \cdots & 0 & 0 \\ 0 & 0 & 0 & A_k(3) & 0 & \cdots & 0 & 0 \\ 0 & 0 & 0 & 0 & A_k(4) & \cdots & 0 & 0 \\ & \cdots & & & \cdots & & & \\ 0 & 0 & 0 & 0 & 0 & \cdots & 0 & A_k(K-1) \\ A_k(K) & 0 & 0 & 0 & 0 & \cdots & 0 & 0 \end{vmatrix},$$

but the subblocks are 40×40 and the order m of the A_k is larger than 2,000. The special features of the coefficient matrices indicate that the random walk on the semi-infinite strip is highly structured. How can such detailed structure be further exploited? A brute force implementation of the standard iterative methods is clearly impractical, but it may be possible to combine the insights gained from the matrix-analytic methods with others, say, aggregation-disaggregation, to obtain accurate algorithms with manageable execution times and storage requirements.

The matrix methods are an excellent testing ground for work on parallel algorithms and for combinatorial algorithms to elucidate further structure of the coefficient matrices. In the past years, they have drawn the attention of persons working on approaches to massive computation. Significant progress in these directions is, therefore, likely. An algorithm to recognize the GI/M/1 and M/G/1 structures, automatically, from model specifications was recently proposed by Berson and Muntz [27]. A kindred combinatorial question is to infer the structure of the matrix G from the incidence matrices corresponding to the A_k. An outline of that problem is found in Neuts [112, pp. 177-179].

More specialized questions are related to combinatorial properties of random walks on the semi-infinite strip. There are well-known connections between the fundamental period of M/G/1 and counts of certain restricted lattice paths. As noted in [112, pp. 170-172], there is a promising relationship between an asymmetric random walk on the integers and matrices of M/G/1 type. Matrix recursions offer a natural formalism for the exploration of the related lattice path counts, and there may be interesting matrix generalizations, for example, of Catalàn numbers.

Starting solutions and bounds. In showing that the minimal nonnegative solution to the nonlinear matrix equations (10.1) and (10.8) for R and G is the solution of greatest interest, we use successive substitutions starting with the zero matrix. For most numerical procedures, that is clearly not the best starting solution. For the stable M/G/1-type models, it is advisable to start with a stochastic matrix $G(0)$. In computing R, there are clear advantages in starting with $R(0) = \eta \mathbf{e}\mathbf{u}$, whenever the Perron-Frobenius eigenvalue η and the corresponding eigenvector $\mathbf{u}$ can be easily computed. Succinct arguments, showing that these are valid starting solutions, are reviewed in Latouche [83], and further generalizations are found in Asmussen [18].

Interesting mathematical problems deal with the characterization of all nonnegative matrices that can be valid starting solutions, i.e., for which successive substitutions or other iterative procedures are guaranteed to converge to the minimal solution of the matrix equations. For the matrix-geometric case, a nontrivial upper bound for R would be useful, that is, a nonnegative matrix $R^\circ \geq R$, with spectral radius < 1, whose computation is easier than that of R. The construction of such a matrix R° is likely to involve probabilistic insight based on the significance of R.

Matrix paradigms for tree structures. Yeung and Sengupta [159] and Takine, Sengupta, and Yeung [153] generalized the theory of matrix analytic methods by considering a discrete-time bivariate Markov chain $\{(X_k, N_k), k \geq 0\}$, in which the values of the X_k are the nodes of a d-ary tree, and N_k takes integer values between 1 and m. They refer to X_k as the *node* and N_k as the auxiliary variable of the Markov chain at time k of which they consider two versions. In the first, the transitions of X_k satisfy a generalized skipfree to the right property (that is the GI/M/1 paradigm) and, in the second, a skipfree to the left property (the M/G/1 paradigm). In both versions, the transitions satisfy a spatial homogeneity property; the transition probability from one node to another depends only on the spatial relationship between the two nodes and not on their specific values. When d is 1, the spatial relationship is just the difference between the values of the two nodes. Their results then reduce to the familiar theory of the GI/M/1 and M/G/1 paradigms developed by Neuts [109,112].

Optimization. Finding the optimal value of a parameter for a tractable queueing model, such as the buffer size K in M/G/1/K, is often a matter of elementary calculus and simple numerical calculations. So far, the simplicity of such problems is stressed only in a few current text books. I have therefore included many problems dealing with simpler optimizations in my forthcoming problem book, Neuts [122]. Applications papers often report optimal parameter values for the models at hand, but these appear to have been found simply by exhaustive computation.

A highly promising area for future research is the integration of optimization into the matrix-analytic methods. In most cases, the procedures will be algorithmic; analytic and structural properties can serve to make the algorithms efficient. For example, all known methods for fitting PH-distributions, see Section 10.2, already have such a flavor.

BIBLIOGRAPHY

[1] Abate, J. and Whitt, W., Numerical inversion of Laplace transforms of probability distributions, *ORSA J. Computing* (1994), (to appear).

[2] Abate, J, Choudhury, G.L. and Whitt, W., Asymptotics for steady-state tail probabilities in structured Markov queueing models, *Stoch. Mod.* **10** (1994), 99-143.

[3] Albores, F.X. and Bocharov, P.P., Two finite queues with relative priority in a single server system with phase-type distributions, *Avtomatika i Telemekhanika* 4 (1993), 96-107.

[4] Aldous, D. and Shepp, L, The least variable phase-type distribution is Erlang, *Stoch. Mod.* **3** (1987), 467-473.

[5] Alfa, A.S. and Chakravarthy, S., Discrete time analysis of two queues with Markovian arrivals and alternating services, Manuscript (1993).

[6] Alfa, A.S., Dolhun, K.L. and Chakravarthy, S., A discrete single-server queue with Markovian arrivals and phase type group services, Manuscript (1993).

[7] Alfa, A.S. and Chakravarthy, S., A discrete queue with the Markovian arrival process and phase-type primary and secondary services, *Stoch. Mod.* **10** (1994), 437-451.

[8] Alfa, A.S., A discrete single-server MAP/PH/1 queue with vacations and exhaustive time-limited service, Manuscript (1994).

[9] Alfa, A.S., Steady state analysis of discrete MAP/PH/1 priority queues, Manuscript (1994).

[10] Alfa, A.S. and Frigui, L., Discrete NT-policy single-server queue with Markovian arrival process and phase-type service, *Euro. J. Oper. Res.* (1995), (to appear).

[11] Alfa, A.S. and Neuts, M.F., Modelling vehicular traffic using the discrete time Markovian arrival process, *Transp. Sci.* (1995), (to appear).

[12] Akimaru, H., Okuda, T. and Nagai, K., A simplified performance evaluation for bursty multiclass traffic in ATM systems, *IEEE Trans. Commun.* **42** (1994), 2078-2083.

[13] Asmussen, S., Risk theory in a Markovian environment, *Scand. Act. J.* (1989), 69-100.

[14] Asmussen, S., Exponential families generated by phase-type distributions and other Markov lifetimes, *Scand. J. Stat.* **16** (1989), 319-334.

[15] Asmussen, S., Ladder heights and the Markov-modulated M/G/1 queue, *Stoch. Proc. Appl.* **37** (1989), 313-326.

[16] Asmussen, S. and Ramaswami, V., Probabilistic interpretation of some duality results for the matrix paradigms in queueing theory, *Stoch. Mod.* **6** (1990), 715-733.

[17] Asmussen, S. and Rolski, T., Computational methods in risk theory: a matrix-algorithmic approach, *Insurance: Math. and Econo.* **10** (1991), 257-274.

[18] Asmussen, S., Phase-type representations in random walk and queueing problems, *Ann. Prob.* **20** (1992), 772-789.

[19] Asmussen, S. and Perry, D., On cycle maxima, first passage problems and extreme value theory for queues, *Stoch. Mod.* **8** (1992), 421-458.

[20] Asmussen, S., Häggström, O., and Nerman, O.., EMPHT - A program for fitting phase-type distributions, In: *Studies in Statistical Quality Control and Reliability*, Mathematical Statistics, Chalmers University and University of Götenborg, Sweden 1992.

[21] Asmussen, S. and Koole, G., Marked point processes as limits of Markovian arrival streams, *J. Appl. Prob.* **30** (1993), 365-372.

[22] Asmussen, S., Nerman, O., and Olsson, M., *Fitting Phase-Type Distributions via the EM Algorithm*, preprint, Chalmers Univ. of Technology, Dept. of Math., Univ. of Göteborg, Sweden 1994.

[23] Asmussen, S. and Bladt, M., Phase-type distributions and risk processes with state-dependent premiums, Manuscript (1994).

[24] Assaf, D. and Levikson, B., Closure of phase-type distributions under operations arising in reliability theory, *Ann. Prob.* **10** (1982), 265-269.

[25] Baiocchi, A., Analysis of the loss probability of the MAP/G/1/K queue: Part I: Asymptotic theory, *Stoch. Mod.* **10** (1994), 867-893.

[26] Baiocchi, A. and Blefari-Melazzi, N., Analysis of the loss probability of the MAP/G/1/K queue: Part II: Approximations and numerical results, *Stoch. Mod.* **10** (1994), 895-925.

[27] Berson, S. and Muntz, R., Detecting block GI/M/1 and block M/G/1 matrices from model specifications, Manuscript (1994).

[28] Blaszczyszyn, B, Frey, A., and Schmidt, V., Light-traffic approximations for Markov-modulated multiserver queues, *Stoch. Mod.* **11** (1995), (to appear).

[29] Blondia, C., The N/G/1 finite capacity queue, *Stoch. Mod.* **5** (1989), 273-294.

[30] Blondia, C., Finite capacity vacation models with nonrenewal input, *J. Appl. Prob.* **28** (1991), 174-197.

[31] Blondia, C., A discrete-time batch Markovian arrival process as *B-ISDN* traffic model, *Belg. J. Oper. Res., Stat., Comput. Sci.* **32** (1992), 3-23.

[32] Bobbio, A. and Trivedi, K.S., Computation of the distribution of the completion time when the work requirement is a *PH* random variable, *Stoch. Mod.* **6** (1990), 133-150.

[33] Bobbio, A. and Telek, M., A benchmark for *PH* estimation algorithms: Results for acyclic-*PH*, *Stoch. Mod.* **10** (1994), 661-677.

[34] Bocharov, P.P. and Naumov, V., Matrix-geometric stationary distribution for the PH/PH/1/r queue, *J. Info. Proc. Cyber.* **22** (1986), 179-186.

[35] Bocharov, P.P. and Litvin, V.G., Ways to analyze and design service systems with phase distributions, *Avtomatika i Telemekhanika* **5** (1986), 5-23 (in Russian).

[36] Bocharov, P.P., Analysis of the queue length and the output flow in single server with finite waiting room and phase type distributions, *Probl. Control Info. Theory* **16** (1987), 211-222.

[37] Bocharov, P.P., Litvin, V.G., and Spesivov, S.S., Performance evaluation of a two-node network with phase-type distribution and two classes of calls,

Avtomatika i Telemekhanika **5** (1989), 169-178.

[38] Bocharov, P.P. and Pavlova, O.L., Matrix-geometric distributions of a queue under the LCFS discipline with interruptions and phase-type distribution, *Avtomatika i Telemekhanika* **9** (1991), 112-122.

[39] Bocharov, P.P. and Pavlova, O.I., Analysis of a queue characterized by a phase-type distribution and a LCFS preemptive resume discipline, *Avtomatika i Telemekhanika* **11** (1992), 83-92.

[40] Bocharov, P.P. Sobre algunas propiedades de las *PH*-distributiones, *Revisita del Seminario de Anseñanza y Titulacion* **IX** No. 75 (1993), 1-16.

[41] Bocharov, P.P. and Nava, S.M., Solucion matricial-geometrica de un sistema de servicio con fases e interrupciones de arribos y de servicios, *Revista del Seminario de Enseñanza y Titulacion* **IX** No. 75 (1993), 17-38.

[42] Bright, L. and Taylor, P.G., Calculating the equilibrium distribution in level dependent quasi-birth-and-death processes, Manuscript (1994).

[43] Chakravarthy, S., A finite capacity dynamic priority queueing model, *Computers Indust. Eng.* **22** (1992), 369-385.

[44] Chakravarthy, S., Analysis of a finite MAP/G/1 queue with group services, *Queueing Sys.* **13** (1993), 385-407.

[45] Chakravarthy, S. and Alfa, A.S., A multiserver queue with Markovian arrivals and group services with thresholds, *Nav. Res. Logist. Quart.* **40** (1993), 811-827.

[46] Chakravarthy, S. and Alfa, A.S., A finite capacity queue with Markovian arrivals and two servers with group services, *J. Appl. Math. Stoch. Anal.* **7** (1994), 161-178.

[47] Chang, C.S. and Nelson, R., Perturbation analysis of the M/M/1 queue in a Markovian environment via the matrix-geometric method, *Stoch. Mod.* **9** (1993), 233-246.

[48] Chemla, J.-P., *Les Lois de Type Phase: Caractérisation et Représentations*, Doctoral Thesis, Lab. d'Automatique de Grenoble, Ecole Nationale Polytech. de Grenoble, France 1994.

[49] Choudhury, G.L. and Whitt, W., Heavy-traffic asymptotic expansions for the asymptotic decay rates in the BMAP/G/1 queue, *Stoch. Mod.* **10** (1994), 445-498.

[50] Choudhury, G.L., Lucantoni, D.M. and Whitt, W., Multidimensional transform inversion with applications to the transient M/G/1 queue, *Ann. Appl. Prob.* **3** (1994), 719-740.

[51] Choudhury, G.L. and Lucantoni, D.M., Numerical computation of the moments of a probability distribution from its transform, *Oper. Res.*, (to appear).

[52] Commault, C. and Chemla, J.P., On dual and minimal phase-type representations, *Stoch. Mod.* **9** (1993), 421-434.

[53] Daigle, J.N. and Lucantoni, D.M., Queueing systems having phase-dependent arrival and service rates, In: *Numerical Solution of Markov Chains*, ed. by W.J. Stewart, Marcel Dekker, New York 1991, 161-202.

[54] Daigle, J.N. and Tang, S.C., The queue length distribution for multiserver discrete time queues with batch Markovian arrivals, *Stoch. Mod.* **8** (1992), 665-683.

[55] Dehon, M. and Latouche, G., A geometric interpretation of the relations between the exponential and generalized Erlang distributions, *Adv. Appl. Prob.* **14** (1982), 885-907.

[56] Doshi, B., Generalizations of the stochastic decomposition results for single server queues with vacations, *Stoch. Mod.* **6** (1990), 307-333.

[57] Dukhovny, A., Multiple roots in some equations of queueing theory, *Stoch. Mod.* **10** (1994), 519-524.

[58] Elwalid, A.I. and Mitra, D., Effective bandwidth of general Markovian traffic sources and admission control of high speed networks, *IEEE/ACM Trans. Networking* **1** (1993), 329-343.

[59] Faddy, M., Compartmental models with phase-type residence time distributions, *Appl. Stoch. Models Data Anal.* **6** (1990), 121-127.

[60] Faddy, M., A structured compartmental model for drug kinetics, *Biometrics* **49** (1993), 243-248.

[61] Faddy, M., Fitting structured phase-type distributions, *Appl. Stoch. Mod. Data Anal.*, (to appear).

[62] Falkenberg, E., On the asymptotic behavior of the stationary distribution of Markov chains of M/G/1 type, *Stoch. Mod.* **10** (1994), 75-97.

[63] Fratini, S., Analysis of a dynamic priority queue, *Stoch. Mod.* **6** (1990), 415-444.

[64] Gail, H.R., Hantler, S.L., and Taylor, B.A., Spectral analysis of M/G/1 and G/M/1 type Markov chains, *Adv. Appl. Prob.* (1995), (to appear).

[65] Gail, H.R., Hantler, S.L., and Taylor, B.A., Solutions of the basic matrix equations for the M/G/1 and G/M/1 Markov chains, *Stoch. Mod.* **10** (1994), 1-43.

[66] Gail, H.R., Hantler, S.L., Konheim, A., and Taylor, B.A., An analysis of a class of telecommunications models, *Performance Evaluation*, special issue on Discrete-Time Markov Chains (1995), (to appear).

[67] Gün, L. and Makowski, A., Matrix-geometric solution for finite capacity queues with phase-type distributions, *Proc. Performance 87*, North Holland (1988), 269-282.

[68] Gün, L., Experimental results on matrix-analytical solution techniques - Extensions and comparisons, *Stoch. Mod.* **5** (1989), 669-682.

[69] Hajek, B., Birth-and-death processes on the integers with phases and general boundaries, *J. Appl. Prob.* **19** (1982), 488-499.

[70] He, Q.M., Differentiability of the matrices in the matrix-analytic method, *Stoch. Mod.* **11** (1995), (to appear).

[71] Huang, J. and Hayes, J.F., A study of the matrix analytic method and its application in performance evaluation of broadband and related system, Report on post doctoral work, Dept. of Elect. and Comput. Eng., Concordia Univ., Montreal Canada 1994.

[72] Johnson, M.A. and Taaffe, M.R., Matching moments to phase distributions: Mixtures of Erlang distributions of common order, *Stoch. Mod.* **5** (1989), 711-743.

[73] Johnson, M.A. and Taaffe, M.R., Matching moments with a class of phase distributions: Nonlinear programming approaches, *Stoch. Mod.* **6** (1990), 259-281.

[74] Johnson, M.A. and Taaffe, M.R., Matching moments with a class of phase distributions: Density function shapes, *Stoch. Mod.* **6** (1990), 283-306.

[75] Johnson, M.A., An empirical study of queueing approximating based on phase-type distributions, *Stoch. Mod.* **9** (1993), 531-561.

[76] Johnson, M.A., Selecting parameters of phase distributions: Combining nonlinear programming, heuristics, and Erlang distributions, *ORSA J. Comput.*

5 (1993), 69-83.

[77] Kao, E.P.C. and Lin, C., Computing the R matrices in matrix-geometric solutions for a class of QBD queues: A phase substitution approach, *Stoch. Mod.* **7** (1991), 629-643.

[78] Kijima, M. and Makimoto, N., Computation of the quasi-stationary distributions in M(n)/GI/1/K and GI/M(n)/1/K queue, *Queueing Sys.* **11** (1992), 255-272.

[79] Kijima, M., Quasi-stationary distributions of single server phase-type queues, *Math. Oper. Res.* **19** (1993), 423-437.

[80] Lang, A., *An Empirical Evaluation of Parameter Estimation Methods for Phase-Type Distributions*, Ph.D. Dissertation, Dept. of Stat., Oregon State Univ., August 1994.

[81] Latouche, G., An exponential semi-Markov process, with applications to queueing theory, *Stoch. Mod.* **1** (1985), 137-169.

[82] Latouche, G., Perturbation analysis of a phase-type queue with weakly correlated arrivals, *Adv. Appl. Prob.* **20** (1988), 896-912.

[83] Latouche, G., Algorithms for infinite Markov chains with repeating columns, In: *Linear Algebra, Markov Chains and Queueing Models*, ed. by C.D. Meyer and R.J. Plemmons, Springer-Verlag, New York 1993, 231-265.

[84] Latouche, G. and Ramaswami, V., A logarithmic reduction algorithm for quasi-birth-and-death processes, *J. Appl. Prob.* **30** (1993), 650-674.

[85] Latouche, G. and Ramaswami, V., An efficient algorithm for the PH/PH/1 queue, Manuscript (1994).

[86] Latouche, G. and Ramaswami, V., Expected passage times in homogeneous quasi-birth-and-death processes, *Stoch. Mod.* **11** (1995), (to appear).

[87] Li, S.Q., Overload control in a finite message storage buffer, *IEEE Trans. Comm.* (1989), 1330-1338.

[88] Liu, D. and Neuts, M.F., Assessing the effecting of bursts of arrivals on the characteristics of a queue, In: *Recent Advances in Statistics and Probability*, Proc. of the 4th Internat. Meeting of Statistics in the Basque Country - IMSIBAC 4, ed. by J. Perez Vilaplana and M.L. Puri, VSP International Science Publishers 1994, 371-387.

[89] Liu, D. and Neuts, M.F., A queueing model for an ATM rate control scheme, *Telecom. Syst.* **2** (1994), 321-348.

[90] Liu, D. and Zhao, Y.Q., Determination of explicit solution for a general class of Markov processes, Manuscript (1994).

[91] Lucantoni, D.M. and Ramaswami, V., Efficient algorithms for solving the nonlinear matrix equations arising in phase-type queues, *Stoch. Mod.* **1** (1985), 29-51.

[92] Lucantoni, D.M., Meier-Hellstern, K.S. and Neuts, M.F., A single-server queue with server vacations and a class of nonrenewal arrival processes, *Adv. Appl. Prob.* **22** (1990), 676-705.

[93] Lucantoni, D.M., New results on the single-server queue with a batch Markovian arrival process, *Stoch. Mod.* **7** (1991), 1-46.

[94] Lucantoni, D.M., The BMAP/G/1 queue: A tutorial, In: *Models and Techniques for Performance Evaluation of Computer and Commun. Systems*, ed. by L. Donatiello and R. Nelson, Springer-Verlag, New York 1993, 330-358.

[95] Lucantoni, D.M., Choudhury, G.L. and Whitt, W., The transient BMAP/G/1 queue, *Stoch. Mod.* **10** (1994), 145-182.

[96] Lucantoni, D.M. and Neuts, M.F., Simpler proofs of some properties of the

fundamental period of the MAP/G/1 queue, *J. Appl. Prob.* **31** (1994), 235-243.

[97] Lucantoni, D.M. and Neuts, M.F., Some steady-state distributions for the MAP/SM/1 queue, *Stoch. Mod.* **10** (1994), 575-598.

[98] Machihara, F., A new approach to the fundamental period of a queue with phase-type Markov renewal arrivals, *Stoch. Mod.* **6** (1990), 551-560.

[99] Maier, R.S., The algebraic construction of phase-type distributions, *Stoch. Mod.* **7** (1991), 573-602.

[100] Makimoto, N., Quasi-stationary distributions in a PH/PH/c queue, *Stoch. Mod.* **9** (1992), 195-212.

[101] Matendo, S.K., Application of Neuts' method to vacation models with bulk arrivals, *Belg. J. Oper. Res., Stat. Comput. Sci.* **31** (1991), 34-48.

[102] Matendo, S.K., A single-server queue with server vacations and a batch Markovian arrival process, *Cahiers du CERO* **35** (1993), 87-114.

[103] Matendo, S.K., Some performance measures for vacation models with a batch Markovian arrival process, *J. Appl. Math. Stoch. Anal.* **7** (1994), 111-124.

[104] Matendo, S.K., Application of Neuts' method to bulk queueing models with vacations, *Recherche Operationnelle- Oper. Res.* **28** (1994), (to appear).

[105] Narayana, S. and Neuts, M.F., The first two moment matrices of the counts for the Markovian arrival process, *Stoch. Mod.* **8** (1992), 459-477.

[106] Nelson, R.D. and Iyer, B.R., Analysis of a replicated data base, *Perf. Eval.* **5** (1985), 133-148.

[107] Nelson, R. A performance evaluation of a general parallel processing model, *Perf. Eval. Rev.* **18** (1990), 13-26.

[108] Nelson, R.D. and Squillante, M.S., Analysis of contention in multiprocessor scheduling, In: *Performance '90*, ed. by P.J.B. King, I. Mitrani, and R.J. Pooley, Elsevier Science Publ. (1990), 391-405.

[109] Neuts, M.F., *Matrix-Geometric Solutions in Stochastic Models: An Algorithm Approach*, The Johns Hopkins University Press, Baltimore 1981.

[110] Neuts, M.F. and Takahasi, Y., Asymptotic behavior of the stationary distributions in the GI/PH/c queue with heterogeneous servers, *Z. f. Wahrscheinlichkeitstheorie* **57** (1982), 441-452.

[111] Neuts, M.F., The caudal characteristic curve of queues, *Adv. Appl. Prob.* **18** (1986), 221-254.

[112] Neuts, M.F., *Structured Stochastic Matrices of* M/G/1 *Type and Their Applications*, Marcel Dekker, Inc., New York 1989.

[113] Neuts, M.F., The fundamental period of the queue with Markov-modulated arrivals, In: *Probability, Stat., and Math.: Papers in Honor of Samuel Karlin*, ed. by T.W. Anderson, K.B. Athreya and D.L. Iglehart, Academic Press 1989, 187-200.

[114] Neuts, M.F. and Rao, B.M., Numerical investigations of a multiserver retrial model, *Queueing Sys.* **7** (1990), 169-190.

[115] Neuts, M.F., On Viterbi's formula for the mean delay in a queue of data packets, *Stoch. Mod.* **6** (1990), 87-98.

[116] Neuts, M.F., The joint distribution of arrivals and departures in quasi-birth-and-death processes, In: *Numerical Solution of Markov Chains*, ed. by W.J. Stewart, Marcel Dekker, New York 1991, 147-159.

[117] Neuts, M.F. and Rao, B.M., On the design of a finite capacity queue with phase-type service and hysteretic control, *Euro. J. Oper. Res.* **62** (1992),

221-240.
[118] Neuts, M.F., Models based on the Markovian arrival process, *IEEE Trans. Comm.* **E75-B** (1992), 1255-1265.
[119] Neuts, M.F., Liu, D., and Surya, N., Local Poissonification of the Markovian arrival process, *Stoch. Mod.* **8** (1992), 87-129.
[120] Neuts, M.F., Two further closure properties of *PH*-distributions, *Asia-Pacific J. Oper. Res.* **9** (1992), 77-85.
[121] Neuts, M.F., The burstiness of point processes, *Stoch. Mod.* **9** (1993), 445-466.
[122] Neuts, M.F., *Algorithmic Probability: A Collection of Problems*, Chapman and Hall, New York 1995, (in press).
[123] Niu, Z.S., Kawai, T., Akimaru, H., and Tadokoro, Y., A unified solution to mixed loss and delay systems with partial preemptive priority, *Trans. IEICE* **J77-BI** (1994), 322-330 (in Japanese).
[124] O'Cinneide, C.A., On nonuniqueness of representations of phase-type distributions, *Stoch. Mod.* **5** (1989), 247-259.
[125] O'Cinneide, C.A., Characterization of phase-type distributions, *Stoch. Mod.* **6** (1990), 1-57.
[126] O'Cinneide, C.A., Phase-type distributions and invariant polytopes, *Adv. Appl. Prob.* **23** (1991), 515-535.
[127] O'Cinneide, C.A., Triangular order of triangular phase-type distributions, *Stoch. Mod.* **9** (1993), 507-529.
[128] Rama Murthy, G., Kim, M., and Coyle, E.J., Equilibrium analysis of skip-free Markov chains - Nonlinear matrix equations, *Stoch. Mod.* **7** (1991), 547-571.
[129] Ramaswami, V., The busy period of queues which have a matrix-geometric steady state probability vector, *Opsearch* **19** (1982), 238-261.
[130] Ramaswami, V. and Lucantoni, D.M., Algorithmic analysis of a dynamic priority queue, In: *Appl. Prob. - Comput. Sci., The Interface*, Vol. **2**, Birkhäuser, Boston 1982, 157-204.
[131] Ramaswami, V. and Lucantoni, D.M., Algorithms for the multiserver queue with phase-type service, *Stoch. Mod.* **1** (1985), 393-417.
[132] Ramaswami, V. and Lucantoni, D.M., Stationary waiting time distribution in queues with phase-type service and in quasi-birth-and-death processes, *Stoch. Mod.* **1** (1985), 125-136.
[133] Ramaswami, V., Independent Markov processes in parallel, *Stoch. Mod.* **1** (1985), 419-432.
[134] Ramaswami, V., Nonlinear matrix equations in applied probability: Solution techniques and open problems, *SIAM Review* **30** (1988), 256-263.
[135] Ramaswami, V., A stable recursive solution for the steady-state vector in Markov chains of M/G/1 type, *Stoch. Mod.* **4** (1988), 183-188.
[136] Ramaswami, V. and Latouche, G., An experimental evaluation of the matrix-geometric method for the GI/PH/1 queue, *Stoch. Mod.* **5** (1989), 629-667.
[137] Ramaswami, V., A duality theorem for the matrix paradigms in queueing theory, *Stoch. Mod.* **6** (1990), 151-161.
[138] Ramaswami, V., From the matrix-geometric to the matrix-exponential, *Queueing Sys.* **6** (1990), 229-260.
[139] Ramaswami, V. and Taylor, P.G., An operator-recurrent approach to product-form networks, Manuscript (1994).

[140] Ramaswami, V. and Taylor, P.G., Some spectral properties of the rate matrices in level dependent quasi-birth-and-death processes with a countable number of phases, Manuscript (1994).

[141] Rydén, T., *Parameter Estimation for Markov Modulated Poisson Processes and Overload Control of SPC Switches*, Doctoral Thesis, Dept. of Math. Stat., Lund Institute of Technology, Lund, Sweden 1993.

[142] Rydén, T., Parameter estimation for Markov modulated Poisson processes, *Stoch. Mod.* **10** (1994), 794-829.

[143] Saito, H., The departure process of an N/G/1 queue, *Perf. Eval.* **11** (1991), 341-351.

[144] Saito, H., *Teletraffic Technologies in ATM Networks*, Artech House, Boston, London 1994.

[145] Schellhaas, H., On Ramaswami's algorithm for the computation of the steady state vector in Markov chains of M/G/1-type, *Stoch. Mod.* **6** (1990), 541-550.

[146] Schellhaas, H., Single-server queues with a batch Markovian arrival process and server vacations, *Oper. Res. Spektrum* **15** (1994), 189-196.

[147] Schmickler, L., MEDA: Mixed Erlang distributions as phase-type representation of empirical distribution functions, *Stoch. Mod.* **8** (1992), 131-156.

[148] Sengupta, B., Markov processes whose steady state distribution is matrix-exponential with an application to the GI/PH/1 queue, *Adv. Appl. Prob.* **21** (1989), 159-180.

[149] Sengupta, B., Phase-type representations for matrix-geometric solutions, *Stoch. Mod.* **6** (1990), 163-167.

[150] Sengupta, B., The semi-Markovian queue: Theory and applications, *Stoch. Mod.* **6** (1990), 383-413.

[151] Stanford, D.A. and Stroiński, K.J., Recursive methods for computing finite-time ruin probabilities for phase-distributed claim sizes, *ASTIN Bulletin* (1994), (to appear).

[152] Takine, T. and Hasegawa, T., The workload in the MAP/G/1 queue with state-dependent services: Its applications to a queue with preemptive resume priority, *Stoch. Mod.* **10** (1994), 183-221.

[153] Takine, T., Sengupta, B., and Yeung, R.W., A generalization of the matrix M/G/1 paradigm for Markov chains with a tree structure, Manuscript (1994).

[154] Tweedie, R.L., Operator-geometric stationary distributions for Markov chains with application to queueing models, *Adv. Appl. Prob.* **14** (1982), 368-391.

[155] Whitt, W., Tail probabilities with statistical multiplexing and effective bandwidths in multiclass queues, *Telecom. Syst.* **2** (1993), 71-107.

[156] Ye, J. and Li, S.Q., Analysis of multimedia traffic queues with finite buffer and overload control - Part I: Algorithm, *Proc. IEEE Infocom'92*, May 1992, 1464-1474.

[157] Ye, J. and Li, S.Q., Analysis of multimedia traffic queues with finite buffer and overload control - Part II: Applications, *Proc. IEEE Infocom'92*, May 1992, 848-859.

[158] Ye, J. and Li, S.Q., Folding algorithm: A computational method for finite *QBD* processes with level-dependent transitions, *IEEE Trans. Comm.* **42**:2 (1994), 625-639.

[159] Yeung, R.W. and Sengupta, B., Matrix product-form solutions for Markov

chains with a tree structure, *Adv. Appl. Prob.* (1994), (to appear).

[160] Zhang, J. and Coyle, E.J., Transient analysis of quasi-birth-and-death processes, *Stoch. Mod.* **5** (1989), 459-496.

Chapter 11

Explicit Wiener-Hopf factorizations for the analysis of multidimensional queues

Jos H.A. de Smit

ABSTRACT This chapter discusses Wiener-Hopf factorizations for multidimensional queues. For the analysis of these models, the classical (scalar) Wiener-Hopf factorization for single-server queues is generalized to the matrix case. First, we present our method for the single-server queue GI/G/1 and derive solutions for the distributions of actual waiting times and virtual waiting times for GI/K_m/1 and K_m/G/1. Next, we consider two classes of multidimensional queues: (*i*) *Single-server semi-Markov queues* and (*ii*) *Multiserver queues with phase-type service times*. For both classes we discuss general results as well as explicit solutions for important special cases.

CONTENTS

11.1	Introduction	293
11.2	Preliminaries	294
11.3	The single-server queue GI/G/1	296
11.4	The single-server semi-Markov queue SM/SM/1	300
11.5	The multiserver GI/PH/s queue (with phase-type service times)	304
11.6	Open problems and research directions	308
	Bibliography	309

11.1 INTRODUCTION

This chapter discusses Wiener-Hopf factorizations for multidimensional queues. For the analysis of these models, the classical (scalar) Wiener-Hopf factorization for the single-server queue GI/G/1 is generalized to the matrix case. We shall discuss two classes of models which give rise to a multidimensional system of Wiener-Hopf-type equations. For important special cases of these classes, explicit Wiener-Hopf factorizations can be obtained, leading to solutions which are both explicit and numerically tractable. These two classes are

(*a*) single-server semi-Markov queues

and

(*b*) multiserver queues with phase-type service times.

Scalar Wiener-Hopf factorizations apply to stochastic processes ($W_n, n = 0, 1, \ldots$) satisfying a relationship of the form

$$W_{n+1} = [W_n + Z_{n+1}]^+, \quad n = 0, 1, \ldots; \tag{$*$}$$

with state space $\mathbb{R}_+$ where ($Z_n, n = 1, 2, \ldots$) is a sequence of i.i.d. random varia-

0-8493-8074-x/95/\$0.00+\$.50

bles which can be both positive or negative. Matrix generalizations apply to vector processes $((W_n, X_n),\ n = 0, 1, \ldots)$, with state space $\mathbb{R}_+ \times \{1, \ldots, N\}$, with finite N, which again satisfy $(*)$ and where $((Z_n, X_n), n = 0, 1, \ldots)$ is a Markov-renewal sequence with state space $\mathbb{R} \times \{1, \ldots, N\}$. This structure can be found in both classes mentioned above. An interesting treatment of the general theory of matrix factorizations for the solution of systems of Wiener-Hopf-type equations can be found in [3].

Since our methods can be considered as matrix generalizations of the scalar case, we first discuss the classical single-server queue GI/G/1. Although the results for this case are not new, we believe that our method of solution is simpler and more powerful than those found in the current literature on this topic. Understanding our analysis requires some elementary background in complex analysis (see e.g., [17]). We shall concentrate on the distributions of actual and virtual waiting times. Using our methods, results for other quantities such as queue lengths both in discrete and continuous time can be obtained (see e.g., [6] and [9]). The derivations, however, are somewhat more technical. It is the aim of this chapter to show how relatively simple methods for single-server queues can be generalized, in a natural way, to multidimensional systems. The class of models, which can be solved this way, has considerable overlap with the class of models which can be solved by matrix geometric methods developed by Neuts (see e.g., [13] and [14]). In Neuts' approach, the basic process is the queue length process at arrival or departure epochs, whereas for our methods the basic process is the actual waiting time process. In both cases, these processes are supplemented by components with finite state space in order to make them Markovian. Using relationships between the distributions of waiting times and those of queue lengths, Neuts obtained results for the waiting time processes whereas we arrive at results for the queue length processes.

We believe, however, that, for the models discussed in this chapter, our methods lead to more efficient numerical algorithms than other methods do. This opinion is based on our own numerical experience, some results of which have been published (see [7] and [15]). Unfortunately, very little has been published about the numerical performance of other algorithmic methods. Consequently, it is difficult to compare different approaches to the same model. The point of view that analytic results in queueing theory are not useful for numerical calculations (see e.g., the preface of [13]) turns out to be unjustified in the models we consider. In fact, there is now much more evidence that many analytic results in queueing theory may lead to efficient and accurate numerical results (see e.g., [1]). The claimed contrast between analytic and algorithmic methods is evidently irrelevant.

The chapter is organized as follows. Section 11.2 contains some preliminaries, which apply to all the subsequent sections. In Section 11.3, we discuss a classical single-server queue. Section 11.4 deals with a single-server semi-Markov queue. A multiserver queue with phase-type service times is analyzed in Section 11.5. Finally, in Section 11.6, we discuss open problems and research directions in this area.

11.2 PRELIMINARIES

The following notations and conventions will be used throughout the chapter. Other notations and conventions for special models are introduced in the corresponding sections and remain valid throughout these sections.

We consider a queueing model in which customers arrive at time epochs $T_0, T_1, \ldots$; with $T_0 = 0$. The interarrival times are denoted by $A_n = T_n - T_{n-1}$, $n = 1, 2, \ldots$; and the service time of the nth customer by B_n, $n = 0, 1, \ldots$. The queueing discipline is first come, first served. We denote the (actual) waiting time of the nth customer by W_n and the virtual waiting time at time t by V_t. The number of customers arriving during the first busy period is denoted by N_1, where

$$N_1 = inf\,\{n > 0 : W_n = 0\},$$

the length of the first busy period by T^*, where

$$T^* = inf\,\{t > 0 : V_t = 0\},$$

and the length of the first busy cycle by T, where

$$T = inf\,\{t > 0 : V_{t-} = 0, V_t > 0\}.$$

Denote $x^+ = max(0, x)$, and $x^- = min(0, x)$. Let $\delta_{i,j}$ be Kronecker's delta, i.e., $\delta_{i,j} = 0$, for $i \neq j$, and $\delta_{j,j} = 1$. I is the identity matrix. By $1(A)$ we shall denote the indicator function of the event A, i.e., $1(A) = 1$ if A occurs and $1(A) = 0$, otherwise.

In the following sections, we apply Rouché's theorem several times in order to determine the number of zero's of some functions in the left half- (complex) plane or the right half-plane. For that purpose, we shall consider the following contours.

Definition 11.1. For $R > \delta \geq 0$, $C^+_{\delta, R}$ is the closed contour consisting of

(a) the part of the line $Re(\phi) = -\delta$, running from

$-\delta + i\sqrt{R^2 - \delta^2}$ to $-\delta - i\sqrt{R^2 - \delta^2}$ and

(b) the part of the circle $|\phi| = R$, running counterclockwise from

$-\delta - i\sqrt{R^2 - \delta^2}$ to $-\delta + i\sqrt{R^2 - \delta^2}$.

$C^-_{\delta, R}$ is the closed contour consisting of

(a) the part of the line $Re(\phi) = -\delta$, running from

$-\delta - i\sqrt{R^2 - \delta^2}$ to $-\delta + i\sqrt{R^2 - \delta^2}$ and

(b) the part of the circle $|\phi| = R$, running counterclockwise from

$-\delta + i\sqrt{R^2 - \delta^2}$ to $-\delta - i\sqrt{R^2 - \delta^2}$.

For the factorization methods we shall need the following properties.

Definition 11.2. We say a function f satisfies property A^+ if $f(\phi)$ is

(i) analytic on $Re(\phi) > 0$,

(ii) continuous and bounded on $Re(\phi) \geq 0$,

and we say that it satisfies property $\tilde{A}^+$ if, in addition, it is

(iii) bounded away from 0 on $Re(\phi) \geq 0$.

We say that a function f satisfies property A^- if $f(\phi)$ is

(i) analytic on $Re(\phi) < 0$,

(ii) continuous and bounded on $Re(\phi) \leq 0$,

and we say that it satisfies property $\tilde{A}^-$ if, in addition, it is

(iii) bounded away from 0 on $Re(\phi) \leq 0$.

We shall also need a matrix generalization of Rouché's theorem which was proved in [6].

Theorem 11.3. *Let* $A(z) = (a_{ij}(z))$ *and* $B(z) = (b_{ij}(z))$ *be complex* $n \times n$ *mat-*

rices, where $B(z)$ is diagonal. The elements a_{ij} and b_{ij}, $1 \leq i \leq n$, $1 \leq j \leq n$, are meromorphic functions in a simply connected region S in which T is the set of all poles of these functions. C is a rectifiable closed Jordan curve in $S-T$. $N_A[N_{A+B}]$ is the number of zeros inside C of $\det B(z)[\det(A(z)+B(z))]$ and $P_B[P_{A+B}]$ the number of poles inside C (zeros and poles of higher order are counted according to this order). If

(i) $|b_{ii}(z)| > \sum_{j=1}^{n} |a_{ij}(z)|$ *on C for all $i = 1,\ldots,n$*

or

(ii) *$A(z)$ is indecomposable on C and $|b_{ii}(z)| \geq \sum_{j=1}^{n} |a_{ij}(z)|$ on C, for all $i = 1,\ldots,n$ with strict inequality for at least one i,*

then on C

$$\det(A(z)+B(z)) \neq 0,$$

$$\det B(z) \neq 0,$$

and

$$N_{A+B} - P_{A+B} = N_B - P_B.$$

11.3 THE SINGLE-SERVER QUEUE GI/G/1

11.3.1 Introduction

Assume that the interarrival times A_n are i.i.d. with common distribution function F and finite expectation α, while the service times B_n are i.i.d. with common distribution function G and finite expectation β. The traffic intensity ρ is defined by $\rho = \beta/\alpha$. The Laplace-Stieltjes transform (L-S-transform) of the interarrival times is denoted by A, i.e.,

$$A(\phi) = \int_0^\infty exp(-\phi x)F(dx), \quad Re(\phi) \geq 0,$$

and the L-S-transform of the service is denoted by B, i.e.,

$$B(\phi) = \int_0^\infty exp(-\phi x)G(dx), \quad Re(\phi) \geq 0.$$

11.3.2 Derivation of the distribution of actual waiting times by Wiener-Hopf factorization

The actual waiting times satisfy the recurrence relation

$$W_{n+1} = [W_n + B_n - A_{n+1}]^+, \quad n = 0,1,\ldots. \tag{11.1}$$

Let for $|r| < 1$, $Re(\phi) \geq 0$,

$$Z_w(r,\phi) = \sum_{n=0}^{\infty} r^n \mathbf{E}_w[exp(-\phi W_n)],$$

where $\mathbf{E}_w[\cdot]$ stands for $\mathbf{E}_{W_0 = w}[\cdot]$. Straightforward calculations show that

$$Z_w(r,\phi)\{1 - rA(-\phi)B(\phi)\} = e^{-\phi w} + V_w(r,\phi), \tag{11.2}$$

for $|r| < 1$, $Re(\phi) = 0$, where

$$V_w(r,\phi) = \sum_{n=0}^{\infty} r^{n+1}\{1 - \mathbf{E}_w[exp(-\phi[W_n + B_n - A_{n+1}]^-)]\}.$$

We immediately see that (11.2) is (the Laplace-Stieltjes transform of) a Wiener-Hopf-type equation. The expression $1 - rA(-\phi)B(\phi)$ is called its symbol (see [3]). Note that for fixed $|r| < 1$, the function $Z_w(r,\phi)$ satisfies A^+ and $V_w(r,\phi)$ satisfies A^-.

It can be shown (see [5]) that for fixed $|r| < 1$, the symbol $1 - rA(-\phi)B(\phi)$ can be factorized, i.e., for $Re(\phi) = 0$,

$$1 - rA(-\phi)B(\phi) = K^+(r,\phi)K^-(r,\phi),$$

where $K^+(r,\phi)$ satisfies $\tilde{A}^+$ and $K^-(r,\phi)$ satisfies $\tilde{A}^-$. Then, from (11.2) we have, for fixed $|r| < 1$ and $Re(\phi) = 0$, that

$$Z_w(r,\phi)K^+(r,\phi) = [e^{-\phi w} + V_w(r,\phi)][K^-(r,\phi)]^{-1}. \tag{11.3}$$

Below, we restrict ourselves to the case $W_0 = w = 0$. For this case, the left-hand side of (11.3) satisfies A^+ and the right-hand side satisfies A^-. We, therefore, have the following solution of (11.2).

Theorem 11.4. *For* $|r| < 1$, $Re(\phi) \geq 0$, *we have*

$$Z_0(r,\phi) = \frac{1}{1-r}K^+(r,0)[K^+(r,\phi)]^{-1}.$$

Proof. The left-hand side of (11.3) satisfies A^+ and the right-hand side satisfies A^-. By analytic continuation, we can define an entire function which is equal to the left-hand side of (11.3) for $Re(\phi) \geq 0$ and equal to the right-hand side for $Re(\phi) \leq 0$. But this entire function is bounded, and hence, by Liouville's theorem, it is a constant. So, for $Re(\phi) \geq 0$,

$$Z_0(r,\phi)K^+(r,\phi) = Z_0(r,0)K^+(r,0),$$

which proves the theorem. □

Equation (11.3) can also be solved for $W_0 = w > 0$, which requires a decomposition step in addition to the factorization applied for the case $W_0 = w = 0$.

If $\rho < 1$ and $K^+(1,\phi) = \lim_{r \uparrow 1} K^+(r,\phi)$ exists, we have, for the L-S-transform of the steady state waiting time, $Z(\phi) = \mathbf{E}[exp(-\phi W)]$, using Abel's theorem

$$Z(\phi) = \lim_{r \uparrow 1}(1-r)Z_0(r,\phi) = K^+(1,0)[K^+(1,\phi)]^{-1}, \quad Re(\phi) \geq 0. \tag{11.4}$$

11.3.3 The virtual waiting time

We easily see that

$$\int_0^{T^*} exp(-\phi V_t)dt$$

$$= \sum_{k=0}^{N_1 - 1} \int_0^{min(A_{k+1}, W_k + B_k)} exp(-\phi(W_k + B_k - u))du. \tag{11.5}$$

Since

$$\int_0^{min(x,y)} exp(-\phi(y-u))du = \frac{1}{\phi}[exp(-\phi(y-x)^+) - exp(-\phi y)],$$

we have from (11.5) and (11.1) that

$$\int_0^{T^*} exp(-\phi V_t)dt$$

$$= \frac{1}{\phi} \sum_{k=0}^{N_1-1} [exp(-\phi W_{k+1}) - exp(-\phi(W_k + B_k))] + T - T^*. \qquad (11.6)$$

Assume that the traffic intensity $\rho < 1$ and that the distribution function of the interarrival times F is non-arithmetic. Then, the steady state virtual waiting time distribution exists. The processes $(W_n, n = 0, 1, \ldots)$ and $(V_t, t \geq 0)$ are both regenerative so that

$$\mathbf{E}[exp(-\phi W)] = \mathbf{E}_0[\sum_{k=0}^{N_1-1} exp(-\phi W_k)]/\mathbf{E}_0[N_1]$$

and

$$\mathbf{E}[exp(-\phi V)] = \mathbf{E}_0[\int_0^T exp(-\phi V_t)]/\mathbf{E}_0[T].$$

Since, from Wald's equation, we have $\mathbf{E}_0[T] = \alpha \mathbf{E}_0[N_1]$ and $\mathbf{E}_0[T^*] = \beta \mathbf{E}_0[N_1]$, taking expectations of both sides of (11.6) yields

$$\mathbf{E}[exp(-\phi V)] = 1 - \rho + Z(\phi)\frac{1 - B(\phi)}{\alpha\phi}, \quad Re(\phi) \geq 0. \qquad (11.7)$$

This result can be found in Cohen [5] and is originally due to Takács.

11.3.4 Examples

The results in this section can also be found in [5].

Example 11.5. *The system* GI/K$_m$/1. For the model GI/K$_m$/1, the L-S-transform of the service times has the form

$$B(\phi) = \frac{B_1(\phi)}{\prod_{i=1}^m (\phi + \mu_i)},$$

where $Re(\mu_i) > 0$, $i = 1, \ldots, m$, and $B_1(\phi)$ is a polynomial of degree $m - 1$ or less. Hence, in this case,

$$1 - rA(-\phi)B(\phi) = \frac{\prod_{i=1}^m (\phi + \mu_i) - rA(-\phi)B_1(\phi)}{\prod_{i=1}^m (\phi + \mu_i)}. \qquad (11.8)$$

We immediately see that for $|r| < 1$ and R large enough $|rA(-\phi)B(\phi)| < 1$ on $C_{0,R}^-$, so that by Rouché's theorem, the numerator of (11.8) has exactly m zeros in the left half-plane $Re(\phi) < 0$, which we denote by $\lambda_1(r), \ldots, \lambda_m(r)$. Consequently, we can write

$$1 - rA(-\phi)B(\phi) = K^-(r, \phi)K^+(r, \phi), \quad Re(\phi) = 0,$$

where

$$K^-(r, \phi) = \frac{\prod_{i=1}^m (\phi + \mu_i) - rA(-\phi)B_1(\phi)}{\prod_{i=1}^m (\phi - \lambda_i(r))}$$

and

$$K^{+}(r,\phi)=\prod_{i=1}^{m}\frac{\phi-\lambda_i(r)}{\phi+\mu_i}.$$

We see that K^{+} satisfies $\widetilde{A}^{+}$ and K^{-} satisfies $\widetilde{A}^{-}$ for fixed $|r|<1$. From Theorem 11.4, we thus have

$$Z_0(r,\phi)=\frac{1}{1-r}\prod_{i=1}^{m}\frac{(\phi+\mu_i)(-\lambda_i(r))}{(\phi-\lambda_i(r))\mu_i},\quad |r|<1,\ Re(\phi)\geq 0. \tag{11.9}$$

Next, for $\rho<1$, we consider the steady state distribution of the waiting times i.e., (cf. (11.4)) we have to study the behavior of the zeros $\lambda_i(r)$ for $r\uparrow 1$. Take $\delta>0$ and consider the contour $C^{-}_{\delta,R}$. Then, for R large enough $|A(-\phi)B(\phi)|<1$ on the semi-circle $|\phi|=R$, $Re(\phi)<0$, whereas on $Re(\phi)=-\delta$,

$$\begin{aligned}|A(-\phi)B(\phi)| &\leq A(\delta)B(-\delta)=(1-\alpha\delta+o(\delta))(1+\beta\delta+o(\delta))\\ &=1-\alpha\delta(1-\rho)+o(\delta),\ \delta\downarrow 0.\end{aligned}$$

Hence, $|A(-\phi)B(\phi)|<1$ on $Re(\phi)=-\delta$ if $o(\delta)/\delta<\alpha(1-\rho)$, which will hold for $\rho<1$ and δ small enough. With the aid of Rouché's theorem, we now see that for $\rho<1$, $1-A(-\phi)B(\phi)$ has exactly m zeros in the left half-plane $Re(\phi)<0$, which we denote by $\lambda_1(1),\ldots,\lambda_m(1)$. Since the $\lambda_i(r)$ are continuous functions in r for $|r|\leq 1$, we may write $\lambda_i(1)=\lim_{r\uparrow 1}\lambda_i(r)$, $i=1,2,\ldots,m$. From (11.4) and (11.9), we conclude that

$$Z(\phi)=\prod_{i=1}^{m}\frac{(\phi+\mu_i)(-\lambda_i(1))}{(\phi-\lambda_i(1))\mu_i},\quad Re(\phi)\geq 0 \tag{11.10}$$

and, due to (11.7),

$$\mathbf{E}[exp(-\phi V)]=\frac{\prod_{i=1}^{m}(\phi+\mu_i)-B_1(\phi)}{\alpha\phi}\prod_{i=1}^{m}\frac{-\lambda_i(1)}{(\phi-\lambda_i(1))\mu_i},\quad Re(\phi)\geq 0.$$

Example 11.6. *The system $K_m/G/1$.* The L-S-transform of the interarrival times for the model $K_m/G/1$ can be written as

$$A(\phi)=\frac{A_1(\phi)}{\prod_{i=1}^{m}(\phi+\lambda_i)},$$

where $Re(\lambda_i)>0$, $i=1,\ldots,m$, and $A_1(\phi)$ is a polynomial of degree $m-1$ or less. So,

$$1-rA(-\phi)B(\phi)=\frac{\prod_{i=1}^{m}(\lambda_i-\phi)-rA_1(-\phi)B(\phi)}{\prod_{i=1}^{m}(\lambda_i-\phi)}. \tag{11.11}$$

Analogous to the case $GI/K_m/1$, we see that, for $|r|<1$, the numerator of (11.11) has exactly m zeros in the right half-plane $Re(\phi)>0$, which we denote by $\mu_1(r),\ldots,\mu_m(r)$, so that

$$1-rA(-\phi)B(\phi)=K^{-}(r,\phi)K^{+}(r,\phi),\quad Re(\phi)=0,$$

with

$$K^{+}(r,\phi)=\frac{\prod_{i=1}^{m}(\lambda_i-\phi)-rA_1(-\phi)B(\phi)}{\prod_{i=1}^{m}(\mu_i(r)-\phi)}$$

and

$$K^{-}(r,\phi)=\prod_{i=1}^{m}\frac{\phi-\mu_i(r)}{\phi-\lambda_i}.$$

Now K^+ satisfies $\widetilde{A}^+$ and K^- satisfies $\widetilde{A}^-$ for fixed $|r| < 1$, and, from Theorem 11.4, we have for $|r| < 1$, $Re(\phi) \geq 0$,

$$Z_0(r,\phi) = \left(\prod_{i=1}^{m} \frac{\mu_i(r) - \phi}{\mu_i(r)}\right) \frac{\prod_{i=1}^{m} \lambda_i}{\prod_{i=1}^{m}(\lambda_i - \phi) - rA_1(-\phi)B(\phi)}. \tag{11.12}$$

For $\rho < 1$, we consider the steady state behavior. Similar to the case of GI/K$_m$/1, we see that for R large enough and δ small enough, $|A(-\phi)B(\phi)| < 1$ on $C^+_{\delta,R}$, which implies that $1 - A(-\phi)B(\phi)$ has exactly m zeros $\mu_1(1), \ldots, \mu_m(1)$ in the right half-plane $Re(\phi) \geq 0$, with $\mu_i(1) = \lim_{r \uparrow 1} \mu_i(r)$, $i = 1, 2, \ldots, m$. One of these zeros, say $\mu_1(1)$, is a simple zero at 0, while $Re(\mu_i(1)) > 0$, for $i = 2, \ldots, m$. Differentiating the identity,

$$1 - rA(-\mu_1(r))B(\mu_1(r)) = 0$$

with respect to r and then letting $r \uparrow 1$, we find that $\mu_1'(1) = 1/(\beta - \alpha)$ so that, with l'Hôpital's rule,

$$Z(\phi) = \phi(\beta - \alpha)\left(\prod_{i=2}^{m} \frac{\mu_i(1) - \phi}{\mu_i(1)}\right) \frac{\prod_{i=1}^{m} \lambda_i}{\prod_{i=1}^{m}(\lambda_i - \phi) - A_1(-\phi)B(\phi)}.$$

11.4. THE SINGLE-SERVER SEMI-MARKOV QUEUE SM/SM/1

11.4.1 Introduction

The single-server semi-Markov queue SM/SM/1 generalizes the queue GI/G/1 by relaxing the assumptions of i.i.d. interarrival times and service times. Let $(Y_n, n = 0, 1, \ldots)$ be an irreducible aperiodic Markov chain with finite state space $\{1, 2, \ldots, N\}$. The stationary distribution of the Markov chain $(Y_n, n = 0, 1, \ldots)$ is denoted by the row vector $\pi = (\pi_1, \ldots, \pi_N)$. Assume that, for all choices of n, x, y and j,

$$\begin{aligned} &\mathbf{P}(A_{n+1} < x, B_n \leq y, Y_{n+1} = j \mid A_1, \ldots, A_n, B_0, \ldots, B_{n-1}, Y_0, \ldots, Y_n) \\ &\quad = \mathbf{P}(A_{n+1} \leq x, B_n \leq y, Y_{n+1} = j \mid Y_n), \end{aligned} \tag{11.13}$$

while the latter conditional probability does not depend on n. We see that the model is completely specified by the functions

$$G_{i,j}(\phi, \psi) = \mathbf{E}[exp(-\phi A_{n+1} - \psi B_n)1(Y_{n+1} = j) \mid Y_n = i],$$

$Re(\phi) \geq 0$, $Re(\psi) \geq 0$, $i = 1, 2, \ldots, N$; $j = 1, 2, \ldots, N$.

Let $G(\phi, \psi)$ be the $N \times N$ matrix with elements $G_{i,j}(\phi, \psi)$.

Let

$$\alpha_{i,j} = \mathbf{E}[A_{n+1}1(Y_{n+1} = j) \mid Y_n = i]$$

and

$$\beta_{i,j} = \mathbf{E}[B_n 1(Y_{n+1} = j) \mid Y_n = i]$$

and let $\bar{\alpha}$ and $\bar{\beta}$ be the $N \times N$-matrices with elements $\alpha_{i,j}$ and $\beta_{i,j}$, respectively. The expected steady state interarrival time is then given by

$$\alpha = \pi \bar{\alpha} \mathbf{1},$$

and the expected steady state service time by

$$\beta = \pi\overline{\beta}\mathbf{1}.$$

The traffic intensity is defined by

$$\rho = \frac{\beta}{\alpha},$$

assuming that both α and β are finite.

For this model, we shall restrict ourselves to the case $W_0 = 0$.

11.4.2 The system of Wiener-Hopf-type equations

In this section, we study the Markov process $((W_n, X_n), n = 0, 1, \ldots)$, with state space $\mathbb{R}_+ \times \{1, \ldots, N\}$, for finite N. Define for $i = 1, \ldots, N$; $j = 1, \ldots, N$; $|r| < 1$ and $Re(\phi) \geq 0$,

$$Z_{i,j}(r,\phi) = \sum_{n=0}^{\infty} r^n \mathbf{E}[exp(-\phi W_n)1(Y_n = j) \mid Y_0 = i]$$

and for $Re(\phi) \leq 0$,

$$V_{i,j}(r,\phi)$$
$$= \sum_{n=0}^{\infty} r^{n+1}\mathbf{E}[(1 - exp(-\phi[W_n + B_n - A_{n+1}]^-))1(Y_{n+1} = j) \mid Y_0 = i].$$

Let $Z(r,\phi)$ and $V(r,\phi)$ be the $N \times N$-matrices with elements $Z_{i,j}(r,\phi)$ and $V_{i,j}(r,\phi)$, respectively. Then we obtain the following system of Wiener-Hopf-type equations

$$Z(r,\phi)\{I - rG(-\phi,\phi)\} = I + V(r,\phi), \tag{11.14}$$

$|r| < 1$, $Re(\phi) = 0$. We see that (11.14) is a straightforward matrix generalization of equation (11.2) for the ordinary single-server queue discussed in Section 11.3. Similar to the scalar case, the matrix function $I - rG(-\phi,\phi)$ is called the symbol. As a consequence the method of solution, given in Section 11.3 for the scalar case, immediately generalizes to the present matrix version. It can be shown that the symbol $I - rG(-\phi,\phi)$ can be factorized (see [2]), i.e., for $Re(\phi) = 0$,

$$I - rG(-\phi,\phi) = K^+(r,\phi)K^-(r,\phi),$$

where (the $N \times N$-matrix) $K^+(r,\phi)$ satisfies $\widetilde{A}^+$ (entrywise) and $K^-(r,\phi)$ satisfies $\widetilde{A}^-$.

Theorem 11.7. *For* $|r| < 1$, $Re(\phi) \geq 0$, *we have*

$$Z(r,\phi) = [I - rG(0,0)]^{-1}K^+(r,0)[K^+(r,\phi)]^{-1}.$$

Proof. See the proof of Theorem 11.4. □

For $\rho < 1$, the vector (W_n, Y_n) converges in distribution to a steady state vector (W, Y). Assume that $K(1,\phi) = \lim_{r\uparrow 1} K(r,\phi)$ exists. Denote

$$Z_j(\phi) = \mathbf{E}[exp(-\phi W)1(Y = j)]$$

and let $Z(\phi)$ be the row vector with elements $Z_j(\phi)$. Then, by Abel's theorem, we have

$$Z(\phi) = \pi K^+(1,0)[K^+(1,\phi)]^{-1}, \quad Re(\phi) \geq 0. \tag{11.15}$$

11.4.3 The virtual waiting time

Expression (11.16) remains valid for the single-server semi-Markov queue. So, analogous to the derivation for the ordinary single-server queue, we have

$$\mathbf{E}[exp(-\phi V)] = 1 - \rho + Z(\phi)\frac{[I - G(0,\phi)]\mathbf{1}}{\alpha\phi}, \quad Re(\phi) \geq 0. \tag{11.16}$$

11.4.4 Examples

The queue SM/SM/1 represents a very broad class of queueing models. This class includes the queue with a Markovian arrival process MAP/SM/1 and the model with a batch Markovian arrival process BMAP/SM/1. See [12] for more details on these models. The analysis leads to solutions for the distributions of actual and virtual waiting times. In the case of single arrivals, there exists a simple relationship between the distributions of waiting times and queue lengths (see e.g. [9]), so that results for queue length distributions, both at arrival instants and in continuous time, can also be obtained. For models with batch arrivals, however, no such relationships exist, so that, with our approach, no results can be found for the queue lengths in BMAP/SM/1. As examples, we mention here two special cases of the SM/SM/1 queue which have been worked out in detail.

Example 11.8. *The* M/G/1 *queue with Markov-modulated arrivals and service times.* A queue M/G/1 with Markov modulated arrivals and service times is described as follows. For more details and proofs, see [15]. Let $X_t, t \geq 0$ be a continuous time Markov chain on $\{1,\ldots,N\}$ with undecomposable generator Q. $Q_{i,j}$, $i = 1,\ldots,N$, $j = 1,\ldots,N$, is the (i,j)th element of Q and $q_i = -Q_{i,i}$. The stationary distribution of $X_t, t \geq 0$ is $p = (p_1,\ldots,p_N)$. When $X_t = i$, customers arrive according to a Poisson process with rate λ_i, and an arriving customer has service time distribution G_i. The stationary arrival rate is denoted by $\lambda = \sum_{i=1}^N p_i\lambda_i$. Given $X_t, t \geq 0$, the service times $B_0, B_1,\ldots$, are independent and also independent of the arrival process. Denote

$$B_i(\phi) = \int_0^\infty exp(-\phi x)G_i(dx)$$

and

$$\beta_i = \int_0^\infty xG_i(dx).$$

For technical reasons, we assume that there exists a $\delta > 0$ such that the $B_i(\phi)$ can be continued analytically to the region $Re(\phi) > -\delta$. Let $\Lambda = diag\{\lambda_1,\ldots,\lambda_N\}$, $Y_n = X_{T_n}$ and $A(\phi)$ the matrix with elements

$$A_{i,j}(\phi) = \mathbf{E}[exp(-\phi A_n)\mathbf{1}(Y_n = j) \mid Y_{n-1} = i].$$

Then, we easily see that

$$A(\phi) = \Lambda[\phi I + \Lambda - Q]^{-1}.$$

We see that the present system is an SM/SM/1 queue with

$$G(\phi,\psi) = B(\psi)A(\phi),$$

where $B(\psi) = diag\{B_1(\psi),\ldots,B_N(\psi)\}$. In order to find the solution of (11.14) for this model, we consider the symbol

$$H(r,\phi) = I - rG(-\phi,\phi) = 1 - rB(\phi)A(-\phi) = L(r,\phi)[-\phi I + \Lambda - Q]^{-1},$$

where

$$L(r,\phi) = -\phi I + (I - rB(\phi))\Lambda - Q.$$

Using Theorem 11.3, we see that the N eigenvalues $\nu_1,\ldots,\nu_N$ of $\Lambda - Q$ all lie in the right half-plane $Re(\phi) > 0$ and that, for fixed $|r| < 1$, the determinant $\det L(r,\phi)$ has exactly N zeros $\mu_1(r),\ldots,\mu_N(r)$ in the right half-plane $Re(\phi) > 0$. Similar to the argument for $K_m/G/1$ in the previous section, we see that $L(1,\phi)$ has exactly $N-1$ zeros in the right half-plane $Re(\phi) > 0$ and a simple pole, say $\mu_1(1)$, at 0. Moreover, we can show that, for $\rho < 1$,

$$\lim_{r\uparrow 1} \frac{1-r}{\mu_1(r)} = \frac{1-\rho}{\lambda}.$$

Let R_i be a right eigenvector of $\Lambda - Q$ corresponding to the eigenvalue ν_i and let R be the matrix of which the R_i are the row vectors. For $i = 1,\ldots,N$, let $D^i(r)$ be a (nonunique) nonzero column vector satisfying

$$H(r,\mu_i(r))D^i(r) = 0,$$

and let $D(r)$ be the $N \times N$ matrix whose ith column is $D^i(r)$. Define the $N \times N$ matrix $S(r)$ by

$$S_{i,j}(r) = \frac{1}{\nu_i - \mu_j(r)} R_i \Lambda D^j(r).$$

Denote $C(r) = S^{-1}(r)R\,\Lambda$, and define the matrix $K(r,\phi)$ by

$$K(r,\phi) = I - D(r)diag\left\{\frac{1}{\phi - \mu_1(r)},\ldots,\frac{1}{\phi - \mu_N(r)}\right\}C(r).$$

We now have the following factorization of the symbol H:

$$H(r,\phi) = K^+(r,\phi)K^-(r,\phi), \quad Re(\phi) = 0,$$

where

(a) $K^+(r,\phi) = H(r,\phi)K(r,\phi)$ satisfies $\tilde{A}^+$ and

(b) $K^-(r,\phi) = K^{-1}(r,\phi)$ satisfies $\tilde{A}^-$.

From Theorem 11.7, we now have the following solution for $|r| < 1$, $Re(\phi) \geq 0$:

$$Z(r,\phi) = K(r,0)K^{-1}(r,\phi)H^{-1}(r,\phi).$$

If $\rho < 1$, we find for the steady state distribution of the waiting times for $Re(\phi) \geq 0$, $\phi \neq 0$,

$$\mathbf{E}[exp(-\phi W)] = Z_0 K(\phi)^{-1} H(\phi)^{-1},$$

where

$$Z_0 = \lim_{r\uparrow 1}(1-r)K(r,0) = \frac{1-\rho}{\lambda} D^1(1)C_1(1),$$

with $D^1(1)$ being the first column of $D(1)$ and $C_1(1)$ the first row of $C(1)$, while for $\phi \neq 0$,

$$K(\phi) = \lim_{r\uparrow 1} K(r,\phi), \quad H(\phi) = \lim_{r\uparrow 1} H(r,\phi).$$

Example 11.9 *A semi-Markov queue with exponential service times.* Our second example is a semi-Markov queue in which the arrival process is a (general) semi-Markov process, and the service time of a customer, who finds the underlying Markov chain in state i, is exponentially distributed with rate μ_i. For more details and proofs see [10]. We denote for $i = 1,\ldots,N$, $j = 1,\ldots,N$, $Re(\phi) \geq 0$,

$$A_{i,j}(\phi) = \mathbf{E}[exp(-\phi A_{n+1})1(Y_{n+1} = j) \mid Y_n = i],$$

and let $A(\phi)$ be the matrix with elements $A_{i,j}(\phi)$ so that

$$G(\phi,\psi) = diag\left\{\frac{\lambda_1}{\psi+\lambda_1},\ldots,\frac{\lambda_N}{\psi+\lambda_N}\right\}A(\phi).$$

Let $H(r,\phi) = I - rG(-\phi,\phi)$. Again using Theorem 11.3, it can be shown that for $|r| < 1$, or $\rho < 1$ and $|r| \leq 1$, for fixed r, the determinant $\det H(r,\phi)$ has exactly N zeros $\mu_1(r),\ldots,\mu_N(r)$ in the left half-plane $Re(\phi) < 0$. Let $B_k(r)$ be a (nonunique) nonzero row vector satisfying

$$B_k(r)H(r,\mu_k(r)) = 0,$$

and let $B(r)$ be the matrix with rows $B_k(r)$. Moreover, define the matrix $L(r)$ with elements $L_{i,j}(r)$ by

$$L_{i,j}(r) = \frac{B_{i,j}(r)}{\lambda_j + \mu_i(r)},$$

for $i = 1,\ldots,N$, $j = 1,\ldots,N$. Let

$$K(r,\phi) = I + L^{-1}(r)diag\left\{\frac{1}{\phi-\mu_1(r)},\ldots,\frac{1}{\phi-\mu_N(r)}\right\}B(r).$$

We now have the following factorization of the symbol H:

$$H(r,\phi) = K^+(r,\phi)K^-(r,\phi), \quad Re(\phi) = 0,$$

where

(a) $K^+(r,\phi) = K^{-1}(r,\phi)$ satisfies $\widetilde{A}^+$ and

(b) $K^-(r,\phi) = K(r,\phi)H(r,\phi)$ satisfies $\widetilde{A}^-$.

From Theorem 11.7 we conclude that, for $|r| < 1$ and $Re(\phi) \geq 0$,

$$Z(r,\phi) = [I - rA(0)]^{-1}K^{-1}(r,0)K(r,\phi).$$

If $\rho < 1$, we find, for the steady state distribution of the waiting times for $Re(\phi) \geq 0$,

$$\mathbf{E}[exp(-\phi W)] = \pi K^{-1}(0)K(\phi),$$

where

$$K(\phi) = \lim_{r \uparrow 1} K(r,\phi).$$

11.5 THE MULTISERVER GI/PH/s QUEUE (WITH PHASE-TYPE SERVICE TIMES)

11.5.1 Introduction

We consider the multiserver queue, GI/PH/s, with phase-type service times. The model is described as follows. The interarrival times A_n, $n = 1, 2, \ldots$ are i.i.d. with common distribution function F and expectation α. The service times B_n, $n = 0, 1, \ldots$ are i.i.d. with common distribution function G, which is a phase-type distribution with irreducible representation (γ, S) of size m (see [13]). The expectation of B_n is denoted by $\beta = -\gamma S^{-1}\mathbf{1}$. The sequences of interarrival times and service times are independent. There are s servers and the service discipline is first come, first served.

Let W_n be the waiting time of the nth customer and $W_{n,i}$, $i = 1, \ldots, s$, the service backlog or workload of the ith server just before the arrival of the nth customer, i.e., if the nth and subsequent customers would not enter the system then the ith server would become idle at time $T_n + W_{n,i}$. Since the queue discipline is first come, first served, we may assume that, in front of each server, there is a separate queue and that an arriving customer joins the queue of the server with the smallest workload. In the case when there are several servers with this smallest workload, the arriving customer will select one of them at random. Hence,

$$W_n = \min_{1 \le i \le s} W_{n,i}.$$

We assume that, at $T_0 = 0$, the 0th customer finds the system empty, i.e., $W_{0,i} = 0$, $i = 1, \ldots, s$. We denote

$$U_{n,i} = W_{n,i} - W_n, \quad i = 1, \ldots, s.$$

The Laplace-Stieltjes transform of F is denoted by A, i.e.,

$$A(\phi) = \int_0^\infty exp(-\phi x) dF(x), \quad Re(\phi) \ge 0.$$

11.5.2 The system of Wiener-Hopf-type equations

All vectors in this and the next subsection, unless indicated otherwise, are m-dimensional row vectors. A vector x has components $x_1, \ldots, x_m$. The inner product of two vectors x and y will be denoted by $xy = x_1y_1 + \ldots + x_my_m$. We write $x \le y$ if $x_1 \le y_1, \ldots, x_m \le y_m$. The vector $(x_1, \ldots, x_{i-1}, x_i \pm 1, x_{i+1}, \ldots, x_m)$ is denoted by $x \pm 1_i$ and the vector $(x_1 \pm y_1, \ldots, x_m \pm y_m)$ by $x \pm y$.

Let R_m^k be the class of vectors which have nonnegative integer components and for which $x\mathbf{1} = x_1 + \ldots + x_m = k$. Obviously R_m^k contains $\binom{m+k-1}{k}$ elements. For brevity, we shall write $c(k)$ instead of $\binom{m+k-1}{k}$.

We again assume that, in front of each server, there is a queue and that an arriving customer lines up to the server with the smallest workload. We see that $U_{n,i}$, if nonzero, is the residual service time at time $T_n + W_n$ of the last customer who joined the queue of server i before T_n. If at time $T_n + W_n -$, the residual service time of this customer is in phase j, $j \in \{1, \ldots, m\}$, we say that at $T_n -$, server i is in phase j. Let $X_{n,j}$ be the number of servers who at $T_n -$ are in phase j; the vector $X_n = (X_{n,1}, \ldots, X_{n,m})$ is called the phase vector of the system at time $T_n -$, and we see that $X_n \in \bigcup_{k=0}^{s-1} R_m^k$. Obviously W_n, X_n, $n = 0, 1, \ldots$, is a vector Markov process. Let X_n^* be the phase vector at T_n, i.e., X_n^* gives the phases at $T_n -$ as well as the initial phase of the nth customer, i.e., $X_n^* = X_n + 1_i$

with probability γ_i. Define

$$V_n = min^{(2)}(U_{n,1},\dots,U_{n,s},B_n),$$

where $min^{(2)}(x_1,\dots,x_k)$ is the smallest but one element of $(x_1,\dots,x_k)$. Of course, $min(U_{n,1},\dots,U_{n,s},B_n) = 0$. All random variables $U_{n,1},\dots,U_{n,s},B_n$ are phase-type with the same matrix T and initial states given by X_n^*. Now the joint distribution of V_n, A_{n+1} and X_{n+1}, given the past of the process, is completely determined by X_n^* and hence by X_n, or $((V_n, A_{n+1}, X_{n+1}),\ n = 0,1,\dots)$ is a Markov-renewal sequence, which is completely described by the functions

$$G_{x,y}(\phi,\psi) = \mathbf{E}[exp(-\phi A_{n+1} - \psi V_n)1(X_{n+1} = y) \mid X_n = x],$$

for $Re(\phi) \geq 0$, $Re(\psi) \geq 0$ and $x, y \in \bigcup_{k=0}^{s-1} R_m^k$. We see that we are now in the same situation as in Section 11.4 so that the solution given in Theorem 11.7 also applies here, assuming that we can find the Wiener-Hopf factors K^+ and K^-. The analysis, which is required to obtain these factors, is in general not easy. Even determining the function G explicitly may be quite hard. We have solved this problem for one important special case, viz., the system with hyper-exponential service times.

11.5.3 The multiserver GI/H$_m$/s queue (with hyperexponential service times)

For the queue GI/H$_m$/s, the service time distribution G is given by

$$G(x) = \begin{cases} \sum_{i=1}^m p_i(1 - exp(-b_i x)), & x \geq 0, \\ 0, & x < 0. \end{cases}$$

More details about this model can be found in [6], [7] and [8]. Assume that at $T_0 = 0$ the system is empty, and define for $|r| < 1$, $Re(\phi) \geq 0$,

$$Z(r;x) = \sum_{n=0}^{\infty} r^n \mathbf{P}(X_n = x), \quad x \in \bigcup_{k=0}^{s-2} R_m^k,$$

$$Z^*(r;x) = \sum_{i=1}^{m} p_i Z(r; x - 1_i), \quad x \in R_m^{s-1},$$

and

$$Z(r,\phi;x) = \sum_{n=0}^{\infty} r^n \mathbf{E}[exp(-\phi W_n)1(X_n = x)], \quad x \in R_m^{s-1}.$$

Let Y_n be the phase vector corresponding to those $U_{n,1},\dots,U_{n,s},B_n$, which are not equal to 0 or V_n, i.e., $Y_n = X_n^* - 1_j$ with probability $x_j b_j / xb$. For $|r| < 1$, $Re(\phi) \geq 0$, we define

$$D(r,\phi;x) = \sum_{n=0}^{\infty} r^{n+1} \mathbf{E}[exp(\phi[W_n + V_n - A_{n+1}]^-)1(Y_n = x)], \quad x \in R_m^{s-1}.$$

Let $Z(r,\phi)$, $D(r,\phi)$ and $Z^*(r)$ be the $c(s-1)$-dimensional column vectors with components $Z(r,\phi;x)$, $D(r,\phi;x)$, and $Z^*(r;x)$, $x \in R_m^{s-1}$, respectively.

The $c(s-1) \times c(s-1)$ matrix H is defined by

$$H_{x,x}(r,\phi) = 1 - r\sum_{j=1}^{m} p_j \frac{(x_j + 1)b_j}{\phi + xb + b_j} A(-\phi)$$

and $$H_{x,x+1_j-1_i}(r,\phi) = -rp_i\frac{(x_j+1)b_j}{\phi+xb+b_j}A(-\phi),\quad i \neq j,\ x_i > 0,$$
$$H_{x,y}(r,\phi) = 0, \quad \text{otherwise.}$$

For integer-valued vectors x, y and z, we write

$$\binom{x}{y;z} = \prod_{i=1}^{m} \frac{x_i!}{y_i! z_i! (x_i - y_i - z_i)!},$$

and we adopt the convention that $\binom{x}{y;z} = 0$ if for some i, $y_i < 0$, $z_i < 0$ or $x_i - y_i - z_i < 0$.

With the above definitions and notations, we have the following system of equations.

Theorem 11.10 *For $x \in R_m^k$, $k = 0, 1, \ldots, s-2$,*

$$Z(r;x) = \delta_{x1,0} + A(xb)\sum_{i=1}^{m} p_i Z(r; y - 1_i)$$

$$+\sum_{l=k+1}^{s-2} \sum_{y \in R_m^l} \sum_{\nu \leq y - x} \binom{y}{x;\nu} (-1)^{\nu 1} A(xb + \nu b) \sum_{i=1}^{m} p_i Z(r; y - 1_i)$$

$$+\sum_{y \in R_m^{s-1}} \sum_{\nu \leq y - x} \binom{y}{x;\nu} (-1)^{\nu 1} D(r, xb + \nu b; y)$$

and, for $x \in R_m^{s-1}$, $Re(\phi) = 0$,

$$H(r,\phi)Z(r,\phi) = rA(-\phi)Z^*(r) + Z(r,0) - D(r,-\phi).$$

This system of equations is solved by factorizing the symbol H. Below we give a sketch of the solution. Using Theorem 11.3 we can show that, under some technical conditions on F and G, which are almost always satisfied, $\det H(r,\phi)$ has, for $|r| < 1$, or $|r| \leq 1$ and $\rho < 1$, exactly $c(s)$ zeros $\mu_x(r)$, $x \in R_m^s$ in the left half-plane $Re(\phi) < 0$. Define the $c(s) \times c(s)$-matrix by

$$J = diag\{\mu_x(r), x \in R_m^s\}.$$

For $y \in R_m^s$, let B_y be a (nonunique) nonzero $c(s-1)$ dimensional vector satisfy ing

$$H(r, \mu_y(r))B_y = 0,$$

and let B be the $c(s-1) \times c(s)$-matrix whose columns are the B_y. Moreover, we define the $c(s) \times c(s)$-matrix L by

$$L_{x,y} = \sum_{\{i \mid x_i > 0\}} p_i B_{x-1_i, y} \frac{1}{\mu_y(r) + xb}, \quad x, y \in R_m^s$$

and the $c(s) \times c(s-1)$-matrix M by

$$M_{x,y} = \sum_{i=1}^{m} p_i \delta_{x, y+1_i}, \quad x \in R_m^s, y \in R_m^{s-1}.$$

For $C = L^{-1}M$, we finally have

$$K(r,\phi) = I + BJ(\phi)C,$$

which gives the factorization

$$H(r,\phi) = H^{-}(r,\phi)H^{+}(r,\phi),$$

where

(*a*) $H^{+}(r,\phi) = K^{-1}(r,\phi)$ satisfies $\widetilde{A}^{+}$ and

(*b*) $H^{-}(r,\phi) = H(r,\phi)K(r,\phi)$ satisfies $\widetilde{A}^{-}$.

We finally have the following theorem.

Theorem 11.11. *Under some technical conditions on F and G, which are almost always satisfied, we have, for* $|r| < 1$, $Re(\phi) \geq 0$,

$$Z(r,\phi) = K(r,\phi)K^{-1}(r,0)Z(r,0).$$

From the first system of equations in Theorem 11.10, we can find $Z(r,0)$. Theorem 11.11 implies that the steady state distribution of the waiting times is a mixture of $c(s)$ exponentials and a concentration at 0.

11.6 OPEN PROBLEMS AND RESEARCH DIRECTIONS

There are several directions in which the methods applied here could be extended, such as

(*i*) other special cases of the class of single-server semi-Markov queues, e.g., the MAP/SM/1 queue;

(*ii*) special cases of the multiserver queue with phase-type service times other than the case of hyperexponential service times, in particular queues with Erlang and hypoexponential distributions;

(*iii*) models which do not belong to the two classes discussed here, e.g., fluid models in queueing theory (see Rogers [16]);

(*iv*) numerical analysis of the time dependent behavior, when the system is initially empty; also in the case when it is initially not empty, numerical inversion of generating functions and Laplace transforms can be applied (see [1]);

(*v*) other quantities than those considered here; in particular, it seems that for the models studied here, Poisson's equation can be solved by the same methods (see Glynn [11] and Bladt and Asmussen [4]).

Other special cases of the SM/SM/1 queue can be treated along the same lines as the examples discussed here. A further investigation of numerical aspects, such as root finding in the complex plane would be of interest. Multiserver queues in which the service time distribution is not (hyper) exponential give rise to difficult problems. These problems are already present in simple special cases like $G/E_2/2$ for which the form of the steady state distributions of waiting and queue lengths is not known. A solution to the factorization problem for these models would give considerable insight into the structure of these steady state distributions. Little work has been done on the time dependent behavior of queues. The expressions for generating functions with respect to the time, which are derived in this chapter, can be used for the study of time dependence by applying numerical inversion.

BIBLIOGRAPHY

[1] Abate, J. and Whitt, W., Numerical inversion of Laplace transforms of probability distributions, *ORSA J. Computing* (1994), (to appear).

[2] Arjas, E., On a fundamental identity in the theory of semi-Markov processes, *Adv. Appl. Prob.* **4** (1972), 271-284.

[3] Bart, H., Gohberg, I., Kaashoek, M.A., *Minimal Factorization of Matrix and Operator Functions*, Birkhäuser Verlag, Basel 1979.

[4] Bladt, M. and Asmussen, S., Poisson's equation for queues driven by a Markovian point process, *Queueing Sys.* **17** (1994).

[5] Cohen, J.W., *The Single-Server Queue*, North Holland, Amsterdam 1969.

[6] De Smit, J.H.A., The queue GI/M/s with customers of different types or the queue $GI/H_m/s$, *Adv. Appl. Prob.* **15** (1983), 392-419.

[7] De Smit, J.H.A., A numerical solution for the multiserver queue with hyperexponential service times, *Oper. Res. Letters* **2** (1983), 217-225.

[8] De Smit, J.H.A., The queue $GI/H_m/s$ in continuous time, *J. Appl. Prob.* **22** (1985), 214-222.

[9] De Smit, J.H.A., The single-server semi-Markov queue, *Stoch. Proc. Appl.* **22** (1986), 37-50.

[10] De Smit, J.H.A. and Regterschot, G.J.K., A semi-Markov queue with exponential service times, In: *Proc. Internat. Symp. on Semi-Markov processes*, ed. by J. Janssen, Plenum, New York (1986), 369-382.

[11] Glynn, P.W., Poisson's equation for the recurrent M/G/1 queue, *Adv. Appl. Prob.* (1994), (to appear).

[12] Lucantoni, D.M., New results on the single-server queue with a batch Markovian arrival process, *Stoch. Mod.* **7** (1991), 1-46.

[13] Neuts, M.F., *Matrix Geometric Solutions for Stochastic Models*, John Hopkins University Press, Baltimore 1981.

[14] Neuts, M.F., *Structured Stochastic Matrices of* M/G/1 *Type and Their Applications*, Marcel Dekker, New York 1989.

[15] Regterschot, G.J.K. and De Smit, J.H.A., The queue M/G/1 with Markov-modulated arrivals and services, *Math. Oper. Res.* **11** (1986), 465-483.

[16] Rogers, L.C.G., Fluid models in queueing theory and Wiener-Hopf factorization of Markov chains, *Ann. Appl. Prob.* **4** (1994), 390-413.

[17] Silverman, R.A., *Complex analysis with applications*, Prentice Hall, Englewood Cliffs, NJ 1974.

Chapter 12

Applications of singular perturbation methods in queueing[1]

Charles Knessl and Charles Tier

ABSTRACT A survey is presented describing the application of singular perturbation techniques to queueing systems. The goal is to compute performance measures by constructing approximate solutions to specific problems involving either the Kolmogorov forward or backward equation which contain a small parameter. These techniques are particularly useful on problems for which exact solutions are not available. Four different classes of problems are surveyed: (*i*) state-dependent queues; (*ii*) systems with a processor-sharing server; (*iii*) queueing networks; (*iv*) time dependent behavior. For each class, an illustrative example is presented along with the direction of current research.

CONTENTS

12.1	Introduction	311
12.2	State-dependent queues	315
12.3	Processor-shared queues	324
12.4	Networks	326
12.5	Transient behavior of queues	329
12.6	Open problems and research directions	332
	Bibliography	332

12.1 INTRODUCTION

In analyzing queueing models, one would like to compute certain performance measures, such as the steady state queue length distribution, transient queue length distribution, mean length of a busy period, unfinished work distribution, sojourn time distribution, time for the queue to reach some specified number, etc. For a specific model, these quantities may all be characterized as solutions to certain equations. Thus, computing the performance measures amounts to solving these equations together with appropriate boundary/initial conditions. Given a Markov process $X(t)$, the transition probability density $p(x,t\,|\,x_0,t_0) = \mathbf{P}[X(t) = x\,|\,X(t_0) = x_0]$ satisfies the forward and backward Kolmogorov equations, which we write in an abstract form as

$$\frac{\partial p}{\partial t} = L_{x,t}p \quad (t > t_0), \quad p(x,t_0^+\,|\,x_0,t_0) = \delta(x - x_0) \tag{12.1}$$

$$-\frac{\partial p}{\partial t_0} = L^*_{x,t_0}p \quad (t_0 < t), \quad p(x,t_0\,|\,x_0,t_0^-) = \delta(x - x_0). \tag{12.2}$$

---

[1]This research was supported in part by NSF Grant DMS-93-00136 and DOE Grant DE-FG02-93ER25168.

0-8493-8074-x/95/$0.00+$.50

Here, L is a linear operator that involves the variable x and time t, and L^* is its adjoint. The δ in (12.1)-(12.2) is the Kronecker delta, if the state space is discrete, and the Dirac delta, if the state space is continuous. The precise forms of L and L^* depend on whether we are looking at a discrete model (such as the number $N(t)$ of customers in an M/M/1 queue) or a continuous model (such as the unfinished work $U(t)$ in an M/G/1 queue), or some combination of these. For example, in considering the joint queue length distribution in a network of Markovian queues, L is generally a multidimensional difference operator. If the model is space and time homogeneous, which occurs say if the arrival and service rates are constant, then L is a constant coefficient operator. However, the form of L is generally different near the boundaries of the state space, so that inherent to the problems (12.1)-(12.2) are complicated sets of boundary conditions. These make it difficult to obtain *simple, exact* solutions to (12.1)-(12.2) for all but the simplest of queueing models.

For the unfinished work process $U(t)$, the operator L is an integro-differential operator since the process has nonlocal transitions. The difficulty in solving (12.1)-(12.2) has led to the introduction of approximations. A popular tool in queueing theory is the use of diffusion approximations. Here, one replaces the original process $X(t)$ by a diffusion process $\widetilde{X}(t)$ (time may need to be scaled). Then, computing $\widetilde{p}$, the transition density for $\widetilde{X}$, involves solving (12.1)-(12.2) with L now being a partial differential operator of second order. This may itself be a difficult task since the boundary conditions associated with the approximate problem are frequently still complicated, especially for models in more than one space dimension.

We have thus far discussed (12.1)-(12.2) in terms of Markovian models. For models with general interarrival time and/or service time distributions, the processes of interest are no longer Markovian but may be imbedded in a (higher-dimensional) Markov process by using the method of supplementary variables (see [8]). For the new process, we can again obtain (12.1)-(12.2) except that L now depends on (x, y), with y being the vector of supplementary variables, and is usually more complicated than the L associated with exponential arrivals/service. Even for non-Markovian models, it is still generally easy to derive the appropriate problems (12.1)-(12.2); the difficult task is the solution of the equations.

In the queueing literature, exact solutions to (12.1)-(12.2) are generally available only for steady state ($t \to \infty$) problems in one and two dimensions and for time-dependent problems in one dimension. For example, the steady state distribution of $N(t)$ for the M/M/1 queue has a very simple form, but the form for the transient distribution is a complicated expression involving an infinite sum of Bessel functions. For the M/G/1 queue, there is an explicit expression for the Laplace transform of the steady state distribution of $N(t)$ (or $U(t)$), though the transform cannot be (analytically) inverted for general service time densities. The time dependent distribution is very complicated. Its double transform (over space and time) can be characterized in terms of the solution to a functional equation, but there is no hope of inverting the transform. For the GI/G/1 model, solving for even the steady state distribution of $U(t)$ is equivalent to solving a Wiener-Hopf integral equation with a general kernel. Solution of this problem can be expressed in terms of two complex contour integrals, one for inverting a Laplace transform and the other for the analytical solution of the Wiener-Hopf problem. This shows that even for one-dimensional models, one cannot obtain simple analytic expressions for the various performance measures. For problems in more than one dimension, the situation is even worse. Jackson networks, or more generally, "product-form" networks, are an important class of multidimensional models for which one can explicitly obtain the steady state queue length distribution. However, obtaining time-dependent information for these models is much harder. Even for Jackson networks, it is difficult to analyze the busy period and various other "first-passage time" problems. Solutions to non-product form networks are complicated even in two dimensions under Markovian assumptions. Using transforms and function-theoretical arguments, these models ([6]) may be reduced to solving certain classic problems in the theory of singular integral equations, such as Dirichlet and Riemann-Hilbert problems. However, one is again left with inverting a two-dimensional transform, which itself is often characterized in a form that is not particularly explicit.

From the above (brief and incomplete) summary of exactly solvable models, it is clear that approximations must play a major role in the analysis of queues. Here we examine a set of methods called "asymptotic and singular perturbation techniques". Their role in queueing theory is basically twofold. First, they can be used to simplify exact solutions when these are available. Since the exact solutions discussed above are extremely complicated, it is useful to evaluate these expressions in certain limiting cases, in order to gain more insight into the qualitative structure of the particular model. Asymptotic formulas often clearly show the dependence of the solutions on the various variables/parameters in the problem, whereas the full exact expressions may be difficult to interpret in terms of the underlying model. Of course, an asymptotic formula can never contain as much quantitative (numerical) information as an exact answer, but it can provide reasonably accurate numerical results at a greatly reduced computational cost. Also, a queueing model is itself an approximation to a physical system. Thus, we believe that obtaining qualitative information is just as important as obtaining accurate numerical values.

A second, and we believe more important, aspect of using perturbation methods is to make progress on problems for which exact solutions are not available. Since it is likely that most models in queueing theory (and indeed in any other area of applied mathematics) will never be solved exactly, obtaining useful approximations is very important. What do we mean by "useful"? The two main criteria for the usefulness of an asymptotic approximation are (*i*) its numerical accuracy and (*ii*) its ability to make transparent qualitative properties of the solution. It is desirable to have both (*i*) and (*ii*), but it is much better to have either (*i*) or (*ii*) than to have nothing. The verification of (*i*) can only be obtained by comparing the approximation to exact results or to numerical approximations, assuming the latter are reliable. Deciding on (*ii*) is somewhat harder as what looks complicated to one person may look simple to another. We believe that, if an asymptotic result can be expressed in terms of elementary functions (sines, exponentials, Gaussians) or well-studied special functions (Bessel, parabolic cylinder), it can be called "simple" enough to be useful. In deciding whether an asymptotic answer is simple, it is also appropriate to view this result in terms of the exact answer, and how complicated it is, or would be (if it could be obtained at all).

Using asymptotics to simplify exact expressions usually involves the approximate evaluation of sums and integrals, using ideas such as Laplace's method, the method of steepest descent (saddle point method), the Euler-MacLaurin formula, Poisson summation, integral representations of sums, etc. These have been used to good effect in queueing theory; examples are [49], where the authors obtained asymptotic expansions for product-form networks for large populations sizes, and [54], where integral representations and subsequent asymptotics were used on processor-sharing models. Of course, these methods assume that one has an exact (and sufficiently explicit) representation for the quantity to be evaluated.

Many other asymptotic methods exist which may be applied directly to the equation(s) satisfied by the given performance measure. They do not rely on having an exact solution, so that they are clearly applicable to a much wider class of problems. These methods include the *WKB* method, "boundary layer" techniques, the ray method, and the method of matched asymptotic expansions. They are called "singular perturbation" techniques (general references are [1, 15]). To see what is meant by singular, consider a function $f(x;\varepsilon)$ which depends on the variable(s) x and an additional (small) parameter ε. If f is an analytic function of ε, i.e., if the series $\sum_{j=0}^{\infty} f_j(x)\varepsilon^j$ converges for $|\varepsilon|$ sufficiently small, then such a series is called a "regular" perturbation series. If either the series diverges (but is still asymptotic for $\varepsilon \to 0$) or the expansion involves a more complicated asymptotic sequence than powers of ε, then the perturbation series is called "singular". Most interesting problems in applied mathematics are of singular perturbation type, and this, we believe, is also true in queueing theory. An important example of a singular perturbation series is the *WKB* form $e^{-\phi(x)/\varepsilon}\sum_{j=0}^{\infty} A_j(x)\varepsilon^j$, which we shall show arises naturally even in very elementary queueing models. Such a series is generally divergent, but even the leading term $(A_0(x)e^{-\phi(x)/\varepsilon})$ is usually an excellent approximation to the quantity that is to be computed.

Another important feature of singular perturbation methods is that such problems tend to contain several scales, which must be treated separately. This leads to several different asymptotic expansions which must be related to one another, and this is usually done by the "asymptotic matching principle". For example, if we consider the m-server M/M/m queue in the limit $m = \varepsilon^{-1} \to \infty$, it is necessary to construct different asymptotic expansions for $N(t) = O(m)$ and $N(t) = O(1)$, which corresponds, respectively, to having a finite fraction of the servers occupied and to having just a few occupied servers.

The methods discussed above have usually been developed in the context of second order equations (ODEs and PDEs). Also, an individual method was usually introduced and developed in the context of a particular scientific application. The *WKB* method was first used as an approximation tool for solving the Schrödinger equation in quantum mechanics; the ray method was developed to solve problems in high frequency wave propagation; boundary layer ideas were first used in the study of viscous flow past obstacles. From our point of view, however, we consider these methods as mathematical techniques for approximately solving equations. They are useful for analyzing queueing models for which the operators L and L^* in (12.1)-(12.2) depend upon a small parameter, call it ε. The size of ε can be used to simplify the equations in a systematic way, and thus to obtain explicit, approximate formulas. Typically, ε measures the reciprocal of the size the system; e.g., $\varepsilon = m^{-1}$ where m is the number of servers in an m server queue, or ε may be the inverse of the customer population in a large, closed Jackson network. The most commonly used asymptotic approximations in queueing theory are light traffic and heavy traffic. For these, ε may be taken as the arrival rate and the difference between arrival and service rates, respectively.

For most models in queueing theory, (12.1)-(12.2) involve difference equations, integral equations, delay equations and various combinations of these. This is because $N(t)$ takes on discrete values and $U(t)$ has jumps (non-local transitions) at arrival times. For these types of problems, singular perturbation methods are not as well developed as they are for ordinary and partial differential equations. Thus, developing methods for solving (12.1)-(12.2) will also enhance the scope of singular perturbation techniques, as they can now be applied to integral and other equations, and these arise naturally in many fields of science and engineering. Our primary goal is to compute solutions to specific problems that arise in queueing theory and in other stochastic models. However, the mathematical methodology should also be useful in other areas.

In the sections that follow, we will apply perturbation methods to some specific queueing models. We show that they are useful for several different classes of problems. In Section 12.2, we consider queues which have state-dependent parameters. State-dependent (and time-dependent) queues lead to an operator L in (12.1)-(12.2) that is not "constant coefficient", and such problems are difficult (or impossible) to tackle using transform methods. In Section 12.3, we consider several queues which have a processor-sharing server. We shall obtain approximations to the sojourn time through such systems. In Section 12.4, we consider queueing networks. For product-form networks, there exist explicit expressions for the joint, steady-state queue length distribution. We show how to compute this, asymptotically, from a recursion which is satisfied by the normalization constant (partition function). We also show, in Section 12.4, how to use perturbation methods to analyze bottlenecks in networks. For the simplest Jackson network consisting of two M/M/1 queues in tandem, we compute the time until the network population becomes large and then, the time needed to settle back to its equilibrium state. For these first passage time problems, there seem to be no exact expressions available. In Section 12.5, we show how to use the ray method to compute the time-dependent behavior of queueing models. As an example, we consider the Erlang loss model (M/M/m/m queue).

12.2 STATE-DEPENDENT QUEUES

We shall first consider the classical repairman problem, which corresponds to the finite source (finite population) M/M/1 queue. Then we extend our results to systems which have a general service time distribution. We also consider an M/G/1 queue characterized by the unfinished work $U(t)$, and which has state-dependent arrivals and service.

12.2.1 Repairman problems

Denote the service rate by μ_0, and let $p_n(t) = \mathbf{P}\{N(t) = n\}$ with $p_n = p_n(\infty)$. If M is the total number of customers that are in the population, then this model corresponds to a queue with a state-dependent arrival rate equal to $\lambda(M - N(t))$. The steady state balance equations are

$$(\mu_0 + \lambda(M-n))p_n = \lambda(M+1-n)p_{n-1} + \mu_0 p_{n+1}; \quad 1 \le n \le M \tag{12.3}$$

$$\lambda M p_0 = \mu_0 p_1 \tag{12.4}$$

with $p_{M+1} \equiv 0$. Solving (12.3)-(12.4) and normalizing the probabilities, we easily obtain

$$p_n = p_0 \frac{M!}{(M-n)!}\left(\frac{\lambda}{\mu_0}\right)^n, \quad p_0 = 1 \Big/ \sum_{n=0}^{M} \frac{M!}{(M-n)!}\left(\frac{\lambda}{\mu_0}\right)^n. \tag{12.5}$$

Now consider the asymptotic limit $M \to \infty$, $\mu_0 = \mu M = O(M)$ which corresponds to systems with many customers and a fast server. Letting $\rho = \lambda/\mu = \lambda M/\mu_0 = O(1)$ and approximating $M!$ and $(M-n)!$ by Stirling's formula, we obtain

$$p_n \sim \frac{p_0}{\sqrt{1-\xi}} e^{-M\phi(\xi)}, \quad \xi = \frac{n}{M} \tag{12.6}$$

$$\phi(\xi) = \xi \log\left(\frac{1}{\rho}\right) + \xi + (1-\xi)\log(1-\xi).$$

This approximation is valid for all $\xi \in [0,1)$ as $M \to \infty$. It ceases to be valid when $\xi \approx 1$ (i.e., $M - n = O(1)$), but, by then, p_n is exponentially small. Next we use the integral representation

$$1/p_0 = \frac{M}{\rho} \int_0^\infty \exp\left(-M\left[\frac{x}{\rho} - \log(1+x)\right]\right) dx \tag{12.7}$$

to obtain the expansion for p_0. The integral in (12.7) is a Laplace type integral whose asymptotic expansion is easily obtained and is different for $\rho > 1$, $\rho = 1$ and $\rho < 1$. We have

$$\begin{aligned} p_0 &\sim 1 - \rho; \quad \rho < 1 \\ &\sim \sqrt{2/(\pi M)}, \quad \rho = 1 \\ &\sim \frac{1}{\sqrt{2\pi M}} \exp\left(M\left[1 - \frac{1}{\rho} + \log\left(\frac{1}{\rho}\right)\right]\right), \quad \rho > 1. \end{aligned} \tag{12.8}$$

We could also obtain a result for $\rho \approx 1$, valid for $\rho - 1 = O(M^{-1/2})$, which will asymptotically match between first and third formulas in (12.8). From (12.6) and (12.8), we observe that p_n is peaked at $\xi = 0$ if $\rho \le 1$ and at $\xi = 1 - \rho^{-1}$ if $\rho \ge 1$. Also, the probability, that the system is empty, has the asymptotic orders of magnitude $O(1)$, $O(M^{-1/2})$, and $O(M^{-1/2}e^{-cM})$ (i.e., $-c = 1 - \rho^{-1} - \log \rho < 0$) when $\rho < 1$, $\rho = 1$, and $\rho > 1$, respectively. This shows that the asymptotic structure of this problem is very sensitive to the value of ρ.

Now we present an alternate approach to the asymptotics, which uses only (12.3). We scale $n = M\xi$ with $p_n = P(\xi)$ to get

$$(1 + \rho(1-\xi))P(\xi) = \rho(1 - \xi + \varepsilon)P(\xi - \varepsilon) + P(\xi + \varepsilon) \tag{12.9}$$

where $\varepsilon \equiv M^{-1}$. This is a difference equation with small differences which resembles a singularly perturbed ODE, as can be seen by expanding the right side of (12.9) for small ε. We seek solutions of (12.9) in the WKB form

$$P(\xi) = C(\varepsilon)e^{-\phi(\xi)/\varepsilon}[A_0(\xi) + \varepsilon A_1(\xi) + \ldots]. \tag{12.10}$$

The constant $C(\varepsilon)$ will be determined by normalization and we set $A_0(\xi) = A(\xi)$. Using (12.10) in (12.9) and expanding for small ε, we obtain, at the first two orders, the equations

$$1 + \rho(1-\xi) = \rho(1-\xi)e^{\phi'(\xi)} + e^{-\phi'(\xi)} \tag{12.11}$$

and

$$\begin{aligned} 0 = {} & \rho e^{\phi'(\xi)}[A(\xi) - (1-\xi)(A'(\xi) + \tfrac{1}{2}\phi''(\xi)A(\xi))] \\ & + e^{-\phi'(\xi)}[A'(\xi) - \tfrac{1}{2}\phi''(\xi)A(\xi)]. \end{aligned} \tag{12.12}$$

Equation (12.11) is a nonlinear ODE for $\phi(\xi)$, but it is very easy to solve since it is a quadratic equation for $e^{\phi'(\xi)}$. One root of this quadratic is clearly $\phi' = 0$ and the other is

$$\phi'(\xi) = -\log \rho - \log(1-\xi)$$

which integrates to

$$\phi(\xi) = -\xi \log \rho + \xi + (1-\xi)\log(1-\xi), \tag{12.13}$$

where we have chosen $\phi(0) = 0$, since $\phi(0)$ can be incorporated into the constant $C(\varepsilon)$ in (12.10). With (12.13), (12.12) is a *linear, first order* ODE for $A(\xi)$. It is easily solved (up to a multiplicative constant) to yield

$$A(\xi) = (1-\xi)^{-1/2}. \tag{12.14}$$

The solution $\phi' = 0$ must be rejected since it would lead to a nonintegrable $A(\xi)$. Now, we have the approximation $P(\xi) \sim C(\varepsilon)Ae^{-\phi/\varepsilon}$, and it remains only to determine the constant $C(\varepsilon)$, which can be done from the normalization

$$C(\varepsilon)\sum_{n=0}^{1/\varepsilon} e^{-\phi(\varepsilon n)/\varepsilon}[A(\varepsilon n) + \varepsilon A_1(\varepsilon n) + \ldots] = 1.$$

This sum may be approximated for $\varepsilon \to 0$ using Laplace's method for sums and the Euler-MacLaurin formula. Using (12.13)-(12.14), we would find that $C(\epsilon)(\sim p_0)$ is again asymptotically given by (12.8) for the 3 cases of ρ.

Expressions (12.13)-(12.14) agree precisely with (12.6), which we obtained using the exact result and Stirling's formula. To obtain corrections to this leading order approximation, we could use the full Stirling series to approximate $M!$ and $(M-n)!$ in (12.5) and then, obtain the full asymptotic series for the integral in (12.7) using Laplace's method. Alternately, we could continue the expansion of (12.9) using (12.10). The correction terms $A_j(\xi)$, for $j \geq 1$, will satisfy linear ODEs of the same form as (12.12), and these are easy to solve.

What have we gained from our direct approach of using (12.3) (or (12.9)) instead of the full solution? Equation (12.3) is a linear second order difference equation, whereas (12.11) is nonlinear and first order, and (12.12) is linear and first order. It turns out to be slightly easier to solve (12.11)-(12.12) than (12.3). The simplification is not so dramatic for this simple model, but the advantages of the direct approach will be much more apparent when we deal with general service time distributions, transient problems, and problems in more than one dimension. It turns out that it is frequently much easier to solve PDEs of the first order than it is to solve PDEs or difference equations of second order. This is true even if the first order problems are nonlinear.

Next we consider the same model but with a general service time distribution, whose density we denote by $\tilde{b}(x)dx = \mathbf{P}\{\text{service time } \in (x, x+dx)\}$. We allow $\tilde{b}$ to be a delta function so that this analysis also applies to M/D/1 models. Now $N(t)$ is no longer Markov, so we consider the process $(N(t), Y(t))$, where the supplementary variable $Y(t)$ measures the elapsed service time of the customer presently being served. We let

$$p_n(y,t)dy = \mathbf{P}\{N(t) = n, Y(t) \in (y, y+dy)\}, \quad n \geq 1$$

$$p_0(t) = \mathbf{P}\{N(t) = 0\}$$

and denote the steady-state limits by $p_n(y), p_0$; and the marginal (steady-state) queue length distribution by $p_n = \int_0^\infty p_n(y)dy \quad (n \geq 1)$. The balance equations are now

$$p_n'(y) = \lambda(M-n+1)p_{n-1}(y) - [\lambda(M-n) + \widetilde{\mu}(y)]p_n(y); \quad 2 \leq n \leq M \tag{12.15}$$

$$p_n(0) = \int_0^\infty p_{n+1}(y)\widetilde{\mu}(y)dy; \quad 2 \leq n \leq M-1 \tag{12.16}$$

where $\widetilde{\mu}(y) = \widetilde{b}(y) / \int_y^\infty \widetilde{b}(x)dx$ is the service rate, conditioned on the elapsed service time. The boundary conditions turn out to be

$$p_1'(y) = -[\lambda(M-1) + \widetilde{\mu}(y)]p_1(y) \tag{12.17}$$

$$\lambda M p_0 = \int_0^\infty p_1(y)\widetilde{\mu}(y)dy \tag{12.18}$$

$$p_1(0) = \lambda M p_0 + \int_0^\infty p_2(y)\widetilde{\mu}(y)dy \tag{12.19}$$

$$p_M(0) = 0 \tag{12.20}$$

$$p_0 + \sum_{n=1}^{M} \int_0^\infty p_n(y)dy = 1. \tag{12.21}$$

To solve this system asymptotically, we again assume that M is large and that service times tend to be small. To make the latter more precise we write the service density in the scaled form $\widetilde{b}(x) = Mb(Mx)$ so that the mean service time is $O(M^{-1})$. Setting $\widetilde{\mu}(y) = M\mu(My)$, we introduce into (12.15)-(12.21) the scaled variables $\xi = n/M$, $\eta = My$, $\varepsilon = M^{-1}$ with $p_n(y)dy = P(\xi,\eta)d\eta$ and obtain the scaled problem

$$P_\eta(\xi,\eta) = \lambda(1-\xi+\varepsilon)P(\xi-\varepsilon,\eta) - [\lambda(1-\xi) + \mu(\eta)]P(\xi,\eta) \tag{12.22}$$

$$P(\xi,0) = \int_0^\infty P(\xi+\varepsilon,\eta)\mu(\eta)d\eta. \tag{12.23}$$

For the moment we ignore the boundary conditions (12.17)-(12.21).

When service times were exponential, we needed two expansions for $P(\xi)$, valid on the respective scales $0 \leq \xi < 1$ and $1-\xi = O(\varepsilon)$ $(M-n = O(1))$. For the present model, it is necessary to consider 3 scales: (i) $0 < \xi < 1$, (ii) $\xi = O(\epsilon)$, $(n = O(1))$, and (iii) $1-\xi = O(\varepsilon)(M-n = O(1))$. The third scale is again unimportant, since the distribution is exponentially small there for any value of

$$\rho \equiv \lambda M \int_0^\infty x\widetilde{b}(x)dx = \lambda \int_0^\infty xb(x)dx \equiv \frac{\lambda}{\mu} = O(1).$$

We proceed to asymptotically solve the problem on scales (i) and (ii), and then relate the two expansions by asymptotic matching. After we have covered the entire state space, there will remain one undetermined constant, and this is obtained by normalization (12.21).

When $0 < \xi < 1$ and $M \to \infty$, we set

$$P(\xi,\eta) = e^{-\int_0^\eta \mu(z)dz} C(\varepsilon)\exp\left\{-\left[\frac{1}{\varepsilon}\phi_0(\xi,\eta) + \phi_1(\xi,\eta) + \varepsilon\phi_2(\xi,\eta) + \ldots\right]\right\}. \tag{12.24}$$

The first factor is included for convenience. The form (12.24) is asymptotically equivalent to (12.10), if we identify $A_0 = e^{-\phi_1}$, $A_1 = -A_0\phi_2$, etc. We shall only compute the leading term in (12.24), which means that we must compute ϕ_0 *and* ϕ_1. Using (12.24) in (12.22) and expanding for $\varepsilon \to 0$, we obtain at the first two orders

$$-\phi_{0,\eta} = 0 \tag{12.25}$$

$$-\phi_{1,\eta} = \lambda(1-\xi)(e^{\phi_{0'}\xi} - 1). \tag{12.26}$$

Thus ϕ_0 is independent of η, and we write $\phi_0 = \phi_0(\xi)$. Then the right side of (12.26) depends on ξ only, which we denote by $K(\xi)$, and then

$$\phi_1(\xi,\eta) = -\eta K(\xi) + L(\xi), \tag{12.27}$$

where K, L are as yet undetermined. Using (12.24) with (12.27) in (12.23), we obtain, at leading order, the following equation for K

$$1 + \frac{K(\xi)}{\lambda(1-\xi)} = \int_0^\infty e^{\eta K(\xi)} b(\eta) d\eta, \tag{12.28}$$

which is a transcendental equation that involves the *scaled* service density $b(\cdot)$. Explicit solutions to (12.28) can be obtained for exponential E_2 and H_2 servers. In the general case, (12.28) is easily solved numerically, and a convexity argument shows that (12.28) has a unique nonzero solution.

To determine L (and hence ϕ_1), we must examine the third term in the expansion of (12.22), and then use the result in (12.23). Omitting the details, the final result is a linear ODE for L:

$$\begin{aligned}[1 - \lambda(1-\xi)I_1(\xi)]L'(\xi)& \\ = \tfrac{1}{2}\phi_0''(\xi) + \lambda[1 - \tfrac{1}{2}(1-\xi)\phi_0''(\xi)]I_1(\xi) &- \tfrac{1}{2}K'(\xi)\lambda(1-\xi)I_2(\xi)\end{aligned} \tag{12.29}$$

and

$$I_j(\xi) = \int_0^\infty \eta^j e^{\eta K(\xi)} b(\eta) d\eta.$$

Since ϕ_0, K, I_1, I_2 are known via (12.28), (12.29) is easily integrated. After some algebra, we find that the leading term in (12.24) is

$$\begin{aligned} P(\xi,\eta) &\sim e^{-\int_0^\eta \mu(z)dz} C(\varepsilon) e^{-\phi_0(\xi)/\varepsilon} e^{\eta K(\xi)} e^{-L(\xi)} \\ &= C(\varepsilon) e^{-\int_0^\eta \mu(z)dz} e^{\eta K(\xi)} \\ &\quad \times \left| \frac{(1-\xi)K'(\xi)}{\lambda(1-\xi) + K(\xi)} \right|^{1/2} \exp\left\{ -\frac{1}{\varepsilon} \int_0^\xi \log\left[1 + \frac{K(z)}{\lambda(1-z)}\right] dz \right\}. \end{aligned} \tag{12.30}$$

This expansion is not valid for $\xi = O(\varepsilon)$ $(n = O(1))$ since it does not satisfy the boundary equations (12.17)-(12.20).

To obtain an appropriate expansion for $n = O(1)$, we go back to the discrete space variable n and consider (12.15)-(12.16) for $n = O(1)$. Since $M \to \infty$, the upper boundary disappears. If we expand for $n = O(1)$

$$p_n(y) = D(\varepsilon)[Q_n(\eta) + \varepsilon Q_n^{(1)}(\eta) + \varepsilon^2 Q_n^{(2)}(\eta) + \ldots].$$

Then, at the leading order, we get

$$Q_n'(\eta) = \lambda Q_{n-1}(\eta) - [\lambda + \mu(\eta)]Q_n(\eta), \quad n \geq 2, \tag{12.31}$$

and

$$Q_n(0) = \int_0^\infty Q_{n+1}(\eta)\mu(\eta)d\eta; \quad n \geq 2. \tag{12.32}$$

Now, we must consider the boundary equations (12.17)-(12.19), which imply that

$$Q_1'(\eta) = -[\lambda + \mu(\eta)]Q_1(\eta), \tag{12.33}$$

$$\lambda Q_0 = \int_0^\infty Q_1(\eta)\mu(\eta)d\eta,$$

and

$$Q_1(0) = \lambda Q_0 + \int_0^\infty Q_2(\eta)\mu(\eta)d\eta.$$

The problem (12.31)-(12.33) is simpler than the original problem (12.15)-(12.20) in two respects. First, the domain of (12.31)-(12.33) is $n \geq 0$, so that we have a problem on an infinite interval rather than the finite interval $0 \leq n \leq M$. Second, for $n = O(1)$, we can, to leading order, approximate quantities, such as $\lambda(M-n) \approx \lambda M$, so that (12.31)-(12.33) is basically a "constant coefficient" problem in n, which is easy to solve using transforms (generating functions). If $\rho < 1$, then the equations (12.31)-(12.33) are precisely those satisfied by the steady-state probabilities in the standard (∞ population) M/G/1 queue. If $\rho > 1$, the M/G/1 model is transient so that $Q_n(\eta)$ can no longer be interpreted probabilistically. It is simply an approximation to the finite source model valid on the scale $n = O(1)$. The correction terms $Q_n^{(j)}(\eta)$ $(j \geq 1)$ will satisfy inhomogeneous versions of the problem (12.31)-(12.33), and these can also be solved using generating functions. For the leading term we obtain

$$Q_n(\eta) = \frac{\lambda Q_0}{2\pi i}\left\{\int_C \frac{(s-1)e^{\lambda(s-1)\eta}}{s^n[s - \widehat{b}(\lambda - \lambda s)]}ds\right\} e^{-\int_0^\eta \mu(z)dz}. \tag{12.34}$$

Here $\widehat{b}(s) = \int_0^\infty e^{-sz}b(z)dz$ and C is a small loop about $s = 0$ in the complex plane.

Now, the expansions for $n = O(1)$ and $0 < \xi = n/M < 1$ contain the, hitherto, undetermined constants $D(\varepsilon)$, $C(\varepsilon)$. One of these can be determined by normalization, but we need one additional condition. This is obtained by requiring that the two expansions "asymptotically match". This means that they should agree on an intermediate scale where $\varepsilon \ll \xi \ll 1$, which corresponds to $n \to \infty$, $n/M \to 0$. Symbolically the matching condition may be written as

$$C(\varepsilon)\exp\left[-\frac{1}{\varepsilon}\sum_{k=0}^\infty \varepsilon^k \phi_k(\xi,\eta)\right]\Bigg|_{\xi \ll 1} \sim D(\varepsilon)\sum_{k=0}^\infty \varepsilon^k Q_n^{(k)}(\eta)\Bigg|_{n \gg 1} e^{\int_0^\eta \mu(z)dz} \tag{12.35}$$

and must hold to all orders in ε. The left side of (12.35) is evaluated by expanding (12.30) as $\xi \to 0$. To leading order, this gives

$$C(\varepsilon)e^{\eta K(0)}\left|\frac{K'(0)}{\lambda + K(0)}\right|^{1/2}\left(1 + \frac{K(0)}{\lambda}\right)^{-n}, \tag{12.36}$$

where we have used $\xi = \varepsilon n$. From (12.34), the large n behavior is determined by the singularity of the integrand that is closest to $s = 0$. This is a simple pole at

$s = S^*$, which satisfies $S^* > 1$ ($S^* < 1$) according as $\rho < 1$ ($\rho > 1$). Computing the residue at this pole, we see that, to leading order, the right side of (12.35) becomes

$$D(\varepsilon)\lambda Q_0 \frac{(S^*-1)e^{\lambda(S^*-1)\eta}}{\lambda \int_0^\infty z e^{\lambda(S^*-1)z} b(z)dz - 1}(S^*)^{-n}. \tag{12.37}$$

From the definitions of $K(\xi)$ and S^*, we have

$$1 + \frac{K(0)}{\lambda} = S^*,$$

so that the forms of (12.36) and (12.37) agree, and the constants C, D are related by

$$C(\varepsilon) = D(\varepsilon)Q_0 \mid K'(0)[\lambda + K(0)] \mid^{1/2}. \tag{12.38}$$

Our final step is to determine $C(\varepsilon)$ (or $D(\varepsilon)$) from normalization (12.21). This requires that we asymptotically evaluate the sum in (12.21) for $\varepsilon \to 0$ ($M \to \infty$). The expansion of the sum depends on whether $\rho < 1$ or $\rho > 1$. Below, we summarize our final results for the marginal probabilities $p_n = \int_0^\infty p_n(y)dy$

(*a*) $0 < \xi < 1$

$$p_n = P(\xi) \sim \frac{C(\varepsilon)}{\lambda} \left| \frac{K'(\xi)}{\lambda(1-\xi) + K(\xi)} \right|^{1/2} \frac{e^{-M\phi(\xi)}}{\sqrt{1-\xi}},$$

$$\phi(\xi) = \int_0^\xi \log\left[1 + \frac{K(z)}{\lambda(1-z)}\right] dz,$$

$$1 + \frac{K(\xi)}{\lambda(1-\xi)} = \int_0^\infty e^{\eta K(\xi)} b(\eta) d\eta,$$

$$C(\varepsilon) \sim (1-\rho) \mid K'(0)[\lambda + K(0)] \mid^{1/2}, \ \left(\rho = \tfrac{\lambda}{\mu} < 1\right),$$

$$C(\varepsilon) \sim \lambda \sqrt{\frac{2}{\pi M}} \ (\rho = 1),$$

and

$$C(\epsilon) \sim \sqrt{\frac{\lambda\mu}{2\pi M}} \exp\left[M\phi\left(1 - \tfrac{1}{\rho}\right)\right] \ (\rho > 1).$$

(*b*) $\xi = O(M^{-1})$ i.e., $n = O(1)$,

$$p_n \sim \frac{1}{2\pi i}\left\{ \int_C \frac{-1 + \widehat{b}(\lambda - \lambda s)}{s - \widehat{b}(\lambda - \lambda s)} s^{-n} ds \right\} \frac{C(\varepsilon)}{\mid K'(0)[\lambda + K(0)] \mid^{1/2}}, \quad n \geq 1,$$

and

$$p_0 \sim \frac{C(\varepsilon)}{\mid K'(0)[\lambda + K(0)] \mid^{1/2}}.$$

We remark that, if $b(z) = \mu e^{-\mu z}$, then, $K(\xi) = \mu - \lambda(1-\xi)$; (*a*) and (*b*) reduce to (12.6) with (12.8). When $\rho \geq 1$, we have

$$K(1-\rho^{-1}) = 0 \text{ and } K'(1-\rho^{-1}) = (2\lambda)/(\mu^2 m_2),$$

where m_2 is the second moment of the (scaled) service time density $b(\cdot)$. Thus, if we expand our approximation for $\rho > 1$ for ξ close to $\xi^* \equiv 1 - \rho^{-1}$, we obtain the Gaussian form

$$p_n \approx \sqrt{\frac{\lambda}{\pi \mu^3 m_2 M}} \exp\left[-\frac{\lambda M}{\mu^3 m_2}(\xi - \xi^*)^2 \right], \tag{12.39}$$

which corresponds to the "diffusion approximation" to the process $N(t)$. Expression (12.39) is only valid for $\xi - \xi^* = O(M^{-1/2})$, whereas that in (a) is valid for all ξ except $\xi \approx 0$ (where (b) is valid) and $\xi \approx 1$. Thus, the results obtained by the WKB approach are much more uniform than those from a diffusion approximation. They are also more accurate numerically. In Figure 1, we give graphs of the exact probabilities (p_n^{EXA}), the WKB approximation (p_n^{WKB}), and the diffusion approximation (p_n^{DIFF}) when $\rho = 3.0$ and $M = 20$ for $b(z) = \mu e^{-\mu z}$. In this case it is easy to obtain the exact answer and demonstrate the accuracy of the WKB approximation. Let us review the procedure we have used. We started from a complicated systems ((12.15)-(12.21)) of differential-difference equations with global transitions in the y variable and linear (non-constant) coefficients in n. Then, using perturbation methods, we reduced the problem to solving a set of simpler equations. To obtain the WKB approximation we need to solve only the transcendental equation (12.28) and, then, a sequence of linear first order ODEs to get the functions ϕ_j for $j \geq 1$. To get the "boundary layer" approximation valid for $n = O(1)$, it was necessary to solve the "constant coefficient version" of (12.15)-(12.19), which is easy to do using transforms. Using this approach, we could also treat more complicated queues, e.g., ones with a general state-dependent arrival rate $\widetilde{\lambda}(n)$. For such problems, transform methods are not applicable.

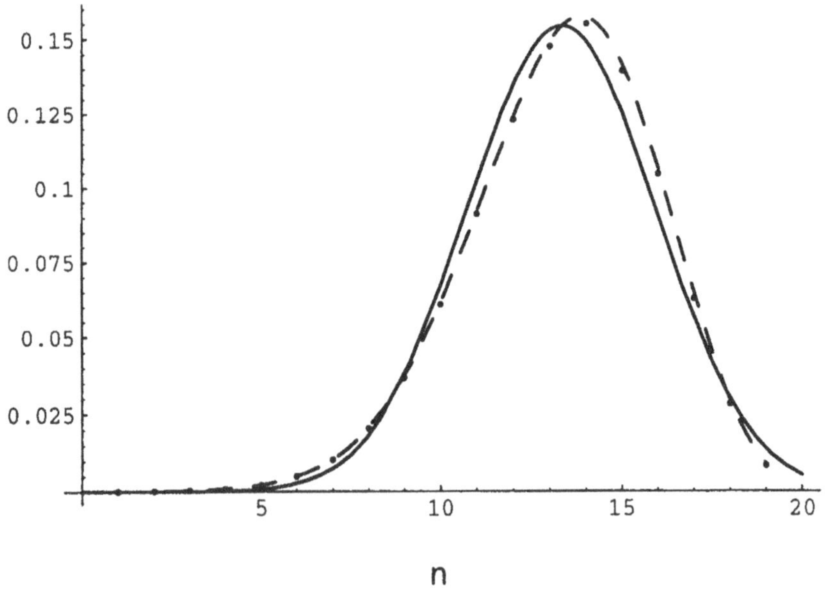

Figure 1: Graphs of $p_n^{EXA} = \ldots$, $p_n^{WKB} = - - -$, and $p_n^{DIFF} = -$, with $\rho = 3.0$ and $M = 20$.

The details of the analysis presented here appear in [28]. These results were extended to the finite source M/G/2 queue in [33] and to problems with more general state-dependent parameters in [25], and a similar analysis was done for the GI/M/m queue in [20].

12.2.2 State-dependent queues described by the unfinished work

Consider an M/G/1 model with an arrival rate that depends on the workload, i.e., $\lambda = \widetilde{\lambda}(U(t))$. Also we consider a state-dependent service density that is allowed to depend on the value of $U(t)$ when that particular customer entered the system. Hence, $\mathbf{P}\{\text{service time } \in (x, x+dx) \mid U(t^*) = w\} = \widetilde{b}(x,w)dx$ where t^* is the arrival time of the customer that requests x units of service. The forward equation for this state-dependent model is a generalization of the equation for the distribution of the unfinished work formulated by Takács (see equation (2) in [23]). We assume that the arrivals are fast and service times (jumps in $U(t)$) are small, so that we write $\widetilde{\lambda}$, $\widetilde{b}$ in the scaled forms

$$\widetilde{\lambda}(w) = \tfrac{1}{\varepsilon}\lambda(w) \text{ and } \widetilde{b}(x,w) = \tfrac{1}{\varepsilon}\, b\left(\tfrac{x}{\varepsilon}, w\right),$$

where ε may be defined, say, as $(\widetilde{\lambda}(0))^{-1}$. Denoting by $p(w,t)$ the unfinished work density for $w > 0$ and by $A(t)$ the probability that the system is empty, the forward equation is

$$p_t = p_w - \tfrac{1}{\varepsilon}\lambda(w)p$$

$$+\tfrac{1}{\varepsilon}\int_0^{w/\varepsilon} p(w-\varepsilon z)\lambda(w-\varepsilon z)b(z, w-\varepsilon z)dz + \frac{A(t)\lambda(0)}{\varepsilon^2} b\left(\tfrac{w}{\varepsilon}, 0\right), \tag{12.40}$$

with the boundary condition

$$A'(t) = -\frac{\lambda(0)}{\varepsilon}A(t) + p(0,t).$$

Also, if $\tau(w)$ is the mean residual busy period (i.e., time to empty the system given $U(0) = w$), it satisfies the backward equation

$$-\tau_w(w) - \tfrac{1}{\varepsilon}\lambda(w)\tau(w) + \tfrac{1}{\varepsilon}\lambda(w)\int_0^\infty \tau(w+\varepsilon z)b(z,w)dz = -1, \tag{12.41}$$

with $\tau(0) = 0$. It is also of interest to compute the time needed for the workload $U(t)$ to exceed a certain value, call it K. Letting $n(w)$ be the mean time for $U(t)$ to exceed K given $U(0) = w$, it satisfies

$$-n_w(w) - \tfrac{1}{\varepsilon}\lambda(w)n(w) + \tfrac{1}{\varepsilon}\lambda(w)\int_0^{(K-w)/\varepsilon} n(w+\varepsilon z)b(z,w)dz = -1, \; 0 < w < K,$$

and

$$-\tfrac{1}{\varepsilon}\lambda(0)n(0) + \tfrac{1}{\varepsilon}\lambda(0)\int_0^{K/\varepsilon} n(\varepsilon z)b(z,0)dz = -1, \tag{12.42}$$

with $n(w) = 0$ for $w \geq K$.

We have analyzed the problems (12.40)-(12.42) using perturbation methods in [23]. Again, the WKB method and asymptotic matching proved useful in analyzing this integro-differential equation, which cannot be solved exactly. Finite capacity queues, in which $U(t)$ is not allowed to exceed a given level, were analyzed in

[24, 26]. In [31, 32] we analyzed Markov-modulated state-dependent queues, in [30] we obtained the distribution of the maximum value of $U(t)$ during a busy period, and in [29], we obtained the (time-dependent) busy period distribution for the model described here. State-dependent GI/G/1 queues were considered in [35].

We close this section by mentioning the work of Keller on time-dependent queues ([14]) that is similar in approach to that discussed here. Time-dependent queues also lead to nonconstant coefficient equations, where L and L^* in (12.1)-(12.2) now explicitly involve time. State-dependent and time-dependent queues are qualitatively very different, but the analysis of each usually involves several different (space/time) scales, and asymptotic matching allows us to relate the various scales to one another.

12.3 PROCESSOR-SHARED QUEUES

Processor-shared (PS) queues are used to model time-sharing computer systems. In some respects their analysis is more difficult than that for FIFO queues. For PS queues, one wishes to compute the *sojourn time*. This is the time period from when a "tagged" customer enters the system, to when that customer leaves (after obtaining the required service).

Consider a Markovian, state-dependent PS model which has (i) a single server, (ii) a state-dependent arrival rate $=\lambda_n$, and (iii) a state-dependent service rate $=\mu_n$, where the rates are conditioned on $N(t)=n$. We denote the sojourn time by R and its conditional density by

$$p_n(t,x)dt = \mathbf{P}\{R \in (t, t+dt) \mid N(0^-)=n, X=x\}. \tag{12.43}$$

Here, X is the tagged customer's required service, and n is the number of customers already in the system as the tagged customer arrives, which is assumed to occur at $t=0$.

The density satisfies the evolution equation

$$\begin{aligned}\frac{\partial}{\partial t}p_n(t,x) = & -\frac{1}{n+1}\frac{\mu_{n+1}}{\mu_0}\frac{\partial}{\partial x}\, p_n(t,x) + \lambda_{n+1}[p_{n+1}(t,x)-p_n(t,x)] \\ & + \frac{n}{n+1}\mu_{n+1}[p_{n-1}(t,x)-p_n(t,x)];\ n=0,1,\ldots \end{aligned} \tag{12.44}$$

For $x>0$, the initial condition is $p_n(t,0)=\delta(t)$ $(n\geq 0)$, since a customer with zero service does not spend any time in the system according to the PS discipline. There are several important special cases of this model which we summarize below as models A-F:

model A: $\lambda_n=\lambda;\ \mu_n=\mu$

model B: $\lambda_n=\lambda(M-n);\ \ \mu_n=\mu$

model C: $\lambda_n=\begin{cases}\lambda, & n<K;\\ 0, & n\geq K\end{cases}\quad \mu_n=\mu$

model D: λ_n general; $\mu_n=\mu$

model E: $\lambda_n=\lambda(M-n);\ \ \mu_n=\dfrac{\mu_0}{1+\omega n}$

model F: $\lambda_n=\begin{cases}\lambda, & n<K;\ \mu_n=\dfrac{\mu_0}{1+\omega n}.\\ 0, & n\geq K\end{cases}$

The exact solution to model A was obtained in [5], and asymptotic properties of the sojourn time were studied in [54]. Model B corresponds to a closed network consisting of a PS-server in series with a set of M terminals (IS node). Model C is a finite capacity model where, at most, K customers are allowed into the system (with additional arrivals lost). Model E again corresponds to the closed network of model B with a state-dependent server. The parameter ω has been used to model the "switching time", which is the time needed for the server to switch between individual customers. Model F has a finite capacity and also considers the switching times.

Sometimes it may not be possible to compute the full conditional density $p_n(t,x)$. Hence we define

$$(a)\ \mathbf{E}[R^j \mid n,x] = \int_0^\infty t^j p_n(t,x)dt;\ j = 1,2,\ldots,$$

$$(b)\ p(t,x) = \sum_n p_n(t,x)\pi_n,$$

$$(c)\ \mathbf{E}[R^j \mid x] = \int_0^\infty t^j p(t,x)dt;\ j = 1,2,\ldots,$$

$$(d)\ p(t) = \int_0^\infty \mu e^{-\mu x} p(t,x)dx, \text{ and}$$

$$(e)\ \mathbf{E}[R^j] = \int_0^\infty t^j p(t)dt;\ j = 1,2,\ldots,$$

which denote, respectively, the conditional sojourn time moments, the distribution conditioned only on the service request, the moments conditioned only on the service request, the unconditional sojourn time distribution, and the unconditional moments. Here, π_n is the (steady-state) probability that the tagged customer finds n others at the arrival instant. By multiplying (12.44) by t^j and integrating with respect to t over $[0,\infty)$, we can obtain recursive equations for the conditional moments. The first two have the form $\mathcal{L}_{n,x}\mathbf{E}[R \mid n,x] = -1, \mathcal{L}_{n,x}\mathbf{E}[R^2 \mid n,x] = -2\mathbf{E}[R \mid n,x]$, where the operator $\mathcal{L}$ is that in the right side of (12.44).

The various models B-F were, in recent years, analyzed using singular perturbation methods. For model B, (d) was computed in [55], (c) was computed in [56], and (b) was computed in [57]. The asymptotic limit assumes that $M \to \infty$ with $\lambda M/\mu$ fixed, and the structure of the problem depends on whether $\lambda M/\mu < 1$ (normal usage), $\lambda M/\mu = 1 + O(M^{-1/2})$ (heavy usage), or $\lambda M/\mu > 1$ (very heavy usage). Some of these results have been extended to systems with multiple customer classes in [53, 58, 59]. For the finite capacity model C, (a) was computed in [16], (d)-(e) were computed in [22], and (b) was considered in [64]. The asymptotics assume that $K \to \infty$ and depend on whether $\lambda/\mu < 1$, $\lambda/\mu = 1 + O(K^{-1})$, $\lambda/\mu = 1 + O(K^{-1/2})$, or $\lambda/\mu > 1$. For model D, the case of a general (but smooth) λ_n was considered in [34]. (a) and (c) were computed for a certain scaling of λ_n and μ. Note that model D contains model B as a special case. The closed model E with switching times was analyzed in [2], where the authors computed asymptotic approximations to (d). In [41, 42], (c) was computed under various assumptions: in the asymptotic limit $M \to \infty$, $\mu_0 = O(M)$, $\omega = O(M^{-1})$. The size of the "switching time" parameter ω significantly affects the asymptotics, and this shows that the introduction of ω is an important consideration in modeling real systems. For model F, the conditional moments (c) were computed asymptotically in [63] for $K \to \infty$, $\omega = O(K^{-1})$ and several cases of the parameters (ρ, β), where $\rho = \lambda/\mu_0$, $\beta = K\omega$.

We also mention some work on the GI/M/1-PS model, which generalizes model A by allowing for a general (renewal) input. Approximations for (a) and (d) were obtained in [21], and (b) was analyzed in [65]. For the GI/M/1-PS model with finite capacity K (which generalizes model C), (c) was computed in [68], and

(d) appears in [69]. For this model, π_n is not known explicitly, but asymptotic approximations for $K \to \infty$ have been obtained in [70].

There are basically two asymptotic approaches that have been developed to analyze problems of the form (12.44). The first is to use a generating function to transform over n and then, analyze the resulting equation, which may be an ODE, PDE or a functional equation, depending on the specific model and on the quantity that is to be computed. A second approach is to analyze (12.44) directly after some appropriate scalings. The first approach was used in [2, 55-58], and the second was used in [16, 34, 41-42, 63-64]. Of course, where both were applicable, they yield identical results, as is shown in [22].

In the next section, we also discuss related work on sojourn times in networks, which may contain PS nodes.

12.4 NETWORKS

Next we consider networks of queues. We will discuss (i) asymptotic expansions for the partition function for large product form networks, (ii) the time needed for large queue lengths to build up, (iii) bottleneck analysis, (iv) sojourn times in networks with overtaking, and (v) approximations for nonproduct form networks.

12.4.1 Partition functions

To illustrate the basic ideas and results, we consider a closed BCMP network which has a single class (chain) consisting of M customers, a single IS (infinite server) node, and K single-server nodes with constant service rates. For this model, the steady-state queue length distribution is well-known to have the product form

$$p(n_1, \ldots, n_K) = \frac{1}{G(M,K)} \frac{M!}{(M - n_1 - n_2 - \ldots - n_K)!} \rho_1^{n_1} \ldots \rho_K^{n_K}. \tag{12.45}$$

Where n_j is the number of customers in node j, ρ_j is the relative utilization, and G is the partition function (normalizing constant),

$$G(M,K) = \sum_{n_1 + \ldots + n_K \leq M} \frac{M!}{(M - n_1 - n_2 - \ldots - n_K)!} \rho_1^{n_1} \ldots \rho_K^{n_K}. \tag{12.46}$$

The performance measures, such as mean queue lengths and throughputs, can be easily calculated from G. The numerical evaluation of G is, however, difficult for M and/or K large. Various ideas have been introduced to simplify the calculation of G, such as computational algorithms (see [3, 7, 48]) and asymptotic expansions using integral representation (see [46, 49, 50, 52]).

In [39], we developed an asymptotic approach to treat networks where *both* the population M and the number of nodes K are large. It has been shown in [3] that G can be computed from the recursion

$$h(m,k) = h(m,k-1) + m\rho_k h(m-1,k), \quad 1 \leq m \leq M, 1 \leq k \leq K, \tag{12.47}$$

$$h(0,k) = 1, \quad 1 \leq k \leq K,$$

and

$$h(m,0) = 1, \quad 1 \leq m \leq M,$$

with $G(M,K) = h(M,K)$. We have obtained asymptotic expansions for G using the ray method and asymptotic matching, applied to the difference equation

(12.47). The asymptotics depend on the relative sizes of the ρ_k. For example, if each of the ρ_k is $O(M^{-1})$ and the utilizations are not much different from one another, our final result for G is

$$G(M,K) \approx \frac{e^{M[U_0 - 1 - \log U_0]}}{\left\{\prod_{k=1}^{K}(1-U_k)\right\}[U_0 + U_0 \sum_{k=1}^{K} \frac{\rho_k}{(1-U_k)^2}]^{1/2}}, \tag{12.48}$$

where $U_0 < 1$ is determined by

$$M = MU_0 + \sum_{k=1}^{K} \frac{U_k}{1-U_k}, \tag{12.49}$$

with $U_k = M\rho_k U_0 < 1$. Equation (12.49) is the fixed-population mean (FPM) approximation in which the closed network is replaced by an equivalent open network, but with the network population constrained to be M. The unknown constant U_0 in (12.49) represents the utilization in the IS node of this open network. We show in [39] that (12.48)-(12.49), as well as the results derived under other assumptions on ρ_k, are very accurate numerically. We have extended this analysis to multiclass networks with all single-server nodes [40] and to multiclass networks with IS node(s) [51].

12.4.2 Buildup of large queue lengths

The steady-state queue length distribution is easy to obtain for tandem Jackson networks consisting of d M/M/1 queues in series. However, it proves much harder to analyze the busy period or to compute the time until the network population becomes large. Recently, we have computed asymptotic approximations to the mean time until the total population reaches M for $M \to \infty$. In a stable network with $\rho_j = \lambda/\mu_j < 1$ $(1 \le j \le d)$, this is a rare event and the mean time grows exponentially with M. When $d = 2$ our final results take the form ([43])

$$T \sim \frac{1}{\lambda} \frac{\rho_> - \rho_<}{(1-\rho_<)(1-\rho_>)^2} \rho_>^{-M}, \quad \rho_> > \rho_<, \tag{12.50}$$

and

$$T \sim \frac{1}{\lambda} \frac{1}{(1-\rho)^3} \frac{a}{\sinh(a/\rho)} \frac{1}{M} \rho^{-M}, \quad \rho_1 \approx \rho_2.$$

Here $T = E[\tilde{T} \mid N_1(0) = N_2(0) = 0]$, $\tilde{T} = min\{t: N_1(t) + N_2(t) = M\}$, $N_j(t) =$ queue length in node $j, \rho_> = max\ \{\rho_1, \rho_2\}$, $\rho_< = min\{\rho_1, \rho_2\}$, $\rho = (\rho_1 + \rho_2)/2$, and $a = (\rho_1 - \rho)M = (\rho - \rho_2)M$. The second formula in (12.50) applies to the limit $M \to \infty$ with $|\rho_2 - \rho_1| = O(M^{-1})$, and $a = 0$ corresponds to the case of identical nodes. The result (12.50) and the analogous formulae for $d > 2$ were obtained by applying singular perturbation methods to a d-dimensional recursion equation. Using such techniques, we have also computed the time until the network population is no longer large, given a large initial value in [44].

12.4.3 Bottleneck analysis

Mean value analysis (MVA) computes *mean* performance measures for product-form networks, without computing the partition function (see [60]). To illustrate the procedure, consider a single class, closed network with population size $= M$ and K single-server nodes. Let $N_i(m)$ be the mean queue length at node i for a network with population size $= m$. Then MVA corresponds to the nonlinear iteration

$$N_i(m) = \frac{mD_i[1 + N_i(m-1)]}{\sum_{k=1}^{K} D_k[1 + N_k(m-1)]}, \quad 1 \le i \le K, \ 1 \le m \le M, \qquad (12.51)$$

where D_i is the load at node i and the initial condition is $N_i(0) = 0$. A "bottleneck" node is one where D_i is largest. Since (12.51) may be difficult to compute if M is very large, approximations have been developed that assume a unique bottleneck ([61, 62]). In particular, [62] develops a perturbation method which scales (12.51) and expands $N_i(m)$ in powers of $\varepsilon \equiv M^{-1}$. This approximation method also applies to multiclass networks. However, it works poorly when there are 2 or more bottleneck nodes. This is because the perturbation expansion becomes invalid if several of the D_i are nearly equal. Recently, we have shown how to treat such nonuniformities. For example, assume that nodes 1 and 2 are the bottlenecks with $D_1 \approx D_2 > D_j (j \ge 3)$. Then, scaling $z = m/M = \varepsilon m$, $h_i(z) = N_i(m)/M$ $(i = 1, 2)$ leads, to leading order in ε, to the ODE

$$zDH_1'(z) + (D - 2az)H_1(z) + 2aH_1^2(z) = zD, \quad H_1(0) = 0, \qquad (12.52)$$

where $H_1(z)$ is the limit of $h_1(z)$ as $\varepsilon \to 0$, $D = (D_1 + D_2)/2$, $a = (D_1 - D)M = (D - D_2)M$, and $h_2(z) \sim H_2(z) = z - H_1(z)$ as $\varepsilon \to 0$. (12.52) is a Ricatti equation whose solution is

$$H_1(z) = -\frac{D}{2a} + \frac{z}{1 - e^{-2az/D}}. \qquad (12.53)$$

The approximation, based on (12.53), leads to accurate numerical results, regardless of the relative sizes of D_1, D_2, and contains the single-node bottleneck approximations as limiting cases.

We have extended this approach to networks with many bottleneck nodes. There, it is necessary to solve a system of nonlinear ODEs of the type (12.52), but this can, nevertheless, be done explicitly. We can also treat multiclass networks near "switch-points", which correspond to regions where the bottleneck configuration changes ([45]).

12.4.4 Networks with overtaking

It is of interest to compute the sojourn time through a network of queues, as a customer goes through the network along a specified path. This can be easily computed for (paths in) product-form networks which are "overtake-free". This means that, along the specified path, customers cannot overtake others that are initially in front of them. Most networks are not overtake-free as service disciplines, such as processor-sharing and also network topologies, allow customers to overtake one another. The simplest examples of networks with overtaking were studied in [4, 10], but no explicit expressions could be obtained for the sojourn time distribution. In [38], we considered a two-node network with an M/M/1 PS node in series with a FIFO node. Asymptotic approximations to the sojourn time moments were obtained, where it was assumed that, as the tagged customer entered the PS node, the number of customers in at least one of the nodes was large. These results were shown to be in excellent agreement with simulations.

In [36, 37], we considered networks with overtaking in the heavy traffic limit. In [37], we analyzed the sojourn time distribution for tandem PS-FIFO and PS-PS queues, assuming that $\rho_i = \lambda/\mu_i \uparrow 1$ $(i = 1, 2)$. A similar analysis was done in [36] for a 3-node network where overtaking was caused by the network topology. In each case, we derived three-term asymptotic approximations to the sojourn time density, which gave simple quantitative measures of the effects of overtaking. Our approach involved using generating functions and then analyzing the resulting functional equations using singular perturbation ideas (e.g., scaling, matching).

12.4.5 Nonproduct form networks

The analysis of these problems is difficult even for two coupled (Markovian) queues. Classic examples of such problems are the shortest queue problem in [12], the fork-join model in [11], and two parallel queues where one server helps the other during periods when one of the queues is empty in [47]. Exact solutions are available ([6, 9]) for some of these problems, but their analytic complexity makes them difficult to interpret.

Multidimensional Markovian networks correspond to solving problems of the type (12.1)-(12.2) with L, L^* being the multidimensional difference operators with complicated boundary conditions. Diffusion approximations have been formulated for queueing networks ([13]), but their solution again involves analyzing (12.1)-(12.2). Now, L is a partial-differential operator of second order, but the "oblique-derivative" boundary conditions make these problems difficult.

We have developed a singular perturbation approach for obtaining the tails of the distributions for multidimensional models. Such an approach can be applied either to difference equations ([27, 38, 44]) or to PDEs ([18, 19]). For many of these problems, there is no natural large or small parameter, so that it seems that the only way to obtain simple formulas is to assume that space and/or time is large. So far, this approach has been applied only to two-dimensional models, but we believe that these ideas can also be used on higher dimensional networks.

12.5 TRANSIENT BEHAVIOR OF QUEUES

It is generally much harder to solve for the transient distribution of a stochastic model than it is to obtain the steady-state distribution. We show how to use perturbation methods to simplify the former task. As an example, we consider the Erlang loss model (the M/M/m/m queue), which is important in the analysis of blocking in teletraffic.

Let $N(t)$ be the number of occupied servers, each with rate μ, and let λ_0 be the arrival rate. Then $p_n(t) = \mathbf{P}\{N(t) = n\}$ satisfies

$$p_n'(t) = \lambda_0 p_{n-1}(t) + \mu(n+1)p_n(t) - (\lambda_0 + \mu n)p_n(t), \quad 1 \le n \le m-1, \tag{12.54}$$

$$p_0'(t) = \mu p_1(t) - \lambda_0 p_0(t), \tag{12.55}$$

$$p_m'(t) = \lambda_0 p_{m-1}(t) - \mu m p_m(t), \tag{12.56}$$

and

$$p_n(0) = \delta_{n,n_0}, \quad 0 \le n_0 \le m, \tag{12.57}$$

where we assume that n_0 servers are occupied at $t = 0$. We assume that the number of servers m is large and that the arrival rate λ_0 is also large. Thus, we scale $\lambda_0 = \lambda m$ with $\lambda = O(1)$ as $m \to \infty$. Letting $\xi = n/m$, $p_n(t) = P(\xi, t)$ and $\varepsilon = m^{-1}$, the scaled version of (12.54) is

$$\varepsilon P_t(\xi,t) = \lambda P(\xi-\varepsilon,t) + \mu(\xi+\varepsilon)P(\xi+\varepsilon,t) - (\lambda+\mu\xi)P(\xi,t). \qquad (12.58)$$

As shown in [17, 66], the asymptotic analysis depends on the size of $\rho = \lambda/\mu = \lambda_0/(\mu m)$, and we must consider $(i)\rho < 1$, $(ii)\ \rho > 1$, and $(iii)\ \rho - 1 = O(m^{-1/2})$. Also, the problem is very sensitive to the initial condition n_0, and we must consider the three cases $(a)\ n_0 = O(1)$, $(b)\ 0 < \xi_0 = n_0/m < 1$, and $(c)\ m - n_0 = O(1)$ which correspond, respectively, to starting the process with only a few occupied servers, a fixed fraction of the servers occupied, and with almost all the servers occupied. Thus, there are nine cases that must be analyzed. In addition, within a particular case of (ρ, n_0), it is necessary to analyze several regions in the (n,t) (or (ξ,t)) plane. All these scales are treated in [17] if $\rho < 1$ and in [66] if $\rho > 1$ or $\rho \approx 1$. Here, we give a few of the main results.

Assume, first, that $\rho < 1$ and $0 < \xi_0 < 1$. We, first, analyze (12.58) for short times $t = \varepsilon\tau = O(\varepsilon)$ and localize space near the initial condition by setting $k = (\xi - \xi_0)/\varepsilon = n - n_0$. To leading order, we obtain

$$p_n(t) \sim e^{-(\lambda+\mu\xi_0)\tau}\left(\frac{\lambda}{\mu\xi_0}\right)^{k/2} I_k(2\sqrt{\lambda\mu\xi_0}\tau),\ \tau = \frac{t}{\varepsilon},\ k = \frac{\xi-\xi_0}{\varepsilon}, \qquad (12.59)$$

where $I_k(\cdot)$ is the modified Bessel function. Relation (12.59) corresponds to a free space birth-death process with birth rate λ and death rate $\mu\xi_0$.

For $t > 0$ and $0 < \xi < 1$, we obtain the approximation to $p_n(t)$ using the *ray method*. We set

$$p_n(t) = P(\xi,t) = C(\varepsilon)e^{-\phi(\xi,t)/\varepsilon}[A(\xi,t) + \varepsilon A_1(\xi,t) + \ldots] \qquad (12.60)$$

in (12.58) and find that ϕ and A satisfy the PDEs,

$$-\phi_t = \lambda(e^{\phi_\xi} - 1) + \mu\xi(e^{-\phi_\xi} - 1) \qquad (12.61)$$

and

$$A_t + (\lambda e^{\phi_\xi} - \mu\xi e^{-\phi_\xi})A_\xi = [\mu e^{-\phi_\xi} - \tfrac{1}{2}\phi_{\xi\xi}(\lambda e^{\phi_\xi} + \mu\xi e^{-\phi_\xi})]A. \qquad (12.62)$$

These are analogous to the "eiconal" and "transport" equations of geometrical optics. They can be solved using standard methods for PDEs. Both are first order equations, though (12.61) is nonlinear. To specify uniquely the functions ϕ and A, we must match the ray approximation to the short time approximation (12.59). This leads to the final results

$$\phi(\xi,t) = \xi\log z_0 - \rho(z_0 - 1)e^{-\mu t} - \xi_0\log(1 - e^{-\mu t} + z_0 e^{-\mu t}),$$

$$z_0(\xi,t) = \frac{1}{2}\left\{\frac{\xi-\xi_0}{\rho(1-e^{-\mu t})} + 1 - e^{\mu t} + \sqrt{\left(\frac{\xi_0-\xi}{\rho(1-e^{-\mu t})} + e^{\mu t} - 1\right)^2 + \frac{4\xi e^{\mu t}}{\rho}}\right\},$$

$$A(\xi,t) = \frac{1}{\sqrt{2\pi}}\left[\xi - \xi_0\left(\frac{z_0}{e^{\mu t} - 1 + z_0}\right)^2\right]^{-1/2}, \qquad (12.63)$$

and $C(\varepsilon) = \sqrt{\varepsilon}$. This approximation becomes invalid near the boundaries $\xi = 0, 1$. There, other expansions must be constructed (see [17, 66]). The latter boundary region is especially important if one wants an accurate approximation to the blocking probability $p_m(t)$. The leading order approximation to this is

$$p_m(t) \sim \sqrt{\varepsilon}A(1,t)e^{-\phi(1,t)/\varepsilon}\left[1 + \frac{1-\rho z_0(1,t)}{1-z_0(1,t)}\frac{1}{\rho z_0(1,t)}\right]. \qquad (12.64)$$

Note that this is different than setting $\xi = 1 (n = m)$ in (12.60), and again, indicates the importance of treating the various scales inherent to the problem.

Now we examine the steady-state limit of the ray approximation by letting $t \to \infty$. From (12.63), we have $A(\xi,\infty) = (2\pi\xi)^{-1/2}$ and $\phi(\xi,\infty) = \xi\log\xi + \rho - \xi - \xi\log\rho$. The exact stationary distribution is

$$p_n(\infty) = \frac{(\rho m)^n}{n!} p_0(\infty) \quad \text{and} \quad p_0(\infty) = 1/\sum_{\ell=0}^{m} \frac{(\rho m)^\ell}{\ell!}. \tag{12.65}$$

If $\rho < 1$, we obtain $p_0(\infty) \sim e^{-\rho m}$ as $m \to \infty$. Then expanding $n!$ by Stirling's formula, we see that $p_n(\infty) \sim (2\pi n)^{-1/2}(\rho m/n)^n e^n e^{-\rho m}$, which is the same as the limit of the ray expansion as $t \to \infty$ (recall that $m = \varepsilon^{-1}$, $\xi = n/m = \varepsilon n$).

When $\rho < 1$, the blocking probability is exponentially small for all times t (cf. (12.64)). Now we consider more heavily loaded systems which have $\rho > 1$ (again taking $0 < \xi_0 < 1$). For this case, the main result(s) for $p_n(t)$ in [66] are (away from boundary and initial layers)

$$\begin{aligned} p_n(t) &\sim \sqrt{\varepsilon} A(\xi,t) e^{-\phi(\xi,t)/\varepsilon}, \quad t < T(\xi) \\ &\sim \left(1 - \rho^{-1}\right)\xi^{-1/2} e^{-\psi(\xi)/\varepsilon}, \quad t > T(\xi), \end{aligned} \tag{12.66}$$

where $\psi(\xi) = 1 + \log\rho + \xi\log\xi - \xi - \xi\log\rho$ and ϕ, A are as in (12.63). The second formula in (12.66) is precisely the expansion of $p_n(\infty)$, as when $\rho > 1$ we have $p_0(\infty) \sim (1 - \rho^{-1}) m! (\rho m)^{-m}$. In (12.66), $T(\xi)$ is defined implicitly be the relation

$$\psi(\xi) = \phi(\xi, T(\xi)), \tag{12.67}$$

which defines a curve in the (ξ, t) plane which passes through the point $\xi = 1$, $\mu t = \log\left[(\rho - \xi_0)/(\rho - 1)\right]$. The curve $t = T(\xi)$ may be viewed as a "front" above which $(t > T(\xi))$ the process forgets the initial condition and settles to its steady-state behavior. Below the front $(t < T(\xi))$, transient effects are still important.

Thus, our analysis of $\rho < 1$ and $\rho > 1$ reveals two mechanisms by which a process approaches its steady state behavior. If $\rho < 1$, the ray approximation depended on time, for all $t < \infty$, and approached the steady state smoothly as $t \to \infty$. When $\rho > 1$, there was a sharp transition to the equilibrium distribution at $t = T(\xi)$. The case $\rho \approx 1$ is more complicated and is treated in detail in [66].

The blocking probability $p_m(t)$ now $(\rho > 1,\ 0 < \xi_0 < 1)$ has the asymptotic expansions

(a) $\mu t < \log\left(\dfrac{\rho - \xi_0}{\rho - 1}\right) = \mu T(1)$

$$p_m(t) \sim \sqrt{\varepsilon} A(1,t) e^{-\phi(1,t)/\varepsilon}\left[1 + \frac{1 - \rho z_0(1,t)}{1 - z_0(1,t)} \frac{1}{\rho z_0(1,t)}\right], \tag{12.68}$$

(b) $\mu t - \log\left(\dfrac{\rho - \xi_0}{\rho - 1}\right) = \dfrac{\sqrt{\varepsilon}}{\rho - 1}\Delta = O(\sqrt{\varepsilon})$,

$$p_m(t) \sim \left(1 - \frac{1}{\rho}\right)\frac{1}{\sqrt{2\pi}}\int_{-\Delta/\sqrt{\beta}}^{\infty} e^{-u^2/2} du, \quad \beta = 1 - \frac{\xi_0(\rho - 1)^2}{(\rho - \xi_0)^2}, \text{ and}$$

(c) $\mu t > \log\left(\dfrac{\rho - \xi_0}{\rho - 1}\right)$,

$$p_m(t) \sim p_m(\infty) \sim \left(1 - \frac{1}{\rho}\right).$$

This shows that, for times $t < T(1)$, the blocking probability is exponentially small and settles to its equilibrium value as t passes through the critical time $T(1)$. Note also that if $\rho > 1$, $z_0(1, T(1)) = 1$, so that (12.68) becomes infinite (and is thus invalid) as $t \uparrow T(1)$.

We believe that the type of structure discussed here (e.g., the front at $t = T(\xi)$) is canonical to many applied probability models. It has also been observed in the finite capacity M/M/1 queue and in the infinite capacity M/M/1 queue, if we take the initial condition $N(0) = n_0$ to be large ([67]). These ideas should also apply to models in more than one space dimension; there, the PDEs (12.61)-(12.62) will involve time t and, say, 2 space variables ξ_1, ξ_2. They will still be of the general form (12.61)-(12.62) and can be readily solved.

Finally, we note that, in [17, 66], detailed numerical comparisons are presented, which show that the various asymptotic approximations are in very good agreement with the exact (numerical) values. This is true even for modest size systems which have, say, $m = 10$. Since this type of asymptotics reveals much about the qualitative structure of the models, our results achieve both of the goals outlined in Section 12.1.

12.6 OPEN PROBLEMS AND RESEARCH DIRECTIONS

We have indicated how to use singular perturbation methods to analyze a variety of queueing models. We have shown that such methods may be used to obtain asymptotic information for models with general interarrival/service time distributions, state-dependent and time-dependent queues, large product-form networks, non-product form networks, and analyzing time-dependent behavior. Except for the work on large product-form networks, most of our work has involved computing steady-state behavior for models in one or two dimensions and transient behavior for one-dimensional models. We expect that these techniques may also be used on models in three or more dimensions. For nonproduct form networks, very little is known for models in more than two dimensions, even under Markovian assumptions on the arrival and service processes. We believe that the "geometric optics" approach outlined in [18, 19, 66, 67] should be useful for computing, asymptotically, steady-state probabilities in more than two dimensions and, also, transient probabilities in more than one dimension. It should be possible to treat multidimensional non-Markovian models using these methods. We are presently investigating such possibilities.

We have shown that the methods discussed are general enough to be used on many specific problems in queueing theory. At the same time, our results show that they capture the subtle structure of the individual model and, thus, lead to better understanding of the differences between models. A good applied mathematics technique should be applicable to a large class of problems, yet it should also be able to lead to an in-depth understanding of specific problems. In this article, we hope that we have demonstrated that perturbation methods satisfy both of these essential criteria.

BIBLIOGRAPHY

[1] Bender, C.M. and Orszag, S.A., *Advanced Mathematical Methods for Scientists and Engineers*, McGraw-Hill, New York 1978.

[2] Bersani, A. and Sciaretta, C., Asymptotic analysis for a closed processor-sharing system with switching times, *SIAM J. Appl. Math.* **51** (1991), 525-541.

[3] Buzen, J.P., Computational algorithms for closed queueing networks with exponential servers, *Commun. ACM* **16** (1973), 527-531.

[4] Coffman Jr., E.G., Fayolle, G., and Mitrani, I., Sojourn times in a tandem queue with overtaking: reduction to a boundary value problem, *Stoch. Mod.* **2** (1986), 43-65.

[5] Coffman Jr., E.G., Muntz, R.R. and Trotter, H., Waiting time distributions for processor-sharing systems, *J. ACM* **17** (1970), 123-130.

[6] Cohen, W. and Boxma, O.J., *Boundary Value Problems in Queueing Systems Analysis*, North-Holland, Amsterdam 1983.

[7] Conway, A.E. and Georganas, N.D., RECAL- A new efficient algorithm for the exact analysis of multiple-chain closed queueing networks, *J. ACM* **33** (1986), 768-791.

[8] Cox, D.R., The analysis of non-Markovian stochastic processes by the inclusion of supplementary variables, *Proc. Camb. Phil. Soc.* (Math. and Phys. Sci.) **51** (1955), 433-441.

[9] Fayolle, G., On functional equations for one or two complex variables arising in the analysis of stochastic models, In: *Mathematical Computer Performance and Reliability*, ed. by G. Iazeolla, P.J. Curtois, and A. Hordijk, North-Holland, Amsterdam (1984), 55-76.

[10] Fayolle, G., Iasnogorodski, R., and Mitrani, I., The distribution of sojourn times in a queueing network with overtaking: reduction to a boundary value problem, In: *Performance 1983*, ed. by A.K. Agrawala and S.K. Tripathi, North-Holland, Amsterdam (1983), 477-486.

[11] Flatto, L. and Hahn, S., Two parallel queues created by arrivals with two demands I, *SIAM J. Appl. Math.* **44** (1984), 1041-1054.

[12] Flatto, L. and McKean, H., Two queues in parallel, *Comm. Pure Appl. Math.* **30** (1977), 255-263.

[13] Harrison, M.J. and Reiman, M.I., On the distribution of multidimensional reflected Brownian motion, *SIAM J. Appl. Math.* **41** (1981), 345-361.

[14] Keller, J.B., Time-dependent queues, *SIAM Rev.* **24** (1982), 401-412.

[15] Kevorkian, J. and Cole, J.D., *Perturbation Methods in Applied Mathematics*, Springer-Verlag, Berlin, New York 1981.

[16] Knessl, C., On finite capacity processor shared queues, *SIAM J. Appl. Math.* **50** (1990), 264-287.

[17] Knessl, C., On the transient behavior of the M/M/m/m loss model, *Stoch. Mod.* **6** (1990), 749-776.

[18] Knessl, C., Diffusion approximation to a fork and join queueing model, *SIAM J. Appl. Math.* **51** (1991), 160-171.

[19] Knessl, C., Diffusion approximation to two parallel queues with processor sharing, *IEEE Trans. Auto. Control* **36** (1991), 1356-1367.

[20] Knessl, C., The *WKB* approximation to the G/M/m queue, *SIAM J. Appl. Math.* **51** (1991), 1119-1133.

[21] Knessl, C., Asymptotic approximations for the GI/M/1 queue with processor sharing service, *Stoch. Mod.* **8** (1992), 1-34.

[22] Knessl, C., On the sojourn time distribution in a finite capacity processor shared queue, *J. ACM* **40** (1993), 1238-1301.

[23] Knessl, C., Matkowsky, B.J., Schuss, Z. and Tier, C., Asymptotic analysis

of a state-dependent M/G/1 queueing system, *SIAM J. Appl. Math.* **46** (1986), 483-505.

[24] Knessl, C., Matkowsky, B.J., Schuss, Z. and Tier, C., A finite capacity single-server queue with customer loss, *Stoch. Mod.* **2** (1986), 97-121.

[25] Knessl, C., Matkowsky, B.J., Schuss, Z. and Tier, C., On the performance of state-dependent single-server queues, *SIAM J. Appl. Math.* **46** (1986), 657-697.

[26] Knessl, C., Matkowsky, B.J., Schuss, Z. and Tier, C., System crash in a finite capacity M/G/1 queue, *Stoch. Mod.* **2** (1986), 171-201.

[27] Knessl, C., Matkowsky, B.J., Schuss, Z. and Tier, C., Two parallel queues with dynamic routing, *IEEE Trans. Comm.* **34** (1986), 1170-1175.

[28] Knessl, C., Matkowsky, B.J., Schuss, Z. and Tier, C., Asymptotic expansions for a closed multiple access system, *SIAM J. Comp.* **16** (1987), 378-398.

[29] Knessl, C., Matkowsky, B.J., Schuss, Z. and Tier, C., Busy period distribution in state-dependent queues, *Queueing Sys.* **2** (1987), 285-305.

[30] Knessl, C., Matkowsky, B.J., Schuss, Z. and Tier, C., Distribution of the maximum buffer content during a busy period for state-dependent M/G/1 queues, *Stoch. Mod.* **3** (1987), 191-226.

[31] Knessl, C., Matkowsky, B.J., Schuss, Z. and Tier, C., A Markov-modulated M/G/1 queue I: Stationary distribution, *Queueing Sys.* **1** (1987), 355-374.

[32] Knessl, C., Matkowsky, B.J., Schuss, Z. and Tier, C., A Markov-modulated M/G/1 queue II: Busy period and time for buffer overflow, *Queueing Sys.* **1** (1987), 375-397.

[33] Knessl, C., Matkowsky, B.J., Schuss, Z. and Tier, C., The two repairmen problem: a finite source M/G/2 queue, *SIAM J. Appl. Math.* **47** (1987), 367-397.

[34] Knessl, C., Matkowsky, B.J., Schuss, Z. and Tier, C., Response times in processor-shared queues with state-dependent arrival rates, *Stoch. Mod.* **5** (1989), 83-113.

[35] Knessl, C., Matkowsky, B.J., Schuss, Z. and Tier, C., A state-dependent GI/G/1 queue, *Euro. J. Appl. Math.* **5** (1994), 217-241.

[36] Knessl, C. and Morrison, J.A., Heavy traffic analysis of the sojourn time in a three node Jackson network with overtaking, *Queueing Sys.* **8** (1991), 165-182.

[37] Knessl, C. and Morrison, J.A., Heavy traffic analysis of the sojourn time in tandem queues with overtaking, *SIAM J. Appl. Math.* **51** (1991), 1740-1763.

[38] Knessl, C. and Tier, C., Approximations to the moments of the sojourn time in a tandem queue with overtaking, *Stoch. Mod.* **6** (1990), 499-524.

[39] Knessl, C. and Tier, C., Asymptotic expansions for large closed queueing networks, *J. ACM* **37** (1990), 144-174.

[40] Knessl, C. and Tier, C., Asymptotic expansions for large closed queueing networks with multiple job classes, *IEEE Trans. Computers* **41** (1992), 480-488.

[41] Knessl, C. and Tier, C., A processor-shared queue which models switching times: normal usage, *SIAM J. Appl. Math.* **52** (1992), 883-899.

[42] Knessl, C. and Tier, C., A processor-shared queue which models switching times: heavy usage asymptotics, *SIAM J. Appl. Math.* **54** (1994), 854-875.

[43] Knessl, C. and Tier, C., Asymptotic properties of first passage times for tandem Jackson networks I: buildup of large queue lengths., *Stoch. Mod.*, (to

appear).

[44] Knessl, C. and Tier, C., Asymptotic properties of first passage times for tandem Jackson networks II: time to empty the system, preprint.

[45] Knessl, C. and Tier, C., Asymptotic bottleneck analysis in single and multiclass networks, in preparation.

[46] Kogan, Y., Another approach to asymptotic expansions for large closed queueing networks, *Oper. Res. Letters* **11** (1992), 317-321.

[47] Konheim, A.G., Meilijson, I. and Melkman, A., Processor-sharing of two parallel lines, *J. Appl. Prob.* **18** (1981), 952-956.

[48] McKenna, J., Extensions and applications of RECAL in the solution of closed product-form queueing networks, *Stoch. Mod.* **4** (1988), 235-276.

[49] McKenna, J. and Mitra, D., Integral representations and asymptotic expansions for closed Markovian queueing networks: normal usage, *Bell Syst. Tech. J.* **61** (1982), 661-683.

[50] McKenna, J. and Mitra, D., Asymptotic expansions and integral representations of moments of queue lengths in closed Markovian networks, *J. ACM* **31** (1984), 346-360.

[51] Mei, J.D. and Tier, C., Asymptotic approximations for a queueing network with multiple classes, *SIAM J. Appl. Math.* **54** (1994), 1147-1180.

[52] Mitra, D. and McKenna, J., Asymptotic expansions for closed Markovian networks with state-dependent service rates, *J. ACM* **33** (1986), 568-592.

[53] Mitra, D. and Morrison, J.A., Asymptotic expansions of moments of the waiting time in closed and open processor-sharing systems with multiple job classes, *Adv. Appl. Prob.* **15** (1983), 813-839.

[54] Morrison, J.A., Response-time distribution for a processor-sharing system, *SIAM J. Appl. Math.* **45** (1985), 152-167.

[55] Morrison, J.A., Asymptotic analysis of the waiting-time distribution for a large closed processor-sharing system, *SIAM J. Appl. Math.* **46** (1986), 140-170.

[56] Morrison, J.A., Moments of the conditioned waiting time in a large closed processor-sharing system, *Stoch. Mod.* **2** (1986), 293-321.

[57] Morrison, J.A., Conditioned response-time distribution for a large closed processor-sharing system in very heavy usage, *SIAM J. Appl. Math.* **47** (1987), 1117-1129.

[58] Morrison, J.A., Conditioned response-time distribution for a large closed processor-sharing system with multiple classes in very heavy usage, *SIAM J. Appl. Math.* **48** (1988), 1493-1509.

[59] Morrison, J.A. and Mitra, D., Heavy-usage asymptotic expansions for the waiting time in closed processor-sharing systems with multiple classes, *Adv. Appl. Prob.* **17** (1985), 163-185.

[60] Reiser, M. and Lavenberg, S.S., Mean-value analysis of closed multichain queueing networks, *J. ACM* **27** (1980), 313-322.

[61] Schweitzer, P., A fixed-point approximation to product-form networks with large population, presented at Second ORSA Telecommunications Conf., Boca Raton, Florida, March 1992.

[62] Schweitzer, P., Serazzi, G. and Broglia, M., A survey of bottleneck analysis in closed networks of queues, In: *Performance Evaluation of Computer and Commun. Systems*, ed. by L. Donatiello and R. Nelson, Lecture Notes in Computer Science, No. **729**, Springer-Verlag, Berlin (1993), 491-508.

[63] Tan, X. and Knessl, C., A finite capacity PS queue which models switching

times, *SIAM J. Appl. Math.* **53** (1993), 491-554.

[64] Tan, X. and Knessl, C., Sojourn time distribution in some processor-shared queues, *Euro. J. Appl. Math.* **4** (1993), 437-448.

[65] Tan, X., Yang, Y. and Knessl, C., The conditional sojourn time distribution in the GI/M/1 processor-sharing queue in heavy traffic, *Queueing Sys.* **14** (1993), 99-109.

[66] Xie, S. and Knessl, C., On the transient behavior of the Erlang loss model: heavy-usage asymptotics, *SIAM J. Appl. Math.* **53** (1993), 555-559.

[67] Xie, S. and Knessl, C., On the transient behavior of the M/M/1 and M/M/1-K queues, *Studies Appl. Math.* **88** (1993), 191-240.

[68] Yang, Y. and Knessl, C., Conditional sojourn time moments in the finite capacity GI/M/1 queue with processor-sharing service, *SIAM J. Appl. Math.* **53** (1993), 1132-1193.

[69] Yang, Y. and Knessl, C., The unconditional sojourn time distribution in the GI/M/1-K queue with processor-sharing service, *Studies Appl. Math.* **93** (1994), 29-91.

[70] Yang, Y. and Knessl, C., Heavy traffic asymptotics of the queue length in the GI/M/1-K queue, preprint.

Chapter 13

The spectral expansion solution method for Markov processes on lattice strips

Isi Mitrani

ABSTRACT A large class of two-dimensional Markov models, whose state space is a lattice strip, can be solved efficiently by means of spectral expansion. The equilibrium distribution of a stochastic process of this type is obtained in terms of the eigenvalues and eigenvectors of a certain matrix polynomial. This method is described in some detail, and examples of its application are presented. The relation between spectral expansion and other existing approaches, such as generating functions or the matrix-geometric method, is discussed. A few possible directions for further research are mentioned.

CONTENTS

13.1	Introduction	337
13.2	The models and their solution	338
13.3	Examples	343
13.4	Spectral expansion and matrix-geometric solutions	346
13.5	Spectral expansion and generating functions	349
13.6	Open problems and future research directions	351
	Bibliography	351

13.1 INTRODUCTION

There are many systems whose behavior is usefully modeled by two-dimensional Markov processes on lattice strips. The system state is described by a pair of integer random variables, I and J; one of these has a finite range, and the other can take any nonnegative value. Often, these models are cast in the framework of a Markov-modulated queue. Demands (jobs) arrive into an unbounded queue and are attended to by a service facility which has a finite number of operational modes. The instantaneous service rate of that facility may depend on its operational mode and, to a limited extent, on the number of jobs present. The instantaneous job arrival rate and the transition rate from one operational mode to another are subject to similar dependencies. In particular, a job arrival, or a service completion, may trigger a change of mode. The pair of random variables, (I, J), where I indicates the operation mode and J is the number of jobs in the system, is then a process on a lattice strip (see, for example, Prabhu and Zhu [18]).

Here, we shall concentrate on an important sub-class of the above processes, characterized by the following properties:

(i) there is a threshold M, such that the instantaneous transition rates out of state (i, j) do not depend on j when $j \geq M$, and

0-8493-8074-x/95/\$0.00+\$.50

(*ii*) the jumps of the random variable J are bounded.
In other words, all transition rates may depend on the operation mode; they may also depend on the queue size when the latter is below a certain value; jobs may arrive and/or depart in fixed or variable batches, provided that those are limited in size.

When the jumps of the random variable J are of size 1, the process is said to be of the *Quasi-Birth-and-Death* type (the term *skipfree* is also used, e.g., in Latouche et al. [9]).

There are three ways of solving such models exactly. Perhaps, the most widely used one is the *matrix-geometric* method [14]. This approach relies on determining the minimal positive solution, R, of a nonlinear matrix equation; the equilibrium distribution is then expressed in terms of powers of R.

The second method uses generating functions to solve the set of balance equations. A number of unknown probabilities which appear in the equations for those generating functions are determined by exploiting the singularities of the coefficient matrix. A comprehensive treatment of that approach, in the context of a discrete-time process with an M/G/1 structure, is presented in Gail et al. [4].

The third (and arguably best) method is the subject of this chapter. It is called *spectral expansion* and is based on expressing the equilibrium distribution of the process in terms of the eigenvalues and left eigenvectors of a certain matrix polynomial. The idea of the spectral expansion solution method has been known for some time (e.g., see Neuts [14]), but there are very few examples of its application in the queueing theory and performance evaluation literature (some instances where that solution has proved useful are reported in Elwalid et al. [2], and Mitrani and Mitra [12]; a more detailed treatment, including numerical results, is presented in Mitrani and Chakka [11].

The spectral expansion solution method and several examples of its application are described in Sections 13.2 and 13.3, respectively. Some observations concerning the efficiency of the solution, compared to that of the matrix-geometric method, are presented in Section 13.4, while Section 13.5 discusses the relation between spectral expansion and generating functions. Possible directions of further research are explored in Section 13.6.

13.2 THE MODELS AND THEIR SOLUTION

Using the terminology of the Markov-modulated queue, let $I(t)$ and $J(t)$ be the random variables representing the operation mode at time t, and the number of jobs in the system at time t, respectively. It is assumed that $X = \{[I(t), J(t)]; t \geq 0\}$ is an irreducible Markov process with state space $\{0, 1, \ldots, N\} \times \{0, 1, \ldots\}$. To start with, suppose that the process is of the Quasi-Birth-and-Death type, i.e., that jobs arrive and depart singly (that restriction will be relaxed later). The evolution of that process proceeds according to the following set of possible transitions:

(*a*) From state (i, j) to state (k, j) $(0 \leq i, k \leq N; i \neq k; j \geq 0)$,
(*b*) From state (i, j) to state $(k, j+1)$ $(0 \leq i, k \leq N; j \geq 0)$, and
(*c*) From state (i, j) to state $(k, j-1)$ $(0 \leq i, k \leq N; j \geq 1)$.

The instantaneous transition rate matrices associated with (*a*), (*b*), and (*c*) are denoted by A_j, B_j, and C_j, respectively (the main diagonal of A_j is zero by definition; also, $C_0 = 0$ by definition). There is a threshold, M $(M \geq 1)$, such that those matrices do not depend on j when $j \geq M$. In other words,

$$A_j = A,\ B_j = B, \text{ and } C_j = C,\ j \geq M. \tag{13.1}$$

Transitions (a) correspond to changes in the operation mode. Those of type (b) represent a job arrival coinciding with a change in the operation mode. If arrivals are not accompanied by such changes, then the matrices B and B_j are diagonal. Similarly, a transition of type (c) represents a job departure coinciding with a change in the operation mode. Again, if such coincidences do not occur, then the matrices C and C_j are diagonal.

The requirement that all transition rates cease to depend on the size of the job queue beyond a certain threshold is not too restrictive. It enables the consideration of models where the arrival, service, and mode transition rates depend on the current operation mode. On the other hand, it is difficult to think of applications where those rates would depend on the number of jobs waiting to begin execution. Note that we impose no limit on the magnitude of the threshold M, although it must be pointed out that the larger M is, the greater the complexity of the solution.

As well as the matrices A_j, B_j, and C_j, it is convenient to define the diagonal matrices D_j^A, D_j^B, and D_j^C, whose ith diagonal element is the ith row sum of A_j, B_j, and C_j, respectively. Those row sums are the total rates at which the process X leaves state (i, j), due to (a) changes in the operation mode, (b) job arrivals (perhaps accompanied by such a change), and (c) job departures (ditto), respectively. The j-independent versions of these diagonal matrices are denoted by D^A, D^B, and D^C, respectively.

The object of the analysis is to determine the joint steady-state distribution of the operation mode and the number of jobs in the system:

$$p_{i,j} = \lim_{t \to \infty} \mathrm{P}(I(t) = i, J(t) = j); i = 0, 1, \ldots, N; j = 0, 1 \ldots \tag{13.2}$$

That distribution exists for an irreducible Markov process if, and only if, the corresponding set of balance equations has a unique normalizable solution.

The balance equations satisfied by the probabilities $p_{i,j}$ can be written in terms of the following row vectors, representing the states with j jobs in the system:

$$\mathbf{v}_j = (p_{0,j}, p_{1,j}, \ldots, p_{N,j});\ j = 0, 1, \ldots \tag{13.3}$$

Those equations are:

$$\mathbf{v}_j[D_j^A + D_j^B + D_j^C] = \mathbf{v}_{j-1}B_{j-1} + \mathbf{v}_jA_j + \mathbf{v}_{j+1}C_{j+1}; j = 0, 1, \ldots, M, \tag{13.4}$$

and $$\mathbf{v}_j[D^A + D^B + D^C] = \mathbf{v}_{j-1}B + \mathbf{v}_jA + \mathbf{v}_{j+1}C;\ j = M+1, M+2, \ldots \tag{13.5}$$

In addition, all probabilities must sum up to 1:

$$\sum_{j=0}^{\infty} \mathbf{v}_j\mathbf{e} = 1, \tag{13.6}$$

where $\mathbf{e}$ is a column vector with $N+1$ elements, all of which are equal to 1.

The first step is to find the general solution of equation (13.5), whose coefficients do not depend on j. That equation can be rewritten in the form

$$\mathbf{v}_jQ_0 + \mathbf{v}_{j+1}Q_1 + \mathbf{v}_{j+2}Q_2 = \mathbf{0};\ j = M, M+1, \ldots, \tag{13.7}$$

where $Q_0 = B$, $Q_1 = A - D^A - D^B - D^C$, and $C_2 = C$. This is a homogeneous vector difference equation of order 2, with constant coefficients. Associated with it is the characteristic matrix polynomial, $Q(\lambda)$, defined as

$$Q(\lambda) = Q_0 + Q_1\lambda + Q_2\lambda^2. \tag{13.8}$$

Denote by λ_k and ψ_k the *generalized eigenvalues* and corresponding *generalized left row eigenvectors* of $Q(\lambda)$. In other words, these are quantities which satisfy

$$det[Q(\lambda_k)] = 0; \;\; \psi_k Q(\lambda_k) = \mathbf{0}; \;\; k = 1, 2, \ldots, d, \tag{13.9}$$

where $d = degree\{det[Q(\lambda)]\}$. In what follows, the qualification *generalized* will be omitted.

The above eigenvalues do not have to be simple, but it is assumed that if one of them has multiplicity m, then it also has m linearly independent left eigenvectors. This has invariably been observed to be the case in practice. So, the numbering in (13.9) is such that each eigenvalue is counted according to its multiplicity.

It is readily seen, by direct substitution, that, for every eigenvalue λ_k and corresponding left eigenvector ψ_k, the sequence

$$\{\psi_k \lambda_k^j; \; j = M, M+1, \ldots\}, \tag{13.10}$$

is a solution of equation (13.7). By combining multiple eigenvalues with each of their independent eigenvectors, one thus obtains a total of d linearly independent solutions. On the other hand, it is known (e.g., see [5]), that the dimensionality of the solution space of equation (13.7) is exactly d. Therefore, any solution of (13.7) can be expressed as a linear combination of the d solutions (13.10):

$$\mathbf{v}_j = \sum_{k=1}^{d} x_k \psi_k \lambda_k^j; \quad j = M, M+1, \ldots, \tag{13.11}$$

where x_k $(k = 1, 2, \ldots, d)$, are arbitrary (complex) constants.

However, the only solutions that are of interest in the present context are those which can be normalized to become probability distributions. Hence, it is necessary to select, from the set (13.11), those sequences for which the series $\sum \mathbf{v}_j \mathbf{e}$ converges. This requirement implies that if $|\lambda_k| \geq 1$, then the corresponding coefficient x_k must be 0. Therefore, the following is true:

Proposition 13.1. *Suppose that c of the eigenvalues of $Q(\lambda)$ are strictly inside the unit disk (each counted according to its multiplicity), while the others are on the circumference or outside. Order them so that $|\lambda_k| < 1$ for $k = 1, 2, \ldots, c$. The corresponding independent eigenvectors are $\psi_1, \psi_2, \ldots, \psi_c$. Then, any normalizable solution of equation (13.7) can be expressed as*

$$\mathbf{v}_j = \sum_{k=1}^{c} x_k \psi_k \lambda_k^j; \quad j = M, M+1, \ldots, \tag{13.12}$$

where x_k $(k = 1, 2, \ldots, c)$, are some constants.

Expression (13.12) is referred to as the *spectral expansion* of the vectors $\mathbf{v}_j$. The coefficients of that expansion, x_k, are yet to be determined.

Note that if there are nonreal eigenvalues in the unit disk, then they appear in complex-conjugate pairs. The corresponding eigenvectors are also complex-conjugate. The same must be true for the appropriate pairs of constants x_k, in order that the right-hand side of (13.12) be real. To ensure that it is also positive, it seems that the real parts of λ_k, ψ_k and x_k should be positive. Indeed, that has always been found to be the case.

So far, expressions have been obtained for the vectors $\mathbf{v}_M, \mathbf{v}_{M+1}, \ldots$; these contain c unknown constants. Now, it is time to consider equations (13.4), for $j = 0, 1, \ldots, M$. This is a set of $(M+1)(N+1)$ linear equations with $M(N+1)$ unknown probabilities (the vectors $\mathbf{v}_j$ for $j = 0, 1, \ldots, M-1$), plus the constants x_k. However, only $(M+1)(N+1) - 1$ of these equations are linearly independent, since the generator matrix of the Markov process is singular. On the other hand, an additional independent equation is provided by (13.6).

In order that this set of linearly independent equations has a unique solution, the number of unknowns must be equal to the number of equations, i.e., $(M+1)(N+1) = M(N+1) + c$, or $c = N+1$. This observation, together with the fact that an irreducible Markov process has a steady-state distribution, if, and only if, its balance and normalizing equations have a unique solution, implies

Proposition 13.2. *The condition $c = N+1$ (the number of eigenvalues of $Q(\lambda)$ strictly inside the unit disk is equal to the number of operation modes), is necessary and sufficient for the ergodicity of the Markov process X.*

In summary, the spectral expansion solution procedure consists of the following steps:

(1) Compute the eigenvalues λ_k and the corresponding left eigenvectors $\boldsymbol{\psi}_k$, of $Q(\lambda)$. If $c \neq N+1$, then stop; a steady-state distribution does not exist.
(2) Solve the finite set of linear equations (13.4) and (13.6), with $\mathbf{v}_M$ and $\mathbf{v}_{M+1}$ given by (13.12), to determine the constants x_k and the vectors $\mathbf{v}_j$ for $j < M$.
(3) Use the obtained solution in order to determine various moments, marginal probabilities, percentiles and other system performance measures that may be of interest.

Careful attention should be paid to step 1. The "brute force" approach, which relies on first evaluating the scalar polynomial $det[Q(\lambda)]$, then finding its roots, is very inefficient for large N and is therefore not recommended. An alternative, which is preferable in most cases, is to reduce the quadratic eigenvalue-eigenvector problem

$$\boldsymbol{\psi}[Q_0 + Q_1\lambda + Q_2\lambda^2] = \mathbf{0}, \tag{13.13}$$

to a linear one of the form $\mathbf{y}Q = \lambda\mathbf{y}$, where Q is a matrix whose dimensions are twice as large as those of Q_0, Q_1 and Q_2. The latter problem is normally solved by applying various transformation techniques. Efficient routines for that purpose are available in most numerical packages.

This linearization can be achieved quite easily if the matrix $C = Q_2$ is non-singular. Indeed, after multiplying (13.13) on the right by Q_2^{-1}, it becomes

$$\boldsymbol{\psi}[R_0 + R_1\lambda + I\lambda^2] = \mathbf{0}, \tag{13.14}$$

where $R_0 = Q_0C^{-1}$, $R_1 = Q_1C^{-1}$ and I is the identity matrix. By introducing the vector $\boldsymbol{\varphi} = \lambda\boldsymbol{\psi}$, equation (13.14) can be rewritten in the equivalent linear form

$$[\boldsymbol{\psi}, \boldsymbol{\varphi}]\begin{bmatrix} 0 & -R_0 \\ I & -R_1 \end{bmatrix} = \lambda[\boldsymbol{\psi}, \boldsymbol{\varphi}]. \tag{13.15}$$

If C is singular, then the desired result is achieved by first making a change of variable, $\gamma = (\theta + \lambda)/(\theta - \lambda)$ or $\lambda = \theta(\gamma - 1)/(\gamma + 1)$ (see [7]). To get the coefficient of γ^2 to be non-singular, the value of θ should be chosen so that the matrix $S = \theta^2 Q_2 + \theta Q_1 + Q_0$ is nonsingular. In other words, θ can have any value which is not an eigenvalue of $Q(\lambda)$. Having made that change of variable, multiplying the resulting equation by S^{-1} on the right reduces it to the form (13.14).

The computational demands of step 2 may be high if the threshold M is large. However, if the matrices B_j $(j = 0, 1, \ldots, M-1)$ are nonsingular (which is often the case in practice), then the vectors $\mathbf{v}_{M-1}, \mathbf{v}_{M-2}, \ldots, \mathbf{v}_0$ can be expressed in terms of $\mathbf{v}_M$ and $\mathbf{v}_{M+1}$, with the aid of equations (13.4) for $j = M, M-1, \ldots, 1$. One is then left with equations (13.4) for $j = 0$, plus (13.6) (a total of $N+1$ independent linear equations), for the $N+1$ unknowns x_k.

Consider now a more general Markov process $X = \{[I(t), J(t)]; t \geq 0\}$ on the same state space $\{0, 1, \ldots, N\} \times \{0, 1, \ldots\}$. The variable J may jump by arbitrary, but bounded amounts, in either direction. In other words, the allowable transitions are:

(a) From state (i, j) to state (k, j) $(0 \leq i, k \leq N; i \neq k)$,

(b) From state (i, j) to state $(k, j+s)$ $(0 \leq i, k \leq N; 1 \leq s \leq r_1; r_1 \geq 1)$, and

(c) From state (i, j) to state $(k, j-s)$ $(0 \leq i, k \leq N; 1 \leq s \leq r_2; r_2 \geq 1)$,

provided, of course, that the source and destination states are valid.

If $r_1 = 1$, then this is (a special case of) a process of the G/M/1 type; similarly, if $r_2 = 1$, then it is (a special case of) a process of the M/G/1 type. Obviously, the Quasi-Birth-and-Death process is both.

Denote, by A_j, $B_j(s)$, and $C_j(s)$, the transition rate matrices associated with (a), (b), and (c), respectively. There is a threshold M, such that

$$A_j = A,\ B_j(s) = B(s), \text{ and } C_j(s) = C(s),\ j \geq M. \tag{13.16}$$

Defining again the diagonal matrices D^A, $D^{B(s)}$, and $D^{C(s)}$, whose ith diagonal element is equal to the ith row sum of $A, B(s)$, and $C(s)$, respectively, the balance equations for $j \geq M + r_1$ can be written in a form analogous to (13.5):

$$\mathbf{v}_j[D^A + \sum_{s=1}^{r_1} D^{B(s)} + \sum_{s=1}^{r_2} D^{C(s)}] = \sum_{s=1}^{r_1} \mathbf{v}_{j-s} B(s) + \mathbf{v}_j A + \sum_{s=1}^{r_2} \mathbf{v}_{j+s} C(s). \tag{13.17}$$

Similar equations, involving A_j, $B_j(s)$, and $C_j(s)$, together with the corresponding diagonal matrices, can be written for $j \leq M + r_1$.

As before, (13.17) can be rewritten as a vector difference equation, this time, of order $r = r_1 + r_2$ with constant coefficients:

$$\sum_{\ell=0}^{r} \mathbf{v}_{j+\ell} Q_\ell = \mathbf{0},\quad j \geq M. \tag{13.18}$$

Here, $Q_\ell = B_{r_1 - \ell}$ for $\ell = 0, 1, \ldots, r_1 - 1$,

$$Q_{r_1} = A - D^A - \sum_{s=1}^{r_1} D^{B(s)} - \sum_{s=1}^{r_2} D^{C(s)},$$

and $Q_\ell = C_{\ell - r_1}$ for $\ell = r_1 + 1, r_1 + 2, \ldots, r_1 + r_2$. The corresponding matrix polynomial is

$$Q(\lambda) = \sum_{\ell=0}^{r} Q_\ell \lambda^\ell. \tag{13.19}$$

From here on, the development proceeds as before. The normalizable solution of equation (13.18) is of the form

$$\mathbf{v}_j = \sum_{k=1}^{c} x_k \boldsymbol{\psi}_k \lambda_k^j;\quad j = M, M+1, \ldots, \tag{13.20}$$

where λ_k are the eigenvalues of $Q(\lambda)$ in the interior of the unit disk, $\boldsymbol{\psi}_k$ are the corresponding left eigenvectors, and x_k are arbitrary constants $(k = 1, 2, \ldots, c)$. The latter are determined with the aid of the state-dependent balance equations and the normalizing equation.

For computation purposes, the polynomial eigenvalue-eigenvector problem of degree r can be transformed into a linear one. For example, suppose that Q_r is nonsingular and multiply (13.18) on the right by Q_r^{-1}. This leads to the problem

$$\psi[\sum_{\ell=0}^{r-1} R_\ell \lambda^\ell + I\lambda^r] = 0, \tag{13.21}$$

where $R_\ell = Q_\ell Q_r^{-1}$. Introducing the vectors $\varphi_\ell = \lambda^\ell \psi$, $\ell = 1,2,\ldots,r-1$, one obtains the equivalent linear form

$$[\psi,\varphi_1,\ldots,\varphi_{r-1}]\begin{bmatrix} 0 & & & -R_0 \\ I & 0 & & -R_1 \\ & \ddots & \ddots & \\ & & I & -R_{r-1} \end{bmatrix} = \lambda[\psi,\varphi_1,\ldots,\varphi_{r-1}].$$

As in the quadratic case, if Q_r is singular then the canonical form (13.21) can be achieved by a change of variable, $\gamma = (\theta+\lambda)/(\theta-\lambda)$. The value of θ can be chose arbitrarily, provided that it is not an eigenvalue of $Q(\lambda)$.

13.3 EXAMPLES

A system that fits naturally in the framework of the Markov-modulated queue is one where a Poisson stream of jobs (arrival rate σ) is attended to by N identical parallel servers (service rate μ), each of which is subject to independent breakdowns and repairs (rates ξ and η, respectively). This model, illustrated in Figure 1, has been analyzed by means of generating functions in [10] and by the matrix-geometric method in [16].

The two variables describing the system state are the number of operative servers, I, and the number of jobs present, J. The instantaneous transitions out of state (i,j) are to state $(i+1,j)$ with rate $(N-i)\eta$, state $(i-1,j)$ with rate $i\xi$, state $(i,j+1)$ with rate σ, and state $(i,j-1)$ with rate $min(i,j)\mu$. Hence, this is a Quasi-Birth-and-Death process. The threshold, beyond which the transition rates cease to depend on the number of jobs in the system, is N. The matrices A, B, and C which appear in equation (13.5) are given by

$$A = \begin{bmatrix} 0 & N\eta & & & \\ \xi & 0 & (N-1)\eta & & \\ & 2\xi & 0 & \ddots & \\ & & \ddots & \ddots & \eta \\ & & & N\xi & 0 \end{bmatrix}, \quad C = \begin{bmatrix} 0 & & & & \\ & \mu & & & \\ & & 2\mu & & \\ & & & \ddots & \\ & & & & N\mu \end{bmatrix},$$

$$B = \sigma I_{N+1}, \tag{13.22}$$

where I_{N+1} is the identity matrix of order $N+1$.

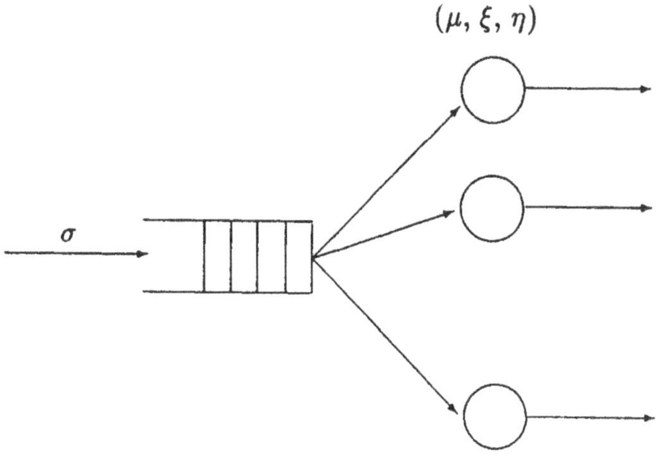

Figure 1: The M/M/N queue with breakdowns and repairs.

When $j < N$, the only dependency of j is in the matrix C_j. Its ith diagonal element is then equal to $min(i, j)\mu$.

The processing capacity of the system, which is defined as the average number of operative servers, is equal to

$$\mathbf{E}[I]) = \frac{N\eta}{\xi + \eta}.$$

The process is ergodic if, and only if, the offered load is less than the processing capacity:

$$\frac{\sigma}{\mu} < \frac{N\eta}{\xi + \eta}. \tag{13.23}$$

The spectral expansion solution yields performance measures such as average response time, average queue size, and percentiles of the queue size distribution (both unconditional and conditioned upon the number of operative servers). Also, it can be used to provide simple estimates in the heavy traffic limit. The important observation, in this connection (see [13]), is that, when the system approaches saturation, expansion (13.12) is dominated by the eigenvalue with the largest modulus inside the unit disk, λ_{N+1}. That eigenvalue is always real (it is in fact, the Perron-Frobenius eigenvalue of Neuts' matrix R, [14]). Thus, when the left-hand side of (13.23) is close to the right-hand side, the average number of jobs in the system is approximately equal to $\lambda_{N+1}/(1 - \lambda_{N+1})$.

A numerical solution displaying this phenomenon is shown in Table 1. The following parameters are fixed: $N = 10$, $\mu = 1$, $\xi = 0.05$, and $\eta = 0.1$. The loading factor is varied by increasing the arrival rate, σ. The saturation point is reached, in this case, when $\sigma = 6.666666\ldots$. The table gives the exact values of $\mathbf{E}[J]$, computed via the spectral expansion method, together with the approximation obtained by ignoring all eigenvalues except the dominant one.

Table 1

σ	$E[J]$	$\lambda_{N+1}/(1-\lambda_{N+1})$
6.0	26.920	31.870
6.2	40.429	45.742
6.4	74.643	80.325
6.6	315.993	322.045
6.63	579.581	585.692
6.66	3215.897	3222.065

Various generalizations of the model in Figure 1, e.g., allowing the job arrival rate to depend on the number of operative servers or the servers to break down and be repaired in batches, can be handled by changing the matrices in (13.22) appropriately.

The second example involves blocking. Consider a network of two nodes in tandem (Figure 2). Jobs arrive into the first node in a Poisson stream with rate σ, and join an unbounded queue. After completing service at node 1 (exponentially distributed with mean $1/\mu$), they attempt to go to node 2, where there is a finite buffer with room for a maximum of N jobs. If that transfer is impossible because the buffer is full, the job remains at node 1, preventing its server from starting a new service until the completion of the current service at node 2 (exponentially distributed with mean $1/\nu$). This model was studied using generating functions in [8] and, before that, in [15].

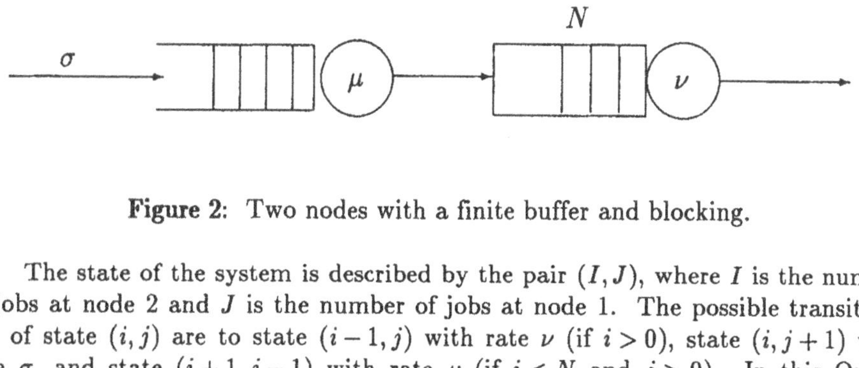

Figure 2: Two nodes with a finite buffer and blocking.

The state of the system is described by the pair (I, J), where I is the number of jobs at node 2 and J is the number of jobs at node 1. The possible transitions out of state (i, j) are to state $(i-1, j)$ with rate ν (if $i > 0$), state $(i, j+1)$ with rate σ, and state $(i+1, j-1)$ with rate μ (if $i < N$ and $j > 0$). In this Quasi-Birth-and-Death process, there are no dependencies on j, apart from the fact that server 1 is idle when $j = 0$. The matrices A, B and C which appear in equation (13.5) are given by

$$A = \begin{bmatrix} 0 & & & \\ \nu & 0 & & \\ & \ddots & \ddots & \\ & & \nu & 0 \end{bmatrix}; \; B = \sigma I_{N+1}; \; C = \begin{bmatrix} 0 & \mu & & \\ & \ddots & \ddots & \\ & & 0 & \mu \\ & & & 0 \end{bmatrix}. \tag{13.24}$$

When the matrix polynomial $Q(\lambda)$ is tri-diagonal, as in the case in the above two examples, it is often possible to prove that all its eigenvalues are real, distinct and nonnegative (e.g., see [8, 10]). However, more general transition structures can (and do) exhibit complex eigenvalues.

The last example is concerned with the execution of task graphs on multiprocessors. For the purpose of this discussion, suppose that a "job" contains one or more processing stages in sequence. A processing stage can be either "serial", i.e., a single task or "parallel", consisting of several independent tasks (see Figure 3). The next stage of a job cannot start until the previous one has completed (and, in the case of a parallel stage, that means all tasks have completed). Different tasks can be executed on different processors, subject to the precedence constraints (see [3]).

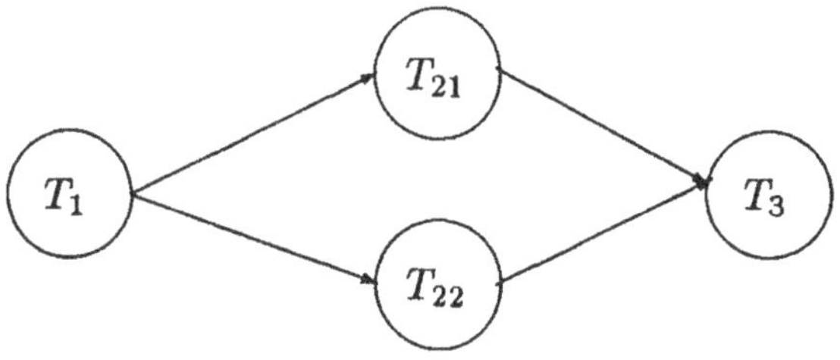

Figure 3: A task graph with three stages: serial, parallel, serial.

Assuming that jobs arrive in a Poisson stream and task execution times are distributed exponentially (perhaps with different parameters for different stages in the graph), the process state can be described by the pair (I, J), where I denotes the particular assignment of tasks on processors and J is the number of jobs in the system. The range of values for I depends on the task graph structure and on the number of available processors. For example, suppose that each job consists of three serial stages, (T_1, T_2, T_3), and there are 2 identical processors. Then as long as $J \geq 2$, two jobs are being executed and there are 6 possibilities for the pair (*task on one processor, task on the other processor*): (T_1, T_1), (T_1, T_2), (T_1, T_3), (T_2, T_2), (T_2, T_3), (T_3, T_3). The values of I are $0, \ldots, 5$.

The construction of matrices A and C for this model can become quite complicated, especially when parallel stages are involved. A systematic algorithm for performing that construction is discussed in [3].

13.4 SPECTRAL EXPANSION AND MATRIX-GEOMETRIC SOLUTIONS

The models described in Section 13.2 can also be solved by the matrix-geometric method [14]. In the case of Quasi-Birth-and-Death processes described by equations (13.4) and (13.5), the vectors $\mathbf{v}_j$, $j \geq M$, are expressed in the form

$$\mathbf{v}_j = \mathbf{v}_M R^{j-M}, \tag{13.25}$$

where R is the minimal non-negative solution of the quadratic matrix equation corresponding to the difference equation (13.7):

$$Q_0 + RQ_1 + R^2 Q_2 = 0. \tag{13.26}$$

Remark 13.3. Equations (13.26) and (13.13) are closely related. Indeed, suppose that λ and ψ are an eigenvalue and a left eigenvector of the matrix R, i.e., they satisfy $\psi R = \lambda \psi$. It is readily verified that, if R satisfies equation (13.26), then λ and ψ satisfy (13.13). Conversely, if all eigenvalues and eigenvectors of R satisfy (13.13), then R satisfies (13.26).

The vector $\mathbf{v}_M$, appearing in the right-hand side of (13.25), and the remaining unknown probabilities, are determined from the balance equations for $j \leq M$, together with the normalizing equation. This part of the solution is very similar to step 2 of the spectral expansion procedure (summary in Section 13.2), and has the same order of complexity.

In the more general case of arbitrary but bounded jumps of the random variable J, the matrix-geometric solution is still of the form (13.25), but the matrix R is obtained as the minimal nonnegative solution of a matrix equation of a higher order. More precisely, if the state-independent balance equations are of the form (13.18), then the equation for R is:

$$\sum_{\ell=0}^{r} R^{\ell} Q_{\ell} = 0. \tag{13.27}$$

These equations are usually solved by iterations. For example, [14] recommends the following procedure for the solution of the quadratic equation (13.26):

$$R_{n+1} = [-Q_0 - R_n^2 Q_2] Q_1^{-1}, \tag{13.28}$$

starting with $R_0 = 0$. The number of iterations depends on the model parameters and the desired accuracy.

Thus, the main operational difference between the two approaches is that spectral expansion involves the computation of eigenvalues and eigenvectors of a given matrix, whereas the matrix-geometric method requires the determination of an unknown matrix by solving a matrix equation. The impact of that difference can be evaluated by performing a series of experiments where the same models are solved by both methods. This was done, in [11], in the context of a multiprocessor system with generalized breakdowns and repairs.

It should be emphasized that both solution methods are exact in principle. That is, if the eigenvalues and eigenvectors, on the one hand, and the matrix R, on the other, are determined exactly, and if the respective sets of linear equations are solved exactly, then, in both cases, the values of the resulting probabilities and performance measures would be correct. However, in practice, both the accuracy of the results and the complexity of the solution depend on many factors and vary between the methods.

The overall computational complexity of determining all eigenvalues and eigenvectors of an $N \times N$ matrix is $O(N^3)$. Moreover, that complexity is, more or less, independent of the parameters of the model. The accuracy of the spectral expansion solution is limited only by the precision of the numerical operations. Using double precision arithmetic, the performance measures computed by this method are correct to about 10 significant digits, i.e., they are exact for all practical purposes.

If the matrix R is computed approximately by an iterative procedure such as (13.28), then the criterion for termination would normally include some tradeoff between computation time and accuracy. The situation is illustrated in Table 2, where a 10-server model is solved by both methods. The procedure (13.28) is terminated when

$$\max_{i,j} | R_{n+1}(i,j) - R_n(i,j) | < \epsilon,$$

for a given value of ϵ. The performance measure, that is computed, is the average queue size, $\mathbf{E}[J]$. Of course, the value, obtained by spectral expansion (column 4), does not depend on ϵ. The offered load is varied by increasing the job arrival rate, σ (the system saturates when $\sigma = 6.6666\ldots$).

The table confirms that, when ϵ decreases, the matrix-geometric solution approaches the solution by spectral expansion. However, it is important to observe that the accuracy of R is not related, in any obvious way, to the accuracy of E[J]. Thus, taking $\epsilon = 10^{-6}$ yields an answer whose relative error is 0.0004% when $\sigma = 3$, 0.06% when $\sigma = 6$, and 6.3% when $\sigma = 6.6$. Another important aspect of the table is that, for a given ϵ, the number of iterations required to compute R increases with σ.

Thus, a comparison between the run times of the spectral expansion and the matrix-geometric algorithms shows that the former is independent of the parameters (and the question of accuracy does not arise), while the latter is influenced both by the offered load and the desired accuracy. For a heavily loaded system, the matrix-geometric run time may be many orders of magnitude larger than that of the spectral expansion. The extra time is almost entirely taken up by the iterative procedure for calculating the matrix R. It seems clear that, even when the accuracy requirements are modest, the computational demands of that procedure have a vertical asymptote at the saturation point.

It can be argued that a different way of computing R may yield better results. A number of alternatives have been compared in [6], and improvements have, indeed, been observed. However, the reported results indicate that all iterative approaches suffer from the drawbacks exhibited in table 1, namely that their accuracy and efficiency decrease when the load increases. Perhaps, the fastest existing method for computing R is not by iterations but via its eigenvalues and eigenvectors (this was done in [1]). However, if that method is adopted, then one might as well use spectral expansion directly and avoid the matrix R altogether.

Table 2: Trade-off between accuracy and complexity

$\sigma = 3.0$				
ϵ	iterations	$E[J]$ (mat. geom.)	$E[J]$ (spect. exp.)	% difference
10^{-3}	29	5.175072879090	5.199710420277	0.4738252555
10^{-6}	93	5.199686527032	5.199710420277	0.0004595110
10^{-9}	158	5.199710396767	5.199710420277	0.0000004521
10^{-12}	223	5.199710420254	5.199710420277	0.0000000004
$\sigma = 6.0$				
ϵ	iterations	$E[J]$ (mat. geom.)	$E[J]$ (spect. exp.)	% difference
10^{-3}	77	35.0382555318	50.4058205572	30.4876795091
10^{-6}	670	50.3726795846	50.4058205572	0.0657483051
10^{-9}	1334	50.4057872349	50.4058205572	0.0000661080
10^{-12}	1999	50.4058205241	50.4058205572	0.0000000657
$\sigma = 6.6$				
ϵ	iterations	$E[J]$ (mat. geom.)	$E[J]$ (spect. exp.)	% difference
10^{-3}	77	58.19477818	540.46702456	89.23250160
10^{-6}	2636	506.34712584	540.46702456	6.31303986
10^{-9}	9174	540.42821366	540.46702456	0.00718099
10^{-12}	15836	540.46698572	540.46702456	0.00000719

Using different starting points for the iterations (rather than $R_0 = 0$), does not bring significant benefits and may produce incorrect results. The procedure could diverge or it could, conceivably, converge to a nonminimal solution of (13.26).

The second step in both algorithms (solving a set of linear equations) is where emerging numerical problems begin to manifest themselves. The observable symptom of such problems is usually an ill-conditioned coefficient matrix. It should be pointed out that neither the spectral expansion nor the matrix-geometric algorithm is immune to ill-conditioned matrices, especially when the size of the problem (as measured by the number of operation modes, $N + 1$), is large. It is a reasonable guess that, within the constraints of double precision arithmetic, neither method can accurately handle models where the random variable I takes more than about 100 possible values. Of course, that range can be enlarged considerably by using extended precision, but then the choice of programming language and/or compiler would be restricted.

13.5 SPECTRAL EXPANSION AND GENERATING FUNCTIONS

Consider a Markov process on a strip and suppose, for simplicity, that its balance equations are given by (13.4) and (13.5) (the more general form (13.17) is treated similarly). These equations can be solved by introducing the vector generating function, $\mathbf{g}(z) = [g_0(z), g_1(z), \ldots, g_N(z)]$, whose ith element represents all states with operation mode i- and j-independent transition rates:

$$\mathbf{g}(z) = \sum_{j=M}^{\infty} \mathbf{v}_j z^{j-M}. \tag{13.29}$$

Using the form (13.7) of (13.5), multiplying the jth equation by z^{j-M} and summing, yields the following equation for $\mathbf{g}(z)$:

$$\mathbf{g}(z)S(z) = \mathbf{b}(z), \tag{13.30}$$

where

$$S(z) = Q_0 + \frac{1}{z}Q_1 + \frac{1}{z^2}Q_2$$

and

$$\mathbf{b}(z) = \mathbf{v}_M[\tfrac{1}{z}Q_1 + \tfrac{1}{z^2}Q_2] + \mathbf{v}_{M+1}\tfrac{1}{z}Q_2.$$

The ith element of $\mathbf{g}(z)$, $g_i(z)$, is given by Cramer's formula:

$$g_i(z) = \frac{D_i(z)}{D(z)}, \quad i = 0, 1, \ldots, N, \tag{13.31}$$

where $D(z)$ is the determinant of $S(z)$ and $D_i(z)$ is the determinant of the matrix obtained from $S(z)$ by replacing its ith row with the vector $\mathbf{b}(z)$.

So, the generating functions are expressed in terms of the unknown probability vector $\mathbf{b}(z)$, including all elements of $\mathbf{v}_M$, together with those elements of $\mathbf{v}_{M+1}$ which are not eliminated by Q_2. Hence, expressions (13.31) depend on $N + 1 + K$ unknown probabilities, where K is the number of nonzero columns in the matrix Q_2. In addition, the vectors $\mathbf{v}_j$ for $j = 0, 1, \ldots, M - 1$ remain to be determined, bringing the total to $(M + 1)(N + 1) + K$ unknown probabilities. The balance equations (13.4), for $j = 0, 1, \ldots, M$, plus the normalizing equation (13.6), provide $(M+1)(N+1)+1$ relations among them. Therefore, in order to determine all the unknowns, $K-1$ extra equations are required.

The additional equations are provided by the observation that $g_i(z)$, being probability generating functions, are analytic in the interior of the unit circle, $|z|<1$, and continuous on the boundary. Therefore, if the polynomial $D(z)$ has a root in that region, say z_0, then $D_i(z)$ must vanish at z_0 for all i. A requirement of the form $D_i(z_0)=0$ provides a linear equation for the unknown probabilities. Moreover, if z_0 is a root of multiplicity m, then $m-1$ derivatives of $D_i(z)$ must vanish at z_0, yielding $m-1$ further equations.

On the other hand, since the columns of $S(z_0)$ are linearly independent, it follows that, if $D_i(z)$, or one of its derivatives, vanishes at z_0 for a particular index i, then, it does so for all other indices. In other words, if one of the generating functions in (13.31) is analytic at z_0, then they all are.

Thus, every root of $D(z)$, which is either inside or on the unit circle, provides as many linear equations involving the unknown probabilities as is its multiplicity (a pair of complex-conjugate roots provides two equations with real coefficients). Consequently, the determination of the unknown probabilities is made possible by finding $K-1$ such roots, counting each according to its multiplicity. It should be mentioned that the root at $z=1$, which is normally present, does not provide an independent equation; the fact, that $D_i(1)=0$ would follow from the other equations.

From the above summary it may appear that the generating function method is quite different from that of the spectral expansion. In fact, the two are very closely related. The first point to note, in this connection, is that the matrix polynomials $Q(\lambda)$ and $S(z)$, appearing in (13.8) and (13.30), respectively, differ only by a change of variable: $S(z)=Q(1/z)$. Hence, an eigenvalue of $Q(\lambda)$ *inside* the unit disk is the reciprocal of an eigenvalue of $S(z)$ *outside* the unit disk and vice versa.

Next, consider the function $g_i(z)$ given by (13.31). This ratio of two polynomials is completely determined by its poles and respective residues. In other words, if $z_1, z_2, \ldots, z_c$ are the roots of $D(z)$ *strictly outside* the unit disk (assume for simplicity that they are distinct), then $g_i(z)$ has the form

$$g_i(z)=\sum_{k=1}^{c}\frac{x_{i,k}z_k}{z_k-z}, \tag{13.32}$$

where $x_{i,k}$ are some constants. Expanding the right-hand side of (13.32) in powers of z and remembering the definition of $g_i(z)$, leads to an expression for the probability vector $\mathbf{v}_j$:

$$\mathbf{v}_j=\sum_{k=1}^{c}\psi_k\left(\frac{1}{z_k}\right)^j;\quad j=M,M+1,\ldots, \tag{13.33}$$

where $\psi_k=z_k^M[x_{0,k},x_{1,k},\ldots,x_{N,k}]$.

The argument has, thus, arrived at an expression very similar to the spectral expansion (13.12). The reciprocals $1/z_k$ are precisely the eigenvalues λ_k. The only difference is that, whereas the coefficients in the expansion (13.12) are obtained by finding the left eigenvectors of $Q(\lambda)$ and then solving a set of simultaneous linear equations, those in (13.33) require the determination of a rational function (by solving a set of simultaneous linear equations) and, then, computing the residues of its poles.

It seems clear from the above discussion that, for the models of interest here, there is very little point in introducing and determining generating functions. Probabilities and other performance measures can be computed directly through spectral expansion. Of course, the situation is different in cases where spectral expansion is not readily applicable, e.g., for processes of type M/G/1 or G/M/1. There, the generating function route is certainly important (see [4]).

13.6 OPEN PROBLEMS AND FUTURE RESEARCH DIRECTIONS

The spectral expansion solution method has been shown to be versatile, readily implementable, and efficient. A strong case can be made for using it, whenever possible, in preference to other existing methods. It can offer considerable advantages in speed without undue penalties in accuracy or numerical stability.

As well as exact solutions, spectral expansion provides a good basis for obtaining asymptotic and approximate recitals. An example of this, in the context of heavy traffic limits, was given in Section 13.3. Other possibilities for approximations exist in cases where a two-dimensional Markov process (infinite in both directions) can be truncated to a strip without losing too much accuracy. Such truncation was used successfully in [3].

The general problem of how to use spectral expansion in order to find approximate solutions to large problems (considerably more than 100 operation modes), certainly deserves further study. One possible approach is to restrict the expansion to the K largest eigenvalues and corresponding eigenvectors, for some K greater than 1, but much smaller than N. Then there would be only K coefficients to determine. When there are fewer unknowns than equations, the latter cannot, in general, be satisfied. However, suppose that some of the balance equations are ignored, and the rest are used to determine the coefficients of the restricted expansion. How good is that approximation?

Another topic for future work is the application of spectral expansion to finite-state models (e.g., a Markov-modulated queue with finite, but large waiting room). It seems that the only major change in the procedure would be to include all eigenvalues in the expansion, not just the ones in the interior of the unit disk. Normalization is always possible. Care should be taken, however, to ensure that the number of unknowns is equal to that of available equations.

Perhaps the most important direction of further research in this area concerns the extension of the spectral expansion method to models with unbounded jumps in the number of jobs, J. It seems likely, for example, that such an extension can be achieved at least for some processes of block-M/G/1 or block-G/M/1 type (see [4]).

BIBLIOGRAPHY

[1] Daigle, J.N. and Lucantoni, D.M., Queueing systems having phase-dependent arrival and service rates, In: *Numerical Solutions of Markov Chains* (ed. by W.J. Stewart), Marcel Dekker 1991.

[2] Elwalid, A.I., Mitra, D. and Stern, T.E., Statistical multiplexing of Markov-modulated sources: Theory and computational algorithms, *Int. Teletraffic Congress* (1991).

[3] Ettl, M. and Mitrani, I., Applying spectral expansion in evaluating the performance of multiprocessor systems, In: *CWI Tracts* (ed. by O. Boxma and

G. Koole) 1994.

[4] Gail, H.R., Hantler, S.L. and Taylor, B.A., Spectral analysis of M/G/1 type Markov chains, *RC17765* IBM Research Division (1992).

[5] Gohberg, I., Lancaster, P. and Rodman, L, *Matrix Polynomials*, Academic Press 1982.

[6] Gun, L., Experimental results on matrix-analytical solution techniques-extensions and comparisons, *Stoch. Mod.* **5**:4 (1989), 669-682.

[7] Jennings, A., *Matrix Computations for Engineers and Scientists*, John Wiley, New York 1977.

[8] Konheim, A.G. and Reiser, M., A queueing model with finite waiting room and blocking, *JACM* **23**:2 (1976), 328-341.

[9] Latouche, G., Jacobs, P.A. and Gaver, D.P., Finite Markov chain models skipfree in one direction, *Nav. Res. Log. Quart.* **31** (1984), 571-588.

[10] Mitrani, I. and Avi-Itzhak, B., A many-server queue with service interruptions, *Oper. Res.* **16**:3 (1968), 628-638.

[11] Mitrani, I. and Chakka, R., Spectral expansion solution for a class of Markov models: Application and comparison with the matrix-geometric method, *Performance Eval.* (1995), (to appear).

[12] Mitrani, I. and Mitra, D., A spectral expansion method for random walks on semi-infinite strips, *IMACS Symp. on Iterative Methods in Linear Algebra*, Brussels (1991).

[13] Mitrani, I. and Puhalskii, A., Limiting results for multiprocessor systems with breakdowns and repairs, *Queueing Sys.* **14** (1993), 293-311.

[14] Neuts, M.F., *Matrix Geometric Solutions in Stochastic Models*, John Hopkins University Press 1981.

[15] Neuts, M.F., Two queues in series with a finite intermediate waiting room, *J. Appl. Prob.* **5** (1968), 123-142.

[16] Neuts, M.F. and Lucantoni, D.M., A Markovian queue with N servers subject to breakdowns and repairs, *Mgt. Sci.* **25** (1979), 849-861.

[17] Sengupta, B., A queue with service interruptions in an alternating Markovian environment, *Oper. Res.* **38** (1990), 308-318.

[18] Prabhu, N.U. and Zhu, Y., Markov-modulated queueing systems, *Queueing Sys.* (1989), 215-246.

Chapter 14

Applications of vector Riemann boundary value problems to analysis of queueing systems

Alexander Dukhovny

ABSTRACT Nonhomogeneous vector Riemann boundary value problems are introduced as a method of studying the generating functions of the steady-state queue-length probabilities. This method is applied to Markov-controlled random walks with finite perturbations that arise in analyses of semi-Markov queues with limited state-dependence. The solutions obtained often involve complex-analytic factorization components of certain matrix functions, but, in some cases, the solution is shown to depend only on selected elements of these components. Methods of finding such components are discussed, depending on the structure of the matrix functions and complex-analytic properties of their elements. In particular, a procedure is introduced to reduce the dimension of the problem and, in the case of special tridiagonal structures, to reduce the problem to scalar factorization. The technique introduced is applied to several special cases of semi-Markov queues, such as "$\pm$" controlled and "queue-increase" controlled systems, and to examples of limited state-dependence in semi-Markov queues such as "warm-up" and "Bailey-type" dependence with additional processing. Potential applications and implementation problems are discussed.

CONTENTS

14.1	Introduction	353
14.2	Initial equations and examples of applications	355
14.3	Banach spaces of Laurent series and operators	357
14.4	Markov-controlled random walks	359
14.5	Finding factorization components	362
14.6	Markov-controlled random walks with finite perturbations	368
14.7	Potential applications of RBVPs and implementation problems	373
	Bibliography	374

14.1 INTRODUCTION

A fundamental identity that appeared in Spitzer [30], almost forty years ago, gave rise to a special methodology in which stochastic processes arising in semi-Markov queueing systems were linked to complex-analytic factorization of scalar or matrix functions of the type $I - rH$, $|r| < 1$. Here, I is the identity matrix and H is a matrix function based on Laplace transforms and generating functions (GFs) of the system's elements (interarrival and service times and arrival and service group

0-8493-8074-x/95/$0.00+$.50

sizes). By construction, the factors were invertible and complex-analytic in complementary regions of the Hilbert boundary value problem for $I - rH$ on the boundary between the regions. Solving such a problem by any method, one could then find the factors and probabilistic elements related to them.

Among the early papers on this subject were, for example, Miller [26], Kemperman [21], Keilson [20], Çinlar [8,9], and Presman [29]. Many new results were obtained with this technique in the 1970's in works such as Arjas [1,2], Borovkov [5], Dagsvik [11], Malyshev [25], Neuts [27], and many other studies. Further development of this method was complicated by the fact that the factorization components were difficult to find, especially in matrix cases related to steady-state problems. In the course of this chapter, we will refer to some of the later publications; we do not intend, though, to provide a complete bibliography.

In queueing systems with state-dependence, where the distributions of the system's elements may depend on the queue-length, one cannot make up a matrix function H that would contain all information about the system. At the same time, in the case of limited state-dependence, when the distributions of the system's elements are the same when the queue-length exceeds a certain threshold, some transitions of the system have those same special features, that were used in the methods based on fundamental identities. To utilize these features we propose a method which is, in a way, opposite to the methods based on fundamental identities. We transform the equilibrium equations for imbedded Markov chains into Riemann boundary value problems (RBVPs) for the GFs of the steady-state probabilities.

In Section 14.2, following a brief general description of RBVPs, we consider the equilibrium equations for a two-dimensional Markov chain with a finite-valued second component and rewrite them in terms of the generating vector functions of the steady-state probabilities. An assumption about transitions of the first component is then introduced that makes the chain a so-called Markov-controlled random walk with finite perturbations. It turns out that this assumption holds, for example, in such systems with group arrival and service, such as $\mathrm{MAP}^X/\mathrm{G}^Y/1$ systems and "queue-increase" controlled and "$\pm$" controlled $\mathrm{M}^X/\mathrm{G}^Y/1$ and $\mathrm{GI}^X/\mathrm{M}^Y/1$ systems with limited state-dependence.

In Section 14.3, similar to Kingman [22] and Takács [31], we define certain Banach spaces of vector-series of complex argument and several important operators in them. In Section 14.4 we introduce and solve a vector RBVP for Markov-controlled random walks; the solution is given in terms of factorization components. We show that, in some cases, the solution depends only on selected elements of these components.

Methods of finding factorization components of the matrix-coefficient, depending on its structure and complex-analytic properties of its elements, are discussed in Section 14.5. A 2×2 problem, corresponding to so-called "$\pm$" controlled systems, is reduced to scalar factorization. For general dimensions, a reduction process is introduced that decreases the matrix size, given certain properties of the problem. An important special case is considered in which the reduction process reduces the problem to scalar factorization, the case of a tridiagonal generating matrix function of jumps with some special properties of elements.

In Section 14.6, we construct and solve a vector RBVP for Markov-controlled random walks with finite perturbations (thereby expressing the GFs of the steady-state probabilities in terms of a finite number of unknown constants), set a system of linear equations for the constants, and prove its nondegeneracy. Several special cases are then studied by this method, such as so-called "Bailey-type" systems, systems with "warm-up", and others. Potential applications of RBVPs and difficulties associated with the technique of RBVPs are discussed in Section 14.7.

14.2 INITIAL EQUATIONS AND EXAMPLES OF APPLICATIONS

A typical scalar RBVP is as follows. For a contour Γ on the complex plane (for example, Γ: $|z| = 1$), find functions $\phi^+(z)$ and $\phi^-(z)$ complex-analytic, respectively, inside and outside Γ (for example, $\Gamma^+ : |z| < 1$ and $\Gamma^- : |z| > 1$) and continuous on Γ, such that on Γ

$$\phi^+(z)f(z) = \phi^-(z) + g(z), \tag{14.1}$$

where $f(z)$ and $g(z)$ are continuous functions defined on Γ. (We will also denote $\bar{\Gamma}^+ : |z| \leq 1$ and $\bar{\Gamma}^- : |z| \geq 1$.) A homogeneous RBVP, where $g(z) \equiv 0$, is often referred to in the literature as the Hilbert problem. The number of solutions of problem (14.1) depends on the complex-analytic properties of $f(z)$ and $g(z)$. We refer the reader to such texts as Gakhov [18] and Litvinchuk and Spitkovsky [23] for the theory of RBVPs. In particular, suppose there exist functions $r^+(z)$ and $r^-(z)$, such that they and $r^+(z)^{-1}$ and $r^-(z)^{-1}$ are analytic, respectively, in Γ^+ and Γ^- and continuous on Γ, and

$$f(z) = r^+(z)^{-1} r^-(z)^{-1}, \quad z \in \Gamma. \tag{14.2}$$

($r^+(z)$ and $r^-(z)^{-1}$ form a solution of the Hilbert problem for $f(z)$ on Γ.) Now, the RBVP (14.1) has a unique solution (up to an arbitrary constant C). To find this solution, we multiply (14.1) by $r^-(z)$ and rewrite it, using (14.2), as

$$\phi^+(z)r^+(z)^{-1} - g^+(z) = \phi^-(z)r^-(z) - g^-(z), \tag{14.3}$$

where

$$g^{+(-)}(z) = \frac{1}{2\pi i}\int_\Gamma \frac{g(t)r^-(t)}{z-t}dt, \text{ and } z \in \Gamma^+ \ (z \in \Gamma^-), \tag{14.4}$$

are Cauchy-integral analytic extensions of $g(z)r^-(z)$. Since both sides of (14.3) are analytic, respectively, in Γ^+ and Γ^-, by Liouville's theorem, they must be identically equal to some constant C, which yields

$$\phi^+(z) = r^+(z)[C + g^+(z)], \tag{14.5}$$

and

$$\phi^-(z) = r^-(z)^{-1}[C + g^-(z)]. \tag{14.6}$$

In case of scalar RBVPs, it is known (see, e.g., Gakhov [18]) that, if $f(z)$ does not vanish anywhere on Γ and

$$\operatorname{Ind}_\Gamma f(z) = (2\pi)^{-1}\Delta_\Gamma \operatorname{Arg} f(z) = 0, \tag{14.7}$$

then, up to an arbitrary multiplicative constant (that does not matter in (14.5) and (14.6)), $r^+(z)$ and $r^-(z)$ are given by

$$r^+(z) = \exp\{-\frac{1}{2\pi i}\int_\Gamma \frac{\ln f(t)}{z-t}dt\}, \ z \in \Gamma^+, \tag{14.8}$$

and

$$r^-(z) = \exp\{\frac{1}{2\pi i}\int_\Gamma \frac{\ln f(t)}{z-t}dt\}, \ z \in \Gamma^-. \tag{14.9}$$

Studying semi-Markov queues with limited state-dependence, we deal with vector RBVPs, where $\phi^+(z)$, $\phi^-(z)$ and $g(z)$ are vector functions and $f(z)$ is a matrix function. These RBVPs follow from the equilibrium equations for the Markov chains associated with the systems in study.

Consider a Markov chain $\{\xi_k, \eta_k\}$, $k = \overline{0,\infty}$, $\xi_k \in \overline{0,\infty}$, $\eta_k \in \overline{1,n}$, the first component of which will have the meaning of the queue length at selected moments of time t_k (for example, post-completion instants for M/G/1 type queues or pre-arrival instants for GI/M/1 type), while the second will be interpreted as the control component at these moments. Denote by

$$A_{ir}^{js} = \mathbf{P}\{(\xi_{k+1}, \eta_{k+1}) = (j,s) \mid (\xi_k, \eta_k) = (i,r)\} \tag{14.10}$$

the transition probabilities of the chain,

$$A_i^j \text{ is an } (n \times n) \text{ matrix with elements } A_{ir}^{js}, \tag{14.11}$$

$$A_{ir}^s(z) = \sum_{j=0}^{\infty} A_{ir}^{js} z^j, \; z \in \bar{\Gamma}^+, \tag{14.12}$$

and $A_i(z)$ is an $(n \times n)$ matrix function with elements $A_{ir}^s(z)$. Let p_{ir} be the steady-state probabilities of this chain (if they exist) and denote

$$\vec{P}_i = (p_{i1}, p_{i2}, \ldots, p_{in}), \; \vec{P}(z) = \sum_{i=0}^{\infty} \vec{P}_i z^i, \; z \in \bar{\Gamma}^+.$$

The classical equilibrium equations for the steady-state probabilities

$$p_{js} = \sum_{i=0}^{\infty} \sum_{r=1}^{n} p_{ir} \; A_{ir}^{js}, \; j = \overline{0,\infty}, \; s = \overline{1,n}, \tag{14.13}$$

and

$$\sum_{i=0}^{\infty} \sum_{r=1}^{n} p_{ir} = 1, \tag{14.14}$$

can be easily rewritten in vector form

$$\vec{P}_j = \sum_{i=0}^{\infty} \vec{P}_i A_i^j, \tag{14.15}$$

and

$$\sum_{i=0}^{\infty} \vec{P}_i \vec{d} = 1, \tag{14.16}$$

where $\vec{d} = (1,1,\ldots,1)^T$, which, in turn, leads to

$$\vec{P}(z) = \sum_{i=0}^{\infty} \vec{P}_i A_i(z), \; z \in \bar{\Gamma}^+, \tag{14.17}$$

and

$$\vec{P}(1)\vec{d} = 1. \tag{14.18}$$

Relations (14.17) and (14.18) will be a starting point for our method. We will assume that, for some $N \geq 0$, transitions of ξ_k follow the relation:

$$\xi_{k+1} = max\{0, \xi_k + \gamma\}, \text{ if } \xi_k \geq N, \tag{14.19}$$

where the distribution of γ depends on the transition of control component η_k. A chain with property (14.19) will be called a Markov-controlled random walk with finite perturbations; here, we will study its steady-state probabilities based on Dukhovny [13, 14]. Under (14.19), the transition probabilities A_{ir}^{js}, $i \geq N$, $j > 0$, depend on the difference $j - i$. Such chains belong to the class of "block-structured

Markov chains with repeating rows" studied in Grassmann and Heyman [19] by the block-elimination method in cases of so-called "left-banded", "right-banded", and "banded" transition matrices. These terms mean that there exist some integers m_1 and m_2 such that A_{ir}^{js}, $i \geq N$, $j > 0$, are zeros when $j - i < -m_1$ or $j - i > m_2$, or both. Markov chains under these assumptions were also extensively studied by matrix-analytic methods originated by Neuts. The method of vector RBVPs, that we propose here, is based on (14.19) and does not assume "bandedness".

In queueing applications, γ is typically equal to $\alpha - \beta$. In M/G/1-type systems, α is the number of arrivals during the period (t_k, t_{k+1}) and β is the service group size. In GI/M/1 type systems, α is the arrival group size and β is the number of customers that can be potentially served during the period (t_k, t_{k+1}). In particular, relation (14.19) holds for imbedded Markov chains in the following systems:

1. MAP^X/G^Y/1 system, where groups of customers arrive at the server according to the Markov arrival process; arrival and service group sizes are arbitrarily distributed; η is the parameter that controls the arrival process; the distributions of arrival and service group sizes may also depend on transitions of η. See e.g., Neuts [28] for a description of MAP processes. Also, in the case of $Y \equiv 1$, the system is studied by the matrix-analytic methods in Lucantoni [24] under the name of BMAP/G/1.

2. "queue-increase controlled" M^X/G^Y/1 system with a nonwaiting server (see Dukhovny [16]). At post-completion instants t_k, the system selects its operating mode (distributions of its elements) depending on transitions of the control variable η by the following rule: for every possible value r of η there is a set of control intervals $\{\sigma_{rs},\ s = \overline{1,n}\}$, partitioning the set of all integers. If $\eta_k = r$, then the distribution of the service period $t_{k+1} - t_k$ (and, therefore, the distribution of α) is determined by r as well as the distribution of the service group size β. If $\gamma \in \sigma_{rs}$, η switches from r to s.

3. Similarly, one can consider a GI^X/M^Y/1 system with "queue-increase" control at prearrival instants t_k, where $\eta_k = r$ determines the distributions of the forthcoming interarrival period and the arrival group size.

4. Systems M^X/G^Y/1 and GI^X/M^Y/1 with so-called " $\pm$ " control: the system switches its mode η (that determines the distributions of the system's elements) from 1 to 2, if over the period (t_k, t_{k+1}) the queue-length has increased, or from 2 to 1, if the queue-length has decreased; otherwise the system preserves its current mode.

14.3 BANACH SPACES OF LAURENT SERIES AND OPERATORS

We consider several Banach spaces of formal Laurent series in powers of a complex variable z. Denote by W the Wiener algebra of Laurent series with absolutely summable coefficients:

$$W = \{f(z) = \sum_{i=-\infty}^{\infty} |f_i z^i| \ \|f\|_W = \sum_{i=-\infty}^{\infty} |f_i| < \infty\}.$$

For each element $f(z)$ of W, its sum-function (which will also be denoted by $f(z)$) converges absolutely on Γ and is continuous there. There is an obvious one-to-one correspondence between elements in W and their sum-functions on Γ generated by

the relation
$$f_j = \frac{1}{2\pi i}\int_{\Gamma} z^{-j-1} f(z)dz.$$
If two elements of W are equal, values of their sum-functions are equal wherever the series converge. However, equality of two sum-functions in regions that do not completely contain Γ does not imply equality of their Laurent expansions. In what follows, all equalities are equalities in W.

Introduce subspaces W^+ and W^- of W consisting of series with only positive or only nonpositive powers of z; W^0 will denote the subspace of constants. $\forall f^+(z) \in W^+$ the sum-function of $f(z)$ is analytic in Γ^+ and $f^+(0) = 0$. $\forall f^-(z) \in W^-$ its sum-function is analytic in Γ^-. We introduce several projection operators on W:

$$T^+ f(z) = \sum_{i=1}^{\infty} f_i z^i, \tag{14.20}$$

$$T^- f(z) = \sum_{i=-\infty}^{0} f_i z^i, \tag{14.21}$$

$$S f(z) = \sum_{i=-\infty}^{\infty} f_i = f(1), \tag{14.22}$$

and
$$D_i f(z) = f_i. \tag{14.23}$$

Every element of W can obviously be uniquely partitioned into the sum of two elements from W^+ and W^-:

$$f(z) = T^+ f(z) + T^- f(z). \tag{14.24}$$

This partition also gives a unique representation of the sum-function of $f(z)$ on Γ as a sum of two functions: one analytic in Γ^+, continuous on Γ and vanishing at zero; the other analytic in Γ^- and continuous on Γ. This means that once two such functions are found, whether by means of the Cauchy-type integral or otherwise, their Maclaurin expansions in Γ^+ and Γ^-, respectively, add up to $f(z)$.

Based on W, we define the space V^nW consisting of vector-rows
$$\vec{f}(z) = [f_1(z), \ldots, f_n(z)],\ f_i(z) \in W,\quad i \in \overline{1,n},$$
$$\| \vec{f}(z) \|_{VW} = \sum_{r=1}^{n} \| f_r(z) \|_W,$$
and the space M^nW of $n \times n$ matrices $F(z)$ with the norm
$$\| F(z) \|_{MW} = \max_r \sum_{s=1}^{n} \| f_{rs}(z) \|_W,$$
where elements $f_{rs}(z) \in W$. Taking elements from, respectively, W^+, W^-, W^0, we define subspaces V^nW^+, V^nW^-, V^nW^0 and M^nW^+, M^nW^-, M^nW^0. Definitions of operators T^+, T^-, S, and D_i are extended to these spaces elementwise; partition (14.24) holds in both V^nW and M^nW.

The product of two Laurent series is defined by means of convolution of their coefficients:

$$f(z)g(z) = \sum_{i=-\infty}^{\infty} f_i z^i \sum_{i=-\infty}^{\infty} g_i z^i = \sum_{i=-\infty}^{\infty} z^i \sum_{j=-\infty}^{\infty} f_j g_{i-j},$$

given the absolute convergence of $\sum_{j=-\infty}^{\infty} f_j g_{i-j}$, $\forall i$. When both series belong to W, their product also does and the sum-function of the product is equal to the product of sum-functions of the factors.

Let $e(z) = \sum_{i=-\infty}^{0} z^i$. The lemma below readily follows from the definition of products of series.

Lemma 14.1. *Let $f(z) \in W$ be such that*

$$\sum_{i=-\infty}^{\infty} |if_i| < \infty, \; f(1) = \sum_{i=-\infty}^{\infty} f_i = 0.$$

Then $f(z)e(z) \in W$ and

$$Sf(z)e(z) = \sum_{i=-\infty}^{\infty} if_i = f'(1). \tag{14.25}$$

The operators, introduced above, have a clear probabilistic meaning. Let $h(z)$ be the GF of an integer random variable γ. Obviously, $h(z) \in W$, and

$$T^+h(z) = \mathbf{E}[z^\gamma I_{\{\gamma > 0\}}], \quad T^-h(z) = \mathbf{E}[z^\gamma I_{\{\gamma \le 0\}}],$$

$$ST^+h(z) = \mathbf{P}\{\gamma > 0\}, \; ST^-h(z) = \mathbf{P}\{\gamma \le 0\}, \; \text{ and } \; D_i h(z) = \mathbf{P}\{\gamma = i\},$$

where I_A is the indicator of event A.

14.4 MARKOV-CONTROLLED RANDOM WALKS

Consider a homogeneous Markov chain $\{\xi_k, \eta_k\}$, $k = \overline{0,\infty}$, $\xi_k \in \overline{0,\infty}$, and $\eta_k \in \overline{1,n}$, that satisfies (14.19) for $N = 0$:

$$A_{ir}^{js} = h_{rs}^{j-i}, \quad j > 0, \tag{14.26}$$

and

$$A_{ij}^{0s} = \sum_{j=-\infty}^{-i} h_{rs}^{j}, \quad \forall i, \; r, s. \tag{14.27}$$

Such a Markov chain will be called a Markov-controlled random walk. In light of (14.20)-(14.23), we can rewrite (14.26) and (14.27) as

$$A_{ir}^{s}(z) = [T^+ + ST^-]z^i H_{rs}(z), \tag{14.28}$$

or, in matrix form,

$$A_i(z) = [T^+ + ST^-]z^i H(z), \tag{14.29}$$

where

$$H_{rs}(z) = \sum_{j=-\infty}^{\infty} h_{rs}^j z^j, \quad r, s \in \overline{1,n}, \tag{14.30}$$

$H(z)$ is the $n \times n$ matrix with components $H_{rs}(z)$. By definition, $H(z) \in M^n W$ and $H(1)$ is the transition matrix of the Markov chain $\{\eta_k\}$. We additionally assume that:

1. The chain $\{\xi_k, \eta_k\}$ is irreducible and aperiodic.
2. $\text{Det}[I - H(z)]$ is an aperiodic series.
3. The series

$$\sum_{j=-\infty}^{\infty} |j| \, h_{rs}^j < \infty, \quad r, s \in \overline{1,n}. \tag{14.31}$$

These assumptions are often met in applications; they will be referred to as "natural." It follows from the first assumption that the chain $\{\eta_k\}$ has the steady-state distribution $\vec{\pi} = [\pi_1, \ldots, \pi_n]$,

$$\vec{\pi} = \vec{\pi} H(1). \tag{14.32}$$

Denote by $\mathbf{E}(z)$ the matrix that differs from I by only the first column elements all of which are equal to $e(z)$.

Lemma 14.2. *Under natural assumptions,*

$$F(z) = [I - H(z)]\mathbf{E}[z] \in M^n W.$$

The proof follows from Lemma 14.1 and (14.31).

Theorem 14.3. (Dukhovny [13]) *Under natural assumptions, the chain $\{\xi_k, \eta_k\}$ has a stationary distribution if, and only if,*

$$\delta = -\sum_{r=1}^{n} \pi_r \sum_{s=1}^{n} H'_{rs}(1) > 0, \tag{14.33}$$

or, equivalently,

$$\det F(1) = \frac{d}{dz} \det [I - H(z)] \Big|_{z=1} > 0. \tag{14.34}$$

For the proof of Theorem 14.3 in the general case, see Dukhovny [13]. Note that $-\delta$ is the expected stationary jump of ξ_k.

Theorem 14.4. (Dukhovny [13]) *Under natural assumptions, if* (14.33) *holds, then*

$$\operatorname{Ind}_\Gamma \det F(z) = \frac{1}{2\pi} \Delta_\Gamma \operatorname{Arg} \det F(z) = 0; \tag{14.35}$$

$$\textit{all partial indices of } F(z) \textit{ are zeros;} \tag{14.36}$$

there exist invertible matrices $R^+(z)$ and $R^-(z)$ such that:

$$R^+(z)^{-1},\ R^+(z) \in M^n W^+ \oplus M^n W^0, \tag{14.37}$$

$$R^-(z)^{-1},\quad R^-(z) \in M^n W^-, \tag{14.38}$$

$$R^+(0) = I, \tag{14.39}$$

and

$$F(z) = R^+(z)^{-1} R^-(z)^{-1}. \tag{14.40}$$

The proof of Theorem 14.4 is given in Dukhovny [13] and is based on Theorem 5.11 of Litvinchuk and Spitkovsky [23].

Based on Theorems 14.3 and 14.4, we can now set and solve a matrix RBVP for the vector GF of the steady-state probabilities of the chain $\{\xi_k, \eta_k\}$.

Theorem 14.5. (Dukhovny [13]) *Under natural assumptions, if* (14.33) *holds, then*

$$\vec{P}(z) = \vec{\pi} R^+(1)^{-1} R^+(z). \tag{14.41}$$

Proof. By definition, $\vec{P}(z) \in V^n W$. Applying (14.29) in (14.17), we obtain

$$\vec{P}(z) = [T^+ + ST^-]\vec{P}(z)H(z),$$

and, by (14.24), we can rewrite it as

$$\vec{P}(z)F(z) = \vec{\Phi}^-(z), \tag{14.42}$$

where

$$\vec{\Phi}^-(z) = \{[ST^- - T^-]\vec{P}(z)H(z)\}\mathbf{E}[z]. \tag{14.43}$$

By Lemma 14.2, $\vec{P}(z)F(z) \in V^nW$, so on the strength of (14.42) $\vec{\Phi}^-(z) \in V^nW$. By construction, $T^-\vec{P}(z)H(z) \in V^nW^-$; therefore, by (14.43), $\vec{\Phi}^-(z) \in V^nW^-$. Now (14.42) becomes a matrix RBVP for $\vec{P}(z) \in V^nW^+ \oplus V^nW^0$ and $\vec{\Phi}^-(z)$. To solve this problem, we multiply both sides of (14.42) by $R^-(z)$ and use (14.40):

$$\vec{P}(z)R^+(z)^{-1} = \vec{\Phi}^-(z)R^-(z).$$

Applying T^+ to both sides, we easily obtain

$$\vec{P}(z)R^+(z)^{-1} - \vec{P}_0 = 0,$$

from which we obtain

$$\vec{P}(z) = \vec{P}_0R^+(z). \tag{14.44}$$

Taking into account the obvious fact that $\vec{P}(1) = \vec{\pi}$, we obtain that

$$\vec{P}_0(z) = \vec{\pi}R^+(1)^{-1}, \tag{14.45}$$

which transforms (14.44) into (14.41). □

Corollary 14.6. (14.41) *can also be written as*

$$\vec{P}(z) = \delta\vec{e}_1R^-(1)R^+(z), \tag{14.46}$$

where δ *is defined by* (14.33), $\vec{e}_1 = (1,0,\dots,0)$.

The proof follows from (14.41) and the following lemma.

Lemma 14.7. *Under natural assumptions, if* (14.33) *holds, then*

$$\vec{\pi}R^+(1)^{-1} = \delta\vec{e}_1R^-(1) \tag{14.47}$$

and

$$\delta = [\vec{e}_1F(1)^{-1}\vec{d}]^{-1}. \tag{14.48}$$

Proof. By construction, $F(1)$ differs from $I - H(1)$ only by the elements of the first column which are equal to $-\sum_{s=1}^{n} H'_{rs}(1)$, $r \in \overline{1,n}$. Using (14.32), we obtain that

$$\vec{\pi}F(1) = \delta\vec{e}_1. \tag{14.49}$$

By (14.40), (14.49) yields (14.47) and (14.48) (note that $\vec{\pi}\vec{d} = 1$). □

The steady-state probability of no customers in the system or in the waiting line can be obtained from (14.45) and (14.47).

Corollary 14.8.

$$\vec{P}_0 = \delta\vec{e}_1R^-(1), \tag{14.50}$$

$$\lim_{k\to\infty} \mathrm{P}\{\xi_k = 0\} = \vec{P}_0\vec{d} = \delta\vec{e}_1R^-(1)\vec{d}. \tag{14.51}$$

It turns out that sometimes one does not have to know all elements of matrices $R^+(z)$ and $R^-(z)$.

Corollary 14.9. *If the only element of the first row of* $R^-(z)$, *that is not identically zero, is the first element* $R^-_{11}(z)$, *then*

$$\vec{P}(z) = \delta\frac{R^-_{11}(1)}{R^-_{11}(z)\det F(z)}\vec{e}_1 \operatorname{Adj} F(z). \tag{14.52}$$

Proof. We rewrite (14.46) on the strength of (14.40) as

$$\vec{P}(z) = \delta\vec{e}_1R^-(1)R^-(z)^{-1}F(z)^{-1}. \tag{14.53}$$

Under the assumptions made, the only nonzero elements of first rows of matrices $R^-(1)$ and $R^-(z)^{-1}$ are $R^-_{11}(1)$ and $1/R^-_{11}(z)$, respectively, so

$$\vec{e}_1 R^-(1)R^-(z)^{-1} = R^-_{11}(1)R^-_{11}(z)^{-1}\vec{e}_1.$$

Now (14.53) yields (14.52). □

Remark 14.10. Note that $\vec{e}_1 \operatorname{Adj} F(z)$ is the first row of $\operatorname{Adj}[I - H(z)]$, since $F(z)$ and $[I - H(z)]$ differ by the first column only.

14.5 FINDING FACTORIZATION COMPONENTS

Knowing matrices $R^+(z)$ and $R^-(z)$ is the integral part of the approach developed in the previous section. To our knowledge, there is no universal method of finding these matrices. However, in many realistic cases, additional assumptions on the complex-analytic properties of $H(z)$ or on its structure (or both) make this problem solvable.

Theorem 14.11. *If the sum-function of* $H(z)$ *is meromorphic in* Γ^+, $R^-(z)$ *is the Laurent series of a rational matrix function; if the sum-function of* $H(z)$ *is meromorphic in* Γ^-, $R^+(z)$ *is the Laurent series of a rational matrix function.*

In both cases, elements of these rational matrix functions can be found in terms of poles and degeneracy points of $F(z)$ in Γ^+ and Γ^-, respectively. The other factor, $R^+(z)$ or $R^-(z)$, respectively, is then found from (14.40). For the proof of this theorem and particular steps of this process, we refer the reader to Litvinchuk and Spitkovsky [23]. The following result is instrumental in such an approach.

Lemma 14.12. *Under natural assumptions, if the sum-function of* $H(z)$ *is meromorphic in* $\Gamma^+(\Gamma^-)$ *and the ergodicity condition* (14.33) *holds true, then the number of roots of* $\det F(z)$ *in* $\Gamma^+(\Gamma^-)$ *is equal to the number of its poles there (counting with multiplicities).*

The proof immediately follows from (14.35).

In the scalar case, the poles and roots of $F(z)$ completely specify the factorization components. Consider, for example, the $GI^X/M^Y/1$ system with a never idle server at pre-arrival instants. Here, $H(z) = a(z)G(\mu - \mu b(1/z))$, where $a(z)$ and $b(z)$ are the GFs of the arrival and service group sizes, respectively. $G(\cdot)$ is the Laplace transform of the interarrival time and μ is the service rate. If $a(z)$ is meromorphic in Γ^- with m poles α_i^{-1}, then, by Lemma 14.12, $F(z)$ should have m roots β_i^{-1} there (some roots and poles may be repeating). Now, one can directly verify that

$$R^+(z) = \prod_{i=1}^{m}(1-\alpha_i z)\prod_{i=1}^{m}(1-\beta_i z)^{-1}$$

and $R^-(z)$, defined by (14.40), satisfy (14.37)-(14.40); so from (14.41) we obtain

$$P(z) = \prod_{i=1}^{m}(1-\beta_i)(1-\beta_i z)^{-1}\prod_{i=1}^{m}(1-\alpha_i)^{-1}(1-\alpha_i z).$$

In particular, when the arrival group size is m, $a(z) \equiv z^m$, $\alpha_i = 0$, $i = \overline{1,m}$, and the previous formula reduces to (see also Chaudhry and Templeton [6]):

$$P(z) = \prod_{i=1}^{m}(1-\beta_i)(1-\beta_i z)^{-1}.$$

Similarly, in the case of an $M^X/G^Y/1$ system with a never idle server considered at post-completion instants, $H(z) = b(1/z)G(\lambda - \lambda a(z))$, where $G(\cdot)$ is the Laplace-Stieltjes transform of the service time distribution, λ is the arrival rate, and $a(z)$ and $b(z)$ are the GFs of arrival and service group sizes. If $b(1/z)$ is meromorphic in Γ^+ with m poles β_i, then, by Lemma 14.12, $F(z)$ should have m roots α_i in Γ^+ (some of these roots and poles may be repeating). By the definition of $F(z)$, one of these roots, say α_m, is zero. Up to a multiplicative constant,

$$R^-(z) = \prod_{i=1}^{m}(z-\beta_i)\prod_{i=1}^{m}(z-\alpha_i)^{-1},$$

and, from (14.40) and (14.41), we obtain that

$$P(z) = \frac{F(1)\prod_{i=1}^{m}(1-\beta_i)}{\prod_{i=1}^{m}(1-\alpha_i)}\,\frac{\prod_{i=1}^{m}(z-\alpha_i)}{F(z)\prod_{i=1}^{m}(z-\beta_i)}.$$

In particular, when the service group size is m, $b(z) \equiv z^m$, $\beta_i = 0$, $i = \overline{1,m}$, and the previous formula yields the classical result

$$P(z) = \frac{(m-K'(1))(z-1)\prod_{i=1}^{m-1}(z-\alpha_i)}{(z^m - K(z))\prod_{i=1}^{m-1}(1-\alpha_i)},$$

where $K(z) = G(\lambda - \lambda a(z))$ and α_i, $i = \overline{1, m-1}$, are the roots of $z^m - K(z)$ in Γ^+.

In the matrix case, finding factorization components is very difficult even under the assumption that $H(z)$ is meromorphic in $\Gamma^+(\Gamma^-)$ (see the discussion in Section 14.7). In this section, we consider situations when the factorization process can be simplified by certain assumptions on the structure of $H(z)$ and complex-analytic properties of some of its elements.

Theorem 14.13. *Let $n = 2$ and assume that $H_{12}(z)$, $H_{22}(z) \in W^+ \oplus W^0$. If the ergodicity condition* (14.33) *and natural assumptions hold true, then*

$$R^+(z) = \tilde{R}^+(0)^{-1}\tilde{R}^+(z), \tag{14.54}$$

$$\tilde{R}^+(z) = \begin{bmatrix} 1 & 0 \\ -T^+\psi(z) & 1 \end{bmatrix} \times \begin{bmatrix} r^+(z) & 0 \\ 0 & 1 \end{bmatrix} \times \begin{bmatrix} 1-H_{22}(z) & H_{12}(z) \\ 0 & \dfrac{1}{1-H_{22}(z)} \end{bmatrix}, \tag{14.55}$$

and

$$R^-(z) = \begin{bmatrix} r^-(z) & 0 \\ 0 & 1 \end{bmatrix} \times \begin{bmatrix} 1 & 0 \\ -T^-\psi(z) & 1 \end{bmatrix} \times \tilde{R}^+(0), \tag{14.56}$$

where

$$r^{\pm}(z) = \exp\{-T^{\pm}\ln\det F(z)\} \tag{14.57}$$

and

$$\psi(z) = \frac{r^-(z)[1-H_{11}(z)-H_{12}(z)]e(z)}{1-H_{22}(z)}. \tag{14.58}$$

Proof. Under natural assumptions, $\|H_{22}(z)\|_W < 1$; so under the assumptions of the theorem, $[1-H_{22}(z)]^{-1} \in W^+$, and, therefore, $\psi(z) \in W$. Now one can directly verify that $R^{\pm}(z)$, defined by (14.54)-(14.58), satisfy (14.37)-(14.40).□

The special form of $R^{\pm}(z)$ in Theorem 14.13 leads to substantial simplifications in the formulas for the steady-sate probabilities.

Theorem 14.14. *Under the assumptions of Theorem* 14.13,

$$\vec{P}(z) = \delta r^{-}(1) r^{+}(z)[1 - H_{22}(z), H_{12}(z)], \tag{14.59}$$

$$\vec{\pi} = \left[\frac{1 - H_{22}(1)}{1 - H_{22}(1) + H_{12}(1)}, \frac{H_{12}(1)}{1 - H_{22}(1) + H_{12}(1)}\right], \tag{14.60}$$

and

$$\vec{P}(0) = \delta r^{-}(1)[1 - H_{22}(0), H_{12}(0)]. \tag{14.61}$$

Proof. To obtain (14.59), we apply (14.54)-(14.56) in (14.46). When $z = 1$, (14.59) yields (14.60); when $z = 0$ we get (14.61) (note that by (14.57) $r^{+}(0) = 1$). □

Let us consider, for example, a 2×2 case that appears in the analysis of an $(\mathrm{M}^X/\mathrm{G}^Y/1)^{\pm}$ queueing system, where the GFs of service group sizes in modes 1, 2 are $b_1(z) \equiv z$, $b_2(z) \equiv (1-q)z(1-qz)^{-1}$. Here we have

$$H_{11}(z) = \frac{k_{10}}{z} + k_{11}, \; H_{12}(z) = \sum_{j=2}^{\infty} k_{1j} z^{j-1} = \frac{K_1(z) - k_{10} - k_{11}z}{z},$$
$$H_{21}(z) = (1-q)\frac{K_2(q)}{z-q}, \text{ and } H_{22}(z) = (1-q)\frac{K_2(z) - K_2(q)}{z-q},$$

where $K_r(z) = \sum_{i=0}^{\infty} k_{ri} z^i$, $r = 1, 2$. From (14.60), we obtain that

$$\vec{\pi} = (1 - k_{10} - k_{11} + K_2(q))^{-1}[K_2(q), 1 - k_{10} - k_{11}].$$

The ergodicity condition (14.33) reduces to

$$\delta = \pi_1[1 - K_1'(1)] + \pi_2[(1-q)^{-1} - K_2'(1)] > 0.$$

Matrix $F(z)$ can now be written as

$$F(z) = \begin{bmatrix} \dfrac{z - K_1(z)}{z-1} & -\dfrac{K_1(z) - k_{10} - k_{11}z}{z} \\ \dfrac{[z - q - (1-q)K_2(z)]z}{(z-q)(z-1)} & 1 - (1-q)\dfrac{K_2(z) - K_2(q)}{z-q} \end{bmatrix}.$$

One can see that the only pole of $f(z) \equiv \det F(z)$ in Γ^{+} is $z = q$. By Lemma 14.12, $f(z)$ has exactly one root there, say, θ (observe that $\theta \neq 0$). From (14.57), we find

$$r^{+}(z) = \frac{qf(0)(z-\theta)}{\theta f(z)(z-q)}, \text{ and } r^{-}(z) = \frac{\theta(z-q)}{qf(0)(z-\theta)},$$

so (14.59) yields

$$\vec{P}(z) = \delta \frac{(1-q)(z-\theta)}{(1-\theta)(z-q)f(z)}[1 - H_{22}(z), \; H_{12}(z)]. \tag{14.62}$$

When $n > 2$, under assumptions similar to those in Theorem 14.13, it is possible to reduce the dimension of the problem by the reduction process outlined by the following propositions.

Introduce a Markov-controlled random walk $\{\overset{*}{\xi}_k, \overset{*}{\eta}_k\}$, $\overset{*}{\eta}_k \in \overline{1, n-1}$ with the generating matrix function of jumps $\overset{*}{H}(z)$ the elements of which are

$$\overset{*}{H}_{rs}(z) = H_{rs}(z) + \frac{H_{rn}(z)H_{ns}(z)}{1 - H_{nn}(z)},\ r,s = \overline{1,n-1}. \tag{14.63}$$

We will refer to this random walk as "reduced" and use (*) to denote related functions and matrices.

Lemma 14.15. *Under natural assumptions on $H(z)$, $\overset{*}{H}(z)$ given by (14.63) is indeed a generating matrix function.*

Proof. Under natural assumptions, $\| H_{nn}(z) \|_W < 1$ (otherwise $H_{nn}(1) = 1$ would make $H(1)$ reducible), so $[1 - H_{nn}(z)]^{-1} \in W$ and all coefficients of its Laurent expansion are nonnegative. Therefore, the same is true for $\overset{*}{H}_{rs}(z)$ defined by (14.63). Also, by (14.63),

$$\sum_{s=1}^{n-1} \overset{*}{H}_{rs}(1) = \sum_{s=1}^{n-1} H_{rs}(1) + \frac{H_{rn}(1)}{1 - H_{nn}(1)} \sum_{s=1}^{n-1} H_{ns}(1) = 1,\ r \in \overline{1,n-1},$$

which completes the proof of the lemma. □

Lemma 14.16. *The natural assumptions hold for $\overset{*}{H}(z)$ if they hold for $H(z)$.*

The proof readily follows from (14.63).

Theorem 14.17. *If (14.33) holds, then the reduced chain is also ergodic and functions $\overset{*}{F}(z), \overset{*}{R}{}^{\pm}(z)$ satisfy (14.34)-(14.40) with $\overset{*}{I}$ being the identity matrix of size $n-1$ and n replaced by $n-1$.*

Proof. It follows from (14.63) that

$$\det \overset{*}{F}(z) = \det F(z)[1 - H_{nn}(z)]^{-1}, \tag{14.64}$$

so (14.34) holds for $\det \overset{*}{F}(1)$ if it holds for $\det F(1)$. Now, we can apply Theorem 14.4 to the reduced chain. □

Theorem 14.18. *Suppose $H_{rn}(z) \in W^{+} \oplus W^{0}$, $r = \overline{1,n}$. Then,*

$$R^{+}(z) = \widetilde{R}^{+}(0)^{-1}\widetilde{R}^{+}(z), \tag{14.65}$$

$$\widetilde{R}^{+}(z) = \begin{bmatrix} \overset{*}{I} & \vec{0}^T \\ -T^{+}\vec{\psi}(z) & 1 \end{bmatrix} \times \begin{bmatrix} \overset{*}{R}{}^{+}(z) & \vec{0}^T \\ \vec{0} & 1 \end{bmatrix} \times \begin{bmatrix} \overset{*}{I} & \vec{\phi}(z)^T \\ \vec{0} & [1 - H_{nn}(z)]^{-1} \end{bmatrix} \tag{14.66}$$

and

$$R^{-}(z) = \begin{bmatrix} \overset{*}{R}{}^{-}(z) & \vec{0}^T \\ \vec{0} & 1 \end{bmatrix} \times \begin{bmatrix} \overset{*}{I} & \vec{0}^T \\ -T^{-}\vec{\psi}(z) & 1 \end{bmatrix} \times \widetilde{R}^{+}(0), \tag{14.67}$$

where $\vec{0}$ is the $(n-1)$-dimensional vector-row of zeros, $\vec{\phi}(z)$ is the $(n-1)$-dimensional vector-row with the elements $H_{rn}(z)/[1 - H_{nn}(z)]$, $r = \overline{1,n-1}$, and $\vec{\psi}(z)$ is the $(n-1)$-dimensional vector-row that results from the following procedure: take the first $n-1$ elements of the nth row of $F(z)$, divide all of them by $1 - H_{nn}(z)$, and multiply this row from the right by $\overset{*}{R}{}^{-}(z)$.

Proof. Under the assumptions of the theorem, $\overset{*}{R}{}^{\pm}(z)$ satisfy (14.37) and (14.38) (with $n-1$ in place of n) and $[1 - H_{nn}(z)]^{-1} \in W^{+} \oplus W^{0}$, so by construction $\vec{\phi}(z) \in V^{n-1}W^{+} \oplus V^{n-1}W^{0}$, $\vec{\psi}(z) \in V^{n-1}W$.

Therefore, looking at the products in (14.65)-(14.67) and their inverses, one can see that (14.65) and (14.67) yield $R^{\pm}(z)$ satisfying (14.37) and (14.38). (14.40) follows from (14.65)-(14.67) by direct multiplication; (14.39) follows from (14.65). Since by Theorem 14.4, matrix functions $R^{\pm}(z)$ satisfying (14.37)-

(14.40) are unique, this completes the proof of Theorem 14.18. □

The reduction process replaces the problem of finding complex-analytic factorization components of an $n \times n$ matrix-series $F(z)$ by the similar problem for an $(n-1) \times (n-1)$ matrix-series $\overset{*}{F}(z)$. Theorem 14.21, below, shows that sometimes this process can go all the way down to factorization of scalar functions.

Corollary 14.19. *GFs of the steady-state probabilities of the chains* $\{\xi_k, \eta_k\}$ *and* $\{\overset{*}{\xi}_k, \overset{*}{\eta}_k\}$ *are related by*

$$\vec{P}(z) = \frac{\delta}{\overset{*}{\delta}}[\overset{*}{\vec{P}}(z), \overset{*}{\vec{P}}(z)\vec{\phi}(z)^T]. \tag{14.68}$$

Proof. Applying (14.65)-(14.67) in (14.46), we obtain (note that $\vec{e}_1 = [\overset{*}{\vec{e}}_1, 0]$)

$$\vec{P}(z) = \delta[\overset{*}{\vec{e}}_1 \overset{*}{R}^-(1)\overset{*}{R}^+(z), 0] \times \begin{bmatrix} \overset{*}{I} & \vec{\phi}(z)^T \\ \vec{0} & [1 - H_{nn}(z)]^{-1} \end{bmatrix},$$

and, using (14.46) for the reduced chain, we arrive at (14.68). □

Corollary 14.20. *GFs of the steady-state probabilities of the chains* $\{\xi_k, \eta_k\}$ *and* $\{\overset{*}{\xi}_k, \overset{*}{\eta}_k\}$ *are related by*

$$\frac{1}{\delta}\vec{P}(z)F(z) = \frac{1}{\overset{*}{\delta}}[\overset{*}{\vec{P}}(z)\overset{*}{F}(z), 0]. \tag{14.69}$$

Proof. Rewrite (14.46) using (14.40) as

$$\vec{P}(z) = \delta\vec{e}_1 R^-(1)R^-(z)^{-1}F(z)^{-1}. \tag{14.70}$$

Applying (14.67) in (14.70), we obtain

$$\vec{P}(z) = \delta[\overset{*}{\vec{e}}_1 \overset{*}{R}^-(1)\overset{*}{R}^-(z)^{-1}, 0]F(z)^{-1},$$

and use again (14.70), this time for the reduced chain. □

Theorem 14.21. *Let* $H(z)$ *be a tridiagonal matrix, the elements of which satisfy the following relations:*

$$H_{rr}(z) \in W^+ \oplus W^0, \quad \forall r > 1, \tag{14.71}$$

$$H_{r\,r+1}(z) \in W^+ \oplus W^0, \quad \forall r < n, \tag{14.72}$$

and

$$H_{r\,r-1}(z)H_{r-1\,r}(z) \in W^+ \oplus W^0, \quad \forall r \in \overline{3,n}. \tag{14.73}$$

Then $\overset{*}{H}(z)$*, defined by* (14.63), *also satisfies* (14.71) – (14.73).

Proof. Since $H(z)$ is tridiagonal, it follows from (14.63) and (14.71)-(14.73) that

$$\overset{*}{H}_{rs}(z) = H_{rs}(z), \quad \forall r,s \in \overline{1,n-1}, \ (r,s) \neq (n-1, n-1) \tag{14.74}$$

and

$$\overset{*}{H}_{n-1\,n-1}(z) = H_{n-1\,n-1}(z) + \frac{H_{n-1\,n}(z)H_{n\,n-1}(z)}{1 - H_{nn}(z)}. \tag{14.75}$$

Using (14.71) for $r = n-1$ and $r = n$, and (14.73) for $r = n$ in (14.75), we conclude that $\overset{*}{H}_{n-1\,n-1}(z) \in W^+ \oplus W^0$. By (14.74), all other $\overset{*}{H}_{rs}(z)$ also satisfy (14.71)-(14.73).

□

The importance of Theorem 14.21 is in the obvious fact that, under its assumptions, the reduction process can be continued after each reduction until n becomes 1. It is remarkable, though, that, in order to find the steady-state probabilities, we do not have to find all the intermediate matrices suggested by Theorem 14.18.

Theorem 14.22. *Under the assumption of Theorem* 14.21,

$$\vec{P}(z) = \delta r^{-}(1) r^{+}(z) \vec{e}_1 \operatorname{Adj} F(z), \tag{14.76}$$

where

$$r^{\pm}(z) = \exp\{-T^{\pm} \ln \det F(z)\}.$$

Proof. It follows from (14.69), as we apply it at each of the $n-1$ steps of the reduction process, that

$$\frac{1}{\delta}\vec{P}(z)F(z) = \frac{1}{\delta^{(1)}} P^{(1)}(z) F^{(1)}(z) \vec{e}_1, \tag{14.77}$$

where we use the superscript (k) to indicate quantities corresponding to the $(n-k)$th reduction step. Indeed, after the $(n-1)$st step, vectors and matrices become scalars.

To continue the proof, we need the following lemma.

Lemma 14.23. *In the reduction process defined by* (14.63),

$$\det F(z) = F^{(1)}(z) \prod_{r=2}^{n} [1 - H_{rr}^{(r)}(z)] \tag{14.78}$$

$$= e(z) \prod_{r=1}^{n} [1 - H_{rr}^{(r)}(z)]. \tag{14.79}$$

Proof. Applying (14.64) $n-1$ times for $n-1$ reduction steps, we obtain (14.78), which, in turn, yields (14.79) by the definition of $F^{(1)}(z)$. □

Under the assumptions of Theorem 14.21,

$$[1 - H_{rr}^{(r)}(z)], [1 - H_{rr}^{(r)}(z)]^{-1} \in W^{+} \oplus W^{0}, \quad \forall r > 1. \tag{14.80}$$

Now we compare complex-analytic factorizations (see (14.39), (14.40))

$$\det F(z) = [r^{-}(z) r^{+}(z)]^{-1}$$

and

$$F^{(1)}(z) = [R^{-(1)}(z) R^{+(1)}(z)]^{-1},$$

and conclude on the strength of (14.78) and (14.80) that

$$\frac{r^{-}(z)}{r^{-}(1)} = \frac{R^{-(1)}(z)}{R^{-(1)}(1)} \tag{14.81}$$

and

$$\frac{r^{+}(z)}{r^{+}(1)} = \frac{R^{+(1)}(z) \prod_{r=2}^{n} [1 - H_{rr}^{(r)}(1)]}{R^{+(1)}(1) \prod_{r=2}^{n} [1 - H_{rr}^{(r)}(z)]}. \tag{14.82}$$

From (14.48), for $n = 1$, we obtain that

$$\delta^{(1)} = F^{(1)}(1).$$

Now we apply (14.46), for $n = 1$, in (14.77), using (14.78), (14.81), (14.82) along with the classical formula $F(z)^{-1} = [\det F(z)]^{-1} \operatorname{Adj} F(z)$ and arrive at (14.76). □

Let us consider, for example, a 3×3 case which arises in the M/G/1-type bulk system with "queue-increase" control. The service group size is 3. Three service modes are available; the increase control intervals σ_{rs} are: $[-3, 2]$, $[3, \infty)$, $\emptyset$, in mode 1; $[-3, -1]$, $[0, 2]$, $[3, \infty)$, in mode 2; and $\emptyset$, $[-3, -1]$, $[0, \infty)$, in mode 3. Denote by $K_r(z)$ the GF of the number of arrivals during the service period in the mode $r = 1,2,3$, and let

$$H_{rs}(z) = \sum_{j \in \sigma_{rs}} z^j D_j z^{-3} K_r(z), \quad r,s = 1,2,3.$$

One can see that $H(z)$ satisfies the assumptions of Theorem 14.21, so we can use (14.76) to find the steady-state probabilities. Analyzing $\det F(z)$, we find that the only pole it has in Γ^+ is $z = 0$ with multiplicity 2. By Lemma 14.12, $\det F(z)$ should have two roots (possibly equal) in Γ^+, say θ_1 and θ_2. It now follows that

$$\frac{r^-(z)}{r^-(1)} = \frac{z^2(1-\theta_1)(1-\theta_2)}{(z-\theta_1)(z-\theta_2)},$$

and (14.76) yields

$$\vec{P}(z) = \delta \frac{(z-\theta_1)(z-\theta_2)}{z^2(1-\theta_1)(1-\theta_2)\det F(z)} \vec{e}_1 \operatorname{Adj} F(z). \tag{14.83}$$

14.6 MARKOV-CONTROLLED RANDOM WALKS WITH FINITE PERTURBATIONS

In this section, we consider a homogeneous Markov chain $\{\xi_k, \eta_k\}$, $k = \overline{0,\infty}$, $\xi_k \in \overline{0,\infty}$, $\eta_k \in \overline{1,n}$, that satisfies (14.19) for $N > 0$. Now relations (14.26)-(14.29) hold only for $i \geq N$. Such a Markov chain will be referred to as a Markov-controlled random walk with finite perturbations. In the scalar case, it was studied by the method of an RBVP in Dukhovny [17]. In addition to natural assumptions on $H(z)$, we will assume that

$$\sum_{j=0}^{\infty} j A_{ir}^{js} < \infty, \; i < N \tag{14.84}$$

and that the chain $\{\xi_k, \eta_k\}$ is irreducible and aperiodic.

Theorem 14.24. (Dukhovny [14]) *Under the above assumptions, the chain* $\{\xi_k, \eta_k\}$ *has a stationary distribution if, and only if,* (14.33) *holds.*

(See [14] for the proof of this theorem in the general case.)

Given (14.33), we, once, again will represent the vector GF of the steady-state probabilities as a solution of a vector RBVP in V^nW. The starting point is, again, relation (14.17) that now assumes the form

$$\vec{P}(z) = \sum_{i=0}^{N-1} \vec{P}_i A_i(z) + \sum_{i=N}^{\infty} [T^+ + ST^-] \vec{P}_i z^i H(z).$$

Using the projection properties of operators $T^{\pm}$, we can rewrite this as

$$\vec{P}(z)[I - H(z)] = \sum_{i=0}^{N-1} \vec{P}_i z^i [I - H(z)] + \sum_{i=0}^{N-1} \vec{P}_i [A_i(z) - z^i I] + \vec{\Phi}^-(1) - \vec{\Phi}^-(z),$$

and, then, multiply from the right by $\mathbf{E}[z]$ to obtain

$$\vec{P}(z)F(z) = \vec{\phi}^-(z) + \sum_{i=0}^{N-1} \vec{P}_i z^i F(z) + \sum_{i=0}^{N-1} \vec{P}_i [A_i(z) - z^i I]\mathbf{E}[z], \tag{14.85}$$

where $\vec{\Phi}^-(z) = \sum_{i=N}^{\infty} T^- \vec{P}_i z^i H(z)$, $\vec{\phi}^-(z) = [\vec{\Phi}^-(1) - \vec{\Phi}^-(z)]\mathbf{E}[z]$.

Relation (14.85), as we prove below, is a vector RBVP in the space V^nW. Solving this problem gives us $\vec{P}(z)$ in terms of probabilities p_{ir}, $i < N$.

Theorem 14.25. (Dukhovny [14]) *Under the assumptions made, and if the ergodicity condition* (14.33) *holds true, then*

$$\vec{P}(z) = \sum_{i=0}^{N-1} \vec{P}_i z^i + \sum_{i=0}^{N-1} \vec{P}_i \{T^+[A_i(z) - z^i I]\mathbf{E}(z)R^-(z)\}R^+(z). \tag{14.86}$$

Proof. First of all, we show that $[A_i(z) - z^i I]\mathbf{E}[z] \in M^n W$. This follows from Lemma 14.1 on the strength of assumption (14.84) and the obvious fact that $A_i(1)$ should be stochastic. Now, it follows from (14.85) itself that $\vec{\phi}^-(z) \in V^n W$ and, therefore, by its definition, $\vec{\phi}^-(z) \in V^n W^-$.

Thus, (14.85) becomes a vector RBVP in $V^n W$ for $\vec{P}(z)$ and $\vec{\phi}^-(z)$. Now we multiply both sides of (14.85) by $R^-(z)$ from the right and use (14.40) to get

$$[\vec{P}(z) - \sum_{i=0}^{N-1} \vec{P}_i z^i]R^+(z)^{-1} = \vec{\phi}^-(z)R^-(z)$$
$$+ \sum_{i=0}^{N-1} \vec{P}_i[A_i(z) - z^i I]\mathbf{E}[z]R^-(z). \tag{14.87}$$

Similar to the proof of Theorem 14.5, we apply T^+ to (14.87), multiply both sides by $R^+(z)$ from the right, and obtain (14.86). □

Theorem 14.26. (Dukhovny [14]) *Under the assumptions of Theorem* 14.25, *the remaining probabilities* p_{ir}, $i < N$ *and a constant* c *form a unique solution of the system of linear equations*

$$\sum_{i=0}^{N-1} \vec{P}_i D_j[A_i(z) - z^i I]\mathbf{E}[z]R^-(z) = \vec{0},\ 0 < j < N, \tag{14.88}$$

$$\sum_{i=0}^{N-1} \vec{P}_i ST^-[A_i(z) - z^i I]\mathbf{E}[z]R^-(z) = -c\vec{e}_1 R^-(1), \tag{14.89}$$

and

$$\sum_{i=0}^{N-1} \vec{P}_i S[A_i(z) - z^i I]\mathbf{E}[z]F(1)^{-1}\vec{d}^T + \sum_{i=0}^{N-1} \vec{P}_i \vec{d}^T = 1 - c/\delta. \tag{14.90}$$

Proof. To prove (14.88), we rewrite (14.87) as

$$\sum_{i=N}^{\infty} \vec{P}_i z^i R^+(z)^{-1} = \vec{\phi}^-(z)R^-(z) + \sum_{i=0}^{N-1} \vec{P}_i[A_i(z) - z^i I]\mathbf{E}[z]R^-(z). \tag{14.91}$$

The left-hand side of (14.91) evidently contains z only in powers greater than or equal to N; $\vec{\phi}^-(z)R^-(z)$ does not contain positive powers of z. Now (14.88) follows upon applying operators D_j, $0 < j < N$ to both sides of (14.91). To obtain (14.89), we apply ST^- to both sides of (14.91) and use the fact that

$$ST^-\vec{\phi}^-(z)R^-(z) = S\vec{\phi}^-(z)R^-(1).$$

It follows from the definition of $\vec{\phi}^-(z)$ that only the first element of $S\vec{\phi}^-(z)$ differs from 0. Denoting this element by c, we obtain (14.89) from

$$S\vec{\phi}^-(z) = c\vec{e}_1. \tag{14.92}$$

Finally, to show that (14.90) should be satisfied, we apply operator S to both sides of (14.85), then multiply both sides from the right by $F(1)^{-1}\vec{d}^T$ and use (14.92), again, along with (14.48) of Lemma 14.7.

Suppose, now, that, in addition to p_{ir}, $i < N$ and c, the system (14.88)-(14.90) has one more solution, say $\tilde{p}_{ir}$ and $\tilde{c}$. Using this solution on the right-hand side of (14.86), we would obtain, there, a vector-series $\tilde{\tilde{\vec{P}}}(z) \in V^nW^+ \oplus V^nW^0$. The Laurent expansion coefficients $\tilde{p}_{ir}$, $i = \overline{0,\infty}$, $r = \overline{1,n}$ of this vector, on the strength of (14.86)-(14.90), would satisfy (14.17) and (14.18) and, therefore, would make a new absolutely summable solution of transition equations (14.13), (14.14). According to Chung Kai-Lai [7], there can be only one such solution. This contradiction completes the proof of Theorem 14.26. □

An obvious application of Theorems 14.24-14.26 arises in the case of the simplest form of state-dependence known as the "warm-up" discipline, when distributions of the system's elements differ from the usual ones if, at t_k, the system is empty ($\xi_k = 0$).

Corollary 14.27. *Under the assumptions of Theorem* 14.25, *when* $N = 1$, *it follows from Theorems* 14.25 *and* 14.26 *that*

$$\vec{P}(z) = \vec{P}_0 + \vec{P}_0\{T^+[A_0(z) - I]\mathrm{E}[z]R^-(z)\}R^+(z), \tag{14.93}$$

where

$$\vec{P}_0 = c\vec{e}_1R^-(1)\{ST^-[I - A_0(z)]\mathrm{E}[z]R^-(z)\}^{-1} \tag{14.94}$$

and

$$c^{-1} = \vec{e}_1R^-(1)\{ST^-[I - A_0(z)]\mathrm{E}[z]R^-(z)\}^{-1} \times \{I + \{ST^+[A_0(z) - I]\mathrm{E}[z]R^-(z)\}R^+(1)\}\vec{d}. \tag{14.95}$$

In certain special types of state-dependence, in which $A_i(z)$, $i < N$, are related to $H(z)$, the system (14.88)-(14.90) can be completely solved. It happens, for example, in the case below, known in the literature as the "Bailey-type" system, where $A_i(z) = A_N(z)$, $i < N$. Such a structure appears, in particular, in the transition matrix of the Markov chain which is the queue length in an $\mathrm{M}^{(X)}/\mathrm{G}^{(N)}/1$ system with a nonwaiting server considered at postcompletion moments of time:

$$A_i(z) = K(z),\ i < N,\ A_i(z) = z^{i-N}K(z),\ \ i \geq N,$$

where $K(z)$ is the GF of the number of arrivals during a service period. In the classical case, there is no control component η, and $H(z)$ is simply $z^{-N}K(z)$. In the theorem below, we consider the "Bailey-type" system with the control component.

Theorem 14.28. *Under the assumptions of Theorem* 14.25, *if*

$$A_i(z) = [T^+ + ST^-]z^NH(z),\ \ i < N, \tag{14.96}$$

then

$$\vec{P}(z) = \vec{\pi}R^+(1)^{-1}[T^+ + ST^-]z^NR^+(z)H(z). \tag{14.97}$$

Proof. It is possible to derive (14.97) as a corollary of Theorems 14.25 and 14.26, substituting (14.96) for $A_i(z)$. However, it is much simpler to use (14.96) along with (14.29) for $i \geq N$ directly in (14.17) and obtain, after some transformations,

$$z^N\tilde{\vec{P}}(z)[I - H(z)] = \sum_{i=0}^{N-1}\vec{P}_i(z^N - z^i) + \vec{\Phi}^-(1) - \vec{\Phi}^-(z), \tag{14.98}$$

where

$$\tilde{\vec{P}}(z) = \sum_{i=0}^{N-1}\vec{P}_i + \sum_{i=N}^{\infty}\vec{P}_iz^i \quad \text{and} \quad \vec{\Phi}^-(z) = \sum_{i=0}^{N-1}T^-\vec{P}_iz^NH(z) + \sum_{i=N}^{\infty}T^-\vec{P}_iz^iH(z).$$

At $z=1$, it follows from (14.98) that $\widetilde{P}(1)=\vec{\pi}$. Denote $\vec{\phi}^-(z)=[\vec{\Phi}^-(1)-\vec{\Phi}^-(z)]\mathbf{E}[z]$, multiply (14.98) by $\mathbf{E}[z]$, and divide it by z^N:

$$\widetilde{P}(z)F(z)=z^{-N}\vec{\phi}^-(z)+\sum_{i=0}^{N-1}\vec{P}_i(1-z^{i-N})\mathbf{E}[z]. \tag{14.99}$$

Similar to (14.85), (14.99) is an RBVP in V^nW. By construction

$$z^{-N}\vec{\phi}^-(z)+\sum_{i=0}^{N-1}\vec{P}_i(1-z^{i-N})\mathbf{E}[z]\in V^nW^-;$$

so, following the reasoning of Theorem 14.5, we obtain that

$$\widetilde{P}(z)=\vec{\pi}R^+(1)^{-1}R^+(z). \tag{14.100}$$

By the definition of $\widetilde{P}(z)$, it follows from (14.100) that

$$\vec{P}(z)=\sum_{i=0}^{N-1}\vec{P}_i(z^i-z^N)+\vec{\pi}z^NR^+(1)^{-1}R^+(z). \tag{14.101}$$

Now we use (14.100) in (14.99), apply operator T^++ST^- to both sides and use the result to transform (14.101) to (14.97). □

Remark 14.29. Due to (14.47), we can use $\delta\vec{e}_1R^-(1)$ in (14.97) in place of $\vec{\pi}R^+(1)^{-1}$.

Corollary 14.30. *Under the assumptions of Theorem* 14.28, *if* $N=1$, *then*

$$\vec{P}(z)=\vec{\pi}R^+(1)^{-1}[T^++ST^-]zR^+(z)H(z), \tag{14.102}$$

Remark 14.31. The classical M/G/1 Kendall's formula

$$P(z)=(1-K'(1))\frac{K(z)(z-1)}{z-K(z)}$$

follows from (14.102) when $n=1$, $H(z)=z^{-1}K(z)$.

In some M/G/1-type systems state-dependence means that an independent additional processing phase is added to the usual service time if $\xi_k<N$. Denote by $B(z)$ the generating matrix function of the number of additional arrivals during this time, conditioned on the concurrent transitions of η. Combining this pattern of state-dependence with the "Bailey-type" pattern we obtain the following result.

Theorem 14.32. *Under the assumption of Theorem* 14.25, *if*

$$A_i(z)=[T^++ST^-]z^NB(z)H(z),\quad i<N, \tag{14.103}$$

where matrix series $B(z)=\sum_{j=0}^{\infty}B_jz^j$ *is such that matrices* $B_j\geq 0$, $\sum_{j=0}^{\infty}jB_j$ *converges, and* $B(1)$ *is stochastic, then*

$$\vec{P}(z)=\vec{Q}[T^++ST^-]z^N\psi^+(z)R^+(z)H(z), \tag{14.104}$$

where $\vec{Q}\equiv\vec{P}_1+\vec{P}_2+\ldots+\vec{P}_N$ *is found from*

$$\vec{Q}\{I-B(1)+\psi^+(1)R^+(1)[I-H(1)]\}=\vec{0}, \tag{14.105}$$

and the normalizing condition

$$\vec{Q}\psi^+(1)R^+(1)\vec{d}=1 \tag{14.106}$$

and

$$\psi^+(z)=[D_0+T^+][B(z)-z^{-1}I]\mathbf{E}[z]R^-(z). \tag{14.107}$$

Proof. Following the line of proof of Theorem 14.28, we apply (14.103) and

(14.29) for $i \geq N$ in (14.17) and transform the result to

$$z^N \tilde{P}(z)[I - H(z)] = \vec{Q} z^N B(z) - \sum_{i=0}^{N-1} \vec{P}_i z^i + \vec{\Phi}^-(1) - \vec{\Phi}^-(z), \qquad (14.108)$$

where

$$\vec{\Phi}^-(z) = \sum_{i=0}^{N-1} T^- \vec{P}_i z^N B(z) H(z) + \sum_{i=N}^{\infty} T^- \vec{P}_i z^i H(z)$$

and

$$\tilde{P}(z) = \vec{Q} B(z) + \sum_{i=N}^{\infty} \vec{P}_i z^{i-N}.$$

Denote $\vec{\phi}^-(z) = [\vec{\Phi}^-(1) - \vec{\Phi}^-(z)]\mathbf{E}[z]$, multiply (14.108) by $\mathbf{E}[z]$, and divide it by z^N:

$$\tilde{P}(z)F(z) = z^{-N}\vec{\phi}^-(z) + \vec{Q}[B(z) - z^{-1}I]\mathbf{E}[z] + \sum_{i=0}^{N-1} \vec{P}_i (z^{-1} - z^{i-N})\mathbf{E}[z]. \qquad (14.109)$$

Like (14.98), (14.109) is an RBVP in V^nW. By construction

$$z^{-N}\vec{\phi}^-(z) + \sum_{i=0}^{N-1} \vec{P}_i (z^{-1} - z^{i-N})\mathbf{E}[z] \in V^nW^-.$$

Also, by Lemma 14.1 and by the definition of $B(z)$,

$$\vec{Q}[B(z) - z^{-1}I]\mathbf{E}[z] \in V^nW^+ \oplus V^nW^0.$$

Now we multiply (14.109) by $R^-(z)$ from the right, use (14.40), and apply operator $[D_0 + T^+]$ to obtain

$$\tilde{P}(z)R^+(z)^{-1} = \vec{Q}\psi^+(z). \qquad (14.110)$$

Using (14.110) in (14.108), we apply operator $T^+ + ST^-$ to both sides of (14.108):

$$\vec{Q} z^N B(z) - \sum_{i=0}^{N-1} \vec{P}_i z^i = [T^+ + ST^-]\vec{Q} z^N \psi^+(z) R^+(z)[I - H(z)]. \qquad (14.111)$$

By the definition of $\tilde{P}(z)$, $\vec{P}(z) = z^N \tilde{P}(z) - \vec{Q} z^N B(z) + \sum_{i=0}^{N-1} \vec{P}_i z^i$, and using here (14.110) and (14.111), we obtain (14.104). (14.105) follows from (14.108) at $z = 1$ by virtue of (14.110). The normalizing condition (14.106) follows from (14.104) at $z = 1$ upon multiplication of both sides from the right by $\vec{d}$. □

If the customers arriving during the additional time phase are rejected by the system, then $B(z)$ becomes a constant matrix.

Corollary 14.33. *If $B(z)$ is a constant matrix B, then*

$$\vec{P}(z) = \vec{Q}\psi[T^+ + ST^-]z^N R^+(z) H(z),$$

$$\vec{Q}\{I - B + \psi R^+(1)[1 - H(1)]\} = \vec{0},$$

and

$$\vec{Q}\psi R^+(1)\vec{d} = 1,$$

where $\psi = B\mathbf{E}[\infty]R^-(\infty)$.

Proof. In this case, $[B - z^{-1}I]\mathbf{E}[z]R^-(z) \in V^nW^-$ and it follows from (14.107) that $\psi^+(z) \equiv \psi = B\mathbf{E}[\infty]R^-(\infty)$. □

If, on the other hand, no changes in the control component η are allowed during the additional time phase, then $B(z)$ becomes $b(z)I = \sum_{j=0}^{\infty} b_j z^j I$.

Corollary 14.34. *If $B(z) = b(z)I$, then $\vec{P}(z)$ is given by* (14.104) *where*

$$\vec{Q}\psi^{+}(1)R^{+}(1) = \vec{\pi}$$

and

$$\psi^{+}(z) = [D_0 + T^{+}][b(z) - z^{-1}]\mathbf{E}[z]R^{-}(z).$$

Proof. Under the assumptions made, $B(1) = b(1)I = I$ (since $B(1)$ is stochastic). From (14.105) and (14.106), now follows the formula for $\vec{Q}$. The formula for $\psi^{+}(z)$ follows from (14.107). □

14.7 POTENTIAL APPLICATIONS OF RBVPS AND IMPLEMENTATION PROBLEMS

The RBVPs constructed in this chapter for semi-Markov queues with limited state-dependence were ultimately based on property (14.19). Three major problems arise when one uses the formulas of the previous sections to find solutions of RBVPs: finding the factorization components of $F(z)$ that satisfy (14.37)-(14.40), applying operators T^{+} and T^{-} in formulas (14.86) and (14.88)-(14.90), and solving the system of equations given by (14.88)-(14.90). In the examples studied in this work, we made special assumptions on the structure of $H(z)$, on the complex-analytic properties of its elements, and on the nature of state-dependence. However, the subclasses of queues that satisfy these assumptions are very broad and the distributions of systems' elements can be further specified to obtain more specific results. Also, several other disciplines of state-dependence in bulk systems can be suggested for possible application of the method of RBVPs.

1. One idea involves M/G/1-type systems with preliminary accumulation of customers, in which no service is offered until the number of customers in the queue reaches a certain level. The level may be fixed or random and its distribution may also depend on the queue-length and the value of the control parameter η at the last completion instant.

2. Another version of the same idea arises in systems with server vacations, the number and durations of which may depend on the queue-length and η.

3. Additional processing phases could be offered if the queue-length is below N; the number of phases may depend on $N - \xi_k$ and η_k.

Similar disciplines of state-dependence can be considered for GI/M/1-type systems. However, the corresponding imbedded Markov chains do not fall into the class of Markov-controlled random walks with finite perturbations. One of the reasons is that accumulation and vacation periods may encompass several arrivals. Nevertheless, our experience shows that RBVPs can also be successfully applied in these cases.

In Dukhovny [15], the method of RBVPs was applied to bulk priority systems. In this paper, the multivariate GF of the steady-state probabilities was found by solving a sequence of univariate RBVPs. (The paper was also presented later at the ORSA/TIMS conference in San Francisco, 1992.) Further analysis has shown that this method can be extended to the case when distributions of the system's elements depend on the composition of the queue.

Multivariate RBVPs are often used to study random walks arising in the theory of parallel queues. Such random walks are essentially multidimensional, which makes the methods used to analyze these RBVPs very different from those proposed here. On this subject, we refer the reader to Cohen and Boxma [10] and also papers by such authors as Fayolle, Malyshev, and Yasnogorodsky.

As we mentioned before, finding factorization components for $F(z)$ in the matrix case is a very difficult task. One approach is to reduce this problem to the scalar factorization using specific features of $H(z)$, as in the case of the tridiagonal $H(z)$ studied in Section 14.5. Theorem 14.11 gives rise to the approach based on $H(z)$ being meromorphic on Γ^+ or Γ^-. Suppose, for example, that $H(z)$ (and, therefore, $F(z)$) is meromorphic in Γ^+. By Theorem 14.11, $R^-(z)$ has to be a rational matrix function. Rewrite (14.40) as

$$R^+(z) = [F(z)R^-(z)]^{-1}. \tag{14.112}$$

Under the assumptions made, (14.112) leads to the equality of two sum-functions everywhere in Γ^+. Since $R^+(z)$ is analytic in Γ^+, so also must be the right-hand side of (14.124), which leads to some obvious results.

Lemma 14.35. *If the sum-function of* $H(z)$ *is meromorphic in* Γ^+ *and the ergodicity condition* (14.33) *and natural assumptions hold, then:*

1. *The roots and poles of* $\det R^-(z)$ *in* Γ^+ *are, respectively, the poles and roots of* $\det F(z)$ *there.*
2. *For every* $z_0 \in \Gamma^+$ *which is a pole of any element of* $F(z)$ *or* $F(z)^{-1}$, $\lim\limits_{z \to z_0} F(z)R^-(z)$ *exists and is invertible.*

For each particular $H(z)$ one can use Lemma 14.35 to set equations for the elements of $R^-(z)$.

De Smit [12] used a similar approach in finding factorization components for a Hilbert problem on the imaginary half-line. The author considers some special cases of a single-server semi-Markov queue, in particular, a semi-Markov modulated M/G/1 queue.

Several technical problems arise in this approach: (1) finding all the roots and poles mentioned in Lemma 14.35, (2) setting and solving equations for the elements of $R^-(z)$, and (3) stability with respect to inevitable errors involved in statistically estimated elements of the queueing system in study.

The number of poles and roots of $\det F(z)$ usually depends on the distributions of the service and arrival group sizes. The example of an $(M^X/G^Y/1)^{\pm}$ system solved by Theorem 14.13 (see (14.62)) shows that in some cases this number can be made small by a suitable model for the service group size. Note that assuming for $b_2(z)$ a finite distribution with a large upper bound in this example would result in a large number of poles and roots of $\det F(z)$ in Γ^+. At the same time, solving this system by matrix-analytic or block-elimination methods would lead to large blocks.

BIBLIOGRAPHY

[1] Arjas, E., On the fundamental identity in the theory of semi-Markov processes, *Adv. Appl. Prob.* **4** (1972), 258-270.

[2] Arjas, E., On the use of a fundamental identity in the theory of semi-Markov queues, *Adv. Appl. Prob.* **4** (1972), 271-284.

[3] Arndt, K., On the distribution of the supremum of a random variable on a Markov chain, In: *Adv. in Prob. Theory, Limit Theorems and Related Problems, Optim. Software* (ed. by A.A. Borovkov), New York 1984.

[4] Bertsimas, D.J., Keilson, J., Nakasato, D. and Zhang, H., Transient and

busy period analysis of the GI/G/1 queue as a Hilbert factorization problem, *J. Appl. Prob.* **28** (1991), 873-885.

[5] Borovkov, A.A., *Stochastic Processes in Queueing Theory*, Springer-Verlag, Berlin 1976.

[6] Chaudhry, M.L. and Templeton, J.D.C., *A First Course in Bulk Queues*, Wiley, New York 1983.

[7] Chung, Kai-Lai, *Markov Chains with Stationary Transition Probabilities*, Springer-Verlag, New York 1967.

[8] Çinlar, E., Time dependence of queues with semi-Markov service, *J. Appl. Prob.* **4** (1967), 356-364.

[9] Çinlar, E., Queues with semi-Markov arrivals, *J. Appl. Prob.* **4** (1967), 365-379.

[10] Cohen, J.W. and Boxma, O.J., *Boundary Value Problems in Queueing Systems Analysis*, North-Holland, New York 1983.

[11] Dagsvik, J., The general bulk queue as a matrix factorization problem, part 1, *Adv. Appl. Prob.* **7** (1975), 636-646, 647-655.

[12] De Smit, J.H.A., The single server semi-Markov queue, *Stoch. Proc. Appl.* **22** (1986), 37-50.

[13] Dukhovny, A.M., One-dimensional random walk that depends on a parameter, *Dokl. Akad. Nauk Ukrain. SSR, ser.* **A**:3 (1982), 9-12.

[14] Dukhovny, A.M., Random walk depending on a parameter with a distortion in a finite region, *Dokl. Akad. Nauk Ukrain. SSR, ser.* **A**:6 (1984), 8-11.

[15] Dukhovny, A.M., Priority systems with batch arrival and batch servicing, *Dokl. Akad. Nauk Ukrain. SSR, ser.* **A**:11 (1986), 67-69.

[16] Dukhovny, A.M., Combined feedback in queueing systems with bulk arrival and batch service, In: *Asymptotic Methods in Analysis of Stochastic Models*, Institute of Math. of the Ukrain. Academy of Sciences, Kiev (1987), 46-54.

[17] Dukhovny, A.M., Markov chains with quasi-toeplitz transition matrix, *J. Appl. Math. Simul.* **2**:1 (1989), 71-82.

[18] Gakhov, F.D., *Boundary Value Problems*, Fizmatgiz, Moscow 1977 (in Russian).

[19] Grassmann, W.K. and Heyman, D.P., Equilibrium distribution of block-structured Markov chains with repeating rows, *J. Appl. Prob.* **27** (1990), 557-576.

[20] Keilson, J., General bulk queue as a Hilbert problem, *J. Roy. Statist. Soc.* **24** (1962), 344-358.

[21] Kemperman, J.H.B., *The Passage Problem for a Stationary Markov Chain*, Univ. of Chicago Press, Chicago 1961.

[22] Kingman, J.F.C., *On the Algebra of Queues*, Methuen, London 1966.

[23] Litvinchuk, G.S. and Spitkovsky, I.M., *Factorization of Measurable Matrix Functions*, Birkhauser Verlag, Basel-Boston 1987.

[24] Lucantoni, D.M., New results on the single-server queue with a batch Markovian arrival process, *Stoch. Mod.* **7** (1991), 1-46.

[25] Malyshev, V.A., On homogeneous random walks on the product of a finite set and a half-line, *Veroyatn. Mat. Issledov.* **41** (1972), 5-14.

[26] Miller, H.D., A matrix factorization problem in the theory of random variables defined on a finite Markov chain, *Proc. Camb. Phil. Soc.* **58** (1962), 268-285.

[27] Neuts, M.F., Some explicit formulas for the steady-state behavior of the queue with semi-Markov service times, *Adv. Appl. Prob.* **9** (1977), 141-157.

[28] Neuts, M.F., A versatile Markovian point process, *J. Appl. Prob.* **16** (1979), 764-779.

[29] Presman, E.L., Factorization methods and a boundary value problem for sums of random variables given on a Markov chain, *Izv. Akad. Nauk SSR Ser. Mat.* **33** (1969), 861-900.

[30] Spitzer, F., A combinatorial lemma and its applications to probability theory, *Trans. Amer. Math. Soc.* **82** (1956), 323-339.

[31] Takács, L., A Banach space of matrix functions and its application in the theory of queues, *Sankhya, Ser* A **38** (1976), 201-211.

[32] Teghem, J., Loris-Teghem, J. and Lambotte, J.P., *Modèles d'attente* M/G/1 *et* GI/M/1 *à arrivées et services en groupes*, Lecture Notes in Operations Research and Mathematical Economics **8**, Springer-Verlag, New York 1969.

Part III

Approximation, Estimates, and Simulation of Queues

Chapter 15

Light-traffic approximation in queues and related stochastic models

Bartłomiej Błaszczyszyn[1], Tomasz Rolski[2], and Volker Schmidt

ABSTRACT Light-traffic approximations in queues refer to results on the asymptotic behavior of queueing characteristics when arrivals are getting sparse and, in the limit, there is no arrival. In the present chapter, we discuss several light-traffic schemes which were introduced and studied by different authors. We analyze light-traffic results with respect to methods, models, and the order of approximation. Moreover, we show how the idea of light-traffic approximation can be utilized in risk and dam models. We also discuss open problems and areas for future research.

CONTENTS

15.1	Introduction	379
15.2	Basic notions and theories	381
15.3	Approximations for Poisson-driven and other Markovian systems	384
15.4	First-order approximations for queues with more general input	389
15.5	Higher order approximations	395
15.6	Analyticity	401
15.7	Discussion and further research problems	402
	Acknowledgements	403
	Bibliography	403

15.1 INTRODUCTION

Recently, a good deal of effort has been expended to develop alternatives to simulations of stochastic models. In this chapter, we survey light-traffic approximations (LTA for short). A survey of other traffic models may be found in Jagerman *et al.* [40]. Roughly speaking, by LTA, we mean results on the asymptotic behavior of characteristics when arrivals are getting sparse and, in the limit, there is no arrival. In other words, we look for systems under conditions opposite to heavy traffic. The heavy-traffic approximations mostly rely on distributional limit theorems of probability theory, and they feature some invariance or robustness regarding input conditions. It turns out, however, that LTA rely essentially on structural properties of the input and, in particular, on the light-traffic scheme (LTS), e.g., how clusters of arrivals are formed in an otherwise sparse input. For example, it is observed that, under certain regularity conditions in single-server queues, clusters of

[1]Supported in part by the Deutsche Forschungsgemeinschaft.

[2]Supported in part by KBN under grant 2 1023 91 01.

0-8493-8074-x/95/$0.00+$.50

of two arrivals play a dominating role, while in c-server queues, the same holds for clusters of $c+1$ arrivals. In this chapter, our goal is to systematize LTA with respect to *methods*, *models*, and the *order of approximation*. We discuss these issues on the basis of over 70 references, most of them recent papers.

For each ξ, let $N^{(\xi)}$ describe input data such as arrival epochs, service times, etc., which, in this chapter, are modeled by marked point processes. Suppose, moreover, that, when $\xi \to \xi_0$ (usually $\xi_0 = 0$ or $\xi_0 = \infty$), the considered queueing system approaches light-traffic conditions. Let the queueing characteristic of interest be given as a functional $\psi(N^{(\xi)})$ of $N^{(\xi)}$. We wish to approximate the expectation $\mathbf{E}[\psi(N^{(\xi)})]$ by studying its asymptotic behavior for $\xi \to \xi_0$. Possible types of approximations are $a_1\xi^{\alpha_1}$ or, more generally, $a_1\xi^{\alpha_1} + \ldots + a_k\xi^{\alpha_k}$, where $\alpha_1, \ldots, \alpha_k$ are nonnegative numbers and $a_1, \ldots, a_k$ are nonzero coefficients. The number of terms in the approximation gives its order. Most of the earlier results in queueing literature are approximations of order one, and we survey some of them in Section 15.4. In particular, we discuss two simple LTSs proposed in Daley and Rolski [27]. In the first case, interarrival times are dilated by a positive factor γ, $\gamma \to \infty$ (called *γ-dilation*). In the other case, arrivals are independently thinned with retention probability π, $\pi \to 0$ (called *π-thinning*). Typical for the π-thinning scheme are polynomial approximations and, in this context, we study systems driven by a Markov-modulated (MM) input. In this case, one is interested in derivatives of $\mathbf{E}[\psi(N^{(\pi)})]$ at $\pi = 0+$ or in results of the form $\mathbf{E}[\psi(N^{(\pi)})] = a_0 + a_1\pi^1 \ldots + a_n\pi^n + o(\pi^n)$, or in the (stronger) analyticity property, i.e., the infinite series-representation $\mathbf{E}[\psi(N^{(\pi)})] = \sum_{j=0}^{\infty} a_j\pi^j$ for $0 \le \pi < \pi_0$ for certain π_0. For the γ-dilation scheme, polynomial approximations are not all that typical, except for some special cases, e.g., Poisson-driven systems where π-thinning and γ-dilation lead to the same LTA. Surprisingly, in general, these two schemes lead to different LTA. This circumstance might be the reason, at least at first sight, why the theory of LTA was widely neglected in the sixties and seventies when classical queueing theory was established, although there were some pioneering papers as Beneš [9] and Bloomfield and Cox [11]. What we know now, and we will aim to show to the reader, is that the key to understanding the hierarchy of LTA is the modern theory of point processes. Because of that, we briefly review some aspects of this theory in Sections 15.2 and 15.5, and in particular, the notion of Palm probabilities and factorial moment measures.

We observe that the principle of LTA has been successfully employed not only in queueing but also in dam and risk processes. These aspects of LTA are discussed in this chapter as well, e.g., in connection with approximation formulas for ruin probabilities. Analogous results can be derived for functionals of sparse marked point processes in d-dimensional Euclidean space. These lead to approximation formulas for models of stochastic geometry and will be reported in a separate paper.

One of the ideas in deriving LTA for queueing characteristics was that the LTA can be combined with some heavy-traffic approximations to provide a base for interpolation so as to get good approximations for a wide range of parameters. We do not discuss the details of this interpolation method, but refer to such chapters like [25, 51, 54, 67] and to the references therein. Another problem domain related to LTA is sensitivity analysis for stochastic models. It concerns, in particular, the investigation of derivatives of the mapping $\xi \to \mathbf{E}[\psi(N^{(\xi)})]$ at some $\xi = \xi_0$, where $0 < \xi_0 < \infty$. In some cases of Poisson-driven systems, sensitivity analysis can be carried out by using results of a corresponding light-traffic analysis. For systems with non-Poisson input, new methods have to be developed, e.g., in the spirit of Brémaud and Vázquez-Abad [23], Brémaud and Gong [22], or Konstantopoulos and Zazanis [44, 45]. For relationships between light-traffic analysis and the rare event method, we refer to [47] and Kovalenko's chapter [48] in this book.

The present chapter is organized in the following way. In Sections 15.2.1 to 15.2.3, we review some basic notion of the theory of stationary marked point processes, such as Palm probabilities and Campbell's formula, which will be useful in later sections. Some examples of point processes are also discussed. In Section 15.2.4, the two basic LTS's are introduced: γ-dilation and π-thinning. Because we consider LTA for different types of queueing systems, in Section 15.2.5, we briefly recall Kendall's notation for queueing systems. Approximation formulas for Poisson-driven and other Markov-type systems are given in Section 15.3. Here, we also discuss several methods used in this context, such as simplification of balance equations and ladder-height distributions under conditions of light traffic. In Section 15.3.3, a method for calculating light-traffic derivatives with respect to the intensity of an underlying stationary Poisson process is reviewed; a generalization of this method to arbitrary marked point processes is presented in Section 15.5. In Section 15.3.4, we describe a likelihood-ratio approach to LTA for Poisson-driven systems. First-order approximations for queues with more general input are discussed in Section 15.4. In particular, a recurrence-equation method for LTA is presented which, when combined with a so-called "sandwich inequality" and a multivariate version of the Abelian lemma, gives useful approximations for single-server queues, multiserver queues, and queues in series. It turns out that, in some cases, the LTA of workload do not depend on the clustering properties of arrival epochs, whereas the LTA of (actual) waiting time do. In Section 15.5, LTA of higher order are considered which are derived by using an expansion method with respect to factorial moment measures of the underlying marked point process. When the form of these moment measures is not overly complicated (e.g., for Markov-modulated Poisson processes), this method leads to polynomial approximations, at least in the π-thinning scheme. Finally, in Section 15.6, analyticity properties of queueing characteristics are considered, i.e., the question regarding under which conditions these characteristics can be expressed by an infinite polynomial series is raised.

15.2 BASIC NOTIONS AND THEORIES

15.2.1 Random marked point processes

In this chapter, we consider queueing systems whose input (offered traffic) is modeled by a random marked point process (m.p.p.); consequently, we proceed to recall some basic notions from the theory of point processes. We consider m.p.p.'s on the real line $\mathbb{R}$ with marks in a space $\mathbb{K}$, in particular, $\mathbb{K} = \mathbb{R}$ or $\mathbb{K} = \mathbb{R}^d$, but we mention that the theory is valid for point processes in quite general Polish spaces. For further details, we refer the reader to the monographs of Baccelli and Brémaud [7], Daley and Vere-Jones [30], Kallenberg [41], Karr [43], König and Schmidt [46] and Sigman [62]. Readers who are primarily interested in getting a first insight into point process theory are referred to the survey in Serfozo [60].

A realization of an m.p.p. is a sequence of arrival epochs $\{t_i\}$ and marks

$\{m_i\}$ such that the ith arrival is associated with the mark $m_i \in \mathbb{K}$. It is customary to represent a realization of an m.p.p. as a (counting) point measure

$$\mu = \sum_{i=l}^{k} \epsilon_{(t_i, m_i)}, \tag{15.1}$$

where $-\infty \le l \le k \le \infty$, $\{t_i\}$ is an increasing sequence $\ldots \le t_{-1} \le t_0 \le 0 < t_1 \le t_2 \le \ldots$, in $\mathbb{R}$ without accumulation points, $m_i \in \mathbb{K}$, and $\epsilon_{(t,m)}$ is the point mass at (t,m), defined by

$$\epsilon_{(t,m)}(D \times K) = \begin{cases} 1 & \text{if } (t,m) \in D \times K \\ 0 & \text{if } (t,m) \notin D \times K. \end{cases}$$

We refer to the t_i's as points and the m_i's as marks. Let $\mathcal{M}$ denote the space of all realizations of a random marked point process and let $\mathcal{M}_s$ be the set of measures, $\mu \in \mathcal{M}$ such that $\mu(\{t\} \times \mathbb{K})$ is either 0 or 1 for each t (its elements are *simple* point measures). By o, we denote the null measure, representing an input with no arrivals (ie., $o(\mathbb{R} \times \mathbb{K}) = 0$).

We denote by $\mathcal{B}(\mathcal{M})$ the minimal σ-algebra of subsets of $\mathcal{M}$, which renders the functions $\mathcal{M} \ni \mu \to \mu(E)$ measurable for all bounded $E \in \mathcal{B}(\mathbb{R} \times \mathbb{K})$, where $\mathcal{B}(\mathbb{R} \times \mathbb{K})$ is the Borel σ-algebra of $\mathbb{R} \times \mathbb{K}$. For an underlying probability space $(\Omega, \mathcal{F}, \mathbf{P})$ and $(\mathcal{M}, \mathcal{B}(\mathcal{M}))$, a *marked point process* is a mapping $\omega \to N(\omega) \in \mathcal{M}$ from the sample space Ω to the measurable state space $\mathcal{M}$. If not otherwise stated, we identify $(\Omega, \mathcal{F})$ with $(\mathcal{M}, \mathcal{B}(\mathcal{M}))$ under the identity mapping $\mathcal{M} \ni \mu \to N(\mu) = \mu$. Thus, every probability measure $\mathbf{P}$ on $(\mathcal{M}, \mathcal{B}(\mathcal{M}))$ defines a *random marked point process* $(N, \mathbf{P})$ (usually we omit the word *random* and call it simply m.p.p.). We say that an m.p.p. $(N, \mathbf{P})$ is simple if $\mathbf{P}(N \in \mathcal{M}_s) = 1$. All m.p.p.s to be considered in the following are assumed simple. Sometimes, we do not consider the space of marks but a (non-marked) point process (p.p.) only.

15.2.2 Stationarity and Palm distributions

Models with time-invariant characteristics play a special role in the analysis of systems driven by a stochastic input. To formalize this idea, one usually defines a parametric family of mappings $\theta_t: \mathcal{M} \mapsto \mathcal{M}$, $t \in \mathbb{R}$, such that, for each $\mu \in \mathcal{M}$, $D \in \mathcal{B}(\mathbb{R})$ and $K \in \mathcal{B}(\mathbb{K})$,

$$(\theta_t \mu)(D \times K) = \mu\left((D+t) \times K\right),$$

where $D + t = \{s + t : s \in D\}$. The marked point process $(N, \mathbf{P})$ is said to be *stationary* if

$$\mathbf{P} \circ \theta_t = \mathbf{P} \quad \text{for each } t \in \mathbb{R}.$$

Note that, for a stationary m.p.p. $(N, \mathbf{P})$, the (first factorial) moment measure $M^{(1)}$, defined by $M^{(1)}(D \times K) = \mathbf{E}[N(D \times K)]$, is θ_t-invariant and hence,

$$M^{(1)}(dt \times \mathbb{K}) = \lambda dt.$$

The constant $\lambda = M^{(1)}([0,1] \times \mathbb{K})$ is called the *intensity* of $(N, \mathbf{P})$. Now, assume that λ is positive and finite. Then, the probability distribution on $(\mathcal{M}, \mathcal{B}(\mathcal{M}))$

$$\mathbf{P}^0(\Gamma) = \frac{1}{\lambda}\mathbf{E}\left[\int_0^1 1(\theta_t N \in \Gamma) N(dt \times \mathbb{K})\right], \quad \Gamma \in \mathcal{B}(\mathcal{M}) \tag{15.2}$$

is called the *Palm distribution* of $(N, \mathbf{P})$, where $\mathbf{1}(\cdot)$ is the indicator function equal 1 if true and 0 if false. It can be interpreted as a conditional distribution, given that the origin is an arrival point. The most important property of $(N, \mathbf{P}^0)$ is the random-time invariance property

$$\mathbf{P}^0 \circ \theta_{t_n} = \mathbf{P}^0,$$

where the t_n are as in (15.1), that is, $\mathbf{P}^0$ is invariant with respect to pointwise shifting in time. Finally, we remark that (15.2) implies the so-called *Campbell formula*. For any measurable $f\colon \mathbb{R}\times\mathbb{K}\times\mathcal{M}\to\mathbb{R}_+$,

$$\mathbf{E}\left[\int_{\mathbb{R}\times\mathbb{K}} f(t,\theta_t N)N(dt\times\mathbb{K})\right] = \lambda\mathbf{E}^0\left[\int_{\mathbb{R}\times\mathbb{K}} f(t,N)dt\right],$$

where $\mathbf{E}^0$ denotes the expectation operator under $\mathbf{P}^0$ (for details see also Schmidt and Serfozo [59] in this book).

15.2.3 Examples

We now give some important examples of the point processes on the real line.

Example 15.1. [Poisson Point Process] A (nonmarked) point process $(N, \mathbf{P})$ on $\mathbb{R}$ is called a *Poisson p.p.* if, for each finite sequence $D_1,\ldots,D_n \in \mathcal{B}(\mathbb{R})$ of bounded Borel sets, the random variables $N(D_1),\ldots,N(D_n)$ are independent and Poisson distributed. If, additionally, $N(D)$ has mean $\lambda\,|\,D\,|$ for any $D\in\mathcal{B}(\mathbb{R})$ where $|\,D\,|$ denotes the Lebesgue measure of D, then the Poisson p.p. is stationary and λ is its intensity.

Example 15.2. [Markov-Modulated M.P.P.] Let $\{X(t); t\geq 0\}$ be an irreducible homogeneous Markov process which takes its values in the finite set $\{1,\ldots,\ell\}$ and is governed by an intensity matrix $\mathbf{\Lambda}$. We say that the process $\{X(t)\}$ defines a random environment. Since the state space of $\{X(t)\}$ is finite, there exists a unique stationary initial distribution of $\{X(t)\}$ denoted by $\boldsymbol{\pi} = (\pi_1,\ldots,\pi_\ell)$, i.e, $\boldsymbol{\pi}\mathbf{\Lambda} = \mathbf{0}$. Now let $\boldsymbol{\lambda} = (\lambda_1,\ldots,\lambda_\ell)$ be a vector of nonnegative numbers and $B = (B_1,\ldots,B_\ell)$ a vector whose components are distributions on $\mathbb{K}$. If the environment is in state i (note that the realizations of $\{X(t)\}$ can be taken to be piecewise constant), then points occur according to a stationary Poisson process with intensity λ_i, and the corresponding marks are drawn from distribution B_i, independently of everything else. We call such an m.p.p. *Markov modulated* (MM). Formally, $(N, \mathbf{P})$ can be defined as a *Cox* process (i.e., as a mixture of Poisson processes) with stochastic intensity measure on $\mathbb{R}\times\mathbb{K}$ of the form

$$\lambda_{X(t)}B_{X(t)}(dm)dt.$$

Example 15.3. [Renewal Process] A *renewal process* $(N = \sum_{i=-\infty}^{\infty}\epsilon_{t_i}, \mathbf{P})$ is given by a sequence of independent identically distributed (i.i.d.) random variables $\{T_i = t_{i+1} - t_i\colon i\in\mathbb{Z}\backslash\{0\}\}$, (where $\mathbb{Z} = \{\ldots,-1,0,1,\ldots\}$), with common distribution function F and any initial vector $(-t_0, t_1)$ independent of the $\{T_i\}$. In the stationary case, $\mathbf{P}(-t_0 > x, t_1 > y) = \lambda\int_{x+y}^{\infty}[1-F(v)]dv$ with $\lambda = 1/\int_0^{\infty}[1-F(v)]dv$.

15.2.4. Light-traffic schemes

We consider stochastic models in light-traffic conditions induced by a certain LTS.

This means that the model is driven by a parameterized family $N^{(\xi)}$ of m.p.p.'s and that, in some sense, we have the "null convergence" $N^{(\xi)} \to o$ if $\xi \to \xi_0$, where $o \in \mathcal{M}$ is the *null measure*, that is, points of $N^{(\xi)}$ are getting ever more rare according to a certain scheme. Light-traffic theorems describe the asymptotic behavior of the model driven by $N^{(\xi)}$ as $\xi \to \xi_0$ resulting in LTA.

We consider the following two basic LTSs:

- γ-dilation: N is an m.p.p. and its time scale is dilated by an increasing factor γ, i.e.,
$$N^{(\gamma)} = \sum_i \epsilon_{(\gamma t_i, m_i)} \quad \text{and} \quad \gamma \to \infty,$$
- π-thinning: N is an m.p.p. and, if it has a point at t, then this point is retained with probability π independently of everything else. Otherwise, the point and its associated mark are deleted, i.e.,
$$N^{(\pi)} = \sum_i U_i^{(\pi)} \epsilon_{(t_i, m_i)} \quad \text{and} \quad \pi \to 0,$$
where $\{U_i^{(\pi)}: i \in \mathbb{Z}\}$ is a sequence of i.i.d. r.v.'s with $\mathbf{P}(U_i = 1) = \pi = 1 - \mathbf{P}(U_i = 0)$. Moreover, N and $\{U_i: i \in \mathbb{Z}\}$ are independent.

Note that, for a stationary Poisson process, γ-dilation is equivalent to π-thinning with the retention probability $\pi = 1/\gamma$ $(\gamma \geq 1)$. Moreover, this property characterizes the Poisson process.

15.2.5. Kendall's notation

Throughout the chapter, we use Kendall's notation for queueing systems, e.g., GI/GI/c stands for a stable system with i.i.d. interarrival times and i.i.d. service times which are independent of each other, and c is the number of servers $(1 \leq c \leq \infty)$. We write G/GI/c, if service times are i.i.d. and independent from the arrival process, and GI/G/c, if interarrival times are i.i.d. and independent of service times. Let T and S be generic interarrival times and service times with distributions A and B, respectively. We denote the actual waiting time (delay) in the system by W, the workload (or virtual waiting time) by V, and the queue length by L. We will append superscripts (γ), (π), (ξ), etc. to all the quantities T, S, W, V, L which vary in an LTS.

Throughout the chapter we use the notation $\boldsymbol{x} = (x_1, \ldots, x_k)$ for $\mathbb{R}^k$-valued or $\mathbb{K}^k$-valued vectors.

15.3 APPROXIMATIONS FOR POISSON-DRIVEN AND OTHER MARKOVIAN SYSTEMS

The simplest case of a stochastic model with Markovian structure assumes Poisson arrivals and exponentially distributed service times. Typical extensions consider Markov-modulated arrivals (see Example 15.2) and so called *phase-type* (PH) *distributions* of service times. A distribution B on $\mathbb{R}_+ = [0, \infty)$ is said to be PH *with representation* $(\boldsymbol{\alpha}, \boldsymbol{T}, d)$ if there is a homogeneous Markov process on the state space $\{0, 1, \ldots, d\}$, with intensity matrix

$$\left(\begin{array}{c|c} 0 & \mathbf{0} \\ \hline t_0 & T \end{array}\right) = \begin{array}{c} 0 \\ 1 \\ 2 \\ \vdots \\ d \end{array} \overset{\begin{array}{ccccc} 0 & 1 & 2 & \dots & d \end{array}}{\left(\begin{array}{c|cccc} 0 & 0 & 0 & \dots & 0 \\ \hline t_{10} & t_{11} & t_{12} & \dots & t_{1d} \\ t_{20} & t_{21} & t_{22} & \dots & t_{2d} \\ \vdots & \vdots & \vdots & \ddots & \vdots \\ t_{d0} & t_{d1} & t_{d2} & \dots & t_{dd} \end{array}\right)},$$

state 0 being the absorbing state, such that B is the distribution of the time to absorption in 0, given the initial distribution

$$\left(1 - \sum_{i=1}^{d} \alpha_i, \boldsymbol{\alpha}\right), \quad \text{where } \boldsymbol{\alpha} = (a_1, \dots, \alpha_d).$$

We assume that the states $1,\dots,d$ in T are all transient, so that absorption in state 0 from any initial state is certain.

A Markovian structure usually renders stochastic models more tractable. This is also the case for light-traffic conditions. In this section, we present some LTA which are essentially based on Markov-type assumptions. We also present results concerning stochastic models which, although not strictly Markovian, can be viewed in a way as driven by a Poisson process. In other LTA, phase-type methods also play an important role, as general constants in approximation formulas become more analytically tractable under the assumption of PH-distributed service times; see also Sections 15.4 and 15.5.

15.3.1 Simplification of balance equations

A queueing system with MM arrivals and PH-distributed service times can be described by a continuous-time Markov process. The first successful LTA were obtained for such Markov processes, for which global balance equations can be written down, but they are often too complex to be solved explicitly. However, sending the traffic intensity to zero simplifies these equations, leading to asymptotic results. Burman and Smith [24] studied in this way, e.g., the continuous-time Markov process describing the number of customers in the M/PH$_i$/c queue, where the service time at the ith server has a PH-representation $(\boldsymbol{\alpha}^i, \boldsymbol{T}^i, d^i)$. The equilibrium-excess service time τ^i at the ith server is a PH-distribution and can be expressed in terms of Kronecker calculus; see [24, 50] and, for the Kronecker calculus, [35]. Similarly, the distribution of the minimum, $\tau = \min_{1,\dots,c} \tau^i$ of independent τ^i's can also be expressed by Kronecker calculus; see e.g., [24, 50]. In [24], the following LTA are derived.

Theorem 15.4. *For the stationary number of customers* $L^{(\lambda)}$ *in an* M/PH$_i$/c *queue,*

$$\lim_{\lambda \to 0} \lambda^{-(c+1)} \mathbf{E}[L^{(\lambda)}] = \frac{\mathbf{E}[\tau]}{c! \prod_{j=1}^{c} \mu_j},$$

where λ *is the arrival intensity, and* μ_j *is the expected service time for the* j*th server.*

A similar approach was applied to the MM/PH$_i$/1 queue in Burman and

Smith [25].

15.3.2. Simplification of ladder-height distributions

The idea of simplifying a general representation formula of ladder-height distributions was used in Asmussen [2] to obtain LTA for the distribution of stationary waiting times in the GI/GI/1 queue in light-traffic regime. This idea is based on random-walk techniques. In a related paper, Daley and Rolski [26] studied conditions in an LTS, under which the ratio between the mean waiting time $\mathbf{E}[W^{(\gamma)}]$ and $\mathbf{E}[(S-\gamma T)_+]$ in the GI/GI/1 queue tends to 1. Following this paper, Asmussen [2] sought conditions under which $W^{(\xi)}$ and $(S^{(\xi)}-T^{(\xi)})$ are *equivalent* in light traffic, where the following definition has been used: For $\xi \to \xi_0$, two families, $\{X^{(\xi)}\}$ and $\{Y^{(\xi)}\}$, of nonnegative random variables are *asymptotically conditionally equivalent* (ACE) (or referred to in [2] as *distributionally light-traffic equivalent*), if

$$\mathbf{P}(X^{(\xi)}>0)\to 0, \quad \mathbf{P}(Y^{(\xi)}>0)\to 0, \quad \frac{\mathbf{P}(X^{(\xi)}>0)}{\mathbf{P}(Y^{(\xi)}>0)}\to 1 \tag{15.3}$$

and the total variation distance,

$$\begin{aligned}&\| \mathbf{P}(X^{(\xi)}\in\cdot\mid X^{(\xi)}>0)-\mathbf{P}(Y^{(\xi)}\in\cdot\mid Y^{(\xi)}>0)\| \\ &=2\sup_{B\in\mathfrak{B}(\mathbb{R})}\left|\mathbf{P}(X^{(\xi)}\in B\mid X^{(\xi)}>0)-\mathbf{P}(Y^{(\xi)}\in B\mid Y^{(\xi)}>0)\right|\end{aligned}$$

converges to 0 as $\xi \to \xi_0$. Moreover, we say that families are ACE *of order p* if, for moments of any order $q \le p$,

$$\frac{\mathbf{E}\left[X^{(\xi)}\right]^q}{\mathbf{E}\left[Y^{(\xi)}\right]^q}\to 1. \tag{15.4}$$

The key to study the stationary waiting time in GI/GI/1 queues is the random walk generated by the sequence $\{S_i-T_i \;\; i\le 0\}$ for which the *ascending ladder epoch* τ_+ and the *ascending ladder height* R_+ are defined by

$$\tau_+ = min\{n: \sum_{i=0}^{n}(S_{-i}-T_{-i})>0\}, \;\; R_+=\sum_{i=0}^{\tau_+}(S_{-i}-T_{-i}).$$

Assume that the following condition $\mathfrak{LT}(p)$ is fulfilled: For all $q\le p$ let

$$\lim_{u\uparrow\infty}\limsup_{\xi\to\xi_0}\frac{\mathbf{E}\left[\left[S^{(\xi)}-T^{(\xi)}\right]^{q+1};S^{(\xi)}-T^{(\xi)}>u\right]}{\mathbf{E}\left[\left[S^{(\xi)}-T^{(\xi)}\right]^{q};S^{(\xi)}-T^{(\xi)}>0\right]}=0. \tag{15.5}$$

The following result was proved in [2].

Lemma 15.5. *Assume that* $S^{(\xi)}-T^{(\xi)}\xrightarrow{\mathfrak{D}}-\infty$, *and that condition* $\mathfrak{LT}(0)$ *holds. Then the ascending ladder height,* $R_+^{(\xi)}$, *of the random walk generated by* $\{S_i^{(\xi)}-T_i^{(\xi)} \;\; i\le 0\}$ *and* $(S^{(\xi)}-T^{(\xi)})_+$ *are ACE. If condition* $\mathfrak{LT}(p)$ *holds for some* $p>0$, *then* $R_+^{(\xi)}$ *and* $(S^{(\xi)}-T^{(\xi)})_+$ *are ACE of order p.*

The result of the above lemma carries out to the waiting time, $W^{(\xi)}$, and $(S^{(\xi)}-T^{(\xi)})_+$ because the distribution of $W^{(\xi)}$ can be expressed as a geometric sum of ladder height distributions, that is,

$$\mathbf{P}(W^{(\xi)}\in D)=(1-\|G_+^{(\xi)}(\cdot)\|)\sum_{n=0}^{\infty}(G_+^{(\xi)})^{*n}(D), \tag{15.6}$$

where $G_+^{(\xi)}$ is the distribution of the ladder height $R_+^{(\xi)}$. In particular, for the case of PH-distributed service times, the following result has been proved in Asmussen [2].

Theorem 15.6. *The stationary waiting times $W^{(\xi)}$ in GI/PH/1 queue, such that $T^{(\xi)} \to \infty$ in distribution and service times have a fixed PH distribution with representation $(\boldsymbol{\alpha}, \boldsymbol{T}, d)$, are ACE with respect to a family of PH-distributed random variables with representations $(\boldsymbol{\nu}^{(\xi)}, \boldsymbol{T}, d)$, where*

$$\boldsymbol{\nu}^{(\xi)} = \int_0^\infty \boldsymbol{\alpha} e^{\boldsymbol{T} t} A^{(\xi)}(dt),$$

and $A^{(\xi)}$ is the distribution function of $T^{(\xi)}$.

15.3.3. Poisson-intensity derivatives

Several LTA are considered complex stochastic systems which can be seen as driven by a Poisson p.p. A general method for investigating characteristics of such systems in light traffic was proposed in Reiman and Simon [53], where these characteristics have the form of expected values of some functionals of a Poisson p.p. In particular, a real function f of the form

$$f(\lambda) = \mathbf{E}[\psi(N^{(\lambda)})]$$

is considered, where $N^{(\lambda)}$ is a homogeneous Poisson p.p. with intensity λ and ψ is a functional of $N^{(\lambda)}$ which satisfies a certain admissibility condition (see [53] for details). The light-traffic information is obtained from $f(0)$ and from the derivatives $\frac{d^n}{d\lambda^n} f(\lambda)\Big|_{\lambda=0}$ $(n \geq 1)$. It appears that the determination of the nth derivative of f involves consideration of some function $\psi^{(n)}_{t_1,\ldots,t_n}$ which is constructed from ψ on sample paths with at most n arrivals, occurring, say at the epochs $t_1, \ldots, t_n$. Then,

$$\frac{d^n}{d\lambda^n} f(\lambda)\Big|_{\lambda=0} = \int_{-\infty}^{\infty} \cdots \int_{-\infty}^{\infty} \psi^{(n)}_{t_1,\ldots,t_n} dt_1 \ldots dt_n, \tag{15.7}$$

where $\psi^{(n)}_{t_1,\ldots,t_n} = \psi^{(n)}_{t_1,\ldots,t_n}(o)$ are defined similarly to the marked case in (15.23).

The proof of (15.7) proposed in [53] is essentially based on properties of Poisson p.p.'s and involves an interchange of limits, whose justification requires a great deal of effort. However, a similar result can be obtained for more general point processes when applying Palm calculus. This was done in [6] and [12] (see Section 15.5.1 below). In a similar vein, particular LTA for the distribution and moments of sojourn time in an open Markovian queueing system with different classes of customers were derived in Simon [65]. For the sake of simplicity, we present here an example of such a result for a system with one class of customers only.

Theorem 15.7. *For the mth moment of the stationary sojourn time, $\overline{W}^{(\lambda)}$, in an open queueing system with Poisson arrivals of intensity λ*

$$\frac{d}{d\lambda}\mathbf{E}[\overline{W}^{(\lambda)}]^m\Big|_{\lambda=0} = (-1)^{m+1} m!$$
$$\times\left(\eta Q_1^{-1}\Theta Q_2^{-m} e - \eta Q_1^{-1} e Q_1^{-m} e + \sum_{i=0}^{m-1}\left(\eta Q_1^{-(m-i)}\Theta Q_2^{-(i+1)} e - \eta Q_1^{-(m+1)} e\right)\right),$$

where Q_1, Q_2 are the generators of the Markov processes describing the behavior of the system with, precisely, one and two customers, respectively η is the initial *probability vector corresponding to* Q_1, *e is the column vector of* 1*'s, and* Θ *is the transition matrix from the states of* Q_1 *to those* Q_2.

Similar arguments (combined with the likelihood-ratio method as in Section 15.3.4) were used in Wang and Wolff [66] where the asymptotic behavior of a well-known approximation for first-moment performance measures is investigated in light traffic for the M/GI/c queue. In Fleming [31], the light-traffic derivative, $\frac{d}{d\lambda}\mathbf{E}[\bar{W}^{(\lambda)}]^m\Big|_{\lambda=0}$, has been calculated for the mth moment of the stationary sojourn time, $\bar{W}^{(\lambda)}$, in the M/GI/1 queue with the round-robin service discipline.

Another example, where the representation formula (15.7) has been used in light-traffic analysis of Poisson-driven queues, is given in Kroese and Schmidt [49]. Consider the following continuous polling system: Customers arrive according to a homogeneous Poisson p.p. with intensity λ and wait on a circle in order to be served by a single-server. The server is *greedy*, in the sense that he always moves (with constant speed α^{-1}) towards the nearest customer on the circle. The customers are served, according to an arbitrary service time distribution B, in the order in which they are encountered by the server. In [49], the following second-order Taylor-expansion has been found for the stationary mean queue length, $\mathbf{E}[L^{(\lambda)}]$, and for the stationary mean work-load, $\mathbf{E}[V^{(\lambda)}]$.

Theorem 15.8. *Assume that the fifth moment of service times is finite. Then,*

$$\mathbf{E}[L^{(\lambda)}] = \lambda\left(e_1 + \frac{\alpha}{4}\right) + \lambda^2\left(\frac{e_2}{2} + \frac{\alpha e_1}{4} + \frac{\alpha^2}{48}\right) + O(\lambda^3)$$

and

$$\mathbf{E}[V^{(\lambda)}] = \lambda\left(\frac{e_2}{2} + \frac{e_1\alpha}{4}\right) + \lambda^2\left(\frac{e_1e_2}{2} + \frac{\alpha e_1^2}{4} + \frac{\alpha^2 e_1}{48}\right) + O(\lambda^3),$$

where $e_i = \int x^i dB(x)$.

In Frey and Schmidt [33], formula (15.7) has been applied in order to derive a Taylor-series expansion for (finite-horizon) multivariate characteristics of the continuous-time risk model with compound Poisson input (see also Sections 15.3.2 and 15.5.2). Namely, the joint probability $\mathbf{P}(\tau_+ < t,\ R_- \le x,\ R_+ \le y)$ of ruin time τ_+, surplus R_- just before ruin and deficit R_+ at ruin time τ_+, is considered as a function of the arrival intensity λ. It is shown in [33] that the nth derivative $f^{(n)}_{t,u,x,y}(0)$ at $\lambda = 0$ of the function $f_{t,u,x,y}(\lambda) = \mathbf{P}(\tau_+ < t, R_- \le x, R_+ \le y)$, where u is the initial reserve, can be given by the following recursion formula.

Theorem 15.9. *For each* $n \ge 1$, $0 \le u$, $t < \infty$, $0 < x$, $y \le \infty$,

$$\frac{f^{(n)}_{t,u,x,y}(0)}{n!} = \frac{f^{(n)}_{u,x,y}(0)}{n!} - \sum_{k=1}^{n} q^{(n-k,k)}_{t,u,x,y}$$

where the quantities $q^{(n,k)}_{t,u,x,y}$ *are given recursively by*

$$q^{(n,k)}_{t,u,x,y} = \int_0^t \left(\int_0^{u+s} q^{(n-1,k)}_{t-s,u+s-z,x,y} dB(z) - q^{(n-1,k)}_{t-s,u+s,x,y} \right) ds,$$

$q^{(0,k)}_{t,u,x,y} = \frac{f^{(k)}_{u+t,x,y}(0)}{k!}$ *and* $\frac{f^{(n)}_{u,x,y}(0)}{n!} = F_{x,y} * G^{*(n-1)}(u)$ *with* $F_{x,y}(v) = \int_v^{max\{x,v\}} (B(s+y) - B(s))ds$, $G(u) = \int_0^u (1 - B(s))ds$.

Remark 15.10. It would be interesting to determine the coefficients $f^{(n)}_{t,u,x,y}(0)/n!$ for more general inputs. In a forthcoming paper ([13]), this problem will be investigated in the case of a Markov-modulated input.

15.3.4. The likelihood-ratio method

Another approach to LTA for systems driven by a homogeneous Poisson p.p. was proposed in Reiman and Weiss [55]. It is based on the following idea. Suppose that N is a stationary Poisson process on $[0,\infty)$ with intensity λ. In this context, it is convenient to consider the canonical probability space $(\mathcal{M}, \mathcal{B}(\mathcal{M}), \mathbf{P}^\lambda)$ (see Section 15.2.1), where $\mathbf{P}^\lambda$ is the distribution of N. Denote by $\mathcal{F}_t$ the history of N up to time t. For fixed $\lambda' > \lambda > 0$ and any given $\mathbf{P}^{\lambda'}$-a.s. finite stopping time T, the Radon-Nikodym derivative $d\mathbf{P}^\lambda/d\mathbf{P}^{\lambda'}$ on $\mathcal{F}_T$ can be represented (see e.g., Brémaud [20]) as a likelihood ratio given by

$$\frac{d\mathbf{P}^\lambda}{d\mathbf{P}^{\lambda'}}(\mu) = e^{(\lambda' - \lambda)T(\mu)}\Big(\frac{\lambda}{\lambda'}\Big)^{\mu([0,T(\mu)))}.$$

Thus, for a $\mathcal{F}_T$-measurable functional ψ, we have

$$\mathbf{E}^\lambda[\psi(N)] = \mathbf{E}^{\lambda'}\left[\psi(N)e^{(\lambda - \lambda')T}\Big(\frac{\lambda}{\lambda'}\Big)^{N([0,T))}\right],$$

where $\mathbf{E}^\lambda$ and $\mathbf{E}^{\lambda'}$ denote the expectation operators corresponding to the measures $\mathbf{P}^\lambda$ and $\mathbf{P}^{\lambda'}$, respectively. Performing formal differentiation and letting $\lambda \to 0$, we obtain the following LTA:

$$\lim_{\lambda \to 0}\frac{d^k}{d\lambda^k}\mathbf{E}^\lambda[\psi(N)] \tag{15.8}$$

$$= \mathbf{E}^{\lambda'}\left[\psi(N)(\lambda')^{-N([0,T))}e^{\lambda' T}\frac{k!}{(k - N([0,T)))!}(-T)^{k - N([0,T))}\mathbf{1}(N([0,T)) \le k)\right].$$

To justify the interchange of the limits in (15.8), some additional assumptions are made in [55], which, unfortunately, are of a rather technical nature.

Similar techniques were considered in Zazanis [71], when finding conditions for analyticity of characteristics of Poisson-driven stochastic systems represented as functions of the arrival intensity. This approach is also used in sensitivity analysis which seeks derivatives and analyticity conditions for positive (nonzero) intensities. In light-traffic analysis, some additional technical assumptions, such as boundedness of T, are usually required (see also Section 15.6).

15.4. FIRST-ORDER APPROXIMATIONS FOR QUEUES WITH MORE GENERAL INPUT

In this section, we consider some standard queueing systems in light traffic, including single-server and multiserver queues and queues in series, with general renewal or general stationary input. Various characteristics of such models are studied, mainly in γ-dilation and π-thinning LTS's. The survey of particular results is preceded by a sketch of the main tools used in the analysis.

15.4.1. A recurrence-equation method

Many models of queues, dams, etc. can be represented as dynamical systems driven by a realization $\mu \in \mathcal{M}$ of a certain m.p.p. N, where the quantity of interest is described by a real-valued process $\{W(t): t \in \mathbb{R}\} = \{W(t,\mu): t \in \mathbb{R}\}$, which satisfies a *recurrence equation* of the form

$$W_{n+1}(\mu) \equiv W(t_{n+1},\mu) = h(W_n(\mu), T_n(\mu), m_n(\mu)), \quad n \in \mathbb{Z}, \tag{15.9}$$

for some function $h: \mathbb{R} \times \mathbb{R}_+ \times \mathbb{K} \to \mathbb{R}$, where the $t_n(\mu) = t_n$ and $m_n(\mu) = m_n$ are as in (15.1) and $T_n(\mu) = t_{n+1}(\mu) - t_n(\mu)$. Suppose that the underlying m.p.p. is $(N, \mathbf{P}^0)$, where $\mathbf{P}^0$ is the Palm distribution corresponding to a stationary m.p.p. $(N, \mathbf{P})$. Then, we can suppose that all $W_n(N)$ $(n \in \mathbb{Z})$ have the same distribution, i.e.,

$$W_1(N) = h(W_0(N), T_0(N), m_0(N)) \overset{\mathbf{P}^0}{=} W_0(N),$$

where $\overset{\mathbf{P}^0}{=}$ means that random variables on both the side, defined on $(\mathcal{M}, \mathfrak{B}(\mathcal{M}), P^0)$, have the same distributions. The existence and uniqueness of such stationary solutions of (15.9) has been studied for various stochastic models, which are surveyed in, for example, [7, 19, 32]. Equation (15.9) is the basis of the proof technique for the light-traffic approximations proposed by Daley and Rolski [26, 27, 28, 29] and Blaszczyszyn and Rolski [15]. Roughly speaking, when $N^{(\xi)} \to o$, in a sense, then

$$W_0(N^{(\xi)}) \overset{\mathbf{P}^0}{=} h(W_0(N^{(\xi)}), T_0^{(\xi)}, m_0^{(\xi)}) \approx h(0, T_0^{(\xi)}, m_0^{(\xi)}),$$

where $T_0^{(\xi)} = T_0(N^{(\xi)})$, $m_0^{(\xi)} = m_0(N^{(\xi)})$. Thus, it remains to justify " $\approx$ " and to study the behavior of $h(0, T_0^{(\xi)}, m_0^{(\xi)})$ when $\xi \to \xi_0$.

To justify " $\approx$ " a *sandwich inequality* is used in [26, 27, 28, 29] where one of the bounds of $h(W_0(N^{(\xi)}), T_0^{(\xi)}, m_0^{(\xi)})$ is $h(0, T_0^{(\xi)}, m_0^{(\xi)})$. This requires a kind of monotonicity property of the function h as well as of $W_0(N^{(\cdot)})$. For the evaluation of the light-traffic limit, e.g., of the quantity $\mathbf{E}^0[h(0, T_0^{(\xi)}, m_0^{(\xi)})]$, this limit is represented in the form (when $T_0^{(\xi)}$ is independent of $m_0^{(\xi)}$, with d.f. $A^{(\xi)}$)

$$\int_{\mathbb{R}_+} \mathbf{E}^0[h(0, t, m_0^{(\xi)}) A^{(\xi)}(dt)]$$

and then, an Abelian lemma (see Lemma 15.11) is used in the case of the γ-dilation LTS. In some cases, e.g., for the waiting time in multiserver queues or for the departure process in a series of queues, one has to take into account more than one interarrival time T_n in the recursive equation (15.9) in order to obtain proper approximations. In that case, the following multivariate version of the Abelian lemma, proved in Daley and Rolski [28], is needed. Fix $\alpha > 0$ and denote by $S_\alpha \equiv S_\alpha(C_A)$ the class of nondecreasing right-continuous functions A on $\mathbb{R}_+$ for which $A(0_+) = 0$ and $\lim_{t \to 0_+} A(t)/t^\alpha = C_A$ for some finite positive constant C_A. In this chapter, A is the distribution function of a nonnegative random variable. Note that, in particular, the exponential distribution with parameter λ belongs to $S_1(\lambda)$, and the gamma distribution with parameters (a, λ) belongs to $S_a(\lambda^a/a)$.

Lemma 15.11. *Let $j \in \mathbb{N} = \{1, 2, \ldots\}$ and let $f: \mathbb{R}^j_+ \to \mathbb{R}_+$ be a componentwise nonincreasing function with $f(0) < \infty$ and $f(\boldsymbol{x}) \downarrow 0$ for $\max_{1 \le i \le j} x_i \to \infty$. If $A \in S_\alpha(C_A)$ for some $0 < C_A < \infty$, then,*

$$\gamma^{j\alpha} \int \cdots \int_{\mathbb{R}^j_+} f(\gamma \boldsymbol{x}) dA(x_1) \cdots dA(x_j) \to (\alpha C_A)^j \int \cdots \int_{\mathbb{R}^j_+} (x_1 \cdots x_j)^{\alpha - 1} f(\boldsymbol{x}) dx_1 \cdots dx_j$$

for $\gamma \to \infty$, whenever the integrals on both sides are finite.

15.4.2. Single-server queues

We begin with waiting time approximations for single-server queues.

Theorem 15.12. *Let $f(x) \leq C + Dx^{\beta}$, where $\beta \geq 1$, and consider the* GI/GI/1 *queue, with an interarrival time distribution A and generic service time S, approaching light-traffic conditions by γ-dilation, with $A \in S_{\alpha}$ for some $\alpha > 0$ and $\mathbf{E}^0[S^{\beta+\alpha+1}] < \infty$. Then, for the stationary waiting time $W^{(\gamma)}$, we have*

$$\lim_{\gamma \to \infty} \gamma^{\alpha} \mathbf{E}^0[f(W^{(\gamma)})] = \alpha C_A \int_0^{\infty} t^{\alpha-1} \mathbf{E}^0[f(S-t)dt.]$$

The proof of Theorem 15.12 given in [26] (see also [27]) is based on the stochastic recurrence equation

$$W \overset{\mathbf{P}^0}{=} (W + S - T)_+.$$

The following result obtained in [29] relaxes the assumption of independent interarrival times.

Theorem 15.13. *In a stationary metrically transitive* G/GI/1 *queue, the stationary waiting time $W^{(\pi)}$ in the π-thinning LTS satisfies*

$$\lim_{\pi \to 0} \pi^{-1} \mathbf{P}^0(W^{(\pi)} > x) = \mathbf{E}^0[H_-(S-x)], \tag{15.10}$$

provided that $\mathbf{E}^0[W^{(\pi)}] < \infty$, where $H_-(u) = \sum_{i=1}^{\infty} \mathbf{P}^0(T_1 + \cdots + T_i < u)$ is the left-continuous version of the (generalized) distribution function of the first factorial moment measure of the Palm arrival process $(N, \mathbf{P}^0)$. Moreover,

$$\lim_{\pi \to 0} \pi^{-1} \mathbf{E}^0[W^{(\pi)}[= \mathbf{E}^0\left[\int_0^S H_-(u)du\right], \tag{15.11}$$

provided that $\mathbf{E}^0[(W^{(\pi)})^2] < \infty$.

Remark 15.14. The proof of Theorem 15.13 requires some nontrivial calculation, and it is not clear how to justify the following extension: Under what conditions do we have

$$\lim_{\pi \to 0} \pi^{-1} \mathbf{E}^0[f(W^{(\pi)})] = \mathbf{E}^0\left[\int_0^S f(s-u)dH_-(u)\right] \tag{15.12}$$

when $f(x) \leq C + Dx^{\beta}$ for some $\beta > 1$?

In the special case of Cox/GI/1 queues, the function H_- can be computed. By *Cox* arrivals, we mean that customers arrive according to a doubly stochastic Poisson process. We suppose that its intensity process $\{\lambda(t)\}$ is stationary and that the (expected) arrival intensity $\lambda = \mathbf{E}[\lambda(0)]$ is positive and finite. Then

$$dH_-(t) = \frac{\mathbf{E}[\lambda(t)\lambda(0)]}{\lambda} dt,$$

provided the covariance $\mathbf{E}[\lambda(t)\lambda(0)]$ is finite. For queues with periodic Poisson input, formulas (15.10) and (15.11) were derived in [27]. For Markov-modulated queues with i.i.d. PH-distributed service times, these LTA were found in [25] and, for MM/G/1 (with non-independent service times), in [16] (see also Section 15.5.2).

It follows from Theorem 15.13 that, under the π-thinning LTS, waiting-time limits reflect any local clustering or clumping behavior of the arrival process, whereas this is not the case for the first order approximation of workload. We illustrate this following an idea of Sigman [61] (see also Daley and Rolski [29], Kroese and Schmidt [49]). Let a single-server queue under general light-traffic conditions have interarrival times $\{T_n^{(\xi)}\}$ and service times $\{S_n^{(\xi)}\}$ fulfilling:

$(C1)$ For each ξ, let $\mathbf{E}^0[T^{(\xi)}] = 1/\lambda^{(\xi)}$, $\mathbf{E}^0[S^{(\xi)}] = 1/\mu^{(\xi)}$ and $\rho^{(\xi)} = \lambda^{(\xi)}/\mu^{(\xi)}$ such

that $0 < \lambda^{(\xi)} < \infty$, $0 < \mu^{(\xi)} < \infty$ and $0 < \rho^{(\xi)} < 1$.
$(C2)$ $T^{(\xi)} \to \infty$ as $\xi \to \infty$.
$(C3)$ $S^{(\xi)} \xrightarrow{\mathfrak{D}} S$ as $\xi \to \infty$, where S has mean $1/\mu > 0$ and $\mu^{(\xi)} \to \mu$.
$(C4)$ For some $\xi_0 > 0$, the generic sojourn time $W^{(\xi)} + S^{(\xi)}$ is dominated in distribution by $W^{(\xi_0)} + S^{(\xi_0)}$ for each $\xi \geq \xi_0$.

Theorem 15.15. *Consider the* G/G/1 *queue in an LTS satisfying conditions* $(C1)$ *to* $(C4)$. *Then, for the workload* $V^{(\xi)}$ *and for any measurable function* $f: \mathbb{R}_+ \to \mathbb{R}_+$ *such that* $\mathbf{E}[(V^{(\xi)})^\beta] < \infty$ *and* $f(x) \leq C + Dx^\beta$ *for some* $\beta > 1$, *we have*

$$\lim_{\xi \to \infty} \left(\lambda^{(\xi)}\right)^{-1} \mathbf{E}[f(V^{(\xi)})] = \int_0^\infty \mathbf{E}^0\left[\int_0^S f(S-t)dt\right]. \tag{15.13}$$

The *proof* is based on Campbell's formula (see Section 15.2.2):

$$\mathbf{E}[f(V^{(\xi)})] = \lambda^{(\xi)}\left(\int_0^\infty f(x)(1 - F^{(\xi)}_{W+S}(x))ds - \int_0^\infty f(x)(1 - F^{(\xi)}_W(x))dx\right),$$

where $F^{(\xi)}_{W+S}$ and $F^{(\xi)}_W$ are the (Palm) stationary distribution functions of $W^{(\xi)} + S^{(\xi)}$ and $W^{(\xi)}$, respectively. From $(C2)$ and $(C3)$ we have $W^{(\xi)} \xrightarrow{\mathfrak{D}} 0$ and $W^{(\xi)} + S^{(\xi)} \xrightarrow{\mathfrak{D}} S$, which completes the proof in view of $(C4)$.

15.4.3. Multiserver queues

Several approaches have been investigated in order to establish approximation formulas for stationary characteristics of multiserver queues. Most of the earlier papers assume that the arrival epochs form a homogeneous Poisson point process; see Boxma, Cohen and Huffels [18] and Burman and Smith [24] and also related papers by Blanc [10], Wang and Wolff [66], Yechiali [70]. In Daley and Rolski [28], light-traffic approximations for waiting-time characteristics are given for multiserver queues with independent, but not necessarily exponentially distributed interarrival times (i.e., the arrival epochs form a renewal point process which is not constrained to be Poisson). Their proof of the following theorem is based on a *multicustomer* recurrence equation (analogous to that considered in Section 15.4.1) and a multidimensional version of the Abelian Lemma (see Lemma 15.11).

Theorem 15.16. *Consider a* GI/GI/c *queue for which the interarrival time d.f.* A *is in* $S_\alpha(C_A)$ *for some positive finite* α *and* C_A. *Let* S *be the generic service time and* $g: \mathbb{R}_+ \to \mathbb{R}_+$ *a monotone nondecreasing function with* $g(0) = 0$. *Furthermore, let* $j \in \{1, \ldots, c\}$ *and let* W_{c+1-j} *be the* $(c+1-j)$*th component of the Kiefer-Wolfowitz vector of waiting times.*
(i) *Approaching light-traffic conditions by* γ*-dilation we have*

$$\lim_{\gamma \to \infty} \gamma^{\alpha c} \mathbf{E}^0\left[g\left(W^{(\gamma)}_{c+1-j}\right)\right] = (\alpha C_A)^j C^g_j(\alpha) \quad \textit{and} \tag{15.14}$$

$$C^g_j(\alpha) = \int_{\mathbb{R}^j_+} \cdots \int x_1^{\alpha-1} \cdots x_j^{\alpha-1} \mathbf{E}^0\left[g\left(\min_{1 \leq i \leq j}\{(S_i - x_{[i,j]})_+\}\right)\right] dx_1 \ldots dx_j, \tag{15.15}$$

provided that for some $\gamma_0 \in (0, \infty)$,

$$\int \cdots \int_{\mathbb{R}^j_+} (x_1 \ldots x_j)^{\alpha - 1} \mathbf{E}^0\left[g\left(W^{(\gamma_0)}_{c+1-j} + S_1 - x_{[1,j]}\right)_+\right] dx_1 \ldots dx_j < \infty, \tag{15.16}$$

where $x_{[i,j]} = x_i + x_{i+1} + \ldots + x_j$.

(*ii*) *Approaching light-traffic conditions by* π*-thinning*

$$\lim_{\pi \to 0} \pi^{-1} \int \cdots \int_{\mathbb{R}^j_+} \mathbf{E}^0\left[g\left(W^{(\pi)}_{c+1-j}\right)\right]$$

$$= \int \cdots \int_{\mathbb{R}^j_+} \mathbf{E}^0\left[g\left(\min_{1 \le i \le j}\{(S_i - x_{[i,j]})_+\}\right)\right] dH(x_1)\cdots dH(x_j) \tag{15.17}$$

provided that, for some positive π',

$$\int \cdots \int_{\mathbb{R}^j_+} \mathbf{E}^0\left[g\left(\min_{1 \le i \le j}\{(W^{(\pi')}_{c+1-j} + S_i - x_{[i,j]})_+\}\right)\right] dH(x_1)\cdots dH(x_j) < \infty,$$

where $H = \sum_{i=1}^{\infty} A^{*i}$ *is the (nondelayed) renewal function.*

Moreover, it is proved in [28] that for a GI/GI/c queue, the waiting times $W^{(\xi)}$ and the r.v.'s $\min_{0 \le i \le c-1}\{(S_i^{(\xi)} - T^{(\xi)}_{[i,c-1]})_+\}$, where $T^{(\xi)}_{[i,j]} = T_i^{(\xi)} + T_{i+1}^{(\xi)} + \cdots + T_j^{(\xi)}$, are ACE both for γ-dilation and π-thinning LTS's.

Remark 15.17. For multiserver queues it would be interesting to have LTA similar to those developed in Asmussen [2] for GI/GI/1. Notice, however, that we cannot obtain this directly from [2] because, for multiserver queues, we do not have a representation of waiting times via ladder-height distributions. Nothing is known for dependent interarrival times besides the recent results of Błaszczyszyn et al. [14] who worked out multiserver queues with Markov-modulated input.

15.4.4. Queues in series

We now consider a series of d single-server queues GI/GI/1→···→GI/1, such that, at the first queue, arrivals occur at the epochs of a recurrent point process; specifically, the interarrival times $\{T_n: n = 0, \pm 1, \ldots\}$ are assumed to constitute a renewal process, which is independent of the service times $\{\boldsymbol{S}_n\} \equiv \{(S_{i,n}, i = 1, \ldots, d): n = 0, \pm 1, \ldots\}$. Moreover, $\{\boldsymbol{S}_n\}$ is assumed to be a sequence of i.i.d. nonnegative random vectors. Here $S_{i,n}$ denotes the service time of the nth customer at the ith station. Customers proceed from one station to the next and are served at each station in order of arrival. Denote by $W_{i,n}$ the waiting time of the nth customer at the ith station. The investigation of tandem queues with Poisson arrivals in light traffic was originated in Wolff [68] (see also [37]) in connection with studies of the optimal order of servers for tandem queues. Another related paper is Yamazaki and Ito [69]. The characteristics of interest were the waiting time at a given station and the total waiting time (response time) of a customer in the system. More general results in this field appear in Błaszczyszyn and Rolski [15]. They proved the following approximations for the vector of stationary waiting times at successive nodes $\boldsymbol{S}$ denotes the generic service time vector for an arrival and its independent copy by $\widehat{\boldsymbol{S}}$.

Theorem 15.18. (*i*) *In* γ*-dilated* GI/GI/1→···→GI/1 *queueing systems for which the interarrival time d.f. is in* $\boldsymbol{S}_\alpha(C_A)$ *for some* $0 < \alpha < \infty$, *we have*

$$\lim_{\gamma\to\infty}\gamma^{\alpha}\mathbf{E}^0[g(W_1^{(\gamma)},\ldots,W_d^{(\gamma)})]$$

$$=C_A\sum_{i=1}^{d}\mathbf{E}^0\left[\int_0^{r_i}(p_i-t)^{\alpha}g(0,\ldots,0,dt,r_{i+1},\ldots,r_d)\right],$$

where $g:\mathbb{R}^d\to\mathbb{R}$ is a coordinatewise monotone, nondecreasing function for which $g(0,\ldots,0)=0$, provided that for some $\gamma_0>0$ and each $i=1,\ldots,d$,

$$\mathbf{E}^0\left[\int_0^{\bar{r}_i}(\bar{p}_i-t)^{\alpha}g(0,\ldots,0,dt,\bar{r}_{i+1},\ldots,\bar{r}_d)\right]<\infty,$$

where $\bar{p}_i=(\bar{P}_i)_+$, $\bar{r}_i=(\bar{R}_i)_+$ with

$$\bar{P}_i=\bar{P}_i^{(\gamma_0)}=W_1^{(\gamma_0)}+S_1+\sum_{k=2}^{i}(W_k^{(\gamma_0)}+S_k-\widehat{S}_{k-1}),$$

$$\bar{R}_i=\bar{R}_i^{(\gamma_0)}=\begin{cases}\bar{P}_1 & (i=1),\\ \min_{2\le j\le i}\left\{\sum_{k=j}^{i}(W_k^{(\gamma_0)}+S_k-\widehat{S}_{k-1})\right\} & (i=2,\ldots,d),\end{cases}$$

and the p_i, r_i are analogously defined with $\gamma_0=0$ (and hence $W_k^{(\gamma_0)}=0$).

(*ii*) In a GI/GI/1→···→GI/1 *queueing system in which the arrival process is subject to π-thinning, we have, for any coordinatewise non-decreasing function g: $\mathbb{R}_+^d\to\mathbb{R}$ with $g(0,\ldots,0)=0$,*

$$\lim_{\pi\to 0}\pi^{-1}\mathbf{E}^0\left[g\left(W_1^{(\pi)},\ldots,W_d^{(\pi)}\right)\right]$$

$$=\sum_{i=1}^{d}\mathbf{E}^0\left[\int_0^{r_i}H(p_i-t)g(0,\ldots,0,dt,r_{i+1},\ldots,r_d)\right],$$

provided that for some π' in $(0,1]$ and each $i=1,\ldots,d$

$$\mathbf{E}^0\left[\int_0^{\bar{r}_i}H(\bar{p}_i-t)g(0,\ldots,0,dt,\bar{r}_{i+1},\ldots,\bar{r}_d)\right]<\infty,$$

where $H=\sum_{i=1}^{\infty}A^{*i}$ is the renewal function, and the $\bar{p}_i$, $\bar{r}_i$, p_i, r_i are defined as in part (i) with $W_i^{(\pi')}$ replacing $W_i^{(\gamma_0)}$.

Again, the proof technique is the recurrence-equation method explained in Section 15.4.1. The same method, with more than one interarrival time in the recurrence equation, allows the study of the departure process from GI/GI/1→··· →GI/1. An interesting feature of the sequence of interdeparture times is that this sequence, both in γ-dilation and π-thinning LTS's, is *asymptotically one dependent* (see [15] for definition).

Remark 15.19. It would be interesting to work out queues in series with Markov-modulated input.

15.5. HIGHER ORDER APPROXIMATIONS

In this section we present more accurate LTA, that is, approximations consisting of more terms. For Poisson-driven systems, some results are presented in Section 15.3.3. In the next section, we discuss a general technique allowing higher order LTA. Another approach is studied in Borovkov [17], where formulas for the first two coefficients of the Taylor expansion of γ-dilated systems, are given explicitly.

15.5.1. Factorial moment expansion

For a realization $N(\mu)=\mu$ of an m.p.p. and for $k\in\mathbb{N}$, define the point measure $N^{(k)}$ on $(\mathbb{R}\times\mathbb{K})^k$ by

$$N^{(k)}=\sum\epsilon_{(t_{i_1},m_{i_1},\ldots,t_{i_k},m_{i_k})}$$

where the summation extends over all k-tuples of atoms (t_i,m_i) of μ such that the points t_i are distinct. Moreover, we adopt the following notation for $(\mathbb{R}\times\mathbb{K})^k$-valued vectors: $(\boldsymbol{x},\boldsymbol{s})=(x_1,s_1,\ldots,x_k,s_k)$.

Definition 15.20. The factorial moment measure $M^{(k)}$ of order k of $(N,\mathbf{P})$ is given by

$$M^{(k)}(d(\boldsymbol{x},\boldsymbol{s}))=\mathbf{E}\Big[N^{(k)}(d(\boldsymbol{x},\boldsymbol{s}))\Big],\quad \text{for } k=1,2,\ldots.$$

We now define the notion of higher order Palm distributions of an m.p.p. $(N,\mathbf{P})$ which can be interpreted as conditional distributions of $(N,\mathbf{P})$, given that $(N,\mathbf{P})$ has points (and marks) at finitely many specified locations. We, first, introduce certain auxiliary measures on $(\mathbb{R}\times\mathbb{K})^k\times\mathcal{M}$.

Definition 15.21. For each k, the measure $C^{(k)}$, given by

$$C^{(k)}(D\times\Gamma)=\mathbf{E}\left[\int_D \mathbb{1}\left(N-\sum_{i=1}^{k}\epsilon_{(x_i,s_i)}\in\Gamma\right)N^{(k)}(d(\boldsymbol{x},\boldsymbol{s}))\right],\tag{15.18}$$

where $D\in\mathcal{B}((\mathbb{R}\times\mathbb{K})^k)$ and $\Gamma\in\mathcal{B}(\mathcal{M})$, is said to be the k-fold (reduced) Campbell measure of $(N,\mathbf{P})$.

Note that the Campbell measure $C^{(k)}$ can be seen as a refinement of the factorial moment measure $M^{(k)}$ because $C^{(k)}(D\times\mathcal{M})=M^{(k)}(D)$.

Definition 15.22. The reduced k-fold Palm distributions $\mathbf{P}^k_{\boldsymbol{x},\boldsymbol{s}}$ of $(N,\mathbf{P})$ (parametrized by $(\boldsymbol{x},\boldsymbol{s})$) are given by the kernel $\mathbf{P}^k:(\mathbb{R}\times\mathbb{K})^k\times\mathcal{B}(\mathcal{M})\to\mathbb{R}_+$, defined by

$$\mathbf{P}^k_{\boldsymbol{x},\boldsymbol{s}}(\Gamma)=\frac{dC^{(k)}(\cdot\times\Gamma)}{dC^{(k)}(\cdot\times\mathcal{M})}(\boldsymbol{x},\boldsymbol{s}).$$

In principle, the construction of higher-order Palm distributions is simple: For each $\Gamma\in\mathcal{B}(\mathcal{M})$ we have $C^{(k)}(\cdot\times\Gamma)\ll C^{(k)}(\cdot\times\mathcal{M})$, i.e., there exists a Radon-Nikodym derivative $\mathbf{P}^k_{\boldsymbol{x},\boldsymbol{s}}(\Gamma)$. However, one must then ensure that $\mathbf{P}^k_{\boldsymbol{x},\boldsymbol{s}}$ is a probability measure for $M^{(k)}$-almost every $(\boldsymbol{x},\boldsymbol{s})$; see e.g., Kallenberg [41] or König and Schmidt [46] for details of this idea which goes back to Ryll-Nardzewski [58].

Various performance characteristics in queueing and other stochastic models can be obtained by taking the expectation $\mathbf{E}[\psi(N)]$ of a functional ψ of some m.p.p. $(N,\mathbf{P})$. We assume that $\psi:\mathcal{M}\to\mathbb{R}$ is measurable and that $\psi(N)$ is integrable with respect to $\mathbf{P}$. Baccelli and Brémaud [6], referring to a result of Reiman and Simon [53], proposed one-term LTA of such characteristics based on Campbell's formula for general stationary m.p.p.'s (cf. Section 15.2.2). Brémaud

[21] also suggested application of k-fold Palm distributions to higher order LTA. This program was carried out by Błaszczyszyn [12], who developed a general theory of *factorial moment expansion* (FME) for such functionals, which proves to be useful in light-traffic analysis of queues, particularly in deriving higher order approximations (see [8, 13, 14, 16, 33, 49]). The theory shows that the expectation $\mathbf{E}[\psi(N)]$ can be expressed as a sum of certain kernels (depending on the form of the functional ψ) integrated with respect to factorial moment measures of $(N, \mathbf{P})$ with a remainder, which is the integral of a functional with respect to a higher order Campbell measure.

For any functional ψ on $\mathcal{M}$, define the *k-th left FME kernel* $\psi^{(k)}$ as a functional on $(\mathbb{R}\times\mathbb{K})^k\times\mathcal{M}$, given by the k-fold composition

$$\psi^{(1)}_{x,s}(\mu)=\psi_{x,s}(\mu)=\psi(\mu\,|_x+\epsilon_{(x,s)})-\psi(\mu\,|_x) \tag{15.19}$$

and

$$\psi^{(k)}_{\mathbf{x},\mathbf{s}}(\mu)=\Big(\ldots(\psi_{x_1,s_1})_{x_2,s_2}\cdots\Big)_{x_k,s_k}(\mu) \quad k>1, \tag{15.20}$$

where $\mu\,|_x$ is the restriction of $\mu\in\mathcal{M}$ to the subset $(-\infty,x)\times\mathbb{K}$, i.e.,

$$\mu\,|_x(D)=\mu(D\cap(-\infty,x)\times\mathbb{K});\quad D\in\mathcal{B}(\mathbb{R}\times\mathbb{K}).$$

Similarly, the *k-th right FME kernel* $\psi_{(k)}$ is the functional on $(\mathbb{R}\times\mathbb{K})^k\times\mathcal{M}$ given by

$$\psi^{x,s}_{(1)}(\mu)=\psi^{x,s}(\mu)=\psi(\mu\,|^x+\epsilon_{(x,s)})-\psi(\mu\,|^x) \tag{15.21}$$

and

$$\psi^{\mathbf{x},\mathbf{s}}_{(k)}(\mu)=\Big(\ldots(\psi^{x_1,s_1})^{x_2,s_2}\ldots\Big)^{x_k,s_k}(\mu) \quad k>1, \tag{15.22}$$

where $\mu\,|^x$ is given by

$$\mu\,|^x(D)=\mu(D\cap(x,\infty)\times\mathbb{K});\quad D\in\mathcal{B}(\mathbb{R}\times\mathbb{K}).$$

Note that the FME kernels $\psi^{(k)}_{\mathbf{x},\mathbf{s}}$ and $\psi^{\mathbf{x},\mathbf{s}}_{(k)}$ can be written in the form

$$\psi^{(k)}_{\mathbf{x},\mathbf{s}}(\mu)=\begin{cases}\displaystyle\sum_{j=0}^{k}(-1)^{k-j}\sum_{\pi\in\{\binom{k}{j}\}}\psi(\mu\,|_{x_k}+\sum_{i\in\pi}\epsilon_{(x_i,s_i)}) & \text{if } x_k<\ldots<x_1,\\ 0 & \text{otherwise,}\end{cases} \tag{15.23}$$

and

$$\psi^{\mathbf{x},\mathbf{s}}_{(k)}(\mu)=\begin{cases}\displaystyle\sum_{j=0}^{k}(-1)^{k-j}\sum_{\pi\in\{\binom{k}{j}\}}\psi(\mu\,|^{x_k}+\sum_{i\in\pi}\epsilon_{(x_i,s_i)}) & \text{if } x_1<\ldots<x_k,\\ 0 & \text{otherwise,}\end{cases}$$

where $\left\{\binom{k}{j}\right\}$ denotes the collection of all subsets of $\{1,\ldots,k\}$ consisting of j elements. The proof of (15.23) proceeds by induction with respect to k. Furthermore, observe that, for the null measure $\mu=o\in\mathcal{M}$, both the left and right kernels coincide, i.e.,

$$\psi^{(k)}_{x_1,s_1,\ldots,x_k,s_k}(o)=\psi^{x_k,s_k,\ldots,x_1,s_1}_{(k)}(o),$$

and we call these functions *null kernels.*

For univariate point processes, $\psi^{(k)}_{\mathbf{x}}$ was introduced in Błaszczyszyn [12].

However, the function $\psi^{(k)}_{x_1,\ldots,x_k}(o)$ appeared for the first time in Reiman and Simon [53], which mainly investigated the case of univariate Poisson point processes. For independently marked Poisson processes, the notion of $\psi^{(k)}_{\boldsymbol{x},\boldsymbol{s}}$ has first been used in Kroese and Schmidt [49].

The functional ψ on $\mathcal{M}$ is said to be *left continuous at infinities* if

$$\lim_{x\to-\infty}\psi(\mu\,|\,_x+\nu)=\psi(\nu) \quad\text{and}\quad \lim_{x\to\infty}\psi(\mu\,|\,_x)=\psi(\mu)$$

and *right-continuous at infinities* if

$$\lim_{x\to-\infty}\psi(\mu\,|\,^x)=\psi(\mu) \quad\text{and}\quad \lim_{x\to\infty}\psi(\mu\,|\,^x+\nu)=\psi(\nu)$$

for every $\mu,\nu\in\mathcal{M}$, such that $\nu(\mathbb{R}\times\mathbb{K})<\infty$.

We now present an extension of Theorem 3.2 in Błaszczyszyn [12] to the case of m.p.p.'s (see also [14, 16, 49]).

Theorem 15.23. *Let ψ be a left-continuous at infinities functional of $(N,\mathbf{P})$. If*

$$\int\limits_{(\mathbb{R}\times\mathbb{K})^i}\int\limits_{\mathcal{M}}\left|\psi^{(i)}_{\boldsymbol{x},\boldsymbol{s}}(\mu)\right|\mathbf{P}^i_{\boldsymbol{x},\boldsymbol{s}}(d\mu)M^{(i)}(d(\boldsymbol{x},\boldsymbol{s}))<\infty$$

for all $i=1,\ldots,n$, then

$$\mathbf{E}[\psi(N)]=\psi(o)+\sum_{i=1}^{n-1}\int\limits_{(\mathbb{R}\times\mathbb{K})^i}\psi^{(i)}_{\boldsymbol{x},\boldsymbol{s}}(o)M^{(i)}(d(\boldsymbol{x},\boldsymbol{s})) \tag{15.24}$$

$$+\int\limits_{(\mathbb{R}\times\mathbb{K})^n}\int\limits_{\mathcal{M}}\psi^{(n)}_{\boldsymbol{x},\boldsymbol{s}}(\mu)\mathbf{P}^n_{\boldsymbol{x},\boldsymbol{s}}(d\mu)M^{(n)}(d(\boldsymbol{x},\boldsymbol{s})).$$

If ψ is a right-continuous at infinities functional of $(N,\mathbf{P})$ and

$$\int\limits_{(\mathbb{R}\times\mathbb{K})^i}\int\limits_{\mathcal{M}}\left|\psi^{\boldsymbol{x},\boldsymbol{s}}_{(i)}(\mu)\right|\mathbf{P}^i_{\boldsymbol{x},\boldsymbol{s}}(d\mu)M^{(i)}(d(\boldsymbol{x},\boldsymbol{s}))<\infty$$

for all $i=1,\ldots,n$, then

$$\mathbf{E}[\psi(N)]=\psi(o)+\sum_{i=1}^{n-1}\int\limits_{(\mathbb{R}\times\mathbb{K})^i}\psi^{\boldsymbol{x},\boldsymbol{s}}_{(i)}(o)M^{(i)}(d(\boldsymbol{x},\boldsymbol{s})) \tag{15.25}$$

$$+\int\limits_{(\mathbb{R}\times\mathbb{K})^n}\int\limits_{\mathcal{M}}\psi^{\boldsymbol{x},\boldsymbol{s}}_{(n)}(\mu)\mathbf{P}^n_{\boldsymbol{x},\boldsymbol{s}}(d\mu)M^{(n)}(d(\boldsymbol{x},\boldsymbol{s})).$$

The factorial moment expansions (15.24) and (15.25) yield useful approximations of queuing characteristics under quite general light-traffic conditions (see Sections 15.5.2 and 15.5.3). In particular, this is the case when the factorial moment measures $M^{(k)}$ of $(N,\mathbf{P})$ have a relatively uncomplicated form. For the examples of point processes already considered in Section 15.2.3, the following holds.

Example 15.24. [Poisson Point Process] For a stationary Poisson process on $\mathbb{R}$ with intensity λ,

$$M^{(k)}(d(x_1,\ldots,x_k))=\lambda^k dx_1\ldots dx_k. \tag{15.26}$$

Furthermore, for a (marked) Poisson process on $\mathbb{R}\times\mathbb{K}$ with intensity measure $\lambda(x)B_x(ds)dx$,

$$M^{(k)}(d(\boldsymbol{x},\boldsymbol{s})) = \lambda(x_1)\ldots\lambda(x_k)B_{x_1}(ds_1)\ldots B_{x_k}(ds_k)dx_1\ldots dx_k. \qquad (15.27)$$

Example 15.25. [Markov Modulated M.P.P.] The kth factorial moment measure $M^{(k)}$ of a Markov-modulated m.p.p. is given by

$$M^{(k)}(d(\boldsymbol{x},\boldsymbol{s})) = \boldsymbol{\pi}S(ds_{(k)})e^{\boldsymbol{\Lambda}(x_{(k-1)}-x_{(k)})}S(ds_{(k-1)})e^{\boldsymbol{\Lambda}(x_{(k-2)}-x_{(k-1)})}\ldots$$
$$S(ds_{(2)})e^{\boldsymbol{\Lambda}(x_{(1)}-x_{(2)})}S(ds_{(1)})\boldsymbol{e}dx_1dx_2\ldots dx_k \qquad (15.28)$$

where

$$S(K) = (\lambda_i B_i(K)\delta_{ij})$$

is the diagonal matrix with ith diagonal element $\lambda_i B_i(K)$ and, by $(x_{(1)},\ldots,x_{(k)})$, we mean a permutation of $(x_1,\ldots,x_k)$ such that the components $x_{(k)}, x_{(k-1)},\ldots, x_{(1)}$ are arranged in ascending order. $(s_{(1)},\ldots,s_{(k)})$ denotes the corresponding permutation of $(s_1,\ldots,s_k)$, and $\boldsymbol{e}$ is a column vector with all entries equal to 1. By $e^{\boldsymbol{\Lambda}t} = \sum_{k=0}^{\infty}(\boldsymbol{\Lambda}t)^k/k!$, we denote the matrix exponential function.

Example 15.26. [Renewal Process] The kth factorial moment measure has the form

$$M^{(k)}(d(x_1,\ldots,x_k)) = H(dx_{(k-1)}-x_{(k)})\ldots H(dx_{(1)}-x_{(2)})\lambda dx_{(k)}, \qquad (15.29)$$

where $H(\cdot)$ is the renewal function $H(u) = \sum_{n=1}^{\infty}F^{*n}(u)$.

When ψ is driven by a family of m.p.p.'s $(N^{(\xi)},\mathbf{P})$, then (15.24) and (15.25) yield expansions of $\mathbf{E}[\psi(N^{(\xi)})]$ in ξ. Note that the leading terms of this expansion depend on ξ through $(M^{(\xi)})^{(i)}$ only and the last term also through $(\mathbf{P}^{(\xi)})^n_{\boldsymbol{x},\boldsymbol{s}}$. The main idea of the FME-approach to light-traffic approximations is to show that the remainder term in (15.24) and (15.25), respectively, has a lesser order of magnitude (as $\xi \to \xi_0$) than the other terms, and hence can be neglected.

An illustrative example is the Markov-modulated m.p.p. Following [14] and [16], we introduce the following parameterization according to the notation used in Examples 15.2 and 15.25. Suppose that $\lambda_i = \lambda\lambda_i'$ and that

$$\sum_{i=1}^{\ell}\pi_i\lambda_i' = 1.$$

Thus the (mean) arrival intensity is λ. Fix the λ_i''s, B_i's and $\boldsymbol{\Lambda}$, and let $\lambda \to 0$. We write $\mathbf{P}^{\lambda}$, if the underlying Markov-modulated m.p.p. is specified by the $\lambda\lambda_i'$, B_i, and $\boldsymbol{\Lambda}$. From Theorem 15.23 and from (15.28), we get the representation

$$\mathbf{E}^{\lambda}[\psi(N)] = a_0 + \lambda a_1 + \cdots + \lambda^{n-1}a_{n-1} + \lambda^n a_n(\lambda)$$

where $a_0,\ldots,a_{n-1}$ are constants. Now the following results may be sought:

$$a_n(\lambda) = O(1), \text{ for } \lambda \to 0, \qquad (15.30)$$

and, in this case,

$$f(\lambda) = a_0 + \ldots + \lambda^{n-1}a_{n-1} + O(\lambda^n) \qquad (15.31)$$

or

$$a_n(\lambda) \to a_n, \text{ for } \lambda \to 0, \qquad (15.32)$$

and, in this case,

$$f(\lambda) = a_0 + \ldots + \lambda^n a_n + o(\lambda^n). \qquad (15.33)$$

In the case of (15.31), we say that we have a *weak approximation of order n* and, in the case of (15.33), a *strong approximation of order n*.

Remark 15.27. Although we usually expect a_n in (15.32) to be just $\psi^{(n)}_{x,s}(o)$, integrated with respect to the nth factorial moment measure of $(N, \mathbf{P}^\lambda)$, verifying (15.30) and (15.32) may be problematic, and it would be desirable to find convenient conditions. To this end, one can choose left or right kernels as the integrand of the remainder integral.

It should be noted that the above parameterization with respect to λ coincides, in general, with π-thinning for which we also obtain a polynomial form in π of the expansion. This, however, is not true for general LTSs, as evidenced by the γ-dilation case. Borovkov [17] studies conditions for the Taylor expansion in the LTS via γ-dilation.

15.5.2. Risk processes

We present here an approximation of the ruin function for risk processes in a Markov-modulated environment, because risk theory and queueing theory are closely related and, in particular, analyzing the ruin function of a risk process is equivalent to analyzing the distribution function of waiting time in a single-server queue (with time-reversed input; see also Sections 15.3.2 and 15.3.3). Let $u > 0$ be fixed and let

$$\psi_u = 1\left(\sup_{v>0} R(v) > u\right), \tag{15.34}$$

where the claim surplus process $R(v) = R(u, \mu)$ aggregates claim sizes in the interval $(0, v]$ and moves between jumps (at arrival epochs) according to a certain premium rate function $p(x)$ which depends on the current reserve. That is

$$R(t, \mu) = \int_{\mathbb{R}_+} m\mu((0,t] \times dm) - \int_0^t p(u - R(s, \mu))ds.$$

Then $\phi(u, \lambda) = \mathbf{E}^\lambda[\psi_u(N)]$ is the *ruin function* (of the initial reserve u). Suppose that the premium rate function is uniformly bounded away from 0; that is, $p(x) \geq \delta > 0$. For the case that the claims arrive according to a Markov-modulated m.m.p., we now present a strong approximation of order 2 of the ruin function which has been derived by Błaszczyszyn and Rolski [16].

Theorem 15.28. *If* $\int_0^\infty x^3 B_i(dx) < \infty$ *for all* $i = 1, \ldots, \ell$, *then,*

$$\phi(u, \lambda) = \lambda \sum_{i=1}^{\ell} \pi_i \lambda'_i \int_u^\infty \frac{1 - B_i(v)}{p(v)} dv$$

$$+ \lambda^2 \sum_{j_1, j_2 = 1}^{\ell} \pi_{j_2} \lambda'_{j_2} \lambda'_{j_1} \left\{ 1\,(j_1 = j_2) \int_0^\infty \int_{(u-t)_+}^\infty \frac{1 - B_{j_2}(t)}{p(s+t)} \, \frac{1 - B_{j_1}(s)}{p(s)} ds dt \right.$$

$$+ \int_0^\infty \int_{(u-t)_+}^\infty \int_{(u-t)_+}^s \frac{1 - B_{j_2}(t)}{p(v+t)} \, \frac{1 - B_{j_1}(s)}{p(s)} \, \frac{1}{p(v)} \left(\Lambda e^{\Lambda \int_v^s \frac{dr}{p(r)}} \right)_{j_2 j_1} dv\, ds\, dt$$

$$\left. - \int_u^\infty \int_0^\infty \frac{1 - B_{j_2}(t)}{p(t)} \, \frac{1 - B_{j_1}(s)}{p(s)} \left(e^{\Lambda \int_0^s \frac{dr}{p(r)}} \right)_{j_2 j_1} ds\, dt \right\} + o(\lambda^2).$$

The coefficients of this expansion are the appropriate integrals of first and second null FME-kernels with respect to first-order and second-order factorial moment measures. To justify the approximation, the second left FME-kernel has to be considered in the remainder.

Remark 15.29. Note that the results of this section, in view of [5], are also relevant to a dam with release rate depending on the present context x.

15.5.3. MM/G/c queues

Błaszczyszyn, Frey and Schmidt [14] derived Taylor-series expansions, with respect to the arrival intensity, for several characteristics of multi-server MM/G/c queues (such as the work load vector and the total work load), where the arrival epochs formed a stationary Markov-modulated process and the distribution of service times might depend on the actual state of the Markov-modulated environment.

Consider the parametrized family $(N, \mathbf{P}^\lambda)$ of Markov-modulated m.p.p.'s, as introduced in Section 15.5.1. For the dth component of the work load vector, the following result has been proved in [14].

Theorem 15.30. *Let $f: \mathbb{R}_+ \to \mathbb{R}_+$ be a nondecreasing function, such that $f(0) = 0$ and $f(x) \leq C + Dx^\alpha$ for some $C, D, \alpha \geq 0$. If $\int_0^\infty s^{\alpha + max(k,2)} B_i(ds) < \infty$ for each $i = 1, \ldots, \ell$, then for the dth component of the stationary work load vector in the* MM/G/c *queue we have*

$$\mathbf{E}^\lambda[f(V_d(N))]$$
$$= \sum_{i=c+1-d}^{k} \lambda^i(-1)^{i+d-c-1} \binom{i-1}{i+d-c-1} \int_{(\mathbb{R}_- \times \mathbb{R}_+)^i} f(\xi_{(i)}^{x,s}) M^{(i)}(d(x,s)) + o(\lambda^k)$$

for each $k \in \{c+1-d, \ldots, c\}$, $d \in \{1, \ldots, c\}$, where $\xi_{(i)}^{x,s}$ is given by

$$\xi_{(i)}^{x,s} = \min_{1 \leq i \leq k} (s_i + x_i)_+ \mathbf{1}(x_1 < \ldots < x_i < 0)$$

and $M^{(i)}$ is given by (15.28).

This is a strong approximation of order k (see (15.33)) which follows from (15.25), using right FME-kernels, since the null kernels of the work load vector are given by

$$V_{(i)}^{x,s}(o) = \left(\overbrace{0, \ldots, 0}^{c-i}, \xi_{(i)}^{x,s}, -\binom{i-1}{1} \xi_{(i)}^{x,s}, \ldots, (-1)^{i-2} \binom{i-1}{i-2} \xi_{(i)}^{x,s}, (-1)^{i-1} \xi_{(i)}^{x,s} \right).$$

The above approximation formula is then obtained by neglecting the remainder. Moreover, in [14], it has been shown that the integrals appearing in Theorem 15.30 can be analytically evaluated in particular cases, e.g., for PH-distributed service times. The following result pertains to the latter case.

Theorem 15.31. *Let the B_i be phase-type distributions with representations $(\boldsymbol{\alpha}_i, \boldsymbol{T}_i, d_i)$ for $i = 1, \ldots, l$. Then,*

$$\int (\xi_{(k)}^{x,s})^n M^{(k)}(d(x,s))$$
$$= (-1)^{n+k} n! \sum_{j_1, \ldots, j_k = 1}^{\ell} \pi_{j_i} \left(\prod_{i=1}^{k} \lambda_{j_i} \right) \tag{15.35}$$
$$\left(\left(\ldots (\alpha_{j_1} \left[(\boldsymbol{\Lambda} \oplus \boldsymbol{T}_{j_1})^{-1} \right]_{j_1 j_2} \otimes \boldsymbol{\alpha}_{j_2}) [(\boldsymbol{\Lambda} \oplus \boldsymbol{T}_{j_1} \oplus \boldsymbol{T}_{j_2})^{-1}]_{j_2 j_3} \right.\right.$$

$$\ldots \otimes \boldsymbol{\alpha}_{j_{k-1}}\Big)\left[\Big(\boldsymbol{\Lambda} \oplus \bigoplus_{i=1}^{k-1} \boldsymbol{T}_{j_i}\Big)^{-1}\right]_{j_{k-1}j_k} \otimes \boldsymbol{\alpha}_{j_k}\Big)\Big(\bigoplus_{i=1}^{k} \boldsymbol{T}_{j_i}\Big)^{-1-n} e_{j_1\ldots j_k},$$

where $e_{j_1\ldots j_k}$ *is the appropriate column vector of* 1*'s,* $\oplus$ *and* $\otimes$ *denote the Kronecker sum and product, respectively, and* $[\boldsymbol{\Lambda} \oplus \boldsymbol{A}]_{jj'}$ *is the* (jj')*th block of size* $\boldsymbol{A}$ *of the matrix* $\boldsymbol{\Lambda} \oplus \boldsymbol{A}$.

Remark 15.32. It would be useful to study other performance characteristics of the MM/G/c queue, such as the actual waiting time or the queue size.

15.6. ANALYTICITY

The validity of polynomial approximations depends on analyticity properties, i.e., whether the function of interest can be expressed as an infinite polynomial series, and, in queueing context, which part of the stability interval of the underlying queueing system is the region of convergence of this series expansion. On this region, an arbitrarily long part of the infinite series can be used in order to obtain sufficiently good approximations. An early idea on expanding probabilities of a Markov process is due to Beneš [9]. He expanded the ratio p_x/p_0 in a Taylor series with respect to Poisson arrival intensity λ, where $\{p_x: x \in \mathbb{E}\}$ is the stationary initial distribution of a Markov process describing a Markovian queueing system with blocking. The coefficients of the series are derived by sample-path and combinatorial arguments. Analyticity for a general class of functionals $\psi(N)$ of a stationary Poisson p.p. N, represented as functions of its intensity λ, is studied in Zazanis [71], where the following result is obtained through the likelihood-ratio method (see also Section 15.3.4).

Theorem 15.33. *Suppose that* $T: \mathcal{M} \to \mathbb{R}_+$ *is a bounded* $(\mathcal{F}_t)$*-stopping time and that* ψ *is an* $\mathcal{F}_T$*-measurable functional of the Poisson p.p.* $(N, \mathbf{P}^\lambda)$*, such that* $\mathbf{E}^\lambda[\psi(N)] < \infty$ *for each* $\lambda \in [0, \lambda_0)$ *and for some* $\lambda_0 \le \infty$. *Then* $f(\lambda) = \mathbf{E}^\lambda[\psi(N)]$ *is an analytic function on* $[0, \lambda_0)$.

Unfortunately, this result is of rather limited interest for light-traffic approximation of stationary queueing characteristics, since usually they are infinite-horizon functionals. A weaker condition appears in [71] for analyticity on intervals which are bounded away from 0. Another approach to analyticity of expected stationary waiting and sojourn times in GI/GI/1 queues is given in Hu [39]. A recursive method for calculating the coefficients of the series considered in Theorem 15.34 is given in Gong and Hu [34].

Theorem 15.34. *Assume that the interarrival time d.f. in the* GI/GI/1 *queue is an analytic function on* $\mathbb{R}_+$*, that* S *is a positive r.v. which has all moments finite, and that, moreover,* $\mathbf{E}[S^n] < n!C^n$ *for some constant* $C < \infty$ *and for all* $n \ge 1$. *Then, the expectations of stationary waiting and sojourn times in the queue with generic service time* θS *are analytic functions of* θ *at* $\theta = 0$.

For general functionals of π-thinned arrival processes, the FME method considered in Section 15.5 can be used. Namely, it follows from (15.24) that $\mathbf{E}[\psi(N^{(\pi)})]$ is an analytical function of π at $\pi = 0+$ with the power-series representation

$$\mathbf{E}[\psi(N^{(\pi)})] = \psi(o) + \sum_{n=1}^{\infty} \pi^n \int_{(\mathbb{R} \times \mathbb{K})^n} \psi^{(n)}_{\boldsymbol{x}, \boldsymbol{s}}(o) M^{(n)}(d(\boldsymbol{x}, \boldsymbol{s}))$$

which is convergent whenever

$$\int_{(\mathbb{R}\times\mathbb{K})^n} \int_{\mathcal{M}} \psi^{(n)}_{x,s}(\mu)\left(\mathrm{P}^{(\pi)}\right)^n_{x,s}(d\mu)M^{(n)}(d(x,s)) \to 0 \text{ for } n\to\infty.$$

An example of this approach appears in Błaszczyszyn, Frey and Schmidt [14] (see also Błaszczyszyn and Rolski [16]) where the following conditions for analyticity of the work load vector in multiserver queues have been derived.

Theorem 15.35. *The expected value of each component of the stationary work load vector in the* MM/G/c *queue is an analytic function of the arrival intensity* λ *at* $\lambda = 0+$, *and its power series is convergent whenever* $\lambda < 1/(\lambda_{max}\int_0^\infty sB(ds))$ *and*

$$\lim_{n\to\infty}\frac{(2\lambda)^{n-1}}{(n-1)!}\int_{\mathbb{R}^n}\left(\sum_{i=1}^{n}s_i\right)^{n+1}B(ds_1)\ldots B(ds_n)=0,$$

where $\lambda_{max} = \max_{1\leq i\leq \ell}\lambda_i$.

Remark 15.36. An open problem, which still puzzles, is to compare the analyticity conditions considered in [9, 14, 39, 71] and to interpret the coefficients given there.

15.7. DISCUSSION AND FURTHER RESEARCH PROBLEMS

Although an expansion (like Taylorian and others) is one of the well-known methods of approximation, the idea of expanding functionals of m.p.p.'s is recent, and there are not very many examples of its application to particular stochastic models. Besides the results, which are called, in this chapter, "of the first order", and which are good only for relatively low traffic, there exist results mainly for Poisson-driven or Markov-modulated queues. These results however, reported, e.g., in [14, 16, 49], seem to indicate an improvement in the quality of the approximation, and they show that the second or third order LTA are good for moderate traffics, too.

One of the areas for future research would be to consider other stochastic models driven by a Poisson or Markov-modulated input, and get for them explicit LTA of an arbitrary order. The only published results for more complex queueing networks are of Markovian networks, as [53]. Forthcoming papers in this area are [8, 13, 33], where Taylor-series expansions are derived for stationary state variables of Poisson-driven (max +)-linear systems and for finite horizon characteristics of risk processes. This leads to the evolution of FME kernels of appropriate functionals.

Another possible direction is to investigate π-thinning of other input processes. For systems driven by a renewal process, e.g., the only available LTA are the first order. In this case, factorial moment measures have to be determined. For both general systems and inputs, we expect that packets for symbolic computations as *Mathematica* or *Maple* are useful in order to obtain particular coefficients of the expansion. We wish also to point out that, besides the coefficients to be computed, from the mathematical point of view, an interesting problem is to show that the remainder is of a suitable order of accuracy. This might be, in our experience, quite a challenging task and would require individual tricks.

Although the factorial moment expansion theorem is valid for general point processes, some preliminary derivations show that it just gives polynomial LTA for the π-thinning LTS. No studies, except for the first order LTA and some special cases made in [17], have been undertaken for other LTSs.

Most of the results surveyed in this chapter deal with continuous performance characteristics such as work load and actual waiting time. It would be worthwhile to derive (using the presented techniques) LTA e.g., for loss probabilities in various models with blocking.

ACKNOWLEDGEMENTS

We thank Jewgeni Dshalalow, Benjamin Melamed and Gary Russell for their careful reading of the manuscript.

BIBLIOGRAPHY

[1] Asmussen, S., Ladder heights and the Markov-modulated M/G/1 queues, *Stoch. Proc. Appl.* **37** (1991), 313-326.

[2] Asmussen, S., Light traffic equivalence in single-server queues, *Ann. Appl Prob.* **2** (1992), 555-574.

[3] Asmussen, S., *Ruin Probabilities*, World Scientific Publishers, Singapore 1995/1996.

[4] Asmussen, S. and Bladt, M., Phase-type distributions and risk processes with state-dependent premiums, preprint R92-2022, University of Aalborg (1992).

[5] Asmussen, S. and Schock Petersen, S., Ruin probabilities expressed in terms of storage processes, *Adv. Appl. Prob.* **20** (1988), 913-916.

[6] Baccelli, F. and Brémaud, P., Virtual customers in sensitivity and light traffic analysis via Campbell's formula for point processes, *Adv. Appl. Prob.* **25** (1993), 221-224.

[7] Baccelli, F. and Brémaud, P., *Elements of Queueing Theory*, Springer-Verlag, Berlin 1994.

[8] Baccelli, F. and Schmidt, V., Taylor expansions for Poisson-driven (max, +)-linear systems, working paper (1994).

[9] Beneš, V., *Mathematical Theory of Connecting Networks and Telephone Traffic*, Academic Press, New York 1965.

[10] Blanc, J.P.C., The power-series algorithm applied to the shortest-queue model, *Oper. Res.* **40** (1992), 157-167.

[11] Bloomfield, P. and Cox, D.R., A low traffic approximation for queues, *J. Appl. Prob.* **9** (1972), 832-840.

[12] Błaszczyszyn, B., Factorial moment expansion for stochastic systems, *Stoch. Proc. Appl.* **45** (1995), (to appear).

[13] Błaszczyszyn, B., Frey, A., Rolski, T. and Schmidt, V., Taylor-series expansion for finite-horizon characteristics of Markov-modulated risk processes, working paper, (1994).

[14] Błaszczyszyn, B., Frey, A. and Schmidt, V., Light-traffic approximations for Markov-modulated multiserver queues, *Stoch. Mod.* **11** (1995), (to appear).

[15] Błaszczyszyn, B. and Rolski, T., Queues in series in light traffic, *Ann. Appl. Prob.* **3** (1993), 881-896.
[16] Błaszczyszyn, B. and Rolski, T., Expansions for Markov-modulated systems and approximations of ruin probability, *J. Appl. Prob.* **33** (1996), preprint 2236, INRIA.
[17] Borovkov, A.A., Asymptotic expansions for functionals of dilation of point processes, preprint, (1994).
[18] Boxma, O.J., Cohen, J.W. and Huffels, N., Approximations of the mean waiting time in an M/G/s queueing system, *Oper. Res.* **27** (1979), 1115-1127.
[19] Brandt, A., Franken, P. and Lisek, B., *Stationary Stochastic Models*, John Wiley, Chichester 1990.
[20] Brémaud, P., *Point Processes and Queues*, Springer-Verlag, New York 1981.
[21] Brémaud, P., Private communication 1992.
[22] Brémaud, P. and Gong, W.-B., Derivatives of likelihood ratios and smoothed perturbation analysis for the routing problem, *Trans. Simu.* **3** (1993), 134-161.
[23] Brémaud, P. and Vázquez-Abad, F.J., On the pathwise computation of derivatives with respect to the rate of a point process: The phantom RPA method, *Queueing Sys.* **10** (1992), 249-270.
[24] Burman, D.Y. and Smith, D.R., A light-traffic theorem for multiserver queues, *Math. Oper. Res.* **8** (1983), 15-25.
[25] Burman, D.Y. and Smith, D.R., An asymptotic analysis of queueing systems with Markov modulated arrivals, *Oper. Res.* **34** (1986), 105-119.
[26] Daley, D.J. and Rolski, T., A light-traffic approximation for a single-server queue, *Math. Oper. Res.* **9** (1984), 624-628.
[27] Daley, D.J. and Rolski, T., Light-traffic approximations in queues, *Math. Oper. Res.* **16** (1991), 57-71.
[28] Daley, D.J. and Rolski, T., Light-traffic approximations in many-server queues, *Adv. Appl. Prob.* **24** (1992), 202-218.
[29] Daley, D.J. and Rolski, T., Light-traffic approximations in general stationary single-server queues, *Stoch. Proc. Appl.* **49** (1994), 141-158.
[30] Daley, D.J. and Vere-Jones, D., *An Introduction to the Theory of Point Processes*, Springer-Verlag, New York 1988.
[31] Fleming, P.J., An approximate analysis of sojourn times in the M/G/1 queue with round-robin service discipline, *Bell Lab. Tech. J.* **63** (1984), 1521-1535.
[32] Franken, P., König, D., Arndt, U. and Schmidt, V., *Queues and Point Processes*, John Wiley, Chichester 1982.
[33] Frey, A. and Schmidt, V., Taylor-series expansion for multivariate characteristics of risk processes, preprint, University of Ulm (1994).
[34] Gong, W.B. and Hu, J.Q., The MacLaurin series for the GI/G/1 queue, *J. Appl. Prob.* **29** (1992), 176-184.
[35] Graham, A., *Kronecker Products and Matrix Calculus with Applications*, Ellis Horwood, Chichester 1981.
[36] Grandell, J., *Doubly Stochastic Poisson Processes*, Lecture Notes in Mathematics **529**, Springer-Verlag, Berlin 1976.
[37] Greenberg, B.S. and Wolff, R.W., Optimal order of servers for tandem queues in light traffic, *Mgt. Sci.* **34** (1988), 500-508.

[38] Harrison, J.M. and Resnick, S.I., The recurrence classification of risk and storage processes, *Math. Oper. Res.* **3** (1978), 57-66.

[39] Hu, J.Q., Analyticity of single-server queues in light traffic, *Queueing Sys.* (1994), (to appear).

[40] Jagerman, D.L., Melamed, B. and Willinger, W., Stochastic modeling of traffic processes, In: *Frontiers in Queueing: Models and Applications in Science and Engineering*, ed. by J.H. Dshalalow, CRC Press, Boca Raton, Florida 1996.

[41] Kallenberg, O., *Random Measures*, 3rd ed., Akademie-Verlag, Berlin and Academic Press, New York 1983.

[42] Karr, A.F., Palm distributions of point processes and their applications to statistical inference, *Cont. Math* **80** (1988), 331-358.

[43] Karr, A.F., *Point Processes and Their Statistical Inference*, 2nd ed., Marcel Dekker, New York 1991.

[44] Konstantopoulos, P. and Zazanis, M.A., Sensitivity analysis for stationary and ergodic queues, *Adv. Appl. Prob.* **24** (1992), 738-750.

[45] Konstantopoulos, P. and Zazanis, M.A., Sensitivity analysis for stationary and ergodic queues: Additional results, *Adv. Appl. Prob.* **26** (1994), 556-560.

[46] König, D. and Schmidt, V., *Random Point Processes*, B.G. Teubner-Verlag, Stuttgart, 1992 (in German).

[47] Kovalenko, I.N., Rare events in queueing systems - a survey, *Queueing Sys.* **16** (1994), 1-50.

[48] Kovalenko, I.N., Approximation of queues via small parameter method, as Chapter 19 in this book, 481-506.

[49] Kroese, D.P. and Schmidt, V., Light-traffic analysis for queues with spatially distributed arrivals, *Math. Oper. Res.* (1995), to appear.

[50] Neuts, M.F., *Matrix-Geometric Solutions in Stochastic Models*, Johns Hopkins University Press, Beltimore, Maryland 1981.

[51] Reiman, M.I. and Simon, B., An interpolation approximation for queueing systems with Poisson input, *Oper. Res.* **36** (1988), 454-469.

[52] Reiman, M.I. and Simon, B., Light traffic limits of sojourn time distributions of Markovian queueing networks, *Stoch. Mod.* **4** (1988), 191-233.

[53] Reiman, M.I. and Simon, B., Open queueing systems in light traffic, *Math. Oper. Res.* **14** (1989), 26-59.

[54] Reiman, M.I., Simon, B. and Willie, J.S., Simterpolation: A simulation based interpolation approximation, *Oper. Res.* **40** (1992), 706-723.

[55] Reiman, M.I. and Weiss, A., Light-traffic derivations via likelihood ratios, *IEEE Trans. Inform. Theory* **35** (1989), 648-654.

[56] Rolski, T., Approximation of performance characteristics in periodic Poisson queues, In: *Queueing and Related Models*, ed. by U.N. Bhat and I.V. Basava, Claredon Press, Oxford (1992), 285-298.

[57] Rudemo, M., Point processes generated by transitions of Markov chains, *Adv. Appl. Prob.* **5** (1973), 262-282.

[58] Ryll-Nardzewski, C., Remarks on processes of calls, *Proc. 4th Berkeley Symp. Math.: Statist. Probab.* **2** (1961), 455-465.

[59] Schmidt, V. and Serfozo, R.F., Campbell's formula and applications to queueing, as Chapter 8 in this book, 225-242.

[60] Serfozo, R.F., Point processes, In: *Handbooks in OR & MS*, ed. by D.P. Heyman and M.J. Sobel, vol. **2**, Stochastic Models, North Holland, Amster-

dam (1990), 1-93.

[61] Sigman, K., Light traffic for work load in queues, *Queueing Sys.* **11** (1992), 429-442.

[62] Sigman, K., *Stationary Marked Point Processes*, Chapman and Hall, London 1994.

[63] Sigman, K., Light traffic for work load and virtual delay in split and match queues, preprint, Columbia University, New York (1994).

[64] Sigman, K. and Yamazaki, G., Heavy and light traffic in fluid models with burst arrivals, *Ann. Inst. Stat. Math.* **45** (1993), 1-7.

[65] Simon, B., Calculating light-traffic limits for sojourn times in open Markovian queueing systems, *Stoch. Mod.* **9** (1993), 213-231.

[66] Wang, C.-L. and Wolff, R.W., Light-traffic approximations for the M/G/c queue, preprint, University of California, Berkeley (1993).

[67] Whitt, W., An interpolation approximation for the mean work load in a GI/G/1 queue, *Oper. Res.* **37** (1989), 936-952.

[68] Wolff, R.W., Tandem queues with dependent service times in light traffic, *Oper. Res.* **30** (1982), 619-635.

[69] Yamazaki, G. and Ito, H., Optimal order for two servers in tandem, *Ann. Inst. Stat. Math.* (1994), (to appear).

[70] Yechiali, U., On the relative waiting times in GI/M/s and GI/M/1 queueing systems, *Opnl. Res. Quart.* **28** (1977), 325-337.

[71] Zazanis, M.A., Analyticity of Poisson-driven stochastic systems, *Adv. Appl. Prob.* **24** (1992), 532-541.

Chapter 16

Quantitative estimates in queueing

Vladimir V. Kalashnikov

ABSTRACT We discuss a problem of obtaining quantitative bounds in the theory of queues. We assume that these bounds are tight in order to employ them in the analysis of real systems. Basic attention is paid to the evaluation of the rate of convergence to the steady state and comparison bounds. We consider the two problems for queueing models described as wide sense regenerative processes. It is shown that their efficient solution can be found with the aid of such tools as coupling and crossing. We also discuss various ways of obtaining tight quantitative bounds.

CONTENTS

16.1	Reasoning	407
16.2	Regenerative processes	408
16.3	Convergence rate to the steady state	413
16.4	Comparison of queues	419
16.5	Concluding remarks and open problems	425
	Bibliography	427

16.1 REASONING

Long ago, while discussing possible ways of highlighting physical phenomena, Ernest Rutherford claimed, "Qualitative is poor quantitative." In this chapter, we discuss some problems in queueing theory associated with this quotation.

Let us regard a queueing model as a mapping $\mathbf{F}: \mathfrak{X} \to \mathfrak{Y}$. It is not necessary to give more details about this mapping, but it would be useful to imagine inter-arrival and service times (or probability distributions defining them), service disciplines, etc., as input $X \in \mathfrak{X}$ and waiting times, queue lengths, busy periods (or their distributions), etc., as output $Y \in \mathfrak{Y}$. When treating a queueing model as a mapping $\mathbf{F}$, we mean that the output Y can be defined as $Y = \mathbf{F}(X)$, but this does not mean necessarily that we know how to do this in an explicit from.

Imposing additional restrictions allows one to use analytic investigations or qualitative ("poor quantitative") analysis. It should be noted that much attention has been paid to the problem of deriving explicit analytic formulas of various queueing characteristics. The Pollaczek-Khintchine formula, Little's formula, and Jackson's multiplicative formula are perhaps the most famous of such results. In all those cases, one singles out a specific subset $\mathfrak{X}^* \subset \mathfrak{X}$ reflecting the imposed restrictions. For example, let $\mathfrak{X}^*$ be a subset of single server queues with Poisson input process, infinite waiting room, and arbitrary distributed service times (independent for different customers) under the ergodicity condition. Then it is possible

0-8493-8074-x/95/$0.00+$.50

to find an explicit dependence $Y = \mathbf{F}(X)$ (in terms of the Laplace transform), given that Y is the distribution function (d.f.) of the queue length in the steady state. Namely, this was rendered by Pollaczek and Khintchine. Explicit formulas of those kinds may not be satisfactory for the following reasons. First, their domains can be extremely restrictive. Second, they usually yield implicit results (e.g., in terms of the Laplace transform), and this requires elaboration of additional numerical methods in accordance with the desired accuracy, rate of convergence, etc. Such numerical methods, being quantitative in essence, can be employed to a narrow set of queueing models because they are based on explicit analytical formulas.

Consider the following example, quantitative in form but qualitative in essence. Let $p_k(t)$ be the probability that, in a specific queueing model, the queue length is equal to k at time t. Suppose that we can prove that there exists the limit, $p_k = \lim_{t\to\infty} p_k(t)$, and that we can calculate p_k. In order for this assertion to be useful in applications, one needs to know the rate of convergence in this limiting relation. Derivation of the rate of convergence is more difficult than the proof of the existence of the steady state. It is important to realize that such a relation as

$$| p_k(t) - p_k | = O\left(\frac{1}{t^\alpha}\right), \text{ for some } \alpha > 0,$$

or, equivalently,

$$| p_k(t) - p_k | \leq \frac{c}{t^\alpha}, \text{ for some } \alpha > 0 \text{ and } c > 0,$$

adds almost nothing from a practical standpoint because a virtual usage of the result above depends strongly on the values of c and α, but we assert only their existence. Let us now assume that we overcame this problem. Then, there arises a new question inspired by the following circumstances. Suppose probabilities p_k were calculated under some restrictions in $\mathfrak{S}^*$ imposed on input data (e.g., under the assumption that customers enter the system in accordance with the Poisson process). But this is only an approximation of the real situation. Usually, there are at least two reasons to make such assumptions. First, we are only able to derive analytic formulas of certain characteristics. Second, existing statistical data confirms our assumption. In both cases, a real input does not agree exactly with our assumptions. Hence, one needs to find *comparison* or *continuity bounds* displaying intervals which "*real*" probabilities (corresponding to real input data unknown to us) belong to. In fact, this can be regarded as a "quantitative analysis" of queueing models.

Note that simulation does not enlarge much the subset $\mathfrak{S}^*$, provided that we do not reduce a simulation process to generating paths of models but treat simulation as a mathematically correct data processing.

16.2 REGENERATIVE PROCESSES

Some important problems of quantitative analysis of queueing models can be solved in rather general terms. We consider here convergence rate and continuity problems for so-called *wide sense regenerative processes.* This allows us to apply such results to a variety of queueing models which can be described in terms of such regenerative processes. This concept goes back to the classic work of Smith, [25] but its wide usage started after works of Thorisson [29] and Asmussen [2]. We recommend [15] for further references and discussions.

There are continuous and discrete time cases, under the assumption that the set T of possible times is either $[0,\infty)$ or $\{0,1,\ldots\}$. However, for now, we will treat only the more difficult continuous time case. Suppose $Z = \{Z(t,\omega)_{t\in T}\}$ is a random process having the state space $(\mathfrak{Z},\mathfrak{B})$ and right continuous paths. Evidently, for discrete time, the requirement of the right continuity is unnecessary. Let $S = (S_n)_{n\geq 0}$ be an increasing sequence of nonnegative finite random times. Let us call instants S_n *renewals* or *regeneration times* and define:

$$N(t) = \#\{n{:}\, S_n \leq t,\ n \geq 0\}, \tag{16.1}$$

the number of renewals (including S_0) which occur in $[0,t]$;

$$N_-(t) = \#\{n{:}\ S_n < t,\ n \geq 0\}, \tag{16.2}$$

the number of renewals (including S_0), occurring in $[0,t)$;

$$\theta_t(Z,S) = \Big((Z(t+s))_{s\geq 0}, (S_{n_-(t)+k} - t)_{k\geq 0}\Big), \tag{16.3}$$

the shifted pair (Z,S) on time t onward;

and
$$W_n = S_n - S_{n-1},\ n \geq 1, \tag{16.4}$$

inter-regeneration times.

The notations above were introduced in [30]. They are convenient for the definition of regenerative processes.

Definition 16.1. A random pair (Z,S) is called a *wide sense regenerative process* (or simply *regenerative process*) if, for all $n \geq 0$,
(*i*) $\theta_{S_n}(Z,S)$ are identically distributed;
(*ii*) $\theta_{S_n}(Z,S)$ does not depend on $(S_0,\ldots,S_n)$.

It follows from Definition 16.1, that S_n, $n \geq 0$, forms a *renewal process*; that is, all W_n, $n \geq 1$, are independent identically distributed random variables (i.i.d. r.v.'s), and they are all independent of S_0. Regeneration times $S_n, n \geq 0$, split the process (Z,S) on the *delay*

$$D = (Z,S)_0^{S_0} \equiv \Big((Z(t))_{t<S_0}, S_0\Big) \tag{16.5}$$

and the *sequence of cycles*

$$C_n = (\theta_{S_{n-1}}(Z,S))_0^{W_n} \equiv (\theta_{S_{n-1}}(Z,S)(t),\ 0 \leq t < W_n),\ n \geq 1. \tag{16.6}$$

Obviously, all cycles are identically distributed random elements but they can be dependent. Here lies the basic difference between Definition 16.1 and Smith's definition of a *classic sense regenerative process* (see [26]) where $\theta_{S_n}(z,S)$ was supposed to be independent of the prehistory $((z(t))_{t<S_n}, S_0,\ldots,S_n)$. Let us call a regenerative process *positive recurrent* if $\mathbf{E}[W_1] = \mu < \infty$.

Example 16.2. *The* GI/GI/1/∞ *model.* Let e_k be the interarrival time between kth and $(k+1)$th customers, s_k be the service time, and w_k be the waiting time of the kth customer, $k \geq 1$. Suppose customers enter the server in the order of their arrivals. Then, the Lindley equation relates recursively waiting times of successive customers (see [14]):

$$w_{k+1} = (w_k + s_k - e_k)_+,\quad k \geq 1, \tag{16.7}$$

where $(\cdot)_+ = max(0,\cdot)$. Let $w_1 = 0$ and

$$S_0 = 1, \quad S_{n+1} = min\{k: w_k = 0, \; k > S_n\}, \quad n \geq 1. \tag{16.8}$$

Set $w = (w_1, w_2, \ldots)$. Letters GI in the notation of the model mean that both sequences $s = (s_1, s_2, \ldots)$ and $e = (e_1, e_2, \ldots)$ are independent and consist of i.i.d.r.v.'s. It is well-known that (w, S) is a regenerative process (even, a classic sense regenerative process). This process is positive recurrent under the *underload condition* $\mathbf{E}[e_1] > \mathbf{E}[s_1]$, where $\mathbf{E}[e_1] = \infty$ may hold.

Example 16.3. *The* GI/GI/N/∞ *model.* Consider a queueing model consisting of N parallel identical servers. Let e and s have the same meaning as in Example 16.2; customers enter the servers in order of their arrival, and both sequences e and s are independent and consist of independent and identically distributed random variables (i.i.d.r.v.). A server takes a customer from the line as soon as it becomes free. If the queue is empty at that instant, the server becomes idle. If a customer arrives and reaches at least one idle server, it immediately goes for service to any idle server. Let us assign to the kth customer, the Kiefer-Wolfowitz vector (see [14]):

$$w_k = (w_k(1), \ldots, w_k(N)),$$

where $w_k(j)$, $1 \leq j \leq N$, is the waiting time of the kth customer until the nearest time when at least j servers become free of customers which arrived earlier than the kth customer (i.e., with numbers $1, 2, \ldots, k-1$), $0 \leq w_k(1) \leq \ldots \leq w_k(N)$. Let $\mathbf{i} = (1, 0, \ldots, 0)$, $\mathbf{1} = (1, \ldots, 1)$ and let operator $\mathbf{R} = \mathbf{R}(w)$ reorder coordinates of the vector w in an ascending order (if some coordinates are equal to each other, their order is insignificant).

Assume that the first customer arrives at the idle system: $w_1 = (0, \ldots, 0)$. Describe the dynamics of $(w_n)_{n \geq 1}$ by the Kiefer-Wolfowitz equation:

$$w_{n+1} = \mathbf{R}(w_n + s_n \mathbf{i} - e_n \mathbf{1})_+, \quad n \geq 1. \tag{16.9}$$

Define the sequence

$$S_0 = inf\{n: w_n = (0, \ldots, 0)\},$$

$$S_{i+1} = inf\{n: w_n = (0, \ldots, 0), n > S_i\}, \quad i \geq 0,$$

which is a renewal process; $(w, (S_0, S_1, S_2, \ldots))$ is a classic sense regenerative process. The positive recurrence takes place if the following two conditions hold:

$$\mathbf{E}[s_1] < N\mathbf{E}[e_1] \tag{16.10}$$

and

$$\mathbf{P}(s_1 < e_1) > 0 \tag{16.11}$$

(see [14]). It stands to reason that (16.10) yields (16.11) if $N = 1$. But, for $N > 1$, the relation (16.11) is not necessarily true if (16.10) is satisfied. Additional restriction (16.11) can be omitted but the resulting process is wide sense regenerative in this case. Let us give the corresponding construction skipping exact statements (we refer the reader to [14, 15] for details) but explaining its meaning.

Assume that $S_0 = 0$ and $m \geq 1$ is a fixed integer. Define successive epochs S_n, $n \geq 1$, such that $S_{n+1} - S_n \geq m$ and each S_n possesses the property that the "influence" of all customers occupying the system at time $S_n - m$ vanishes by the moment S_n in the sense that all values w_k, $k \geq S_n$, do not depend on the values s_k and e_k, $k \leq S_n - m$. It can be proved that, under (16.10), such a sequence can be constructed, and process (w, S) is wide sense regenerative and positive recurrent.

Similar constructions can also be suggested for other queueing models (for example, for a multiphase model, see [14]) as well as for continuous time queueing processes (see [5, 15]). For instance, the queue length continuous time process $Q(t)$ in GI/GI/N/∞ can be embedded into a wide sense regenerative process (Z, S), where $Z(t) = (Q(t), s_{n(t)})$, $n(t)$ is the number of customers that entered the systems within $[0, t]$, and sequence S is constructed in a specific way but very similar to the discrete time case (see [15, Section 1.4, Example 3]). For our purposes, it is sufficient to know that such a process (Z, S) exists. A multiserver model (in both discrete and continuous time) is used, in this text, as *Drosophila melanogaster* because most features can be illustrated with the help of this model. The results referring to this model are collected into Propositions appearing in Examples.

Interregeneration time W_1 is an important characteristic of a regenerative process. Denote by

$$F(x) = \mathbf{P}(W_1 \leq x), \quad x \geq 0, \tag{16.12}$$

the d.f. of W_1. So far, we mentioned only one property of F; namely,

$$\mu \equiv \mathbf{E}[W_1] = \int_0^\infty x dF(x) < \infty, \tag{16.13}$$

that was regarded as positive recurrence of the underlying regenerative process. Let us now *quantify* this property as follows.

Definition 16.4. Let a nonnegative function G defined on $[0, \infty)$ and a constant $\overline{g} > 0$ be fixed. We say that regenerative process (Z, S) belongs to the class $\mathcal{M}(G, \overline{g})$ if

$$\mathbf{E}[G(W_1)] \leq \overline{g}. \tag{16.14}$$

It is important that, for *every* regenerative process (Z, S) from $\mathcal{M}(G, \overline{g})$, one has *the same* estimate (16.14) of interregeneration times or more precisely, their G-moments.

The following three classes of functions G are of importance for us:

$$\mathcal{E} = \{G: G(x) = \exp(\lambda x), \ \lambda > 0\}, \tag{16.15}$$

$$\mathcal{P} = \{G: G(x) = x^s, s \geq 1\}, \tag{16.16}$$

and

$$\Theta_c = \left\{G: g(x) \equiv \frac{dG(x)}{dx} \text{ is concave}, \lim_{x \to \infty} g(x) = \infty\right\}. \tag{16.17}$$

Obviously, inequality (16.14) singles out a class of regenerative processes with uniformly bounded exponential moments if $G \in \mathcal{E}$, and power moments of some order $s \geq 1$ if $G \in \mathcal{P}$. We now give an additional definition and theorem (see [10, 14, 15, 21]) which we state for interregenerative times though it is valid for any family of r.v.'s.

Definition 16.5. A family $\{X\}$ of real r.v.'s is called *uniformly integrable* if

$$\lim_{x \to \infty} \sup_{\{X\}} \int_x^\infty \mathbf{P}(|X| > u) du = 0. \tag{16.18}$$

Note that

$$\lim_{x \to \infty} \int_x^\infty \mathbf{P}(|X|) > u) du = 0$$

if, and only if, $\mathbf{E}[|X|] < \infty$, for a *fixed* r.v. X. Definition 16.5 requires the uniformity of this property within class $\{X\}$.

Theorem 16.6. *A family $\mathcal{W}$ of interregeneration times "induced" by a family of regenerative processes $\{(Z,S)\}$ is uniformly integrable if, and only if, there exists a function $G \in \Theta_c$ and a constant $\bar{g} > 0$ such that* (16.14) *holds true for every r.v. $W_1 \in \mathcal{W}$.*

Theorem 16.6 allows one to verify the uniform integrability (qualitative, by definition) by quantitative relation (16.14).

As we have mentioned, the property that a regenerative process belongs to a definite class $\mathcal{M}(G,\bar{g})$ is crucial, and it is necessary to have criteria to provide this property. The problem lies in the following. Any queueing model (for example, our *Drosophila melanogaster*) is described in terms of input parameters (the d.f. of interarrival times and the d.f. of service times). After embedding this model into a regenerative process, it is necessary to be able to express properties inherent in the regenerative process in terms of the input parameters.

Example 16.7. *Quantification of the positive recurrence property.* Consider the GI/GI/N/∞ model which is underloaded, that is, (16.10) holds true (possibly, $\mathbf{E}[e_1] = \infty$). It follows that there exists such positive numbers α and Δ for which

$$\mathbf{E}[N min(\alpha, e_1) - s_1] = \Delta > 0, \tag{16.19}$$

but for each model these constants may be different. It is worth emphasizing that constants Δ and α define, in fact, a *subclass* of multiserver models meeting (16.19). In this subclass, ergodicity is uniform in a certain sense, and constants Δ and α measure the uniformity. In particular, Δ may be regarded as a distance from the ergodicity boundary while α measures a "compactness" of possible d.f.'s of interarrival times (the compactness of d.f.'s of service times is implied by the existence and boundedness of mean values of s_1).

Suppose, first, that the discrete time regenerative process (Z,S) from Example 16.3 contains the Kiefer-Wolfowitz waiting time process as a component. The assertions below are proved in [14], Section 5.3.

Proposition 16.8. *Given* (19) *and* $\mathbf{E}[G(s_1)] \le \beta_G < \infty$ *for* $G \in \mathcal{P} \cup \Theta_c$, *there exists a constant* $\bar{g} = \bar{g}(G, \beta_G, \Delta, \alpha)$ *such that* $(Z,S) \in \mathcal{M}(G,\bar{g})$.

Proposition 16.9. *Given* (19) *and* $\mathbf{E}[G(s_1)] \le \beta_G < \infty$ *for* $G(x) = \exp(\lambda x)$, $\lambda > 0$, *there exist constants* $\bar{g} = \bar{g}(\lambda, \beta_G, \Delta, \alpha)$ *and* $0 < \lambda' = \lambda'(\lambda, \beta_G, \Delta, \alpha) \le \lambda$ *such that* $(Z,S) \in \mathcal{M}(G',\bar{g})$, *where* $G'(x) = \exp(\lambda' x)$.

Now let the regenerative process (Z,S), mentioned in Example 16.3, be continuous time and contain the queue length process $Q(t)$ as a component.

Proposition 16.10. *Propositions* 16.8 *and* 16.9 *remain valid in the continuous time case if one additionally requires that* $\mathbf{E}[G(e_1)] \le \alpha_G < \infty$ *in both cases. Constant* α_G *must be added to the list of the constants which* $\bar{g}$ *depends on.*

The necessity of the additional restriction in the continuous time case can be explained by the fact that, in that case, interregeneration times consist of sums of interarrival times and, therefore, $\mathbf{E}[G(W_1)] \ge \mathbf{E}[G(e_1)]$. In this sense, inequality $\mathbf{E}[G(e_1)] < \infty$ is necessary for the correctness of Proposition 16.10. This is not true for the discrete time case because there the time n is associated with the number of incoming customers and does not depend on interarrival times.

In both cases, constant $\bar{g}$ *can be evaluated explicitly* in terms of the involved parameters. This means that we are able to evaluate moments of interregeneration times *uniformly* over classes of queueing models, and the boundaries of those classes can be expressed in terms of Δ, β_G, and α. Moreover, it is possible to prove that corresponding values of $\bar{g}$ have the correct order along $\Delta \to 0$. A common tool for this is the *test functions method* (see [10, 14, 15, 22]) although other methods can also be applied.

Definitely, analytical bounds can be improved and such an improvement is an important problem in the investigation of queueing models. In the subsequent sections, we discuss possible applications of simulation for this purpose. Somewhat similar results hold for other queueing models. (See [14].)

16.3 CONVERGENCE RATE TO THE STEADY STATE

Suppose output Y of a queueing model is a random process $Y = (Y_t)_{t \geq 0}$, and we are interested whether Y has the steady state. More precisely, we seek a subset $\mathfrak{X}^* \subset \mathfrak{X}$ for which the relation $X \in \mathfrak{X}^*$ implies that $Y = \mathbf{F}(X)$ has the limiting distribution

$$Q(\cdot) = \lim_{t \to \infty} Q_t(\cdot), \tag{16.20}$$

where

$$Q_t(\cdot) = \mathbf{P}(Y_t \in \cdot). \tag{16.21}$$

In queueing theory, such a subset $\mathfrak{X}^*$ is usually defined by ergodicity conditions similar to (16.10) (see [2, 3, 14]). However, it is clear that these conditions are insufficient if we are interested in quantitative estimates of the rate of convergence to the steady state. Let us show how this important problem can be solved for regenerative processes. For this, suppose that the process Y is embedded into a wide sense regenerative process (Z, S). Let us take $Y_t \equiv \theta_t(Z, S)$. Then,

$$Q_t(\cdot) = \mathbf{P}(\theta_t(Z, S) \in \cdot). \tag{16.22}$$

Additional concepts help us resolve our problem. Sometimes, it is convenient not to pay attention to delays and regard regenerative processes with identically distributed cycles and different (in general) delays as belonging to the same class. First of all, given a regenerative process (Z, S), we can define a *zero-delayed process* (Z^0, S^0) associated with (Z, S):

$$(Z^0, S^0) = \theta_{S_0}(Z, S). \tag{16.23}$$

Now let us consider a more general concept; see [30].

Definition 16.11. A pair $(\widetilde{Z}, \widetilde{S})$ is a *version* of regenerative process (Z, S) if $(\widetilde{Z}, \widetilde{S})$ is a regenerative process and

$$\theta_{\widetilde{S}_0}(\widetilde{Z}, \widetilde{S}) \stackrel{d}{=} (Z^0, S^0), \tag{16.24}$$

where $\stackrel{d}{=}$ stands for equality in distribution.

Definition 16.12. A regenerative process (Z, S) is called *stationary* if

$$\theta_t(Z, S) \stackrel{d}{=} (Z, S) \tag{16.25}$$

for all $t \geq 0$.

Evidently, if the regenerative process (Z, S) has a stationary version (Z^*, S^*), then, every distribution (corresponding to each $t \geq 0$)

$$Q_t^*(\cdot) = \mathbf{P}(\theta_t(Z^*, S^*) \in \cdot) \tag{16.26}$$

coincides with the steady state distribution Q of (Z, S) (given that Q exists):

$$Q_t^*(\cdot) \equiv Q(\cdot). \tag{16.27}$$

Therefore, *the problem of evaluation of the rate of convergence of* Q_t *to* Q *can be replaced by the problem of comparison of the two nonstationary distributions* $Q_t(\cdot)$ *and* $Q_t^*(\cdot)$. The second problem is more convenient to solve. To do this, one must first state the existence of a stationary version. The following theorem was proved by Thorisson (see [30]) though its roots can be traced back to C. Palm.

Theorem 16.13. *Let* (Z,S) *be a wide sense regenerative process. Then a stationary version* (Z^*,S^*) *of* (Z,S) *exists if, and only if, the mean inter-regeneration time* μ *is finite:* $\mu<\infty$. *The stationary version* (Z^*,S^*) *is unique in the sense that its distribution*

$$Q^*(\cdot)\equiv \mathbf{P}\big((Z^*,S^*)\in\cdot\big)\equiv \mathbf{P}(\theta_t(Z^*,S^*)\in\cdot)\equiv Q_t^*(\cdot),\quad t\geq 0$$

is defined uniquely.

Henceforth, we suppose that all underlying regenerative processes have stationary versions or, by Theorem 16.13, their interregeneration times have a finite mean value. In order to compare stationary process (Z^*,S^*) with (Z,S), we use the *coupling* concept; see [20, 29, 31, 32]. Here we use only one of the possible versions of this concept, referring the reader to [20] and [32] for other variants. The basic idea is to couple (Z,S) with (Z^*,S^*) at some moment τ (possibly, random) so that $\theta_t(Z,S)=\theta_t(Z^*,S^*)$ for $t\geq\tau$. But this involves some formal difficulties. For example, processes (Z,S) and (Z^*,S^*) may be defined even on different probability spaces. In addition, the desired equality $\theta_t(Z,S)=\theta_t(Z^*,S^*)$ is too restrictive in some cases (for example, for wide sense regenerative processes). The following definitions are intended to overcome these difficulties.

Definition 16.14. A pair of processes $(\widehat{Z},\widehat{S})$ and $(\widehat{Z}^*,\widehat{S}^*)$ is a *coupling* of regenerative processes (Z,S) and (Z^*,S^*) if $(\widehat{Z},\widehat{S})\overset{d}{=}(Z,S)$ and $(\widehat{Z}^*,\widehat{S}^*)\overset{d}{=}(Z^*,S^*)$.

Definition 16.15. Let a pair of processes $(\widehat{Z},\widehat{S})$ and $(\widehat{Z}^*,\widehat{S}^*)$ be a coupling of regenerative processes (Z,S) and (Z^*,S^*), and let τ and τ^* be two nonnegative random times such that

$$\big((\widehat{Z},\widehat{S},\tau)\big)\overset{d}{=}\big((\widehat{Z}^*,\widehat{S}^*),\tau^*\big). \tag{16.28}$$

Then r.v.'s $\tau\overset{d}{=}\tau^*$ are called *weak* (or *distributional*) *coupling times*.

The distributional coupling concept, being applied to wide sense regenerative processes, implies the following statement containing the desired rate of convergence in terms of the coupling time. But, before stating this, let us introduce a natural comparison measure between Q_t and $Q\equiv Q_t^*$, namely, the *total variation distance*:

$$Var(Q_t,Q)=sup\{\,|\,Q_t(B)-Q(B)\,|\,\}, \tag{16.29}$$

where *sup* is taken over all measurable subsets B of the space of possible "values" of (Z,S), each being identified with the element

$$((Z(t)_{t\in T},(S_0,S_1,\ldots)).$$

Theorem 16.16. *For any regenerative process* (Z,S), *the following rate of convergence is valid:*

$$\mathrm{Var}(Q_t,Q)\leq \mathbf{P}(\tau>t), \tag{16.30}$$

where τ *is a distributional coupling time.*

The proof of this theorem can be found in [20, 29, 32]. Let us discuss this statement. First of all, in order for this to make any sense, a stationary version of (Z,S) must exist and, therefore, by Theorem 16.13, $\mu<\infty$. Secondly, inequality

(16.30) holds for *any* coupling of (Z,S) and its stationary version. But it has not been clear until now what additional restrictions must be imposed on (Z,S) in order for a coupling to exist in the sense that $\tau<\infty$ a.s. Below, we give conditions ensuring existence and evaluate the probability on the right-hand side of (16.30). Third, given a coupling exists, a natural question arises about the existence of the *maximal* coupling, such that inequality (16.30) is actually an *identity for all* times t. It turns out that the maximal coupling exists under non-restrictive conditions imposed on the underlying state space. We refer the reader to [1, 6, 7, 8, 28] for further discussions.

One can see that we reduced the initial problem of evaluating the rate of convergence to another problem that seems to be much more complicated because it requires a construction of *probabilistic copies* of (Z,S) and its stationary version (Z^*,S^*) that couple at some random time τ. Our goal is to show that coupling arguments are constructive.

It can be shown that a *stationary renewal process* S^* is defined by the two d.f.'s

$$\mathrm{P}(W_1^*\le x)\equiv \mathrm{P}(W_1\le x)=F(x), \tag{16.31}$$

and

$$\mathrm{P}(S_0^*\le x)=\frac{1}{\mu}\int_0^x(1-F(u))du. \tag{16.32}$$

Suppose for simplicity that (Z,S) is a zero-delayed process, that is, $S_0=0$. It turns out that we can consider a coupling of embedded renewal processes instead of a coupling of regenerative processes in order to evaluate τ (see [15, 20, 29, 32]).

Definition 16.17. A pair $(\widehat{S},\widehat{S}^*)$, defined on a common probability space, is a *coupling* of renewal processes S and S^* if $\widehat{S}\overset{d}{=}S$ and $\widehat{S}^*\overset{d}{=}S^*$.

Definition 16.18. Let $(\widehat{S},\widehat{S}^*)$ be a coupling of renewal processes S and S^* and let κ and κ^* be two nonnegative random indices such that

$$(\widehat{S}_\kappa,\widehat{S}_{\kappa+1},\ldots)=(\widehat{S}^*_{\kappa^*},\widehat{S}^*_{\kappa^*+1},\ldots)\ \text{a.s.} \tag{16.33}$$

Then r.v.

$$\tau=\widehat{S}_\kappa=\widehat{S}^*_{\kappa^*} \tag{16.34}$$

is called the *coupling time* of S and S^*.

Let us compare Definitions 16.15 and 16.18. A remarkable difference is that Definition 16.15 requires the equality in distribution (see (16.28)) of "coupled" process), while Definition 16.18 requires the equality with probability (see (16.33)). This difference can be explained mainly by the fact that Definition 16.15 deals with wide sense regenerative processes while Definition 16.18 deals with much simpler renewal processes. Note that it is possible (and reasonable) to replace equality in distribution (16.28) with a similar equality with probability if one considers classic sense regenerative processes (when their cycles C_n are i.i.d.). However, maybe the most important fact is that Definition 16.18 enables one to "construct" coupling $(\widehat{S},\widehat{S}^*)$ explicitly and thus, with the help of the following theorem (see [15, 16]), evaluate coupling time for regenerative processes.

Theorem 16.19. *If there exists a coupling* $(\widehat{S},\widehat{S}^*)$ *of renewal processes* S *and* S^*, *then there exists a coupling* $((\widehat{Z},\widehat{S}),(\widehat{Z}^*,\widehat{S}^*))$ *of regenerative processes* (Z,S) *and* (Z^*,S^*) *with the same coupling time.*

In Definition 16.4, we introduced the class $\mathcal{M}(G,\overline{g})$ of regenerative processes.

Since the definition uses interregeneration times only, one can apply this to renewal processes. Class $\mathcal{M}(G,\bar{g})$ quantifies the positive recurrence property of renewal processes and, by the way, guarantees, in accordance with Theorem 16.13, that (Z,S) has a stationary version. For evaluation of coupling times, we need to quantify arithmetic properties of interregeneration (or interrenewal) times. Recall, first, the well-known concepts associated with d.f.'s of non-negative r.v.'s. Let F be such a generic d.f.

Definition 16.20.

(*i*) A point $x \geq 0$ is a *point of increase* for F if $F(x) - F(x-\epsilon) > 0$ for an $\epsilon > 0$.

(*ii*) A d.f. F is called *lattice with span* $d > 0$ if all points of increase of F belong to the set $\{0, d, 2d, \ldots\}$ and if d is the largest number satisfying this property. F is *nonlattice* if its points of increase do not belong to a set of this form.

(*iii*) A d.f. F is called *spread-out* if, for an integer $m > 0$, the m-fold convolution F_*^m has a nonzero, absolutely continuous component with respect to the Lebesgue measure, i.e., the d.f. F_*^m has the representation

$$F_*^m(x) = \int_0^x f_m(u)du + G_m(x),$$

where f_m is a density function, $0 < \int_0^\infty f_m(u)du \leq 1$, and G_m is a singular d.f.,

$$G_m(\infty) = 1 - \int_0^\infty f_m(u)du < 1.$$

Let $f = \{f_n\}_{n \geq 1}$ be a probability distribution of a positive integer-valued r.v.

Definition 16.21. Distribution f is called *periodic with period* d if

$$GCD\{n: f_n > 0\} = d,$$

where GCD stands for the *greatest common divisor*. If $d = 1$, then distribution f is called *aperiodic*.

The following definition *quantifies* the aperiodicity property (see [9, 10, 15]).

Definition 16.22. Given integer N_0 and real $\alpha > 0$, distribution f belongs to class $\mathbf{U}_a(N_0, \alpha)$ if

$$GCD\{n: f_n \geq \alpha, n \leq N_0\} = 1.$$

Evidently, any distribution belongs to class $\mathbf{U}_a(N_0, \alpha)$ for some N_0 and α if, and only if, it is aperiodic. In this sense, each set $\mathbf{U}_a(N_0, \alpha)$ can be regarded a class of *uniformly aperiodic distributions* where N_0 and α measure the uniformity, in a way.

Now we introduce, similarly, a class of *uniformly spread-out distributions.*

Definition 16.23. Given $0 \leq a_* < a^*$, $\alpha > 0$ and integer $m > 0$, d.f. F belongs to class $\mathbf{U}_s(a_*, a^*, m, \alpha)$ of uniformly spread-out distributions if there exists a natural $l \leq m$ such that the absolutely continuous component f_l of the l-fold convolution F_*^l exists and satisfies the relation

$$\sup_{a_* \leq x \leq a^*} f_l(u) \geq \alpha.$$

It can be easily proved that d.f. F is spread-out if, and only if, there exist a_*, a^*, m, α, such that $F \in \mathbf{U}_s(a_*, a^*, m, \alpha)$. Therefore, parameters a_*, a^*, m, α can be viewed as those measuring the uniformity property.

Now it is possible to formulate basic results concerning bounds of coupling times. In order to do this, let us consider discrete and continuous cases simultaneously. For brevity, we introduce class $\mathbf{U}(\gamma)$ in the following way:

- $\mathbf{U}(\gamma) = \mathbf{U}_a(N_0, \alpha)$ and $\gamma = (N_0, \alpha)$ in the discrete time case;
- $\mathbf{U}(\gamma) = \mathbf{U}_s(a_*, a^*, m, \alpha)$ and $\gamma = (a_*, a^*, m, \alpha)$ in the continuous time case.

This notation allows us not to distinguish between discrete and continuous time cases.

Theorem 16.24. *Let renewal process $S \in \mathcal{M}(G, \overline{g})$ and $F \in \mathbf{U}(\gamma)$. Then there exists a coupling of S with its stationary version S^* with coupling time τ that satisfies the relations*

(i) $\mathbf{E}[g(\tau)] < g^*$ *if* $G \in \mathcal{P} \cup \Theta_c$ *(where $g(x) = dG(x)/dx$);*

(ii) $\mathbf{E}[\exp(\lambda'\tau)] < g^*$ *if* $G(x) = \exp(\lambda x)$, $\lambda > 0$, $\lambda' \leq \lambda$.

Constants g^* (different in the two cases, in general) and λ' in (i) and (ii) depend on the involved parameters only and can be written in a closed form.

The proof of this theorem can be found in [15, 20]. It follows from (16.30) that the quantitative estimates exhibited in Theorem 16.24 allow us to evaluate the rate of convergence of distributions Q_t to Q (when $t \to \infty$) since they allow us to evaluate the distribution of the coupling time τ. Note that the coupling time indicated in Theorem 16.24 can be applied to evaluation of the rate of convergence for *any* regenerative process (Z, S) having the embedded renewal process S (cf. Theorem 16.19). Condition $S \in \mathcal{M}(G, \overline{g})$ implies that $\mu = \mathbf{E}[W_1] < \infty$. Hence by Theorem 16.13, that process (Z, S) has a stationary version.

Since g^* can be evaluated in an explicit form, one can bound the d.f. of τ with the help of Chebyshev's inequality

$$\mathbf{P}(\tau > x) \leq \frac{g^*}{g(x)}. \tag{16.35}$$

Inequality (16.35) cannot be regarded as a good approximation because of at least two reasons: too low accuracy (in general) of Chebyshev's inequality and too rough bound of g^*. Therefore, Theorem 16.24 together with an analytical estimate of g^* can be viewed as only a *qualitative* result implying the rate of convergence to the steady state for (Z, S).

Example 16.25. *Convergence rates for the* GI/GI/N/∞ *model.* The following propositions are direct consequences of Example 16.7, inequalities (16.30), (16.35) and Theorem 16.24. Consider first the discrete time model described by the Kiefer-Wolfowitz relations (16.9). Let $W_n(x) = \mathbf{P}(w_n \leq x)$ be the d.f. of waiting time vector w_n. In this case, condition $F \in \mathbf{U}(\gamma)$ is satisfied automatically, where γ is defined by "input parameters" of the model.

Proposition 16.26. *Given* (19) *and* $\mathbf{E}[G(s_1)] \leq \beta_G < \infty$ *for* $G \in \mathcal{P} \cup \Theta_c$, *there exists the limiting distribution* $W(x) = \lim_{n \to \infty} W_n(x)$ *and a constant* $g^* = g^*(G, \beta_G, \Delta, \alpha)$ *such that*

$$\mathrm{Var}(W_n, W) \equiv \frac{1}{2} \int_0^\infty |\, d(W_n - W)(x)\,| \leq \frac{g^*}{g(n)}.$$

Proposition 16.27. *Given* (16.19) *and* $\mathbf{E}[G(s_1)] \leq \beta_G < \infty$ *for* $G(x) = \exp(\lambda x)$, $\lambda > 0$, *there exist constants* $g^* = g^*(\lambda, \beta_G, \Delta, \alpha)$ *and* $0 < \lambda' = \lambda'(\lambda, \beta_G, \Delta, \alpha) \leq \lambda$ *such that*

$$Var(W_n, W) \leq g^* \exp(-\lambda' n).$$

Consider now a queue length process $Q(t)$ evolving in continuous time. Let

$$A(x) = \mathbf{P}(e_a \leq x),$$

and

$$p_k(t) = \mathbf{P}(Q(t) = k).$$

Proposition 16.28. *Suppose $A \in \mathbb{U}(\gamma)$ for some γ and relation* (16.19) *holds.*

(*i*) *If $\mathbf{E}[G(e_1)] \leq \alpha_G < \infty$ and $\mathbf{E}[G(s_1)] \leq \beta_G < \infty$ for $G \in \mathcal{P} \cup \Theta_c$, there exists the limiting distribution $p_k = \lim_{t \to \infty} p_k(t)$ and a constant $g^* = g^*(G, \alpha_G, \beta_G, \Delta, \alpha, \gamma)$ such that*

$$Var(p(t), p) \equiv \frac{1}{2} \sum_{k=0}^{\infty} | p_k(t) - p_k | \leq \frac{g^*}{g(t)}.$$

(*ii*) *If $\mathbf{E}[G(e_1)] \leq \alpha_G < \infty$ and $\mathbf{E}[G(s_1)] \leq \beta_G < \infty$ for $G(x) = \exp(\lambda x)$, $\lambda > 0$, there exist constants $g^* = g^*(\lambda, \alpha_G, \beta_G, \Delta, \alpha, \gamma)$ and $0 < \lambda' = \lambda'(\lambda, \alpha_G, \beta_G, \Delta, \alpha, \gamma) \leq \lambda$ such that*

$$Var(p(t), p) \leq g^* \exp(-\lambda' t).$$

In order to proceed to a quantitative result which can be employed in real applications, let us view Theorem 16.24 as an assertion claiming the *existence* of τ only. However, we evaluate d.f. $\mathbf{P}(\tau \leq x)$ or moments of τ using the constructive character in the proof of Theorem 16.24. Details of such an approach can be found in [13, 15], and here we only outline its main stages. The constructiveness in the proof of Theorem 16.24 consists of an explicit specification of a coupling $(\widehat{S}, \widehat{S}^*)$ given S. Of course, this coupling is not maximal, but it is "fair" in the sense of the statement of Theorem 16.24. More importantly, this coupling $(\widehat{S}, \widehat{S}^*)$ can be specified in an *algorithmic form*, that is, one can construct Monte-Carlo algorithms (using methods proposed by Kalashnikov in [13, 15] and Lindvall in [19, 20]) producing successive renewals of the two processes $\widehat{S}$ and $\widehat{S}^*$ which are *dependent* in general. This suggests the availability of appropriate estimators for evaluating any standard statistical data of τ (histogram, sample moment, etc.) when running such an algorithm. It follows that the probability $\mathbf{P}(\tau > x)$ or expectation $\mathbf{E}[G(\tau)]$ can be evaluated statistically and thus, drawbacks of the bounds from Theorem 16.24 do not appear.

Consider some ways of implementation of this approach. Let d.f. F of inter-renewal times be known and satisfy Theorem 16.24 (in order to ensure the correctness of the constructions). Then, using mentioned Monte-Carlo algorithms, one can specify $(\widehat{S}, \widehat{S}^*)$ and thus, obtain a sample value τ_1 of the coupling time. Repetition of this procedure m times leads to m sample values $\tau_1, \ldots, \tau_m$. Then, employing any appropriate statistical estimator, one can evaluate either $\mathbf{P}(\tau > x)$ or $\mathbf{E}[G(\tau)]$. In order to evaluate the real efficiency of such an approach, two programs (according to methods contained in [13, 15, 19, 20]) were written in Turbo Pascal 5.0. These programs require approximately the same time for obtaining sample values of coupling times. Specifically, it takes about 5 sec. per 1000 sample values on IBM-PC/XT. The time mentioned is quite reasonable and enables one to estimate empirical d.f. of the coupling time and use this for evaluating the rate of convergence.

The above approach works if F is known. But this is not the typical case. More often, one is aware of general properties of F only (such as those contained in Propositions 16.8 through 16.28) obtained from other input data. In this case, these general properties can be employed to ensure that F satisfies Theorem 16.24. Because F is unknown, let us use simulation of generating paths of the underlying regenerative process (Z, S) and obtain sample values of interregeneration times.

Suppose, for simplicity, that F has a density f. First, we construct an estimate $\widehat{f}$ of f based on the observed values of its i.i.d. interregeneration times with the help of any statistical density estimator. Then we use one of the aforementioned algorithms (replacing f by $\widehat{f}$) to obtain the sample values of coupling times. It should be noted that those algorithms have the consistency property in the sense that replacing f with $\widehat{f}$ leads to "close" distributions of coupling times. Doing this, we arrive at quite good estimates of the rate of convergence, and their accuracy can be evaluated with the help of standard statistical arguments.

In the theory of queues, the proposed approach makes it possible to obtain accurate quantitative bounds. For simulation, it allows one to evaluate such characteristics as the rate of convergence to the steady state in an *indirect way*.

16.4 COMPARISON OF QUEUES

In the theory of queues, we often have to compare probability characteristics of two or more queueing models taken from a definite set. To illustrate this, let us consider continuity and approximation problems. In both cases we describe a queueing model as a mapping $\mathbf{F}: \mathfrak{X} \to \mathfrak{Y}$.

We have mentioned that the explicit form $Y = \mathbf{F}(X)$ can be found in a few cases. Should this form be known, one must remember that model $\mathbf{F}: \mathfrak{X} \to \mathfrak{Y}$ is just an approximation of the reality, which means that we do not know exact values of inputs X, and we do not know exact form of mapping $\mathbf{F}$. The deviation of a model from the reality can be explained by our desire to make the model as simple as possible and by the fact that the input data result from statistical estimations. In any case, it is necessary to know whether "small deviations" of both X and $\mathbf{F}$ lead to "small deviations" of outputs Y. To be more precise, the problem (which can be called the *continuity problem*) consists of the following. Consider another queueing model $\mathbf{F}': \mathfrak{X} \to \mathfrak{Y}$. Suppose that $\mathbf{F}'$ is close (in some sense) to $\mathbf{F}$ and input data X' are close to X. If, under these assumptions, output data $Y' = \mathbf{F}'(X')$ are close to $Y = \mathbf{F}(X)$, then model $\mathbf{F}$ possesses a continuity property. Of course, in order to convert these words into a correct mathematical statement, one must introduce convergence concepts in spaces $\mathfrak{X}$ and $\mathfrak{Y}$ and define a convergence of $\mathbf{F}'$ to $\mathbf{F}$. In order to obtain quantitative continuity bounds, it is necessary to metrize these convergences. Let us emphasize the following feature of the continuity analysis. First, an origin of "perturbed values" of $\mathbf{F}'$ and X' is a shortage of information about the reality. Hence, we can choose neither $\mathbf{F}'$ nor X' in an appropriate form but we may know that X' belongs to a specific subset $\mathfrak{X}^* \subset \mathfrak{X}$ of input data and $\mathbf{F}'$ belongs to a definite neighborhood of $\mathbf{F}$. Continuity of queues (using different concepts of convergence) were considered in [3, 10, 12-16, 18, 24, 27, 33].

Now let us consider another situation when we cannot analyze model $Y = \mathbf{F}(X)$ but must approximate this by another model $Y' = \mathbf{F}'(X')$. For example, we can use the hyper-Erlang approximation for d.f.'s governing the queueing model in the hopes of finding some characteristics of the model in a closed form. There arise questions about the accuracy of the approximation and about the possibility of choosing $X^{(n)}$ and $\mathbf{F}^{(n)}$, $n \geq 1$, from the desired set, which converge, in a way, to X and $\mathbf{F}$ and imply the convergence to $Y^{(n)}$ of the unknown output Y. The basic difference between approximation and continuity problems can be explained as follows. In the continuity analysis, it is impossible to choose the "perturbed model" $\mathbf{F}'$ because, as we have mentioned, the only information we have is that the perturbed initial data X' belongs to set $\mathfrak{X}^*$. When solving an approximation problem, one can choose $X^{(n)}$ and $\mathbf{F}^{(n)}$ arbitrarily, provided that $Y^{(n)}$ can be calculated (either explicitly or numerically or by simulation) and the calculation process is simple enough. There are many works where one approximates a queuing model by another one and the approximation is based on the fact that respective governing d.f.'s are close enough. There are a few works which discuss the consequences of such an approximation. We refer the reader to [11,17].

A general frame for the mentioned problems and further references can be found in [10,12,14,15,17]. The basic component of these problems is a comparison of two processes which can be called *nonperturbed* and *perturbed*. Let us examine this problem and suppose that we must compare two wide sense regenerative processes (Z,S) and (Z',S'). The basic idea is to make assumptions on the studied processes that permit one to construct so-called crossing times which are simultaneous regeneration times for another pair of regenerative processes (called a *crossing*). Each element of the pair coincides in distribution with one of the initial processes. One can see that the concept of crossing is a generalization of the concept of coupling. The difference between them is due to the fact that (Z',S') *is not* a version of (Z,S). Provided that intercrossing times have moments higher than of the first order, the problem of uniform-in-time comparison of regenerative processes is reduced (using renewal-type arguments) to the evaluation of comparison estimates over finite horizons only.

We now introduce necessary notions starting from embedded renewal processes S and S'. Let us assume, for simplicity, that the two processes are without delays, that is $S_0 = S'_0 = 0$. Similar to coupling, we define the concept of crossing.

Definition 16.29. A pair $(\bar{S},\bar{S}')$ defined on a common probability space is a *crossing* of renewal processes S and S' if $\bar{S} \stackrel{d}{=} S$ and $\bar{S}' \stackrel{d}{=} S'$.

Notice that Definitions 16.17 and 16.29 are quite similar. The difference between them reveals itself in the following definition.

Define three sequences $(\kappa(j))_{j\ge 0}$, $(\kappa'(j))_{j\ge 0}$, and $(\sigma(j))_{j\ge 0}$ as follows:

$$\kappa(0) = \kappa'(0) = 0,$$

$$\kappa(j+1) = min\{n: \bar{S}_n = \bar{S}'_k \text{ for some } k > \kappa'(j), n > \kappa(j)\}, \quad j \ge 0,$$

$$\kappa'(j+1) = min\{n: \bar{S}'_n = \bar{S}_\kappa \text{ for some } k > \kappa(j),\ n > \kappa'(j)\}, \quad j \ge 0,$$

$$\sigma(j) = \bar{S}_{\kappa(j)} = \bar{S}'_{\kappa'(j)}, \quad j \ge 0,$$

and denote by $\sigma = (\sigma(0),\sigma(1),\ldots)$ the collection of times, which are simultaneous regeneration epochs for both $\bar{S}$ and $\bar{S}'$. It is clear that $\sigma(0) = 0$, but the existence and finiteness of other times $\sigma(1)$, $\sigma(2)$, ... depend on the joint distribution of $\bar{S}$ and $\bar{S}'$ in general. Let, for any $i \ge 0$,

$$\theta_{\sigma(i)}(\bar{S},\bar{S}') = \Big((\bar{S}_{\kappa(i)+k} - \bar{S}_{\kappa(i)})_{k\ge 0}, (\bar{S}'_{\kappa'(i)+k} - \bar{S}'_{\kappa'(i)})_{k\ge 0}\Big) \tag{16.36}$$

be the shift of $(\bar{S},\bar{S}')$, and

$$\psi(i) = \Big(\kappa(i), (\bar{S}_k)_{0\le k\le \kappa(i)}, \kappa'(i), (\bar{S}'_k)_{0\le k\le \kappa'(i)}\Big) \tag{16.37}$$

be the corresponding "prehistory" of $(\bar{S},\bar{S}')$.

Definition 16.30. Given a crossing $(\bar{S},\bar{S}')$ of zero-delayed renewal processes S and S', the sequence $(\sigma(k))_{k\ge 0}$ is called a sequence of *crossing times* if $\theta_{\sigma(i)}(\bar{S},\bar{S}')$ does not depend on $\psi(i)$ for any $i \ge 0$ provided that $\sigma(i) < \infty$.

The following lemma shows that, if we are able to construct a crossing

$(\bar{S},\bar{S}')$ with $\sigma(1)<\infty$, then it is possible to take $\sigma(n)<\infty$ for all $n\geq 1$ without loss of generality; see [15].

Lemma 16.31. *If there exists a crossing $(\bar{S},\bar{S}')$ such that $\mathbf{P}(\sigma(1)<\infty)=1$, then there exists another crossing $(\tilde{S},\tilde{S}')$, for which the sequence of crossing times $(\tilde{\sigma}(i))_{i\geq 0}$ forms a renewal process, and*

$$\tilde{\sigma}(i+1)-\tilde{\sigma}(i)\stackrel{d}{=}\sigma(1).$$

The following theorem uses the same notations as Theorem 16.24 and generalizes it. Let $F(x)=\mathbf{P}(W_1\leq x)$ and $F'(x)=\mathbf{P}(W_1'\leq x)$ be d.f.'s of inter-renewal times of renewal processes S and S', respectively.

Theorem 16.32. *If $S\in\mathcal{M}(G,\bar{g})$, $S'\in\mathcal{M}(G,\bar{g})$, $F\in\mathbf{U}(\gamma)$, and $F'\in\mathbf{U}(\gamma)$, for given $G,\bar{g},\gamma$, then there exists a crossing of S and S' such that the following relations hold:*

(*i*) $\mathbf{E}[G(\sigma(1))]<g^*$ *if* $G\in\mathcal{P}\cup\Theta_c$;

(*ii*) $\mathbf{E}[\exp(\lambda'\sigma(1))]<g^*$ *if* $G(x)=\exp(\lambda x)$, $\lambda>0$.

Constants g^* (different in the two cases, in general) and λ' in (i) and (ii) depend on the involved parameters only and can be written in a closed form.

Let us pay attention to the following peculiarities of the statement of Theorem 6. First, in contrast to Theorem 16.24, there exists a G-moment of intercrossing time (recall that we could guarantee the existence g-moments of coupling times in Theorem 16.24 but not G-moments, in general). This can be explained by the fact that both S and S' are zero-delayed renewal processes, while S^*, in Theorem 16.24, is a stationary renewal process having the delay S_0^* defined by (16.32). It is well known that

$$\mathbf{E}[G(W_1)]<\infty\Leftrightarrow\mathbf{E}[g(S_0^*)]<\infty$$

and, moreover, the following equality holds

$$\mu\mathbf{E}[g(S_0^*)]\equiv\mathbf{E}[G(W_1)].$$

Second, the estimates indicated in Theorem 16.32 do not depend on S and S' but only on the involved parameters defining classes $\mathcal{M}(G,\bar{g})$ and $\mathbf{U}(\gamma)$.

We now generalize the concept of crossing on wide sense regenerative processes.

Definition 16.33. The collection $((\bar{Z},\bar{S}),(\bar{Z}',\bar{S}'))$ is called a *crossing* of (Z,S) and (Z',S'), if

$$(\bar{Z},\bar{S})\stackrel{d}{=}(Z,S),\quad(\bar{Z}',\bar{S}')\stackrel{d}{=}(Z',S'),\tag{16.38}$$

where $(\bar{S},\bar{S}')$ is a crossing of renewal processes of S and S' and shift $\theta_{\sigma(i)}((\bar{Z},\bar{S}),(\bar{Z}',\bar{S}'))$ does not depend on prehistory $\psi(i)$ for any $i\geq 0$, provided $\sigma(i)<\infty$, where $\psi(i)$ is defined by (16.37).

The following theorem discloses an important fact that a crossing of regenerative processes can be treated as a regenerative process; see [15].

Theorem 16.34. *If a crossing $(\bar{S},\bar{S}')$ of renewal processes S and S' exists, then there exists a crossing $((\bar{Z},\bar{S}),(\bar{Z}',\bar{S}'))$ of regenerative processes (Z,S) and (Z',S') with the same crossing times and $(\bar{Z},\bar{Z}',\sigma)$ is a wide sense regenerative process.*

Thus, for constructing a crossing of regenerative processes, we can consider their embedded renewal processes only.

Example 16.35. *Crossing times for the* GI/GI/N/∞ *model.* Propositions applied to this example are similar to those from Example 16.25. We preserve basic notations used in Example 10.25, but suppose that now we consider *two* models (perturbed and nonperturbed). Propositions 16.36 and 16.37 deal with the discrete time models described by the Kiefer-Wolfowitz relations while Proposition 16.38 deals with continuous time models.

Proposition 16.36. *Given both models are subjected to the conditions of Proposition 16.26 with the same parameters $G, \beta_G, \Delta, \alpha$, there exists a crossing of regenerative processes describing the two models having the crossing time $\sigma(1)$ and a constant $g^* = g^*(G, \beta_G, \Delta, \alpha)$ such that*

$$\mathbf{E}[G(\sigma(1))] \le g^*.$$

Proposition 16.37. *Given that both models are subjected to the conditions of Proposition 16.27 with the same parameters $\lambda, \beta_G, \Delta, \alpha$, there exists a crossing of regenerative processes describing the two models having the crossing time $\sigma(1)$ and a constant $g^* = g^*(\lambda, \beta_G, \Delta, \alpha)$ such that*

$$\mathbf{E}[\exp(\lambda' \sigma(1))] \le g^*.$$

Proposition 16.38. *Let both models be subjected to the conditions of Proposition 16.28 with the same parameters $G, \alpha_G, \beta_G, \Delta, \alpha, \gamma$.*

(*i*) *If $G \in \mathfrak{P} \cup \Theta_c$, then there exists a crossing of regenerative processes describing the two models having the crossing time $\sigma(1)$ and a constant $g^* = g^*(G, \alpha_G, \beta_G, \Delta, \alpha, \gamma)$ such that*

$$\mathbf{E}[G(\sigma(1))] \le g^*.$$

(*ii*) *If $G(x) = \exp(\lambda x)$, $\lambda > 0$, then there exists a crossing of regenerative processes describing the two models having the crossing time $\sigma(1)$ and constants $g^* = g(\lambda, \alpha_G, \beta_G, \Delta, \alpha, \gamma)$ and $0 < \lambda' = \lambda'(\lambda, \alpha_G, \beta_G, \Delta, \alpha, \gamma) \le \lambda$ such that*

$$\mathbf{E}[\exp(\lambda' \sigma(1))] \le g^*.$$

In all cases, constants g^* and λ' can be written in an explicit form.

We now consider a comparison problem for regenerative processes. For this, take the following comparison measure. Let state space $\mathfrak{Z}$ of regenerative processes under comparison be a complete separable space endowed with a metric h. In applications, $\mathfrak{Z}$ often represents a usual Euclidean space. Let $\mathfrak{F}_{BL}$ be a set of real functions φ defined on $\mathfrak{Z}$ such that $|\varphi(z) - \varphi(z')| \le h(z, z')$ and $|\varphi| \le 1$. Suppose Z and Z' are two r.v.'s taking values in $\mathfrak{Z}$. Define the distance between these r.v.'s (or, more exact, between their distributions) by the following formula:

$$\zeta_{BL}(Z, Z') = sup\{ |\mathbf{E}[\varphi(Z)] - \mathbf{E}[\varphi(Z')]| : \varphi \in \mathfrak{F}_{BL}\}. \tag{16.39}$$

It can be shown (see [4, 17]) that metric ζ_{BL} thus defined induces a *weak convergence* in the space of distributions of r.v.'s, taking values from $\mathfrak{Z}$. We do not consider other possible metrics which can be employed for comparison of regenerative processes, referring the reader to [14-16].

In what follows, we are interested in a so-called *uniform-in-time* comparison estimate of two regenerative processes (Z, S) and (Z', S'), namely, we evaluate the following quantity

$$\epsilon(Z, Z') \equiv \sup_t \zeta_{BL}(Z_t, Z'_t). \tag{16.40}$$

The following statement, proven in [15], explains a usage of the crossing concept.

Lemma 16.39. *Let there exist, for two zero-delayed regenerative processes (Z,S) and (Z',S'), a crossing with a sequence $\sigma=(0,\sigma(1),\sigma(2),\ldots)$ of crossing times such that*

$$0 < c_1 \le \mathbf{E}[R(\sigma(1))] \le c_2 < \infty \tag{16.41}$$

for some monotone positive function R, $\lim_{x\to\infty}(R(x)/x)=\infty$. Then,

$$\epsilon(Z,Z') \le \min_T \left(\sup_{u\le T} \zeta_{BL}(Z_u, Z'_u) + c\left(\frac{1}{r(T)} + \frac{1}{R(T)}\right)\right), \tag{16.42}$$

where

$$r(T) = \inf_{u>T} \frac{R(u)}{u}$$

and the constant c depends only on c_1 and c_2.

Let us discuss possible consequences of Lemma 16.39. Assume that, instead of a pair (Z,S) and (Z',S') of regenerative processes, we have a collection of such pairs, each pair belonging to the same subset indicated in Theorem 16.32 and defined by a triple $(G,\overline{g},\gamma)$. Then one can choose R and c_2, such that $\mathbf{E}[R(\sigma(1))] \le c_2$ for *any pair* from this subset, by Theorem 16.32. For example, if $G\in\mathfrak{P}\cup\Theta_c$, then it is possible to take $R=G$ and $c_2=g^*$ (see Theorem 16.32). The existence of c_1 is a technical requirement, and it is implied by the condition that $F\in\mathbf{U}(\gamma)$ and $F'\in\mathbf{U}(\gamma)$. The existence of c_1, which is a "common" constant for all regenerative processes under consideration means that no interregeneration times approach zero in probability. If this is the case, then the term $c(1/r(T)+1/R(T))$ in (16.42) depends on the triple $(G,\overline{g},\gamma)$ only. Note that this term in (16.42) tends to zero when $T\to\infty$. Suppose that there exists a sequence of pairs $(Z(n),S(n))$ and $(Z'(n),S'(n))$, such that

$$\lim_{n\to\infty} \sup_{u\le T} \zeta_{BL}(Z_u(n), Z'_u(n)) = 0 \tag{16.43}$$

for each *fixed* T. Relation (16.43) can be called *finite-time* continuity property. This property takes place for a variety of queueing models (see [14, 17]). It follows from Lemma 16.39 that, under such assumptions,

$$\lim_{n\to\infty} \epsilon(Z(n), Z'(n)) = 0,$$

i.e., *uniform-in-time continuity* property holds, where

$$\epsilon_T(Z,Z') = \sup_{u\le T} \zeta_{BL}(Z_u, Z'_u). \tag{16.44}$$

Theorem 16.40. *Let regenerative process (Z,S) and (Z',S') be such that all assumptions of Theorem 16.32 hold. Then,*

$$\epsilon(Z,Z') \le \min_T\left(\epsilon_T(Z,Z') + \frac{c}{h(T)}\right), \tag{16.45}$$

where $h(t)\equiv g(t)=dG(t)/dt$ if $G\in\mathfrak{P}\cup\Theta_c$, and $h(t)=\exp(\lambda' t)$ if $G(x)=\exp(\lambda x)$, $\lambda'\le\lambda$. Constants c in (16.45) and λ' depend on triple G, $\overline{g},\gamma$ only and can be written in a closed form.

Theorem 16.41. *Given the assumptions of Theorem 8 and $\epsilon_T(Z,Z')\le\delta T$, $\delta>0$, the following comparison bounds are true.*

(*i*) *If $G(x)=x^s$, $s>1$, then*

$$\epsilon(Z,Z') \le c\left(\frac{s}{s-1}\right)\delta^{(s-1)/s}. \tag{16.46}$$

(*ii*) *If $G(x)\in\Theta_c$, then*

$$\epsilon(Z,Z') \leq c\delta G^{-1}\left(\frac{1}{\delta}\right). \tag{16.47}$$

(*iii*) *If* $G(x) = \exp(\lambda x)$, $\lambda > 0$, *then*

$$\epsilon(Z,Z') \leq -c\delta \ln \delta. \tag{16.48}$$

Constant $c > 0$ *(different in the relations above) can be written in an explicit form in terms of the involved parameters.*

Requirement $\epsilon_T(Z,Z') \leq \delta T$ from Theorem 16.41 often holds true in the theory of queues (see [14, 17]).

Example 16.42. *Finite horizons comparison bounds.* Let us compare two GI/GI/N/∞ models over finite horizons. All notations referring to the perturbed model are the same as in Example 16.25, but they are labeled with primes. We consider discrete and continuous time models separately.

Proposition 16.43. (Discrete time case). *Let the two models be subjected to the conditions of Propositions* 16.26 *or* 16.27 *with the same parameters* G, β_G, Δ, *and* α. *Then there exists a constant*

$$L = L(G, \beta_G, \Delta, \alpha) > 0$$

such that

$$\zeta_{BL}(w_n, w_n') \leq (L+n)\delta,$$

where $n = 0$ *is regarded to be a simultaneous regeneration time for the two models and*

$$\delta = \zeta_{BL}(e_1, e_1') + \zeta_{BL}(s_1, s_1').$$

Proposition 16.44. (Continuous time case). *Let both models be subjected to the conditions of Proposition* 16.28 *with the same parameters* $G, \alpha_G, \beta_G, \Delta, \alpha$ *and* γ. *Then there exists a constant*

$$L = L(G, \alpha_G, \beta_G, \Delta, \alpha, \gamma) > 0$$

such that

$$Var(Q(t), Q'(t)) = \frac{1}{2}\zeta_{BL}(Q(t), Q'(t)) \leq (L+t)\delta,$$

where

$$\delta = Var(e_1, e_1') + Var(s_1, s_1').$$

Constant L *from Propositions* 16.43 *and* 16.44 *can be written in a closed form.*

The difference between definitions of δ in Propositions 16.43 and 16.44 can be explained by the fact that the weak convergence of e_1' to e_1 need not imply the convergence of $Q'(t)$ to $Q(t)$ for some fixed t (for instance, for discontinuity points of d.f. A). In fact, bounds exposed in Propositions 16.43 and 16.44 can be obtained *without* requiring ergodicity conditions for the regenerative processes under comparison (see [14]). But we must assume that initial states for the two processes are the same (for instance, that both systems are idle at the very beginning).

Let us now combine results presented in Examples 16.4 through 16.6 and Theorem 16.41 in order to obtain uniform-in-time comparison bounds for a pair of the GI/GI/N/∞ models.

Example 16.45. *Uniform-in-time comparison bounds.* We now state a few propositions referring to the uniform-in-time comparison of two regenerative models, by considering discrete and continuous time cases separately.

Proposition 16.46. *Let the two discrete time models be subjected to the conditions of Proposition* 16.26 *with the same parameters* G, β_G, Δ, *and* α. *Then,*

$$\sup_n \zeta_{BL}(w_n, w'_n) \le L\delta + c\delta G^{-1}\left(\frac{1}{\delta}\right),$$

where L *and* δ *are taken from Proposition* 16.43 *and* c *is taken from* (16.46) *or* (16.47).

Proposition 16.47. *Let both discrete time models be subjected to the conditions of Proposition* 16.27 *with the same parameters* λ, β_G, Δ, *and* α. *Then,*

$$\sup_n \zeta_{BL}(w_n, w'_n) \le L\delta + c\delta \ln \delta,$$

where L *and* δ *are taken from Proposition* 16.43 *and* c *is taken from* (16.48).

Proposition 16.48. *Let both continuous time models be subjected to the conditions of Proposition* 16.28 (*i*) *with the same parameters* $G, \alpha_G, \beta_G, \Delta, \alpha$, *and* γ. *Then,*

$$Var(Q(t), Q'(t)) \le L\delta + c\delta G^{-1}\left(\frac{1}{\delta}\right),$$

where L *and* δ *are taken from Proposition* 16.44 *and* c *is taken from* (16.46) *or* (16.47).

If the models are subjected to the conditions of Proposition 16.28 (*ii*) with the *same* parameters $\lambda, \alpha_G, \beta_G, \Delta, \alpha, \gamma$, then,

$$Var(Q(t), Q'(t)) \le L\delta + c\delta \ln \delta,$$

where L and δ are taken from Proposition 16.44 and c is taken from (16.48).

16.5 CONCLUDING REMARKS AND OPEN PROBLEMS

We discussed only two topics related to quantification in the theory of queues and suggested methods of evaluation of important characteristics, such as convergence rate to the steady state and comparison estimates. In some aspects, these methods were efficient: they enable one to obtain tight bounds of the desired characteristics not only with the help of analytical methods but with the help of simulation, thus broadening the domain of the methods. Let us pay attention to an unusual way of employing simulation tools. Namely, we suggested so-called *directed simulation experiments* in which we simulate, not the process under investigation, but the coupling of two renewal processes seeking their coupling time. Having done this, we are able to obtain a bound on the rate of convergence with the help of coupling inequality (16.30). Therefore, in such an approach, mathematics is used for the study of interconnections between different characteristics of the processes and for the design of corresponding simulation experiments. Consequently, simulation is used for evaluating some auxiliary values (such as coupling time). Respective estimates are well grounded. The author considers these results as examples of ways to combine mathematical and simulation tools for the analysis of queueing and other models. Apparently, further development of such methods requires elaborating new concepts to solving new problems (as coupling and crossing fitted to convergence rate and comparison problems) and, of course, designing new simulation experiments. In order to corroborate the prospect of this approach, let us mention estimates of probability characteristics of rare events which were obtained in a similar way (see [15]).

Those readers who are less motivated by simulation can work on new quantitative bounds by analytical tools. Problems of interest would be such as generalization of existing results, use of different metrics (implying not only weak convergence but convergence of moments), investigation of broader classes of random processes (not only regenerative), and also new setups.

Let us discuss some worthy open problems:

(1) There are various ways of constructing a regenerative process that our model is embedded into. Because of this, there arises a question about the construction having the shortest interregeneration time W_1 (in an appropriate probabilistic sense) or at least an interregeneration time that is short enough. Such a construction may allow shortening simulation time of regenerative processes in order to evaluate their stationary characteristics and the rate of convergence to the steady state. In addition, this may make it possible to improve some analytical estimates expressed in terms of interregeneration times (such as estimates given in Theorems 16.24, 16.32, 16.40, 16.41, and Propositions 16.8, 16.9, 16.26-16.38, 16.44-16.48.)

(2) Examples show that embedding a model into a regenerative process may be accompanied by additional constructions (such as splitting), and these constructions may not have a clear interpretation in terms of the model (see [14, 23]). In turn, this may lead to obstacles in the application of elaborated "regenerative" methods. Hence, an important problem is to build "observable" regeneration times. This means that we must have a way to determine regeneration epochs observing the "prehistory" of the model. We refer readers to [15] for further discussions of the "observability" notion. In fact, each "unobservable" regeneration epoch can be converted into an "observable" epoch by equipping the initial regeneration process with supplementary variables. The problem is how to do this in the most reasonable way.

(3) Let us discuss some problems associated with comparison bounds. One can see that a crucial step in obtaining them is the evaluation of crossing times. Formally, the construction of crossing times is very similar to that of coupling times, and, therefore, one can apply an approach similar to that from Section 16.3, but with one modification. Namely, when considering the coupling, we dealt with a *single* regenerative process, and the coupling, time was defined in terms of a *single* d.f. F. Crossing time depends on the pair of d.f.'s F and F'. We can regard "non-disturbed" d.f. F to be known, but "disturbed" d.f. F' can be known only in an approximation problem when we choose it by ourselves. In the continuity problem, it remains unknown to us. Consequently, it is impossible to find bounds of crossing times for *all* values of F'. If (Z,S) is close, in a way, to (Z',S'), then it is reasonable to expect that the two processes have close d.f.'s of interregeneration times, and, therefore, the crossing time must be close in distribution to interregeneration time of (Z,S) (though, the weak convergence of Z' to Z does not necessarily imply the weak convergence of S' to S). One can prove, however, that, for some standard queueing models, it is possible to replace crossing time by inter-regeneration time in comparison estimates (see [3, 14]). It is a plausible conjecture that the crossing time, under nonrestrictive conditions, is close in distribution to interregenerative time of (Z,S). Given this conjecture, it is reasonable to evaluate the closeness between the indicated r.v.'s that enables employing d.f.'s of interregeneration times instead of d.f.'s of crossing times in final formulas.

(4) Another important problem is the evaluation of finite horizons "discrepancies" between processes (Z,S) and (Z',S'). As we have mentioned, properties of

probability metrics play an important role in this (see [14, 17]). But, in addition, features of specific queueing models affect the evaluation process. In order to understand this, it is sufficient to compare estimates above for discrete and continuous time models. Therefore, an important problem is a classification of queueing models in accordance with different types of infinite times comparison bounds.

(5) The evaluation of various distributions in queueing models, when estimated distributions of "governing" r.v.'s (such as interarrival and service times) are used in simulation, is of exceptional importance. Such an evaluation is widely used in simulation. An approach to the solution of this problem is suggested in [24]. In particular, it was shown that goodness of fit in terms of Kolmogorov-Smirnov distance may not provide satisfactory estimates of d.f.'s of participating r.v.'s while the empirical d.f.'s do satisfy necessary requirements. This problem closely relates to continuity estimates, and this may lead to a discovery of new statistical procedures to be employed in the theory of queues and simulation of queueing systems.

BIBLIOGRAPHY

[1] Aldous, D. and Thorisson, H., Shift-coupling, *Stoch. Proc. Appl.* **44** (1993), 1-14.

[2] Asmussen, S., *Applied Probability and Queues*, John Wiley, Chichester 1987.

[3] Borovkov, A., *Asymptotic Methods in Queueing Theory*, John Wiley, Chichester 1984.

[4] Dudley, R., *Real Analysis and Probability*, Wadsworth & Brooks/Cole, Pacific Grove 1989.

[5] Foss, S. and Kalashnikov, V., Regeneration and renovation in queues, *Queueing Sys.* **8** (1991), 211-224.

[6] Goldstein, S., Maximal coupling, *Z. Wahrscheinlichkeitstheorie verw. Geb.* **46** (1979), 193-204.

[7] Griffeath, D., A maximal coupling for Markov chains, *Z. Wahrscheinlichkeitstheorie verw. Geb.* **31** (1975), 95-106.

[8] Griffeath, D., Coupling methods for Markov processes, Studies in Probability and Ergodic Theory, *Adv. Math. Suppl. Studies* **2** (1978), 1-43.

[9] Kalashnikov, V., An uniform estimate of the rate of convergence in the discrete time renewal theorem, *Prob. Theory Appl.* **22**:2 (1977), 399-403.

[10] Kalashnikov, V., *Qualitative Analysis of the Behavior of Complex Systems by the Test Functions Method*, Nauka, Moscow (in Russian) 1978.

[11] Kalashnikov, V., Approximation of stochastic models, *Semi-Markov Models: Theory and Applications*, Plenum Press, New York (1986), 319-336.

[12] Kalashnikov, V., Regenerative queueing processes and their qualitative and quantitative analysis, *Queueing Sys.* **6** (1990), 113-136.

[13] Kalashnikov, V., Statistical estimates of transient periods for regenerative models by the coupling method, *Math. Meth. Statistics* **1** (1992), 39-48.

[14] Kalashnikov, V., *Mathematical Methods in Queueing Theory*, Kluwer Acad. Publ., Dordrecht 1994.

[15] Kalashnikov, V., *Topics on Regenerative Processes*, CRC Press, Boca Raton 1994.

[16] Kalashnikov, V., Crossing and comparison of regenerative processes, *Acta*

Appl. Math. **34** (1994), 151-172.

[17] Kalashnikov, V. and Rachev, S., *Mathematical Methods for Construction of Queueing Models*, Wadsworth & Brooks/Cole, Pacific Grove 1990.

[18] Kennedy, D., The continuity of the single-server queue, *J. Appl. Prob.* **9**:2 (1972), 370-381.

[19] Linvall, T., On coupling of renewal process with use of failure rates, *Stoch. Proc. Appl.* **22** (1986), 1-15.

[20] Lindvall, T., *Lectures on the Coupling Method*, John Wiley, New York 1992.

[21] Meyer, P., *Probability and Potentials*, Blaisdell Publ. Co., Waltham 1966.

[22] Meyn, S. and Tweedie, R., *Markov Chains and Stochastic Stability*, Springer-Verlag, New York 1993.

[23] Nummelin, E., *General Irreducible Markov Chains and Nonnegative Operators*, Cambridge University Press, Cambridge 1984.

[24] Sharma, V., On the problem of estimation in queueing systems, *Queueing Sys.* (1995), (to appear).

[25] Smith, W., Regenerative stochastic processes, *Proc. Roy. Soc. Edinb.* **A64** (1955), 9-48.

[26] Smith, W., Renewal theory and its ramifications, *J. Roy. Stat. Soc.* **B20** (1958), 243-302.

[27] Stoyan, D., *Comparison Methods for Queues and Other Stochastic Models*, John Wiley, Chichester 1983.

[28] Sverchkov, M. and Smirnov, S., Maximal coupling on D-valued processes, *Sov. Math. Dokl.* **41** (1990), 352-354.

[29] Thorisson, H., The coupling of regenerative processes, *Adv. Appl. Prob.* **15** (1983), 531-561.

[30] Thorisson, H., Construction of a stationary regenerative process, *Stoch. Proc. Appl.* **42**:2 (1992), 237-253.

[31] Thorisson, H., *Coupling Methods in Probability Theory*, Science Institute, Univ. of Iceland, preprint RH-18-92 (1992).

[32] Thorisson, H., Coupling and convergence of random elements, processes and regenerative processes, *Acta Appl. Math.* **34** (1994), 85-107.

[33] Whitt, W., The continuity of queues, *Adv. Appl. Prob.* **6**:1 (1974), 175-183.

Chapter 17

Steady state rare events simulation in queueing models and its complexity properties[1]

Søren Asmussen and Reuven Y. Rubinstein

ABSTRACT This chapter gives an overview of rare events simulation via importance sampling in queueing models, as well as some new results and proofs in the area. Special attention is given to complexity properties of the estimators, exponential change of measure, steady-state implementation via Sigmund duality, and robustness under perturbation of the optimal set of parameters. Open problems and further research are discussed as well.

CONTENTS

17.1	Introduction	429
17.2	Preliminaries	431
17.3	ECM for simple random walks	434
17.4	OECM	436
17.5	ECM for more general models	440
17.6	Steady-state rare events simulation	443
17.7	Robustness of the OECM	448
17.8	Examples	450
17.9	Further research and open problems	455
17.10	Appendix: Proof of an auxiliary random walk result	457
	Acknowledgement	459
	Bibliography	459

17.1 INTRODUCTION

Estimation of probabilities $\alpha = \mathrm{P}(A)$ of rare events A is crucial for many modern systems like telecommunication networks and, in particular, for asynchronous transfer mode (ATM) multiplexers in broadband integrated switching digital networks.

Analytical and even "good" asymptotical expressions for such rare event probabilities ($\alpha \leq 10^{-6}$) are only available for a very small class of systems. Hence,typically, one has to resort to simulation. Unfortunately, estimation of rare event probabilities under the original measures, that is, by the crude Monte Carlo method (CMC), is very time consuming and, thus, extremely costly. Instead, a method based on changing the underlying distribution, called *importance sampling* (IS), is typically used [44].

In the past decade, IS has been applied to a variety of problems arising in es-

[1]This work was supported by the Technion V.P.R. Fund - B.R.L. Bloomfield Industrial Management R.F.

0-8493-8074-x/95/$0.00+$.50

timating rare events of queueing systems, (see e.g., [2,4,13,20,21,22,28,30,42,45,46, 55]). The main idea of IS approach is to make the occurrence of rare events more frequent or, in other words, to speed up the simulation of rare events under the new probability measure. Then, in order to obtain an unbiased estimator of the desired rare event, the simulated events (under the new probability measure) are weighted by the likelihood ratio (LR; also called the Radon-Nikodym derivative).

The main goal of IS is to select a simulation distribution which minimizes the computational cost (simulation time) of the estimator, subject to a required accuracy (width of the confidence interval).

It is well known that choosing the simulation distribution as the original distribution, given that the rare event has occurred, is not only optimal but, in fact, leads to an estimator with zero variance. Unfortunately, implementation of such optimal IS distribution is, typically, unfeasible for the following reasons. First, it explicitly depends on α, the unknown quantity that we are trying to estimate. If, in fact, α was known, there would be no need to run the simulation experiment at all. Second, even with known α, it would, typically, be very difficult to sample from such a conditional distribution in a simple way.

Therefore, the discussion of optimality of IS has to be put in different terms. One way is to look for the optimal simulation distribution within a restricted class of distributions, say, within the *same parametric family* of distributions, and, then, minimize the computational cost w.r.t. the parameters of the distribution rather than w.r.t. the distribution itself. For reasonably simple systems like the GI/G/1 queue, it has turned out that such parametrization, and, in particular, the so-called *exponential change of measure* (ECM), leads to very convincing results. Much effort has been devoted to determine which of the parameters θ governing the ECM is the optimal one, say θ^*, to extend this *optimal exponential change of measure* (OECM) to more complex queueing systems and to find conditions under which the OECM is optimal, also, in a broader IS setting.

The theoretical framework in which we study rare events simulation in this chapter is based on complexity theory and was introduced in Kriman and Rubinstein [31]. According to [31], the IS estimators are classified either as *polynomial time* or as *exponential time* ones. Specifically, we present necessary and sufficient conditions in various settings under which the IS estimators are polynomial time. We show that the OECM plays the key rôle for generating such polynomial time estimators. This is the first of the main themes to be discussed in this chapter.

The second main theme is steady-state rare events simulation (in contrast to the study of transient phenomena with which most papers in the literature deal). Here, the rare event is typically of the form $A = \{L > x\}$ where L is the sample performance of the underlying system and x some large number. We distinguish the following three approaches. The first one, proposed by Siegmund [52] and further developed by Asmussen [4], relates the original rare event queueing problem to a ruin problem associated with a random walk, which is then treated by ECM; various approaches to OECM have been proposed and are discussed in Section 17.5.

Note that this approach cannot be generalized, say, to the multiserver case (see discussions in [28, 45]). However, we propose a generalization based on duality of Markov processes, originating from Siegmund [53] and developed more generally by Asmussen [7] (in this book), which has a scope far beyond random walk like models (one example we treat in detail concerns finite buffer problems) and is discussed in detail in a simulation context for the first time here. The second approach for estimating rare events is based on standard regenerative simulation (see e.g., [33, 5]). Under the assumption that the IS distribution comes from the same parametric family as the original one, Asmussen, Rubinstein, and Wang [11] calculated explicitly the variance for the M/M/1 queue, and showed empirically for more complex queueing models that, in order to obtain variance reduction with the regenerative LR estimator relative to the CMC estimator, one has to choose the IS densities such that the associated traffic intensity ρ_0 (under the IS) will be moderately larger than the original traffic intensity ρ. We prove rigorously that, even under the optimal simulation distribution, the regenerative LR estimators in [5] have *exponential time* complexity. The third approach, introduced independently by several authors (e.g., [18, 26, 41, 44]), is based on the so-called *switching regenerative* (SR) simulation. The idea is to use, initially, at each regenerative cycle, a simulation distribution that leads to quick occurrence of a rare event (say, the first buffer overflow in the GI/GI/1/b queue), and, then, use the original distribution for the rest of the regenerative cycle. Some empirical studies with this technique are given, for example, in [26], [41], and [44]. It was proved in [31] that this estimator is polynomial time.

The last main theme is *robustness* of the OECM in the sense that we find how much one can perturb this optimal value, say θ^*, such that the estimator still leads to polynomial time estimators, so that the method leads to dramatic variance reduction (compared to the CMC method) and can be used in practice.

The chapter is organized as follows. The necessary notation and definitions are introduced in Section 17.2. In Section 17.3, we study the ECM in its simplest form in the framework of random walk problems; in particular, the discussion applies to the GI/G/1 queue and classical ruin problems. In Section 17.4, we look at various approaches to the OECM, in particular, *large deviations* (LD) theory and conditional limit theorems, and present a simple proof of a basic optimality result proved in its greatest generality in [32]. Section 17.5 deals with generalizations of the ECM. The two main subjects are Markov additive processes, generalizing simple random walks and covering some main cases like Markovian fluids, a variety of Markov-modulated systems, and local ECM as introduced in [16]; for further applications of ECM which go beyond simple random walks but are not treated here, see e.g., Sadowsky [45]-[47] and Glasserman and Liu [23]. In Section 17.6, we study polynomial and exponential time algorithms for steady-state rare events simulation in the framework of the three approaches outlined above. The robustness results are in Section 17.7 with an auxiliary result of some independent interest on first passage times deferred to the Appendix. Section 17.8 contains various examples, ranging from elementary ones of primarily didactic nature to two complicated and realistic case studies, OECM for a fluid model with alternating renewal on-off sources, and Markov-modulated queues with rejection; these examples are at the frontier of the current state of the area; in fact, they even involve new ideas. Finally, in Section 17.9, we discuss open problems and future research directions.

17.2 PRELIMINARIES

17.2.1 Importance sampling

Assume that the rare event A is defined on the probability space $(\Omega, \mathcal{F}, \mathbf{P})$, and let Y be the identity mapping $\Omega \rightarrow \Omega$. A CMC experiment then amounts to generating

N i.i.d. replications $Y(1),\dots,Y(N)$ of Y (i.e., selecting a point in Ω at random N times) and using the estimator

$$\hat{\alpha}_N = \frac{1}{N}\sum_{n=1}^{N} I(n),$$

where $I(n) = I(Y(n) \in A)$. In the IS, one chooses a different probability measure $\boldsymbol{Q}$ on $(\Omega, \mathfrak{F})$, which is absolutely continuous w.r.t $\mathbf{P}$ on A (i.e., the Radon-Nikodym derivative $W = W(Y) = d\mathbf{P}/d\boldsymbol{Q}$ exists on A), simulates $Y(1),\dots,Y(N)$ from the underlying distribution $\boldsymbol{Q}$, and uses the estimator

$$\bar{\alpha}_N = \frac{1}{N}\sum_{n=1}^{N} I(n)W(n), \tag{17.1}$$

where $W(n) = W(Y(n))$.

Example 17.1. Let $(\Omega, \mathfrak{F})$ be $\mathbf{R}^{k+1}$ equipped with the Borel σ-algebra. Then $Y = (Y_0,\dots,Y_k)$, where $Y_i(y_0,\dots,y_k) = y_i$. Assume that $Y_0,\dots,Y_k$ are i.i.d. with common distribution F, i.e., $\mathbf{P} = F^{\otimes k}$. A natural choice of $\boldsymbol{Q}$ is $\boldsymbol{Q} = G^{\otimes k}$, where G is a distribution equivalent to F, and, then,

$$W(Y) = \frac{dF}{dG}(Y_1)\dots\frac{dF}{dG}(Y_k).$$

If F has a density $f(z) = f_{\theta_0}(z)$ belonging to a parametric family $(f_\theta)_{\theta \in \Theta}$, one often takes G as the distribution with density f_{θ_1} for some $\theta_1 \neq \theta_0$; then, $dF/dG(Y_i) = f_{\theta_0}(Y_i)/f_{\theta_1}(Y_i)$.

Example 17.2. In the setting of real-valued discrete time stochastic processes, the most straightforward choice is sequence space $\Omega = \mathbf{R}^{\{0,1,2,\dots\}}$. Assume that $\mathbf{P}, \boldsymbol{Q}$ correspond to $Y_0, Y_1, \dots$ being i.i.d. with marginals F, resp. $G \neq F$. Then, if y is such that $F(y) \neq G(y)$, the events

$$\left\{\lim \sum_{0}^{N} I(Y_n \le y)/N = F(y)\right\}, \quad \left\{\lim \sum_{0}^{N} I(Y_n \ge y)/N = G(y)\right\}$$

are mutually exclusive and have probability 1 w.r.t. $\mathbf{P}$, resp. $\boldsymbol{Q}$.

This observation indicates that, in virtually all interesting cases, any distribution $\boldsymbol{Q} \neq \mathbf{P}$ fails to satisfy the relevant absolute continuity condition when considering an infinite realization of a stochastic process. However, if we restrict ourselves to segments prior to stopping times, we get something interesting. Formally, let $\Omega_\infty = E^{\{0,1,2,\dots\}}$ where $(E, \mathcal{E})$ is some measurable space, let $\mathfrak{F}_\infty = \mathcal{E}^{\{0,1,2,\dots\}}$ be the product σ-field, and let $\mathbf{P}, \boldsymbol{Q}$ be two different distributions of a stochastic process $Y_0, Y_1, \dots$ on $(\Omega_\infty, \mathfrak{F}_\infty)$. Assume that, for each n, the restrictions of $\mathbf{P}, \boldsymbol{Q}$ to $\mathfrak{F}_n = \sigma(Y_0,\dots,Y_n) = \mathcal{E}^{n+1}$ are equivalent with likelihood ratio W_n. Then, if τ is a stopping time which is finite a.s. w.r.t. $\boldsymbol{Q}$, one has

$$\mathbf{P}G = \mathbf{E}_{\boldsymbol{Q}}[W_\tau; G], \tag{17.2}$$

when $G \in \mathfrak{F}_\tau$, $G \subseteq \{\tau < \infty\}$.

For a concrete example, assume that E is finite or countable and that Y_0, $Y_1, \dots$ is a Markov chain with transition probabilities p_{ij}, q_{ij} w.r.t. $\mathbf{P}$, resp. $\boldsymbol{Q}$, and the same initial probabilities $\mathbf{P}(Y_0 = i) = \boldsymbol{Q}(Y_0 = i)$. Then,

$$W_n = \frac{p_{Y_0 Y_1}}{q_{Y_0 Y_1}} \frac{p_{Y_1 Y_2}}{q_{Y_1 Y_2}} \dots \frac{p_{Y_{n-1} Y_n}}{q_{Y_{n-1} Y_n}}.$$

In the rest of the chapter, the rare event $A = A(x)$ depends on a parameter x, such that $\alpha(x) = \mathbf{P}(A(x)) \to 0$ as $x \to \infty$.

17.2.2 Framework for complexity analysis

Let $\bar{\alpha}_N(x) = (Z_1(x) + \ldots + Z_N(x))/N$ be an estimator of $\alpha(x)$ based upon N i.i.d. replications of some r.v. $Z(x)$ satisfying $\mathbf{E}[Z(x)] = \alpha(x)$; the main example is IS where $Z(x) = W(Y)I(Y \in A(x))$.

Following [31], we say that $\bar{\alpha}_N$ is an (ϵ, δ)-accurate estimator of $\alpha(x)$, ($0 < \epsilon$, $\delta < 1$) if

$$\mathbf{P}(\,|\,\bar{\alpha}_N - \alpha(x)\,|\, < \epsilon\alpha(x)) > 1 - \delta. \tag{17.3}$$

Let $\kappa(x) = Var Z(x)/(\mathbf{E}[Z(x)])^2$ be the square coefficient of variation (SCV) of $Z(x)$. By the Central Limit Theorem (CLT), we have that

$$N \approx \gamma\kappa(x),$$

where $\gamma = \Phi^{-1}(1 - \delta/2)^2\epsilon^{-2}$ and Φ denotes the standard normal c.d.f. That is, the sample size for an (ϵ, δ)-accurate estimator is proportional to the SCV.

Let $T(x)$ be the computational cost (expected CPU time) for generating $Z(x)$. Of course, this concept is not *a priori* rigorous, but, in most examples, there is a natural mathematical definition; in the setting of Example 17.1, one would take $T(x)$ as some constant independent of x and, in Example 17.2, as the expected value of the stopping time $\tau = \tau(x)$.

Definition 17.3. An IS estimator is called a *polynomial time* one if (for a given pair of (ϵ, δ)) (17.3) is guaranteed by a sample of size $N = N(x)$ such that $N(x)T(x) = O(\,|\log\alpha(x)\,|^p)$ for some $p < \infty$. An IS estimator, for which (17.3) requires that $N(x)T(x)$ is of order at least $\alpha(x)^{-q}$ for some $q > 0$, is called an *exponential time* estimator.

For related definitions of time complexity in the deterministic setting, see e.g., [54]. In the stochastic setting, optimality criteria related to the polynomial property appear, e.g., in [26, 24].

Note that, most often in the literature, one neglects taking $T(x)$ into account. Whereas, in most concrete examples, this is unimportant because $T(x)$ does note dependent too much on x, we believe that a general definition not involving $T(x)$ runs into logical difficulties. To see this, consider a new estimator $Z^*(x)$ obtained by averaging $N(x)$ replications of $Z(x)$; taking $N(x)$ large enough produces an estimator with any desired degree of precision.

Example 17.4. Consider the CMC so that $Z(x) = I(Y \in A(x))$. Then, $T(x)$ does not depend on x and

$$\mathbf{E}[Z^2(x)] = \alpha(x)(1 - \alpha(x)) \approx \alpha(x).$$

Thus, (17.3) requires $N(x)$ to be of order at least $\alpha(x)^{-1}$, so that *the CMC estimator is exponential time.*

The following results give conditions for an IS estimator to be polynomial/exponential time and is the typical setup for the rest of the chapter (the proof is trivial and omitted):

Proposition 17.5. *Assume that*

(*i*) $\alpha(x) \approx Ce^{-\gamma x}$ *for some constants* $C, \gamma \in (0, \infty)$, *and*

(*ii*) $T(x) \approx xT_0$ *for some* T_0.

Then, the IS estimator is polynomial time, provided

(*iii*) $\mathbf{E}[Z^2(x)] = O(x^q e^{-2\gamma x})$ *with* $q < \infty$

and exponential time if

(*iv*) $\liminf_{x\to\infty} e^{-\delta x}\mathbf{E}[Z^2(x)/\alpha(x)^2] > 0$ *for some* $\delta > 0$.

Note that (i) is no restriction since it can be obtained by replacing x by $g(x)$ for some function g. In queueing, the stationary distributions typically have exponential tails and thus (i) holds. These facts motivate the terms polynomial and exponential time. Of course, there are estimators which are neither polynomial nor exponential time, e.g., the ones with $N(x)T(x) \approx e^{|\log \alpha(x)|^{\epsilon}}$ with $0 < \epsilon < 1$, but the estimators commonly met in practice are polynomial or exponential time. One can obviously weaken (ii) to $T(x) = O(x^p)$, for the IS estimator to be polynomial time, and to $\liminf_{x\to\infty} T(x)/x^p > 0$ (for some $p > 0$), for the IS estimator to be exponential time.

Example 17.6. Let Y be exponential with rate $\delta = \delta_0$ and $A(x) = I((x,\infty))$, and assume that the IS distribution $\boldsymbol{Q}$ is such that Y is exponential with rate $\delta_1 = \delta_1(x)$. Here, $\alpha(x) = e^{-\delta x}$ and, again, $T(x)$ does not depend on x. Furthermore,

$$\mathbf{E}_{\delta}[Z^2(x)] = \mathbf{E}_{\delta_1}[W^2; Y > x] = \mathbf{E}_{\delta_1}\left[\left(\frac{\delta}{\delta_1}\right)^2 e^{-2(\delta-\delta_1)Y}; Y > x\right]$$

$$= \frac{\delta^2}{\delta_1(2\delta - \delta_1)} e^{-(2\delta-\delta_1)x},$$

where the subscripts δ and $\delta_1$mean that the expectations are taken with respect to exponential p.d.f. with rates δ and δ_1, respectively.

The optimal value (the minimizer of $\mathbf{E}_{\delta_1}[\cdot]$ with respect to δ_1) satisfies $\delta_1^* \approx 1/x$; the corresponding estimator fulfills (iii) (with $q = 1$) and is, therefore, polynomial time. More generally, the expression for $\mathbf{E}[Z^2(x)]$ easily shows that the necessary and sufficient condition for the IS estimator to be polynomial time is

$$\frac{C_1}{x^p} \le \delta_1(x) \le C_2 \frac{\log x}{x}$$

for all large x and some constants p, C_1, C_2. Similarly, the IS estimator is exponential time if, and only if, for all large x, $\delta_1(x)$ belongs to a region of the form

$$(0, e^{-\eta x}) \cup (\beta, \infty),$$

for some constants $\eta, \beta > 0$.

17.3 ECM FOR SIMPLE RANDOM WALKS

Let

$$S_n = Y_1 + \ldots + Y_n \tag{17.4}$$

be a random walk with increment distribution $F(x) = \mathbf{P}(Y_k \le x)$ and the corresponding moment and cumulant generating functions (m.g.f. and c.g.f.) given as

$$\widehat{F}[s] = \mathbf{E}[e^{sY_k}] = \int_{-\infty}^{\infty} e^{sx} F(dx) \text{ and } \kappa(s) = \log \widehat{F}[s],$$

respectively. We assume, throughout, that the mean drive $\mu = \mathbf{E}[Y_k] = \kappa'(0)$ is negative. The exponential change of measure (ECM) corresponds to one where the c.d.f. F is replaced by the distribution with density $e^{\theta y}/\widehat{F}[\theta]$, i.e., with c.d.f.

$$F_\theta(x) = \frac{1}{\widehat{F}[\theta]} \int_{-\infty}^{x} e^{\theta y} F(dy).$$

In this case, we have

$$\widehat{F}_\theta[s] = \frac{\widehat{F}[s+\theta]}{\widehat{F}[\theta]}, \quad \kappa_\theta(s) = \log \widehat{F}_\theta[s] = \kappa(\theta + s) - \kappa(\theta).$$

The likelihood ratio is $W_n(F \mid F_\theta) = W_n(F_0 \mid F_\theta)$, where

$$W_n(F_{\theta_1} \mid F_{\theta_2}) = \exp\{(\theta_1 - \theta_2)S_n - n(\kappa(\theta_1) - \kappa(\theta_2))\}.$$

Of special importance is the case $\theta = \theta^*$, where θ^* is the (unique) solution > 0 of the equation

$$\widehat{F}[\theta] = 1 \tag{17.5}$$

or, equivalently, $\kappa(\theta) = 0$. Another important value is θ_0, the point at which $\kappa(s)$ attains its minimum: if $\theta > \theta_0$, the mean drift $\kappa'_\theta(0)$ corresponding to the IS distribution is positive rather than negative as for the CMC method (this is one of the key points in the approach). The situation is illustrated in Figure 1 (the roles of $\theta(y)$ and $I(y)$ are explained in Section 17.4)

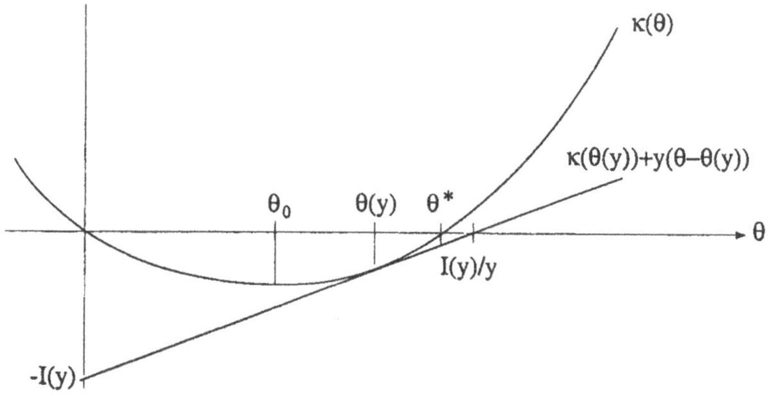

Figure 1

The classical example, where the ECM comes up, is first passage time probabilities; see Siegmund [52], Asmussen [4] for some early examples: the rare event is $A(x) = \{\tau(x) < \infty\}$, where $\tau(x) = inf\{n > 0: S_n > x\}$ with x large, and the corresponding first passage time probability is $\alpha(x) = \mathbf{P}(\tau(x) < \infty)$. Besides its intrinsic interest (sequential analysis in [52] and insurance risk in [4]), $\alpha(x)$ is also of basic importance in queueing because of the formula

$$\mathbf{P}(L \geq x) = \mathbf{P}(M \geq x) = \mathbf{P}(\tau(x) < \infty), \tag{17.6}$$

where L is the steady-state waiting time, $M = max_{n=0,1,\ldots} S_n$, and S_n, is a random walk with the Y_n being the independent difference between the nth service time and the nth interarrival time, cf. [5] Ch. III.7.

ECM for random walks is not only of interest in simulation but also for a variety of other topics surveyed in [5] Ch. XII. The reader who is a novice in the area may want to look at the elementary examples in Section 17.8.1 before proceeding to the study of the IS problem which we now present.

We choose $\theta \geq \theta_0$ (in order to have $\tau(x) < \infty$ a.s. w.r.t. to the IS distribution), and our estimator of $\alpha(x)$ is

$$\begin{aligned} Z(x) &= W_{\tau(x)}(F \mid F_\theta) = \exp\{-\theta S_{\tau(x)} + \tau(x)\kappa(\theta)\} \\ &= \exp\{-\theta x\}\exp\{-\theta\xi(x) + \tau(x)\kappa(\theta)\}, \end{aligned}$$

where $\xi(x) = S_{\tau(x)} - x$ is the overshoot. A fundamental result states that the optimal IS distribution corresponds to a random walk with increment distribution F_{θ^*}:

Theorem 17.7. *Consider a random walk, and let* $A(x) = \{\tau(x) < \infty\}$. *Let* $\mathbf{Q}$ *be a probability measure corresponding to* $Y_1, Y_2, \ldots$ *being i.i.d. with common distribution* G. *Then, the IS estimator is polynomial time if, and only if,* $G = F_{\theta^*}$.

Proof. We prove sufficiency here only and defer necessity to Section 17.4.1 If $G = F_{\theta^*}$, then

$$\mathbf{E}[Z^2(x)] = \mathbf{E}_{F_{\theta^*}}[e^{-2\theta^* x}e^{-2\xi(x)}] \le e^{-2\theta^* x} = O(\alpha(x)^2).$$

Here, we used the standard formula

$$\alpha(x) \approx ce^{-\theta^* x}, \quad x \to \infty, \quad \text{where } c = \lim_{x\to\infty} \mathbf{E}_{F_{\theta^*}}[e^{-\theta^*\xi(x)}] \tag{17.7}$$

for the last equality. Now just apply Proposition 17.5. □

As a variant of Theorem 17.7, we consider the problem of estimating the probability that the waiting time L will exceed some quantity (level) x within a cycle. In the simple random walk setting, the cycle is $C = inf\{n = 1, 2, \ldots : S_n \le 0\}$, the rare event is $A(x) = \{\tau(x) < C\}$, and the IS estimator is based upon

$$Z(x) = I(\tau(x) < C)W_{\tau(x)}.$$

Corollary 17.8. *The IS estimator for the probability* $\alpha(x) = \mathbf{P}(\tau(x) < C)$ *is polynomial time if, and only if,* $G = F_{\theta^*}$.

Proof. For sufficiency, let $G = F_{\theta^*}$. Then,

$$\mathbf{E}[Z^2(x)] = \mathbf{E}_{F_{\theta^*}}[e^{-2\theta^* x}e^{-2\xi(x)}I(\tau(x) < C)] \le e^{2\theta^* x} = O(\alpha(x)^2).$$

Necessity follows by a similar trivial variant of the proof, to be given in Section 17.4, of the necessity part of Theorem 17.7. □

17.4 OECM

Throughout this section, we consider the random walk setting of Section 17.3 and address the question of why the OECM given by Theorem 17.7 is, in fact, optimal. Various approaches have been suggested and are surveyed in the subsections below.

17.4.1 Optimality calculations for simple random walks

Assume that the IS is given by changing F to G (say) and keeping the random walk structure, that is, the independence of the Y_i. No simple inequality is known saying that, for a fixed x, the minimal variance is obtained by taking $G = F_{\theta^*}$. Instead, the results are asymptotical. Siegmund [52] basically showed that $G = F_{\theta^*}$ is asymptotically optimal as $x \to \infty$, provided that G is chosen from the exponential family (F_θ) (his setting is a two-barrier problem and slightly different, but the calculation is easily adapted to the present setting). Lehtonen and Nyrhinen [32] extended this result to G being optimal in almost the whole class of distributions G equivalent to F. Thus, our Theorem 17.7 is basically equivalent to [32]. Finally, Asmussen [4] showed that $G = F_{\theta^*}$ is asymptotically optimal in the exponential family (F_θ) when the limit is not $x \to \infty$ but a diffusion (heavy traffic) limit (again, the setting is slightly different by specializing to classical insurance risk models in continuous time, but the proof is easily adapted). This result confirms empirical findings that the choice $G = F_{\theta^*}$ works well not only for large x, but also for small or moderate x. Mathematically, [52] uses explicit LR calculations for exponential families and [32] uses LD techniques.

We now present an approach which is more elementary than [32] by essentially using only the information inequality (which in turn is just a special case of Jensen's inequality):

Proof of necessity in Theorem 17.7. Assume that the IS distribution is G and that G and F $(G \neq F_{\theta^*})$ are equivalent in the Radon-Nikodym sense. Then, (cf. Example 17.1)

$$Z(x) = W_{\tau(x)}(F \mid G) = \frac{dF}{dG}(Y_1)\ldots\frac{dF}{dG}(Y_{\tau(x)}).$$

By the chain rule for Radon-Nikodym derivatives,

$$\begin{aligned}\mathbf{E}_G[Z^2(x)] &= \mathbf{E}_G[W^2_{\tau(x)}(F \mid G)] = \mathbf{E}_G[W^2_{\tau(x)}(F \mid F_{\theta^*})W^2_{\tau(x)}(F_{\theta^*} \mid G)] \\ &= \mathbf{E}_{F_{\theta^*}}[W^2_{\tau(x)}(F \mid F_{\theta^*})W_{\tau(x)}(F_{\theta^*} \mid G)] = \mathbf{E}_{F_{\theta^*}}[\exp\{K_1 + \ldots + K_{\tau(x)}\}],\end{aligned}$$

where

$$K_i = \log\left(\frac{dF_{\theta^*}}{dG}(Y_i)\left(\frac{dF}{dF_{\theta^*}}(Y_i)\right)^2\right) = -\log\frac{dG}{dF_{\theta^*}}(Y_i) - 2\theta^* Y_i.$$

Here,

$$\mathbf{E}_{F_{\theta^*}}[K_i] = \epsilon' - 2\theta^*\mathbf{E}_{F_{\theta^*}}[Y_i] = \epsilon' - 2\theta^*\kappa'(\theta^*),$$

where

$$\epsilon' = -\mathbf{E}_{F_{\theta^*}}\left[\log\frac{dG}{dF_{\theta^*}}(Y_i)\right] > 0,$$

by the information inequality. Since $K_1, K_2, \ldots$ are i.i.d., Jensen's inequality and Wald's identity yield

$$\mathbf{E}_G[Z^2(x)] \geq \exp\left\{\mathbf{E}_{F_{\theta^*}}[K_1 + \ldots + K_{\tau(x)}]\right\} = \exp\{\mathbf{E}_{F_{\theta^*}}[\tau(x)(\epsilon' - 2\theta^*\kappa'(\theta^*))]\}.$$

Since $\mathbf{E}_{F_{\theta^*}}[\tau(x)/x] \to 1/\kappa'(\theta^*)$, it, thus, follows (using (17.7) that, for $0 < \epsilon < \epsilon'/\kappa'(\theta^*)$,

$$\limsup_{x\to\infty}\frac{\mathbf{E}_G[Z^2(x)]}{\alpha(x)^2 e^{\epsilon x}} = \limsup_{x\to\infty}\frac{\mathbf{E}_G[Z^2(x)]}{c^2 e^{-2\theta^* x + \epsilon x}} \geq \limsup_{x\to\infty}\frac{e^{-2\theta^* x}}{c^2 e^{-2\theta^* x}} = \frac{1}{c^2} > 0,$$

which completes the proof. □

17.4.2 Conditioned limit theorems

The optimal (but unfeasible) change of measures for the IS is the conditional distribution $\mathbf{P}(\cdot \mid A(x))$ given $A(x)$. Therefore, an obvious way to look for a good IS distribution is to try to find a simple asymptotic description of this conditional distribution and to simulate, using this asymptotic description.

For the random walk setting in Section 17.3, where $A(x) = \{\tau(x) < \infty\}$, it turns out that, an asymptotic description of $\mathbf{P}(\cdot \mid A(x))$ is available. The results state roughly that, up to time $\tau(x)$, the random walk behaves as if it changes increment distribution from F to F_{θ^*}, which is precisely the type of behavior needed to infer (at least heuristically) the optimality of θ^*. Several precise statements supporting this informal description were given by Asmussen [3]. As a main example, consider the one-dimensional empirical distribution of the Y_i. Define

$$\widehat{F}^{(n)}(y) = \frac{1}{n}\sum_{i=1}^{n} I(Y_i \le y).$$

Proposition 17.9. *As* $x \to \infty$,

$$\mathbf{P}\left(\| \widehat{F}^{(\tau(x))} - F_{\theta^*} \| > \epsilon \mid \tau(x) < \infty \right) \to 0, \tag{17.8}$$

where $\| \cdot \|$ *denote the supremum norm.*

Proof. The l.h.s. of (17.8) is equal to

$$\frac{\mathbf{P}\left(\| \widehat{F}^{(\tau(x))} - F_{\theta^*} \| > \epsilon ; \tau(x) < \infty \right)}{\mathbf{P}(\tau(x) < \infty)}$$

$$= \frac{\mathbf{E}_{\theta^*}\left[e^{-S_{\tau(x)}}; \| \widehat{F}^{(\tau(x))} - F_{\theta^*} \| > \epsilon, \tau(x) < \infty \right]}{\mathbf{E}_{\theta^*}\left[e^{-S_{\tau(x)}}; \tau(x) < \infty \right]}$$

$$= \frac{\mathbf{E}_{\theta^*}\left[e^{-\xi_{\tau(x)}}; \| \widehat{F}^{(\tau(x))} - F_{\theta^*} \| > \epsilon \right]}{\mathbf{E}_{\theta^*}\left[e^{-\xi(\tau(x))} \right]}$$

$$\le \frac{\mathbf{P}_{\theta^*}\left[\| \widehat{F}^{(\tau(x))} - F_{\theta^*} \| > \epsilon \right]}{\mathbf{E}_{\theta^*}\left[e^{-x(\tau(x))} \right]} \tag{17.9}$$

(using $S_{\tau(x)} = x + \xi_{\tau(x)}$ and $\tau(x) < \infty$ $\mathbf{P}_{\theta^*}$-a.s.). By the Glivenko-Cantelli theorem, $\| \widehat{F}^{(n)} - F_{\theta^*} \| \to 0$ $\mathbf{P}_{\theta^*}$-a.s. as $n \to \infty$. Taking $n = \tau(x)$, it follows that the numerator in (17.9) tends to zero. As in (17.7), the denominator has a limit in $(0, \infty)$, and the proof is complete. □

The results of [3] are, in fact, somewhat more general by allowing inference, also, on the dependency structure in the conditional limit. For example, it is straightforward to show that

$$\frac{1}{\tau(x)} \sum_{i=1}^{\tau(x)} I(Y_i \le y_1, \ldots, Y_{i+k-1} \le y_k) \to F_{\theta^*}(y_1) \ldots F_{\theta^*}(y_k)$$

in $\mathbf{P}(\cdot \mid \tau(x) < \infty)$-probability. Perhaps, the most convincing indication that the Y_i are asymptotically conditionally independent is the fact that variance constants, coming up in conditional approximations by Brownian motion and Brownian bridge, are the same as in the unconditional F_{θ^*}-random walk. See [3] for more detail.

17.4.3 LD

The LD approach to OECM has several variants. We give one involving the concept of *optimal path* which may be seen as an alternative approach to conditioned limit theorems. Some relevant references on this and other LD points of view in simulation are Cottrell et al. [16], Parekh and Walrand [42], Frater et al. [14], Sadowsky and Bucklew [48], and Sadowsky [45]-[47]. For LD theory, in general, we refer to [13, 17, 51].

We first introduce the function

$$I(y) = \sup_{\theta \in \Theta} (\theta y - \kappa(\theta))$$

which, in the literature, goes under names like the LD *rate function*, the *Legendre transform*, the *Legendre-Fenchel transform*, the *Cramér transform* etc. (sometimes the sign is reversed). For simplicity, we assume that the interval Θ is open. Then $\kappa(\theta)/\mid\theta\mid\uparrow\infty$ as we approach the boundary of Θ. This shows that the *sup* is attained for some $\theta(y)$ satisfying $\kappa'(\theta(y)) = y$, so that $I(y) = \theta(y)y - \kappa(\theta(y))$. It is, then, not too hard to show that I is nonnegative, convex, and attains its minimum 0 for $y = \kappa'(0)$. The crucial fact for OECM is:

Lemma 17.10. $\min_{0<y<\infty} \frac{I(y)}{y}$ *is attained for* $y^* = \kappa'(\theta^*)$.

Proof. Obviously, $I(y)/y \to \infty$ as $y\downarrow 0$, and (by convexity) $I(y)/y$ is non-decreasing for large y, so that the minimum is attained. By straightforward different-iation, we get

$$I'(y) = \theta'(y)y + \theta(y) - \theta'(y)\kappa'(\theta(y)) = \theta(y),$$

$$\frac{d}{dy}\frac{I(y)}{y} = \frac{yI'(y) - I(y)}{y^2} = \frac{y\theta(y) - \theta(y)y + \kappa(\theta(y))}{y^2} = \frac{\kappa(\theta(y))}{y^2}.$$

Putting the last expression equal to 0 yields $\theta(y^*) = \theta^*$ (since we look at minimum values for $y > 0$ only, $\theta(y^*) = 0$ is excluded), from which we immediately get $y^* = \kappa'(\theta^*)$. □

One of the main themes of LD theory is to give estimates of the probability that the random walk (or some more general process) follows an atypical path. In the random walk setting, this means that $S^{(n)}(\cdot)$ follows a path different from the one $\varphi_0(t) = \mu t$ given by the law of large numbers where $S^{(n)}(t) = S_{[nt]}/n$, $0 \le t \le 1$ (here $[\cdot] =$ integer part). The LD results state that, under appropriate regularity conditions,

$$\mathrm{P}(S^{(n)}(\cdot) \in \mathcal{S}) \approx \exp\left\{ -n \inf_{\varphi \in \mathcal{S}} \int_0^1 I(\varphi'(t))dt \right\} \tag{17.10}$$

for suitable subsets $\mathcal{S}$ of continuous paths with $\varphi_0 \notin \mathcal{S}$. In many examples, there is a single path φ^* for which the minimum is attained, and this is the *optimal path.*

In order to understand how the random walk reaches the high level x, we perform the optimization not only over φ but also over n. We then write x in the form $x = ny$, and let $\mathcal{S}$ be the set of continuous functions on $[0,1]$ with $\varphi(0) = 0$, $\varphi(1) = y$. Then, (17.10) takes the form

$$\mathrm{P}(S_n \approx x) \approx \exp\left\{ -\frac{x}{y} \inf_{\varphi \in \mathcal{S}} \int_0^1 I(\varphi'(t))dt \right\}. \tag{17.11}$$

By Jensen's inequality and the convexity of I,

$$\int_0^1 I(\varphi'(t))dt \ge I\left(\int_0^1 \varphi'(t)dt \right) = I(y),$$

with equality if and only $\varphi(t) = ty$. Hence for fixed n,

$$\mathbf{P}(S_n \sim x) \sim \exp\left\{-x\frac{I(y)}{y}\right\}$$

Viewing x as fixed and taking into account that minimizing over n is the same as minimizing over y, we obtain, by Lemma 17.10, that the minimizer is $y^* = \kappa'(\theta^*)$. In conclusion, if x is large, the most likely way in which the random walk can cross level x is by crossing at time

$$\tau(x) = n = \frac{x}{y} = \frac{x}{\kappa'(\theta^*)}$$

and by moving linearly at rate $y = \kappa'(\theta^*)$ up to that time. But *this is precisely the same way as that in which the random walk with increment distribution* F_{θ^*} *crosses level* x, which motivates that the conditional distribution, given the rare event $\tau(x) < \infty$, is the one of the random walks with increment distribution F_{θ^*}.

We return to a discussion of the LD method in Section 17.9.

17.5 ECM FOR MORE GENERAL MODELS

17.5.1 Local ECM

Assume that S_n is a Markov chain on the real line rather than a random walk. For the purpose of evaluating first passage probabilities like $\mathbf{P}_y(\tau(x) < C)$ with

$$\tau(x) = inf\{n: S_n > x\}, 0 \le y < x, C = inf\{n: S_n \le 0\}, \mathbf{P}_x = \mathbf{P}(\cdot \mid S_0 = x),$$

Cottrell et al. [16] suggested performing the OECM *locally*. This means that, if $F_x(y) = \mathbf{P}_x(X_1 - x \le y)$, one computes $\theta^*(x)$ as solution of $\widehat{F}_x[\theta^*(x)] = 1$ and simulates using the $F_{x:\theta^*(x)}(dy) = e^{\theta^*(x)y}F_x(dy)$, instead of the F_x, to govern the transitions of S_n. The LR, is then,

$$W_n = \prod_{k=1}^{n} e^{\theta^*(S_{n-1})X_k}.$$

Optimality properties of the procedure are obtained in [16] and involve the concept of slow Markov walks from LD theory, i.e., a scaling of the increments by ϵ and the limit $\epsilon\downarrow 0$.

So far, the method has not had many practical implementations; Heidelberger [26] gives some relevant references, to which we add, here, Asmussen and Nielsen [10]. However, this limited experience shows that the method can be highly efficient, and, in particular, we believe that it is worthwhile to combine it with other ideas like Markov-modulation and duality to be exploited later in this chapter. For such an example, see Section 17.8.3.

17.5.2 Markov-modulated models

Consider, first, the discrete time case, and let $J = \{J_n\}_{n=0,1,2,\ldots}$ be an irreducible Markov chain with a finite state space E. A Markov additive process (MAP) $\{S_n\}_{n=0,1,2,\ldots}$ is an extension of random walks, defined as $S_n = Y_1 + \ldots + Y_n$ where the Y_n are conditionally independent, given J, such that the distribution of Y_n is $H^{(ij)}$ and the transition matrix $\mathbf{P} = (p_{ij})_{i,j \in E}$ of J or, equivalently, $F^{(ij)} = p_{ij}H^{(ij)}$; note that

$$F^{(ij)}(\infty) = p_{ij},\ F^{(ij)}(y) = \mathbf{P}_i(X_1 \le y, J_1 = j).$$

Of models where discrete time MAPs play an important role, we mention in particular Markov chains with transition matrices of GI/M/1 or M/G/1 type, see Neuts [38] or [5] Ch. X.4.

The generalization of the m.g.f. is the $E \times E$ matrix $\widehat{\boldsymbol{F}}[\theta]$ with ijth element $\widehat{F}^{(ij)}[\theta]$, and as generalization of the cumulant g.f., one can take $\kappa(\theta)$ where $e^{\kappa(\theta)}$ is the Perron-Frobenius eigenvalue of $\widehat{\boldsymbol{F}}[\theta]$; denote the corresponding right eigen vector by $\boldsymbol{h}^{(\theta)} = \left(h_i^{(\theta)}\right)_{i \in E}$, i.e., $\widehat{\boldsymbol{F}}[\theta]\boldsymbol{h}^{(\theta)} = e^{\kappa(\theta)}\boldsymbol{h}^{(\theta)}$. The ECM corresponding to θ is given by

$$\widetilde{\boldsymbol{P}} = e^{-\kappa(\theta)}\boldsymbol{\Delta}^{-1}_{\boldsymbol{h}^{(\theta)}}\widehat{\boldsymbol{F}}[\theta]\boldsymbol{\Delta}_{\boldsymbol{h}^{(\theta)}} \quad \text{and} \quad \widetilde{H}^{(ij)}(dx) = \frac{e^{\theta x}}{\widehat{H}^{(ij)}[\theta]}H^{(ij)}(dx)$$

and the OECM by taking $\theta = \theta^*$ where θ^* is the solution of $\kappa(\theta^*) = 0$. Here $\boldsymbol{\Delta}_{\boldsymbol{h}^{(\theta)}}$ is the diagonal matrix with the $h_i^{(\theta)}$ on the diagonal. The likelihood ratio is

$$W_n(\boldsymbol{P} \mid \widetilde{\boldsymbol{P}}) = \frac{h^{(\theta)}(J_0)}{h^{(\theta)}(J_n)}e^{-\theta S_n + n\kappa(\theta)}. \tag{17.12}$$

In continuous time, a MAP with an underlying finite Markov process $\{J_t\}_{t \ge 0}$ has a simple description, cf. e.g., Neveu [40]. The clue for understanding is the structure of a continuous time random walk (a process with stationary independent increments) as the independent sum of a deterministic drift, a Brownian component, and a pure jump (Levy) process; see e.g., [5] Ch. III.8. Let the intensity matrix of $\{J_t\}$ be $\boldsymbol{\Lambda} = (\lambda_{ij})_{i,j \in E}$. On an interval $[t, t+s)$ where $J_t \equiv i$, the MAP $\{S_t\}$, then, evolves like a process with stationary independent increments with the drive μ_i, the variance σ_i^2 of the Brownian component, and the Levy measure $\nu_i(dx)$ depending on i. In addition, a jump of $\{J_t\}$ from i to $j \ne i$ has probability q_{ij} of giving rise to a jump of $\{S_t\}$ at the same time, the distribution of which then has some distribution $B^{(ij)}$.

Let $\widehat{\boldsymbol{F}}_t[\theta]$ be the matrix with ijth element $\mathbf{E}_i\left[e^{\theta S_t}; J_t = j\right]$. It is easy to see that $\widehat{\boldsymbol{F}}_t[\theta] = e^{tG[\theta]}$, where

$$G^{(ij)}[\theta] = \begin{cases} q_{ij}\lambda_{ij}\widehat{B}^{(ij)}[\theta] + (1 - q_{ij})\lambda_{ij}, & i \ne j \\ \lambda_{ii} + \mu_i\theta + \frac{1}{2}\sigma_i^2\theta^2 + \displaystyle\int_{-\infty}^{\infty}(e^{\theta x} - 1)\nu_i(dx), & i = j. \end{cases}$$

We define $\kappa(\theta)$ as the dominant eigenvalue of $G[\theta]$ and $\boldsymbol{h}^{(\theta)}$ as the corresponding right eigenvector. Equivalently, $e^{t\kappa(\theta)}$ is the Perron-Frobenious eigenvalue of $\widehat{\boldsymbol{F}}_t[\theta]$ and $\boldsymbol{h}^{(\theta)}$ is the right eigenvector. The ECM corresponding to θ is then given by

$$\widetilde{\boldsymbol{\Lambda}} = \boldsymbol{\Delta}^{-1}_{\boldsymbol{h}^{(\theta)}}G[\theta]\boldsymbol{\Delta}_{\boldsymbol{h}^{(\theta)}} - \kappa(\theta)\boldsymbol{I},\ \widetilde{\mu}_i = \mu_i + \theta\sigma_i^2,\ \widetilde{\sigma}_i^2 = \sigma_i^2,$$

$$\nu_i(dx) = e^{\theta x}\nu_i(dx),\ \widetilde{q}_{ij} = \frac{q_{ij}\widehat{B}^{(ij)}[\theta]}{1 + q_{ij}(\widehat{B}^{(ij)}[\theta] - 1)},\ \widetilde{B}^{(ij)}(dx) = \frac{e^{\theta x}}{\widehat{B}^{(ij)}[\theta]}B^{(ij)}(dx).$$

In particular, the expression for $\widetilde{\boldsymbol{\Lambda}}$ means

$$\widetilde{\lambda}_{ij} = \frac{h_j^{(\theta)}}{h_i^{(\theta)}}\lambda_{ij}\left[1 + q_{ij}(\widehat{B}_{ij}[\theta] - 1)\right],\ i \neq j$$

(the diagonal elements are determined by $\lambda_{ii} = -\sum_{j \neq i}\lambda_{ij}$), and, if $\nu_i(dx)$ is compound Poisson, $\nu_i(dx) = \beta_i B_i(dx)$ with $\beta_i < \infty$ and B_i a probability measure, then, also, $\widetilde{\nu}_i(dx)$ is compound Poisson with

$$\widetilde{\beta}_i = \beta_i \widehat{B}_i[\theta], \quad \widetilde{B}_i(dx) = \frac{e^{\theta x}}{\widehat{B}_i[\theta]}B_i(dx). \tag{17.13}$$

The likelihood ratio on $[0, T]$ is just (17.12) with n replaced by T.

Example 17.11. If all $\sigma_i^2 = 0$, $\nu_i = 0$, $q_{ij} = 0$, we have a process with piecewise linear sample paths with slope μ_i when $J_t = i$. This process (or rather its reflected version, cf. Example 17.14) is a *Markovian fluid*, a process of considerable current interest because of its relevance for ATM technology; for some recent references, see Asmussen [8] and Rogers [43]. The ECM just means to replace λ_{ij} with $\widetilde{\lambda}_{ij} = h_j^{(\theta)}\lambda_{ij}/h_i^{(\theta)}$ for $i \neq j$ and is, therefore, another Markovian fluid. For some special structures like independent sources, the eigenvalue problem determining the OECM (i.e., θ^*) can be reduced quite a lot by using the concept of *effective bandwidth*; see the case study in Section 17.8.2.

Example 17.12. Assume that all $\sigma_i^2 = 0$, $\mu_i = -1$, $q_{ij} = 0$, and that $\nu_i = \beta_i B_i(dx)$ corresponds to the compound Poisson case with the B_i concentrated on $(0, \infty)$. This process (or rather its reflected version) corresponds to the workload process in the Markov-modulated M/G/1 queue with arrival intensity β_i and service time distribution B_i of the customer arriving when $J_t = i$, and the ECM just means to replace these parameters by the ones given by (17.13).

Example 17.13. Assume that $\sigma_i^2 = 0$, $\mu_i = 0$, $\nu_i = 0$, $q_{ij} = 0$ for all i and that $B^{(ij)}$ is concentrated on $\{0, 1\}$ for all i, j. Then the MAP is a counting process, in fact, the same as the Markovian point process introduced by Neuts [37] and increasingly popular as a modeling tool.

ECM for Markov additive processes goes back to a series of papers by Keilson and Wishart and others in the sixties, e.g., [27]. To our knowledge, the first use of the concept in simulation is Asmussen [6]. Further recent references are Bucklew [13], Bucklew et al. [14], Lehtonen and Nyrhinen [33], and Chang et al. [15]. Again, some of the most interesting applications are in combination with duality ideas, which for infinite buffer problems just means time reversal.

Example 17.14. Let S_t be the MAP described in Example 17.11. Then, the fluid model of interest is

$$V_t = S_t - \min_{0 \leq u \leq t} S_u.$$

Define the cycle as $C = inf\{t > 0: S_t = 0\}$ (this definition is only interesting if $J_0 = i$ with $\mu_0 > 0$), and assume that the rare event $A(x)$ is the event $\{sup_{0 < t < C} V_t \geq x\}$ of buffer overflow within the cycle. Then, (noting that $S_t = V_t$ for $t < C$), we can just perform the simulation by performing OECM for the MAP $\{S_t\}$ and running it until it hits either x or 0. If, instead, $A(x) + \{V \geq x\}$ is defined in terms of the steady state, we first note the well-known representation $V \stackrel{\mathfrak{D}}{=} max_{0 \leq t < \infty} S_t^*$ where $\{S_t^*\}$ is the MAP we obtain by time-reversing $\{J_t\}$ (replacing λ_j by $\widetilde{\lambda}_{ij} = \pi_j\lambda_{ij}/\pi_i$, where π is the stationary distribution), leaving $\tau^*(x) = inf\{t: S_t^* \geq x\}$ and the simulation is performed by running $\{S_t^*\}$, using the OECM, until it hits x. Similar remarks apply to Example 17.13.

The approach can, to some extent, be generalized beyond finite E. An example is given in Section 17.8. Note, however, that if E is infinite, a MAP may be quite complicated (an example is provided by the local time of a diffusion) and that the existence of dominant eigenvalues for the relevant integral operator does not always hold.

17.6 STEADY-STATE RARE EVENTS SIMULATION

Consider the situation where the rare event is $A(x) = \{L > x\}$, with L being an r.v. having the limiting stationary distribution of some stochastic sequence $\{L_n\}$. Even if $A(x)$ is not rare, the problem of estimating $\alpha(x) = \mathbf{P}A(x)$ from simulation is nontrivial because it is not obvious how to generate an r.v. with the same distribution as L (in fact, this is generally impossible, cf. [9]) or even an r.v. $Z(x)$ with $\mathbf{E}[Z(x)] = \alpha(x)$. For the GI/G/1 queue, a solution to this problem is provided by (17.6), but, in general, it is not obvious at first sight how to express steady-state probabilities in terms of r.v.'s which can be simulated in finite time.

As for methods developed to deal with steady-state rare events simulation, the regenerative method plays a prominent role. Let $\{L_n\}$ be regenerative with generic cycle length C; then,

$$\alpha(x) = \frac{\alpha_1(x)}{\alpha_0}, \text{ where } \alpha_1(x) = \mathbf{E}[Z_1(x)], \qquad Z_1(x) = \sum_{i=1}^{C} I(L_i > x),$$
$$\alpha_0 = \mathbf{E}[Z_0], \quad Z_0 = C. \tag{17.14}$$

Standard regenerative simulation amounts to simulating N replications of the random vector $(Z_1(x), Z_0)$ and estimating the mean by

$$\bar{\alpha}(x) = \frac{\bar{\alpha}_1(x)}{\bar{\alpha}_0}, \text{ where } \bar{\alpha}_1(x) = \frac{1}{N}\sum_{j=1}^{N} Z_{1j}(x), \bar{\alpha}_0 = \frac{1}{N}\sum_{j=1}^{N} Z_{0j}. \tag{17.15}$$

Under standard conditions, cf. [44], $\bar{\alpha}_1/\bar{\alpha}_0$ is asymptotically distributed normal with mean α_1/α_0 and variance

$$\frac{1}{N}\left(\frac{1}{\alpha_0^2}\sigma_{11} + \frac{\alpha_1^2}{\alpha_0^4}\sigma_{00} - \frac{2\alpha_1}{\alpha_0^3}\sigma_{01}\right). \tag{17.16}$$

Here, $(\sigma_{ij})_{i,j=0,1}$ denotes the covariance matrix of $(Z_1(x), Z_0)$.

We now survey two approaches on how to combine the regenerative estimator (17.15) with IS for rare events estimation and (in somewhat more detail) a method based upon a generalization of (17.6). This generalization was first given in Siegmund [53] and, then, further extended in Asmussen [7]. Its potential for simulation analysis is discussed in detail, here, for the first time.

17.6.1 Standard IS regenerative estimator

Assume that the regenerative process is driven by an i.i.d. sequence $Y_1, Y_2, \ldots$ with common distribution F of the Y_i, and let $L_n = L_n(Y_1, \ldots, Y_n)$ be the sample performance, e.g., the steady-state number of customers in a queueing network. Asmussen, Rubinstein, and Wang [11] suggested, instead of (17.15), the following LR estimator:

$$\bar{\alpha}(x) = \frac{\bar{\alpha}_1(x)}{\bar{\alpha}_0}, \text{ where } \bar{\alpha}_1(x) = \frac{1}{N}\sum_{j=1}^{N}\sum_{i=1}^{C_j} W_{ji} I(L_{ji} > x), \tag{17.17}$$

$$\bar{\alpha}_0 = \frac{1}{N}\sum_{j=1}^{N}\sum_{i=1}^{C_j} W_{ji}, \tag{17.17a}$$

and

$$W_{ji} = g(Y_{j1})\ldots g(Y_{ji}), \quad \text{where } g = dF/dG. \tag{17.18}$$

From a practical point of view, the finding of [11] is that, in order to obtain substantial variance reduction, one should choose the changed distribution G of the Y_i such that it corresponds to a queue with *moderately larger traffic intensity.*

Nevertheless, the estimator is *exponential time.* To demonstrate this, we return to the simple random walk (GI/G/1) structure of Section 17.3 where L_n is a reflected random walk. Note that the OECM is not feasible here since $\mathbf{E}_{\theta^*}[Y_i] > 0$ so that $\mathbf{P}_{\theta^*}(C < \infty) < 1$ and the simulation may never terminate. For the same reason, one has to impose the constraint $\mathbf{E}_G[Y_i] \le 0$ on the IS distribution G; in fact, the G needs to be even further restricted for the estimator to have *finite variance*, cf. [11].

Proposition 17.15. *Consider the random walk setting of Section* 17.3. *Then, the estimator $\bar{\alpha}(x)$ in* (17.17) *is exponential time.*

Proof. Note, first, that $\alpha(x)$ and $\alpha_1(x)$ are both of order $e^{-\theta^* x}$. Obviously, $Z_1(x) \ge W_{\tau(x)} I(\tau(x) > C)$ so that Corollary 17.8 implies that $\beta(x) = \mathbf{E}_G[Z_1^2(x)/\alpha(x)^2]$ goes to infinity at least as fast as $e^{\epsilon x}$ for some $\epsilon > 0$. Therefore, the first term in (17.16) grows at least like $\alpha(x)^2 e^{\epsilon x}$. Since σ_{00} does not depend on x, the second term is $O(\alpha(x)^2)$, whereas for the last we use the Cauchy-Schwartz inequality to infer that

$$\sigma_{01} = O(\sqrt{\sigma_{11}}) = O\Big(\sqrt{\mathbf{E}_G[Z_1^2(x)]}\Big) = \alpha(x)O(\sqrt{\beta(x)}).$$

Hence, the variance in (17.16) can be written as

$$O(1)\alpha(x)^2\beta(x) + O(\alpha(x)^2) - \alpha(x)^2 O(\sqrt{\beta(x)}),$$

which is of order $\alpha(x)^2\beta(x)$, i.e., at least $\alpha(x)^2 e^{\epsilon x}$. This is the desired conclusion. □

17.6.2 The switching regenerative estimator

The idea behind switching estimators is to change the IS distribution G dynamically within the cycle. The basic example is to use OECM ($G = F_{\theta^*}$) until time $\tau(x)$ (the 'rare event is hit') and then switch off the IS ($G = F$) during the rest cycle; note that this procedure does not run into the difficulty of a possibly nonterminating simulation to which the naive combination of OECM with regenerative simulation is prone, as discussed above.

Switching estimators were introduced independently by several authors (e.g., [18], [26], [41] and [44]). Formally, we define, here, a switching regenerative estimator by the requirement that the IS distribution at time n (the distribution used for generating Y_n) is allowed to be dependent on n and $Y_1, \ldots, Y_{n-1}$. Here, we shall look only at the case where the IS distribution at time n is, say G_1, for $n \le \tau(x)$ and is G_2 for $\tau(x) < n \le C$; for a more general case, see Kriman and Rubinstein [31]. They consider a wide class of IS distributions such that $\mathbf{E}[Y_n] > 0$, $1 \le n \le \tau(x)$, and $\mathbf{E}[Y_n] < 0$, $n \in (\tau(x), C]\backslash I_1$, where I_1 is a finite set of indices. They proved that, for the switching estimator $\bar{\alpha}(x)$ to be polynomial time, it is necessary to use the OECM $G_n = F_{\theta^*}$ for $n \in [1, \tau(x)]\backslash I_2(x)$, where $I_2(x)$ is a set of indices consisting of $o(x)$ elements.

If $g_i = dF/dG_i$, then the LR is

$$W_n = \begin{cases} g_1(Y_1)\ldots g_1(Y_n) & n \le \tau(x) \\ W_{\tau(x)} g_2(Y_{\tau(x)+1})\ldots g_2(Y_n) & \tau(x) < n \le C. \end{cases} \tag{17.19}$$

It follows that the switching estimator $\bar{\alpha}_1(x)$ for $\alpha_1(x)$ based upon N replications of $Z_1(x)$ is defined as in (17.17), only with the changed LR W_n. We assume, without further discussion, that G_1 and G_2 have been chosen such that the *variance is finite.* The following result can be seen as an extension of Corollary 17.8 for the transient case (Corollary 17.8 corresponds to the case, where one switches to a server with infinite service rate after reaching level x).

Proposition 17.16. *Consider the random walk setting for estimating* $\alpha_1(x) = \mathbf{E}[\sum_{t=1}^{C} I_{\{L_t > x\}}]$, *where* L_t *is the waiting time in the* GI/G/1 *queue. Then, regardless of the choice of* G_2, *the switching estimator* $\bar{\alpha}_1(x)$ *for* $\alpha_1(x)$ *in* (17.17), (17.18) *is polynomial time if, and only if,* $G_1 = F_{\theta^*}$.

Proof. Necessity again follows by Corollary 17.8, using

$$Z_1(x) \ge g_1(Y_1)\ldots g_1(Y_{\tau(x)})I(\tau(x) < C).$$

For sufficiency, let

$$Z_1(y,x) = \sum_{i=1}^{\tau_{-(y,x)}} g_2(Y_1)\ldots g_2(Y_n)I(y + S_n > 0),$$

$$Z_3(y) = \sum_{i=1}^{\infty} g_2(Y_1)\ldots g_2(Y_n)I(y + S_n > 0),$$

where $\tau_{-(y,x)}$ is the first passage time of the random walk from $y > 0$ to levels $< -x$. Clearly, $Z_2(y,x)\uparrow Z_3(y)$ as $x\to\infty$, and the assumption of finite variance is easily seen to be equivalent to $\mathbf{E}[Z_3^2(y)] < \infty$ for each y. Further, $\mathbf{E}[Z_3^2(y)] = O(y^2)$, $Z_1(x) = e^{-\theta^* x}e^{-\xi(x)}Z_2(\xi(x),x)$, so that

$$\mathbf{E}[Z_1^2(x)] \le e^{-2\theta^* x}\mathbf{E}[e^{-2\xi(x)}Z_3(\xi(x))] = \alpha(x)^2\mathbf{E}[e^{-2\xi(x)}O(\xi(x)^2)] = O(\alpha(x)^2).$$

□

Now, consider the ratio estimator $\bar{\alpha}_1(x)/\bar{\alpha}_0$ for $\alpha(x) = \alpha_1(x)/\alpha_0$. We have, here, several options for the precise definition. The two most obvious ones to estimate α_0 seem to be (a) from *the same sample path* (governed by $G_1 = F_{\theta^*}$ up to $\tau(x)$ and $G_2 = F$ after that) and (b) from an *independent simulation* by the CMC method so that $Z_0 = C$ and the governing measure is $\mathbf{P}_F$ throughout.

Proposition 17.17. *Consider the random walk setting of Section* 17.3. *Then the switching regenerative estimator* $\bar{\alpha}_1(x)/\bar{\alpha}_0$ *with* $G = F_{\theta^*}$ *is polynomial time in both cases* (a) *and* (b).

Proof. By Proposition 17.16, the variance $\sigma_{11}(x)$ of $\bar{\alpha}_1(x)$ is $O(\alpha(x)^2)$ and, hence, it is sufficient to check that $\sigma_{00} = O(x^p)$ for some p (use (17.16) with $\sigma_{0\ 1}(x)$ estimated by Cauchy-Schwartz). In case (b), the assertion is trivial because σ_{00} does not depend on x (here in fact, $\sigma_{01} = 0$). In case (a), note that

$$A_0 = Z_0^{(1)} + Z_0^{(2)}, \text{ where } Z_0^{(1)} = \sum_{i=1}^{\tau(x)} W_n, \quad Z_0^{(2)} = C - \tau(x).$$

Here $W_n = e^{-\theta^* S_n} \le 1$, so that $Z_0^{(1)} \le \tau(x)$. The proof is, then, easily completed with $p = 2$ by noting that $\mathbb{E}[\tau(y)^2] = O(y^2)$ for a random walk with positive drift (take the $\mathbf{P}_{F_{\theta^*}}$-walk with $y = x$ for $Z_0^{(1)}$ and the $\mathbf{P}_F$-walk with $y = S_{\tau(x)}$ and the sign reversed for $Z_0^{(2)}$). □

17.6.3 Duality between transient and steady-state probabilities

The motivation for the study in this section is the success that formula (17.6) has had for the simulation of steady-state characteristics of the GI/G/1 queue. As a second motivating example, consider a random walk $L_0, L_1, \ldots$ between two reflecting barriers $0, b$, i.e., $L_{n+1} = \Gamma(L_n + Y_n)$ where $Y_1, Y_2, \ldots$ are i.i.d. and Γ is the two-sided reflection operator

$$\Gamma(z) = \begin{cases} 0, & z \le 0 \\ z, & 0 \le z \le b \\ b, & b \le z < \infty. \end{cases}$$

Let $S_n = Y_1 + \ldots + Y_n$ be the unrestricted random walk and define

$$\tau[a, b) = inf\{n = 0, 1, 2, \ldots : S_n < a \text{ or } S_n \ge b\}$$

$(a \le 0, b \ge 0)$. The following representation of the stationary distribution of L is implicit in Lindley [36] and explicit in Siegmund [53]. The result does not appear to have been noticed in queueing theory at all; for a direct proof, see [7].

Proposition 17.18. *For $x \in [0, b]$, $\mathbf{P}(L \ge x) = \mathbf{P}(S_{\tau[x-b, x)} \ge x)$.*

An example of the structure in Proposition 17.18 is provided by $L_n =$ *the queue length just prior to the nth arrival in the* GI/M/1 *queue with waiting room (buffer) of size $b < \infty$.* Here, $Y_i = 1 - Z_i$, where Z_i is distributed as the number of Poisson events (fictitious service events) in an interarrival interval, i.e.,

$$\mathbf{P}(Z_i = j) = \int_0^\infty e^{-\delta t} \frac{(\delta t)^j}{j!} F(dt), \tag{17.20}$$

where δ is the service intensity and F the interarrival distribution. The rare event of interest is $A(b) = \{L = b\} = \{L \ge b\}$, the event of the buffer being full. In order to make the event $\{L = b\}$ rare in the limit $b \to \infty$, we assume $\rho < 1$. The method suggested by combining Proposition 17.18 with the study of the infinite buffer case in Section 17.3 is to perform the OECM for the Y_i, simulate the random walk from this IS distribution until either $\{-1, -2, \ldots\}$ or $\{b, b+1, \ldots\}$ is hit, and let

$$Z(b) = W_{\tau[0b)} I(S_{\tau[0,b)} \ge b). \tag{17.21}$$

An easy calculation, which we omit, shows that the OECM for Y_i is the same as we obtain by first performing the OECM for the interarrival and service time distribution, as in Section 17.8.1, i.e., switch A, δ to $A_{-\theta^*}, \delta^* = \delta - \theta^*$, and just plug these parameters into (17.20). It is also straightforward to modify the proof of Theorem 17.7 to show that this IS estimator is polynomial time and becomes exponential time by any other change of measure in the class of distributions with the Y_i being i.i.d.

An appealing feature of the method is that it allows simulation of $\alpha(b)$ *simultaneously for numerous values of* b, *say* $b_1 < b_2 < \ldots < b_p$, *from a single sample path*: run the simulation until exit from $[0, b_1)$. If $S_{\tau[0,b_1)} < 0$, stop and put

(17.21) equal to 0 for $i = 1,\dots,p$. Otherwise, if $S_{\tau[0,b_1)} \geq b_1$, compute (17.21) for $b = b_1$ by letting $I(\cdot) = 1$ and recording the current LR W, and continue the simulation until exit from $[0, b_2)$. If $S_{\tau[0,b_2)} < 0$, stop and put (17.21) equal to 0 for $i = 2,\dots,p$. Otherwise, compute (17.21) for $b = b_2$, as before, and continue the simulation until exit from $[0, b_3)$. Continuing in this way, the simulation stops upon exit from $[0, b_p)$, and (17.21) has been recorded for all b_i at that time.

The approach to (finite or infinite buffer) queueing probabilities via random walk first passage probabilities may appear special, but a general framework was, in fact, suggested by Siegmund [53]; we survey the discrete time case only but include a continuous-time example (Section 17.8.3). Denote the given process by $L_0, L_1, \dots$ (in the setting of rare events, we are interested in $\alpha(x) = \mathbf{P}(L \geq x)$), and assume it is a Markov chain with state space $[0, \infty)$, which is ergodic (possibly with the stationary distribution concentrated on a reduced absorbing subset of $[0, \infty]$) and *stochastically monotone* in the sense that $\mathbf{P}_y(L_1 \geq x)$ is an increasing function of y. The *dual* process is, then, a Markov chain $R_0, R_1, \dots$ on $[0, \infty]$ with transition function given by

$$\mathbf{P}_x(R_1 \leq y) = \mathbf{P}_y(L_1 \geq x) \tag{17.22}$$

(existence requires the r.h.s. to be right continuous in y), and, if $\tau = inf\{n: R_n = 0\}$, one has

$$\mathbf{P}(L \geq x) = \mathbf{P}_x(\tau < \infty); \tag{17.23}$$

here, in fact $\{0\}$ is an absorbing set for $R_0, R_1, \dots$. The small twists in these definitions necessary to make the setup work will be clear from the following example.

Example 17.19. Let $L_0, L_1, \dots$ be the reflecting random walk $L_{n+1} = \Gamma(L_n + Y_n)$ with some arbitrary initial value $L_0 \in [0, \infty]$. A tedious inspection of all possible cases ($r = 0, 0 < r < b, b \leq r < \infty$, $r = \infty, 0 \leq y \leq r, r < y < \infty$), then, shows that the dual process can be described by the dynamic equation $R_{n+1} = \Delta(R_n, Y_n)$, where

$$\Delta(r, y) = \begin{cases} 0, & r = 0 \text{ or } r - y \leq 0 \\ r - y, & 0 < r \leq b \text{ and } r - y \geq 0 \\ \infty, & r > b. \end{cases}$$

That is, when $0 < x = R_0 \leq b$, $R_0, R_1, \dots$ evolve as $x - S_n$ until either $(-\infty, 0]$ or (b, ∞) is hit. In the first case, the chain is instantaneously set to 0 and absorbed there. In the second, it is set to ∞ in the next step and absorbed there. It follows immediately from this description that the events $\tau < \infty$ and $S_{\tau[x-b,x)} > b$ coincide when $R_0 = x$, and, thus, Proposition 17.18 and equation (17.23) are the same result.

In general, the simulation method we suggest for $\alpha(x)$ is as follows: compute the dual chain R and evaluate $\mathbf{P}_x(\tau < \infty)$ by simulating R, driven by the local OECM discussed in Section 17.5.1, starting from $R_0 = x$ and continuing until 0 is hit. Note that, by ergodicity, the V chain will, typically, have a downwards drift in an overall sense (the size depends in general upon location). The duality relation (17.22), then, indicates that the overall drive of R is positive, i.e., is changed to negative by the local OECM, so that, indeed, τ is finite w.r.t. this IS distribution, and the simulation terminates in finite time.

Seemingly, neither the stochastic monotonicity assumption nor the required right-continuity in (17.22) drastically restricts the usefulness of the setup; much more serious is the Markov property. To this end, an extension of [53] has recently been developed by Asmussen [7] by allowing v to be modulated by a Markov process J. More precisely, it is assumed that $\{(J_n, V_n)\}$ is a Markov chain with state space $E \times [0,\infty)$, such that the marginal distribution of $\{J_n\}$ corresponds to an ergodic Markov process on E (the stationary distribution π then exists) and that $\mathbf{P}_{i,x}(V_t \geq y, J_t = j)$ is a nondecreasing right-continuous function of x for fixed i, j, y (we take E to be finite here though this is not crucial). The dual is, then, a Markov chain $\{(J_n^*, R_n)\}$ with state space $E \times [0,\infty]$, such that (marginally) $\{J_n^*\}$ is just the standard time reversed version of $\{J_n\}$ and that

$$\mathbf{P}_{j,y}(R_1 \leq x, J_1^* = i) = \frac{\pi_i}{\pi_j}\mathbf{P}_{i,x}(V_1 \geq y, J_1 = j),$$
$$\mathbf{P}(V \geq x, J = j) = \pi_j \mathbf{P}_{j,x}(\tau < \infty). \tag{17.24}$$

If $\alpha(x) = \mathbf{P}(V \geq x, J = j)$, it is suggested in [7] to simulate $\alpha(x)$ by computing the dual $\{(J_n^*, R_n)\}$ and evaluate $\mathbf{P}_{j,x}(\tau < \infty)$ by simulating $\{(J_n^*, R_n)\}$ driven by the local Markov-modulated OECM, starting from $J_0^* = j$, $R_0 = x$ and continuing until 0 is hit.

The Markov dependence in this setup seems to widen the scope of the method quite a lot. An obvious example is Markov-modulated queues and fluids with a finite buffer; another one is presented in Section 17.8.3.

17.7 ROBUSTNESS OF THE OECM

We now address the following question: How much can one perturb the parameter θ in the ECM from the optimal value θ^* so that the IS estimator remains polynomial time?

The practical motivation for this question comes from the fact that the OECM may be difficult to compute exactly, say, for a general Markovian fluid of more complicated structure than the one studied in Section 17.5.1. Here is the theoretical answer to the question separately for the random walk and the regenerative settings.

Proposition 17.20. *Consider the random walk setting of Section 17.3. Assume that $\theta = \theta(x)$ varies with x such that $\theta \to \theta^*$ as $x \to \infty$. Then, a necessary and sufficient condition for $Z(x) = W_{\tau(x)}(F \mid F_{\theta(x)})$ to be polynomial time is*

$$\Delta = \Delta(x) = \theta(x) - \theta^* = O\left(\sqrt{\frac{\log x}{x}}\right). \tag{17.25}$$

Proof. The second moment is

$$\begin{aligned}\mathbf{E}_\theta[W_{\tau(x)}(F \mid F_\theta)^2] &= \mathbf{E}_{\theta^*}\left[W_{\tau(x)}(F \mid F_\theta)^2 W_{\tau(x)}(F_\theta \mid F_{\theta^*})\right] \\ &= e^{-2\theta^* x} e^{-\Delta x}\mathbf{E}_{\theta^*}[e^{-(\theta+\theta^*)\xi + \tau\kappa}],\end{aligned}$$

where $\theta = \theta(x)$, $\kappa = \kappa(\theta)$, and so, we are looking for condition

$$e^{-\Delta x}\mathbf{E}_{\theta^*}[e^{-(\theta+\theta^*)\xi + \tau\kappa}] = O(x^p) \tag{17.26}$$

for some $p < \infty$.

To this end, we use Theorem 17.23 of the Appendix to the random walk with governing probability measure $\mathbf{P}_{\theta^*}$ and $\beta(x) = \sqrt{x}\kappa(\theta)$. Then $\mu = \kappa'(\theta^*)$, $\sigma^2 = \kappa''(\theta^*)$. By Taylor expansion, $\kappa(\theta) \approx \Delta\mu$; in particular, the condition $\beta(x) = o(\sqrt{x})$ is fulfilled because of $\Delta \to 0$. It follows that

$$e^{-\Delta x}\mathbf{E}_{\theta^*}[e^{-(\theta+\theta^*)\xi + \tau\kappa}]$$
$$= \exp\{\beta(x)(\tau(x) - x/\mu)/\sqrt{x}\} \cdot \exp\{\beta(x)\sqrt{x}/\mu - \Delta x\}$$
$$= \exp\{\beta(x)^2\kappa''(\theta^*)(1+o(1))/2\kappa''(\theta^*)^3\} \cdot \exp\{x\Delta^2\kappa''(\theta^*)(1+o(1))/2\kappa'(\theta^*)\}$$
$$= \exp\{x\Delta^2(\kappa''(\theta^*)/\kappa'(\theta^*) + o(1))\}.$$

Thus the necessary and sufficient condition is $\Delta^2 x = O(\log x)$. □

Corollary 17.21. *Consider the SR estimator* $\bar{\alpha}_1(x)/\bar{\alpha}_0$. *Assume that* $\theta = \theta(x)$ *varies with* x *such that* $\theta \to \theta^*$. *Then, the estimator is polynomial time, provided* (17.25) *holds.*

Proof. Since $Z_1(x) \le W_{\tau(x)}(F \mid F_\theta)$, it follows immediately from the above proof that $Var Z_1(x) = O(x^p e^{-2\theta^* x})$. The rest of the proof is just as in the proof of Proposition 17.17. □

Notice that Proposition 17.20 and Corollary 17.21 are associated with the transient and the steady-state regimes, respectively.

The following result characterizes the computational cost when the estimator $\bar{\alpha}$ is not polynomial time (the proof follows directly from Theorem 5.1 of [31]).

Proposition 17.22. *Consider an importance sampling distribution* G *(with p.d.f.* g*) such that:*

1. $\mathrm{E}_G[Y_n] > 0$;
2. $\int_{-\infty}^{\infty} g^2(y)/f(y)dy = C < \infty$;
3. *There exists a unique solution* $\theta^{**} > 0$ *of the equation*
$$\int_{-\infty}^{\infty} f^2(y)/g(y)\exp(-\theta^{**}y)dy = 1$$
such that
$$-\infty < \int_{-\infty}^{\infty} yf^2(y)/g(y)\exp(-\theta^{**}y)dy < 0.$$

Then, the simulation cost of the estimator $\bar{\alpha}$ *can be represented in the form*

$$T_G(x)N_G(x) = \exp(zx + o(x)),$$

where $z = \theta^{**} - 2\theta^* > 0$.

Here, z is called the *exponential rate* of the computation cost and can be used as an indicator of "robustness" of the computational cost with respect to the choice of the importance sampling distribution in the sense that the ratio $R \equiv T_G(x)N_G(x)/(T_{F_{\theta^*}}N_{F_{\theta^*}})$ can be approximated rather well by the term $\exp(zx)$. Note that, for $G = F_{\theta^*}$, we have $z = 0$ and the estimator $\alpha(x)$ reduces to polynomial time.

Table 1 presents the computational cost (simulation time) $T_G(x)N_G(x) \equiv T_\theta N_\theta$ corresponding to the relative width 0.1 of the 95% confidence interval for $\alpha(x)$, the ratio R_θ, called the *loss factor*, (approximation of $\exp(zx)$), and the exponential rate z, all as functions of θ (and the relative perturbation $\delta = (\theta - \theta^*)/\theta^*$), while estimating the probability $\alpha(x) = 5.34 \cdot 10^{-10}$, $(x = 40)$ for the M/M/1 queue length process, with $\rho = 0.6$ and service rates $\delta = 1$.

Table 1. The Efficiency of the RS Estimator as a Function of θ for the M/M/1 Queue with $\rho = 0.6$, $\delta = 1$, and $\alpha(x) = 5.34 \cdot 10^{-10}$, $(x = 40)$

θ	δ	$T_\theta N_\theta$	R_θ	$\exp(zx)$	z
0.30	-0.25	$1.20 \cdot 10^6$	4.5	6.1	$4.53 \cdot 10^{-2}$
0.35	-0.125	$5.87 \cdot 10^5$	2.2	1.7	$1.33 \cdot 10^{-2}$
0.40*	0.00	$2.76 \cdot 10^5$	1.0	1.0	0.0
0.43	0.075	$4.27 \cdot 10^5$	1.6	1.3	$1.36 \cdot 10^{-2}$
0.46	0.15	$1.52 \cdot 10^6$	5.5	5.2	$4.13 \cdot 10^{-2}$

More numerical results on such kinds of robustness are given in Kriman and Rubinstein [31]. They found that the SR estimator is robust with respect to small and moderate perturbation of θ in the sense that, for relative perturbations $\delta = |\theta - \theta^*|/\theta^* < 0.2$, one still obtains dramatic variance reduction, at least when $\alpha(x) \geq 10^{-20}$.

17.8 EXAMPLES

17.8.1 ECM for simple random walks and GI/G/1 queues

1. Assume that F is normal with mean μ and variance σ^2. Then $\kappa(s) = \mu s + s^2\sigma^2/2$ and
$$\kappa_\theta(s) = \mu(s+\theta) + (s+\theta)^2\sigma^2/2 = (\mu+\theta)s + s^2\sigma^2/2$$
which shows that F_θ is normal with mean $\mu + \theta$ and variance σ^2. Also, the solutions of $\kappa(\theta^*) = 0$ are 0 and -2μ, so that F_{θ^*} is normal with mean $-\mu$ and variance σ^2.
2. Assume that $Y_k = U_k - T_k$ where the U_k are i.i.d. with distribution B, the T_k are i.i.d. with distribution A, and the sequences $\{U_k\}$, $\{T_k\}$ are independent. Then, $\widehat{F}[s] = \widehat{B}[s]\widehat{A}[-s]$ so that
$$\widehat{F}_\theta[s] = \frac{\widehat{B}[s+\theta]}{\widehat{B}[\theta]}\frac{\widehat{A}[-s-\theta]}{\widehat{A}[-\theta]} = \widehat{B}_\theta[s]\widehat{A}_{-\theta}[-s]$$
which shows that $\widehat{F}_\theta$ is such that the F_θ-distribution of X_k corresponds to U_k, T_k being independent and having distributions B_θ, resp. $A_{-\theta}$. Here are some special cases:
M/M/1 Here, U, T are exponential, say, with rates $\delta > \beta$. Thus
$$\widehat{B}_\theta[s] = \frac{\delta/(\delta - s - \theta)}{\delta/(\delta - \theta)} = \frac{\delta - \theta}{\delta - s - \theta}.$$
That is, the changed service time distribution B_θ is exponential with rate $\delta_\theta =$

$\delta - \theta$. Similarly, it is seen that the changed interarrival time distribution A_θ is exponential with rate $\beta_\theta = \beta + \theta$. Furthermore, (17.5) becomes

$$1 = \mathbf{E}[e^{\theta^* U}] \cdot \mathbf{E}[e^{-\theta^* T}] = \frac{\delta}{\delta - \theta^*} \cdot \frac{\beta}{\beta + \theta^*}$$

with solution $\theta^* = \delta - \beta$. Thus the exponential change of measure given by θ^* means $\delta_{\theta^*} = \beta$, $\beta_{\theta^*} = \delta$. That is, with a popular term it is obtained by interchanging δ and β.

M/E$_2$/1 or **E$_2$/M/1** We mention these examples not so much because of their intrinsic interest but because (17.5) here reduces to a quadratic, hence, is explicitly solvable and *these two examples are more or less the only ones beyond* M/M/1 *where this is possible.*

M/D/1 Here, B is degenerate, say at 1, and A is exponential with rate $\beta < 1$. It is easily seen that B_θ is, again, degenerate at 1 and that A_θ is exponential with rate $\beta + \theta$. Furthermore, (17.5) becomes

$$1 = \mathbf{E}[e^{\theta^* U}] \cdot \mathbf{E}[e^{-\theta^* T}] = e^{\theta^*} \cdot \frac{\beta}{\beta + \theta^*}$$

which is a transcendental equation without an explicit solution and, thus, must be solved numerically.

D/M/1 Here, A is degenerate, say, again, at 1, and B is exponential with rate $\delta > 1$. It is easily seen that A_θ is, again, at 1 and that B_θ is exponential with rate $\delta - \theta$. Furthermore, (17.5) becomes another transcendental equation,

$$1 = \mathbf{E}[e^{\theta^* U}] \cdot \mathbf{E}[e^{-\theta^* T}] = \frac{\delta}{\delta - \theta^*} \cdot e^{-\theta^*}.$$

17.8.2 OECM for a fluid model with alternating renewal on-off sources

Let V_t be the buffer content at time in a fluid ATM models with N i.i.d. sources. Each source alternates between being off and on according to an alternating renewal process with distributions F_0, F_1 of the off-, resp. on-period. An active source feeds fluid into the buffer at rate a, and the maximal output rate of the buffer is c. The buffer capacity is $b < \infty$, and our rare event probability is $\alpha(b) = \mathbf{P}(V = b)$. From a numerical point of view, $\alpha(b)$ is, in principle, computable from the algorithms of Rogers [43] and Asmussen [8], provided F_0, F_1 are both phase-type, say with p_0, p_1 phases, so that the whole system is Markovian with Np_0p_1 states. However, since N is typically very large (500-1000 is typical for the literature), it is clear that this method is impracticable for complexity reasons if more than just a few phases for F_0, F_1 are required. Thus, simulation will typically be the only possibility for nonexponential F_0, F_1. We shall describe an algorithm which appears rather elegant and, at the same time, nontrivial to design by today's standard.

Let $M_t^{(j)}$ be the amount of fluid produced by a source j up to time t, let $B_j(t) \in \{0, 1\}$ be the state of source j at time t, and let $R_j(t)$ be the residual holding time of the state. Further write

$$\boldsymbol{J}(t) = (\boldsymbol{B}(t), \boldsymbol{R}(t)) = (B_j(t), R_j(t))_{j = 1, \ldots, N};$$

a typical state of $\{\boldsymbol{J}(t)\}$ is denoted $(\boldsymbol{y}, \boldsymbol{x})$ where $\boldsymbol{y} = (y_1, \ldots, y_N)$ are the states and $\boldsymbol{x} = (x_1, \ldots, x_N)$ are the holding times. Then, the buffer content is obtained by reflecting the process

$$S_t = \sum_{j=1}^{N} M_t^{(j)} - ct \tag{17.27}$$

at the boundaries $0, b$. We can view (17.27) as an MAP with underlying Markov process $\{(\boldsymbol{B}(t), \boldsymbol{R}(t))\}$, having state space $\{0,1\}^N \times (0,\infty)^N$. It is easy to see that this MAP is time reversible, so that the representation $\alpha(b) = \mathbf{P}(S_{\tau[0,b)} \geq b)$ in Proposition 17.18 is available. Thus, the simulation is performed by running $\{S_t\}$ until either 0 or b is hit. The starting values are $S_0 = 0$ whereas $(\boldsymbol{B}(0), \boldsymbol{R}(0))$ should be chosen with the stationary distribution, meaning that

$$\mathbf{P}(B(0) = i, R(0) \in dx) = \frac{\mu_i}{\mu_1 + \mu_2} \cdot \frac{1 - F_i(x)}{\mu_i} + dx,$$

where μ_i is the mean of F_i. Thus, what remains to describe is only the OECM for the MAP.

The first step is the calculation of exponential moments for the individual source. Let θ, j be fixed, write $M_t = M_t^{(j)}$ be the amount of fluid produced by source j up to time t, and let $K_i(t) = \mathbf{E}_i[e^{\theta M_t}]$, where $i = 1$ $(i = 0)$ means the source starts an on (off) period at time 0. It is easily seen that

$$K_0(t) = 1 - F_0(t) + \int_0^t K_1(t-y)F_0(dy)$$

and

$$K_1(t) = e^{\alpha\theta t}(1 - F_1(t)) + \int_0^t K_0(t-y)e^{\alpha\theta y}F_1(dy).$$

From this, it follows by Markov renewal theory ([5], Ch. X.2) that, for suitable c_0, c_1 (explicitly but canceling in the following), we have $K_i(t) \approx c_i e^{t\nu(\theta)}$, where $\nu(\theta)$ is determined by the equation

$$\widehat{F}_0[-\nu(\theta)]\widehat{F}_1[a\theta - \nu(\theta)] = 1. \tag{17.28}$$

Letting $K_{i,x}(t) = \mathbf{E}_{i,x}[e^{\theta M_t}]$, where i, x means that i is the initial state and x is its residual holding, it follows that, for $t > x$,

$$K_{0,x}(t) = K_1(t-x) \approx g_0(x)e^{t\nu(\theta)}, \quad K_{1,x}(t) = e^{ax}K_0(-x) \approx g_1(x)e^{t\nu(\theta)},$$

where $g_0(x) = c_0 e^{-\nu(\theta)x}$, $g_1(x) = c_1 e^{(a\theta - \nu(\theta))x}$.

We next compute the function $\kappa(\theta)$ for the whole system in order to find the solution θ^* of $\kappa(\theta^*) = 0$ determining the OECM. Clearly,

$$\mathbf{E}_{\boldsymbol{y},\boldsymbol{x}}[e^{\theta S_t}] \approx e^{-c\theta t}\prod_{i=1}^N K_{y_i,x_i}(t) = h_1(\boldsymbol{y},\boldsymbol{x},\theta)e^{(N\nu(\theta) - c\theta)t},$$

where

$$h_1(\boldsymbol{y},\boldsymbol{x},\theta) = \prod_{i=1}^N g_{y_i}(x_i;\theta).$$

Thus $\kappa(\theta) = N\nu(\theta) - c\theta$ and θ^* is determined by $N\nu(\theta^*) - c\theta^* = 0$. In the language of Chang et al. [15], $\nu(\theta^*)$ is the *effective bandwidth* of the individual source.

Write

$$h(\boldsymbol{y},\boldsymbol{x}) = h_1(\boldsymbol{y},\boldsymbol{x},\theta^*) = c_0^{N - |\boldsymbol{y}|}c_1^{|\boldsymbol{y}|}\exp\{a\theta^*|\boldsymbol{xy}| - \nu(\theta^*)|\boldsymbol{x}|\},$$

where $|\boldsymbol{x}| = \sum_1^N x_i$, $|\boldsymbol{xy}| = \sum_1^N x_i y_i$. The OECM is determined by

$$\begin{aligned}&\mathbf{P}^*_{\boldsymbol{y},\boldsymbol{x}}(\boldsymbol{B}(t) \in d\widetilde{\boldsymbol{y}}, \boldsymbol{R}(t) \in d\widetilde{\boldsymbol{x}})\\ &= e^{\theta^* S_t}\frac{h(\widetilde{\boldsymbol{y}},\widetilde{\boldsymbol{x}})}{h(\boldsymbol{y},\boldsymbol{x})}\mathbf{P}_{\boldsymbol{y},\boldsymbol{x}}(\boldsymbol{B}(t) \in d\widetilde{\boldsymbol{y}}, \boldsymbol{R}(t) \in d\widetilde{\boldsymbol{x}})\end{aligned}$$

$$= e^{\theta^* S_t}\left(\frac{c_0}{c_1}\right)^{|\boldsymbol{y}| - |\widetilde{\boldsymbol{y}}|} \exp\{a\theta^*(|\widetilde{\boldsymbol{y}}\widetilde{\boldsymbol{x}}| - |\boldsymbol{x}\boldsymbol{y}|) - \nu(\theta^*)(|\widetilde{\boldsymbol{x}}| - |\boldsymbol{x}|)\}$$

$$\cdot \mathbf{P}_{\boldsymbol{y},\boldsymbol{x}}(\boldsymbol{B}(t) \in d\widetilde{\boldsymbol{y}}, \boldsymbol{R}(t) \in d\widetilde{\boldsymbol{x}}). \tag{17.29}$$

Take t as the minimal x_i, without loss of generality attained for $i = 1$. Assume first $y_1 = 1$. Then, the only possible $(\widetilde{\boldsymbol{y}}, \widetilde{\boldsymbol{x}})$ is of the form

$$\widetilde{y}_i = y_i,\ \widetilde{x}_i = x_i - t_i,\ i = 2,\ldots,N,\ \widetilde{y}_1 = 0$$

and $\widetilde{x}_1$ arbitrary. Noting that

$$|\boldsymbol{y}| = 1 + |\widetilde{\boldsymbol{y}}|,\ |\boldsymbol{x}| = Nt + |\widetilde{\boldsymbol{x}}| - \widetilde{x}_1,\ |\boldsymbol{x}\boldsymbol{y}| = t|\boldsymbol{y}| + |\widetilde{\boldsymbol{y}}\widetilde{\boldsymbol{x}}|,$$

(17.29) takes the form

$$e^{\theta^* a|\boldsymbol{y}|t - c\theta^* t}\frac{c_0}{c_1}\exp\{-a\theta^* t|\boldsymbol{y}| + \nu(\theta^*)(Nt - \widetilde{x}_1)\}F_0(d\widetilde{x}_1)$$
$$= \frac{c_0}{c_1}\exp\{-\nu(\theta^*)\widetilde{x}_1\}F_\theta(d\widetilde{x}_1).$$

Combined with a similar calculation for the case $y_1 = 0$, this shows that *the OECM corresponds to replacing F_0, F_1 by the distributions*

$$F_0^*(dx) = \frac{e^{-\nu(\theta^*)x}}{\widehat{F}_0[-\nu(\theta^*)]}F_\theta(dx) = \frac{e^{-c\theta^*/N}}{\widehat{F}_0[-c\theta^*/N]}F_\theta(dx)$$

and $$F_1^*(dx) = \frac{e^{a\theta^* - \nu(\theta^*)x}}{\widehat{F}_1[a\theta^* - \nu(\theta^*)]}F_1(dx) = \frac{e^{(a - c/N)\theta^* x}}{\widehat{F}_1[(a - c/N)\theta^* x]}F_1(dx).$$

Note that $a - c/N > 0$ except for the trivial case that the buffer is always empty in stationarity. Thus, the OECM makes F_0 stochastically smaller and F_1 stochastically larger.

17.8.3 A Markov-modulated queue with rejection

In this section, we give an example of steady-state simulation which involves both local OECM (Section 17.5.1) combined with Markov-modulation (Section 17.5.2) and the duality ideas of Section 17.6.3.

Assume that $\{V_t\}_{t \geq 0}$ is a birth-death process with rates modulated by $\{J_t\}$. That is, $\{(J_t, V_t)\}_{t \geq 0}$ is a Markov jump process on $E \times \{0,1,2,\ldots\}$ with intensity matrix

$$\begin{pmatrix} \boldsymbol{Q}_0 & \boldsymbol{B}_0 & \boldsymbol{0} & \boldsymbol{0} & \boldsymbol{0} & \cdots \\ \boldsymbol{D}_1 & \boldsymbol{Q}_1 & \boldsymbol{B}_1 & \boldsymbol{0} & \boldsymbol{0} & \cdots \\ \boldsymbol{0} & \boldsymbol{D}_2 & \boldsymbol{Q}_2 & \boldsymbol{B}_2 & \boldsymbol{0} & \cdots \\ \boldsymbol{0} & \boldsymbol{0} & \boldsymbol{D}_3 & \boldsymbol{Q}_3 & \boldsymbol{B}_3 & \cdots \\ & & & & & \ddots \end{pmatrix} \tag{17.30}$$

(the blocks are $E \times E$), where (assuming that $\{J_t\}$ has only two states for simplicity)

$$\boldsymbol{D}_n = \begin{pmatrix} \delta_n^{(1)} & 0 \\ 0 & \delta_n^{(2)} \end{pmatrix},\ n = 1,2,\ldots,\ \boldsymbol{B}_n = \begin{pmatrix} \beta_n^{(1)} & 0 \\ 0 & \beta_n^{(2)} \end{pmatrix},\ n = 0,1,\ldots \tag{17.31}$$

and $$Q_n = Q - D_n - B_n, \text{ where } Q = \begin{pmatrix} q_1 & q_1 \\ q_2 & -q_2 \end{pmatrix}$$

is the intensity matrix for $\{J_t\}$. Thus, $\{V_t\}$ evolved as a birth-death process with birth rates $\beta_n^{(i)}$ and death rates $\delta_n^{(i)}$ when $J_t = i$. We want to simulate $\alpha(x)$, the probability that $V \geq x$ in the steady state. The procedure has two steps: First, determine the dual in the sense of (17.22) (or rather its continuous time analog), and, next, perform the local OECM for this dual process.

Since there are only two states, $\{J_t\}$ is time-reversible, so that $\{J_t^*\}$ has the same Q-matrix, and it follows easily from [7] that $\{(J_t^*, R_t)\}$, again, has an intensity matrix of the form (17.30) but with the B_n, D_n replaced by

$$B_0^{(R)} = D_0^{(R)} = 0,\ B_n^{(R)} = D_n,\ D_n^{(R)} = B_{n-1} (n = 1, 2, \ldots),\ Q_n = Q - B_n - D_n.$$

Now consider the local OECM for the R-process at level n. The corresponding MAP is a pure jump process with the Levy measure $\nu_i = \nu_{i,n}$ concentrated at $\{-1, 1\}$, with mass $\beta_n^{(i,R)}$ at 1 and $\delta_n^{(i,R)}$ at -1. In obvious notation (cf. Section 17.5.2), this means that

$$G_n[\theta] = e^{-\theta} D_n^{(R)} + Q - B_n^{(R)} - D_n^{(R)} + e^{\theta} D_n^{(R)}.$$

Thus $\theta^*(n)$, $h^{(n)}$ are determined by the requirement that the dominant eigenvalue of $G_n[\theta^*]$ is 0 and that $h^{(n)}$ is the corresponding right eigenvector, and the changed MAP process has parameters

$$q_{1,n}^* = \frac{h_2^{(n)}}{h_1^{(n)}} q_1,\quad q_{2,n}^* = \frac{h_1^{(n)}}{h_2^{(n)}} q_2,\quad \nu_{i,n}^*(dx) = e^{\theta^*(n)x} \nu_{i,n}(dx); \tag{17.32}$$

the expression for $\nu_{i,n}^*$ means that, when $m > 0$, the process goes from (i, m) to $(i, m+1)$ with rate $\beta_n^{*(i)}$ and to $(i, m-1)$ with rate $\delta_n^{*(i)}$ where

$$\beta_n^{*(i)} = e^{\theta^*(n)} \beta_{n-1}^{(i)},\quad \delta_n^{*(i)} = e^{-\theta^*(n)} \delta_n^{(i)}. \tag{17.33}$$

Thus, for the simulation we should run $\{(J_t^*, R_t)\}$ starting from $R_0 = x$ with J_0^* having the stationary distribution $\boldsymbol{\pi}$, using the parameters (17.32), (17.33) until a transition of the R component is observed, say, to $Y(1)$ at time σ_1. Then, the same procedure is repeated with x replaced by $Y(1)$ in (17.32), (17.33), and we continue until R hits 0, say, after $\sigma(x)$ transitions of R. The estimator is, then, the likelihood ratio $W_{\sigma(x)}$. To evaluate it, let $I(k)$, $Y(k)$ be the value of J, resp. R, just after the kth transition of R (thus $I(0) = J_0^*$, $Y(0) = x$). Then, the corresponding likelihood ratio is

$$W_n = \prod_{i=1}^{n} \frac{h_{I(i-1)}^{(Y(i-1))}}{h_{I(i)}^{(Y(i-1))}} \exp\{-\theta^*(Y(i-1))(Y(i) - Y(i-1))\}$$

as follows easily by (17.12).

For a concrete numerical example, we took

$$Q = \begin{pmatrix} -5 & 5 \\ 1 & -1 \end{pmatrix},\ \lambda_n^{(1)} = 2 \cdot \rho^n,\ \lambda_n^{(2)} = 0,\ \mu_n^{(1)} = \mu_n^{(2)} = 3.$$

A possible interpretation is that we have an exponential server working at rate 3 (regardless of the environment), interrupted Poisson arrivals at rate 2 in state 1 and rate 0 in state 2, and that an arriving customer seeing n persons in the line actually joins the line w.p. ρ^n. We simulated $\alpha(x,\rho) = \mathbf{P}_\rho(V \geq x)$ for $\rho = 1$, 0.9, 0.8 and $x = 1,\ldots,10$, taking $N = 1000$ replications for each pair (x,ρ). The simulation was programmed in Pascal and the execution time was a few minutes on a microcomputer (Macintosh IIfx). The point estimates and 95% confidence intervals are in Table 2. It is seen that the relative error stays remarkably constant, indicating that the estimator is *polynomial time.*

Table 2. The Point Estimates and 95% Confidence Intervals for $\alpha(x,\rho)$.

x	$\rho = 1.0$	$\rho = 0.9$	$\rho = 0.8$
1	$(1.13 \pm 0.04) \cdot 10^{-1}$	$(1.10 \pm 0.04) \cdot 10^{-1}$	$(1.02 \pm 0.04) \cdot 10^{-1}$
2	$(2.84 \pm 0.10) \cdot 10^{-2}$	$(2.43 \pm 0.09) \cdot 10^{-2}$	$(2.16 \pm 0.09) \cdot 10^{-2}$
3	$(7.76 \pm 0.29) \cdot 10^{-3}$	$(5.18 \pm 0.21) \cdot 10^{-3}$	$(3.61 \pm 0.16) \cdot 10^{-3}$
4	$(2.00 \pm 0.08) \cdot 10^{-3}$	$(1.00 \pm 0.04) \cdot 10^{-3}$	$(4.82 \pm 0.21) \cdot 10^{-4}$
5	$(5.12 \pm 0.19) \cdot 10^{-4}$	$(1.71 \pm 0.07) \cdot 10^{-4}$	$(5.47 \pm 0.25) \cdot 10^{-5}$
6	$(1.35 \pm 0.05) \cdot 10^{-4}$	$(2.87 \pm 0.12) \cdot 10^{-5}$	$(5.23 \pm 0.24) \cdot 10^{-6}$
7	$(3.47 \pm 0.13) \cdot 10^{-5}$	$(4.25 \pm 0.18) \cdot 10^{-6}$	$(3.79 \pm 0.17) \cdot 10^{-7}$
8	$(8.95 \pm 0.33) \cdot 10^{-6}$	$(5.40 \pm 0.23) \cdot 10^{-7}$	$(2.23 \pm 0.10) \cdot 10^{-8}$
9	$(2.39 \pm 0.09) \cdot 10^{-6}$	$(6.79 \pm 0.30) \cdot 10^{-8}$	$(1.09 \pm 0.05) \cdot 10^{-9}$
10	$(6.42 \pm 0.24) \cdot 10^{-7}$	$(7.09 \pm 0.32) \cdot 10^{-9}$	$(4.33 \pm 0.20) \cdot 10^{-11}$

17.9 FURTHER RESEARCH AND OPEN PROBLEMS

The LD approach for rare events plays a dominant role in today's literature so we shall start with a discussion of its pros and cons.

On the negative side, one may note that LD arguments are often somewhat heuristic and that the direct approach might lead to stronger results. E.g., the LD argument in Section 17.4.3 identifies the optimal IS only in terms of its mean, and many limit results are not the strongest possible, involving only logarithmic asymptotics. Also, when getting to more complex (and thereby also more realistic and practically challenging!) models, LD theory leads into complex variational problems which typically do not have an explicit solution. For example, Frater et al. [21] show that finding the OECM for an open Jackson network reduces to the solution of a complex minimax problem. Thus, LD theory allows one to identify explicitly the OECM for simple models *only.*

The strength of the LD approach is its generality and the fact that the mathematical state of the area is very advanced, providing a considerable body of theory to draw upon. The philosophy is that, once it has been understood how to paraphrase the optimality properties of OECM for simple systems in LD language, the generality of LD theory will allow finding suitable IS distributions, also, in more complicated settings. The implementation of this program has just begun and its further development has the potential of being one of the most interesting areas in rare events simulation.

One of the main application areas for rare events simulation is queueing networks. Unfortunately, the state of the art, here, is less satisfying than for the relatively simple models discussed in the rest of the chapter, and we find it fair to say that *no general satisfying method exists.* In particular, methods based on regenerative simulation run into difficulties for large networks (the important case!) because of excessive cycles, and, also, it is not clear at all if the duality ideas in Section 17.6.3 can be generalized to networks.

Even the change of measure is problematic for networks. The parameter vector $\boldsymbol{v} = (v_i)_{i \in E}$ (consisting of arrival rates, parameters for the service time distributions at different nodes, routing probabilities, etc.) is of high dimension and there is no canonical way, like ECM, to change it from $\boldsymbol{v}$ to a reference parameter $\boldsymbol{v}_\theta$, say. One possibility is to find an optimal reference parameter, say $\boldsymbol{v}_\theta^*$, from the solution of a stochastic optimization problem, the optimization problem being associated with minimization of the variance of the rare event estimator with respect to $\boldsymbol{v}_\theta$. In fact, this optimal solution $\boldsymbol{v}_\theta^*$ can be obtained by employing some standard stochastic optimization procedure to find the argmin of the *empirical variance.* The rare event probability $\alpha(x)$ can be, then, estimated from simulation by using $\boldsymbol{v}_\theta^*$ as parameter for the IS.

A modified version of this idea is based on the notion of *bottleneck module.* The bottleneck module is defined as the bottleneck (most congested) queue plus all the associated parameters (a subset $\boldsymbol{v}_B = (v_i)_{i \in B}$ of components of $\boldsymbol{v}$) which have direct impact for fast estimation of the rare event in the network. These associated parameters, called the *important parameters,* need not necessarily be part of the bottleneck queue. In other words, the important parameters might belong to queues different from the bottleneck one. In this modification, the IS is performed w.r.t. the parameters $\boldsymbol{v}_B$ associated with the bottleneck module *alone*, rather than w.r.t. the *entire set* $\boldsymbol{v}$ of the parameters of input distributions, most of which do not contribute to fast estimation of rare events. That is, one, first, determines $\boldsymbol{v}_B^\#$ as a solution to the stochastic optimization problem of minimizing the empirical variance (keeping the v_i, $i \in E \backslash B$ fixed) and, next, estimates the rare event probability $\alpha(x)$ by simulation, using the parameter vector $\boldsymbol{v}^0$ given by

$$v_i^0 = \begin{cases} v_i^\# & i \in B \\ v_i & i \in E \backslash B \end{cases}$$

as parameter for the IS.

Clearly, since the set B is typically *much smaller* than E, the stochastic program determining the parameters for the IS has a reduced dimensionality, so that the method is typically preferable to the one involving changing *all* the v_i. Moreover, we do not need to know a priori either the bottleneck queue or its associated important parameters. This can be done either adaptively via a stochastic approximation algorithm or by using a screening algorithm similar to Algorithm 5.1 in [34].

Notice that procedures of this type have been implemented for static models, like stochastic PERT models, reliability models, and static queueing models. We already have some supporting numerical results for tandem queues. It is our belief that the above approach is suitable for estimating rare events for general queueing networks in both transient and steady-state regimes. Based upon Theorem 3.1 in [34], we also conjecture that it has good complexity properties.

17.10 APPENDIX: PROOF OF AN AUXILIARY RANDOM WALK RESULT

Let $\tau(x) = inf\{n > 0 : S_n > x\}$, $\xi(x) = S_{\tau(x)} - x$ (in particular, $\tau(0)$ is the first ladder epoch and $\xi(0)$ the ladder height), and let $\xi(\infty)$ be the limit in distribution $\xi(x)$ as $x \to \infty$.

Theorem 17.23. *Consider a random walk with mean $\mu = \mathbf{E}[X] > 0$, variance σ^2 and $\mathbf{E}[e^{sX}] < \infty$ for all s in a neighborhood of 0. Then as $x \to \infty$,*

$$\mathbf{E}\left[\exp\left\{\alpha\xi(x) + \beta(x)\frac{\tau(x) - x/\mu}{\sqrt{x}}\right\}\right] = \mathbf{E}\left[\exp\{\alpha\xi(\infty)\}\cdot\exp\left\{\frac{\beta(x)^2\sigma^2}{2\mu^3}(1 + o(1))\right\}\right]$$

for $\alpha \le 0$ and any function $\beta(x) \in (-\infty, \infty)$ with $|\beta(x)| = o(\sqrt{x})$.

For $\beta(x) \equiv \beta, \alpha = 0$, Theorem 17.23 is, of course, just the CLT for $\tau(x)$.

Proof. Let $\delta = \delta_x = \beta(x)/\sqrt{x}$,

$$Z(u) = Z_x(u) = \mathbf{E}[e^{\alpha\xi(u) + \delta\tau(u)}]$$

and

$$z(u) = z_x(u) = \mathbf{E}\left[e^{\alpha(\xi(0) - u) + \delta\tau(0)}; \xi(0) > u\right].$$

Then,

$$\begin{aligned} Z(u) &= z(u) + \int_0^\infty\int_0^u Z(u-y)e^{\alpha t}\mathbf{P}(\tau(0) \in dt, \xi(0) \in dy) \\ &= z(u) + \int_0^u Z(u-y)G(dy), \end{aligned}$$

where

$$G(dy) = G_x(dy) = \mathbf{E}\left[e^{\alpha\tau(0)}; \xi(0) \in dy\right].$$

Choose $\gamma = \gamma_x$ such that

$$\int_0^\infty e^{-\gamma y}G(dy) = \mathbf{E}[e^{\delta\tau(0) - \gamma\xi(0)}] = 1,$$

and let $U = U_x = \sum_0^\infty G^{*n}$ be the renewal measure associated with $e^{-\gamma y}G(dy)$. Note that, since $\delta \to 0$ as $x \to \infty$, also $\gamma \to 0$, and $Z(u)$ has a limit (denoted $Z_\infty(u)$ and equal to $\mathbf{E}[e^{\alpha\xi(u)}]$). Similar remarks apply to z, G, U and other quantities in the following.

By a standard argument from renewal theory ([5] Ch. V.6),

$$e^{-\gamma u}Z(u) = \int_0^u z(u-y)U(dy). \tag{17.34}$$

Furthermore, by Taylor expansion, it follows that, up to $O(\delta^3)$ terms,

$$0 = \delta\mathbf{E}[\tau(0)] - \gamma\mathbf{E}[\xi(0)] + \frac{\delta^2}{2}\mathbf{E}[\tau(0)^2] + \frac{\gamma^2}{2}\mathbf{E}[\xi(0)^2] - \delta\gamma\mathbf{E}[\tau(0)\xi(0)]. \tag{17.35}$$

By Wald's identities for the first and second moment of a stopped sum,

$$\mathbf{E}[\zeta(0)] = \mathbf{E}[S_{\tau()}] = \mu\mathbf{E}[\tau(0)], \quad \mathbf{E}[(\xi(0)^2 - \mu\tau(0))^2] = \sigma^2\mathbf{E}[\tau(0)].$$

This yields, first, $\gamma = \delta/\mu + O(\delta^2)$ and, next, by substitution back in (17.35) that (up to $O(\delta^3)$ terms)

$$0 = \Delta\mathbf{E}[\tau(0)] - \gamma\mathbf{E}[\xi(0)] + \frac{\delta^2}{2\mu^2}\sigma^2\mathbf{E}[\tau(0)], \quad \gamma = \delta\frac{1}{\mu} + \delta^2\frac{\sigma^2}{2\mu^3}.$$

Taking $u = x$, it now follows that

$$\mathbf{E}\left[\exp\left\{\alpha\xi(x) + \beta(x)\frac{\tau(x) - x/\mu}{\sqrt{x}}\right\}\right] = Z(x)e^{-\delta x/\sqrt{x}},$$

$$= e^{\gamma x - \delta x/\mu}\int_0^u z(u-y)U(dy)$$

$$= e^{\delta^2 x(1+o(1))\sigma^2/2\mu^3}\int_0^u z(u-y)U(dy)$$

$$= e^{\beta(x)^2(1+o(1))\sigma^2/2\mu^3}\int_0^x z(x-y)U(dy),$$

and all that remains to prove is that $\int_0^x z(x-y)U(dy) \to \mathbf{E}[\exp\{\alpha\xi(\infty)\}]$; this is essentially a triangular array version of the key renewal theorem since

$$\int_0^u z_\infty(u-y)U_\infty(dy) \to \frac{1}{\mathbf{E}[\xi(0)]}\int_0^\infty z_\infty(u)du = \mathbf{E}[\exp\{\alpha\xi(\infty)\}].$$

In the following, we bound $z_x(u)$ for all large x by a function of the form

$$\overline{z}(u) = \mathbf{E}\left[e^{\alpha(\xi(0)-u) + \overline{\delta}\,\tau(0)}; \xi(0) > u\right],$$

where $\overline{\delta} > 0$ does not depend on x and is so small that $\mathbf{E}[\xi(0)e^{\overline{\delta}\,\tau(0)}] < \infty$.

The measure $U_x(x - dy)$ restricted to $[0, x]$ is bounded and converges vaguely $x \to \infty$ to the restriction of $dy/\mathbf{E}[\xi(0)]$ to $[0, \infty)$. Furthermore, z_x is bounded by $\overline{z}$ and converges a.e. to $z_\infty(u)$. These facts imply that

$$\int_{x-N}^x z_x(x-y)U_x(dy) \to \frac{1}{\mathbf{E}[\xi(0)]}\int_0^N z_\infty(x)dx.$$

Using Lorden's inequality for the renewal function ([5] Ch. V.5) and the fact that the mean and variance of G_x have limits, we can infer that $U(k+1) - U(k) \le c$ uniformly in k and all sufficiently large x. Hence

$$\int_0^{x-N} z_x(x-y)U_x(dy) \le c\sum_{k=N}^\infty \sup_{k \le x \le k+1} \overline{z}(x),$$

which can be made arbitrarily small by taking N large enough. Similarly, $\int_N^\infty z_\infty(x)dx$ can be made arbitrarily small, and the proof is complete. □

ACKNOWLEDGEMENT

We would like to thank Vladimir Kriman from the Technion for several valuable suggestions on the earlier draft of this work.

BIBLIOGRAPHY

[1] Anantharam, V., How large delays build up in a GI/GI/1 queue, *Queueing Sys.* **5** (1988), 345-368.

[2] Anantharam, V., Heidelberger, P. and Tsoucas, P., Analysis of rare events in continuous time Markov chains via time reversal and fluid approximation, *Adv. Appl. Prob.* (1990), (to appear).

[3] Asmussen, S., Conditioned limit theorems relating a random walk to its associate, with applications to risk reserve processes and the GI/G/1 queue, *Adv. Appl. Prob.* **14** (1982), 143-170.

[4] Asmussen, S., Conjugate processes and the simulation of ruin problems, *Stoch. Proc. Appl.* **20** (1985), 213-229.

[5] Asmussen, S., *Applied Probability and Queues*, John Wiley, New York 1987.

[6] Asmussen, S., Risk theory in a Markovian environment, *Scand. Actuarial J.* (1989), 69-100.

[7] Asmussen, S., Stationary distributions via first passage time, as Chapter 3 in this book, 79-102.

[8] Asmussen, S., Stationary distributions for fluid flow models with or without Brownian noise, *Stoch. Mod.* **11** (1994), (to appear).

[9] Asmussen, S., Glynn, P.W. and Thorisson, H., Stationary detection in the initial transient problem, *ACM TOMACS* **2** (1992), 130-157.

[10] Asmussen, S. and Nielsen, H.M., Ruin probabilities via local adjustment coefficients, *J. Appl. Prob.* (1995), (to appear).

[11] Asmussen, S., Rubinstein, R.Y. and Wang, C.L., Regenerative rare events simulation via likelihood ratios, *J. Appl. Prob.* **31** (1994), 797-815.

[12] Bolotin, V.A., Kappel, J.G. and Kuehn, P.J., Teletraffic analysis of ATM systems, *IEEE J. Select. Areas Commun.* **9** (1991), 281-283.

[13] Bucklew, J.A., *Large Deviation Techniques in Decision, Simulation, and Estimation*, John Wiley, New York 1990.

[14] Bucklew, J.A., Ney, P. and Sadowsky, J.S., Monte Carlo simulation and large deviations theory for uniformly recurrent Markov chains, *J. Appl. Prob.* **27** (1990), 44-59.

[15] Chang, C.S., Heidelberger, P., Juneja, S. and Shahabuddin, P., Effective bandwidth and fast simulation of ATM intree networks, Research report, IBM Research Division, T.J. Watson Research Center, New York 1992.

[16] Cottrell, M., Fort, J.C. and Malgouyres, G., Large deviations and rare events in the study of stochastic algorithms, *IEEE Trans. Automat. Control* **AC-28** (1983), 907-918.

[17] Dembo, A. and Zeitouni, O., *Large Deviation Techniques*, Jones and Bartlett, Boston 1993.

[18] Devetsikiotis, M. and Townsend, K.R., A dynamic importance sampling methodology for the efficient estimation of rare events probabilities in regenerative simulations of queueing systems, In: *Proc. of IEEE Globecom '92* (1992), 1290-1297.

[19] Devetsikiotis, M. and Townsend, K.R., Statistical optimization of dynamic importance sampling parameters for efficient simulation of communication networks, (1993), preprint.

[20] Frater, M.R., Lennon, T.M. and Anderson, B.D.O., Optimality efficient estimation of the statistics of rare events in queueing networks, *IEEE Trans. Automat. Contr.* **AC-36** (1991), 1395-1405.

[21] Frater, M.R. and Anderson, B.D.O., Fast estimation of the statistics of excessive backlogs in tandem networks of queues, *Australian Telecom. Res.* **23** (1989), 49-55.

[22] Frater, M.R., Walrand, J. and Anderson, B.D.O., Optimality efficient estimation of the buffer overflow in queues with deterministic service times, *Australian Telecom. Res.* **24** (1990), 1-8.

[23] Glasserman, P. and Liu, T.-W., Rare-event simulation for multistage production-inventory systems, Manuscript, Columbia University (1994).

[24] Glasserman, P. and Kou, S.-G., Analysis of an importance sampling estimator for tandem queues, Manuscript, Columbia University (1994).

[25] Glynn, P.W. and Iglehart, D.L., Importance sampling for stochastic simulations, *Mgt. Sci.* **35** (1989), 1367-1392.

[26] Heidelberger, P., Fast simulation of rare events in queueing and reliability models, IBM Research Report RC 19028, Yorktown Heights, New York. Preliminary version published in *Performance Eval. of Computer and Commun. Syst.*, Springer Lecture Notes in Computer Science **729** (1993), 165-202.

[27] Keilson, J. and Wishart, D.M.G., A central limit theorem for processes defined on a finite Markov chain, *Proc. Cambridge Philos. Soc.* **60** (1964), 547-567.

[28] Kingman, J.F.C., On the algebra of queues, *Ann. Math. Stat.* **3** (1966), 285-326.

[29] Kovalenko, I.N., Rare events in queueing systems - A survey, *Queueing Sys.* **16** (1994), 1-49.

[30] Kriman, V., Sensitivity analysis of GI/GI/m/B queues with respect to buffer size by the score function method, *Stoch. Mod.* (1993), (to appear).

[31] Kriman, V. and Rubinstein, R.Y., Polynomial and exponential time algorithms for estimation of rare events in queueing models, Manuscript, Technion, Haifa (1994).

[32] Lehtonen, T. and Nyrhinen, H., Simulating level crossing probabilities by importance sampling, *Adv. Appl. Prob.* **24** (1992), 858-874.

[33] Lehtonen, T. and Nyrhinen, H., On asymptotically efficient simulation of ruin probabilities in Markovian environment, *Scand. Actuarial J.* (1992), 60-75.

[34] Lieber, D., Rubinstein, R.Y. and Elmakis, D., Quick estimation of rare events in stochastic networks, Manuscript, Technion, Haifa (1994).

[35] Lindley, D., The theory of queues with a single server, *Proc. Cambr. Philos. Soc.* **48** (1952), 277-289.

[36] Lindley, D., Discussion of a paper of C.B. Winsten, *J. Roy. Stat. Soc. Ser.* **B21** (1959), 22-23.

[37] Neuts, M.F., A versatile Markovian point process, *J. Appl. Prob.* **16** (1977), 764-779.

[38] Neuts, M.F., *Matrix-Geometric Solutions in Stochastic Models*, Johns Hopkins University Press, Baltimore 1981.

[39] Neuts, M.F., *Structured Stochastic Matrices of the* M/G/1 *Type and their Applications*, Marcel Dekker, New York 1989.

[40] Neveu, J., Une generalisation des processus à accroissements positifs independants, *Sem. Math. Abh. Hamburg.* (1961).

[41] Nicola, V.F., Shahabuddin, P., Heidelberger, P. and Glynn, P.W., Fast simulation of steady-state availability in non-Markovian highly dependable systems, In: *Proc. of the 20th Intern. Symp. on Fault-Tolerant Computing*, IEEE Computer Society Press (1993), 491-498.

[42] Parekh, S. and Walrand, J., A quick simulation method for excessive backlogs in networks of queues, *IEEE Trans. Automat. Contr.* **AC-34** (1989), 54-66.

[43] Rogers, L.C.G., Fluid models in queueing theory and Wiener-Hopf factorization of Markov chains, *Ann. Appl. Prob.* (1994), (to appear).

[44] Rubinstein, R.Y. and Shapiro, A., *Discrete Event Systems: Sensitivity Analysis and Stochastic Optimization via the Score Function Method*, John Wiley, New York 1993.

[45] Sadowsky, J.S., Large deviations theory and efficient simulation of excessive backlogs in a GI/GI/m queue, *IEEE Trans. Automat. Contr.* **AC-36** (1991), 1383-1394.

[46] Sadowsky, J.S., On the optimality and stability of exponential twisting in Monte Carlo simulation, *IEEE Trans. Info. Theory* **IT-39** (1993), 119-128.

[47] Sadowsky, J.S., Monte Carlo estimation of large deviations probabilities, Manuscript, Arizona State University, Tempe, Arizona (1993).

[48] Sadowsky, J.S. and Bucklew, J.A., On large deviations theory and asymptotically efficient Monte Carlo estimation, *IEEE Trans. Info. Theory* **IT-36** (1990), 579-588.

[49] Sadowsky, J.S. and Szpankowski, W., The probability of large queue length and waiting times in a heterogeneous multiserver queue. Part I: Tight limits, Manuscript, Purdue University, West Lafayette, Louisiana (1992).

[50] Sadowsky, J.S. and Szpankowski, W., The probability of large queue length and waiting times in a heterogeneous multiserver queue. Part II: Positive recurrence and logarithmic limits, Manuscript, Purdue University, West Lafayette, Louisiana (1992).

[51] Schwartz, A. and Weiss, A., Large deviation for performance analysis: queues, communication and computers, Manuscript, Technion IIT, Haifa, Israel (1992).

[52] Siegmund, D., Importance sampling in the Monte Carlo study of sequential tests, *Ann. Stat.* **4** (1976), 673-684.

[53] Siegmund, D., The equivalence of absorbing and reflecting barrier problems for stochastically monotone Markov process, *Ann. Prob.* **4** (1976), 914-924.

[54] Stockmeyer, L.J., Computational complexity, *Handbooks in OR and MS*, **3** (1992).

[55] Tsoucas, P., Rare events in series of queues, *J. Appl. Prob.* **29** (1992), 168-175.

Chapter 18

Piecewise-linear diffusion processes

Sid Browne and Ward Whitt

ABSTRACT Diffusion processes are often regarded as among the more abstruse stochastic processes, but diffusion processes are actually relatively elementary, and, thus, are natural first candidates to consider in queueing applications. To help demonstrate the advantages of diffusion processes, we show that there is a large class of one-dimensional diffusion processes for which it is possible to give convenient explicit expressions for the steady-state distribution, without writing down any partial differential equations or performing any numerical integration. We call these tractable diffusion processes *piecewise linear*; the drift function is *piecewise linear*, while the diffusion coefficient is *piecewise constant*. The explicit expressions for steady-state distributions, in turn, yield explicit expressions for long-run average costs in optimization problems, which can be analyzed with the aid of symbolic mathematics packages. Since diffusion processes have continuous sample paths, approximation is required when they are used to model discrete-valued processes. We discuss strategies for performing this approximation, and we investigate when this approximation is good for the steady-state distribution of birth-and-death processes. We show that the diffusion approximation tends to be good when the difference between the birth and death rates is small compared to the death rates.

CONTENTS

18.1	Introduction and summary	463
18.2	Diffusion approximations	467
18.3	Piecewise-continuous diffusions	470
18.4	Four basic linear diffusion processes	471
18.5	Stochastic comparisons	474
18.6	On the quality of diffusion approximations for BD processes	475
18.7	Optimization	476
18.8	Conclusions and open problems	478
	Acknowledgement	478
	Bibliography	479

18.1 INTRODUCTION AND SUMMARY

In the natural sciences, diffusion processes have long been recognized as relatively simple stochastic processes that can help describe the first-order behavior of important phenomena. This simplicity is illustrated by the relatively quick way that the model is specified in terms of a drift function and a diffusion function (plus boundary behavior, which, here, we will take to be standard). However, the

0-8493-8074-x/95/$0.00+$.50

analysis of diffusion processes can involve some formidable mathematics, which can reduce the appeal, and evidently has impeded applications to queueing problems. Our purpose, here, is to circumvent the formidable mathematics and focus solely on creating the model and obtaining the answer, which here is regarded as the steady-state distribution. From a theoretical standpoint, very little, here, is new. Our goal is to show that diffusion processes are easier to work with than often supposed.

For accessible introductory accounts of diffusion processes, see Glynn [11], Harrison [15], §9.4 and §13.2 of Heyman and Sobel [17], Chapter 15 of Karlin and Taylor [20], and Chapter 7 of Newell [27]. For accessible advanced treatments, see Billingsley [4], Breiman [5], Ethier and Kurtz [7], Karatzas and Shreve [19], and Mandl [25].

A diffusion process is a continuous-time Markov process $\{X(t): t \geq 0\}$ with continuous sample paths. We will consider only real-valued, time-homogeneous diffusion processes. Such a diffusion process is characterized by its *drift function* or *infinitesimal mean*

$$\mu(x) = \lim_{\epsilon \downarrow 0} \mathbf{E}[X(t+\epsilon) - X(t) \mid X(t) = x], \tag{18.1}$$

its *diffusion function* or *infinitesimal variance*

$$\sigma^2(x) = \lim_{\epsilon \downarrow 0} \mathbf{E}[(X(t+\epsilon) - X(t))^2 \mid X(t) = x], \tag{18.2}$$

and its boundary behavior. We assume that the state space is the subinterval (s_0, s_k), where $-\infty \leq s_0 < s_k \leq +\infty$. If the boundaries s_0 and s_k are finite, then we assume that the boundaries are *reflecting*. It is easy to understand what reflecting means by thinking of what happens with an approximating simple random walk; from the boundary, the next step is back into the interior. If the boundary points are not finite, then we assume that they are *inaccessible* (cannot be reached in finite time). The boundary behavior can be subtle, and nonstandard variations can be relevant for applications, e.g., see Harrison and Lemoine [16], Kella and Whitt [22], and Kella and Taksar [21]. However, here we consider only the standard case.

We call the diffusion processes, that we consider, *piecewise-linear diffusions*, because we assume that the drift function $\mu(x)$ is *piecewise-linear* and the diffusion function $\sigma^2(x)$ is *piecewise-constant* in the state x. These piecewise-linear diffusion processes are of interest both as models in their own right and as approximations. The piecewise-linear diffusions can serve as approximations for both nondiffusion processes (e.g., birth-and-death processes, see Section 18.2) and diffusion processes with more general piecewise-continuous drift and diffusion functions. In some of the literature on diffusion processes, it is assumed that the drift function and diffusion coefficient are continuous, e.g., see p. 159 of Karlin and Taylor [20], but this stronger assumption is actually not necessary, as can be seen from pp. 13, 25, 90 of Mandl [25] and other references.

An example of a piecewise-linear diffusion process is the heavy-traffic diffusion approximation for the GI/M/s queue developed by Newell [26], Halachmi and Franta [12], and Halfin and Whitt [13]. This diffusion approximation plays an important role in approximations for the general GI/G/s queue in Whitt [33, 36, 38]. In this diffusion process, the drift is constant when all servers are busy and linear otherwise, while the variance is constant throughout. In the context of this GI/M/s example, our purpose is to show that the steady-state distribution can be immediately written down and understood. For this example, it will become evident that the steady-state distribution of the diffusion process has a density that is a piece of an exponential density connected to a piece of a normal density.

Another example of a piecewise-linear diffusion process occurs in the diffusion approximation for large trunk groups in circuit-switched networks with trunk reservation; see Reiman [28,29]. These papers illustrate optimization applications, which we discuss in Section 18.7. All these examples involve queues with state-dependent arrival and service processes; for more examples of this kind, see Whitt [35] and references cited there. A nonqueueing example is the two-drift skew Brownian motion in the control problem of Beneš, Shepp and Witsenhausen [2]; see §6.5 of Karatzas and Shreve [19].

It should be clear that, when we use a diffusion approximation for a queueing process, we are assuming that we can disregard the detailed discrete behavior of the queueing process. The diffusion approximation tends to be appropriate when the jumps are relatively small compared to the magnitude of the process, which tends to occur under heavy loads. Formally, diffusion approximations can be justified by heavy-traffic limit theorems, in which we consider a sequence of models with an associated sequence of traffic intensities approaching the critical value for stability from below; e.g., see Halfin and Whitt [13].

We now specify, in more detail, what we mean by piecewise linear. We assume that there are $k+1$ real numbers s_i such that $-\infty \leq s_0 < s_1 < \ldots < s_k \leq \infty$. Then, the state space is (s_0, s_k) with $\mu(x) = a_i x + b_i$ and $\sigma^2(x) = \sigma_i^2 > 0$ on the interval (s_{i-1}, s_i), $1 \leq i \leq k$. (Often the variance function can be regarded as constant overall, but we will consider the general case; motivation is given in Section 18.2). As indicated above, if the boundary points s_0 and s_k are finite, then, we assume that they are reflecting. Otherwise, we assume that they are inaccessible. Moreover, if $s_0 = -\infty$, then, we require that $a_1 > 0$ of ($a_1 = 0$ and $b_1 > 0$). Similarly, if $s_k = +\infty$, then we require that $a_k < 0$ or ($a_k = 0$ and $b_k < 0$). From pp. 13, 25, 90 of Mandl [25], these conditions guarantee the existence of a proper steady-state limit (convergence in distribution).

The important point is that the steady-state limit has a density of the form

$$f(x) = p_i f_i(x), \quad s_{i-1} \leq x < s_i, \tag{18.3}$$

where $\sum_{i=1}^{k} p_i = 1$, $\int_{s_{i-1}}^{s_i} f_i(x)dx = 1$, f_i has a known relatively simple form and p_i can be easily computed. Consequently, the steady-state mean is

$$m \equiv \int_{s_0}^{s_k} x f(x) ds = \sum_{i=1}^{k} p_i m_i, \tag{18.4}$$

where m_i is the mean of f_i, and similarly for higher moments. In particular, in Section 18.3, we show that

$$p_i = r_i / \sum_{j=1}^{k} r_j, \qquad 1 \leq i \leq k, \tag{18.5}$$

where $r_1 = 1$ and

$$r_i = \prod_{j=2}^{i} \frac{\sigma_{j-1}^2 f_{j-1}(s_{j-1}-)}{\sigma_j^2 f_j(s_{j-1}+)}, \qquad 2 \leq i \leq k. \tag{18.6}$$

Since the component densities f_i are all continuous, the overall density f is continuous if, and only if, $\sigma_i^2 = \sigma_1^2$ for all i. In all cases, the cumulative distribution function is continuous. Our experience indicates that, for most queueing applications, it is appropriate to have $\sigma_i^2 = \sigma_1^2$ and, thus, a continuous steady-state density f.

For piecewise-linear diffusions with $a_i \leq 0$ for all i, the component densities f_i in (18.3) have a relatively simple form, so that it is easy to calculate the component means m_i (and second moments) and the probability weights p_i without performing any integrations. This makes the characterization attractive as an algorithm when k is large, as well as an insightful representation when k is small. In particular, if $a_i \leq 0$ for all i, then the component densities are all truncated and renormalized pieces of normal, exponential, and uniform densities. The relatively simple form for the steady-state distribution follows quite directly from the general theory, as we indicate in Section 18.4, but it does not seem to be well-known (among nonexperts).

Conceptually, the characterization can be explained by the properties of truncated reversible Markov processes; see §1.6 of Kelly [23]. If the state space of a reversible Markov process is truncated (and given reflecting boundaries), then the truncated process is reversible with a steady-state distribution which is a truncated and renormalized version of the original steady-state distribution, i.e., the truncated steady-state distribution is the conditional steady-state distribution of the unrestricted process given the truncation subset. This property holds for multidimensional reversible Markov processes, but we restrict attention here to real-valued processes. For a multidimensional diffusion process application, see Fendick and Hernandez-Valencia [9]. This truncation property is also a natural *approximation* more generally, e.g., see Whitt [33].

For example, if a diffusion process on the real line behaves like an Ornstein-Uhlenbeck (OU) diffusion process over some subinterval of the state space, then, its steady-state distribution restricted to that subinterval is a truncation and renormalization of the normal steady-state distribution of the full OU process with those parameters. Moreover, by exploiting basic properties of the normal distribution, it is possible to give explicit expressions for the moments of the conditional distribution restricted to this subinterval; see Proposition 18.3 below. These explicit expressions, in turn, help produce closed-form expressions for long-run average costs in optimization problems; see Section 18.7. This makes it possible to tackle the optimization problems with symbolic mathematics packages such as Maple V; see Char et al. [6].

Here is how the rest of the chapter is organized. In Section 18.2, we discuss diffusion approximations for birth-and-death processes and give some examples showing how piecewise-linear diffusions can naturally arise. In Section 18.3, we present the steady-state distribution of a piecewise-continuous diffusion, drawing on the basic theory in Karlin and Taylor [20] and Mandl [25]. In Section 18.4, we present four basic linear diffusion processes whose restrictions will form the pieces of the piecewise-linear diffusion process. In the cases with $a_i \leq 0$, we exhibit the appropriate conditional distribution and its first two moments. In Section 18.5, we establish a stochastic comparison that can be used to show that piecewise-linear diffusions, which serve as approximations for a more general piecewise-continuous diffusion, actually are stochastic bounds. In Section 18.6, we investigate when the simple diffusion approximation for birth-and-death processes, introduced in Section 18.2, should be reasonable. In Section 18.7, we discuss optimization. Finally, we state our conclusions in Section 18.8.

In this chapter, we only consider steady-state distributions. However, it should be noted that diffusion processes can also help us understand transient phenomena, such as arise in simulation experiments; e.g., see Whitt [34, 37].

18.2 DIFFUSION APPROXIMATIONS

We often can obtain a diffusion process as an approximation of another process. In this section, we briefly discuss how.

18.2.1 Diffusion approximations of Birth-and-Death processes

We first discuss approximations of *birth-and-death* (BD) processes. As we indicate in Section 18.6 below, the steady-state distribution of a birth-and-death process is not difficult to calculate directly. However, in some cases, it may be desirable to have the closed-form formulas (18.3)-(18.6), especially when the number k of pieces is small.

We begin by showing how a diffusion process can arise as a limit of a sequence of birth-and-death processes. To express the limiting behavior, let $\lfloor x \rfloor$ be the greatest integer less than or equal to x. For each positive integer n, let $\{B_n(t): t \geq 0\}$ be a birth-and-death process on the integers from $\lfloor c_n + l_n\sqrt{n} \rfloor$ to $\lfloor c_n + u_n\sqrt{n} \rfloor$ with state-dependent birth-and-death rates $\beta_n(j)$ and $\delta_n(j)$, respectively. Let the boundary behavior be the same as assumed for the diffusion processes. Let

$$X_n(t) = \frac{B_n(t) - c_n}{\sqrt{n}}, \quad t \geq 0. \tag{18.7}$$

In the context of (18.7), the drift and diffusion functions of $X_n(t)$ are

$$\begin{aligned} \mu_n(x) &\equiv \lim_{\epsilon \downarrow 0} \mathbf{E}[X_n(t+\epsilon) - X_n(t) \mid X_n(t) = x] \\ &= \frac{\beta_n\left(\lfloor c_n + x\sqrt{n} \rfloor\right) - \delta_n\left(\lfloor c_n + x\sqrt{n} \rfloor\right)}{\sqrt{n}} \end{aligned} \tag{18.8}$$

and

$$\begin{aligned} \sigma_n^2(x) &\equiv \lim_{\epsilon \downarrow 0} \mathbf{E}[(X_n(t+\epsilon) - X_n(t))^2 \mid X_n(t) = x] \\ &= \frac{\beta_n\left(\lfloor c_n + x\sqrt{n} \rfloor\right) + \delta_n\left(\lfloor c_n + x\sqrt{n} \rfloor\right)}{n}. \end{aligned} \tag{18.9}$$

If $l_n \to l$, $u_n \to u$, $\mu_n(x) \to \mu(x)$ and $\sigma_n^2(x) \to \sigma^2(x)$ as $n \to \infty$, then $X_n(t)$ can be said to converge to the diffusion process on (l, u) with drift function $\mu(x)$ and diffusion function $\sigma^2(x)$; see Stone [31] and Iglehart [18]. This convergence is in a strong sense, including the finite-dimensional distributions of the stochastic processes and more, see Billingsley [4], but we will consider only the steady-state distributions. Convergence of the steady-state distributions can be shown directly by a modification of the argument in Section 18.6 below.

Example 18.1. The M/M/s queue. The number of customers in the system in the classical M/M/s queue is a birth-and-death process with birth (arrival) rate $\beta(j) = \beta_0$ and death rate $\delta(j) = \eta\, min\{j, s\}$ in state j, where η is the individual service rate. For states in the interval $[0, s]$, we have $\delta(j) = \eta j$, while, for states in the interval (s, ∞), we have $\delta(j) = \eta s$. Consider a sequence of M/M/s queueing models indexed by n. In model n, let the number of servers be $s_n = n$, let the arrival rate be $\beta_n(j) = n - a\sqrt{n}$ for all j, and let the individual service rate be 1, so that the death rate is $\delta_n(j) = min\{j, n\}$. Then it is natural to let $c_n = n$, so that $l_n = -\sqrt{n}$ and $u_n = +\infty$. Then we have convergence to a diffusion process, as shown in Halfin and Whitt [13].

Of course, in applications, we, typically, have only *one* birth-and-death process. Then, we can form the diffusion approximation by letting $l = l_n$, $u = u_n$, $\mu(x) = \mu_n(x)$ and $\sigma^2(x) = \sigma_n^2(x)$ where $\mu_n(x)$ and $\sigma_n^2(x)$ are defined by (18.6) and (18.7) for some given n, which we take as $n = 1$. Setting $n = 1$ corresponds to simply matching the infinitesimal means and variances. Based on Berger and Whitt [3], §8.5, we suggest refining this direct diffusion approximation by making the state space for the diffusion process $(l - 1/2,\ u + 1/2)$ instead of (l, u). This corresponds to the familiar refinement when a continuous (e.g., the normal) distribution is used to approximate an integer-valued probability distribution; see p. 185 of Feller [8].

Henceforth, here, we will concentrate on the direct approximation for the steady-state distribution of a birth-and-death process based on $n = 1$. We hasten to point out that a user should check whether the accuracy of the approximation is adequate for the intended application. We investigate when the crude direct approximation for the steady-state distribution is reasonable in Section 18.6.

Suppose that the birth-and-death parameters β and δ are both *linear*; i.e., $\beta(j) = \beta_0 + \beta_1 j$ and $\delta(j) = \delta_0 + \delta_1 j$ for $l \le j \le u$. Instead of (18.8) and (18.9), we can use the *linear approximations*

$$\mu(x) \approx \beta_0 + \beta_1 x - \delta_0 - \delta_1 x \tag{18.10}$$

and

$$\sigma^2(x) \approx \beta_0 + \beta_1 x + \delta_0 + \delta_1 x \tag{18.11}$$

for $l - 1/2 \le x \le u + 1/2$. Furthermore, assuming that the process will mostly be in the region of x_0 in which $\mu(x_0) \approx 0$, we can further approximate the variance by

$$\sigma^2(x) \approx \beta_0 + \delta_0 + (\beta_1 + \delta_1)x_0, \tag{18.12}$$

provided that $\mu(x_0) \approx 0$ for some x_0 with $l - 1/2 \le x_0 \le u + 1/2$. Otherwise, we let $\sigma^2(x)$ be either $\sigma^2(l)$ or $\sigma^2(u)$, whichever is closer.

Finally, even when β and δ are not linear, we may be able to produce (18.10) and (18.12) over subintervals by making a piecewise-linear approximation.

Example 18.1 (continued). Returning to the M/M/s queue, we apply (18.10) and (18.12) to obtain $\mu(x) = \beta_0 - \eta x$ and $\sigma^2(x) = 2\beta_0$ over $(-1/2, s + 1/2)$, and $\mu(x) = \beta_0 - \eta s$ over $(s + 1/2, \infty)$. To have constant variance overall, we argue that $\mu(x) \approx 0$ for $x \approx s$, so that $\beta_0 \approx \eta s$; thus, we have the further approximation $\sigma^2(x) = 2\beta_0$ for $x \in (s, \infty)$ as well as for $x \in [0, s]$. The relevant values of x are $s + c\sqrt{s}$ for some constant c. For this example, the exact steady-state distribution of the birth-and-death process combines a truncated Poisson distribution below s with a truncated geometric distribution above s, while the diffusion approximation yields a truncated normal distribution below s and a truncated exponential distribution above s; see Halfin and Whitt [13]. These approximations often tend to be good, as is well-known.

Example 18.2. Secondary servers with a buffer. We now consider an example of a birth-and-death process with three linear regions. There is a service facility with one primary server plus a buffer of capacity c_1. There are s secondary servers that accept overflows from the primary buffer. There is an additional buffer of capacity c_2 to hold arrivals when all servers are busy. The secondary system is costly, so that, whenever space opens up in the primary buffer, a customer in service in the secondary system immediately leaves and enters the primary buffer. With this last feature, the number of customers in the system can be modeled as a birth-and-death process.

Let the arrival rate be constant, so that $\beta(j) = \beta_0$ for all j. The service rate is linear in the three regions

$$\delta(k) = \begin{cases} \eta_1, & 1 \le k \le c_1 + 1 \\ \eta_1 + (k - c_1 - 1)\eta_2, & c_1 + 2 \le k \le c_1 + s \\ \eta_1 + s\eta_2, & c_1 + s + 1 \le k \le c_1 + c_2 + s + 1. \end{cases} \quad (18.13)$$

The resulting direct diffusion approximation has drift function

$$\mu(x) = \begin{cases} \beta_0 - \eta_1, & -1/2 \le x < c_1 + 3/2 \\ \beta_0 - \eta_1 - (x - c_1 - 1)\eta_2, & c_1 + 3/2 \le x < c_1 + s + 3/2 \\ \beta_0 - \eta_1 - s\eta_2, & c_1 + s + 3/2 \le x \le c_1 + c_2 + s + 3/2 \end{cases} \quad (18.14)$$

and diffusion function

$$\sigma^2(x) = \begin{cases} \beta_0 + \eta_1, & -1/2 \le x < c_1 + 3/2 \\ \beta_0 + \eta_1 + (x_0 - c_1 - 1)\eta_2, & c_1 + 3/2 \le x < c_1 + s + 3/2 \\ \beta_0 + \eta_1 + s\eta_2, & c_1 + s + 3/2 \le x \le c_1 + c_2 + s + 3/2 \end{cases} \quad (18.15)$$

provided that

$$\mu(x_0) = \beta_0 - \eta_1 - (x_0 - c_1 - 1)\eta_2 \approx 0. \quad (18.16)$$

for $c_1 + 3/2 \le x_0 \le c_1 + s + 3/2$. If $\mu(x) > 0$ (< 0) for all x in this region, then we can set $\sigma^2(x) = \sigma^2(c_1 + s + 3/2)$ ($\sigma^2(x) = \sigma^2(c_1 + 3/2)$).

Note that (18.15) and (18.16) lead to a piecewise-constant diffusion function. We can further simplify (18.15) by just letting $\sigma^2(x) \approx 2\beta_0$, assuming that $\mu(x) \approx 0$ over the entire range of relevant values.

18.2.2 Diffusion approximations for general integer-valued processes

Diffusion approximations are even more important when the stochastic process being approximated is not a birth-and-death process, because, then, there may be no alternative formula for the steady-state distribution. The crude direct approximation, above, easily generalizes; we just match the infinitesimal means and variances, as in (18.8) and (18.9). However, the infinitesimal means and the variances are often hard to determine. An alternative approach is to match the large-time behavior, as discussed in Whitt [32] and references cited there.

To match the large time behavior, let $\{X(t): t \ge 0\}$ be a given integer-valued stochastic process and let $X_j(t)$ represent the sum of the jumps from state j during the period that X has spent t units of time in state j. To formally define $X_j(t)$, let $T_j(t)$ be the time when X has spent t units of time in state j, defined by setting

$$t = \int_0^{T_j(t)} 1_{\{X(u) = j\}} du, \quad (18.17)$$

where 1_A is the indicator function of the set A. Let J_i be the time of the ith jump of X, and let $N(t)$ be the number of jumps of X in $[0,t]$. Then

$$X_j(t) = \sum_{i=1}^{N(T_j(t))} (X(J_i) - j) \; 1_{\{X(J_i -) = j\}}, \quad t \geq 0. \tag{18.18}$$

Typically, we can only approximately determine $\{X_j(t): t \geq 0\}$, but even an estimate can serve as the basis for the diffusion approximation.

We assume that $\{X_j(t): t \geq 0\}$ obeys a central limit theorem, i.e.,

$$\frac{X_j(t) - \lambda_j t}{\sqrt{\lambda_j c_j^2 t}} \Rightarrow N(0,1) \quad \text{as } t \to \infty, \tag{18.19}$$

where $N(0,1)$ is a standard (zero mean, unit variance) normal random variable and $\Rightarrow$ denotes convergence in distribution. We, then, create the distribution approximation by first setting

$$\mu(j) = \lambda_j \quad \text{and} \quad \sigma^2(j) = \lambda_j c_j^2 \tag{18.20}$$

and then fitting continuous functions to $\mu(j)$ and $\sigma^2(j)$. It is easy to see that this procedure coincides with (18.8) and (18.9) with $n = 1$ when X is a birth-and-death process, but it also applies more generally.

18.2.3 Birth-and-death approximations

Since birth-and-death processes are also relatively easy to work with, we could consider constructing approximating birth-and-death processes instead of approximating diffusion processes. This might be convenient for looking at the time-dependent behavior, e.g., for doing simulation or optimization via Markov programs in the spirit of Kushner and Dupuis [24]. However, it is not as easy to approximate by a birth-and-death process as it is by a diffusion process.

Starting from a diffusion process, we can obtain an approximating birth-and-death process by solving (18.8) and (18.9) for the birth-and-death rate functions β and δ. In particular, we get

$$\beta(j) = \frac{\sigma^2(j) + \mu(j)}{2} \quad \text{and} \quad \delta(j) = \frac{\sigma^2(j) - \mu(j)}{2}. \tag{18.21}$$

Obviously, this birth-and-death construction works only when $\sigma^2(j) \geq \mu(j)$ for all j. When $\sigma^2(j)$ is significantly less than $\mu(j)$, we should not anticipate that a birth-and-death approximation will be good.

We, also, note that piecewise-linear birth-and-death processes can be considered. The geometric, Poisson, and discrete uniform distributions play the role of the exponential, normal, and continuous uniform distributions below. The truncation property holds because the birth-and-death process is also a reversible Markov process.

18.3 PIECEWISE-CONTINUOUS DIFFUSIONS

We, now, exhibit the steady-state distribution for a (time-homogeneous) diffusion with piecewise-continuous drift and diffusion functions $\mu(x)$ and $\sigma^2(x)$, with $\sigma^2(x) > 0$. As before, we use the $k+1$ points s_i and assume that the drift and diffusion coefficients are continuous on (s_{i-1}, s_i) with limits from the left and right at each s_i for each i; see pp. 13, 25 and 90 of Mandl [25]. We, also, assume that the boundary points s_0 and s_k are reflecting, if finite, and inaccessible, if infinite. We assume there is a proper time-dependent distribution which converges to a proper steady-state distribution with density $f(x)$. (For the piecewise-linear case, this follows from the extra structure.) The general theory implies that

$$f(x) = \frac{m(x)}{M(s_k)}, \quad s_0 \leq x \leq s_k, \tag{18.22}$$

where

$$m(x) = \frac{2}{\sigma^2(x)s(x)} \tag{18.23}$$

is the *speed density,*

$$s(x) = exp\left\{ - \int_{\theta}^{x} \frac{2\mu(y)}{\sigma^2(y)} dy \right\} \tag{18.24}$$

is the *scale density* with θ arbitrary satisfying $s_0 < \theta < s_k$, and

$$M(x) = \int_{s_0}^{x} m(y)dy, \quad s_0 \leq x \leq s_k, \tag{18.25}$$

provided that all integrals are finite; see pp. 13, 25, 90 of Mandl [25] and §15.3 and 15.5 of Karlin and Taylor [20].

From (18.22)-(18.25), we see that the density $f(x)$ can easily be calculated by numerical integration. Our object is to obtain more convenient explicit expressions. From (18.24) and (18.25), we see that $s(x)$ and $M(x)$ are continuous on (s_0, s_k), so that m and f are continuous everywhere in the interval (s_0, s_k) except perhaps at the points s_i, $1 \leq i \leq k-1$, where $\sigma^2(x)$ is discontinuous. Indeed, since $\sigma^2(x)$ has positive limits from the left and the right at s_i for each i, $1 \leq i \leq k-1$, so will the density f, and we can relate the right and left limits. In particular,

$$f(s_i+) = \frac{\sigma^2(s_i-)}{\sigma^2(s_i+)} f(s_i-). \tag{18.26}$$

From (18.26) and (18.3), we easily obtain the formula for the probability weights in (18.5).

From (18.22)-(18.25), we also directly deduce that the conditional density, conditioning on a subinterval is $Kf(x)$ for x in this subinterval. Moreover, this conditional density is the steady-state density of the diffusion process obtained by restricting the original diffusion process to this subinterval, using reflecting boundaries at all finite boundary points.

18.4 FOUR BASIC LINEAR DIFFUSION PROCESSES

We construct the component densities f_i in (18.3) from the steady-state densities of four basic diffusion processes.

18.4.1 The Ornstein-Uhlenbeck diffusion process

If

$$\mu(x) = -a(x-m) \quad \text{and} \quad \sigma^2(x) = \sigma^2 > 0 \tag{18.27}$$

for $a > 0$ and $-\infty < x < \infty$, then we have the Ornstein-Uhlenbeck (OU) process, for which the steady-state limit is normally distributed with mean m and variances $\sigma^2/2a$.

Let $N(m, b^2)$ denote a normally distributed random variable with mean m and variance b^2. Let Φ be the cumulative distribution function (cdf) and ϕ the density of $N(0,1)$. If X is the steady-state distribution of the OU process in (18.27) restricted to the interval (s_{i-1}, s_i), then X has the distribution of $N(m, \sigma^2/2a)$ conditioned to be in the interval (s_{i-1}, s_i); i.e., X has the density

$$f(x) = \frac{b^{-1}\phi\left(\frac{x-m}{b}\right)}{\Phi\left(\frac{s_i - m}{b}\right) - \Phi\left(\frac{s_{i-1}-m}{b}\right)}, \quad s_{i-1} < x < s_i, \tag{18.28}$$

where $b^2 = \sigma^2/2a$.

Of course, the cdf Φ appearing in (18.28) involves an integral, but it can be calculated approximately without integrating using rational approximations; see §26.2 of Abramowitz and Stegun [1].

Note that we can easily infer the shape of f from (18.28). For example, f is unimodal; the mode is in the interior of (s_{i-1}, s_i), and, thus, at m, if, and only if, $s_{i-1} < m < s_i$. In general, $f(x)$ increases as x moves toward m.

The following proposition gives the first two moments of x.

Proposition 18.3. *If* $-\infty \le s_{i-1} < s_i \le \infty$, *then,*

$$\mathbf{E}[N(m,b^2) \mid s_{i-1} \le N(m,b^2) \le s_i] = m + b\,\frac{\left[\phi\left(\frac{s_{i-1}-m}{b}\right) - \phi\left(\frac{s_i - m}{b}\right)\right]}{\Phi\left(\frac{s_i - m}{b}\right) - \Phi\left(\frac{s_{i-1}-m}{b}\right)} \tag{18.29}$$

and

$$\begin{aligned}
&\mathbf{E}[N(m,b^2)^2 \mid s_{i-1} \le N(m,b^2) \le s_i] = \\
&\qquad m^2 + 2mb\,\frac{\left[\phi\left(\frac{s_{i-1}-m}{b}\right) - \phi\left(\frac{s_i - m}{b}\right)\right]}{\Phi\left(\frac{s_i - m}{b}\right) - \Phi\left(\frac{s_{i-1}-m}{b}\right)} + b^2 \\
&\qquad + b^2\,\frac{\left[\left(\frac{s_{i-1}-m}{b}\right)\phi\left(\frac{s_{i-1}-m}{b}\right) - \left(\frac{s_i - m}{b}\right)\phi\left(\frac{s_i - m}{b}\right)\right]}{\Phi\left(\frac{s_i - m}{b}\right) - \Phi\left(\frac{s_{i-1}-m}{b}\right)}.
\end{aligned} \tag{18.30}$$

Proof. First note that $x\phi(x) = -\phi'(x)$ for all x, so that

$$\mathbf{E}[N(0,1) \mid s_{i-1} \le N(0,1) \le s_i] = \frac{\phi(s_{i-1}) - \phi(s_i)}{\Phi(s_i) - \Phi(s_{i-1})}.$$

Consequently,

$$\mathbf{E}[N(m,b^2) \mid s_{i-1} \le N(m,b^2) \le s_i] = m + b\mathbf{E}\left[\frac{N(m,b^2) - m}{b} \mid s_{i-1} \le N(m,b^2) \le s_i\right]$$

$$= m + b\mathbf{E}\left[N(0,1) \mid \frac{s_{i-1}-m}{b} \le N(0,1) \le \frac{s_i - m}{b}\right].$$

Next note that $x^2\phi(x) = \phi(x) + \phi''(x)$, so that

$$\mathbf{E}[N(0,1)^2 \mid s_{i-1} \le N(0,1) \le s_i] = 1 + \frac{s_{i-1}\phi(s_{i-1}) - s_i\phi(s_i)}{\Phi(s_i) - \Phi(s_{i-1})}.$$

Consequently,

$$\begin{aligned}\mathbf{E}[N(m,b^2)^2 \mid s_{i-1} \le N(m,b^2) \le s_i] &= m^2 \\ &+ 2mb\mathbf{E}[N(0,1) \mid s_{i-1} \le N(m,b^2) \le s_i] \\ &+ b^2\mathbf{E}[N(0,1)^2 \mid s_{i-1} \le N(m,b^2) \le s_i].\end{aligned}$$

□

18.4.2 Reflected Brownian motion with zero drift

If

$$\mu(x) = 0 \quad \text{and} \quad \sigma^2(x) = \sigma^2 > 0 \tag{18.31}$$

on (s_{i-1}, s_i) for $-\infty < s_{i-1} < s_i < \infty$, then we have the reflected Brownian motion (RBM) process with zero drift, for which the steady-state limit X is uniformly distributed on (s_{i-1}, s_i) with mean $(s_{i-1} + s_i)/2$ and second moment $(s_i^3 - s_{i-1}^3)/3(s_i - s_{i-1})$. The conditional distribution on a subinterval is again uniform with the new endpoints playing the role of s_{i-1} and s_i.

18.4.3 Reflected Brownian motion with drift

If

$$\mu(x) = -a \quad \text{and} \quad \sigma^2(x) = \sigma^2 > 0 \tag{18.32}$$

for $a > 0$ on (s, ∞), then we have RBM with negative drift, for which the steady-state limit is distributed as s plus an exponential with mean $\sigma^2/2a$. This case also covers RBM with positive drift a on $(-\infty, -s)$, say, $\{R(t): t \ge 0\}$, because $\{-R(t): t \ge 0\}$ is then the RBM with negative drift above. Hence, if f and g are the steady-state densities with negative and positive drift, respectively, then $g(-s-x) = f(s+x)$ for $x \ge 0$. Hence, it suffices to focus only on the negative drift case.

It is well-known and easy to see that the conditional distribution of s plus an exponential, given that it is contained in the interval (s_{i-1}, s_i), where $s_{i-1} > s$, is the same as an exponential on $(0, s_i - s_{i-1})$; i.e., the conditional density is

$$f(x) = \frac{\lambda e^{-\lambda(x - s_{i-1})}}{1 - e^{-\lambda(s_i - s_{i-1})}}, \quad s_{i-1} < x < s_i, \tag{18.33}$$

where λ^{-1} is the mean of the exponential random variable; here $\lambda^{-1} = \sigma^2/2a$.

Let X be a random variable with the density f in (18.33). Then elementary calculations yield

$$\mathbf{E}[X] = s_{i-1} + \lambda^{-1}\frac{[1 - \lambda e^{-\lambda(s_i - s_{i-1})}(1 + \lambda(s_i - s_{i-1}))]}{1 - e^{-\lambda(s_i - s_{i-1})}} \tag{18.34}$$

and

$$\mathbf{E}[X^2] = s_{i-1}^2 + \frac{2s_{i-1}\lambda^{-1}[1 - \lambda e^{-\lambda(s_i - s_{i-1})}(1 + \lambda(s_i - s_{i-1}))]}{1 - e^{-\lambda(s_i - s_{i-1})}} \tag{18.35}$$

$$+\lambda^{-2}\frac{\left[1-e^{-\lambda(s_i-s_{i-1})}\left(1+\lambda(s_i-s_{i-1})+\frac{\lambda^2(s_i-s_{i-1})^2}{2}\right)\right]}{1-e^{-\lambda(s_i-s_{i-1})}},$$

where $\lambda^{-1}=\sigma^2/2a$.

18.4.4 Positive linear drift

A relatively difficult case occurs if

$$\mu(x)=a(x-m) \quad \text{and} \quad \sigma^2(x)=\sigma^2>0 \tag{18.36}$$

for $a>0$ and $s_{i-1}<x<s_i$. Then, there is positive linear drift away from m. By partitioning the interval into two subintervals and performing a change of variables, it suffices to consider the case

$$\mu(x)=ax \quad \text{and} \quad \sigma^2(x)=\sigma^2>0 \tag{18.37}$$

on $(0,s)$. However, even (18.37) is difficult. Indeed, no nice explicit form is available for (18.37). In particular, from (18.22)-(18.25), we see that the steady-state density (18.37) is of the form

$$f(x)=Ke^{ax^2/\sigma}, \quad 0\le x\le s, \tag{18.38}$$

and the mean is

$$\mathrm{E}[X]=\frac{\sigma K}{2a}(e^{as^2/\sigma}-1) \tag{18.39}$$

for a constant K such that $\int_0^s f(x)dx=1$. Except for the constant K, the forms of (18.38) and (18.39) are quite simple and thus easily understood. However, K does not have a simple expression. The constant K can be found from Dawson's integral $D(y)\equiv e^{-y^2}\int_0^y e^{x^2}dx$, whose values appear in Table 7.5 of Abramowitz and Stegun [1]. The maximum value is $D(y)=0.541$ occurring at $y=0.924$; see 7.1.17 of Abramowitz and Stegun.

Since the constant K in (18.38) is relatively intractable, if this case is present, then we would resort either to direct numerical integration in the setting of Section 18.3 or approximation of the drift coefficient in (18.36) by piecewise-constant drift coefficients, as in Section 18.4.2 and Section 18.4.3, over several subintervals.

Example 18.4. Insurance fund. We now give a (nonqueueing) example with a positive state-dependent drift. As in Harrison [14], consider an insurance firm with an asset process that is a diffusion with state-dependent drift $\mu(x)=\alpha x$ for positive x where $\alpha>0$ and constant variance function, but let the process have a reflecting barrier at zero instead of the absorbing barrier. Moreover, combine this with DeFinetti's model of an insurance fund as discussed on pages 146-147 of Gerber [10], in which all proceeds above some level b are paid out as dividends. Then, the asset process is a linear diffusion on $[0,b]$ with drift function $\mu(x)=\alpha x$, where $\alpha>0$.

18.5 STOCHASTIC COMPARISONS

Since we may want to approximate a general piecewise-continuous diffusion by a piecewise-linear diffusion, it is useful to have results providing insight into the quality of the approximation.

From Section 18.3, we easily can obtain sufficient conditions for a stochastic comparison. We say that one density f_1 is less than or equal to another f_2 on the same interval (s_0, s_k) in the sense of *likelihood ratio ordering*, and we write $f_1 \leq_{lr} f_2$, if $f_2(x)/f_1(x)$ is nondecreasing in x. A likelihood ratio ordering implies that the distribution determined by f_1 is stochastically less than or equal to the distribution determined by f_2; see Ross [30].

Proposition 18.5. *Consider two piecewise-continuous diffusions on a common interval* (s_0, s_k) *satisfying* (18.22)-(18.25). *If* $\sigma_1^2(x)/\sigma_2^2(x)$ *is nondecreasing in* x *and* $\mu_2(x)/\sigma_2^2(x) \geq \mu_1(x)/\sigma_1^2(x)$ *for all* x, *then* $f_1 \leq_{lr} f_2$.

Proof. Note that $f_2(x)/f_1(x)$ is nondecreasing if, and only if, $\sigma_1^2(x)s_1(x)/\sigma_2^2(x)s_2(x)$ is nondecreasing, by (18.22) and (18.23). Next, by (18.24), $s_1(x)/s_2(x)$ is nondecreasing if, and only if, $\mu_2(x)/\sigma_2^2(x) \geq \mu_1(x)/\sigma_1^2(x)$ for all x. □

Note that the condition in Proposition 18.5 is satisfied if $\sigma_1^2(x) = \sigma_2^2(x)$ and $\mu_1(x) \leq \mu_2(x)$ for all x.

From (18.22)-(18.25), we can also establish continuity results showing that $f_n(x) \to f(x)$ for each x if $\mu_n(x) \to \mu(x)$ and $\sigma_n^2(x) \to \sigma^2(x)$ for each x, plus extra regularity conditions, for a sequence of piecewise-continuous diffusions.

18.6 ON THE QUALITY OF DIFFUSION APPROXIMATIONS FOR BIRTH-AND-DEATH PROCESSES

We, now, investigate when the direct diffusion approximation for birth-and-death processes with $n = 1$ in (18.8) and (18.9) is reasonable for the stationary distribution for the birth-and-death process. For simplicity, we assume that $l > -\infty$. Recall that the steady-state probability mass function for a birth-and-death process is

$$\pi_j = \rho_j / \sum_{i=l}^{u} \rho_i, \quad l \leq j \leq u, \tag{18.40}$$

where $\rho_l = 1$ and

$$\rho_j = \prod_{i=l+1}^{j} (\beta_{i-1}/\delta_i) = \frac{\beta_l}{\delta_j} exp \sum_{i=l+1}^{j-1} log\left(\frac{\beta_i}{\delta_i}\right), \quad l+1 \leq j \leq u. \tag{18.41}$$

To relate (18.41) to the steady-state distribution of the diffusion, we exploit the expansion of the logarithm, i.e.,

$$log(1+x) = x - \frac{x^2}{2} + \frac{x^3}{3} - \cdots. \tag{18.42}$$

From (18.41) and (18.42), we obtain a condition for the diffusion approximation to be good. *The condition is that* $(\beta_i - \delta_i)/\delta_i$ *is suitably small for the* i *of interest.* Assuming this is the case, we have

$$\rho_j \approx \frac{\beta_l}{\delta_j} exp \sum_{i=l+1}^{i-1} \left[\left(\frac{\beta_i - \delta_i}{\delta_i}\right) + O\left(\frac{\beta_i - \delta_i}{\delta_i}\right)^2\right]. \tag{18.43}$$

From (18.8) and (18.9) with $n = 1$, $\beta_l \approx \sigma^2(l)$, $\delta_j \approx \sigma^2(j)/2$, $\delta_i \approx \sigma^2(i)/2$ and

$$\rho_j \approx \frac{2\sigma^2(l)}{\sigma^2(j)} exp \sum_{i=l+1}^{j-1} \frac{2\mu(i)}{\sigma^2(i)}. \tag{18.44}$$

If, in addition, $2\mu(i)/\sigma^2(i)$ is suitably smooth, e.g., linear, then

$$\rho_j \approx \frac{2\sigma^2(l)}{\sigma^2(j)} exp \int_{l+1/2}^{j-1/2} \frac{2\mu(y)}{\sigma^2(y)} dy \tag{18.45}$$

and indeed, by (18.22), (18.23), (18.24), and (18.45),

$$\pi_j \approx \int_{j-1/2}^{j+1/2} f(y)dy \approx f(j), \;\; 1 \le j \le u, \tag{18.46}$$

where f is the diffusion process density in (18.22). Formula (18.46) shows that the steady-state birth-and-death probability mass function values π_j are reasonably approximated by the steady-state diffusion density $f(j)$.

18.7 OPTIMIZATION

It can be rather straightforward to handle costs in a piecewise-linear diffusion process. Suppose a cost is charged to the system at rate $g_i(x)$ per unit time when $x \in [s_{i-1}, s_i)$. Then standard renewal-reward theory tells us that the expected average cost per unit time is

$$\sum_{i=1}^{k} p_i \int_{s_{i-1}}^{s_i} g_i(x) f_i(x) dx. \tag{18.47}$$

We discuss one example of optimization below; see Beneš, Shepp and Witsenhausen [2], Reiman [28, 29] and Kushner and Dupuis [24] for others.

Example 18.2 revisited. Suppose that we consider the secondary service with buffers again. The piecewise-linear diffusion process approximation is given in (18.14) and (18.15). By the results of Sections 18.3 and 18.4, we find that in regions 1 and 3, the stationary distribution is truncated exponential, and in region 2 it is truncated normal. To simplify notation, we will let $\beta_0 - \eta_1 = -\mu \le 0$ and $\beta_0 + \eta_1 = \delta$. We also let $(\beta_0 - \eta_1)/\eta_2 = -\alpha \le 0$ and $\sqrt{\beta_0/\eta_2} = \gamma$. Then, from Section 18.4, we find

$$f_1(x) = \frac{\lambda_1 e^{-\lambda_1(x+1/2)}}{1 - e^{-\lambda_1(c_1+2)}}, \tag{18.48}$$

$$f_2(x) = C(s) \cdot \phi\left(\frac{x + \alpha - (c_1+1)}{\gamma}\right), \tag{18.49}$$

$$f_3(x) = \frac{\lambda_2 e^{-\lambda_2(x - (c_1 + 1 + s + 1/2))}}{1 - e^{-\lambda_2 c_2}}, \tag{18.50}$$

where

$$\lambda_1 = \frac{\delta}{2\mu}, \; \frac{1}{C(s)} = \gamma\left[\Phi\left(\frac{1/2+\alpha}{\gamma}\right) - \Phi\left(\frac{1/2+s+\alpha}{\gamma}\right)\right], \; \lambda_2 = \frac{\delta + s\eta_2}{2(\mu - s\eta_2)}. \tag{18.51}$$

(It follows from Section 18.4.1 that, for the normal part, we have the mean $m = c_1 + 1 - \alpha$.) From (18.5) and (18.6), we also get $r_1 = 1$,

$$r_2 = \frac{\delta\lambda_1 e^{-\lambda_1(c_1+2)}}{(1 - e^{-\lambda_1(c_1+2)}) 2\beta_0 C(s) \phi((\alpha + 1/2)/\gamma)}, \tag{18.52}$$

and
$$r_3 = r_2 \cdot \frac{(1-e^{-\lambda_2 c_2})2\beta_0 C(s)\phi((\alpha+s+1/2)/\gamma)}{(\delta+s\eta_2)\lambda_2}. \tag{18.53}$$

Now we consider optimization problems. Even if we restrict attention to choosing the parameters c_1, s and c_2, there are quite a few possibilities. For example, the s secondary servers could be a given, as would be the number of buffer spaces, $c_1 + c_2$. In this case, the decision problem would be *how to split the buffers* and where to place the secondary servers (if we restrict ourselves to using them as *dedicated* group). In extreme cases, we might want to place all of the buffer spaces in between the single server and the secondary servers (if e.g., $h_1 = h_3 < h_2$), or, in the other extreme (e.g., if $h_2 < min\{h_1, h_3\}$), we may want to place all the servers together at the head of the system, thus effectively working as a (partially) *ranked* M/M/s+1/$c_1 + c_3$ system with a strange cost structure. (In both of these cases, there would only be 2 regions.) We will call this Problem 1.

Alternatively, the buffer spaces as well as their positions might be fixed externally, and the decision variable might simply be how many excess servers, s, to hire within a given budget constraint. We will call this Problem II.

In both cases, since *queueing* occurs only in the regions 1 and 3, costs should be *quadratic* in those 2 regions, and *linear* in the region where service is in *parallel*; i.e., we will take $g_i(x) = h_i \cdot (x - s_{i-1})^2$, $i = 1,3$, and $g_2(x) = h_2 \cdot (x - s_1) \equiv h_2 \cdot (x - (c_1 + 1 + 1/2))$. Let $B(c_1, s, c_2)$ denote the cost function for the system. Then we have

$$\mathbf{E}[B(c_1,s,c_2)] = p_1 h_1 \mathbf{E}[(X_1 - s_0)^2] + p_2 h_2 \mathbf{E}[X_2 - s_1] + p_3 h_3 \mathbf{E}[(X_3 - s_2)^2], \tag{18.54}$$

where X_i has density f_i. These values are then easily obtained from (18.29) and (18.35), yielding

$$\mathbf{E}[(X_1 - s_0)^2] = \frac{1}{4} - \frac{1-\lambda_1 e^{-\lambda_1(c_1+2)}(1+\lambda_1(c_1+2))}{\lambda_1(1-e^{-\lambda_1(c_1+2)})}$$
$$+\frac{1}{4}\,\frac{1-e^{-\lambda_1(c_1+2)}(1+\lambda_1(c_1+2)+\lambda_1^2(c_1+2)^2/2)}{1-e^{-\lambda_1(c_1+2)}}, \tag{18.55}$$

$$\mathbf{E}[X_2 - s_1] = c_1 + 1 - \alpha + \frac{\gamma^2}{C(s)}\left[\phi\left(\frac{\alpha+1/2}{\gamma}\right) - \phi\left(\frac{s+\alpha+1/2}{\gamma}\right)\right], \tag{18.56}$$

and

$$\mathbf{E}[(X_3 - s_2)^2] = (c_1 + 1 + s + 1/2)^2$$
$$+\left(\frac{2(c_1+1+s+1/2)}{\lambda_2}\right)\left(\frac{1-\lambda_2 e^{-\lambda_2 c_2}(1+\lambda_2 c_2)}{1-e^{-\lambda_2 c_2}}\right)$$
$$+\frac{1-e^{-\lambda_2 c_2}(1+\lambda_2 c_2+\lambda_2^2 c_2^2/2)}{\lambda_2^2(1-e^{-\lambda_2 c_2})}. \tag{18.57}$$

It should, of course, be recalled that $p_2 = p_2(c_1, s)$, $p_3 = p_3(c_1, s, c_2)\lambda_2 = \lambda_2(s)$.

Standard numerical optimization techniques can, now, be used to optimize the system. For example, for Problem I, suppose that $c_1 + c_2 = K$; then, let $c_1 = c$, and $c_2 = K - c$ in (18.55)-(18.57), and just optimize $\mathbf{E}[B(c, s, K - c)]$ with respect to c. The two extreme cases referred to above correspond, respectively, to the cases $c = 0$, $c = K$. For Problem II, we would try to maximize x subject to $\mathbf{E}[B(C_1, s, c_2)] \leq l$, where l is our budget per unit time.

For example, we applied the symbolic mathematical package, Maple V, to differentiate $\mathbf{E}[B(c, s, K - c)]$ with respect to c in order to find the optimal solution for Problem I; see Char et al. [6]. Using piecewise-linear diffusion processes together with symbolic mathematics packages seems like a promising approach.

18.8 CONCLUSIONS AND OPEN PROBLEMS

In Sections 18.1, 18.3, and 18.4 we showed that the steady-state distribution of a one-dimensional piecewise-linear diffusion can be expressed conveniently in closed form, in a way that is insightful. It remains to obtain corresponding results for multidimensional diffusions.

In Sections 18.2 and 18.6, we discussed diffusion approximations for birth-and-death processes and other integer-valued processes. It remains to further evaluate the quality of these approximations.

In Sections 18.3 and 18.5, we discussed piecewise-linear diffusion approximations for more general diffusion processes with piecewise-continuous drift and diffusion functions. In Section 18.7, we showed how the piecewise-linear diffusion processes can be used effectively for optimization, especially when combined with a symbolic mathematics package such as Maple V. It remains to exploit the use of symbolic mathematics packages further. Moreover, it remains to develop effective algorithmic methods for solving and optimizing more complicated multidimensional diffusion processes; see Kushner and Dupuis [24] for significant progress in this direction.

Overall, we have tried to support the idea that one-dimensional diffusion processes can be useful for queueing and other applied problems.

ACKNOWLEDGEMENT

We thank Jewgeni Dshalalow for helpful comments on the presentation.

BIBLIOGRAPHY

[1] Abramowitz, M. and Stegun, I.A., *Handbook of Mathematical Functions*, Dover, New York, 1972.

[2] Beneš, V.E., Shepp, L.A., and Witsenhausen, H.S., Some solvable stochastic control problems, *Stochastics* **4** (1980), 39-83.

[3] Berger, A.W. and Whitt, W., The Brownian approximation for rate-control throttles and the G/G/1/C queue, *Discrete Event Dynamic Systems* **2** (1992), 7-60.

[4] Billingsley, P., *Convergence of Probability Measures*, John Wiley, New York 1968.

[5] Breiman, L., *Probability*, Addison-Wesley, Reading, MA 1968.

[6] Char, B.W., Geddes, K.O., Gonnet, G.H., Leong, B.L., Monagan, M.B., and Watt, S.M., *First Leaves: A Tutorial Introduction to Maple V*, Springer-Verlag, New York 1993.

[7] Ethier, S.N. and Kurtz, T.G., *Markov Processes, Characterization and Convergence*, Wiley, New York 1986.

[8] Feller, W., *An Introduction to Probability Theory and its Applications*, 3rd ed., Wiley, New York 1968.

[9] Fendick, K.W. and Hernandez-Valencia, E., A Stochastic Flow Model for an ATM Switch with Shared Buffers and Loss Priorities, AT&T Bell Laboratories, Holmdel, NJ 1992.

[10] Gerber, H.O., *An Introduction to Mathematical Risk Theory*, Heubner Foundation Monograph **8**, Irvin, Chicago 1979.

[11] Glynn, P.W., Diffusion approximations, In: *Stochastic Models* (ed. by D.P. Heyman and M.J. Sobel), North Holland (1990), 145-198.

[12] Halachmi, B. and Franta, W.R., Diffusion approximations to the multiserver queue, *Mgt. Sci.* **24** (1978), 522-529.

[13] Halfin, S. and Whitt, W., Heavy-traffic limits for queues with many exponential servers, *Oper. Res.* **29** (1981), 567-588.

[14] Harrison, J.M., Ruin problems with compounding assets, *Stoch. Proc. Appl.* **5** (1977), 67-79.

[15] Harrison, J.M., *Brownian Motion and Stochastic Flow Systems*, John Wiley, New York 1985.

[16] Harrison, J.M. and Lemoine, A.J., Sticky Brownian motion as the limit of a storage process, *J. Appl. Prob.* **18** (1981), 216-226.

[17] Heyman, D.P. and Sobel, M.J., *Stochastic Models in Operations Research, Vol. I: Stochastic Processes and Operating Characteristics*, McGraw-Hill, New York 1982.

[18] Iglehart, D.L., Limit diffusion approximations for the many-server queue and the repairman problem, *J. Appl. Prob.* **2** (1965), 429-441.

[19] Karatzas, I. and Shreve, S., *Brownian Motion and Stochastic Calculus*, Springer-Verlag, New York, 2nd edition 1991.

[20] Karlin, S. and Taylor, H.M., *A Second Course in Stochastic Processes*, Academic Press, New York 1981.

[21] Kella, O. and Taksar, M., A heavy traffic limit for the cycle counting process in G/G/1; optional interruptions and elastic screen Brownian motion, *Math. Oper. Res.* **19** (1994), 132-151.

[22] Kella, O. and Whitt, W., Diffusion approximations for queues with server vacations, *Adv. Appl. Prob.* **22** (1990), 706-729.

[23] Kelly, F.P., *Reversibility and Stochastic Networks*, Wiley, New York 1979.

[24] Kushner, H.J. and Dupuis, P.J., *Numerical Methods for Stochastic Control Problems in Continuous Time*, Springer-Verlag, New York 1992.

[25] Mandl, P., *Analytic Treatment of One-Dimensional Markov Processes*, Springer-Verlag, New York 1968.

[26] Newell, G.F., *Approximate Stochastic Behavior of n-Server Service Systems with Large n*, Springer-Verlag, New York 1973.

[27] Newell, G.F., *Applications to Queueing Theory*, 2nd ed., Chapman and Hall, London 1982.

[28] Reiman, M.I., Asymptotically optimal trunk reservations for large trunk groups, *Proc. IEEE 28th Conf. on Decision and Control* (1989), 2536-2541.

[29] Reiman, M.I., Optimal trunk reservation for a critically loaded link, *Proc.*

13th Int. Teletraffic Congress, Copenhagen (1991), 247-252.

[30] Ross, S.M., *Stochastic Processes*, Wiley, New York 1982.

[31] Stone, C., Limit theorems for random walks, birth and death processes and diffusion processes, *Ill. J. Math.* **4** (1963), 638-660.

[32] Whitt, W., Refining diffusion approximations for queues, *Oper. Res. Letters* **1** (1982), 165-169.

[33] Whitt, W., Heavy traffic approximations for service systems with blocking, *AT&T Bell Lab. Tech. J.* **63** (1984), 689-708.

[34] Whitt, W., Planning queueing simulations, *Mgt. Sci.* **35** (1989), 1341-1366.

[35] Whitt, W., Queues with service times and interarrival times depending linearly and randomly on waiting times, *Queueing Sys.* **6** (1990), 335-352.

[36] Whitt, W., Understanding the efficiency of multiserver service systems, *Mgt. Sci.* **13** (1992), 708-723.

[37] Whitt, W., Asymptotic formulas for Markov processes with applications to simulation, *Oper. Res.* **40** (1992), 279-291.

[38] Whitt, W., Approximations for the GI/G/m queue, *Prod. Oper. Mgt.* **2** (1993), 141-161.

Chapter 19

Approximation of queues via small parameter method

Igor N. Kovalenko

ABSTRACT The "small parameter method" is recently developed to efficiently solve various queueing problems both analytically and numerically. In this chapter, the author discusses principles of this method applied to busy period parameters. This is also a basis for the derivation of steady state distributions and the distribution of time until the occurrence of a rare event.

CONTENTS

19.1 Main concepts 481
19.2 Examples 483
19.3 Small parameter analysis of the busy period 490
19.4 Rare events in queueing systems 493
19.5 Light-traffic limits and perturbation analysis of queueing systems 498
19.6 Unsolved problems and research directions 501
Acknowledgements 503
Bibliography 503

19.1 MAIN CONCEPTS

Hundreds, if not thousands, of queueing models have been analyzed during the decades elapsed after publishing the pioneering works by A.K. Erlang. Most of these models deal with a few well-known steady state distributions (of queue length, waiting time, busy period and other processes) and their moments. One would be fortunate to have a closed form analytical expression for a desired parameter, but even in such an exceptionally favorable case, evaluation of the expression may be very difficult. For example, the well-known simple and elegant Pollaczek-Khintchine formula

$$\phi(s) = \frac{1-\rho}{1-\frac{\lambda}{s}\left(1-\beta(s)\right)}$$

requires a numerical inversion of the Laplace transform to compute the waiting time distribution. In addition, in most cases, even Laplace transforms or generating functions are not available in closed forms. The small parameter method is a method for overcoming various computational complexities. In many situations, it leads to simple approximations of queuing system parameters. It is necessary to choose an appropriate small parameter describing the problem. Roughly speaking, the investigator of an analyzed system, say, S, should construct a simpler system S_0 with a higher computational efficiency, and he or she should be able to compute a correction to the transfer from S_0 to S.

0-8493-8074-x/95/$0.00+$.50

In an attempt to illustrate convenience of the small parameter method in queueing theory, the author gives illustrative examples. He proposes a preliminary classification of the most widely occurring varieties of small parameter methods. He also develops an approach to verify the convergence and to establish error bounds. However, most attention is paid to the algorithmic aspect.

The small parameter method is used in the following sense. Assume that a queuing system S is defined by:

(*i*) its structure (number of channels, service mechanism, etc.) and

(*ii*) a set of distributions $F^{(k)}(t)$ of some underlying independent random variables $X_n^{(k)}$ (where, for example, $X_n^{(0)}$ is the nth interarrival time and $X_n^{(1)}$ is the nth service time).

Suppose that, in a practical situation, there is stable tendency of one type of random variables to dominate the other, for example, $\mathbf{E}[X_n^{(1)}]/\mathbf{E}[X_n^{(0)}] < 0.1$. In such a situation, if a closed form expression for a desired parameter, say, the mean waiting time $\bar{W}$, is not available, one can consider the following alternative.

Assume that a family of queueing systems S_ϵ differs from S_0 in such a way that $X_n^{(k)}$ is replaced by $Y_n^{(k)}h_k(\epsilon)$, where $Y_n^{(k)}$ are supposed to have ϵ-independent distributions $G^{(k)}(t)$ and $h_k(\epsilon)$ are some functions of a small positive parameter ϵ. Then, we have $\bar{W} = \bar{W}(\epsilon)$, and in many cases there exists a convergent power series $\bar{w} = c_0 + c_1\epsilon + c_2\epsilon^2 + \ldots$ or at least an asymptotical expansion $\bar{w} \sim c_0 + c_1\epsilon + \ldots + c_r\epsilon^r$. Either can be used for approximation by using methods of numerical analysis.

When analyzing time-dependent queueing phenomena, a typical applied problem often suggests an ϵ-dependent time scaling. If an event of interest is very rare, then, instead of the associated counting process $X(t)$ (that is, the number of events within time interval $(0,t)$), it is reasonable to consider

$$Y_\epsilon(t) = X(t/\epsilon^r).$$

In such a case, a two-fold parameter dependence exists: the process $X(t)$ depends on ϵ through the underlying variables $Y_n^{(k)}h_k(\epsilon)$, and, moreover, the time scaling factor ϵ^r is considered.

The small parameter approach is an excellent source of many approximate expressions for numerical analysis. Each of the expressions, in a large variety of cases, can be treated as means of relatively simple functions of random variables $Y_n^{(k)}$. Thus, the simulation method can be regarded as an appropriate numerical approach.

Small parameter algorithms are initially constructed under strong assumptions on underlying distributions $G^{(k)}(t)$. The Cramèr's condition,

$$\exists\delta > 0, \text{ such that } \int e^{\delta\,|\,t\,|}\,dG^{(k)}(t) < \infty,$$

is an obvious assumption to suggest a desired algorithm. But as soon as the algorithm is derived, the problem of an error estimate arises. In a relevant case, a residual term estimate of the form

$$\mathbf{E}\Big[\theta(\{X_n^{(k)}\})\Big]$$

can be derived relative to a restricted number of variables, $X_n^{(k)}$. In this case, we observe that no small parameter is needed since the bound depends on distributions of the original random variables $X_n^{(k)}$ and not on ϵ.

The small parameter transfers its functions to the "small" functional $\mathbf{E}[\theta]$. The situation can be illustrated by the Central Limit Theorem. Indeed, consider a sequence of i.i.d. random variables X_n with a common characteristic function $f(t)$ with $\mathbf{E}[X_n]=0$ and $Var(X_n)=1$. Then, the Central Limit Theorem for the normed sums $(X_1+\ldots+X_n)/\sqrt{n}$ involves a simple small parameter analysis of the function $n\log f(t/\sqrt{n})$, or, equivalently, $\epsilon^{-2}\log f(t\epsilon)$, where $\epsilon = 1/\sqrt{n}$. Moreover, the asymptotic expansion in powers of ϵ can be derived. (See Feller [19].) But, as soon as the triangle scheme is considered (in which $X_1+\ldots+X_n$ is replaced by $X_{n1}+\ldots+X_{nn}$ and X_{nk} is distributed dependent on $n>k$), no small parameter is needed. Indeed, the Lindeberg condition insuring the asymptotical normality for a triangle scheme contains no parameters excluding the functional

$$\sum_k \mathbf{E}\left[Z_{nk}^2 I_{\{|z_k|>\tau\}}\right]$$

for appropriately normed random variables Z_{nk}.

For many cases, however, a simple small parameter approach leading to simple approximations with residual terms of the form $O(\ldots)$ or $o(\ldots)$, is of practical and theoretical importance. In a more typical situation, where no appropriate small functional estimate of an approximation error is available, a small parameter approach seems to be the only one to analyze a system.

Roughly speaking, two large classes of parameters are of interest in the analysis of queueing systems: (*i*) steady state parameters (loss probability, mean queue length, mean waiting time, distribution function of the waiting time, sojourn time, etc.) and (*ii*) parameters connected to the global system behavior expressed as a rule by the occurrence of some rare events within a large time interval. The busy period is just one of the main targets for the analysis of both (*i*) and (*ii*). Indeed, well-known ergodic balance equations allow us to derive steady state distributions given those for the busy period. On the other hand, many exponential type limit theorems and exact and asymptotic bounds are established. They enable one to derive rare events distribution via busy period parameters.

The two outlined interconnections can explain the construction of this chapter. Section 19.3 is devoted to busy period parameters almost trivially linked to the steady state parameters. Then, in Section 19.4, the rare event analysis based on busy period parameters is developed. Section 19.5 surveys other important features of small parameter analysis of queueing systems. Section 19.2 contains some examples in which the specific queueing system is considered, and the small parameter approach is demonstrated. All the results are almost elementary but they suggest substantial generalizations.

19.2 EXAMPLES

19.2.1 An example from reliability theory

Consider an unreliable system, whose behavior can be described by a finite state random process $X(t)$. The state 0 indicates workable states of all the system components, whereas in each of the other states some components fail. Failures of the components, as well as their repairs, obey a Markov process, so that the $X(t)$ is a homogeneous Markov process. Denote by λ_{ij} the transition rate from state i to state j, so that

$$\mathbf{P}\{X(t+dt)=j \mid X(t)=i\} = \lambda_{ij}dt \text{ for } i \neq j$$

and $\lambda_i = -\lambda_{ii} = \sum_{j \neq i} \lambda_{ij}$.

The steady state distribution of the process is the unique solution of the linear algebraic system

$$\lambda_j \pi_j = \sum_{j \neq i} \lambda_{ij} \pi_i \tag{19.1}$$

and

$$\sum \pi_j = 1 \tag{19.2}$$

or, in the matrix form,

$$\pi \Lambda = 0 \tag{19.3}$$

$$\pi \mathbb{I} = 1 \tag{19.4}$$

where $\pi = (\pi_j)$ is the (row) vector of steady state probabilities, $\Lambda = (\lambda_{ij})$ and $\mathbb{I}$ denotes the vector $(1, \ldots, 1)^T$. For a repairable system, the ergodicity property holds true; hence, (19.3-19.4) can be solved in an obvious manner.

However, as it frequently occurs, "when a problem is thought to be solved for a mathematician it only arises for an engineer." In our case, the multidimensionality prevents one from obtaining a numerical solution (let alone an analytical one!) even for systems of moderate complexity. A small parameter approach can improve the situation considerably.

Let us consider $X(t)$ as a regenerative process with the epochs of transitions from state 0 to any other state being regeneration points. A maximal connected interval of nonzero states can be called a "busy period." Then, busy periods alternate with idle periods. For any $j \neq 0$,

$$\pi_j = \frac{\lambda_0 T_j}{1 + \lambda_0 \sum_{i \neq 0} T_i} \tag{19.5}$$

and

$$\pi_0 = \frac{1}{1 + \lambda_0 \sum_{i \neq 0} T_i}, \tag{19.6}$$

where T_j is the mean sojourn time in the state j within a busy period. In many practical cases, it is reasonable to assume failure rates to be comparatively small. In one such situation, for a two-dimensional process $Y(T) = (X(t), \gamma(t))$, the second component can be defined as follows. Within idle periods $\gamma(t) = 0$, whereas each failure implies a jump of the unit magnitude of this function. Thus, $\gamma(t)$ equals the number of failures that occur from the beginning of the current busy period up to time t.

Let $\mathfrak{S}_m$ be the subset of all states of the process $Y(t)$ with the second component equal to m. Thus, $\{Y(t)\} = \mathfrak{S} = \bigcup_{m=0}^{\infty} \mathfrak{S}_m$. We, hereinafter, use the notations

$$|x| = m \text{ for } x \in \mathfrak{S}_m,$$

$$\lambda_{xy} = \lambda_{ij} \text{ for } x = (i, m),\ y = (j, m+1),$$

$$\lambda_{xy} = 0 \text{ otherwise if } x \neq y$$

$$\lambda_x = -\lambda_{xx} = \sum_{y \in \mathfrak{S}_{m+1}} \lambda_{xy}, \quad x \in \mathfrak{S}_m,$$

$$\mu_{xy} = \lambda_{ij} \text{ if } x = (i, m),\ y = (j, m),\ i \neq j \text{ or } x = (i, m),\ m > 0,\ j = 0$$

and
$$\mu_x = -\mu_{xx} = \sum_{y \neq x} \mu_{xy}.$$

Such an asymmetry in the notations can be explained by different roles of λs and μs. Namely, we shall consider λs as small parameters and μs as fixed numbers. The property of repairability consists of the following assumption: For each $x_0 \in \mathfrak{S}_m$, $m \geq 1$, there exists a path $x_0 \to x_1 \to x_2 \to \ldots \to x_l$ such that $\mu_{x_0 x_1} > 0, \ldots, \mu_{x_{l-1} x_l} > 0$ and $x_l = 0$. Following equations (19.5-19.6), we obtain

$$\pi_0^{-1} = 1 + \lambda_0 \sum_{m=1}^{\infty} \sum_{x \in \mathfrak{S}_m} T_x \tag{19.7}$$

and
$$\pi_x = \pi_0 \lambda_0 T_x, \quad x \neq 0, \tag{19.8}$$

where 0 represents the state (0,0), π_x is the ergodic probability of a state $x \in \mathfrak{S}$, and T_x is the sojourn time in state x.

One can write down the equations for π_x in a straightforward way. Indeed,

$$(\mu_x + \lambda_x)\pi_x = \sum_{|y| = m, y \neq x} \mu_{yx}\pi_y + \sum_{|y| = m-1} \lambda_{yx}\pi_y \tag{19.9}$$

as soon as $|x| = m \geq 1$.

Define the vectors $\prod_m = (\pi_x, |x| = m)$ and matrices

$$M_m = (\mu_{xy}, |x| = |y| = m)$$

$$L_m = (\lambda_{xy}, |x| = m, |y| = m+1)$$

and
$$K_m = \text{diag}(\lambda_x, |x| = m).$$

Then, system (19.9) can be rewritten in an elegant matrix form

$$\prod_m (M_m - K_m) + \prod_{m-1} L_{m-1} = 0, \; m \geq 1. \tag{19.10}$$

The condition imposed on the path $x_0 \to \ldots x_l$ implies nonsingularity of M_m. Hence, $M_m - K_m$ is nonsingular for a sufficiently small norm of the matrix K_m being defined as:

$$\| K_m \| = \max_{|x| = m} \lambda_x. \tag{19.11}$$

Moreover, the expansion

$$(M_m - K_m)^{-1} = \sum_{n=0}^{\infty} (M_m^{-1} K_m)^n M_m^{-1} \tag{19.12}$$

converges in the norm. Therefore, an iterative computing algorithm

$$\prod_m = -\prod_{m-1} L_{m-1} (M_m - K_m)^{-1}, \quad m \geq 1 \tag{19.13}$$

can be applied. A comprehensive work with matrix $M_m - K_m$ can be avoided if one uses (19.12), so that only M_m is to be inverted.

Making use of the norm of matrices, we can choose a number of iterations. Even the roughest approximation

$$(M_m - K_m)^{-1} \approx M_m^{-1} \tag{19.14}$$

yields the calculation of the main terms. Indeed, define by R_m the remainder term associated with the mth iteration of scheme (19.13). Then, obviously

$$\| R_m \| \leq \frac{\| K_m \|}{\| M_m \| (\| M_m \| - \| K_m \|)}. \tag{19.15}$$

We can define an approximate algorithm

$$\prod_m = -\prod_{m-1} L_{m-1} M_m^{-1}, \ m \geq 1 \tag{19.16}$$

and

$$\prod_0 = 1. \tag{19.17}$$

Assume that $\| K_m \| = \| L_m \|$ are small parameters of the order ϵ. Then,

$$\prod_m = O(\epsilon^m) \tag{19.18}$$

and the error in its calculation by (19.16-19.17) is of the order $O(\epsilon^{m+1})$. An obvious procedure for the approximation of an arbitrary order is as follows. Keep the terms with $m = 0, 1, \ldots, l-1$ in the series (19.12). Then the algorithm (19.13) leads to estimates of $\prod_m / \pi_0$ with errors of order $O(\epsilon^{m+l})$. After that, we compute the valued of $\prod_0 = \pi_0$ from an approximate normalization condition

$$\sum_{m=0}^{l-1} \sum_{|x|=m} \frac{\pi_x}{\pi_0} = \pi_0^{-1}. \tag{19.19}$$

The values $\pi_x = \frac{\pi_x}{\pi_0}\pi_0$ are to be computed in such a way that they have errors of order $O(\epsilon^{|x|+l})$.

Now, we can return to the original problem of computing π_j. We have

$$\pi_j \approx \sum_{m=0}^{l-1} \pi_{(j,m)}. \tag{19.20}$$

We can observe that all π_j are evaluated by the previously proposed process of controlled errors. An admissible order of errors depends on the problem to be solved. Thus, assume that the steady state unavailability q is to be computed. We have

$$q = \sum_{i \in F} \pi_i, \tag{19.21}$$

where F is the set of the system down-states. If the set F does not include states with less than r failures, then,

$$q = O(\epsilon^r). \tag{19.22}$$

Hence, the relative error in the q computation has the order $O(\epsilon^{l-r})$. Of course, accurate bounds for errors can, also, be obtained on the basis of the operator norm.

The problem of choosing an appropriate value of l is of a practical importance. The greater l, the larger is the set of states involved in the computing processes. Indeed, in a system with N uninterchangeable components, there are $\binom{N}{m}$ states, x, with $|x| = m$, so that, in a real computation, there may be a significant difference between, say, $l = 4$ or $l = 5$. A compromise should be reached between the desired accuracy and the number of system states.

Note that the well-known paper by Moore and Shannon [44] seems to be the first one to suggest a far-reaching, small parameter approach to reliability problems.

19.2.2 Loss probability within a busy period

Consider a GI/G/m/r queueing system with $A(t)$ and $B(t)$ being the interarrival and service time distributions, respectively, and both $a = \int_0^\infty t dA(t)$ and $b = \int_0^\infty t dB(t)$ being finite.

We are interested in the probability q of at least one loss within a busy period. A "busy period" is a time interval within which at least one customer is being served. The solution of the problem is very complicated; it is equivalent to finding a solution of a multidimensional integral equation. A surprisingly considerable simplification can be achieved through the small parameter approach. Assume that

$$A'(t) = \frac{1}{\Gamma(\alpha)} t^{\alpha - 1} e^{-t}, \quad t > 0,\ \alpha > 0 \tag{19.23}$$

(thus $a = \alpha$) and

$$B(t) = B_0\left(\frac{t}{\epsilon}\right), \tag{19.24}$$

where $\epsilon > 0$ is considered to be a small parameter, and $B_0(t)$ is a distribution function with finite moments of sufficiently large orders.

Assume that a busy period is initiated by the arrival of a customer labeled by 1. Denote by z_k the time interval between the arrivals of customers k and $k+1$, and let η_k be the service time of customer k. Let q_0 be the probability of the event A consisting of the following inequalities:

$$\begin{gathered} \eta_1 > S_{m+r} + z_{m-1} + \ldots + z_1 \\ \eta_2 > S_{m+r} + z_{m-1} + \ldots + z_2 \\ \ldots\ldots\ldots\ldots\ldots\ldots\ldots\ldots \\ \eta_{m-1} > S_{m+r} + z_{m-1} \\ \eta_m > S_{m+r}, \end{gathered} \tag{19.25}$$

where

$$S_{m+r} = z_m + z_{m+1} + \ldots + z_{m+r}. \tag{19.26}$$

The event A implies the loss of customer $m+r+1$, thus

$$q \le q_0. \tag{19.27}$$

Hence, q_0 plays a central role in the theory of repairable systems with a fast repair. The most significant contribution to this topic was made by A.D. Solovyev and his disciples; see Solovyev [57]. If $N(t)$ denotes the number of customers in the system at time t, then q_0 is the probability of the monotone passage $1 \to 2 \to \ldots \to m+r+1$ of $N(t)$.

Using (19.25), one obtains

$$\begin{aligned} q_0 = {} & (\Gamma(\alpha))^{-m+1} (\Gamma(r+1)\alpha)^{-1} \epsilon^{(m+r)\alpha} \int_0^\infty \ldots \int_0^\infty x_1^{\alpha-1} x_2^{\alpha-1} \ldots x_{m-1}^{\alpha-1} x_m^{(r+1)\alpha - 1} \\ & \times e^{-\epsilon(x_1 + \ldots + x_m)} \bar{B}_0(x_1 + \ldots + x_m) \bar{B}_0(x_2 + \ldots + x_m) \ldots \bar{B}_0(x_m) dx_1 \ldots dx_m. \end{aligned} \tag{19.28}$$

If

$$\begin{aligned} c = {} & (\Gamma(\alpha))^{-m+1} (\Gamma(r+1)\alpha)^{-1} \int_0^\infty \ldots \int_0^\infty x_1^{\alpha-1} x_2^{\alpha-1} \ldots x_{m-1}^{\alpha-1} x_m^{(r+1)\alpha-1} \\ & \times \bar{B}_0(x_1 + \ldots + x_m) \ldots \bar{B}_0(x_m) dx_1 \ldots dx_m < \infty, \end{aligned} \tag{19.29}$$

then,

$$q_0 \sim c\epsilon^{(m+r)\alpha}, \quad \text{as } \epsilon \to 0. \tag{19.30}$$

In the case of nonmonotone failure within a busy period, at least $m+n+2$ customers are processed during the busy period and, thus,

$$\eta_1 + \ldots + \eta_{m+r+1} > z_1 + \ldots + z_{m+r+1}. \tag{19.31}$$

Hence,

$$q - q_0 \le Pr(\eta_1 + \ldots + \eta_{m+r+1} > z_1 + \ldots + z_{m+r+1})$$

$$\le \frac{1}{\Gamma((m+r+1)\alpha)} \int_0^\infty x^{(m+r+1)\alpha - 1} \tag{19.32}$$

$$\times \bar{B}^{(m+r+1)}(x)dx = \frac{\epsilon^{(m+r+1)\alpha}}{\Gamma((m+r+1)\alpha+1)} \mathbf{E}[(Y_1 + \ldots + Y_{m+r+1})^{(m+r+1)\alpha}],$$

where Y are i.i.d.r.v. with the distribution $B_0(t)$, provided $\mathbf{E}[Y_1^{(m+r+1)\alpha}]$ is finite. We observe that

$$\frac{q}{q_0} = 1 + O(\epsilon^\alpha), \quad \epsilon \to 0. \tag{19.33}$$

Obviously,

$$c \le (\Gamma(\alpha+1))^{-m+1}(\Gamma(r+1)\alpha+1)^{-1}(\mathbf{E}[Y_1^\alpha])^{m-1}\mathbf{E}[Y_1^{(r+1)\alpha}]. \tag{19.34}$$

Therefore, the condition $\mathbf{E}[Y_1^{(m+r+1)\alpha}] < \infty$ implies the finiteness of c.

19.2.3 The distribution of the number of losses in a modified system M/G/m/0

Consider a system M/G/m/0 modified in such a way that the input rate equals λ_k, given that k channels are engaged. The service time distribution function $B(t)$ is considered unchanged and $T = \int_0^\infty t dB(t) < \infty$. Denote by ζ the time of the first loss of a customer in the steady state. We desire to estimate $\mathbf{P}(\zeta > t)$ under the assumption that λ^* can be considered a small parameter where $\lambda^* = min\ (\lambda_0, \ldots, \lambda_{m-1})$. If, moreover, all the λ_k are small, then, the estimation approach does not differ from that introduced in the preceding section. Thus, we consider a nontrivial case when $\lambda_m \to 0$, whereas $\lambda_0, \ldots, \lambda_{m-1}$ are fixed positive constants.

Consider a random process $X(t)$ defined as follows: $X(t) = 1$ if all the channels are engaged at time t and $X(t) = 0$ otherwise. Then, obviously,

$$\mathbf{P}(\zeta > t) = \mathbf{E}\left[\exp\{-\lambda_m \int_0^t X(u)du\} \right]. \tag{19.35}$$

A standard regenerative argument (see, for example, Gnedenko and Kovalenko [25]) ensures the ergodicity of the random process $X(t)$. Thus, we have

$$\mathbf{E}[X(t) \mid X(t_0) = i] \to \pi_m \overset{\text{def}}{=} (\lambda_0, \ldots, \lambda_{m-1}\tau^m/m!)\pi_0 \tag{19.36}$$

with

$$\pi_0^{-1} = \sum_{k=0}^m (\lambda_0, \ldots, \lambda_{k-1}\tau^k)/k!. \tag{19.37}$$

Hence,

$$\int_0^t X(u)du = \pi_m t + R_t, \tag{19.38}$$

where R_t is a random variable satisfying the following condition:

$$\mathbf{E}[R_t^2] = o(t^2), \quad t \to \infty. \tag{19.39}$$

Therefore,

$$\mathbf{P}(\zeta > t) - \exp\{-\lambda_m \pi_m t\} \to 0, \quad \lambda_m \to 0 \tag{19.40}$$

uniformly in $t \geq 0$.

19.2.4 A loss queueing system with time-dependent parameters

Consider an M/M/m/0 queueing system with time-dependent arrival and service rates $\lambda = \lambda(t)$ and $\mu = \mu(t)$. The system behavior can be described by the system of equations

$$\pi_0' + \lambda\pi_0 = \mu\pi_1; \quad \pi_k' + (\lambda + \mu)\pi_k = \lambda\pi_{k-1} + k\mu\pi_{k+1}, \quad 1 \leq k \leq m-1;$$
$$\pi_m' + m\mu\pi_m = \lambda\pi_{m-1}, \tag{19.41}$$

where $\pi_k = \pi_k(t)$. Assume that positive time averages $\bar{\lambda} = \lim_{T \to \infty} \frac{1}{T} \int_0^T \lambda(t)dt$, $\bar{\mu} = \lim_{T \to \infty} \frac{1}{T} \int_0^T \mu(t)dt$ exist. System (19.41) implies the following assertion. If $\bar{\pi}_k = \lim_{T \to \infty} \pi_k(t)$ exists, then,

$$\bar{\pi}_k = (\bar{\lambda}^k / \bar{\mu}^k k!)\pi_0 \tag{19.42}$$

with

$$\bar{\pi}_0^{-1} = \sum_{k=0}^m \bar{\lambda}^k / \mu^k k!. \tag{19.43}$$

From system (19.41), we obtain the following condition:

$$\int_\alpha^{\alpha+\Delta} \lambda(t)dt \Big/ \int_\alpha^{\alpha+\Delta} \mu(t)dt \underset{\alpha \to \infty}{\longrightarrow} \bar{\lambda}/\bar{\mu}. \tag{19.44}$$

Equation (19.44) means that, in the pseudotime $\tau = \int_0^t \mu(u)du/\bar{\mu}$, the birth-and-death process has a state-independent birth rate of the form $\lambda + \omega(\tau)$, with $\int_\alpha^{\alpha+\Delta} \omega(\tau)d\tau \underset{\alpha \to 0}{\longrightarrow} 0$, and death rates $k\bar{\mu}$, $1 \leq k \leq m$. Elementary reasoning involving perturbation implies the sufficiency of condition (19.44) for the convergence of $\pi_k(t)$ to positive limits as $t \to \infty$. However, one can observe interesting limit behavior. To illustrate the situation, assume that

$$\lambda(t) = f(\epsilon t), \quad \mu(t) = g(\epsilon t), \tag{19.45}$$

where f and g are continuous periodic functions.

If they have the same period, say T_0, then one can decompose the real half-line into intervals A_n of length $T/(N\epsilon)$. Within each of them, N is sufficiently large and $\lambda(t)$ and $\mu(t)$ vary so slowly that a distribution close to the steady state holds. Considering all the intervals together allows one to reconstruct the global system behavior:

$$\pi_k((nT_0 + t)/\epsilon) \underset{\epsilon \to 0}{\longrightarrow} \frac{1}{k!}\left(\frac{f(t)}{g(t)}\right)^k \Big/ \sum_{j=0}^m \frac{1}{j!}\left(\frac{f(t)}{g(t)}\right)^j, \quad 0 \leq k \leq m. \tag{19.46}$$

If $f(t)$ and $g(t)$ have incompatible periods T_0 and T_1, then the time averages

$$\bar{\pi}_k = \lim_{T\to\infty} \frac{1}{T} \int_0^T \pi_k(t)dt$$

can be defined by the equation

$$\bar{\pi}_k = \frac{1}{T_0 T_1} \int_0^{T_1} \int_0^{T_0} \pi_k(f(t), g(\tau))dt d\tau \tag{19.47}$$

where $\pi_k(f, g)$ are steady state probabilities for an Erlang system M/M/m/0 with parameters $\lambda = f$ and $\mu = g$.

19.3 SMALL PARAMETER ANALYSIS OF THE BUSY PERIOD

Assume that the behavior of a queueing system is described by a stable regenerative process $X(t)$ with the regeneration points t_n as times originating busy periods $(t_n, t_n + \zeta_n)$. The intervals $(t_n + \zeta_n, t_{n+1})$ are idle periods of the system. The entire interval (t_n, t_{n+1}) is a busy cycle of the system. We note that, in our case, a "busy period" is the period of an engagement of at least one queueing channel, whereas "idle period" is the period within which the system is free of any customers.

Let $\Phi(t_0, t) = \Phi_X(t_0, t)$ be an additional functional of the process $X(t)$, so that $\Phi(t_0, t) = \Phi(t_0, t_1) + \Phi(t_1, t)$ for $t_0 < t_1 < t$ with respect to any path function of $X(t)$. Assume that the functional is invariant of time shifts, so that $\Phi_X(t_0, t_1) = \Phi_Y(t_0 + \tau, t_1 + \tau)$ when $Y(t) = X(t - \tau)$, $t_0 \le t \le t_1$. Under very general additional conditions, the renewal ergodic equation holds:

$$\bar{\Phi} = \lim_{T\to\infty} \frac{1}{T}\Phi(0, T) = \mathbf{E}[\Phi(0, \zeta^*)], \tag{19.48}$$

where $[0, \zeta^*)$ is the busy cycle which starts at time 0. Equation (19.48) includes many special cases concerning sojourn times, number of served customers, and so on. In many cases, the value of Φ over an idle period either equals zero or can easily be computed. Thus only $\mathbf{E}[\Phi(0, \zeta)]$ is of interest, where $(0, \zeta)$ denotes the busy period. Equation (19.48) is of principal importance in queueing theory since busy period parameters are much easier to deal with compared with corresponding parameters of "global" behavior, especially asymptotically and numerically.

In a single-channel queueing system, a busy period starts at the time of a customer's arrival at an idle system and terminates as soon as the system becomes idle again. Thus, a busy period may consist of service times of some customers. Historically, results on busy period parameters were first established for the M/G/1 queue.

Takács' functional equation for the Laplace transform, $g(s)$, of the busy period distribution

$$g(s) = \beta(s + \lambda - \lambda g(s)),$$

where $\beta(s)$ is the Laplace transform of the service time distribution (see Takács [59]), allows one to study exponential tails of the busy period distribution in the case of rational $\beta(s)$ and, also, to obtain some asymptotic expansions and estimates. For example, one can set $g(s) = g(s, \lambda)$, and, thus, get an expansion of the function in a λ-power series.

Prabhu [47] obtained an elegant and very useful formula

$$G_n(x,t) = \int_x^t e^{-\lambda u} \lambda x \frac{(\lambda u)^{n-1}}{n!} dB_n(u-x), \tag{19.49}$$
$$t > x,\ n \geq 0,\ B_n = B^{*(n)}.$$

The function $G_n(x,t)$ is defined as the probability of the following event: A busy period started when the virtual waiting time equal to x will be terminated in time less than t and exactly n customers will be served within that period.

This formula is the key to a small parameter analysis of many parameters connected to the busy period. Consider the following example. Having integrated $G_n(x,\infty)$ over x with the weight $dB(t)$, one obtains the distribution of the number of customers serviced during a busy cycle, namely:

$$p_1 = \beta(\lambda),\ p_2 = -\lambda\beta(\lambda)\beta'(\lambda), \tag{19.50}$$

and so on. Thus, for instance, the probability of at least three serviced customers is

$$1 - p_1 - p_2 \sim \lambda^2(b^2 + \frac{\alpha_2}{2}) \tag{19.51}$$

with $\alpha_2 = \beta''(0)$, and $\alpha_2 < \infty$. Similarly, the estimate

$$1 - p_1 - \ldots - p_r \sim \lambda^r R_r \tag{19.52}$$

holds true, with R_r being a polynomial in the first r moments of the service time distribution, provided the rth moment is finite.

We will highlight some further results related to the busy period. Cohen [16] provided detailed analysis of various busy period random variables. He obtained elegant integral representations for the distributions of the maximum of a busy period in a specified time interval and of the maximal queue length in a busy period. For the latter distribution, the probability that the maximal queue length exceeds a given $x = 2, 3, \ldots$ is given by the expression

$$\frac{1}{2\pi i}\int_{D_\omega} \frac{1-\omega}{\beta(\lambda(1-\omega)) - \omega}\frac{d\omega}{\omega^x} \Big/ \frac{1}{2\pi i}\int_{D_\omega} \frac{1}{\beta(\lambda(1-\omega)) - \omega}\frac{d\omega}{\omega^x}, \tag{19.53}$$

where D_ω is a disc (in the complex plane) centered at $\omega = 0$ with a sufficiently small radius.

Some generalizations of the analytical theory of busy period distributions have been made for a variety of queueing models. A short survey of further results on busy periods is presented in Kovalenko [40]. We cite here only one formula of Stadje [58].

Let τ be the length of a busy period in a GI/G/1 queue, N be the number of customers serviced during this period, and I be the length of the subsequent idle period. Then,

$$\mathbf{E}[z^N e^{-\theta_1\tau - \theta_2 I}]$$
$$= 1 - \exp\left\{ -\sum_{n-1} \frac{z^n}{n} \iint_{v-u\leq 0} e^{-\theta_1 v + \theta_2(v-u)} dA^{*n}(u) \times dB^{*n}(v) \right\}. \tag{19.54}$$

The previous example 19.2.2 demonstrates the main principle of the busy period asymptotical analysis. Many results of such an analysis are reviewed in

Kovalenko [40]. In what follows, we discuss rather general mathematical models of busy period perturbations with a formalism of its asymptotical analysis via a probabilistic interpretation of a small parameter dependent on asymptotic expansion.

Let $[0,\zeta_0)$ be an unperturbed busy period, and let $z_0(t)$ be a random process describing the queueing behavior within it. If no perturbation occurs, then, the busy period terminates at time t_0. But there is a possibility of a perturbation at each time t, $0 \le t < \zeta_0$, with $\lambda(t,z)$ being the perturbation rate at time t, given $z_0(t) = z$. After the perturbation at time t_1, a planned behavior is replaced by a new behavior characterized by a random process $z_1(t)$, $t_1 \le t \le \zeta_1$. Assume that t_{k-1}, ζ_{k-1} and $z_{k-1}(t)$, $t_{k-1} \le t < \zeta_{k-1}$, are given. Then the rate of the kth perturbation equals $\lambda(t,z)$ at time $t, t_{k-1} \le t < \zeta_{k-1}$, given that $z_{k-1}(t) = z$. If the perturbation occurs at time t_k, then, ζ_k and $z_k(t)$, $t_k \le t < \zeta_k$, are chosen "at random," independently of the process behavior before the moment t_k.

Thus our probabilistic model of a busy period can be specified by the following characteristics:

$$\mathfrak{L}[\zeta_0; z_0(t), 0 \le t < \zeta_0], \tag{19.55}$$

$$\mathfrak{L}[\zeta_n; z_n(t), t_n \le t < \zeta_n \mid t_n; z_{n-1}(t) = z], \tag{19.56}$$

and

$$\lambda(t,z). \tag{19.57}$$

Here, the symbol $\mathfrak{L}$ stands for a probability law without obvious measure-theoretical rigor. We note only that the transition distribution law in (19.56) does not depend on n, $n \ge 1$, and that the initial time of a busy period does not affect its behavior.

For simulation purposes, it is convenient to introduce independent random variables ω_n such that

$$\zeta_0 = \tau_0(\omega_0),\ z_0(t) = z_0(t,\omega_0), \tag{19.58}$$

$$\zeta_n = \tau(\omega_n \mid t_n, z_{n-1}(t_n) = z), \tag{19.59}$$

and

$$z_n(t) = z(t, \omega_n \mid t_n, z_{n-1}(t_n) = z), \tag{19.60}$$

where τ_0, τ, z_0 and z are nonrandom functions.

The given model includes many known queueing models. The only example which we mention here is a queueing system perturbed by interruptions of service due to random failures of serving facilities.

Denote by $[0,\zeta)$, the busy period with possible perturbations. Then,

$$[0,\zeta_0) = [0,t_1) \cup [t_1,t_2) \cup \ldots \cup [t_\nu, \zeta_\nu), \tag{19.61}$$

where ν is the number of perturbations within this busy period. From (19.61), we have that

$$\mathbf{E}[\Phi(0,\zeta)] = \mathbf{E}[\Phi(0,t_1 \wedge \zeta_0)] + \mathbf{E}[\Phi(t_1,t_2)] + \ldots + \mathbf{E}[\Phi(t_\nu,\zeta_n)], \tag{19.62}$$

where

$$a \wedge b = min(a,b).$$

Assume that the following conditions are satisfied:

$$\mathbf{E}[\zeta_0] < \infty \tag{19.63}$$

and

$$\mathbf{P}(\zeta_n - \zeta_{n+1} > x \mid t_n, z_{n-1}(t_n) = z) \leq \bar{B}(x), \tag{19.64}$$

for any t_n and z, $\bar{B}(x) = 1 - B(x)$. Assume also that the distribution function $B(x)$ has finite moments and

$$\mid \mathbf{E}[\Phi^r(0, t_1 \wedge \zeta_0)] \mid \leq P_r(t_1 \wedge \zeta_0) \tag{19.65}$$

and

$$\mid \mathbf{E}[\Phi^r(t_n, t_{n+1} \wedge \zeta_n) \mid t_n, z_{n-1}(t_n) = z] \mid \leq P_r(t_{n+1} \wedge \zeta_n - t_n), \tag{19.66}$$

where P_r are some polynomials, $r = 1, 2, \ldots$.

Then, the consecutive terms of (19.62), as it can easily be shown by an appropriate stochastic ordering, form an asymptotic expansion for $\mathbf{E}[\Phi(0,\zeta)]$, provided that $\sup_{t,z} \lambda(t,z) = \epsilon$ is a small parameter. That is, for arbitrary r,

$$\mathbf{E}[\Phi(0,\zeta)] = \mathbf{E}[\Phi(p,\zeta_0) I_{\{\nu = 0\}}] + \sum_{k=1}^{r} \mathbf{E}[\Phi(t_k, t_{k+1} \wedge \zeta_0) I_{\{\nu \geq k\}}] + O(\epsilon^{r+1}), \tag{19.67}$$

where I_A denotes an indicator function of a set A.

The author and his colleagues at V.M. Glushkov Institute of Cybernetics of the National Academy of Sciences of the Ukraine have elaborated some methods for the simulation of the asymptotic expansion (19.67) oriented to the analysis of highly reliable systems (see Kovalenko and Kuznetsov [39]). The variance reduction approach is applied in such a way that the estimate of the probability of a targeted rare event has a bounded relative mean square error as $\epsilon \to 0$; at the same time, a straightforward simulation of an associated queueing system tends to an unbounded relative error under the same conditions.

19.4 RARE EVENTS IN QUEUEING SYSTEMS

19.4.1 The Rényi theorem and its ramifications

Although steady state distributions are of "well-established" practical interest, there are other distributions which deserve attention, such as the time to the occurrence of a rare event (i.e., the loss of a call or the incidence of a very long waiting time).

A general scenario of the asymptotical analysis of these distributions is the following. Let q be the probability of a targeted rare event within a busy cycle and a be the mean length of a busy cycle. Let ν be the number of the first busy cycle within which the event occurred. Then ν is geometrically distributed, and, hence, $q\nu$ is an asymptotically exponential random variable with the unit mean. We can observe that $\nu \to \infty$ in probability, and, hence, the entire length of the first ν busy cycle equals $\nu a + \gamma$, where a random variable γ is small compared to νa. Thus, as $q \to 0$, the time to a rare event is asymptotically exponential with parameter q/a. Historically, the first limit theorem of such a kind is a well-known theorem by Rényi; see Rényi [52]. It is a small parameter theorem because it deals with a geometric sum $\xi_1 + \ldots + \xi_\nu$, where the distribution of each ξ_k is a given function $F(x)$, depending neither on q nor on the occurrence of the rare event. But such a situation takes place rarely. Indeed, as it seems to be clear intuitively, a busy cycle, on which a large queue can be reached, is generally longer than a conventional cycle. Many efforts have been made to generalize Rényi's theorem to a triangle scheme (where $F(x)$ depends on q) and to the search of bounds for the difference between the distribution of $\xi_1+\ldots+\xi_\nu$ and the exponential distribution. Perhaps, one of the most convenient triangle type theorems is as follows. Let $\mathbf{E}[\xi_k]=a$ and $Var(\xi_k)=\sigma^2$. Then the condition

$$\sigma^2 q/a \to 0, \quad q\to 0 \tag{19.68}$$

is sufficient for the asymptotic exponentiality of the "geometric sum" $\xi_1+\ldots+\xi_\nu$ with parameter q/a. For the recent status of the problem, see Kruglov and Korolev [42].

A fruitful study in queueing and reliability applications was rendered by Solovyev [56]. Solovyev considers a regenerative process. An event A can occur in each regeneration cycle with probability q, independently of all other cycles. Let χ_n be the indicator of the event A within the nth cycle, ξ_n the duration of the cycle, $\phi_-(z)=\mathbf{E}[e^{-z\xi_n}\chi_n]$, $\phi_+(z)=\mathbf{E}[e^{-z\xi_n}(1-\chi_n)]$, and $\phi(z)=\phi_-(z)+\phi_+(z)$. Set $\bar{q}=sup\frac{q-\phi_-(z)}{1-\phi(z)}$, and $q_0=max\{q,\bar{q}\}$. The main theorem of the paper is formulated as follows.

Let the distribution of (ξ_n,χ_n) vary in such a way that $q>0$, $q_0\to 0$ and

$$\lim_{q_0\to 0}\int_0^x \frac{t}{q}dF\left(\frac{t}{\gamma}\right)=P(x), \tag{19.69}$$

where $F(t)$ is the distribution function to the duration of a cycle, where $P(x)$ is a nondecreasing function satisfying the conditions

$$P(0)=0, \quad \int_0^\infty \frac{1}{x}dP(x)<\infty. \tag{19.70}$$

(19.69)-(19.70) is a necessary and sufficient condition for the Laplace-Stieltjes transform of the p.d.f. of the time to the first occurrence of event A multiplied by γ, to converge to $(1+w(z))^{-1}$, where

$$w(z)=\int_0^\infty \frac{1}{x}(1-e^{-zx})dP(x). \tag{19.71}$$

The last equation gives a general form of possible asymptotic laws.

The cited paper also gives some sufficient conditions more convenient in applications. Various types of repairable redundant systems in reliability were studied on the basis of the above cited theorem of Solovyev.

Combining this theorem with the analysis of the rare event within a busy period, Solovyev and his followers have established a large variety of exponential type theorems for a rare event occurrence, with a special emphasis on repairable models of the reliability theory. We mention here only a monograph [26] and, two papers in this direction: Gnedenko and Solovyev [27] and Konstantinidis and Solovyev [33]. The former paper treats a large variety of queueing models, whereas the last one presents most recent results. Other approaches were developed by Kalashnikov [31] and Kovalenko [38].

19.4.2 States merging (lumping) for rare events analysis

Complex queueing systems (for example, models of telephone stations) are described by random processes with such a large number of states that no straightforward method exists for solving corresponding systems of linear equations. A few decades ago, in the mathematical and engineering literature on teletraffic analysis, an effort was made to consider lumped states (macrostates) and to derive a system of approximate equations for the macrostates. Such an approach is still important; see Basharin, Bocharov, and Kogan [8], Dobrushin, Kelbert, Rybko, and Sukhov [18]. Mathematicians developed some methods for asymptotic states merging of Markov, semi-Markov, and some other processes. The small parameter method turned out to play the major role in these investigations. It is usually assumed that all the microstates belonging to the same macrostate are mutually communicable whereas transitions from one macrostate to another have small rates. In such a situation, an individual microstate can be ignored and an asymptotic system of equations, concerning macrostates only, can be derived. In an appropriate mathematical model, it is necessary to use slow time scaling to accelerate transitions between macrostates. In such a slow time, the macrostate process is close to a Markov chain. The first order approximation can be expressed in probabilistic terms and based on well-known ergodic principles (see example 19.2.1). A heavy system of singular perturbation methods is needed for the construction of asymptotic expansion of a small parameter. To clarify the situation, consider the following model (Korolyuk, Penev and Turbin [34]).

Let $\xi_\epsilon(t)$ be a Markov chain with continuous time parameter and finite state space $E = \{0, 1, \ldots, m\}$. Let $\eta_\epsilon(n)$ be an embedded Markov chain with transition probabilities

$$p^\epsilon_{ij} = \begin{cases} p_{ij} - \epsilon q_{ij} & \text{if } i, j \in E_0 = \{1, 2, \ldots, m\} \\ \epsilon g_i & \text{if } i \in E_0, j = 0 \\ 1 & \text{if } i = j = 0, \end{cases} \tag{19.72}$$

where

$$g_i = \sum_{j \in E_0} g_{ij}, \quad \sum_{j \in E_0} p_{ij} = 1 \tag{19.73}$$

and $(p_{ij}, i \in E_0, j \in E_0)$ is the transition matrix of an irreducible ergodic chain with ergodic state probabilities ρ_i, $i \in E_0$. Obviously, state 0 is absorbing for the Markov chain $\xi_\epsilon(t)$. Denote, by ζ^ϵ_i, the waiting time in state i, so that

$$\mathbf{P}\{\zeta^\epsilon_i > t\} = e^{-t/q_i}, i \in E_0, \tag{19.74}$$

and, by τ^ϵ_i, the absorption time with initial state i. Set

$$u^\epsilon_i(t) = \mathbf{P}\{\epsilon\tau^\epsilon_i > t\}. \tag{19.75}$$

The Kolmogorov state equations are

$$\epsilon\frac{du^\epsilon_i(t)}{dt} = -\frac{1}{q_i}u^\epsilon_i(t) + \frac{1}{q_i}\sum_{j \in E_0} p^\epsilon_{ij}u^\epsilon_i(t) \tag{19.76}$$

or, in matrix form,

$$\epsilon Q\frac{d}{dt}u^\epsilon(t) = (P - I - \epsilon G)u^\epsilon(t) \tag{19.77}$$

with the initial condition

$$u^{\epsilon}(0) = 1, \tag{19.78}$$

where

$$Q = (\delta_{ij}q_i, i \in E_0, j \in E_0), P = (p_{ij}, i \in E_0, j \in E_0), G = (g_{ij}, i \in E_0, j \in E_0),$$

$1 = (1, \ldots, 1)$. Thus the desired function $u^{\epsilon}(t)$ is a solution of a differential equation with a small parameter factor of the derivative. It turns out that this solution can be expressed as

$$u^{\epsilon}(t) = e^{-\lambda t}(I - \epsilon R)^{-1}1 - (I + \epsilon H_{\tau})^{-1}\exp\{-Q^{-1}(I-P)\tau\}\epsilon R(I - \epsilon R^{-1})1$$

with

$$\lambda = \sum_{i \in E_0} \rho_i g_i \Big/ \sum_{i \in E_0} \rho_i q_i \tag{19.79}$$

and

$$R = R_0(\lambda Q - G), \tag{19.80}$$

where R_0 is the reduced resolvent of the matrix $I - P$ and H_{τ} is an integral operator defined by

$$H_{\tau}f(t) = \int_0^{\tau} \exp\{-Q^{-1}(I-P)(\tau - x)\}Q^{-1}Gf(x)dx \tag{19.81}$$

with $\tau = t/\epsilon$. It can be shown that the solution has the form

$$u^{\epsilon}(t) = e^{-\lambda t}v_{\epsilon} + w_{\epsilon}(\tau), \tag{19.82}$$

where

$$v_{\epsilon} = v_0 + \epsilon v_1 + \epsilon^2 v_2 + \ldots \tag{19.83}$$

does not depend on t and

$$w_{\epsilon}(\tau) = \epsilon w_1(\tau) + \epsilon^2 w_2(\tau) + \ldots. \tag{19.84}$$

v_{ϵ} and w_{ϵ} can be obtained from the equations

$$(I - P + \epsilon(G - \lambda Q))v_{\epsilon} = 0 \tag{19.85}$$

and

$$(Q\frac{d}{dt} + I - P + \epsilon G)w^{\epsilon}(\tau) = 0 \tag{19.86}$$

with the boundary condition $w_{\epsilon}(0) = 1 - v_{\epsilon}$. λ, in (19.79), is the unique eigenvector of the matrix $I - P + \epsilon(G - \lambda Q)$, corresponding to compute v_k and w_k. w_k is a function of the boundary layer type being widely used in asymptotical analysis. Due to this function, the convergence of the absorption time distribution to the exponential distribution is not uniform in a small neighborhood of the origin.

The outlined approach was developed by a group of scientists lead by Korolyuk, Turbin, and Anisimov from Kiev, Ukraine. They generalized cases for finite state Markov chains to discrete space semi-Markov processes and, later on, to arbitrary state space semi-Markov processes and random evolutions, such as queues in a randomly varying environment. A significant contribution to the theory of singular queueing problems was made by Turbin [60].

Let $(\xi_{\epsilon}(t))$ be a family of Markov processes in an arbitrary measurable state space $(X^{(0)}, \mathfrak{B}^{(0)})$. In many practical applications, a dependence of $\xi_{\epsilon}(t)$ on ϵ is defined by its dependence on the infinitesimal operator A_{ϵ} of the semi-group $T_{\epsilon}(t)$ of the Markov process. Turbin considers a dependence of the form

$$A_\epsilon = A_0 - B(\epsilon),\ B(\epsilon) = \sum_{k \geq 1} B_k \epsilon^k, \tag{19.87}$$

where $B(\epsilon)$ can be regarded as a small perturbation, although A_0 and B_k are, in general, unbounded. As $\epsilon = 0$, the state space $X^{(0)}$ admits an ergodic decomposition

$$X^{(0)} = \bigcup_{y \in X^{(0)}} X_y^0, \quad X_y^0 \in \mathfrak{B}^{(0)}, \tag{19.88}$$

where X_y^0 is a minimal closed ergodic class of state y. Let $\phi(x)$, $x \in X^{(0)}$, be a function separating the classes X_y^0 but not states within a given class. An asymptotic phase merging of a Markov process refers to the consideration of the family of processes

$$\xi_\epsilon^{(s)}(t) = \phi(\xi_\epsilon(t/\epsilon^s)) \tag{19.89}$$

and study of its asymptotic behavior as $\epsilon \to 0$. It should be noted that the value of s, under which a proper limiting behavior takes place, is uniquely determined by the infinitesimal operator A_ϵ. In 1969-1970, Korolyuk and Turbin established a connection between the phase merging and problems of matrix inversion perturbed on the spectrum.

It is of interest to find the stochastic limit of $\xi_\epsilon^{(s)}(t)$ as $\epsilon \to 0$. Some operator expansions have been obtained for analytical and numerical solutions for a variety of cases.

For the recent state of the problem, see Korolyuk and Turbin [36] and, in a less rigorous version, Korolyuk [35].

Anisimov [2, 3] introduced a concept of so-called *switching stochastic processes*. It gives an accurate measure for a sequence of transitions of a real system through a given set of random processes. The model is more general than many previous models. For the class of switching processes, some averaging theorems were proved mainly establishing exponential limiting behavior of the distribution of time to a rare event.

19.4.3 Large deviation methods

For many queueing systems, the queueing process can be described by a random walk, and usual system parameters can be treated as its functionals. Such widely considered kinds of functionals give the maximal value of a walking particle coordinates within a given time interval, the time spent by the particle beyond a given level, the number of level crossings, and so on. Considering the random walk with taboo probabilities, one can interpret it in terms of the corresponding busy period parameters; for example, the maximal queue length on the busy period. The theory of random walks (both unbounded and bounded from below) is rich in explicit formulas or (at least) integral representations for a wide class of parameters. Many queueing theorists make use of them for derivations of estimates and bounds for global and local system parameters.

We have already mentioned a wide spectrum of integral representations of queueing system parameters in the monograph by Cohen [16]. Asmussen and Perry [6] have investigated local and global queueing parameters by a unified analytical approach.

For many purposes and, particularly, for rare events simulation, it is convenient to consider the underlying distributions $A(t)$ and $B(t)$ belonging to appropriate parametric families. A family of distributions (F_θ), $\theta \in \Theta$, is called a *conjugate family* if

$$\frac{dF_\theta(x)}{dF_{\theta_0}(x)} = \exp\{(\theta - \theta_0)x - c(\theta_0, \theta)\}, \tag{19.90}$$

and if each θ, for which the RHS of (19.90) defines a distribution for some $c(\theta_0, \theta)$, belongs to Θ. Diverse applications of conjugate families to queueing systems are discussed in the monograph by Asmussen [5]. One of the results consists of an exponential bound for the deviation of the waiting time distribution of the nth customer from its limiting distribution under appropriate analytical conditions.

The large deviation theory was expounded in a monograph by Bucklew [14] in a form convenient for queueing systems parametric analysis. A sequence (S_n) of random variables is said to *satisfy the large deviation principle* if, for some c and any $\epsilon > 0$,

$$\mathrm{P}(\mid S_n - c \mid > \epsilon) \sim K(\epsilon, c, n)\exp\{-nI(\epsilon, c)\} \tag{19.91}$$

as $n \to \infty$, where $I(\epsilon, c)$ is positive for each positive ϵ and c and $K(\epsilon, c, n)$ is a slowly varying function (relative to an exponential function). The derivation of $I(\epsilon, c)$ is presented for the most interesting random sequences (S_n) including random walks, functionals of finite state Markov chains, and diffusion processes (Ventzell-Freidlin theory). A monograph by Keilson [32] deals with the large deviation theory for time-reversible Markov chains. Many queueing models can be described by such random processes.

Anatharam, Heidelberger and Tsoucas [1] also use the time reversal approach for the rare event analysis of continuous time Markov processes. New ideas lead to a considerably simpler analysis. A relevant asymptotic property is based on the following conditions. If it is desired to compute the rate of reaching a far state set S_N, the "reverse way" of the time reversed Markov chain started from S_N consists of a unique sample path. Wide sufficient conditions are established for the asymptotic exponentiality of the time to a rare event. The authors formalized the simulation of rare events.

A theory of large deviations for semimartingales was applied by Liptser [43] to large deviation estimates for parameter N in the environment of a single-server queueing system controlled by strictly stationary input and output sequences. N is a large parameter interpreted as the maximal queue length.

Willekens and Teugels [61] used the Wiener-Hopf expansion for the steady state waiting time distribution in the system M/G/1. With an approximation of its terms, they investigated the long-tailed behavior of this distribution in some typical cases of the subexponential behavior of the service time distribution.

19.5 LIGHT-TRAFFIC LIMITS AND PERTURBATION ANALYSIS OF QUEUEING SYSTEMS

Perturbation analysis (PA) is a very fruitful branch of applications of small parameter methods to system analysis. Ho [28] (see also [29]) defined PA as an analytical technique that calculates the sensitivity of a discrete event dynamical system (DEDS), with respect to system parameters, by analyzing its sample path.

Let us consider a DEDS (in particular, a queueing system) with underlying distributions dependent on parameter θ. Suppose we need to estimate the derivative $\partial \bar{f} / \partial \theta$, where $\bar{f}$ is a system performance parameter. One can model a sample path X and compute $\partial f(X)/\partial\theta$, where $f(X)$ is the value of the performance parameter in a specific realization and, then, construct the estimate

$$\partial \bar{f}/\partial\theta \approx \frac{1}{n}\sum_{k=1}^{n} \partial f(X_k)/\partial\theta \tag{19.92}$$

repeating the procedure for n independent sample paths X_k. The problem is in the consistency of (19.92). So, when analyzing queueing systems, the estimate accounts for only small variations, ignoring possible discontinuities due to parameter variations. A monograph by Glasserman [22] gives a theoretical background for this approach. Fu and Hu [21] applied smoothed PA for the estimation of the loss probability and some other parameters in the queueing system GI/G/1/K. In some papers (Reiman and Weiss, [51]; Rubinstein, [53]; Glynn and Sanders, [24]), the sampling method is applied. Let, for example, $\bar{f} = \mathbf{E}[f(X)]$, where X is a random variable with a density function $p(x,\theta)$. Then, one can construct an estimate of the form

$$\partial \bar{f}/\partial\theta \approx \frac{1}{n}\sum_{k=1}^{n} f(X_k)\partial(\log p(X_k,\theta))/\partial\theta \tag{19.93}$$

which saves us from possible discontinuities but is not always good from the point of view of the variance estimate. The methods elaborated in the framework of PA turned out to be extremely fruitful in light-traffic analysis. Daley and Rolski [17], Brémaud and Vazquez-Abad [13] and other authors investigated the behavior of some queueing systems under light-traffic conditions.

The most general and tractable results are due to Asmussen [4]. Following this paper, consider a stable GI/G/1 queueing system with U, T and W, which stand for a service time, an inter-arrival time, and a steady state waiting time, respectively. As usual, a random walk with increments X_n is considered, with X_n distributed as U-T. The author proposed a random walk following a light-traffic condition for the family $F^{(\gamma)}(x)$ of the distributions of X_n, where γ is a large parameter. Then,

$$LT(p){:}\lim_{z\to\infty}\limsup_{\gamma\to\infty} \int_z^\infty x^{q+1}dF^{(\gamma)}(x) \Big/ \int_0^\infty x^q dF^{(\gamma)}(x) = 0, \quad 0 \le q \le p. \tag{19.94}$$

Two γ-dependent nonnegative random variables $R^{(\gamma)}$ and $S^{(\gamma)}$ are called *distributionally light-traffic equivalent* if both of them tend to zero in probability; the probabilities of them being positive are equal, and the variation distance of their conditional distributions under the condition that $R^{(\gamma)} > 0$ (respectively $S^{(\gamma)} > 0$) vanishes as $\gamma\to\infty$. Asmussen proved that if $X^{(\gamma)} \to -\infty$ in probability as $\gamma\to\infty$ and the condition $LT(0)$ holds, then, the maximum of the random walks is distributionally light-traffic equivalent to $X^{(\gamma)+}$. If the condition $LT(q)$ also holds, then, the moments of both random variables of orders $1, 2, \ldots, p$ are also equal as $\gamma\to\infty$.

Reiman and Simon [50] investigated the following problem. Let $f(\lambda)$ be the performance measure for a queue with Poisson input with rate λ. The aim is to compute the values $f(0)$, $f'(0)$, $f''(0), \ldots$. The main result consists of the following stochastic interpretation of the derivatives. The value of $f(0)$ is obtained by looking at a "tagged" customer in an an otherwise empty system. It turns out that $f^{(n)}(0)$ can be obtained by looking at the tagged customer in a system where, at most, n other customers arrived. The general principle was applied in Reiman and Simon [49] to the investigation of the sojourn time distribution for a wide class of queueing networks.

We would like to point out that the Reiman-Simon effect was discovered and exploited in some much earlier works on reliability theory. In the paper by Kovalenko [37], a multichannel queueing model of a rather general type was proposed. In this model, the service times are of a general form, whereas the streams of customers are Markovian, with transition rates λ_{ij} proportional to a small parameter ϵ. The steady state distribution of the queueing process can be obtained recursively, based on an ϵ-power series expansion.

Pechinkin [46] proposed an effective iterative approach to the small parameter investigation of queueing systems; see also [45]. Baccelli and Brémaud [7] considered a general approach to parameter derivatives of mean values of queueing functionals.

Let N be a Poisson process of intensity λ defined on a real line, and let (T_n) be its sequence of event times; $\ldots < T_{-1} < T_0 < T_1 < \ldots$. Let (Z_n) be an i.i.d. sequence of random marks on a measurable space. The processes (T_n) and (Z_n) are assumed to be independent. Let θ be a functional of $(N, (Z_n))$ of the form

$$\varphi = h(T_0, Z_0, \tau_{-1}, Z_{-1}, \tau_{-2}, Z_{-2}, \ldots), \tag{19.95}$$

where $\tau_n = T_{n+1} - T_n$. We want to compute

$$\lim_{\lambda \to 0} \frac{1}{\lambda} \mathbf{E}_\lambda[\varphi], \tag{19.96}$$

where $\mathbf{E}_\lambda$ denotes the expectation corresponding to a given intensity λ, provided that $\varphi = 0$ as $\lambda = 0$.

Define the functional g by

$$\begin{aligned} g(T_n, Z_n, \tau_{n-1}, Z_{n-1}, \tau_{n-2}, Z_{n-2}, \ldots) &= h(T_n, Z_n, \tau_{n-1}, Z_{n-1}, \ldots) \\ &- h(T_n - \tau_{n-1}, Z_{n-1}, \tau_{n-2}, Z_{n-2}, \ldots). \end{aligned} \tag{19.97}$$

For brevity, we write $g - g(T_n, Z_n, Y_{n-1})$, where $Y_{n-1} = (\tau_{n-1}, Z_{n-1}, \tau_{n-2}, Z_{n-2}, \ldots)$. Also, for $\alpha > 0$, define $Y^\alpha_{n-1} = (\alpha\tau_{n-1}, Z_{n-1}, \alpha\tau_{n-2}, Z_{n-2}, \ldots)$.

Assume that

(1) For some $\alpha_0 > 0$,

$$\int_{-\infty}^{0} \mathbf{E}[\,|\,g(t, Z_0, Y^\alpha_{-1})\,|\,]dt < \infty \tag{19.98}$$

for $\alpha \geq \alpha_0$.

(2) There exists a functional $g_\infty(t, Z_0)$, independent of Y_{-1}, such that

$$\lim_{\alpha \to \infty} \int_{-\infty}^{0} \mathbf{E}_1[\,|\,g(t, Z_0, Y^\alpha_{-1}) - g_\infty(t, Z_0)\,|\,]dt = 0. \tag{19.99}$$

Theorem 1 by Baccelli and Brémaud states that under conditions (1) and (2),

$$\lim_{\lambda \to \infty} \frac{1}{\lambda} \mathbf{E}_\lambda[\varphi] = \int_{-\infty}^{0} \mathbf{E}_1[g_\infty(t, Z_0)]dt. \tag{19.100}$$

In the proof, Campbell's formula (also discussed in Schmidt and Serfozo [54], in this book) is the key tool.

Zazanis [62] has established the analyticity of a queueing system's efficiency with respect to the rate λ of the Poisson input. In an important paper by Glasserman [23], a general scheme for the queueing process of the form

$$W_{n+1} = \phi(W_n, U_n),\ n \geq 0, \tag{19.101}$$

is considered. Here (U_n) is a strictly stationary and ergodic vector control sequence dependent on parameter θ. One can write a recurrence relation for $\partial W_n/\partial\theta$ and use the observed values of the derivative for the estimation of the sensitivity of the waiting time distribution. Stability conditions are established for such an estimate.

Brémaud [12] introduced an estimate based on the coupling method. Thus, two random sequences are simulated up to a moment of their coupling, and the estimate is a functional of the path functions. Cao [15] developed a new approach for performance sensitivity analysis based on the concept of the realization factor of the perturbation. The approach was applied to queueing networks. Some other important papers on light-traffic analysis are by Błaszczyszyn, Rolski and Schmidt [9] (in this book), Sigman [55], and Zhu and Li [63].

In a monograph by Borovkov [10], some approaches were developed for light-traffic analysis. Many results of the asymptotical analysis of reliability are also very close to the concepts and methods of light-traffic analysis; see our survey in Kovalenko [38] and the survey by Reibman, Smith and Trivedi [48]. Other related recent works in this section are by Ho and Cassandras [29] on PA, and Jagerman, et al. [30] on traffic processes.

19.6 UNSOLVED PROBLEMS AND RESEARCH DIRECTIONS

19.6.1 Gathering limit theorems for light-traffic queues

In the well-known monograph [11], Borovkov developed "gathering theory" of queues in a heavy traffic. Let a queue be described by a random walk sequence (W_n) or a random process $(W(t))$. Under some asymptotic conditions, the behavior of the sequence/process is defined mainly by local conditions (i.e., a sort of averaged shift and diffusion coefficients), not by a subtle structure of the queueing system. Thus, theorems established in such a way cover a wide class of queueing processes. We believe that the available techniques can lead to the establishment of some useful gathering type theorems concerning light-traffic situations. Contrary to the heavy traffic, the queueing process behavior under light-traffic condition in a small neighborhood of the origin is essential, whereas large deviations can typically be ignored.

An adequate approach to the averaging is of importance when investigating light-traffic limits. The experience of the highly reliable systems analysis suggests that in many cases, such parameters, as the probability of the system failure on a busy period or the system failure rate, are equivalent to $C\lambda^r$, where λ is a component failure rate considered as a small parameter. It would be worthwhile to generalize such theorems for the case of time dependent parameter $\lambda(t)$. The question is: Can one choose such an averaging process that the result is still valid if one changes λ^r to an appropriate average of $\lambda(t)$? [In some cases one must take the limit of the average $\frac{1}{T}\int_0^T \lambda^r(t)dt$ as $T\to\infty$ where r is defined by the system structure; see [38]: section 5.5].

19.6.2 The insensitivity and ϵ-sensitivity

A great segment of the research was devoted to the insensitivity problem in the queueing theory; we mention here only a monograph [20] for a fundamental exposition of the problem. In the insensitivity conditions, convenient expressions of practical use can usually be derived for the system efficiency parameters. So, the availability of a complex system can be investigated in the framework of the insensitivity method. The following problem can be considered. Let a queueing system S be close to a system S_0 for which the insensitivity conditions are satisfied. Can one compute the efficiency of S, in such a way, that $W(S) = W(S_0) + \Delta W$, where $W(S_0)$ is a known parameter, due to available expressions for the "insensible" system, and can ΔW be computed via a small parameter procedure?

Some useful modifications of the concept can be proposed. An approach worth trying consists of the following. Let us consider a Poisson-driven queueing system with state-dependent and time-dependent input rate, $\lambda_j(z,t)$ for a j type customer. Assume that $\lambda_j(z,t) = \lambda_j^{(0)} + \epsilon_j(z,t)$, where $\lambda_j^{(0)}$ are the parameters for S_0 and $\epsilon_j(z,t)$, which can be regarded as small parameter functions. In general, it should not be too difficult to derive convenient approximate expressions for ΔW in the above situation.

19.6.3 Further elaboration of the rare events theorems

The key point of most of the exponential type theorems concerning rare events probabilities is that a large number ν of renewals occur before the rare event does. However, in practice, it is not the case. Thus, modifications of limit theorems would be of interest as to cover such situations. A possible approach would consist of the development of asymptotic series in a small parameter. The problem must be approached both "from infinity" and "from zero". That is, setting $\epsilon = \mathbf{E}[\nu]$, one can search for an asymptotic expansion in ϵ^{-1} or, alternatively, in ϵ. Each of the methods can be used in an appropriate region of parameters.

19.6.4 Comparisons with the exponential case

Let S_ϵ be a small parameter dependent queueing system and S_ϵ^0 be an associated system which differs from the former one in the following: All the control distributions are replaced by appropriate exponential distributions. Consider the system availability $w(S_\epsilon)$ and the ratio

$$\sigma = \frac{w(S_\epsilon)}{w(S_\epsilon^0)}.$$

The problem consists of the estimation of $\sigma_0 = \liminf_{\epsilon \to 0} \sigma$ and $\sigma_1 = \limsup_{\epsilon \to 0} \sigma$. For example, consider an M/G/m/r queue, S_ϵ, with input rate ϵ and a fixed service time distribution $B(x)$. Stoikova (see [40]) obtained results on estimators for linear and linear-fractional functionals of distributions. They led to some bounds for σ_0 and σ_1 in the class of distributions $B(x)$ with fixed moments $\alpha_1, \alpha_2, \ldots, \alpha_r$ in the case if $w(S)$ is the probability of a customer loss in a busy period.

19.6.5 Small parameter-dependent optimization

A large variety of optimization problems can be considered in the framework of small parameters. Let $S_{\epsilon,u}$ be an ϵ-depending queueing system controlled by a parameter $u_\epsilon \in U$. Assume that $f_{\epsilon,u}$ is a parameter characterizing the system's efficiency. It is worthwhile to investigate the limiting behavior of the solution of an optimization problem

$$f_{\epsilon, u_\epsilon} \Rightarrow max, \quad u_\epsilon \in U.$$

Both regular and singular cases are of practical interest. Some methods were evaluated in [41].

ACKNOWLEDGEMENTS

My work on this chapter would not have been possible without an exchange with leading scientists from some Western schools and an opportunity to use their libraries during the last two years. Therefore, I am happy to thank Søren Asmussen (from Aalborg University), Alessandro Birolini (from the Federal Institute of Technology, Zurich) and Robert Gilchrist (from the University of North London).

Thanks are also due to Jewgeni Dshalalow for his invitation to contribute this chapter and for his difficult editorial duties, and to V.S. Korolyuk for his valuable suggestions.

BIBLIOGRAPHY

[1] Anantharam, V., Heidelberger, P. and Tsoucas, P., *Analysis of Rare Events in Continuous Time Markov Chains via Time Reversal and Fluid Approximation*, Research Report RC 16280 (#71858) 1990.

[2] Anisimov, V.V., Switching processes, *Kibernetica* **4** (1977), 111-115 (in Russian).

[3] Anisimov, V.V., Limit theorems for switching processes and their applications, *Kibernetica* **6** (1978), 108-118 (in Russian).

[4] Asmussen, S., Light-traffic equivalence in single server queues, *Ann. Appl. Prob.* **2**:3 (1992), 555-574.

[5] Asmussen, S., *Applied Probability and Queues*, Wiley, Chichester 1987.

[6] Asmussen, S. and Perry, D., On cycle maxima, first passage problems and extreme value theory for queues, *Stoch. Mod.* **8**:3 (1992), 421-458.

[7] Baccelli, F. and Brémaud, P., Virtual customers in sensitivity and light-traffic analysis via Campbell's formula for point processes, *J. Appl. Prob.* **28**:1 (1991), 1-19.

[8] Basharin, G.P., Bocharov, P.P., and Kogan, Ya.A., *Queueing Analysis in Computing Networks*, Nauka Publ., Moscow 1989 (in Russian).

[9] Błaszczyszyn, B., Rolski, T., and Schmidt, V., Light-traffic approximations in queues and related stochastic models, as Chapter 15 in this book, 379-406.

[10] Borovkov, A.A., *Stochastic Processes in Queueing Theory*, Springer-Verlag, New York, 1976.

[11] Borovkov, A.A., *Asymptotical Methods in Queueing Theory*, Nauka Publ., Moscow 1980.

[12] Brémaud, P., Maximal coupling and rare perturbation sensitivity analysis, *Queueing Sys.* **11** (1992), 307-333.

[13] Brémaud, P. and Vazquez-Abad, F.J., On the pathwise computation of derivatives with respect to the rate of a point process: The phantom RPA method, *Queueing Sys.* **10** (1992), 249-270.

[14] Bucklew, J.A., *Large Deviation Techniques in Decision M Simulations and*

Estimation, John Wiley, Chichester, Brisbane, Toronto, Singapore 1990.

[15] Cao, X.R., A new method of performance sensitivity analysis of non-Markovian queueing networks, *Queueing Sys.* **11** (1992), 313-350.

[16] Cohen, J.W., *The Single Server Queue*, 2nd ed., North Holland, Amsterdam 1992.

[17] Daley, D. and Rolski, T., Light-traffic approximations in queues, *Math. Oper. Res.* **1** (1991), 57-71.

[18] Dobrushin, R.L., Kelbert, M.Ya., Rybko, A.N. and Sukhov, Yu.M., *Qualitative Methods of Queueing Network Theory*, Inst. Problem Peredachi Inform. Acad. Nauk SSSR, Moscow 1986 (in Russian).

[19] Feller, W., *Introduction to Probability Theory and its Applications*, Vol. **2**, Wiley, New York 1966.

[20] Franken, P., König, D., Arndt, U. and Schmidt, V., *Queues and Point Processes*, Wiley, New York 1982.

[21] Fu, M.C. and Hu, J.Q., Smoothed perturbation analysis for queues with finite buffers, *Queueing Sys.* **14** (1993), 57-78.

[22] Glasserman, P., *Gradient Estimation via Infinitesimal Perturbation Analysis*, Kluwer, Boston, Dordrecht, London 1991.

[23] Glasserman, P., Stationary waiting time derivatives, *Queueing Sys.* **12** (1992), 369-390.

[24] Glynn, P., and Sanders, L., Monte Carlo optimization in manufacturing systems: Two new approaches, In: *Proc. ASMECIE Conf.*, Chicago 1986.

[25] Gnedenko, B.V. and Kovalenko, I.N., *Introduction to Queueing Theory*, 2nd ed., Birkhäuser, Boston, Basel, Berlin 1989.

[26] Gnedenko, B.V., Belyaev, Yu.K. and Solovyev, A.D., *Mathematical Methods in Reliability Theory*, Academic Press, New York 1969.

[27] Gnedenko, D.B. and Solovyev, A.D., Estimations of complex repairable systems reliability, *Izv. Akad. Nauk SSSR, Tekhn. Kibern.* **3** (1975), 121-128 (in Russian).

[28] Ho, Y.C., Performance evaluation and perturbation analysis of discrete event dynamical systems, *IEEE Trans. Automatic Contr.* **AC-32** (1987), 563-572.

[29] Ho, Y.C. and Cassandras, C.G., Perturbation analysis for control and optimization of queueing systems: An overview and the state of the art, In: *Frontiers in Queueing: Methods and Applications in Science and Engineering*, ed. by J.H. Dshalalow, CRC Press, Boca Raton 1996.

[30] Jagerman, D.L., Melamed, B. and Willinger, W., Stochastic modeling of traffic processes, In: *Frontiers in Queueing: Methods and Applications in Science and Engineering*, ed. by J.H. Dshalalow, CRC Press, Boca Raton 1996.

[31] Kalashnikov, V.V., Deriving parameters of first failure time by semiregenerative processes analysis, *Problems of Stoch. Models Stab.* **VNIISI** (1990), 21-31 (in Russian).

[32] Keilson, J., *Markov Chain Models- Rarity and Exponentiality*, Springer-Verlag, New York 1979.

[33] Konstantinidis, D.G. and Solovyev, A.D., Uniform estimate of reliability for a complex regenerated system with unlimited number of repair units, *Moscow Univ. Math. Bull.* **46** (1991), 21-24.

[34] Korolyuk, V.S., Penev, I.P. and Turbin, A.F., An asymptotic expansion for the absorbing time of a Markov chain distribution, *Kibernetica* **4** (1973),

133-135 (in Russian).

[35] Korolyuk, V.S., *Stochastic Models of Systems*, Kyiv, "Lybid" 1993 (in Ukrainian).

[36] Korolyuk, V.S. and Turbin, A.E., *Mathematical Foundations of the State Lumping of Large Systems*, Kluwer, Dordrecht, Boston, London 1993.

[37] Kovalenko, I.N., On some topics in the theory of reliability of complex systems, *Kibernetiku na sluzhby Kommunizma* **2**, (1964), 194-205 (in Russian).

[38] Kovalenko, I.N., *Rare Events Analysis in Estimation of Systems Efficiency and Reliability*, Sov. Radio Publ, Moscow 1980 (in Russian).

[39] Kovalenko, I.N. and Kuznetsov, N.Yu., *Methods of High Reliable Systems Account*, Radio i Svyaz, Moscow 1988 (in Russian).

[40] Kovalenko, I.N., Rare events in queueing systems - A survey, *Queueing Sys.* **16** (1994), 1-49.

[41] Kovalenko, I.N. and Nakonechny, A.N., *Approximate Analysis and Optimization of the Reliability*, Naukova Dumka Publ., Kiev 1989 (in Russian).

[42] Kruglov, V.M. and Korolev, Yu. V., *Limit Theorems for Random Sums*, Moscow Univ. Publ., Moscow 1990 (in Russian).

[43] Liptser, R.Sh., Large deviations for a simple closed queueing model, *Queueing Sys.* **14** (1993), 1-31.

[44] Moore, E. and Shannon, C., Reliable circuits using less reliable relays, *J. Franklin Inst.* **3** (1956), 191 and **4** (1956), 281.

[45] Nagonenko, V.A. and Pechinkin, A.V., A light load in a system with inverse order of servicing and a probabilistic priority, *Izv. Akad. Nauk SSSR Tekhn. Kibernet.* **6** (1985), 82-89 (translated as *Soviet J. Comput. Systems Sci.* **23** (1985), 51-58).

[46] Pechinkin, A.V., The analysis of one-server systems with small load, *Izv. Akad. Nauk SSSR, Tekhn. Kibernet.* **3** (1984), 143-148 (in Russian) (translated as *Engrg. Cybernetics* **22** (1985), 129-135).

[47] Prabhu, N.U., *Queues and Inventories*, Wiley, New York 1965.

[48] Reibman, A., Smith, R. and Trivedi, K., Markov and Markov reward model transient analysis: An overview of numerical approaches, *Euro. J. Oper. Res.* **40** (1989), 257-267.

[49] Reiman, M.I. and Simon, B., Light-traffic limits of sojourn time distributions in Markovian queueing networks, *Stoch. Mod.* **4** (1988), 191-233.

[50] Reiman, M.I. and Simon, B., Open queueing systems in light traffic, *Math. Oper. Res.* **14** (1989), 26-59.

[51] Reiman, M.I. and Weiss, A., Sensitivity analysis for simulations via likelihood ratios, Preprint 1986.

[52] Rényi, A., Poisson-folyamat egy jeuẽmzese (A characteristic of the Poisson stream), *Proc. Math. Inst. Hungarian Acad. Sci.* **1**:4 (1956), 563-570.

[53] Rubinstein, R.Y., *Monte Carlo Optimization, Simulation and Sensitivity of Queueing Networks*, Wiley, New York 1986.

[54] Schmidt, V. and Serfozo, R.F., Campbell's formula and applications to queueing, as Chapter 8 in this book, 225-242.

[55] Sigman, K., Light-traffic for workload in queues, *Queueing Sys.* **11** (1992), 429-442.

[56] Solovyev, A.D., Asymptotic behavior of the first occurence time of a rare event in a regenerative process, *Izv. Akad. Nauk SSSR. Tekhn. Kibern* **6** (1971), 79-89 (in Russian).

[57] Solovyev, A.D., Analytic methods of the computation and estimation of the reliability, In: *Voprosy Matematicheskoj Teroii Nadezhnosti*, ed. by B.V. Gnedenko, Radio i Svyaz, Moscow 1983, 9-112 (in Russian).

[58] Stadje, W., A new approach to the distribution of the duration of the busy period for a G/G/1 queueing system, *J. Austral. Math. Soc. Ser. A 48* 1:89; MR 91c: 60133, 1990.

[59] Takács, L., *Introduction to the Theory of Queues*, Oxford University Press, New York 1962.

[60] Turbin, A.F., Investigations of operators and random processes asymptotic phase aggregation, *Avtoref. diss. dokt. fiz. -mat. nauk* (Inst. Math. Acad. Sci. Ukr. SSR, Kiev) 1980 (in Russian).

[61] Willekens, E., and Teugels, J.L., Asymptotic expansions for waiting time probabilities in an M/G/1 queue with long-tailed service times, *Queueing Sys.* **11** (1992), 295-312.

[62] Zazanis, M.A., Analyticity of Poisson-driven stochastic systems, *Adv. Appl. Prob.* **24** (1992), 532-541.

[63] Zhu, Y. and Li, H., The Maclaurin expansion for G/G/1 queue with Markov-modulated arrivals and services, *Queueing Sys.* **14** (1993), 125-134.

Index

Index

A priori (for rate stability conditions) 154
Abelian lemma 390
ACE of order p 386
Actual and virtual waiting time 294
Actual waiting time process 3
Additive component 168
Additive increments 168
Airy function 55
Analyticity 401, 402
Anti-PASTA 199
Aperiodic 416
Approximation by PH-distributions 270
Approximations of queueing system parameters 481
Arrival component 169
Arrival theorem 220
Arrivals see time averages (ASTA) 126, 195
Ascending ladder epoch 386
Ascending ladder height 386
ASTA in Markovian setting 205
ASTA problem 198
ASTA property 126
ASTA under LBA 202
ASTA under lack of dependence 202
ASTA under time reversal 204
ASTA under WLAA 201
Asymptotic behavior 280
Asymptotic deterministic analysis 120
Asymptotic methods 314
Asymptotically conditionally equivalent (ACE) 386
Asymptotically one dependent 394
Asynchronous transfer mode (ATM) 92
Atoms 104
Average stable 152

Bailey-type dependence 353
Ballot theorem 46
Banded transition matrices 357
Batch Markovian arrival process (BMAP) 171
Bernoulli excursion 48
Bernoulli thinning 170
Bias formula 202
Birth-and-death approximation 467
Block-structured Markov chains with repeating rows 356-357
Blocking probabilities 331
Bottleneck analysis 326, 327
Bottleneck module 456
Boundary layer techniques 314
Branching process 65
Brownian excursion 46, 61
Burstiness 271
Busy cycle 156
Busy period 46, 64, 156, 274
Busy period parameters 483
Busy period process 3

Campbell's formula 227, 383
Catalan number 48
Caudal characteristic curve 273
Central limit theorem 48, 470
Characterization of PH-distributions 269
Circuit-switched networks 465
Classic sense regenerative process 409
Closure properties 269
Coloring 167
Combinatorial theorems 46
Comparison 408
Compensator 209
Complex-analytic factorization 353
Compound renewal process 7
Compound phase-type (CPH) arrival processes 177
Compound phase-type Markov renewal process (CPH-MRP) 180
Conditional ASTA 203
Confluent hypergeometric function 55
Continuity bounds 408
Continuity problem 419
Continuous PH-distribution 267
Convergence rate 413
Counting measure 104
Counting process 7
Coupling 414
Coupling of renewal processes 415
Coupling time 415
Covariance formula 126
Cox/GI/1 391
Cox process 170, 383
Crossing 420, 421
Crossing times 420
Cycle maxima 280

D-policy systems 245
Dawson's integral 474
Deficit 388
Diffusion approximation
322, 329, 467, 469
Diffusion function 464
Diffusion process 85, 464
Directed simulation experiment 412
Discrete event dynamical
system (DEDS) 498
Discrete MAPs 266
Discrete PH-distribution 266
Discrete queues 281
Distributionally light-traffic
equivalent 386, 499
D/M/1 451
Doob-Meyer decomposition 209
Drift function 464
DSTA 127
Duality 79, 279, 446
Dynamic priority queue 278

Effective bandwidth 452
Eigenvalue 340, 347
Eigenvector 340, 347
Elementary renewal theorem 122
Embedded Markov chain 8
Embedded Markov renewal
process 265
Erlang, A.K. 4
Erlang's loss formula 6
Erlang's loss system 6
Error estimate 482
Estimation 429
Event-average 132, 196
Exchange formula 230
Exponential change of measure
(ECM) 430
Exponential time 430

Factorial moment expansion
(FME) 396
Factorial moment measure 382, 395
Factorization component 362
Factorization theorem 279
FIFO discipline 142
Finite capacity queue of
M/G/1-type 281
Finite horizons comparison
bound 424
First excess 248
First passage probability 79
First passage time 248
Fitting of PH-distributions 270
Flow control model 278
Fluid model 160
Fluid model with finite or
infinite buffer 80
Folding algorithm 276
Forward recurrence time 127
Fubini's theorem 227
Fundamental period 277

Gamma-dilation 384
Gathering theory 501
General model for random trees 70
Generalized eigenvalue 340
Generating function 53, 353
Geometry of PH-distributions 269
G/G/1 392
G/G/m queue 232
G/GI/1 queue 391
G/M/1 342
GI/GI/∞ 409
GI/GI/1 12, 386, 391, 401
GI/GI/1$\to\cdots\to$GI/1 393
GI/GI/c 392
GI/GI/N/∞ 410
GI/G/1 12, 296, 450
GI/H_m/s 306
GI/K_m/1 298
GI/M/1-type 266, 268
GI/M/s queue 464
GI/PH/1 274, 387
GI/PH/s 304
Global-balance conditions 131
Greatest common divisor 416

$H = \lambda G$ 140, 147
Heavy traffic diffusion
approximation 464
Height of a vertex 67
Hermite polynomial 59
Hilbert problem 355
Homogeneous Markov chain 5

Idle period 62, 156
Importance sampling (IS) 429
Inaccessible boundary 464
Index of the first excess 248

Infinite server queue 161
Infinitesimal generator 176
Infinitesimal matrix 6
Infinitesimal mean 464
Infinitesimal variance 464
Input (primary) process 152
Input-output process 152
Insensitivity 128, 162, 501
Insurance fund 474
Insurance premiums 235
Intensity 382
Interdeparture time 394
Inter-regeneration time 409
Invariant probability measure 5
Inverse-rate formula 128
Inversion formula 228
Irreducible representation 267
Itô's lemma 115

K-fold (reduced) Campbell measure 395
kth left FME kernel 396
kth right FME kernel 396
Key Renewal Theorem 7
Kendall's embedded process 8
Kiefer-Wolfowitz vector of waiting times 392
Killing measure 114
K_m/G/1 299
Kolmogorov equation 311

$L = \lambda W$ 14
LAA 138, 199
Labeled trees 69
Lack of Bias Assumption (LBA) 201, 202
Lack of memory property 167
Ladder height of a random walk 235
Laplace-Stieltjes transform 55
Large deviation methods 497
Large deviation principle 498
Large deviation (LD) theory 431
Lattice 416
Law of large numbers 183
LCFS-PR 128, 131
LD rate function 439
Left-banded transition matrix 357
Left-continuous at infinity 397
Light-traffic approximation (LTA) 280, 379
Light traffic limit 498
Light-traffic scheme (LTS) 379
Likelihood-ratio (LR) 389, 430
Likelihood-ratio ordering 474
Limit distribution 72
Lindley recursion 81
Little's law 225
Little's formula ($L = \lambda W$) 15, 140, 160
Local ECM 440
Local Poissonification 271
Local time 56
Long-run average 123
Loss probability 280
LTS 383

MAP/G/1 278
MAP/SM/1 278
Marked point process 228, 382
Marked random measure 229
Marking 167
Markov-additive process (MAP) 80, 167, 440
Markov-additive property 187
Markov-Bernoulli coloring 191
Markov-Bernoulli marking 191, 192
Markov-Bernoulli recording 189, 190, 191
Markov-Bernoulli thinning 192
Markov chain 5
Markov-controlled random walk with finite perturbations 353
Markov-modulated (MM) 80, 343, 383, 453
Markov-modulated arrivals 170
Markov-modulated arrivals and service times 302
Markov-modulated M/G/1 queues 80
Markov-modulated M/PH/1 queue 80
Markov-modulated M.P.P 383
Markov-modulated Poisson process (MMPP) 170
Markov-modulated queue 337
Markov-modulated queue with rejection 453
Markov process 5, 61
Markov random walk (MRW) 168
Markov renewal process (MRP) 10, 169, 265
Markov renewal theory 282

Markov subordinators 160
Markovian arrival process
(MAP) 266, 267, 271, 302
Markovian setting 213
Martingale 80, 138
Martingale approach to ASTA 209
Martingale central limit theorem 223
Martingale strong law of large
numbers 223
Matched asymptotic expansion 314
Matrix-Analytic Bulletin 281
Matrix-analytic method 265
Matrix-exponential formula for the
steady-state waiting time
distribution 274
Matrix-geometric method
337, 338, 346
Matrix-geometric solution 272
Matrix-geometric solution and
PH-distribution 274
Matrix quadratic equation 275
Mean value analysis 327
Method of Rouché roots 275
Method of supplementary
variables 9, 266
Minimal nonnegative solution
272, 277
M/D/1 451
M/GI/1 queue 388
M/GI/c queue 388
M/G/1 338, 342
M/G/1 structure 277
M/G/1-type 266, 268
M/G/1 with Markov modulated
arrivals 302
$M/E_2/1$ 451
M/M/1 queue in a Markovian
environment 276
MM/G/c 400, 402
MM/PHi/1 queue 385
M/M/N 344
M/M/s queue 467
M/PHi/c queue 385
Moment convergence theorem 55
Monotone stochastic recursions 80
Monte Carlo Method (CMC) 429
Moran dam 84
Multidimensional frequencies 132
Multiserver queue with phase-type
service time 293
Multiserver queues 161
Multiple Rouché roots 275

N-D-policy 256
N-policy systems 244
N-process 171
Neveu's exchange formula 135, 230
Noncommutative Banach algebra 13
Nonhomogeneous vector Riemann
boundary value problem 353
Nonperturbed 420
Normal density function 48
Normal distribution function 48
Null kernels 397
Null measure 384

Operator-geometric theorem 272
Optimal exponential change of
measure (OECM) 430
Optimization 283, 476, 502
Ornstein-Uhlenbeck diffusion
process 466, 471
Output process 3

Palm Calculus 214
Palm distribution 383
Palm formula 215
Palm inversion formula 126, 132, 215
Palm probabilities 225
Palm probability measure 132
Palm transformation formula 126
Papangelou's lemma 214
Partial lack of memory property 1179
Partition functions 326
PASTA 127, 140, 197, 199
PASTA theorem 200
Pathwise analysis 120
Pathwise-uniform-integrability 149
Periodic 416
Perron-Frobenius eigenvalue 273, 282
Perturbation analysis (PA) 276, 498
Perturbation methods 313
Perturbed 420
PH/PH/c queue with
heterogeneous servers 273
PH-distribution 266, 269
Phase-type (PH) distributions 384
Phase-type (PH) renewal process 171
Phase-type service times 304
π-thinning 384

Piecewise continuous diffusions 470
Piecewise-linear birth-and-death processes 470
Piecewise linear diffusions 464
" ± " controlled system 353
Point-average frequency distribution 125
Point of increase 416
Point process 225
Point process of arrivals 171
Poisson point process 383
Poisson random measure 104
Pollaczeck-Khintchine formula 8
Polling system 388
Polynomial time 430
Pretermination index 257
Pretermination level 246, 257
Pretermination processes 256
Pretermination time 257
Primary quantity 154
Priority queue 278
Processor-shared queues 324
Product-form networks 326

QBD process 275
QBDs with finite state spaces 276
Quasi-birth-and-death 338, 342
Quasi-birth-and-death process 275, 346
Quasi-reversible networks 216
Quasi-reversible queues 215
Quasi-stationary distribution 281
Queue-increase controlled system 353
Queue-length process 158
Queueing network 215, 311
Queueing process 3, 45
Queueing system 1
Queues in series 393
Queues with mobile customers 112
Queues with server vacations 244
Queues with vacations 279

Radon-Nikodym derivative 214, 233
Random counting measure 104
Random graphs 45
Random kernel 231
Random measure 103
Random-rooted tree 67, 72
Random walk 80
Random walks between two barriers 80
Rare events 429
Rare events analysis 483
Rate-balance equation 129
Rate conservation laws (RCL) 124, 233
Rate-stable 152
Rational approximations 472
Ray method 314, 330
Rényi theorem 493
Recurrence equation 389
Recurrent process 7
Reduced k-fold Palm distributions 395
Reducibility 277
Reduction process 366
Reflected Brownian motion (RBM) 473
Reflecting boundaries 464
Regeneration times 409
Regenerative process 7, 409
Renewal function 7
Renewal process 7, 383, 409
Renewal reward theorem ($Y = \lambda X$) 119, 122
Renewal-reward theory 476
Renewals 409
Retention 384
Reversed process 127
Riemann boundary value problems (RBVPs) 354
Riemann-Stieltjes sum approach to ASTA 207
Right-banded transition matrices 357
Right-continuous at infinities 397
Risk model 388
Risk process 236, 399
Risk theory 272
(r, R)-quorum system 244
Robustness 431
Rooted trees 46
Row eigenvector 340
Ruin function 399
Ruin probability 79
Ruin time 388

Sample-path Techniques 119
Sandwich inequality 390
Scale density 471
Secondary quantities 154

Secondary recording 107
Secondary servers with a buffer 468
Semi-Markov matrices 282
Semi-Markov process 10
Semi-Markov queues with
limited state-dependence 353
Semi-regenerative process 10, 11
Service cycle 254
Simple 382
Simple MAP of arrivals 169
Simulation 429
Single-server queue 63, 293
Single server semi-Markov queue 293
Singular perturbation techniques 314
Skew Brownian motion 465
Small parameter method 481
SM/G/1 13, 14
SM/M/1 304
SM/SM/1 300
Sojourn time 324
Sojourns of stationary process 234
Source of a MAP 176
Spatial M/G/∞ queue 107
Spatial M/M/1 queue 113
Spatial queue 103
Spectral method 275
Speed density 471
Spread-out 416
Stable process 176
Standard IS regenerative
estimator 444
Starting solutions 283
State-dependent QBDs 276
State-dependent queues 311, 315
States merging for rare events
analysis 494
Stationary 382, 413
Stationary distribution 3, 79
Stationary ergodic 213
Stationary number of customers 385
Stationary processes 225
Stationary queue length 388
Stationary sojourn time 387, 401
Stationary univariate MAPs 171
Stationary waiting time
387, 391, 393, 401
Stationary workload 388
Stationary workload vector 400
Statistical equilibrium 4
Steady state distribution 3
Steady-state rare events
simulation 430
Stochastic comparison 474
Stochastic intensity 139, 210
Stochastic order 474
Stochastic recursions 86
Stochastically monotone
Markov processes 80
Storage process 84
Strictly stationary 150
Strong approximation of order n 399
Strong MAP 173
Strong law of large numbers
(SLLN) for the martingale 210
Superpositioning of independent
processes 170
Surplus 388
Swiss Army formula 230
Switched Poisson processes 186
Switching regenerative (SR)
simulation 431
Symbolic mathematical package
MAPLE V 478
Symmetric random walk 47
Systems with a processor-sharing
server 311
Systems with vector arrivals 246

Takács process (virtual waiting
time process) 3, 9
Takács integro-differential
equations 9
Termination process 252
Test functions method 412
Theory of fluctuations 11
Thinned process 192
Thinning 167
Thorough departure time 145
Tightness 152
Time-average 132, 196
Time-average frequency
distribution 253
Time-dependent behavior 311
Time of ruin 79
Tombstone process 111
Total height of a rooted tree 67
Total variation distance 414
Transform inversion 280
Transient behavior of queue 329
Transient distribution 3

Transition probability matrix (TPM) 5
Travel times in stationary networks 234
Tree structures 283
Triangular PH-distribution 270
Truncated exponential distribution 468
Truncated geometric distribution 468
Truncated normal distribution 468
Truncated Poisson distribution 468
Truncated reversible Markov process 466
Trunk group 465
Trunk reservation 465

Unbiased estimator 430
Uniform integrability 130
Uniform-in-time comparison bounds 424
Uniformly aperiodic distributions 416
Uniformly integrability 411
Uniformly spread-out distribution 416
Univariate MAP of arrivals 169
Unlabeled tree 68

Vacation model 160
Vector difference equation 339, 342
Version of regenerative process 413
Virtual waiting time 3, 9, 297, 302

Waiting time 231, 386
Waiting time distribution 274
Warm-up 353
Weak approximation of order n 399
Weak (distributional) coupling times 414
Wide sense regenerative process 408
Width of a rooted tree 67
Wiener-Hopf factorization 293, 296
WKB method 314
WLAA 200
Workload 232, 392
Workload process 157, 232
Workload vector 402

$Y = \lambda X$ 140
Yule process 5

Zero-delayed process 413